Natural Products and Drug Discovery

Natural Products and Drug Discovery

An Integrated Approach

Edited By

Subhash C. Mandal
Professor
Division of Pharmacognosy
Department of Pharmaceutical Technology
Jadavpur University
Kolkata, India

Vivekananda Mandal
Assistant Professor
Division of Pharmacognosy
Institute of Pharmaceutical Sciences
Guru Ghasidas University (A Central University)
Bilaspur, India

Tetsuya Konishi
Professor Emeritus
Niigata University of Pharmacy & Applied Life Sciences (NUPALS)
Tojima, Akiha-ku, Niigata, Japan
&
Director, Office HALD Food Function Research
Sakai, Nishi-ku, Niigata, Japan

Elsevier
Radarweg 29, PO Box 211, 1000 AE Amsterdam, Netherlands
The Boulevard, Langford Lane, Kidlington, Oxford OX5 1GB, United Kingdom
50 Hampshire Street, 5th Floor, Cambridge, MA 02139, United States

Library of Congress Cataloging-in-Publication Data
A catalog record for this book is available from the Library of Congress

British Library Cataloguing-in-Publication Data
A catalogue record for this book is available from the British Library

ISBN: 978-0-08-102081-4

For information on all Elsevier publications visit our website at
https://www.elsevier.com/books-and-journals

Publisher: Mica Haley
Acquisitions Editor: Anneka Hess
Editorial Project Manager: Michelle W. Fisher
Production Project Manager: Poulouse Joseph
Designer: Matthew Limbert

Typeset by TNQ Books and Journals

Contents

Section II
Leads From Natural Products

7. The Role of Natural Products From Plants in the Development of Anticancer Agents

Danielle Twilley and Namrita Lall

8. Plant Drugs in the Treatment of Osteoporosis

Sudhir Kumar and Rakesh Maurya

9. Phytodrugs and Immunomodulators for the Therapy of Leishmaniasis

C. Benjamin Naman, Ciro M. Gomes, and Gaurav Gupta

10. Natural Products Targeting Inflammation Processes and Multiple Mediators

G. David Lin and Rachel W. Li

11. Biologically Functional Compounds From Mushroom-Forming Fungi

Hirokazu Kawagishi

17. Herb and Drug Interaction

Nilanjan Ghosh, Rituparna C. Ghosh, Anindita Kundu, and Subhash C. Mandal

18. Toxicity Studies Related to Medicinal Plants

Kavimani Subramanian, Divya Sankaramourthy, and Mahalakshmi Gunasekaran

List of Contributors

Sugato Banerjee, Department of Pharmaceutical Sciences and Technology, Birla Institute of Technology, Ranchi, India

Souvik Basak, Dr. B.C. Roy College of Pharmacy and Allied Health Sciences, Durgapur, India

Tuhin K. Biswas, Department of Kayachikitsa (Medicine), J.B. Roy State Ayurvedic Medical College and Hospital, Kolkata, India

Anand K. Chaudhari, Centre of Advanced Study in Botany, Institute of Science, Banaras Hindu University, Varanasi, India

Jihang Chen, Hong Kong University of Science & Technology, Hong Kong SAR, China

Kavi B.S. Chouhan, Institute of Pharmacy, Guru Ghasidas Central University, Bilaspur, India

Somenath Das, Centre of Advanced Study in Botany, Institute of Science, Banaras Hindu University, Varanasi, India

Nawal K. Dubey, Centre of Advanced Study in Botany, Institute of Science, Banaras Hindu University, Varanasi, India

Anirban Dutta, ICAR-Indian Agricultural Research Institute, New Delhi, India

Abhishek K. Dwivedy, Centre of Advanced Study in Botany, Institute of Science, Banaras Hindu University, Varanasi, India

Mostafa Elachouri, Mohammed first University, Oujda, Morocco

Geetanjali, Department of Chemistry, Kirori Mal College, University of Delhi, Delhi, India

Nilanjan Ghosh, Dr. B.C. Roy College of Pharmacy and Allied Health Sciences, Durgapur, India

Rituparna C. Ghosh, Dr. B.C. Roy College of Pharmacy and Allied Health Sciences, Durgapur, India

Ciro M. Gomes, Hospital Universitário de Brasília, Brasília, Brazil

Mahalakshmi Gunasekaran, Department of Pharmacology, Mother Theresa Post Graduate and Research Institute of Health Sciences, Puducherry, India

Gaurav Gupta, Department of Immunology, University of Manitoba, Winnipeg, MB, Canada

Dharmik Joshi, Department of Pharmaceutical Sciences and Technology, Birla Institute of Technology, Ranchi, India

Hirokazu Kawagishi, Shizuoka University, Shizuoka, Japan

Kam M. Ko, Hong Kong University of Science & Technology, Hong Kong SAR, China

Sudhir Kumar, Medicinal and Process Chemistry Division, Central Drug Research Institute, Lucknow, India

Anindita Kundu, Department of Pharmaceutical Technology, Jadavpur University, Kolkata, India

Sukalyan K. Kundu, Department of Pharmacy, Jahangirnagar University, Dhaka, Bangladesh

Namrita Lall, Department of Plant and Soil Sciences, University of Pretoria, Pretoria, South Africa

Pou K. Leong, Hong Kong University of Science & Technology, Hong Kong SAR, China

Rachel W. Li, ANU Medical School, The Australian National University, Canberra, ACT, Australia

G. David Lin, Research School of Chemistry, The Australian National University, Canberra, ACT, Australia

Alex Loukas, Centre for Biodiscovery and Molecular Development of Therapeutics, Australian Institute of Tropical Health and Medicine, James Cook University, QLD, Australia

Anuradha S. Majumdar, Department of Pharmacology, Bombay College of Pharmacy, Mumbai, India

Abhishek Mandal, ICAR-Indian Agricultural Research Institute, New Delhi, India

Vivekananda Mandal, Institute of Pharmacy, Guru Ghasidas Central University, Bilaspur, India

Subhash C. Mandal, Department of Pharmaceutical Technology, Jadavpur University, Kolkata, India

Rakesh Maurya, Medicinal and Process Chemistry Division, Central Drug Research Institute, Lucknow, India

Dhrubojyoti Mukherjee, Department of Pharmaceutical Technology, Jadavpur University, Kolkata, India

C. Benjamin Naman, Center for Marine Biotechnology and Biomedicine, University of California, San Diego, California, United States

Ajay G. Namdeo, Poona College of Pharmacy, Bharati Vidyapeeth Deemed University, Pune, India

Sunday O. Otimenyin, Department of Pharmacology, University of Jos, Plateau State, Nigeria

Partha Palit, Department of Pharmaceutical Sciences, Assam University, Silchar, India

Somdatta Roy, Department of Pharmaceutical Sciences and Technology, Birla Institute of Technology, Ranchi, India

Shubhadeep Roychoudhury, Department of Life Science & Bio-Informatics, Assam University, Silchar, India

Supradip Saha, ICAR-Indian Agricultural Research Institute, New Delhi, India

Divya Sankaramourthy, Department of Pharmacology, Mother Theresa Post Graduate and Research Institute of Health Sciences, Puducherry, India

Ram Singh, Department of Applied Chemistry, Delhi Technological University, Delhi, India

Sahil J. Somani, Department of Pharmacology, RK University, Rajkot, India

Kavimani Subramanian, Department of Pharmacology, Mother Theresa Post Graduate and Research Institute of Health Sciences, Puducherry, India

Roshni Tandey, Institute of Pharmacy, Guru Ghasidas Central University, Bilaspur, India

Danielle Twilley, Department of Plant and Soil Sciences, University of Pretoria, Pretoria, South Africa

Phurpa Wangchuk, Centre for Biodiscovery and Molecular Development of Therapeutics, Australian Institute of Tropical Health and Medicine, James Cook University, QLD, Australia

Sujata Wangkheirakpam, Department of Chemistry, NIT Manipur, Imphal, India

Foreword

I feel genuinely honored in writing this foreword for the book *Natural Products and Drug Discovery: An Integrated Approach*, edited by three excellent scholars from the area of natural products research: Subhash C. Mandal, Vivekananda Mandal, and Tetusya Konishi. Natural products have evidently been one of the major sources of new drugs, and will continue to be so in the years to come. This very reason has prompted a huge body of research exploring natural products for new drugs to combat various ailments. There are several books and excellent review articles available to date covering various areas relating to natural products research, particularly the area of natural products drug discovery, but this book will stand out from the crowd probably because of its inclusive approach to integrating several aspects of natural products drug discovery processes in one book.

This book offers 23 chapters organized in three distinct sections: traditional medicine and drug discovery (six chapters), leads from natural products (nine chapters), and herbal drug research (eight chapters). All these chapters are written by experts from relevant areas of natural products drug discovery.

Natural Products and Drug Discovery: An Integrated Approach integrates several classical and modern aspects of drug discovery, from Chinese traditional medicine to Ayurvedic medicine, as well as modern aspects of drug discovery strategies, e.g., natural products lead discovery, and will act as an outstanding reference book for natural products researchers.

I wholeheartedly recommend this book to all who are interested in natural products drug discovery and related areas.

Professor Satyajit D. Sarker
Editor-in-Chief, Phytochemical Analysis
Director, School of Pharmacy and Biomolecular Sciences
Liverpool John Moores University
Liverpool L3 3AF
United Kingdom

Preface

Natural product research has become the leading force in the drug discovery sector. This fact has been further triggered due to the enormous risk and time involved in the synthetic route of drug discovery. Natural product research, though more complicated due to the complex mixtures involved, still offers a more successful rate when compared to synthetic drug discovery. From ancient histories it becomes evident that traditional medicine (Ayurveda, the Indian traditional system, traditional Chinese medicine, traditional Japanese medicine, etc.) has always been there to reduce the sufferings of human ailments, even before the advent of antibiotics. Today's drug discovery is no longer just a case of trial and error or mere serendipity but rather has become a more programmed and strategized venture. Drug discovery these days has become an integrated approach of modern biology and traditional medicine using a holistic approach. The modern tools of chemistry and biology—in particular, the various "-omics" technologies—now allow scientists to detail the exact nature of the biological effects of natural compounds on the human body, as well as to uncover possible synergies, which hold much promise for the development of new therapies against many devastating diseases. Henceforth, we cannot deny the shift of the scientific community more toward traditional medicines involving complementary and alternative therapies. Well-strategized ethnobotanically inspired natural product research can provide vital leads with the potential for developing them as future drug candidates. Henceforth, this is the perfect time to bring out a book that can act as a fuel to this driving force of drug discovery. This book serves as a "one-stop solution" for all beginners in the field of botanical research leading to drug discovery and is committed to fulfilling the needs of herbal drug researchers. The book is an amalgamation of 23 scientifically crafted chapters prioritized judiciously into three major groups. Through the various chapters, the book acts as a vital support system for natural product researchers where all issues pertaining to drug discovery from botanicals are dealt with under a single umbrella system. The book aims to dig deep into our cultural roots and extract the ancient science of different traditional systems of medicine practiced worldwide to try to integrate ancient knowledge with modern approaches for empowering the drug discovery process. Application of ethnopharmacology in developing preventive and clinical medicine is emphasized upon. On the other hand, the book also amalgamates different strategies and ideologies under one roof, presents a simplified approach of bioassay-guided fractionation and isolation, and showcases important traditional leads that can be explored for future drug discovery. Recent developments in the science of enzyme substrate reactions are highlighted and the role of in vitro techniques is exemplified in the process of drug discovery.

We humbly express our gratitude to our national and international funding agencies and home universities who have supported us in our journey of natural product research. We are also thankful to our peer review team for timely reviewing the manuscripts and providing valuable inputs. Finally, we express our deep gratitude to our family members for their constant support, particularly during the busy days of compiling this book.

Section I

Traditional Medicine and Drug Discovery

Chapter 1

Drug Discovery From *Ayurveda*: Mode of Approach and Applications

Tuhin K. Biswas
Department of Kayachikitsa (Medicine), J.B. Roy State Ayurvedic Medical College and Hospital, Kolkata, India

1. AYUSH AND *AYURVEDA*

Ayurveda, the Indian indigenous system of traditional and complementary medicine (T&CM), is one of the oldest categories of medical sciences, which has existed for thousands of years. It is predicted that *Ayurveda* originated as an outcome of the Indian old classical literature known as *Veda*, particularly from the fourth section *Atharva Veda*. *Ayurveda* is considered a comprehensive system of medical science that emphasizes the promotion of body physiology, prevention of diseases, and pacification of ailments by means of natural procedures. *Ayurveda* is the leading T&CM in India among all other systems commonly practiced in this country, such as the Siddha system of medicine, Unani, Yoga, and Homeopathy. All these systems come under the common umbrella of AYUSH (*Ayurveda*, Yoga, Unani, Siddha, and Homeopathy), which are nourished by the Government of India with a special branch of ministry. *Ayurveda* consists of eight major clinical specialties such as *Kāyachikitsa* (medicine), *Śalya tantra* (surgery), *Śalakyatantra* (diseases of the supraclavicular region), *Kaumārabhrtya* (pediatrics, obstetrics, and gynecology), *Bhūtavidyā* (psychiatry), *Agada tantra* (toxicology), *Rasāyana tantra* (rejuvenation and geriatrics), and *Vājkarana* (aphrodisiac and eugenics). Each and every specialty of *Ayurveda* has a separate vivid description of treatment in different classical texts. AYUSH emphasizes intervention in public health care particularly in the areas of epidemic diseases, geriatric health care, neglected diseases, mental health and cognitive disorders, immunological disorders, anemia and nutritional disorders, maternal and child health, and the study of constitution, temperament, and miasma. The role of the AYUSH system can be used alone or as an add-on treatment in the prevention and

Natural Products and Drug Discovery. https://doi.org/10.1016/B978-0-08-102081-4.00001-0

control of noncommunicable diseases, vector-borne diseases, systemic review and metaanalysis of AYUSH research studies, literary research, and the scientific documentation and development of databases [1]. The mission of AYUSH is the mainstreaming of its branches at all levels in the health care system (primary, secondary, or tertiary), to improve access to and quality of public health delivery, and to focus on the promotion of health and prevention of diseases [2].

2. CHRONOLOGICAL GENESIS OF AYURVEDIC DRUGS FOR THERAPEUTIC APPLICATION

There has been a long history of the genesis of drugs from natural resources in *Ayurveda* since the *Vedic* period (c. 3000 BC). In spite of this there was no systematic therapeutic guideline in the *Veda*; however, two types of therapeutics are described broadly as drug (Ausadha) therapy and natural therapy. A total of 107 medicinal plants are mentioned in the *Veda* under drug therapy, which is available in different seasons [3]. However, no detailed descriptions are mentioned in the *Veda* for specific use of these medicinal plants. The classical description of drug development from plant sources was first scientifically described in classical Ayurvedic texts such as *Charaka Samhita* and *Sushruta Samhita* (c. 1500–1000 BC). There are detailed descriptions of over 700 herbs and their specific therapeutic uses in these two texts [4]. Most of these medicinal plants were used in multiple combinations (polyherbal). A total of 50 groups (*Kasaya*) of medicinal plants, comprising 10 medicinal plants in each group, were described in *Charaka Samhita* [5] according to their therapeutic activities. There were repetitions of some medicinal plants in other groups and nomenclature of some of the medicinal plants was done in multiple names, probably to signify their different actions. These groups of drugs have both preventive and curative aspects of therapeutics. In addition, there is also a description of certain medicinal plants that are described as having health-promoting activities. Apart from the description of groups of drugs, there is also mention of medicinal plants in combined form for the treatment of different diseases such as fever, diarrhea, dyspnea, diabetes, etc. in separate chapters. The categorization of medicinal plants in *Sushruta Samhita* is somehow different when compared with the view of *Charaka Samhita*. Medicinal plants are categorized in 37 groups (*Gana*) in *Sushruta Samhita* [6]. This categorization was done in accordance with their origin or similar morphological characters. However, none of the groups in *Charaka Samhita* or in *Sushruta Samhita* follows the modern botanical system of nomenclature or taxonomy. Likewise, in *Charaka Samhita*, drugs of plant origin are described in *Sushruta Samhita* separately in combined form in different chapters for the treatment of various diseases. There are some similarities in the selection of medicinal plants in groups between *Charaka Samhita* and *Sushruta Samhita* but all plants in all groups categorized by each author aredifferent (Table 1.1). Later on, a more scientific compilation was drawn in another classical text of

TABLE 1.1 Comparative Study of the Group of Medicinal Plants Mentioned in *Charaka Samhita* and *Sushruta Samhita*

SI	*Charaka Samhita* [5]		Corresponding Group (*Gana*) in *Sushruta Samhita* [8]
	Name of Group (*Kasaya*)	Plants Included (10 in Each Group) (Sanskrit and Botanical Names)	
1	*Jeevaneeya* (invigorators)	*Jivaka,*[a] *Risabhaka,*[a] *Meda,*[a] *Mahameda,*[a] *Kakoli,*[a] *Kshirakakoli,*[a] *Mudgaparni* (*Phaseolus trilobus*), *Masaparni* (*Teramnus labialis*), *Jeevanti* (*Leptadenia reticulata*), *Madhuka* (*Glycyrrhiza glabra*)	Kakolyadi
2	*Vrimhaneeya* (nourishing)	*Kshirini* (*Mimusops hexandra*), *Rajaksavaka* (*Euphorbia microphylla*), *Aswagandha* (*Withania somnifera*), *Kakoli,*[a] *Kshirakakoli,*[a] *Vatyayani* (*Sida rhombifolia*), *Bhadradaruni* (*Sida cordifolia*), *Bharadvaji* (*Thespesia lampas*), *Payasya* (*Ipomoea paniculata*), *Risyagandha*[a]	Vidarigandhadi
3	*Lekhaneeya* (antiobese)	*Musta* (*Cyperus rotundus*), *Kustham* (*Saussurea lappa*), *Haridra* (*Curcuma longa*), *Daruharidra* (*Berberis aristata*), *Vacha* (*Acorus calamus*), *Ativisa* (*Aconitum heterophyllum*), *Katurohini* (*Picrorhiza kurroa*), *Chitraka* (*Plumbago zeylanica*), *Chirabilva* (*Pongamia pinnata*), *Haimavati* (*Iris versicolor*)	Mustadi
4	*Bhedaneeya* (cathartic)	*Suvaha* (*Operculina turpethum*), *Arka* (*Calotropis gigantea*), *Urubuka,*[a] *Agnimukhi* (*Gloriosa superba*), *Chitra* (*Baliospermum montanum*), *Chitraka* (*P. zeylanica*), *Chirabilva* (*P. pinnata*), *Sankhini* (*Canscora decussata*), *Sakuladini* (*P. kurroa*), *Swarnakshirini* (*Argemone mexicana*)	Shyamadi
5	*Sandhaneeya* (wound healing)	*Madhuka* (*G. glabra*), *Madhuparni* (*Tinospora cordifolia*), *Prisniparni* (*Uraria picta*), *Ambasthaka* (*Cissampelos pareira*), *Samanga* (*Rubia cordifolia*), *Mocharasa* (*Salmalia malabarica*), *Dhataki* (*Woodfordia fruticosa*), *Lodhra* (*Symplocos racemosa*), *Priyangu* (*Callicarpa macrophylla*), *Katphala* (*Myrica nagi*)	Ambasthadi or Priyangadi
6	*Dipaneeya* (digestive stimulant)	*Pippali* (*Piper longum*), *Pippalimula* (root of *P. longum*), *Cavya* (*Piper chaba*), *Chitraka* (*P. zeylanica*), *Shringavera* (*Zingiber officinalis*), *Amlavetasa* (*Rheum emodi*), *Maricha* (*Piper nigrum*), *Ajamoda* (*Trachyspermum roxburghianum*), *Bhallataka* (*Semecarpus anacardium*), *Hingu* (*Ferula narthex*)	Pippalyadi

Continued

TABLE 1.1 Comparative Study of the Group of Medicinal Plants Mentioned in *Charaka Samhita* and *Sushruta Samhita*—cont'd

	Charaka Samhita [5]		Corresponding Group (*Gana*) in *Sushruta Samhita* [8]
Sl	Name of Group (*Kasaya*)	Plants Included (10 in Each Group) (Sanskrit and Botanical Names)	
7	*Valya* (promotes strength)	*Aindri* (*Citrullus colocynthis*), *Risabhi*,[a] *Atirasa* (*Asparagus racemosus*), *Risyaprokta* (*T. labialis*), *Payasya* (*I. paniculata*), *Aswagandha* (*W. somnifera*), *Sthira* (*Desmodium gangeticum*), *Rohoni* (*P. kurroa*), *Bala* (*S. cordifolia*), *Atibala* (*S. rhombifolia*)	Laghu Panchamula
8	*Varnya* (complexion promoter)	*Chandana* (*Santalum album*), *Tunga* (*Calophyllus inophyllum*), *Padmaka* (*Prunus cerasoides*), *Usira* (*Vetiveria zizanioides*), *Madhuka* (*G. glabra*), *Manjistha* (*R. cordifolia*), *Sariva* (*Hemidesmus indicus*), *Payasya* (*I. paniculata*), *Sita* (white variety of *Cynodon dactylon*), *Lata* (green variety of *C. dactylon*)	Eladi
9	*Kanthya* (voice cleanser)	*Sariva* (*H. indicus*), *Ikshu* (*Saccharum indicum*), *Madhuka* (*G. glabra*), *Pilppali* (*P. longum*), *Draksha* (*Vitis vinifera*), *Vidari* (*I. panniculata*), *Kaitarya* (*M. nagi*), *Hamsapadi* (*Adiantum lunulatum*), *Brihati* (*Solanum indicum*), *Kantikarika* (*Solanum xanthocarpum*)	–
10	*Hridya* (cardiotonic)	*Amra* (*Mangifera indica*), *Amrataka* (*Spondias pinnata*), *Likucha* (*Artocarpus lakoocha*), *Karamarda* (*Carissa carandas*), *Vriksamla* (*Tamarindus indica*), *Amlavetasa* (*R. emodi*), *Kuvala* (*Ziziphus jujuba*), *Dadimba* (*Punica granatum*), *Matulunga* (*Citrus decumana*)	Paruskadi
11	*Triptighna* (contentment promoter)	*Nagara* (*Z. officinalis*), *Chavya* (*P. chaba*), *Chitraka* (*P. zeylanica*), *Vidanga* (*Embelia ribes*), *Murva* (*Clematis triloba*), *Guduchi* (*T. cordifolia*), *Vacha* (*A. calamus*), *Mustaka* (*C. rotundus*), *Pippali* (*P. longum*), *Patola* (*Trichosanthes cucumerina*)	Patoladi
12	Arshoghna (antihemorrhoidal)	*Kutaja* (*Holarrhena antidysenterica*), *Bilva* (*Aegle marmelos*), *Chitraka* (*P. zeylanica*), *Nagara* (*Z. officinalis*), *Ativisa* (*A. heterophyllum*), *Abhaya* (*Terminalia chebula*), *Dhanyaska* (*Fagonia cretica*), *Daruharidra* (*B. aristata*), *Vacha* (*A. calamus*), *Chavya* (*P. chaba*)	Muskakadi

13	*Kusthaghna* (drugs that cure obstinate skin diseases)	*Khadira* (*Acacia catechu*), *Abhaya* (*T. chebula*), *Amalaki* (*Phyllanthus emblica*), *Hridra* (*C. longa*), *Aruskara* (*S. anacardium*), *Saptakarna* (*Alstonia scholaris*), *Aragvadha* (*Cassia fistula*), *Karavira* (*Nerium indicum*), *Vidanga* (*E. ribes*), *Jatipravala* (*Jasminum officinalis*)	Aragvadhadi, Salasaradi, Arkadi, Laksadi
14	*Kandughna* (antipruritic)	*Chandana* (*S. album*), *Nalada* (*Nardostachys jatamansi*), *Kritamala* (*C. fistula*), *Naktamala* (*P. pinnata*), *Nimba* (*Azadirachta indica*), *Kutaja* (*H. antidysenterica*), *Sarsapa* (*Brassica nigra*), *Madhuka* (*G. glabra*), *Daruharidra* (*B. aristata*), *Mustaka* (*C. rotundus*)	Eladi, Aragvadhadi
15	*Krimighna* (antiinfective and anthelmintic)	*Aksiva* (*Moringa oleifera*), *Maricha* (*P. nigrum*), *Gandira* (*Euphorbia antiquorum*), *Kebuka*,[a] *Vidanga* (*E. ribes*), *Nirgundi* (*Vitex negundo*), *Kinihi* (*Achyranthes aspera*), *Svadamstra* (*Tribulus terrestris*), *Vrisaparnika* (*Ipomoea reniformis*), *Akuparnika*[a]	Surasadi, Laksadi
16	*Visaghna* (antitoxin)	*Haridra* (*C. longa*), *Manjistha* (*R. cordifolia*), *Suvaha* (*Pluchea lanceolata*), *Sukshmaila* (*Elettaria cardamomum*), *Palindi*,[a] *Chandana* (*S. album*), *Kataka* (*Strychnos potatorum*), *Sirisa* (*Albizia lebbeck*), *Sindhuvara* (*V. nigundo*), *Slesmataka* (*Cordia dichotoma*)	Rodhradi, Aragvadhadi, Arkadi, Anjanadi
17	*Stanyajana* (galactogogue)	*Virana* (*Vetiveria zizanioides*), *Sali* (*Oryza sativa*), *Sastika* (another variety of *O. sativa*), *Iksuvalika* (*Asteracantha longifolia*), *Darbha* (*Desmostachya bipinnata*), *Kusa* (a variety of *D. bipinnata*), *Kasa* (*Saccharum spontaneum*), *Gundra* (*Saccharum sara*), *Itkata*,[a] *Kattrina* (*Cymbopogon schoenanthus*)	Kakolyadi
18	*Stanyashodhana* (galactopurificatory)	*Patha* (*C. pareira*), *Mahasudha* (*Z. officinalis*), *Suradaru* (*Cedrus deodara*), *Musta* (*C. rotundus*), *Murvi* (*C. triloba*), *Guduchi* (*T. cordifolia*), *Vatsaka* (*H. antidysenterica*), *Kiratikta* (*Swertia chirata*), *Katurohini* (*P. kurroa*), *Sariva* (*H. indicus*)	Mustadi, Vachadi, Haridradi
19	*Shukrajana* (spermatopoietics)	*Jivaka*,[a] *Risabhaka*,[a] *Kakoli*,[a] *Kshirkakoli*,[a] *Mudgaparni* (*P. trilobus*), *Masaparni* (*T. labialis*), *Meda*,[a] *Vriddharuha* (*A. racemosus*), *Jatila* (*N. jatamansi*), *Kulinga*[a]	Kakolyadi
20	*Shukrasodhana* (spermatopurificators)	*Kustha* (*S. lappa*), *Elavaluka* (*Prunus cerasus*), *Katphala* (*M. nagi*), *Samudraphena* (*Sepia officinalis*), *Kadamba* (*Saccharum officinalis*), *Kandeksu* (*S. spontaneum*), *Iksuraka* (*A. longifolia*), *Vasuka* (*Indigofera enneaphylla*), *Usira* (*V. zizanioides*)	Valli Panchamula, Kantaka Panchamula

Continued

TABLE 1.1 Comparative Study of the Group of Medicinal Plants Mentioned in *Charaka Samhita* and *Sushruta Samhita*—cont'd

	Charaka Samhita [5]		Corresponding Group (*Gana*) in *Sushruta Samhita* [8]
Sl	Name of Group (*Kasaya*)	Plants Included (10 in Each Group) (Sanskrit and Botanical Names)	
21	*Snehopaga* (adjuvant of unction)	*Mrdvika* (*V. vinifera*), *Madhuka* (*G. glabra*), *Madhuparni* (*T. cordifolia*), *Meda*,[a] *Vidari* (*I. paniculata*), *Kakoli*, *Kshirkakoli*,[a] *Jivaka*,[a] *Jivanti* (*L. reticulata*), *Salaparni* (*D. gangeticum*)	–
22	*Swedopaga* (adjuvant for fomentation)	*Sobhanjanak* (*M. oleifera*), *Eranda* (*Ricinus communis*), *Arka* (*C. gigantean*), *Vrischika* (white *Boerhavia diffusa*), *Punarnava* (red *B. diffusa*), *Yava* (*Hordeum vulgare*), *Tila* (*Sesamum indicum*), *Kulattha* (*Dolichos biflorus*), *Masa* (*Phaseolus mungo*), *Badara* (*Z. jujuba*)	–
23	*Vamanopaga* (emesis inducer)	*Honey*, *Madhuka* (*G. glabra*), *Kovidara* (red variety of *Bauhinia variegata*), *Karbudara* (white variety of *B. variegata*), *Nipa* (*Anthocephalus indicus*), *Vidula* (*Barringtonia acutangula*), *Bimbi* (*Coccinia indica*), *Sanapuspi* (*Crotalaria verrucosa*), *Sadapuspa* (*C. gigantean*), *Pratyakpuspa* (*A. aspera*)	–
24	*Virechanopaga* (purgative inducer)	*Draksha* (*V. vinifera*), *Kasamarya* (*Gmelina arborea*), *Paruska* (*Grewia asiatica*), *Abhaya* (*T. chebula*), *Amalaki* (*P. emblica*), *Bibhitaka* (*Terminalia bellirica*), *Kuvala* (*Zizyphus sativa*), *Badara* (*Z. jujuba*), *Karkandhu* (*Ziziphus nummularia*), *Pilu* (*Salvadora persica*)	Parusakadi
25	*Asthapanopaga* (enema inducer)	*Trivrita* (*O. turpethum*), *Bilva* (*A. marmelos*), *Pippali* (*P. longum*), *Kustha* (*S. lappa*), *Sarsapa* (*B. nigra*), *Vacha* (*A. calamus*), *Vatsaka* (*H. antidysenterica*), *Satapuspa* (*Foeniculum vulgare*), *Madhuka* (*G. glabra*), *Madana* (*Randia dumetorum*)	–
26	*Anuvasanapaga* (variety of enema inducer)	*Rasna* (*P. lanceolata*), *Suradaru* (*C. deodara*), *Bilva* (*A. marmelos*), *Madana* (*R. dumetorum*), *Satapuspa* (*F. vulgare*), *Vrischira* (white variety of *B. diffusa*), *Punarnava* (red variety of *B. diffusa*), *Svadamstra* (*Tribulus terristris*), *Agnimantha* (*Clerodendrum phlomidis*), *Syonaka* (*Oroxylum indicum*)	–

27	*Shirovirechanopaga* (drugs that induce cleaning of the brain)	*Jyutismati* (*Celastrus paniculatus*), *Ksvaka* (*Centipeda minima*), *Maricha* (*P. nigrum*), *Pippali* (*P. longum*), *Vidanga* (*E. ribes*), *Shigru* (*Moringa oleifera*), *Sarsapa* (*B. nigra*), *Apamarga* (*A. aspera*), *Sveta* (white variety of *Clitoria ternatea*), *Mahasveta* (blue variety of *C. ternatea*)	–
28	*Chardinigaraha* (antiemetic)	*Jambu* (*Syzygium cumini*), *Amra* (*M. indica*), *Matulunga* (*C. decumana*), *Badara* (*Z. jujuba*), *Dadima* (*P. granatum*), *Yava* (*H. vulgare*), *Yastika* (*G. glabra*), *Usira* (*V. zizanioides*), *Mrit* (soil), *Laja* (fried paddy)	Nyagrodhadi
29	*Trishnanigraha* (thirst restrainers)	*Nagara* (*Z. officinalis*), *Dhanvayasaka* (*F. cretica*), *Musta* (*C. rotundus*), *Parpataka* (*Fumaria parviflora*), *Chandana* (*S. album*), *Kiratiktaka* (*S. chirata*), *Guduchi* (*T. cordifolia*), *Hriver* (*Pavonia odorata*), *Dhanyaka* (*Coriandrum sativum*), *Patola* (*T. cucumerina*)	*Guduchyadi, Utpaladi, Sarivadi, Paruskadi*
30	*Hikkanigrah* (hiccough restrainers)	*Sati* (*Hedychium spicatum*), *Puskaramula* (*Inula racemosa*), *Badara* (*Z. jujuba*), *Kantikarika* (*S. xanthocarpum*), *Brihati* (*S. indicum*), *Vriksaruha* (*Dendrophthoe falcata*), *Abhaya* (*T. chebula*), *Pippali* (*P. longum*), *Duralabha* (*F. cretica*), *Kulishrimgi* (*Rhus succedenea*)	*Brihatyadi, Vidaragandhadi*
31	*Purishasamgrahaniya* (bowel binder)	*Priyangu* (*C. macrophylla*), *Ananat* (*H. indicus*), *Amra* (*M. indica*), *Katvanga* (*O. indicum*), *Lodhra* (*S. racemosa*), *Mocharasa* (*S. malabarica*), *Samanga* (*Mimosa pudica*), *Dhataki* (*W. fruticosa*), *Padmaa* (*Clerodendrum serratum*), *Padma* (*Nelumbo nucifera*)	*Rodhradi, Priyangadi, Ambasthadi*
32	*Purisavijaneeya* (bowel color inducer)	*Jambu* (*S. cumini*), *Shallaki* (*Boswellia serrata*), *Kacchura*,[a] *Madhuka* (*Madhuca indica*), *Salmali* (*S. malabarica*), *Shrivestaka* (*Pinus roxburghii*), *Payasya* (*I. paniculata*), *Utpala* (*Nymphaea alba*)	*Nyagrodhadi*
33	*Mutrasamgrahaneeya* (antidiuretic)	*Jambu* (*S. cumini*), *Amra* (*M. indicum*), *Plaksha* (*Ficus lacor*), *Vad* (*Ficus benghalensis*), *Kapitana* (*A. lebbeck*), *Udumbara* (*Ficus racemosa*), *Asvattha* (*Ficus religiosa*), *Bhallataka* (*S. anacardium*), *Asmantaka* (*Bauhinia racemosa*), *Somavalka* (*A. catechu*)	*Nyagrodhadi, Surasaladi*

Continued

TABLE 1.1 Comparative Study of the Group of Medicinal Plants Mentioned in *Charaka Samhita* and *Sushruta Samhita*—cont'd

	Charaka Samhita [5]		
Sl	**Name of Group (*Kasaya*)**	**Plants Included (10 in Each Group) (Sanskrit and Botanical Names)**	**Corresponding Group (*Gana*) in *Sushruta Samhita* [8]**
34	*Mutravirajaneeya* (urinary antiseptic)	*Padma* (*N. nucifera*), *Utpala* (*N. alba*), *Nalina* (variety of *N. nucifera*), *Kumuda* (*N. alba*), *Saugandhika*,[a] *Pundarika* (red variety of *Nymphaea lotus*), *Satapatra* (variety of *N. nucifera*), *Madhuka* (*G. glabra*), *Priyangu* (*C. macrophylla*), *Dhataki* (*W. fruticosa*)	*Utpaladi*
35	*Mutravirechaneeya* (diuretrics)	*Vrikshadani* (*D. falcata*), *Svadamdstra* (*T. terristris*), *Vasuka* (*I. enneaphylla*), *Vasira* (*Gynandropsis gynandra*), *Pasanabheda* (*Bergenia ligulata*), *Darbha* (*D. bipinnata*), *Kusa* (another variety of *D. bipinnata*), *Hasa* (*S. spontaneum*), *Gundra* (*S. sara*), *Itkata*[a]	*Trina Panchamula*, *Virataradi*
36	*Kasaharani* (drugs to control cough)	*Draksha* (*V. vinifera*), *Abhaya* (*T. chebula*), *Amalaki* (*P. emblica*), *Pippali* (*P. longum*), *Duralabha* (*F. cretica*), *Shrimgi* (*Rhus succedanea*), *Kantakarika* (*S. xanthocarpum*), *Vrischira* (white *B. diffusa*), *Punarnava* (red *B. diffusa*), *Tamalaki* (*Phyllanthus niruri*)	*Vidaragandhadi*
37	*Swasahara* (antidyspnea)	*Sati* (*Hedychium spicatum*), *Puskaramula* (*I. racemosa*), *Amlavetasa* (*R. emodi* sp.), *Ela* (*E. cardamomum*), *Hingu* (*F. narthex*), *Aguru* (*Aquilaria agallocha*), *Surasa* (*Ocimum sanctum*), *Tamalaki* (*P. niruri*), *Jivanti* (*L. reticulata*), *Canda*[a]	*Pippyaladi*
38	*Swathuhara* (antiinflammatory)	*Paatalaa* (*Stereospermum suaveolens*), *Agnimantha* (*C. phlomidis*), *Syonaka* (*O. indicum*), *Bilva* (*A. marmelos*), *Kasamarya* (*G. arborea*), *Kantakarika* (*S. xanthocarpum*), *Brihati* (*S. indicum*), *Salaparni* (*D. gangeticum*), *Prisnioarni* (*U. picta*), *Gokshuraka* (*T. terrestris*)	*Dasamula*

39	*Jwarahara* (antipyretic)	*Sariva* (*H. indicus*), sugar, *Patha* (*C. pareira*), *Manjistha* (*R. cordifolia*), *Draksha* (*V. vinifera*), *Pilu* (*S. persica*), *Parusaka* (*G. asiatica*), *Abhaya* (*T. chebula*), *Amalaki* (*P. emblica*), *Vibhitaka* (*T. belerica*)	*Sarivadi, Patoladi, Amalakyadi*
40	*Shramahara* (fatigue dispelling)	*Draksha* (*V. vinifera*), *Kharjura* (*Phoenix sylvestris*), *Priyala* (*Buchanania lanzan*), *Badara* (*Z. jujuba*), *Dadima* (*P. granatum*), *Phalgu* (*Ficus hispida*), *Parusaka* (*G. asiatica*), *Iksu* (*S. officinarum*), *Yava* (*H. vulgare*), *Sastika* (*O. sativa*)	*Paruskadi*
41	*Dahaprashamana* (drugs that control burning sensation)	*Laja* (dried paddy), *Chandana* (*S. album*), *Kasamarya* (*G. arborea*), *Madhuka* (*M. indica*), sugar, *Nilotpala* (*Nymphaea stellata*), *Usira* (*V. zizanioides*), *Sariva* (*H. indicus*), *Guduchi* (*T. cordifolia*), *Hribera* (*P. odorata*)	*Sarivadi, Utpaladi, Anjanadi*
42	*Shitaparamana* (drugs that control chill)	*Tagara* (*Valeriana welchii*), *Aguru* (*A. agallocha*), *Dhanyaka* (*C. sativum*), *Shrimgavera* (*Z. officinalis*), *Bhutika* (*Trachyspermum ammi*), *Vacha* (*A. calamus*), *Kantakari* (*S. xanthocarpum*), *Agnimantha* (*C. phlomidis*), *Syonaka* (*O. indicus*), *Pippali* (*P. longum*)	*Pippyaladi, Surasadi*
43	*Udardaprashamana* (antiurticaria)	*Vidaragandha* (*D. gangeticum*), *Priyala* (*B. lanzan*), *Badara* (*Z. jujuba*), *Khadira* (*A. catechu*), *Saptaparna* (*A. scholaris*), *Asvakarna* (*Dipterocarpus alatus*), *Arjuna* (*Terminalia arjuna*), *Asana* (*Terminalia tomentosa*), *Arimeda* (variety of *A. catechu*)	*Salasaradi*
44	*Angamardaprashamana* (drugs to cure malaise)	*Vidaragandha* (*D. gangeticum*), *Prisniparni* (*U. picta*), *Brihati* (*S. indicum*), *Kantakarika* (*S. xanthocarpum*), *Eranda* (*Ricinus communis*), *Kakoli*,[a] *Chandana* (*S. album*), *Usira* (*V. zizanioides*), *Ela* (*E. cardamomum*), *Madhuka* (*G. glabra*)	*Vidaragandhadi*
45	*Shulaprashamana* (drugs to cure colic pain)	*Pippali* (*P. longum*), *Pippali mula* (root of *P. longum*), *Chabya* (*P. chaba*), *Chitraka* (*P. zeylanica*), *Shrimgavera* (*Z. officinalis*), *Maricha* (*P. nigrum*), *Ajamoda* (*T. roxburghianum*), *Ajagandha* (*G. gynandra*), *Ajaji* (*Cuminum cyminum*), *Gandira* (*Euphorbia antiquorum*)	*Pippyaladi*
46	*Shonitasthapana* (hemostatics)	Honey, *Madhuka* (*G. glabra*), *Rudhira* (*Crocus sativa*), *Mocharasa* (resin of *S. malabarica*), earth pot pieces, *Lodhra* (*S. racemosa*), *Girika* (*Ferrum hematite*), *Priyangu* (*C. macrophylla*), sugar, *Laja* (fried paddy)	*Priyangadi, Anjanadi*

Continued

TABLE 1.1 Comparative Study of the Group of Medicinal Plants Mentioned in *Charaka Samhita* and *Sushruta Samhita*—cont'd

	Charaka Samhita [5]		
Sl	**Name of Group (*Kasaya*)**	**Plants Included (10 in Each Group) (Sanskrit and Botanical Names)**	**Corresponding Group (*Gana*) in *Sushruta Samhita* [8]**
47	*Vedanasthapana* (sedatives)	*Sala* (*Shorea robusta*), *Katphala* (*M. nagi*), *Kadamba* (*A. indicus*), *Padmaka* (*P. cerasoides*), *Tumba* (*Zanthoxylon alatum*), *Mocharasa* (resin of *S. malabarica*), *Sirisha* (*A. lebbeck*), *Vanjual* (*Salix caprea*), *Elavaluka* (*P. cerasus*), *Asoka* (*Saraca indica*)	*Rodhradi*
48	*Samgyasthapana* (restorative of consciousness)	*Hingu* (*F. narthex*), *Kaitarya* (*Murraya koenigii*), *Arimeda* (variety of *A. catechu*), *Vacha* (*A. calamus*), *Koraka* (*Angelica glauca*), *Vayastha* (*Bacopa monnieri*), *Golomi* (variety of *A. calamus*), *Jatila* (*N. jatamansi*), *Palankaasa* (*Commiphora wightii*), *Ashokarohini* (*P. kurroa*)	*Priyangadi*
49	*Prajasthapana* (fertility promoter)	*Aindri* (*C. colocynthis*), *Brahmi* (*Bacopa monnieri*), *Satavirya* (*C. dactylon*), *Sahasravirya* (variety of *C. dactylon*), *Amogha* (*P. emblica*), *Avyatha* (*T. cordifolia*), *Ashiva* (*T. chebula*), *Aristaa* (*P. kurroa*), *Vaatyapuspi* (*S. rhombifolia*), *Visvaksenakanta* (*C. macrophylla*)	*Vidaragandhadi*, *Kakolyadi*
50	*Vayasthapaka* (antiaging)	*Amrita* (*T. cordifolia*), *Abhaya* (*T. chebula*), *Dhatri* (*P. emblica*), *Mukta* (pearl), *Sveta* (variety of *C. ternatea*), *Jivanti* (*L. reticulata*), *Atirasa* (*A. racemosus*), *Mandukaparni* (*Centella asiatica*), *Sthira* (*D. gangetium*), *Punarnava* (*B. diffusa*).	*Kakolyadi*, *Vidaragandhadi*

[a] *Plants are now endangered species and not scientifically evaluated for nomenclature.*

Ayurveda, the *Astanga Hridaya* (c. AD 600). In this text, 10 groups of drugs were designed on the basis of the intensity of diseases, availability of such ingredients, dosage form, and suitability to patients according to their body constitutions (*Prakriti*). The combinations of medicinal plants in some groups can be found in multiple variations. The grouping process of *Astanga Hridaya* [7] is the compilation of both *Charaka Samhita* and *Sushruta Samhita*. In addition to three major texts of *Ayurveda*, namely, *Charaka Samhita*, *Sushruta Samhita*, and *Astanga Hridaya*, descriptions of new medicinal plants were included for specific therapeutic purposes in different classical texts after the period of *Astanga Hridaya*. These texts are known by the name of *Nighantus* (texts dealing with descriptions of drugs of plant origin). During this time various preparations from plant drugs were developed, which are described in *Sharangadhara Samhita*. In this text five basic forms of plant drugs are mentioned, such as fresh juice (*Swarasa*), paste (*Kalka*), cold infusion (*Shita*), hot infusion (*Shrita*), and decoction (*Fanta*). The dose of each type of formulation was described according to the body strength of the patients and virulence of the disease process. In this text, descriptions are available regarding the composition of similar kinds of plant drugs having similar activities and qualities, such as *Trikatu* (*Piper longum*, *Piper nigrum*, and *Zingiber officinalis*) as an appetizer, *Triphala* (*Terminalia chebula*, *Terminalia belerica*, and *Phyllanthus emblica*) as a purgative, *Trisugandhi* (*Elettaria cardamomum*, *Cinnamomum zeylanicum*, and *Cinnamomum tamala*) as a mouth cleanser, etc. [8] In general, it has been reported that a total of 1587 medicinal plants are mentioned in different texts of *Ayurveda*. There are many descriptions of similar medicinal plants in various texts. Individually, 341 medicinal plants are described in *Charaka Samhita*, 395 in *Sushruta Samhita*, and 902 in *Astanga Hridaya*. However, out of these large numbers of medicinal plants described in different texts, many are not found today. Finally, the Ayurvedic Pharmacopoeia of India (API), published by the Government of India, has quoted approximately 395 available medicinal plants. In one study, DNA fingerprinting and sequencing of 347 medicinal plants out of a total of 395 medicinal plants of the API for proper authentication were done using rbcL (ribulose-bisphosphate carboxylase gene) techniques [9].

The introduction of metals and minerals as drugs in *Ayurveda* was classically performed by the great Ayurvedic scientist Nagarjuna around 600 BC. Among many inventions, the most important contribution of Nagarjuna was the special process of purification and preparation of drugs from mercury, iron, zinc, gold, silver, copper, magnesium, etc. Nagarjuna also pioneered the introduction of animal products as drugs in *Ayurveda*, such as conch shell, coral, pearl, etc. Since the introduction of metals, minerals, and animal products in health care, the drugs in *Ayurveda* took on a new shape by combining medicinal plants on the one hand and minerals, metals, and animal products on the other hand for various therapeutic purposes. The basic principles of pharmacodynamics and pharmacokinetics of Ayurvedic drugs depend upon certain factors as described in *Ayurveda*, such as *Rasa* (taste or sense of gustatory organ), *Guna* (qualities of the ingredient), *Veerya* (potency), *Vipaka* (effect of drugs after their assimilation in the body), and *Prabhava* (specific action). Leads from these sources will definitely help to unveil the discovery of new drugs from *Ayurveda*.

3. FUNDAMENTAL PRINCIPLES OF PERSONALIZED MEDICINE, GENETIC STUDY, AND APPLIED ASPECTS OF AYURVEDIC PHARMACODYNAMICS

Ayurvedic science for the diagnosis of disease and therapeutic designs depends upon the fundamental principles of three basic elements: *Vata*, *Pitta*, and *Kapha*. These three in combination are known as *Tridosa* or three physiological units. According to *Ayurveda*, human beings are classified into certain basic categories known as *Prakriti* on the basis of *Tridosa*, which expresses the phenotypic characters of individual variations. In general, *Vata dosa* represents kinetic activity, *Pitta dosa* represents source of energy for digestion and metabolism, and *Kapha dosa* represents potentiality. There are seven basic types of *Prakriti* or phenotypic expressions: three individual examples of *dosa*, three combinations with two *dosas*, and a combination of three *dosas* simultaneously. Characteristic features of seven variations of *Prakriti* or individualized cohorts are identified with different physical and psychological phenomena. The prevalence of diseases in a specific variety of *Prakriti* is also very specific. Susceptibility of diseases and therapeutic modules also depends upon these individualizations. Among the seven types, *Vata*, *Pitta*, and *Kapha* are the three phenotypic extremes that are readily distinguishable. Many of the phenotypic features that distinguish the predominant *Prakriti* types overlap with attributes described for human adaptations. Molecular and genomic evidence is provided for the differences between contrasting constitution types from a genetically homogeneous background. Common variations from a subset of differentially expressed genes are also partitioned differently between the phenotypically stratified *Prakriti* groups. Analysis of EGLN1 variation of gene helped to identify hypoxia as one of the axes and captured the genetic marker attributable to specific constitutions for high-altitude hypoxic adaptation. It has been hypothesized that integration of the comprehensive phenotyping method of *Ayurveda* with genomics might provide scaffolds to connect major axes of variation to an individual's phenomenon. This has been analyzed in a study of a set of ~2800 single nucleotide polymorphisms (SNP), represented in the Indian Genome Variation Consortium panel. Identification of common variations differs between healthy individuals of contrasting *Prakriti* types in genes that govern blood cell traits, hemostasis, metabolism, lipid homeostasis, etc. A study of a group of people showed that hypoxia (EGLN1) to hemostasis (VWF) and red blood cell traits (SPTA1) was found in a *Pitta* constitution-specific manner [10]. EGLN1 is responsible for hypoxia and because it is comparable in *Pitta prakriti*, therapeutic modules can be designed on the basis of *Pitta*-oriented drugs, diets, and regimens. Expression and genetic analysis of healthy individuals phenotyped using the principles of *Ayurveda* could uncover genetic variations that are associated with adaptation to external environments and susceptibility to diseases [11]. There are several factors that may govern the activities of *Vata*, *Pitta*, and *Kapha* such as geographical variations (*Desha*), seasonal variations (*Kala*), familial background (*Kula*), racial traits (*Jati*), age groups (*Vayah*), and individualism (*Pratyatmaniyatah*).

4. CLASSIFICATION OF AYURVEDIC THERAPEUTICS

Ayurvedic therapeutics is aimed mainly at two objectives: preventive (*Swasthsyorjaskara*) and curative (*Artasya roganut*) procedures. One of the important aspects of preventive therapy is promotion of health or promotive therapy. Nowadays, great importance is placed on promotive therapy by various ways and means. These varied groups of therapeutics are described vividly in different classical texts of *Ayurveda* such as *Charaka Samhita* (c. 1500 BC), *Sushruta Samhita* (c. 1000 BC), *Astanga Hridaya* (c. AD 600), *Sharangadhara Samhita* (c. AD 1100), Bhavaprakash Nighantu (c. AD 1400), etc.

4.1 Ayurvedic Treatment for the Promotion of Health

Health promotion is now an important issue in the field of medical science, particularly from the point of view of public health management. Treatment or measures for the promotion of health differs with the preventive aspect of treatment on the basis of its aims and applications. Preventive aspects of therapeutics are aimed at protecting a specific variety or a group of diseases, while treatment for the promotion of health is intended in a broader sense. According to the World Health Organization (WHO), health promotion is defined as the process of enabling people to increase control over and to improve their health. It moves beyond a focus on individual behavior toward a wide range of social and environmental interventions. According to the WHO, this category of therapeutics can be performed by means of communication (raising awareness about healthy behaviors for the general public), education (empowering behavioral change and action through increased knowledge), policies, and social environment [12]. In *Ayurveda*, there are numbers of therapeutics and regimens for the promotion of health. The most important therapeutic measure under this category is *Rasayana* or immunomodulation therapy. In general, the word *Rasayana* is defined as a treatment that helps to increase lifespan, intellect, and a disease-free state, delays the aging process, maintains youthfulness, luster, and complexion, promotes digestive and metabolic activities, tranquilizes activities of the senses and mind, provides body strength, and results in clarity of voice and uniformity of body nutrition [13]. The ultimate goal of *Rasayana* therapy is to improve the normal immune system of the human body. Many groups of drugs are mentioned in *Ayurveda* for *Rasayana* activities for various purposes such as *Medhaya Rasayana* for improvement of intellect, *Ajasrika Rasayana* for improvement of body vitality, etc. Medicinal plants such as *P. emblica*, *Semecarpus anacardium*, *Bacopa monnieri*, *Acorus calamus*, *Withania somnifera*, *Aspharagus racemosus*, etc. are important examples that can be used for various types of activities under this group. In addition, drugs, therapy, and promotion of health can also be maintained by means of maintenance of behavioral regimen. The social impact of an individual definitely reflects his or her health, as mentioned in *Ayurveda* under *Achara Rasayana*. Apart from drug therapy and behavioral regimen maintenance, there are descriptions of regimens to be followed by an individual according to their *Prakriti* predominance throughout daily (*Dinacharya*) or seasonal (*Ritucharya*) variations. Specific diets and activities are advocated in different Ayurvedic texts for the purpose of *Dinacharya* and *Ritucharya* according to the specifications of individual variations.

4.2 Treatment for the Prevention of Disease

The preventive aspect of treatment is an important issue in *Ayurveda*. This aspect of therapeutics is designed specifically for the prevention of certain diseases. Nowadays, the WHO emphasizes this treatment guideline in the domain of public health management. A separate chapter has been devoted to this purpose in *Sushruta Samhita* where a detailed description of therapeutic measures is available by natural ways for the prevention of diseases [14]. Emphasis is given in *Ayurveda* with this type of therapeutics for specific lifestyle disorders. Many drugs of plant origin are described in *Ayurveda* for the prevention of various lifestyle disorders and the most important example is diabetes mellitus. Diabetes mellitus, a major lifestyle disease, is undoubtedly the most challenging public health problem of the 21st century, with a worldwide prevalence of 387 million (8.3%), which is predicted to rise to 592 million by 2035. It has been evidenced that 77% of people with diabetes live in low- and middle-income countries. India, once known as the "diabetes capital of the world," was home to 61.3 million patients with type 2 diabetes mellitus in 2011, with predictions of 101.2 million diabetics by 2030 [15]. Scientists working in the field of diabetes mellitus are now emphasizing more on prevention of the disease, preferably in the stage of prediabetes. Prediabetes can be diagnosed by laboratory findings of an individual having an HbA1C level of 5.7%−6.4%, fasting plasma glucose of 100−125 mg/dL, and an Oral Glucose Tolerance Test (OGTT) of 140−199 mg/dL [16]. These laboratory findings are to be associated with an elevated body mass index at the overweight level and evidence of obesity. Treatments at this stage can resist the devastating situation of fully manifested diabetes mellitus. A number of medicinal plants of Ayurvedic origin have been screened scientifically for the prevention of diabetes and the most important medicinal plant reported among them is *Pterocarpus marsupium* (Vijaysara) [17]. This plant was tested in a multicentric trial method in pharmacological and clinical models and is reported to be a potent agent for the management of prediabetes. However, there is ample scope in *Ayurveda* for more scientific study with medicinal plants described in various texts for the management of prediabetes. Similarly, prevention of cardiac disease has been described in *Ayurveda* and *Terminali arjuna* (Arjuna) is the specific drug that could be able to perform such activities for preventing a massive cardiac attack. The cardioprotective effect of the alcoholic extract of *T. arjuna* an in vivo model of myocardial ischemic reperfusion injury was proved and suggests the potential of Arjuna in the prevention of ischemic heart disease. Preclinical studies in modern medicine suggest that there are strong antioxidant properties of *T. arjuna* and reduction of ischemic perfusion injury [18]. Panchakarma [19] and Yoga therapy play important roles in the prevention of various diseases. "Panchakarma" is a biopurification method similar to emesis (*Vamana*), medicated purgation (*Virechana*), decoctive enema (*Niruha vasti*), nutritive enema (*Anuvasana vasti*), and errhine therapy (*Shira virechana*). Both Panchakarma and Yoga are noninterventional procedural therapies for the prevention of various diseases, particularly stress disorders. Ocular disease such as dry eyes, computer vision syndrome, age-related macular degeneration, glaucoma, and various types of retinopathies are manifested due to lifestyle disorders. Ocular health can be maintained and diseases of the eye can be prevented with time-tested Ayurvedic therapeutics such as Anjana (collyrium), Aschyotana (eye drop), Abhaynga (massage), Panchakarma, etc., which have tremendous potentiality [20]. Apart from these examples, there are descriptions of therapies for the prevention of diseases of joint disorders, liver disorders, etc. in *Ayurveda* and many medicinal plants have already been scientifically screened.

4.3 Curative Management in *Ayurveda*

Curative management depends upon many factors such as body strength of the patient (*Rogi vala*), intensity of diseases (*Roga vala*), capability for assimilation of drugs (*Agni vala*), potency of drugs (*Ousadha vala*), prognosis of diseases (*Sadhyasadhya*), etc. Considering all these factors, drugs are formulated from different sources such as medicinal plants, metals and minerals, and those of animal origin in various dosage forms such as fresh juice (*Swarasa*), paste (*Kalka*), decoction (*Phanta*), cold infusion (*Sheeta*), hot infusion (*Shrita*), powder (*Churna*), calcined drugs (*Bhasma*), etc. The efficacy of different forms of drugs varies from one disease to another and there is specific indication of these dosage forms in accordance with the specificity of diseases or disease states or the involvement and intensity of *dosas* (functional units of the body). Broadly all these drugs can be delivered in three major

TABLE 1.2 Various Dosage Forms of Ayurvedic Drugs

Sl	Liquid Preparations	Semisolid Preparation	Solid Drugs
1	*Swarasa* (fresh juice)	*Kalka* (paste)	*Churna* (powder)
2	*Kwatha* (decoction)	*Avaleha* (linctuses)	*Rasakriya* (concentrated liquid)
3	*Hima* (cold infusion)	*Lepa* (poultice)	*Khanda* (cakes)
4	*Phanta* (hot infusion)	*Sikta taila* (oil sprinkling)	*Gudapaka* (molasses)
5	*Pramathya* (doughy form)	*Malahara* (ointment)	*Guggulu*
6	*Paniya* (liquid preparations)	*Upanaha* (rubbing powder)	*Sattwa* (precipitate)
7	*Usnodaka* (warm water)		*Lavana* (salt preparation)
8	*Ksirapaka* (milk preparations)		*Ayaskriti* (iron formula)
9	*Laksha rasa* (lac juices)		*Masi* (carbon preparation)
10	*Mamsa rasa* (meat juices)		*Kshara* (alkalis)
11	*Mantha* (churning drinks)		*Vati* (pills)
12	*Udaka* (aqueous extraction)		*Varti* (pellets)
13	*Panaka* (fruit juices)		
14	*Arka* (distillation)		
15	*Sarkara* (syrup)		
16	*Sneha* (oil preparation)		
17	*Sandhana* (fermentation)		

forms [21]: liquid, semisolid, and solid (Table 1.2). Each of these forms can be prepared with various natural agents such as plants, minerals, and animals.

4.3.1 Curative Management With Plant Drugs

Plant drugs occupy major ingredients of Ayurvedic drugs. It is reported that in India around 15,000 medicinal plants have been recorded; however, traditional communities use only 7000–7500 plants for curing different diseases.

Medicinal plants are listed in various indigenous systems such as Siddha (600), *Ayurveda* (700), Amchi (600), Unani (700), and Allopathy (30) plant species for different ailments. According to another estimate, 17,000 species of medicinal plants have been recorded, of which nearly 3000 species are used in the medicinal field [22]. In spite of the huge number of medicinal plants mentioned in *Ayurveda* the majority remain scientifically unveiled.

4.3.2 Metals and Minerals for Curative Management

There are a small number of metals and mineral drugs that are used for the management of various diseases. The aim of designing metal drugs, either in single form or in combined form with other metals and minerals or in combination with medicinal plants (herbomineral), is to play an important role in the eradication of diseases. Metal and mineral drugs exhibit immediate action and are beneficial for several acute stages of diseases at the cellular level. The most important advantage of metal drugs is that they can be delivered in microelemental form because they split into finer minute particles during preparation. Preparation of metal and mineral drugs involves a series of steps such as purification of raw materials by treatment with different medicinal plant ingredients (*Shodhana*), trituration (*Bhavana*), oxidation (*Jarana*), combustion for several times (*Putapaka*), and calcination (*Marana*) [23]. These series of chemical changes result in turning macromolecules into nanomolecules so that the drugs can easily reach the target cells. The end point reaction can be classically determined by traditional physical and chemical testing such as filling the creases of fingers (*Rekhapurnata*), floating on water (*Varitaram*), loss of glistening shine (*Nischandrika*), tastelessness (*Niswada*), inability of returning to its original form (*Apunarbhava*), etc. [23].

4.3.3 Animal Products for Curative Management

Besides use of medicinal plants and metal drugs, ingredients from animal parts and products are also used for the management of diseases in Ayurvedic texts. However, application of animal parts and products in clinical practice is very much limited in *Ayurveda*. Use of animal products is nowadays restricted due to wild animal preservation law. In spite of several hurdles, there are still some specific animal products that are used in *Ayurveda* for the treatment of many diseases such as Kapardaka bhasma (calcinated conch shell), Pravala bhasma (calcinated corals), Mukta bhasma (calcinated pearls), etc. It is reported that Kapardaka (*Cyprea moneta*) is composed of high amounts of crystalline calcium carbonate with the presence of several trace elements such as Mg, Al, K, Fe, and Zn [24]. The antiinflammatory and anticancer activities of Kapardaka bhasma, prepared according to Ayurvedic techniques, are also scientifically reported [24].

5. SCIENTIFIC RESEARCH OF *AYURVEDA* FOR DRUG DEVELOPMENT FROM PLANT SOURCES

5.1 *Rasayana* Therapy

Rasayana is a group of drugs that maintains body immunity in general and delays the aging process. In *Ayurveda*, *Rasayana* is described as a therapy that sustains long life, intellect, memory, senses, clarity of voice, and eases digestion and metabolism, etc. The most important action of this group of drugs is adaptogenecity. Much research has been conducted with the plant *Tinospora cordifolia* and it was observed that the plant had an immunomodulatory effect by producing reactive oxygen species [25] as well as activating macrophages [26]. The important chemical constituent of this plant responsible for such activity is sesquiterpene glycoside, called tinocordiside [27] (Fig. 1.1).

5.2 Diabetes Mellitus

In accordance with reports from the WHO, it is predicted that India will be a diabetic country by the year 2030. It is therefore urgent to search for newer molecules from natural sources to combat the disease. Searching for drugs from Ayurvedic leads will be successful by studying in detail about 20 varieties of diseases and their management under the broad heading of *Prameha*. However, to select the specific lead, it is suggested that one needs to be acquainted with the Ayurvedic language of *Prameha* and its effect on the specificity of constituents (*Prakriti* or personalized characters) for the proper implementation of research. It has been reported that more than 25 medicinal plants are described in different Ayurvedic texts for the management of *Prameha* in general and *Madhumeha* (diabetes mellitus) in particular. Most of these drugs are described in the form of polyherbal or herbomineral compositions and research should be directed in that direction. The dosage form of drugs as mentioned in Ayurvedic texts is an important factor, which is to be followed in the same manner for research in this field. Many scientific studies have already been carried out on the basis of the foregoing leads and classical

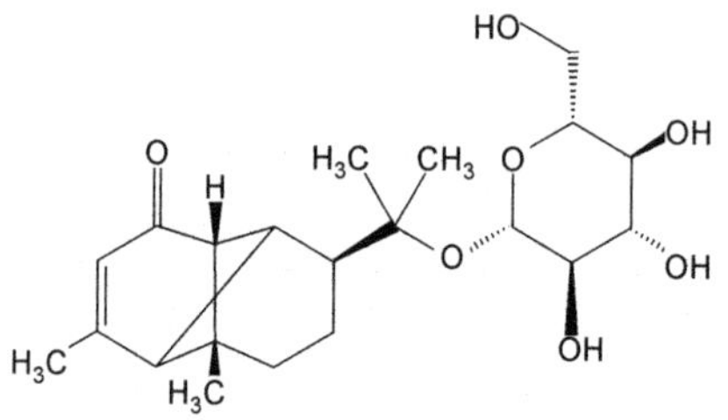

HO
OH
H_3C CH_3 O
O H
O
OH
OH
H_3C
H_3C

FIGURE 1.1 Tinocordiside from *Tinospora cordifolia*.

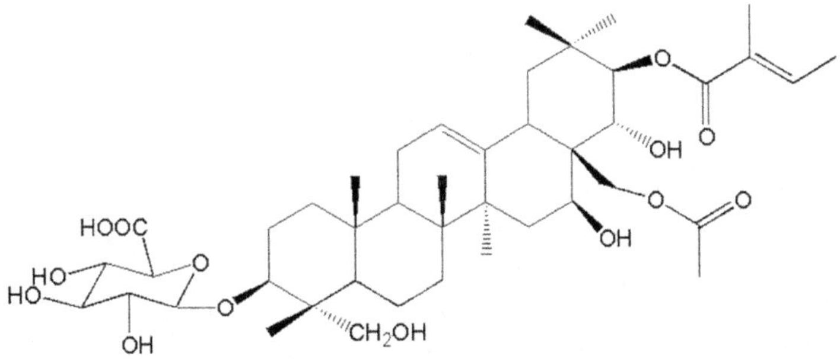

FIGURE 1.2 Gymnemic acid from *Gymnema sylvestre*.

examples are scientific evaluation of *Caesalpinia bonducella* and *Gymnema sylvestre* [28,29]. Both plants are responsible for stimulation of insulin triggering of β-cells of the islets of Langerhans. The antidiabetic property of *G. sylvestre* is the presence of a glycoside gymnemic acid (Fig. 1.2) in this plant. However, it is better to limit the research on diabetes mellitus to natural products of Ayurvedic origin in type 2 diabetes mellitus and not in type 1 diabetes mellitus.

5.3 Wound-Healing Drugs

Research in the field of wound healing is fairly new. Evidence has shown that no direct drug has so far been developed in modern medicine for the successful healing of wounds. Growth factors, hydrogels, etc. are used as healing agents and dressing agents; however, these are not only expensive but may also cause serious adverse effects and are limited to treating certain types of wounds. A group of drugs is described in *Ayurveda* for the management of wounds either from plant sources or from minerals and metals. The lead compounds can be achieved from the description of the treatment of *Vrana* (wounds) as mentioned in classical Ayurvedic texts, particularly *Sushruta Samhita.* Scientific study of *Cynodon dactylon*, *Pterocarpus santalinus*, *Curcuma longa*, etc. for wound-healing activities supports this evidence [30,31]. In spite of previous studies, there is ample scope for research in this field because most of the drugs so far remain untouched.

5.4 Learning, Memory, and Cognitive Disorders

This area of research was selected for three reasons: (1) there is a paucity of modern drugs/agents facilitating acquisition, retention, and retrieval of information and knowledge; (2) with the increasing number of elderly people in the world, the need for drugs to treat cognitive disorders, such as senile dementia and Alzheimer's disease, have acquired special urgency; and (3) *Ayurveda* claims that several plants, the so-called *medhya* plants, possess such activities.

The past two decades have seen tremendous advances in the area of brain physiology, learning, memory, and various brain disorders, and a host of mechanisms at molecular level have been delineated. Synapses—the junctions of nerve cells representing the basic interactive unit of neuronal circuits—constitute the fundamental systemic relationship within the brain. Understanding how this interactive multitude of neuronal circuitry is established initially, and refined continuously throughout life, is fundamental to understanding the molecular basis of learning and memory. At present, an impressive array of chemical entities affecting synapse formation, neuronal differentiation, neurotransmission, nerve growth and repair, and several other functions are recognized. Approximately 50 neurotransmitters belonging to diverse chemical groups have been identified in the brain. Receptors, which are activated by these chemicals, assume special importance in the present context. Specifically, *N*-methyl-D-aspartic acid (NMDA) and γ-aminobutyric acid (GABA) receptors have been implicated in learning and memory [32,33]. It has been further postulated that $GABA_B$ antagonists may enhance memory [32], whereas the NMDA receptor has the ability to mediate synaptic plasticity [33]. Acetylcholine, the first neurotransmitter to be characterized, has a very significant presence in the brain; Winkler et al. [35] determined that acetylcholine is essential for learning and memory. Acetylcholine has been a special target for investigations for almost two decades because its deficit, among other factors, has been held responsible for senile dementia and other degenerative cognitive disorders, including Alzheimer's disease [34]. Major emphasis has focused on acetylcholine. Because the number of acetylcholine receptors declines with advancing age, inhibitors of acetylcholine esterase (AChE), which terminates the action of acetylcholine, have been special targets for development. Some of the Ayurvedic plants studied are reputed to be memory enhancers (*medhya*) and antiaging drugs (*Vayahsthapana*) by standard receptor binding and enzyme inhibition techniques, with the specific aim of identifying any leads based on the foregoing considerations. It was gratifying to see several positive results. Shankhapushpi (leaf) is one of the prime *medhya* plants of *Ayurveda*; it may be useful for neural regeneration and synaptic plasticity. Jatamansi (rhizome) appears to be an excellent candidate for a potential inhibitor of AChE. Haritaki (fruit) is highly prized in *Ayurveda* for antiaging; its extract has displayed several activities. Ashwagandha (root) is another important antiaging plant. This plant was investigated in some detail because its extract showed high affinity for both $GABA_A$ and $GABA_B$ receptors [35].

6. SCIENTIFIC RESEARCH OF *AYURVEDA* FOR DRUG DEVELOPMENT FROM METALS AND MINERALS

Research with Ayurvedic metals and minerals is very much limited. However, there is sufficient potentiality of metal drugs for their therapeutic activity on target cells if it is standardized with modern technological aids. Some of the evidence in this field can prove these facts.

6.1 Drugs From Zinc: *Jasada Bhasma*

In *Ayurveda*, metal-based preparations (*bhasmas*) are indicated for the treatment of several diseases. Standard *Ayurveda* textbooks recommend *Jasada bhasma* (zinc-based *bhasma*) as the treatment of choice for diabetes. Modern medicine also recognizes the important role of zinc in glucose homeostasis. In a study it was observed that single administration of zinc oxide nanoparticles resulted in significant suppression of glucose levels in OGTT carried out in both type 1 and type 2 diabetic rats (~22% and ~30%, respectively). These effects appeared to be more prominent than those obtained with similar doses of *Jasada bhasma* prepared by Ayurvedic methods. After 4 weeks of treatment (1, 3, and 10 mg/kg doses) to diabetic rats, a significant reduction in glucose levels was seen in both nonfasted (~19% and ~29% in type 1 and type 2 diabetic rats, respectively) and fasted (~26% and ~21% in type 1 and type 2 diabetic rats, respectively) state, suggesting multiple mechanisms. Reduction of nonfasted glucose levels can be attributed to insulin secretagogue effects. Reduction of fasted glucose levels may be due to glucagon inhibition, as is reported with zinc. Increased serum insulin levels (~35% and ~70% in type 1 and type 2 diabetic rats, respectively) suggested insulin secretagogue effects. Reduction in serum triglyceride (~48%) and free fatty acid (~41%) levels was also observed after treatment indicating beneficial effects on lipid metabolism. Overall results suggested that zinc oxide nanoparticles were more potent and efficacious than *Jasada bhasma* [36].

6.2 Iron Therapy in *Ayurveda*

Iron has been used in *Ayurveda* widely for a long time for the treatment of various diseases, including iron deficiency anemia. In a study on *Lauha bhasma*, prepared by classical methods of *Ayurveda*, it was observed that as the processes of calcinations are increased, there is incorporation of many essential elements such as Cu, Mg, Zn, etc., but not Fe, which are considered responsible for building the hemoglobin. The same study in animal models showed that the bioavailability of *Lauha bhasma*, prepared by classical Ayurvedic procedures, could have more bioavailability at the cellular level than the synthetic preparation of iron in the form of ferrous sulfate [37]. In an another study it was observed that the X-ray diffractogram of *Lauha bhasma* showed an intense peak at 36 degrees indicating the presence of Fe(III), which conforms to Fe_2O_3 (hematite) as per the PC PDFWIN data. This confirms the presence of Fe_2O_3 in *Lauha bhasma*. In the absence of any prominent peak at 45 degrees, it may be inferred that the amount of free iron is insignificant [38]. The Fourier transform infrared (FTIR) spectrum of *Lauha bhasma* shows a broad band between 3400 and 3500 cm^{-1}, characteristic of υ_{O-H} (stretching vibrations of the O–H bond). The broad band in the region 1700–1650 cm^{-1} is assigned to υ_{CO} of organic constituents. The absorption band near 1620 cm^{-1} may be attributed to the presence of an aromatic ring, while the sharp band at 560 cm^{-1} may be due to the Fe–O bond. All the foregoing FTIR data suggest that *Lauha bhasma* may contain a complex with organic moieties present in the treating agents used in the various stages of its preparation [38].

6.3 Gold Therapy in *Ayurveda*

From ancient times, *Swarnabhasma* (gold ash) has been used in several clinical manifestations, including loss of memory, defective eyesight, infertility, overall body weakness, and incidence of early aging. *Swarnabhasma* has been used by Ayurvedic physicians to treat different diseases such as bronchial asthma, rheumatoid arthritis, diabetes mellitus, nervous disorders, etc. Qualitative analyses of *Swarnabhasma*, prepared after proper purification and calcination as per Ayurvedic pharmacy, indicated that *Swarnabhasma* contained not only gold but also several microelements (Fe, Al, Cu, Zn, Co, Mg, Ca, As, Pb, etc.). Infrared spectroscopy showed that the material was free from any organic compound. The metal content in the *bhasma* was determined by atomic absorption spectrometry. Acute oral administration of *Swarnabhasma* showed no mortality in mice (up to 1 mL/20 g bw of *Swarnabhasma* suspension containing 1 mg of drug). Chronic administration of *Swarnabhasma* also showed no toxicity as judged by serum glutamic pyruvic transaminase, serum glutamic oxaloacetic transaminase, serum creatinine, serum urea level, and histological studies. In an experimental animal model, chronic *Swarnabhasma*-treated animals showed significantly increased superoxide dismutase and catalase activity, two enzymes that reduce free radical concentrations in the body [39].

6.4 Shilajit: A Unique Molecule of *Ayurveda*

Shilajit is used in *Ayurveda* as a remedy for several diseases, particularly chronic diseases. Shilajit is a pale-brown to blackish-brown exudate that oozes from sedimentary rocks worldwide, largely in the Himalayas. It is an important drug of the ancient Ayurvedic materia medica and it is to this day used extensively by Ayurvedic physicians for a variety of diseases. Early Ayurvedic writings from the *Charaka Samhita* describe Shilajit as a cure for all diseases as well as a *Rasayana* (rejuvenator) that promises to increase longevity. It is composed of rock humus, rock minerals, and organic substances that have been compressed by layers of rock mixed with marine organisms and microbial metabolites [40]. In a clinical study with purified Shilajit, it was observed that the compound has potent spermatogenic activity in healthy volunteers. Interestingly, it was also observed that high-performance liquid

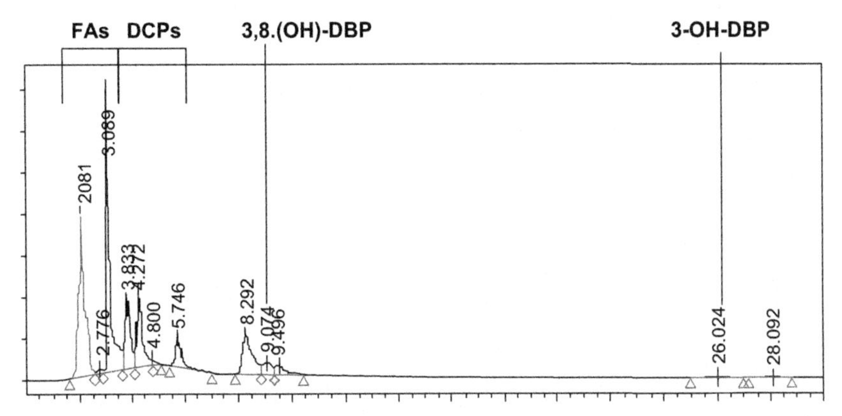

FIGURE 1.3 High-performance liquid chromatography (HPLC) chromatogram of Shilajit showing the presence of dibenzo-α-pyrone. *3,8-(OH)-DBP*, 3,8,hydroxy dibenzi-alpha-pyrone; *3-OH-DPB*, 3, hydroxy dibenzi-alpha-pyrone; *DCPs*, DBP chrmoproteins; *FA*, fulvic acid.

chromatography analysis of purified Shilajit contains a chemical component, dibenzo-α-pyrone (Fig. 1.3), which can stimulate human spermatozogenesis as evidenced from a significant increase in testosterone, dehydroepiandrosterone, follicle-stimulating hormone, and luteinizing hormone [41]. This study supports the earlier clinical study of spermatogenetic activity of purified Shilajit in patients of oligospermia [42].

7. RECOMMENDATION

Ayurveda has been practiced in India for thousands of years and the health care of a wide range of communities has been well maintained with this system. The evidence of Ayurvedic therapeutics is based on observation and application of generations of ancient scholars. Therefore the safety and efficacy aspects of the drugs described in different classical texts of *Ayurveda* may not need further validation. However, apart from this fact, due to the change of global geographical patterns, particularly the ill effect of global warming, the characteristics of natural products may behave in strange ways and may need to be revalidated. Use of drugs from an Ayurvedic source for research and therapeutic applications should therefore be done after thorough scientific evaluation, utilizing traditional leads in classical forms. In addition, selection of any particular therapeutic agent(s) needs interpretation of the Ayurvedic language in a justified way so that proper application can be performed. The therapeutic agent(s) of *Ayurveda* are designed on the basis of their suitability for variations of individualism and drugs should be selected on such a basis. Moreover, most of the drugs in *Ayurveda* are related to multiple compositions, not single forms, and research and applications of such agents should be directed in this way. In the past, the Ministry of AYUSH, the Government of

India, has developed and published good clinical practice [43], which is the hallmark of future applications of research and therapeutic application of Ayurvedic origin. The research portal of the Ministry of AYUSH, Government of India (www.ayushportal.nic.in), may also be helpful in this direction.

REFERENCES

[1] Scheme for Extra Mural Research in Ayurveda, Yoga & Naturopathy, Unani, Siddha, Sowa Rigpa and Homoepathy, Ministry of AYUSH, Government of India, 2015.

[2] Results-framework Documents (RFD) for Department of AYUSH (2014–2015), Ministry of AYUSH, Government of India, 2015.

[3] J. Khandelwal, R.K. Singh, S. Rath, M. Kotecha, Medicinal therapies in Veda, Ayurpharm Int. J. Ayur. Alli. Sci. 3 (2) (2014) 35–40.

[4] B. Patwardhan, A.D.B. Vaidya, M. Chorghade, Ayurveda and natural products drug discovery, Curr. Sci. 86 (6) (2004) 789–799.

[5] Charaka Samhita, Sutrasthana, Sadvirechanasatasriya Adhyaya (chapter describing six hundred compositions) (Chapter IV).

[6] Sushruta Samhita, Sutrasthana, Dravyasamgrahaniya Adhyaya (chapter describing group of drugs) (Chapter 38).

[7] Vagbhatta, Astanga Hridaya Samhita, Sutrasthana, Shodhanadi Gana Samgraha (chapter describing purifying drugs) (Chapter 15).

[8] P.V. Sharma, Introduction to Dravyaguna Vijnana, Chowkhambha Orientalia, Varanasi, 1982, pp. 23–25.

[9] S.L. Vassou, S. Nithaniyal, B. Raju, M. Parani, Creation of reference DNA barcode library and authentication of medicinal plant raw drug used in Ayurvedic medicine, BMC Complement. Altern. Med. 16 (1) (2016) 186–191.

[10] B. Prasher, S. Negi, S. Aggarwal, A.K. Mandal, et al., Whole gene expression and biochemical correlates of extreme constitutional types defined in Ayurveda, J. Translational Med. 6 (48) (2008) 1–12.

[11] S. Agarwal, S. Negi, P. Jha, P.K. Sing, T. Stobdan, M.A.Q. Pasha, S. Ghosh, et al., EGLN1 involvement in high-altitude adaptation revealed through genetic analysis of extreme constitution types defined in Ayurveda, Proc. Natl. Acad. Sci. 107 (44) (2010) 18961–18966.

[12] Participants at the 6th Global Conference on Health Promotion. The Bangkok Charter for Health Promotion in a Globalized World, World Health Organization, Geneva, Switzerland, August 11, 2005.

[13] Charaka Samhita, Vidyotini Hindi commentary, Chikitsasthana, in: sixteenth ed., in: K. Shastri, G. Chaturvedic (Eds.), Pada 1, Verse 7-8, vol. II, Chaukhamba Bharati Academy, 1989, p. p.5 (Chapter 1).

[14] T.A. Jadavji, Sushruta Samhita, Dalhana Commentary, Uttartantra, verse no. 25, Choukhamba Surabharati Prakashan, Edition Reprint, 1994, p. 494 (Chapter 1).

[15] M. Gupta, R. Singh, S.S. Lehl, Diabetes in India: a long way to go, Int. J. Sci. Rep. 1 (1) (2015) 1–2.

[16] G. Sypniewska, How to diagnose the prediabetes? Biochemia Med. 24 (Suppl.) (2014) S31–S34.

[17] N.K. Kumar, V. Muthuswamy, N.K. Ganguly, Initiatives of Indian Council of Medical Research in scientific validation of traditional medicine, Health Administrator, ICMR's Annual Report 1999–2000, XX (1&2), pp. 115–119.

[18] S. Seth, P. Dua, S.K. Moulik, Potential benefits of *Terminalia arjuna* in cardiovascular disease, J. Prev. Cardiol. 3 (1) (2013) 428–432.

[19] K. Divya, J.S. Tripathi, S.K. Tiwari, Utilization of Panchakarma in health care: preventive, nutritive and curative treatment of disease, J. Pharmaceut. Sci. Innov. 2 (5) (2013) 1–5.

[20] P.K. Sahoo, S. Dash, S. Fiaz, Concepts of preventive ophthalmology in Ayurveda, Int. J. Res. Ayurveda Pharm. 7 (2) (2016) 115–119.

[21] N. Arun, K. Vinay, G. Basavaraj, Various dosage forms of Ayurveda, Uniq. J. Ayur. Herbal Med. 2 (4) (2014) 20–23.

[22] A.K. Meena, P. Bansal, S. Kumar, Plants-herbal wealth as a potential source of ayurvedic drugs, Asian J. Trad. Med. 4 (4) (2009) 152–170.

[23] S.P. Rashid, M. Shivashankar, Evaluation of herbo-mineral formulation (bhasma): a review, Int. J. Res. Ayurveda Pharm. 6 (3) (2015) 382–386.

[24] K.M. Krishna, K.K. Singh, A critical review on drug Kapardika (*Cypraea moneta* Linn), Int. Res. J. Pharm. 3 (2) (2012) 8–11.

[25] A. Kapil, S. Sharma, Immunopotantiating compounds from *Tinospora cordifolia*, J. Ethnopharmacol. 58 (1997) 89–98.

[26] P. More, K. Pai, Immunomodulatory effect of *Tinospora cordifolia* (Guduchi) on macrophage activation, Biol. Med. 3 (2) (2011) 134–140.

[27] Devprakash, K.K. Srinivasan, T. Subburaju, S. Gurav, S. Singh, *Tinospora cordifolia*: a review on its ethnobotany, phytochemical and pharmacological profile, Asian J. Biochem. Pharmac. Res. 4 (1) (2011) 291–302.

[28] A.A. Romaiyan, Liu, H. Asare-Anane, C.R. Maity, S.K. Chatterjee, N. Koley, T. Biswas, A.K. Chaterji, G.C. Huang, S.A. Amiel, S.J. Persaudi, P.M. Jonse, A Novel *Gymnema sylvestre* extract stimulates insulin secretion from human islets in vivo and in Vitro, Phytother. Res. 24 (9) (2010) 1370–1376.

[29] S. Chakrabarti, T.K. Biswas, et al., Antidiabetic activity of *Caesalpinia bonducella* F. in chronic type 2 diabetic model in Long Evans rats and evaluation of insulin secretagogue property of its fractions on isolated islets, J. Ethnopharmacol. 97 (2005) 117–122.

[30] T.K. Biswas, L.N. Maity, B. Mukherjee, Wound healing potential of *Pterocarpus santalinus* Linn: a pharmacological evaluation, Int. J. Lower Extremity Wounds 3 (3) (2004) 143–150.

[31] T.K. Biswas, S. Pandit, S. Chakrabarti, S. Banerjee, N. Poyra, T. Seal, Evaluation of *Cynodon dactylon* for wound healing activity, J. Ethnopharmacol. 41 (2016) 378–387.

[32] G.D. Fischbach, Mind and brain, Sci. Am. 267 (1992) 48–57.

[33] J. Willetts, R.L. Balster, J.D. Leander, The behavioral pharmacology of NMDA receptor antagonists, Trends Pharmacol. Sci. 11 (1990) 423–428.

[34] G. Johnson, C.F. Bigge, Recent advances in excitatory amino acid research, Annu. Rep. Med. Chem. 26 (1991) 11–22.

[35] J. Winkler, S.T. Suhr, F.H. Gage, L.J. Thal, L.J. Fisher, Essential role of neocortical acetylcholine in spatial memory, Nature 375 (1995) 484–487.

[36] R.K. Umrani, K.M. Paknikar, Jasada bhasma, a zinc based Ayurvedic preparation: comtemporary evidence of anti-diabetic activity, inspired development of nanomedicine, Evid. Based Complement Alternat. Med. (2015) 1–9. Article ID: 193156.

[37] S. Pandit, T.K. Biswas, P.K. Debnath, A.V. Saha, U. Chowdhury, B.P. Shaw, et al., Chemical and pharmacological evaluation of different ayurvedic preparations of iron, J. Ethnopharmacol. 65 (1999) 149–156.

[38] B. Krishnamacharya, B. Pemiah, S. Krishnaswamy, U.M. Krishnan, S. Sethuraman, R. Sekar, Elucidation of core shell model for lauha bahsma through phyico-chemcial characterization, Int. J. Pharmac Pharmaceutic Sci. 4 (2) (2012) 644–649.

[39] A. Mitra, S. Charabarti, B. Auddy, P.C. Tripathi, S. Sen, A.V. Saha, B. Mukherjee, Evaluation of chemical constituents and free radical scavenging activity of Swarnabhasma (gold ash), – an Ayurvedic drug, J. Ethnophrmacol. 80 (2002) 147–153.

[40] S. Ghosal, The aroma principles of Gomutra and Karpuragandha Shilajit, Indian J. Indg. Med. 11 (1994) 11–14.

[41] S. Pandit, S. Biswas, U. Jana, R.K. De, S.C. Mukhopadhyay, T.K. Biswas, Clinical evaluation of purified shilajit on testosterone levels in healthy volunteer, Int. J. Androlg. 20 (2015) 1–6.

[42] T.K. Biswas, S. Pandit, S. Mondal, S.K. Biswas, U. Jana, T. Ghosh, P.C. Tripathi, P.K. Debnath, R.G. Auddy, A. Auddy, Clinical evaluation of spermatogenic activity of processed shilajit in oligospermia, Int. J. Androlog. 41 (2009) 1–9.

[43] Good Clinical Pracice Guidelines for Clinical Trials of Ayurveda, Sidha and Unani Medicine (GCP-ASU), Department of AYUSH, Government of India, New Delhi, March 2013. www.indianmedicine.nic.in.

Chapter 2

Traditional and Folk Medicine as a Target for Drug Discovery

Sujata Wangkheirakpam
Department of Chemistry, NIT Manipur, Imphal, India

1. INTRODUCTION

Folk and traditional medicine systems are treatises that include a summation of the different medical approaches and practices found at a particular time in different parts of the world. Folk and traditional medicines were the only medicinal knowledge and techniques in olden times to combat many types of ailments before the modern medicine system developed. Altogether, plant, fungus, and animal products along with other minerals have been used in different formulations as prescriptions in the past by the respective healers in different countries. Ayurveda, Siddha medicine, Unani, ancient Iranian medicine, Irani, Islamic medicine, traditional Chinese medicine (TCM), traditional Korean medicine, and traditional African medicine are some of the known traditional medicines. It is well known that natural products have been a highly productive source of leads for drug discovery and development. In India, references to the curative properties of some herbs in the Veda were most probably the earliest records of the use of plants in medicine that later formed the Ayurveda. The works of Charak Samhita and Sushruta Samhita were the two most important works on Indian medicine. In China, TCM is a well-organized and profound traditional medicine system. A major portion of the populations of the world are now using traditional and folk medicines [1]. The reason may be the unavailability of modern drugs and medical facilities, poverty, or strong beliefs in the healing medicines of a particular region and as strategies to escape from undesirable side effects of synthetic drugs. Medicinal plants and other natural products of the regions are being used for the treatment of various ailments by the healing practitioners. Plant sources are well known, namely, tulsi, turmeric, ginger, etc., which are abundant and easily available. Natural products other than plants are also used in traditional medicine systems. Even mushrooms, which are macrofungi, are used in many medicine systems. Two *Auricularia* spp. (family Auriculariaceae), native of

Natural Products and Drug Discovery. https://doi.org/10.1016/B978-0-08-102081-4.00002-2

the state of Manipur, India, have traditional uses for treating diarrhea, dysentery, diabetes, hypertension, constipation, and liver pain. This is recorded in the folk medicine of Manipur, called the "Maibaron." Mushrooms have been studied widely for various bioactive compounds, and the isolation of polysaccharides, phenolics, proteins, etc. has been reported. Many other mushrooms such as *Lentula edodes*, *Grifola frondosa*, and *Tricholoma lobayense* have been reported as having hepatoprotective effects against paracetamol-induced liver injury [2]. Traditional and folk medicines are the knowledge, skills, and practices based on the theories, beliefs, and experiences indigenous to different cultures, whether explicable or not, used in the maintenance of health as well as in the prevention, diagnosis, improvement, or treatment of physical and mental illnesses. According to the World Health Organization (WHO), 80% of the world population is dependent on traditional and folk medicines [3]. However, the WHO clearly states that inappropriate or incomplete use of traditional medicines or practices can have undesirable negative or dangerous effects, because a slight variation in the concentration of the components in a traditional and folk medicine formulation might have adverse consequences. Further research is required to ascertain the efficacy and safety of several of the practices and medicinal plants and other natural products used by traditional medicine systems. These practices are common in many countries of Asia and Africa and the bulk of the population relies on traditional medicine for their primary healthcare needs. When adopted outside of its traditional culture, traditional medicine is often called alternative medicine [3]. Many definitions have been given to define traditional and folk medicines. Traditional medicine may include formalized aspects of folk medicine, the longstanding remedies passed on and practiced by common people from one generation to another. However, folk medicine consists of the healing practices and ideas of body physiology and health preservation known to some in a culture, transmitted informally as general knowledge, and practiced by anyone in the culture having prior experience; these practitioners are known as healers [4]. In some cases, folk medicine may also be referred to as traditional medicine, alternative medicine, indigenous medicine, or natural medicine. These terms are often used as synonyms, and authors give preferences according to certain points they want to emphasize. Most probably, only indigenous medicine and traditional medicine have the same meaning as folk medicine. Alternative and natural medicine refers to modified medicine systems. Here, traditional medicine is a term used to define a collected systematically practiced folk medicine for certain periods of time and well recorded and established for the particular indigenous people of a region. However, the terms traditional and folk medicine are quite similar but difficult to differentiate. The study of herbs dates back 5000 years to the ancient Sumerians, as evident in written records. They described well-established medicinal uses for plants. Indigenous medicine is generally transferred orally through a community, family, and individuals until collected and recorded. Within a given culture, elements of indigenous medicine knowledge may be diffusely known by many, or may be gathered and applied by those in the specific role of healer such as a shaman or midwife. At present a large number of these natural medicine systems are under modern scientific investigation to ascertain their efficacy and safety, and, as mentioned earlier, are directed by the WHO for traditional medicinal plants. The gradual sophistication of phytochemistry, pharmacology, etc. has helped to decipher these plants for a better understanding of their ethnomedical uses. Classic examples of phytochemicals in biology and medicine include Taxol, vincristine, vinblastine, colchicine, artemisinin, etc. It is estimated that the global trade in medicinal plants is US$100 million per year, which is growing at the rate of 10%–15% annually [5]. It is well known that natural products have led to drug discovery. Many modern synthetic drugs have their origins in bioactives from natural products discovered during the course of studying the use of traditional medicinal plants. Paclitaxel is an example of a drug developed from the traditional medicinal plant *Taxus baccata*. Paclitaxel is known by its common name Taxol. Many new modifications have been made from it to produce drugs for cancer therapy. Two main diseases for which therapy is not available are hepatic diseases and dengue, and even for diseases such as diabetes, new antidiabetic drugs need to be found because existing ones have treatment-limiting side effects. The requirement for the development of improved drugs, which will be more efficient and will have fewer or negligible side effects, is in demand along with the search for new drugs for many diseases. For liver disorders and dengue there is no proper therapy available. Some plant extracts are being used for treatment. Silymarin is being used for liver disorders with other synthetic drugs that have adverse side effects. Extracts of *Psidium guajava*, *Azidarachta indica* (neem), *Ocimum sanctum*, *Momordica charantia*, and *Carica papaya* have been used for the treatment of dengue. Any standardized extract or single compound considered for therapeutic use must be evaluated for its ability to act consistently. Investigations are being carried out to discover the key moiety from the available natural sources. Suitable analytical methods for the identification and quantification of components in herbal medicines should be established. To separate and identify the ingredients in herbal medicines, which are part of traditional medicine or which are discovered after further investigation, various current analytical methods and chromatographic techniques such as open column chromatography, gas chromatography-mass spectrometry (GC-MS), liquid chromatography-mass spectrometry (LC-MS), and other spectroscopic techniques have been developed. Spectroscopic techniques such as ultraviolet (UV), infrared (IR), magic-angle spinning, nuclear magnetic resonance (NMR), and X-ray diffraction techniques have been used for the identification of certain major ingredients in the active moieties of the bioactive fraction [6,7]. The new techniques are used to confirm the structure of compounds that are separated and purified. The active moiety can be a single compound or a group of compounds as a fraction. Most of the experiments are carried out using bioassay-guided techniques. However, in the case of the author's experiment with hepatoprotective activity, chemical screening with different standards are performed with advanced techniques such as high-performance liquid chromatography (HPLC). The effect is a single compound in the case of Taxol and a synergistic effect of more than one compound in silymarin. With these techniques, new formulations and compounds have been identified. Science has referred to traditional medicines because they have been developed on a trial and error basis on humans and recorded after successful results. Traditional systems have been successful leads to drug discovery because the systems have been developed after prolonged and repeated experiments on humans by healers. Bioactive compounds have been use as further leads for better drugs through techniques such as combinatorial synthesis, and according to a particular disease and availability, a single compound or extract has been used. There are situations where the synergetic effect works better than a single bioactive compound. There are many traditional and folk medicine systems practiced in different parts of the world.

2. DIFFERENT TRADITIONAL AND FOLK MEDICINES

In different parts of the world, different medicine systems were developed with available natural products, minerals, and cultural beliefs of a particular habitat. Ayurveda, Jamu, Siddha, Sri Lankan, Thai, Vietnamese, Chinese, Japanese, Korean, Mongolian, Tibetan, ancient Egyptian, ancient Greek, ancient Iranian, Byzantine, Irani, Roman, Islamic, Unani, African (Iboga, Muti, Nganga, Hausa, South Africa, Yoruba), American (Aztec, Brazilian, Curandero, Kallawaya, Mapuche, Maya), Australian Bush, and Western European are some known traditional medicines practiced in different corners of the world. Of these, Ayurveda, Siddha, Unani, ancient Iranian medicine, Irani, Islamic medicine, TCM, traditional Korean medicine, and acupuncture are some of the well-established traditional medicines. In spite of advance medical knowledge, traditional and folk medicines are still used for therapy for many ailments in many parts of the world. Medicinal plants and other natural products have provided the basis for great medical systems in human history: Hippocrates, Galen, the great Ayurveda of the Indian subcontinent, TCM, Islamic medical system, and many other cultural traditions (which were often hybrids of the various systems of medicine). All these systems have formed a large part of the respective *Materia Medica*. The world market for plant-derived chemicals for pharmaceuticals, fragrances, flavors, and color ingredients alone exceeds several billion dollars per year. Alternative or complementary medicines are modified traditional and folk medicines. About three-quarters of the world population rely mainly on plants and plant extracts for health care; thus a chemical study of traditionally used medicinal plants is an essential requirement. Some of these traditional medicine systems and alternative medicines are discussed here. There are many medicine systems that need to be documented such as the Maibaron of the Manipuris of India. Traditional medicine systems have been leads to drug discoveries because they are the outcomes of prolonged and repeated trial and error research into natural products singly or with minerals, etc. Recently, traditional systems have been revived through information technology. Online databases, networking, and portals on traditional and folk medicines disseminate knowledge to the whole world. Many folk medicines that are regional are bought together. Of the medicine systems, Ayurveda and TCM were the most eminent. Some of the alternative and lesser known but well-recorded reliable medicine systems are discussed next.

2.1 Ayurveda

Ayurveda is a well-known medicine system followed in India. The word Ayurveda means life-knowledge. The origin of Ayurveda is found in ancient sacred books of the Aryans. It can be traced back to the era of mythology. Ayurveda knowledge was believed to be transferred from the gods to humans through the Hindu God Brahma. He passed it to Prajapati, another king, and then to Atreya Punarvasu. The Charaka Samhita is the first and most important text of Ayurveda. However, the surgical texts are in Sushruta Samhita. Ayurveda was well developed at the time of Gautama Buddha (563–483 BC). Charaka Samhita is ascribed to Atreya Punarvasu. Susruta Samhita is ascribed to Dhanvantari. Susruta along with a number of physicians learned from a god called Dhanvantari. Dhanvantari incarnated himself as Divodasa, a mythical king of Varanasi. Though there are some references to medicines in Rigveda, systematic medicine is found in Atharvaveda. The origins of Ayurveda are also found in Atharvaveda. Some important recordings of Ayurveda are Charaka Samhita, Susruta Samhita, Astanga Samgraha, Astanga Hridaya, Bhela Samita, Sharngadhara Samhita, and Bhavaprakasa. There was also the theory of the Five Elements (Pancha Mahabhutas). Prana, Subtle Fire (tejas), and Essence (ojas) are the quintessential expressions of the "Five Elements" as applied to embodied life. Ayurveda describes *doshas* known as Vata, Pitta, and Kapha. A dosha is a fault or error. Balance of the doshas results in health and imbalance results in disease. Doshas are balanced when they are equal to each other and each human possesses a unique combination of doshas, which define the person's temperament and characteristics. In either case, it is said that each person should modulate their behavior or environment to manipulate the doshas to maintain their natural state. Ayurveda teaches not only about medicine, but also about healthful living, prevention of disease, and personal and social hygiene [8–10]. Ayurveda uses natural products such as plants and animal products, toxic metals, and minerals in its formulations. The roots, leaves, fruits, bark, or seeds of plants such as cardamom, ginger, turmeric, and cinnamon are used for therapy. Animal products such as milk, bones, and gallstones are also used. Fats are prescribed both for consumption and for external use. Ayurveda describes that both oil and tar can be used to stop bleeding. The practice of adding metals, minerals, or gems to herbal preparations in Ayurveda is called *rasa shastra*. There are prescriptions of consumption of minerals such as sulfur, arsenic, lead, copper sulfate, and gold. Toxic heavy metals such as lead, mercury, and arsenic are also used. Ayurveda uses alcoholic beverages called *Madya*. Purified opium is used in eight Ayurvedic preparations and is said to balance the Vata and Kapha doshas and increase the Pitta dosha. Lead, mercury, and arsenic have been detected in a substantial proportion of Indian-manufactured traditional Ayurvedic medicines. Metals may be present due to the practice of rasa shastra (combining herbs with metals, minerals, and gems) [8,9,11]. Adverse reactions to herbs are described in traditional Ayurvedic texts. Making Ayurvedic formulations requires in-depth knowledge of the particular formulation both in quality and quantity. In addition, knowledge of correct dosage is required. Minor variations in dosages would make the prescription ineffective or have adverse effects. There is a communication gap between practitioners of modern medicine and the genuine Ayurveda. For proper documentation of the Ayurveda text in a systematic manner and for further investigation, the Indian government has made policies to preserve and investigate the Ayurveda in depth. The Central Council for Research in Ayurveda and Siddha carries out research through a national network of research institutes. The Traditional Knowledge Digital Library (TKDL) was set up as a repository for formulations of various systems of Indian medicine such as Ayurveda, Unani, and Siddha in 2001. TKDL is a collaborative project between the Council of Scientific and Industrial Research, the Ministry of Science and Technology and the Department of Ayurveda, Yoga and Naturopathy, Unani, Siddha, and Homoeopathy, and the Ministry of Health and Family Welfare. This was done to fight biopiracy and unethical patents. The formulations in the TKDL come from over 100 traditional Indian Ayurveda books [12,13]. With its strong background, Ayurveda is still practiced today and many herbal products based on Ayurveda are in the market and are widely preferred over synthetic products. There are other alternative medicine systems, namely, Siddha and Unani. India, a land of diverse culture and center of many civilizations, still has unexplored folklore healing practices such as the "Maibaron" in the northeastern state of Manipur. It is a well-documented healing system in "Puyas."

Yunani or Unani medicine (tibb Yunani) is the term for Perso-Arabic traditional medicine and was brought by Muslim invaders to India. The Arabic term Yunani means Greek, because the Perso-Arabic system of medicine was, in turn, based on the teachings of the Greek physicians Hippocrates and Galen. It is a combination of Greek medicine and Ayurveda [8,14]. Siddha medicine is a system of traditional medicine that originated in South India. Traditionally, it is taught that the Siddhars laid the foundation for this system of medication. Siddhars were spiritual adepts who possessed the ashta siddhis (the eight supernatural powers). Agastya is considered the first siddha and the guru of all siddhars; the siddha system is believed to have been handed over to him by Murugan, son of Shiva and Parvati. Homeopathy is a system of alternative medicine created in 1796 by Samuel Hahnemann. Ayurveda, Siddha, Unani, Naturopathy, Homeopathy, and Yoga form the Department of Ayush of the Government of India [8,14,15].

2.2 Maibaron

Maibaron is a medicine system of the class of people known as the Meeteis, natives of the state of Manipur, India, to which the author also belongs. Maibaron is in one of the "Puyas," the group of religious books of the Meeteis written in the "Meetei mayek script." It has been described in the book with Bengali script under the Puya group from the original Puyas, under the Puya "Khunungi Houna Lonchat," which describes the culture and tradition of the Meeteis along with the healing system. It is a well-recorded medicine system where the healers are called Maiba and Maibis. In this traditional system, natural products, alcohol, and the umbilical cord were used for treating many ailments along with worshipping and offerings to God. Manipur has been blessed with many medicinal natural products, and it comes under a biodiversity hotspot [16,17]. The author had investigated *Auricularia* spp., which had a traditional use in liver pain, diarrhea, and dysentery as described in the Maibaron. The author extracted chlorogenic acid (CGA), which had antioxidant activity in the ethyl acetate extract of the two species. This might be the key compound for liver healing activity. *Auricularia* spp. were studied for liver healing and antimicrobial properties by the author and results were impactful. In Maibaron, utilization of *Auricularia* spp. has been described for similar therapies. Hence Maibaron could be used for investigations of new drugs as supported by the rich medicinal plants of Manipur [2].

2.3 Traditional Chinese Medicine

TCM is a Chinese medicine system. It is well known all over the world and is quite comparable with the Ayurveda medicine system in India. TCM was developed by trial and error as were other original medicine systems. TCM is largely based on experience and is guided by a holistic concept. In mythology, the emperors Huangdi and Shennong wrote medical manuals (*Huangdi Neijing—The Yellow Emperor's Manual on Corporeal Medicine* and *Shen Nong Ben Cao Jing—Classical Pharmacopeia of the Heavenly Husbandman*) about 5000 years ago. Records of early medical practice in China are found in the *Historical Memoirs* (*Shi Ji*). It is the first book in a series of dynastic records written about 500 BC. Different forms of diagnostic procedures of pulse study, inspection of the tongue, and methods for questioning, as well as the therapeutic modalities of acupuncture, moxibustion, massage, remedial exercise, and the use of plant medicines are described in the book.

The "yin-yang theory" and "Five-Element theory" form the basis that treatment is targeted at correcting an underlying imbalance. The five substances are Qi, Blood (xue), Essence (jing), Spirit (shen), and Fluids (jin ye). The substances form the basis for the development and maintenance of the human body. TCM uses 365 different types of medicinal substances of plant, animal, and mineral nature with studies of their properties and effects in *Materia Medica* (*Shen Nong Ben Cao Jing*) as prescriptions for different diseases. There are more than 6000 available texts of the clinical experiences and evolving theoretical knowledge of Chinese medicine, though many such ancient texts have been lost. Yin-yang (opposites) refers to opposing influences such as positive and negative forces. The Five-Element theory states that everything is maintained in kinetic balance under the movement of Five Elements similar to the Pancha Mahabhutas of Ayurveda. Prescription of herbs based on these theories may comprise a single herb or more commonly a mixture of herbs in differing amounts. The prescribed herbs are classified according to four properties: Zhing, Chen, Zhou, and Shi. The contents of a single herb are complex and most medicines are a mixture of a number of different herbs. For example, the formulation Danggui-Nian-Tong-Tang, which is used for the treatment of acute gouty arthritis, consists of 15 different herbs. Acupuncture and moxibustion were also practiced. Acupuncture is based on inserting needles of various gauges and lengths into the skin at specific acu-points. Moxibustion is based on burning tinder made of Chinese mugwort (*Artemisia argyi* or *Artemisia vulgaris*) next to or on a locus [9,18,19]. The *Encyclopedia of Traditional Chinese Medicines* presents a comprehensive and integrative work on surveying TCM plant sources, chemistry, pharmacology, and medicinal effects and indications in a systematic manner. This encyclopedia has more than 8000 TCM components and a variety of pharmacological data, which are valuable not only for the study of TCM, but also for the development of Western medicine; however, certain data are normalized for fast documenting [20]. *Artemisia annua* is traditionally used to treat fever. It has been found to have antimalarial properties. The antimalarial agent artemisinin and its synthetic derivatives were developed from Qing Hao (*A. annua*). The inventor YuoYou Tou received a Nobel Prize in 2015 for the discovery of this drug by bioassay fractionation [21].

2.4 Traditional Korean Medicine

Traditional Korean medicine refers to the traditional medicine practices developed in Korea. Korean medicine originated in ancient and prehistoric times around 3000 BC. It has a unique origin and history. Korean medicine uses herbal medicine, acupuncture, moxibustion, and aromatherapy for the treatment of diseases. Korean traditional medicine is internationally recognized as a sufficiently reasonable and scientific medical treatment in comparison to Western medicine. The legendary textbook of Korean traditional medicine, *Dong-eui Bo-Gam*, written in the 17th century, was registered to UNESCO Memory of the World in 2009. It was verified by the International Advisory Committee of UNESCO [22].

2.5 African Medicine (Muti in South and Ifa in West)

African traditional medicine is the oldest, and perhaps the most assorted, of all therapeutic systems. Africa is considered to be the cradle of humankind with a rich biological and cultural diversity marked by regional differences in healing practices. The traditional healer typically diagnoses and treats the psychological basis of an illness before prescribing medicines. Mainly medicinal plants are used to treat the symptoms. There are mainly two reasons for the long interest in traditional medicine in Africa:

1. Inadequate access to allopathic medicines and Western forms of treatments. The majority of people in Africa cannot afford access to modern medical care either because it is too costly or because there are no medical service providers.
2. Lack of effective modern medical treatment for some ailments such as malaria and/or HIV/AIDS, which, although global in distribution, disproportionately affect Africa more than any other region in the world.

Africa already contributes nearly 25% to the world trade in biodiversity. In spite of this huge potential and diversity, the African continent has only a few drugs commercialized globally [23]. In due course, Muti and Ifa medicines were developed in different parts of Africa. Muti is practiced in traditional medicine in Southern Africa and as far north as Lake Tanganyika. In Southern Africa, "Muti" is a slang word for medicine in general. Ifa is a religion and system of divination and refers to the verses of the literary corpus known as the *Odu Ifá*. It is practiced in the Americas, West Africa, and the Canary Islands, in the form of a complex religious system [24–26]. Research reports have proved that African herbal medicines had potent bioactives in the treatment of cancer, mainly hypoxides from many plants, including *Hypoxis hemerocallidea* and *Hypoxis colchicifolia* [27].

2.6 Iranian Medicine

Iran has a medicine system known as Iranian medicine. Rhazes and Avicenna were two Persian medical scholars with important contributions to the medicine system. They gave contributions based on two major medical books: *Liber Continens* and *The Canon of Medicine*. They collected and systematically expanded the Greek, Indian, and Persian ancient healing systems with further discoveries. *The Canon of Medicine* was the greatest work of Avicenna, which remained as a standard medical textbook in Europe for nearly 700 years. There was lesser systematic compilation works after the canon of medicine though there were valuable studies after that. It was initially translated into Latin around AD 1150 by Gerard of Cermona from northern Italy and it was printed for the first time in 1473. It consisted of a compilation of five books:

1. *Universalia* (*al-Kulliyyat*) presents general medical principles.
2. *Matrica Medica* (*Mofradat*) deals with simple drugs.
3. The third book is devoted to diseases occurring in particular parts of the body.
4. The fourth book contains general medical topics such as skin rashes, wounds, fractures, luxations, ulcers, and fevers.

Pharmacopoeia (*Qarabadeyn*) deals with compound remedies [28].

Almost all medicine systems discussed used natural products with or without minerals, etc. as prescriptions, which were developed to combat the various health issues in the past, and evolved from myth to history then to written records of the present world. We are using the written documents as guidelines for finding new drugs for experiments, along with important applications for curing ailments though traditional and folk medicines. Suitable recognition of traditional medicine is an important element of national health policies and also has an important basis for new products with significant export potential. Well-recorded medicine systems such as Maibaron discussed earlier should be encouraged for proper recognition and for researching the recorded data. It can be used as a guideline for searching for new drugs required for diseases that currently do not have proper therapeutics or present existing drugs limited by undesirable side effects. Untapped biological resources, combined with knowledge from folk and traditional medicines and technological advances in screening, separation, and synthesis, have to be integrated for new natural product drug discovery ventures. There have been many drugs that have had their origins in natural products in traditional medicine, e.g., Taxol.

3. TAXOL AS A LEAD TO CANCER DRUG DISCOVERY

Cancer is a feared disease for which increasing numbers of medicines are becoming available every year. Because of its complicacy it is very difficult to control cancer in its late stages. So, cancer therapy is a big challenge and specific drugs for different types of cancer need to be developed. The number of cancer cases is growing even though many drugs have been discovered. Modern scientific investigations for new drug discovery have relied on traditional and folk medicines as a guide to their work because they are outcomes of trial and error studies over a long period of time. In fact, many drugs have been developed by tracking their traditional use. In cancer chemotherapy, some 67% of the effective drugs may be traced to natural origins. Many of these drugs are comprehensively reviewed in the recent volumes of anticancer

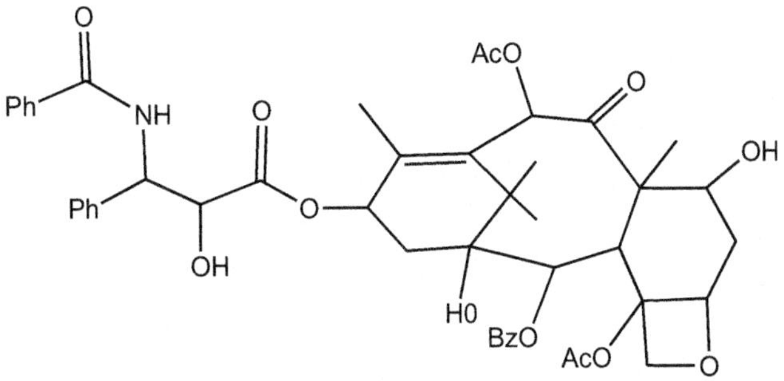

FIGURE 2.1 Paclitaxel structure.

agents from natural products. Examples of such compounds are the vinca alkaloids vinblastine and vincristine from *Caranthus roseus* and paclitaxel (Taxol) (Fig. 2.1) from *Taxus brevifolia*. There are many compounds, such as camptothecin from the bark of *Camptotheca acuminate*, which are precursors to the semisynthetic drugs topotecan (Hycamptin) and irinotecan (Camptosar). Taxol (paclitaxel) is a natural product extracted initially from the bark of the Pacific yew tree *T. brevifolia* but lately it has been increasingly isolated from *T. baccata*. *T. baccata* is a good example of the source of an anticancer drug. *T. baccata* is in demand for Taxol content for its anticancer properties. At present, Taxol and its derivatives are synthesized to meet the growing requirements of anticancer drugs. Therapies may be of synergistic action of different components in the plants and in some cases only a single component is potent [1]. Taxol as a single compound is bioactive. Herein Taxol discovery is discussed for illustrating the discovery of a drug from traditional medicine. In 1021, *T. baccata*, the European yew tree, was documented for phytotherapy in *The Canon of Medicine* for cardiac remedy [29]. *T. baccata* is further used as a traditional medicine treatment for breast and ovarian cancer in the Central Himalayas [30]. Investigations on the species by a research team reported that certain compounds found in the bark of yew have anticancer properties [31]. Later the compound paclitaxel (Taxol) was isolated from the bark of the plant. This illustrates the evolution of a drug from traditional and folk medicines. To discuss at length the schematic presentation given in Fig. 2.2, the route to the discovery of a drug that has its origin in traditional medicine starts with its traditional use recorded in a particular system in the form of books or nowadays databases and portals. *T. baccata* has a traditional medicinal use and in the Central Himalayas it has been used as a treatment for cancer and as a cardiac remedy. The fractionation was performed according to the results of the bioassay. The fraction with the positive bioassay results were further fractionated so as to find out the synergistic bioactives or a single bioactive. The extracts were tested in animal models and then in clinical trials. The precursors of the chemotherapy drug paclitaxel (Taxol) can be obtained from

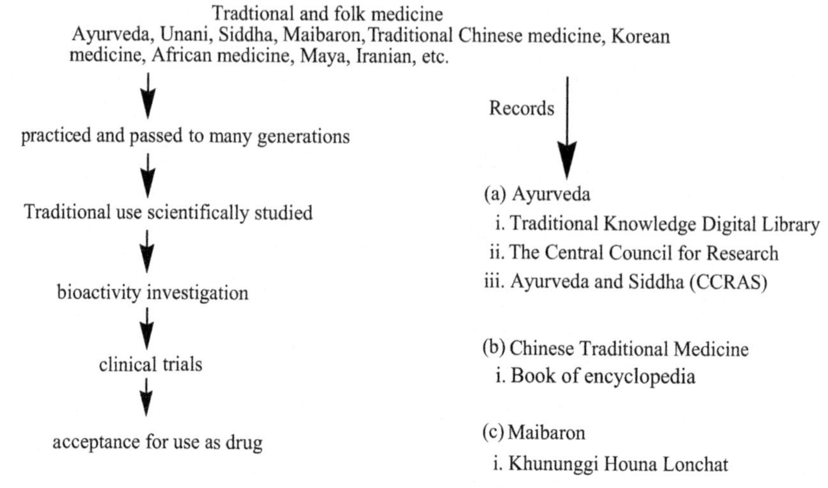

FIGURE 2.2 Schematic presentation of the evolution of a drug. CCRAS, *Central Council for Research in Ayurveda and Siddha.*

the bark of the plant. *T. baccata* was preferred because it is a much more renewable source than the bark of the Pacific yew (*T. brevifolia*) from which the precursors were initially isolated. The active fraction was analyzed and bioactive paclitaxel (Taxol) was isolated. This was the first linking stage in developing the anticancer drug Taxol from folk and traditional medicines. Then, it had to undergo clinical trials in rat models and subsequently to higher models in the single compound form. Then it was further trialed in humans. It was approved by a competent authority for the treatment of ovarian cancer, and further studies involving breast, colon, and lung cancers have shown promising results. After these types of pharmacological and chemical investigations, Taxol was isolated and identified. The success of the clinical trials made Taxol an anticancer drug. The drugs, though bioactive, may have a number of shortcomings, e.g., insolubility or undesirable side effects. So, further research was carried out to synthesize and develop more efficient derivatives of Taxol. One drawback of paclitaxel is its very limited water solubility. To overcome this and to improve its efficiency, a number of research teams prepared water-soluble derivatives, e.g., smart taxanes, which are water-soluble paclitaxel derivatives developed for targeted therapy of cancer. The first tumor-directed derivative of paclitaxel was designed and reported to promote the pharmacological efficacy of Taxol by synthesizing a water-soluble tumor-recognizing conjugate. A 7-amino acid synthetic peptide, BBN [7–13], which binds to the cell surface bombesin/gastrin-releasing peptide, was developed [32]. Furthermore, a more efficient form, silver nanoparticles with Taxol, has been synthesized. As the size reduces to the nanorange the surface area increases and it magnifies many bioactive properties. Multifunctional silver nanoparticles were synthesized with *T. baccata* as reducing agent and stabilizer. The silver nanoparticles possess potential for biomedical properties over the anticancer properties. The biosynthesized silver nanoparticles may have potential as multifunctional nanoparticles for different applications such as diagnosis, targeting, drug delivery, and imaging of the tumors simultaneously during the therapy of cancer cells. Thus the efficiency of anticancer drugs is improved for use in biomedical applications too. The anticancer activity of the silver nanoparticles of taxol had higher anticancer activity from the organic taxol [33]. Taxol derivatives, their preparation, and the pharmaceutical compositions containing them were protected in the US patent US 4814470. New Taxol derivatives of the formula with different substituents and their isomers, which were also useful antitumor agents, were synthesized. New formulations were developed that contained a Taxol derivative combined with one or more pharmaceutically acceptable, inert, or physiologically active diluents [34]. A semisynthetic analog of paclitaxel, Taxotere, has also been found to have good antitumor activity in animal models. Taxotere is currently undergoing clinical trials in Europe and the United States. Amino acid derivatives and a new formulation with the new derivatives have been protected as US patent US5489589. The compositions were of water-soluble amino acid derivatives of paclitaxel. The derivatives can be used as antitumor agents. Taxol has been shown to have excellent antitumor activity in in vivo animal models. It has a peculiar mode of action that triggers an abnormal polymerization of tubulin and disruption of mitosis in cancer cells. It has recently been approved for the treatment of ovarian cancer. Studies involving breast, colon, and lung cancers have shown promising results [35]. It has now been found that the derivative has an activity significantly greater than that of Taxol. With the advent of combinatorial synthesis, new derivatives for better pharmacological efficacy have been developed further, and the quest for new drugs has been accelerated.

Though Taxol was a big discovery for cancer treatment there have been many drugs that use traditional medicine as targets. The discovery of artemisinin, an antimalarial drug from CTM, was also a breakthrough in drug discovery. The route to the discovery of artemisinin was short compared with those of many other phytochemical discoveries in drug development. Artemisinin with its unique sesquiterpene lactone was created by phytochemical evolution. Many other therapeutic compounds have been developed from TCM. Arsenic, an ancient drug used in Chinese medicine, has been found to be an effective and relatively safe drug in the treatment of acute promyelocytic leukemia (APL) from clinical studies. Arsenic trioxide is now considered to be the first-line treatment for APL. Huperzine A, an effective agent for the treatment of memory dysfunction, is derived from the Chinese medicinal herb *Huperzia serrata*. A derivative of huperzine A is now undergoing clinical trials in Europe and the United States for the treatment of Alzheimer's disease. The molecules chuangxiongol and paeoniflorin derived from Chinese medicinal products have been tested for their efficacy in preventing restenosis after percutaneous coronary intervention. Evidence from clinical trials supporting the therapeutic value of related strategies from Chinese medicine aimed at activating blood circulation has been obtained in the treatment of ischemic diseases and in the management of myocardial ischemia/reperfusion injury. The joint application of exercise (to increase the shear stress of blood flow) with extracts from shenlian, another Chinese medicine, shows promise for the prevention of atherosclerosis. Salvianolic acid B, a compound from the root of *Salvia miltiorrhiza*, in combination with increased shear stress, has been identified as a potential mechanism to account for the effect of the functions of endothelial cells [21].

Though there are many drugs for the treatment of existing diseases they have undesirable side effects and some new diseases do not have proper drugs for their cure. There are many life-threatening diseases worldwide. So, there is a great need for phytochemical and pharmacological studies to produce remedies for either a single bioactive or an extract with more than one compound. Even though scientists are continuously working to discover efficient drugs for treating prevalent diseases, new diseases and microbial infections are constantly manifesting themselves. Liver diseases and dengue are two life-threatening diseases for which proper therapies are waiting discovery. Despite the quest for traditional medicinal plants as targets, drugs and therapies for liver disorders and dengue are at the beginning of their phytochemical evolution.

4. DEMAND FOR DRUGS FOR LIVER DISORDERS (HEPATIC DISEASE)

Liver disorders and hepatic diseases are becoming very common with the introduction of adulterants in foods and the consumption of excess alcohol and other drugs. However, drugs for therapy are limited. Other liver diseases are caused by viral infection and some might be of genetic origin. Alcohol overconsumption can lead to several forms of liver disease, including alcoholic hepatitis, fatty liver disease, cirrhosis, and liver cancer. In the advanced stages of alcoholic liver disease, fat builds up in the liver's cells due to increased creation of triglycerides and fatty acids, with a decreased inability to break down fatty acids. If not controlled in time the disease can lead to liver inflammation from the excess fat in the liver. Scarring in the liver often occurs as the body attempts to heal, and extensive scarring can lead to the development of cirrhosis, which is the formation of fibroids instead of liver cells in more advanced stages of the disease. In this case, hepatocellular carcinoma or liver cancer may develop. The challenge is for medical science to thoroughly research the discovery of new drugs to combat liver disorders. There are no proper drugs for the cure of hepatic diseases, and there are many types of hepatic disorders with different causes. Fascioliasis, hepatitis, alcoholic liver disease, fatty liver disease, etc. are just some of the liver diseasesthat remain serious health issues. Their management is a difficult task for modern medicine to tackle without proper therapies. Hepatic diseases are caused by xenobiotics, alcohol consumption, malnutrition, infection, anemia, and certain medications [36]. Paracetamol (acetaminophen or *N*-acetyl-*p*-aminophenol) is a commonly used analgesic and antipyretic drug. At high doses, it is attributed to hepatotoxicity due to the formation of the toxic, highly reactive metabolite *n*-acetyl parabenzoquineimine. Conventional or synthetic drugs used in the treatment of liver diseases are inadequate and can have serious adverse effects. So, there is a global demand for studies of traditional medicinal plants as targets for liver treatment. Many natural products of herbal origin are in use for the treatment of liver ailments [37]. Hepatic problems have caused a significant number of liver transplants and deaths. However, available pharmacotherapeutic options for liver diseases are very limited. So, the development of new effective drugs is in great demand [38]. Silymarin is used for the cure for liver disorders. Silymarin is a unique flavonoid complex containing silybin, silydianin, and silychrisin, which is derived from the milk thistle plant *S. marianum*. Silymarin exerts membrane-stabilizing and antioxidant activity and promotes hepatocyte regeneration.

Silymarin has a hepatoprotective effect in chronic liver diseases caused by oxidative stress. This may be due to alcoholic and nonalcoholic fatty liver diseases, drugs, or chemically induced hepatic toxicity. Silymarin also reduces inflammatory reaction and inhibits fibrogenesis in the liver. These results have been established by experimental and clinical trials. According to open studies, the long-term administration of silymarin significantly increased the survival time of patients with alcohol-induced liver cirrhosis. Some studies showed that silymarin can significantly reduce tumor cell proliferation, angiogenesis, and insulin resistance. The chemopreventive effect of silymarin on hepatocellular carcinoma or liver cancer has been established in several studies using in vitro and in vivo methods. It can exert a beneficial effect on the balance of cell survival and apoptosis by interfering with cytokines. In some neoplastic diseases, silymarin can be administered as an adjuvant therapy as well [39,40].

The unique phytochemicals from milk thistle, silybin, silydianin and silychrisin, have been the subject of many years of research into their hepatoprotective and other properties. Milk thistle gets its name from the white markings on the leaves, its milky white sap, and its traditional use by nursing mothers to increase milk. It is best known for its use as a liver protection and decongestant, which can be traced back to the Greeks and Pliny the Elder (AD 23–79), who recorded that it was excellent for getting rid of bile. The famous English herbalist Culpepper (1616–54) used milk thistle to cleanse the liver and spleen, and to treat jaundice and gallstones [41]. Silymarin is marketed in the form of syrup and tablets with many trade names. It is widely used for many liver diseases. Many extracts from plants and macrofungi have been reported to have hepatoprotective activity. Studies on *Auricularia* by the author have given positive results. *Auricularia* is a genus comprising edible macrofungi that grow on fresh wood or decaying tree trunks. They are found in hilly, swampy forests. According to folklore the use of *Acaulospora delicata, Auricularia auricular*, and *Auricularia polytricha* for healing liver has been given in the literature of the Maibaron of the Meetei of Manipur, India. The discovery of a high content of the highly antioxidant compound CGA in mushrooms could be the key compound responsible liver healing. CGA is highly bioavailable in humans and possesses antiinflammatory activities. Traditional use by the Meeteis of the documented species in the Maibaron is liver healing. Thus the species were used as a target for our research on screening for the highly antioxidant compound CGA. An HPLC method for the preliminary screening of traditional medicinal use and intraspecies comparison of *A. delicata* and *A. polytricha* was reported regarding the qualitative and quantitative analysis of CGA. Though the concentration was different, CGA was present in both species. CGA was estimated as 0.11% in *A. polytricha* and 0.02% in *A. delicata* (Table 2.1). Elemental composition was studied by energy dispersive X-ray analysis. K was 3.84% in *A. delicata* and 2.16% in *A. polytricha*, respectively. Ca was higher in *A. polytricha* (1.79%), while it was only 0.55% in *A. delicata* (Table 2.2). The promising liver-healing lead from traditional use and the presence of the superantioxidant CGA could be correlated. Phytochemical and pharmacology research on liver healing is ongoing by the author and her research group. Maibaron can be a good guideline for other medicine systems for drug discovery for other ailments too [2]. The author used another method of chemical screening instead of bioassay-guided fractionation, which is tedious and costly. This is a new aspect that is ongoing and has been taken up as an experiment for in vivo hepatoprotective activity involving animal sacrifice. To minimize animal sacrifice for in vivo experiments, chemical screening with standards has been used rather than bioassay-guided fractionation. Screening with chemicals and in-vitro antioxidant activity are done prior to in vivo experiments. The effect might be in synergy with the minerals too because in many medicine systems minerals are used with natural products.

TABLE 2.1 Concentration of Chlorogenic Acids (CGAs) in *Auricularia polytricha* and *Acaulospora delicata* by High-Performance Liquid Chromatography

	A. polytricha	*A. delicata*
Concentration of CGA	0.11%	0.02%

TABLE 2.2 The Elemental Content of *Auricularia polytricha* and *Acaulospora delicata* With Energy Dispersive X-Ray Analysis (EDX)

Element	Wt% *A. polytricha*	Wt% *A. delicata*
C	46.65	55.41
N	6.75	8.18
O	31.65	24.78
Na	0.26	0.08
Mg	0.82	0.72
Al	1.67	1.18
Si	6.23	3.51
P	0.80	0.53
S	0.12	0.30
Cl	0.05	0.12
K	2.16	3.84
Ca	1.79	0.55
Fe	1.06	0.80
Total	100	100

Until now liver treatment is mainly performed using silymarin with its synergistic components silybin, silydianin, and silychrisin. Conventional or synthetic drugs are also used in the treatment of liver diseases, but these are insufficient for successful treatment of liver diseases. They also have serious adverse effects, so the future search requirement is for a drug that can heal the liver with minimum side effects.

5. DEMAND FOR DRUGS FOR DENGUE

Dengue fever (DF) is a disease where the infected person suffers from a high fever and the blood platelet count decreases rapidly. If it is not treated promptly it proves to be fatal. Unfortunately, there is no drug to combat this viral infection. A blood transfusion to maintain platelet count and drugs such as antipyretics and medicinal plant extracts are administered to fight the virus. Here there are two contributors: the mosquito and the virus. DF is caused by the arthropod-borne flavivirus called dengue virus (DENV). The virus is transmitted by the *Aedes aegypti* mosquito. There are four antigenically related but distinct virus serotypes (DENV-1, 2, 3, and 4) belonging to the genus *Flavivirus* in the Flaviviridae family, which causes DF. With a shrinking world caused by better transport and communication systems, an increasing human population, and urbanization, favorable conditions have been created for the mosquito vector *A. aegypti* to spread the virus to new areas, causing major epidemics. Even though dengue problems have caused a significant number of deaths and problems worldwide, available pharmacotherapeutic options for dengue are very limited and the development of new effective drugs is in great demand, as it is in the case of liver diseases. Complications in the development of a dengue vaccine and unavailability of synthetic drugs have encouraged the development of a herbal-based antiviral preparation that is safer and less harmful than synthetic drugs. As done in the past, targets for the discovery of a medicine for dengue have been from folk medicines. Traditional medicines such as *P. guajava* leaf extract, *A. indica* (neem), tea prepared by using *O. sanctum* boiled leaves, methanolic extract of *M. charantia*, and *C. papaya* leaf extract have been used for the treatment for dengue. Until now plant extracts are successfully combating dengue with other synthetic medicines. However, there are still no standardized plant extracts available as there are for liver disorders such as silymarin. Investigations are being carried out to discover the key moiety from the available natural sources. Dengue can be controlled by controlling both its vector and the virus. The development of a dengue vaccine is complicated by the antibody-dependent enhancement effect. Thus the development of a natural product-based antiviral preparation promises a more potential alternative in combating dengue disease. There are currently no specific treatments for DF. Only a standard treatment for the management of fever is given, i.e., nursing care, fluid balance, electrolytes, and blood-clotting parameters. Much research has been done so far to plan for controlling dengue and developing formulations and drugs for controlling the virus. Herbal treatments from plants and mushrooms and compounds, which can control the virus and vector, are being assessed and screened because a drug to cure dengue needs to be discovered. And, as the saying goes, prevention is better than cure in controlling the unwanted epidemic of dengue. Some of the plants used and their modes of usage are quite simple. *P. guajava* (guava), *A. indica* (neem), *O. sanctum* (tulsi), *M. charantia* (bitter gourd), and *C. papaya* (papaya) leaf have been used in the treatment of DF. These are plants that are easily available in India and can be used for controlling the virus. The plants and mushrooms that have been used in combating dengue to treat the virus must (1) target —and control of virus and (2) targeting −and control the vector to stop the epidemic; these are discussed next. The extracts from plants as well as mushrooms and single compounds separated and purified were effective.

5.1 Targeting —and Controlling the Virus

5.1.1 Plant Sources

Psidium guava, *Alternanthera philoxeroides*, *Andrographis paniculata*, *A. indica*, *Boesenbergia rotunda*, *C. papaya*, *Cladogynos orientalis*, *Cladosiphon okamuranus*, *Cryptonemia crenulata*, *Cymbopogon citratus*, *Euphorbia hirta*, *Flagellaria indica*, *Gymnogongrus griffithsiae*, *Gymnogongrus torulosus*, *Hippophae rhamnoides*, *Houttuynia cordata*, *Leucaena leucocephala*, *Lippia alba* and *Lippia citriodora*, *Meristiella gelidium*, *Mimosa scabrella*, *M. charantia*, *O. sanctum*, *Piper retrofractum*, *P. guajava*, *Quercus lusitanica*, *Rhizophora apiculata*, *Tephrosia crassifolia*, *Tephrosia madrensis* and *Tephrosia viridiflora*, *Uncaria tomentosa*, and *Zostera marina* were found to be effective for DENV from different investigations. Of the shortlisted plants *P. guajava* leaf extract has been tested in vitro and showed to inhibit the growth of DENV. Tea, which is traditionally prepared by using *O. sanctum* boiled leaves, acts as a preventive medicament against DF. The maximal nontoxic dose of the methanolic extract of *M. charantia* against Vero E6 cells was investigated in vitro. The methanolic extract of *M. charantia* showed an inhibitory effect on DENV-1 by antiviral assay based on cytopathic effects. *C. papaya* leaf has been used traditionally in the treatment of DF. While papaya leaf extract inhibits the bacterial infection that causes the fever, tawa-tawa extract prevents bleeding. In addition, unpublished research has found that *P. guava* leaves are a good way to increase platelets, thus helping to avoid bleeding. The in vitro and in vivo inhibitory potential of aqueous extract of *A. indica* (neem) leaves on the replication of DENV-2 was evaluated [42]. *C. kamuranus*, *L. leucocephala*, *M. scabrella*, *T. madrensis*, *C. crenulata*, *G. torulosus*, *H. cordata*, *M. gelidium*, *B. rotunda*, and *Z. marina* were also reported to have antiviral activity [43]. The methanolic extract of *Vitis cinerea* leaves, the ethanol extract of *Tridax procumbens* stems, and the ethanol extract of *S phaeralcea angustifolia* had activity against the virus [44].

5.1.2 Compound Sources

Many compounds have been screened to be active against dengue. The cyclohexenylchalcone derivatives of *B. rotunda*, 4-hydroxypanduratin A and panduratin A, showed good competitive inhibitory activities toward DENV-2 NS3 protease [45]. A sulfated polysaccharide called fucoidan from *C. kamuranus* was found to potentially inhibit DENV-2 infection [46]. The sulfated polysaccharides from *C. crenulata*, i.e., galactan, were selective inhibitors of DENV-2 multiplication in Vero cells [47]. Galactomannans extracted from seeds of *L. leucocephala* have demonstrated activity against yellow fever virus and DENV-1 both in vitro and in vivo [48]. The flavonoids isolated from *T. madrensis*, glabranine and 7-*O*-methyl-glabranine, exert strong inhibitory effects on DENV replication in LLC-MK2 cells [49]. Fucoidan, galactomanan, glabranine, galactan, hyperoside, kappa carrageenan, zosteric acid, and myrsellinol were also compounds that were effective for DF [43]. Furthermore, in other research, rutin and quercetin, which are strong antioxidants, displayed relevant antiviral activity [50].

In another review, many compounds have been reviewed that have low toxicity to humans and possess antiviral activity. Fucoidans are a group of polysaccharides that contain considerable percentages of L-fucose and sulfate ester groups, sulfated fucose, and its derivatives. Several flavonoids have been screened for dengue antiviral activity, including compounds isolated from Mexican *Tephrosia* species glabranine and 7-*O*-methylglabranine. Emetine is a compound belonging to the ipecacuanha alkaloids. Its dihydrochloride was identified by Low and coworkers as a compound displaying potent DENV antiviral activity. Terpenoids and polycyclic quinones such as hypericin, tetrabromohypericin, and gymnochrome B were also effective against DF [51].

5.1.3 Mushroom Sources

In folk medicine, mushrooms have also been used as medicine; these were also screened to discover their antiviral properties. US patent US 20060171958 A1 describes a formulation having unique antiviral properties prepared from medicinal mushroom mycelium extracts and derivatives. The compositions are derived from *Fomitopsis*, *Piptoporus*, *Ganoderma*, and blends of medicinal mushroom species and are useful in preventing and treating viruses, including *Orthopoxvirus*, influenza, avian influenza, Venezuelan equine encephalitis, yellow fever, West Nile, dengue, New World and Old World arenaviruses, hantavirus, Rift Valley fever, sandfly fever, SARS, rhinovirus, and other viruses [52].

5.1.4 Formulations

There has been no formulation to date to combat DF, though efforts continue. Micro Labs Limited, a Bengaluru-based pharmaceutical company, launched Caripill, made from *C. papaya* leaf extracts, to help increase platelet count in patients suffering from dengue. Approved by the scientific and regulatory authority, the pill does not have any side effects. While the conventional treatment involves platelet transfusion, a dose of Caripill of 1100 mg can be taken three times a day for 5 days to increase platelet count. Caripill syrup with 275 mg/5 mL in different dosages is safe for children too [53]. However, further improved formulations and drugs need to be discovered.

All these results need to be further investigated with clinical trials for an effective drug to combat DF.

5.2 Targeting —and Controlling the Vector Mosquito

Another way of controlling dengue is by controlling the vector spread and stopping any unwanted epidemic caused by the *A. aegypti* mosquito. Spreading of the mosquito is controlled by killing at the larva stage. Larvicidal activity is studied with different extracts of different medicinal plants. Larvicidal activity of *A. aegypti* was exhibited by *Glutaren ghas* bark extract against *Mangifera indica*, *Melanochyla fasciculiflora*, and *Anacardium occidentale* [54]. Larvicidal effects of leaf and stem/bark extracts of *Jatropha curcas*, *Citrus grandis*, and *Tinospora rumphii* were tested on the larvae of the dengue vector *A. aegypti*. The extracts also showed larvicidal properties [55]. This study was undertaken to evaluate larvicidal efficacy of *Vitex negundo*, *Clerodendrum inerme*, and *Gliricidia sepium* individually, and synergistic activities were combined with *Pongamia glabra* seed extract against the early fourth instar vector mosquito *A. aegypti*. The results of the study revealed that *V. negundo* and *C. inerme* plant leaves individually and in combination with extracts of *P. glabra* can be used as a potent source of natural mosquito larvicidal agent [56]. The crude extracts of fruits of *Sapindus mukorossi* and leaves of *Cestrum nocturnum*, *Cestrum diurnum*, and *Asclepias curassavica* were tested under laboratory conditions against *A. aegypti*, a vector of dengue and chikungunya, for their larvicidal properties. Bioassay experiments carried out with crude alcoholic extracts of *S. mukorossi*, *C. nocturnum*, *C. diurnum*, and *A. curassavica* showed larvicidal properties [57].

Basidiomycete macrofungi *Cyptotrama asprata* is a mushroom that shows strong larvicidal activity against the mosquito *A. aegypti*. From submerged cultures of *C. asprata*, a secondary metabolite (oxiran-2-yl)methylpentanoate was isolated, which was responsible for its larvicidal activity against *A. aegypti* [58].

Altogether, many crude extracts and compounds have been discovered, e.g., extracts from *P. guajava*, *A. indica*, *O. sanctum*, and *M. charantia*. *M. charantia* has been effective in treating dengue and *C. papaya* leaf has been used traditionally in the treatment of fever. These are plants that are easily available in India and can be used for controlling the virus. In parallel, for controlling the vector, crude alcoholic extracts of *S. mukorossi*, *C. nocturnum*, *C. diurnum*, and *A. curassavica*, which have larvicidal properties can be used to control the spread of the fever. To control the spread of dengue both the virus and the vector need to be controlled. Finally, more effective pharmaceutically acceptable drugs or formulations need to be developed to study the life cycle of the vector and enzymes involved. However, reliable standardized therapy is not available for dengue and currently the symptoms are combated as a treatment for fever and replenishing blood loss. Newly discovered therapies are in great demand, which can be either a single bioactive or possess synergistic effects of many components in a fraction. Though there are many new extracts with activity against DF, there have been no new drugs that can combat the symptoms of dengue. Some extracts and compounds were effective against the fever and some increased platelet count. However, there is no standardized drug for combating the virus. Therefore the targets are under investigation but further bioactivity screening and chemical isolation need to be performed to produce an effective drug.

In the past, traditional medicine had been a lead to drug discovery. The examples cited previously were Taxol, artemisinin, and silymarin. *A. delicata* in Maibaron is under further investigation as a target for liver therapy. Using traditional and folk medicines as targets, medicines have been discovered that can be a single drug or group of bioactives acting together for combating diseases. As discussed before, in the case of Taxol and artemisinin, it is a single compound and in the case of silymarin it is the synergistic effect of silybin, silydianin, and silychrisin. The discoveries have come about from detailed and repeated pharmacological investigations along with chemical analysis. Extractions using different polarity solvents, different temperatures, fermentation, Soxhlet extraction, etc. have been carried out to discover the optimized methods. However, the best extraction method often is the one that follows traditional knowledge. Bioassay-guided fractionation is performed in most of these cases. However, the author used chemical screening followed by bioassay to save time, chemicals, lessen animal sacrifice, and to be cost-effective. The active extract is fractionated for compound identification with open column chromatography or by preparative HPLC. Identification of the bioactives can be done using UV, IR, NMR, mass spectroscopy, and other chromatographic techniques. Screening is done by fractionating with different modes of extractions. A further crystallographic method with X-ray diffraction of single crystals is used for structural elucidation confirmation because it provides very reliable data regarding the structure of bioactives. For complex essential oils and nonpolar formulations, hyphenated techniques, which are very advanced such as thin layer chromatography-direct analysis in real time (TLC-DART), HPTLC, GC-MS, LC-MS, LC-MS-MS, etc. are used for the evaluation of constituents. The biological activities have to be investigated further.

In *Ficus pomifera* the author's research reports the presence of appreciable amounts of phenols and tannins in the plant leaves. Furthermore, triterpenoids were isolated with open column chromatography and structure confirmed with X-ray crystallography. Triterpenoids have been proved to be cytotoxic in previous research reports. The terpenoids were in the nanorange and had luminescence too. There is literature on the identification and quantization of active compounds with HPLC and LC-MS from natural products. CGA was identified and estimated with HPLC from *A. delicata* and *A. polytricha* by the author. A novel method of identification and estimation without using standards of unknown compounds with the liquid chromatography—electrospray ionization—mass spectrometry method was developed for the first time and validated in *F. pomifera*. A tentative structure of a new triterpenoid was discovered with IR, NMR, and mass spectra and the quantity was estimated. The molecular formula of the compound was assigned as $C_{27}H_{42}O_2$ with molecular mass of 398.5. The mass spectra of the compound exhibited a molecular ion peak at *m/z* 398.5 corresponding to the molecular formula $C_{27}H_{42}O_2$. Accounting for the peak area of the compound, the concentration of the triterpenoid was calculated as 8.64 ppm in 2000 ppm of the extract [2,6,7]. Validated analytical and manufacturing processes, process control, quality control, pharmacological test systems, and well-designed clinical studies are essential primary symbols for successful separation and identification of bioactives and active fractions.

After the isolation and identification of the new bioactives from the traditional medicinal plants, more efficient derivatives with desired solubility are synthesized. The synthesis methods are now further improved for finding active products in less time in a systematic way with combinatorial synthesis. With this a database is also created by picking up active constituents in less time with the use of software based on genetic algorithms, and efficient bioactives are generated.

Over time, many new diseases have emerged and scientists have been challenged from time to time to fight these diseases. At such times the only guide to research is traditional and folk medicines. Scientists have taken traditional and folk medicines as references because they are aware that such data are valuable and dependable. As described earlier, in the past such prescriptions were compiled by the healers by repeated trial and error. The results were compiled after their direct use on humans because in the past there were no developed scientific investigations and therefore standardized medicines. Positive results were compiled and the healers were authorized to practice traditional medicine according to the symptoms. Upon successful results and repeated practice the medicines were recorded for future use. Nowadays in modern research before reaching human trials, drugs or formulations have to undergo many clinical trials, which are performed on rat models and higher animals. As a result of successful trials, drugs can only be used on humans after approval by a competent authority. Old medicine systems are still flourishing in places such as Africa where there is limited supply of modern medicines and fewer medical facilities. Even in the most developed countries, when diseases such as dengue spread we still have to rely on the traditional and folk medicines. These medicine systems have paved the way for discovering new drugs to combat various types of diseases.

6. CONCLUSION

Over the years, humankind has developed many natural products from a number of different civilizations. Evidence of the use of natural products for various ailments by our ancestors has shown the in-depth relationship we have with Mother Nature. With the development of the human way of life, medicines have also evolved. These medicines have been formulated in the correct doses and specific medicines for specific ailments have been planned. Until now, many diseases have had no specific pharmacotherapeutic options as cited earlier, particularly for liver disorders and dengue. For those diseases for which therapies have been developed but not standardized, most are from medicinal plants and other natural products such as mushrooms as used in traditional and folk medicines. The trial and error use and healing practices using plants and other natural products have been passed down from generation to generation. These recorded medicine systems are the result of many experimental trials on humans so they can be used as a guide to search for new therapeutic bioactives for diseases for which no proper therapy has been developed. Even though these therapies were performed without background scientific knowledge, the data are valuable and can be used as leads for new drugs because of their proven track record. Ayurveda and TCM have been scientifically investigated to discover the active moiety and mechanism of action of the formulation of a particular phytochemical. Medicine systems such as the Maibaron, which have been practiced from a long time and are well documented, should be given recognition. This can increase our medicine system database for targets for drug discovery. Presently, from such studies many drugs have been developed, and in the search for new medicines to combat certain ailments, science has been dependent on traditional and folk medicines for answers. As humans evolved, medicine has also developed to treat most ailments. Taxol is an illustrative example of drug discovery by scientific investigation from *T. baccata*, which has a traditional use in cancer cure especially in the Central Himalayas. Silymarin from the milk thistle plant was traditionally used as a liver protectant and decongestant. There were reports on the use of milk thistle to cleanse the liver and spleen and to treat jaundice and gallstones. Here, many compounds act in synergy to produce the desired effect. The potential use of higher plants as a source of new drugs is still poorly explored. From the whole plant kingdom, only a small percentage has been investigated phytochemically and an even smaller percentage has been properly studied in terms of their pharmacological properties, including extensive screening. Research into isolated plant constituents is of great importance for the development of bioactive substances from ethnic medicine. With the emergence of the latest technologies and enhanced knowledge of isolated plant constituents, characterization, and analytical tools, a number of compounds have been efficiently isolated from potential plants and have contributed greatly to drug discovery from ethnic plants. Separating a medicinal herb into its constituents sometimes cannot explain exactly the way it works in its natural form. The whole herb might be worth more than the sum of the activities of its single components. A plant contains hundreds of chemical constituents that interact in a complex way to produce therapeutic effects. The detailed mechanism by which a particular medicinal plant is effective might not be understood, even though its medicinal benefits are well established. Advanced analytical tools such as TLC-DART, HPTLC, HPLC, LC-MS/MS, etc., several plant extracts, and their formulations maybe standardized in a better way to enhance the bioactivity of the ethnic medicines. The advanced analytical techniques along with biotechnological interventions, nowadays called chemical biology, will be a major tool in the development of new therapies and bioactive molecules. Thus with the new techniques, traditional and folk medicines can be used safely as targets to drug discovery. These systems were developed after repeated trials on humans, which have been practiced for a long period of time from generation to generation, from mythology to history to the documented data of today. The medicine systems are legacies for health and healing from the ancient world for drug discovery to the present world. The legacies have to be tapped for active drugs and herbal formulations with synergistic effects.

REFERENCES

[1] W.D. Sujata, L.S. Warjeet, Studies on the uses of some plants for medicinal and dyeing properties, Int. J. Chem. 5 (1) (2016) 93–102.

[2] W.D. Sujata, D.D. Joshi, Screening and comparison of two edible macrofungi of *Auricularia* Spp, Curr. Sci. 112 (3) (2017) 460–463.

[3] Traditional Medicine: Definitions, World Health Organization (WHO) (2000). www.who.int>medicines>areas>defininition.

[4] A. Deepak, S. Anshu, Indigenous Herbal Medicines: Tribal Formulations and Traditional Herbal Practices, Aavishkar Publishers Distributor, Jaipur, India, 2008, ISBN 978-81-7910-252-7, p. 440.

[5] D.D. Joshi, R. Prasad, W. Sujata, Medicinal plants for microbial control, Enviromental microbiol. (2015) 45–59. I.K. International Pvt. Ltd. S-25, Green Park Extension Uphaar Cinema Market New Delhi-110 016, India.

[6] W. Sujata, L.S. Warjeet, HPLC open column chromatography and LC-ESI-MS separation patterns and a new method of estimation of phytoconstituent from HEIBAM (*Ficus pomifera* Wall), Acta Chromatogr. 28 (1) (2016) 5–17.

[7] W. Sujata, L.S. Warjeet, G.S. Jiten, L.S. Surendrajit, W. Amey, Cytotoxic triterpenoids from *Ficus pomifera* wall, Indian J. Chem. Sect. B 15B (2015) 676–681. Highlighted by publication as article in Cheminform Abstract Wiley.

[8] S. Robert, Ayurveda: Life, Health and Longevity, Ayurvedic Press, Alburquerque, New Mexico, 2004.

[9] S. Robert, L. Arnie, Tao and Dharma, Chinese Medicine and Ayurveda, Lotus Press, PO Box 325, Twin Lakes, WI 53181, USA, 2005.

[10] V. Narayanaswamy, Origin and development of Ayurveda: (A Brief History). US National Library of Medicine National Institutes of Health, Anc. Sci. Life 1 (1) (1981) 1–7.

[11] R.B. Saper, R.S. Phillips, A. Sehgal, N. Khouri, R.B. Davis, J. Paquin, V. Thuppil, S.N. Kales, Lead, mercury, and arsenic in US and Indian manufactured medicines sold via the internet, JAMA 300 (8) (2008) 915–923.

[12] About CCRAS. Central Council for Research in Ayurveda and Siddha. Department of AYUSH, Ministry Of Health and Family Welfare. www.ccras.nic.in.

[13] Traditional Knowledge Digital Library (Government of India). https:www.tkdl.res.in.

[14] H.S. Zillur Rahman, Unani medicine in India: its origin and fundamental concepts by Hakim Syed Zillur Rahman, in: B.V. Subbarayappa (Ed.), History of Science, Philosophy and Culture in Indian Civilization, Part 2 (Medicine and Life Sciences in India), vol. IV, Centre for Studies in Civilizations, Project of History of Indian Science, Philosophy and Culture, New Delhi, 2001, pp. 298–325.

[15] Siddha. Indian systems of medicine. www.ism.kerala.gov.in.

[16] N. Sanajaoba, Manipur Past and Present, first ed., vol. 2, K M Mittal for Mittal Publication A-110, Mohan garden, New Delhi-110059, 1991, pp. 45–47.

[17] N. Khelchandra, Th M. Pandit, Khunungi Houna Lonchat, second ed., Th Sanatomba, L Shushila, Th Phulachand for Padma Printers, Paona Bazar, Imphal Manipur, 2006, pp. 50–52.

[18] P.C.A. Kam, S. Liew, Traditional Chinese herbal medicine and anaesthesia, Anaesthesia 57 (2002) 1083–1089.

[19] G. Asaf, The Evolution of Chinese Medicine Song Dynasty, Routledge, Taylor and Francis Group, London and New York, 2008, pp. 960–1200.

[20] Z. Jiaju, X. Guirong, Y. Xinjian, Encyclopedia of Molecular Structures, Pharmacological Activities, Natural Sources and Applications Traditional Chinese Medicines, vol. 6, Springer Heidelberg, Dordrecht, London, New York Library of Congress Springer-Verlag Berlin Heidelberg, 2011.

[21] T. Youyou, The discovery of artemisinin (qinghaosu) and gifts from Chinese medicine, Nat. Med. 17 (10) (2011) 1217–1220.

[22] K.B. Kim, H.J. Park, D.H. Song, Self health diagnosis system for Korean traditional medicine with enhanced ART2, Adv. Sci. Technol. Lett. 33 (2013) 16–19.

[23] M.M. Fawzi, Traditional medicines in Africa: an appraisal of ten potent African medicinal plants, Evid. Based Complement. Altern. Med. (2013) 1–14, 617459.

[24] Traditional African Healing, African Code.

[25] Ifa: The Religion of the Yoruba Peoples. ReligiousTolerance.org. www.religioustolerance.org.

[26] E.A. Adedoja, Sixteen Major Books in Odu Ifa From Iile Ife.

[27] M. Edward, C. Curtiss, S. Dugald, K. Izzy, African herbal medicines in the treatment of HIV: Hypoxis and Sutherlandia. An overview of evidence and pharmacology, Nutr. J. 4 (19) (2005) 1–6.

[28] M.D. Mohammad-Hossein Azizi, History of ancient medicine in Iran. The otorhinolaryngologic concepts as viewed by Rhazes and Avicenna, Arch. Iranian Med. 10 (4) (2007) 552–555.

[29] Y. Tekol, The medieval physician Avicenna used an herbal calcium channel blocker, *Taxus baccata* L. Phytother. Res. 21 (7) (2007) 701–702.

[30] Asia Medicinal Plants Database. agris.fao.org/taxus baccata.

[31] National Non-Food Crops Centre, "Yew". www.nnfcc.co.uk.

[32] A. Safavy, K.P. Raisch, M.B. Khazaeli, D.J. Buchsbaum, J.A. Bonner, Paclitaxel derivatives for targeted therapy of cancer: toward the development of smart taxanes, J. Med. Chem. 42 (23) (1999) 4919–4924.

[33] A.K. Abolghasem, B. Abdol-Khalegh, H.Z.E. Sayyed, R.K. Ahmad, R. Amir, Green synthesis of anisotropic silver nanoparticles with potent anticancer activity using *Taxus baccata* extract, RSC Adv. 4 (2014) 61394–61403.

[34] C. Michel, G. Daniel, Gueritte-Voegelein Francoise, P. Pierre, Taxol Derivatives, Their Preparation and Pharmaceutical Compositions Containing Them US 4814470 A, 1989.

[35] D.W. Mark, F.K. John, Amino Acid Derivatives of Paclitaxel US 5489589 A, 1996.

[36] M. Mroueh, Y. Saab, R. Rizkallah, Hepatoprotective activity of *Centaurium erythraea* on acetaminophen-induced hepatotoxicity in rats, Phytother. Res. 18 (2004) 431–433.

[37] S.K. Mitra, S.J. Seshadri, M.V. Venkataranganna, S. Gopumadhavan, V. Udupa, D.N.K. Sarma, Effect of HD-03-a herbal formulation in galactosamine-induced hepatopathy in rats, Indian J. Physiol. Pharmacol. 44 (2000) 82–86.

[38] A.J. Akindele, K.O. Ezenwanebe, C.C. Anunobi, O.O. Adeyemi, Hepatoprotective and in vivo antioxidant effects of *Byrsocarpus coccineus* Schum. and Thonn. (Connaraceae), J. Ethnopharmacol. 129 (2010) 46–52.

[39] J. Feher, G. Lengyel, Silymarin in the prevention and treatment of liver diseases and primary liver cancer, Curr. Pharm. Biotechnol. 13 (1) (2012) 210–217.

[40] A. Presser, Pharmacist's Guide to Medicinal Herbs, Smart Publications, Petaluma, CA, 2000, pp. 259–260.

[41] H. Wagner, et al., The chemistry of silymarin (silybin), the active principle of the fruits of *Silybium marianum*, Arzneim-Forsch Drug Res. 18 (1968) 688–696 found in Silymarin: A Potent Antioxidant, Liver Protector, and Anti-Cancer Agent www.smart-publications.com/articles/silymarin-a-potent-antioxidant-liver-protector-and-anti-cancer-agent.

[42] L. Abd, K. Siti, Y. Harisun, M.Z. Razauden, Potential anti-dengue medicinal plants: a review, J. Nat. Med. 67 (4) (2013) 677–689.

[43] S. ShasankSekhar, D. Debasmita, Research and reviews, J. Pharmacogn. Phytochem. 1 (2) (2013) 5–9.

[44] H.A. Rothan, M. Zulqarnain, Y.A. Ammar, E.C. Tan, N.A. Rahman, R. Yusof, Screening of antiviral activities in medicinal plants extracts against dengue virus using dengue NS2B-NS3 protease assay, Trop. Biomed. 31 (2) (2014) 286–296.

[45] T.S. Kiat, R. Pippen, R. Yusof, H. Ibrahim, N. Khalid, N.A. Rahman, Inhibitory activity of cyclohexenylchalcone derivatives and flavonoids of fingerroot, *Boesenbergia rotunda* (L.), towards dengue-2 virus NS3 protease, Bioorg. Med. Chem. Lett. 16 (2006) 3337–3340.

[46] K.I.P.J. Hidari, N. Takahashi, M. Arihara, M. Nagaoka, K. Morita, T. Suzuki, Structure and anti-Dengue virus activity of sulfated polysaccharide from marine alga, Biochem. Biophys. Res. Commun. 376 (2008) 91–95.

[47] L.B. Talarico, R.G.M. Zibetti, M.D. Noseda, M.E.R. Duarte, E.B. Damonte, P.C.S. Faria, C.A. Pujol, The antiviral activity of sulfated polysaccharides against Dengue virus is dependent on virus serotype and host cells, Antivir. Res. 66 (2005) 103–110.

[48] L. Ono, W. Wollinger, I.M. Rocco, T.L.M. Coimbra, P.A.J. Gorin, M.R. Sierakowski, In vitro and in vivo antiviral properties of sulfated galactomannans against yellow fever virus (BeH111 strain) and dengue 1 virus (Hawaii strain), Antivir. Res. 60 (2003) 201–208.

[49] I. Sanchez, F.G. Garibay, J. Taboada, B.H. Ruiz, Antiviral effect of flavonoids on the Dengue virus, Phytother. Res. 14 (2000) 89–92.

[50] A.R.A. Silva, S.M. Morais, M.M. M Marques, D.M. Lima, S.C.C. Santos, R.R. Almeida, I.G.P. Vieira, M.I.F. Guedes, Antiviral activities of extracts and phenolic components of two *Spondias* species against dengue virus, J. Venomous Anim. Toxin. Incl. Trop. Dis. 17 (4) (2011) 406–413.

[51] R.T. Robson, L.P. Wagner, A.F. Costa da Silveira Oliveira, M. da Silva, A.S. de Oliveira Adalberto, L. da Silva Milene, C. da Silva Cynthia, O. de Paula Sergio, Natural products as source of potential dengue antivirals, Molecules 19 (2014) 8151–8176.

[52] Pual Statemets, Antiviral Activity from Medicinal Mushrooms, US20060171958 A1, 2006.

[53] Microlabs launches caripill. www.caripillmicro.com/.

[54] Y. Ali, F.Z. Wan, Lethal response of the dengue vectors to the plant extracts from family Anacardiaceae, Asian Pac. J. Trop. Biomed. 5 (10) (2015) 812–818.

[55] P.M. Gutierrez Jr., A.N. Antepuesto, B.A.L. Eugenio, M.F.L. Santos, Larvicidal activity of selected plant extracts against the dengue vector *Aedes aegypti* mosquito, Int. Res. J. Biol. Sci. 3 (4) (2014) 23–32.

[56] S.R. Yankanchi, V.Y. Omkar, S.J. Ganesh, Synergistic and individual efficacy of certain plant extracts against dengue vector mosquito, *Aedes aegypti*, J. Biopest. 7 (1) (2014) 22–28.

[57] K.P. Ravi, R. Chinnalalaiah, Medicinal plants used in dengue treatment: an overview, Int. J. Chem. Nat. Sci. 2 (1) (2014) 70–76.

[58] M.N. Eric, W.N. Alice, O.O. Josiah, K.C. Peter, Larvicidal activity of (oxiran-2-yl) methylpentanoate extracted from mushroom *Cyptotrama asprata* against mosquito *Aedes aegypti*, Int. J. Biol. Chem. Sci. 3 (6) (2009) 1203–1211.

Chapter 3

Bioactivity-Guided Phytofractions: An Emerging Natural Drug Discovery Tool for Safe and Effective Disease Management

Partha Palit
Department of Pharmaceutical Sciences, Assam University, Silchar, India

1. INTRODUCTION

Natural product-based drugs make up a substantial proportion of the pharmaceutical market, predominantly in the therapeutic areas of infectious diseases, oncology, and other lifestyle disorders. The basic motto of any drug development program so far has been to design selective ligands (drugs) that act on single selective disease targets to achieve highly effective and safe drugs with negligible side effects. Even though this approach has been successful for many diseases, there has still been a noteworthy decline in the number of new drug candidates being introduced into clinical practice over the past few decades. The severe shortfall that the pharmaceutical industries are facing is due primarily to the postmarketing failures of hit drugs [1,2]. Many analysts postulate that the current capital-intensive model—"the one drug to fit all" strategy—will be flawed in the future and that a new "less investment, more drugs" model is necessary for further scientific growth. It is now well established that many diseases are multifactorial in nature and there are often multiple ways or alternate routes that may be switched on in response to the inhibition of a specific target [3]. This in turn puts resistant cells or resistant organisms under the specific pressure of a targeted agent, resulting in drug resistance and clinical failure of the drug. Drugs designed to act against individual molecular targets cannot usually fight multifactorial diseases such as cancer or diseases that affect multiple tissues or cell types such as diabetes and immunoinflammatory diseases. Combination drugs that affect multiple

Natural Products and Drug Discovery. https://doi.org/10.1016/B978-0-08-102081-4.00003-4

targets concurrently are better at controlling complex disease systems and are less prone to drug resistance. This multicomponent therapy shapes the basis of phytotherapy or phytomedicine where the holistic therapeutic effect arises because of complex positive (synergistic) or negative (antagonistic) interactions between different components of a cocktail. In this approach, multicomponent therapy is supposed to be advantageous for multifactorial diseases, instead of a single "block-blaster drug" that might better explain the state of affairs. The different interactions between various components might engage the protection of an active substance from decomposition by enzymes, modification of transport across membranes of cells or organelles, avoidance of multidrug resistance (MDR) mechanisms, and multidrug targeting concepts, among others [3–5].

Bioassay-guided fractionation of crude herbal extracts is a very promising and rational approach in any natural drug development program. It means the systematic separation of extracted components based on the differences in their physicochemical properties, and assesses pharmacological activity, followed by the next round of separation and evaluation. It helps to recognize and separate the precise mixture of disease-healing components from crude extracts. It also knocks out the unwanted, inert phytocompounds from active therapeutic natural drug candidates.

In connection with safe, effective, therapeutic, and prophylactic management, and predictable low side effects of herbal medicine, bioassay-guided fractionation of crude herbal extracts is in great demand in developing as well as developed countries for primary and/or daily health care. It is not only used for medicinal purposes, but also applied for herbal cosmetics, food preparation, and nutraceuticals development. Utilization and application of herb-derived natural coloring agents, biofuels, biofertilizers, biopesticides, and bioadsorbents play an important role in developing modern technology-oriented civilization [6,7].

Ginseng, *Withania somnifera*, *Ginkgo biloba*, St. John's wort, *Ocimum sanctum*, and gulancha have been utilized as adaptogens, antidiabetic agents, and antidepressants. There are many polyherbal formulations, such as triphala churna, muci-bael, health-boosting tonics, and skin ointments containing *Aloe vera*, *Arnica*, *Calendula*, *Haritaki*, Amalaki, fenugreek, *Medicago sativa*, *Euphorbia hirta*, etc., that can be found on the market. In 2017 the Nobel Prize for Medicine was awarded to three scientists for their pioneering discovery of artemisinin from herbal plants (*Artemisia annua*) for the effective treatment and management of roundworm and malaria. Different herb-derived nutraceuticals such as prebiotics and probiotics, vitamins, hormones, amino acids, and polyunsaturated fatty acids help us to fight against and prevent infectious diseases and aging processes. Herbal nutraceuticals provide metabolomes as prophylactic or preventive agents, in contrast to drugs, which are active chemical substances used for preprotection or to treat an illness. Numerous herbs such as *Achyranthes aspera*, mentha, banana, and *Emblica officinalis* are applied to remove polluting and toxic heavy metals such as ion and arsenic from foods, wastewaters, and drinking water, due to the presence of adsorbent phytochemicals. Dry neem seed extract and curcuminoids give 100% inhibition of mycelial growth. Curcuminoids, the major coloring constituents of *Curcuma longa* (turmeric) rhizome powder, comprise three related curcumins (I–III). The use of citronellal oil for controlling pests and insects is remarkable. *Vitex negundo* is used as an herbal pesticide. Natural pure coloring matters extracted from numerous herbs such as *Hibiscus rosa-sinensis*, marigold, *C. longa*, tomato, henna, and jatropha are used in different pharmaceutical medicines, food preparations, and biocompatible industrial products.

Herbs are usually used as crude phytoextracts for the previously cited purposes (worldwidescience.org). Crude herbal extracts are delivered as ointments, powders, and polyherbal liquid extracts in dosage form. They comprise multiple phytoconstituents responsible for eliciting numerous pharmaceutical and medicinal benefits. However, earlier reports suggest that crude phytoconstituents extracted from highly polar solvent systems such as hydroethanol or methanol contain versatile types of bioactive primary and secondary metabolites (amino acids, proteins, steroids and terpenoids, alkaloids, flavonoids, tannins, etc.) [8]. However, biological investigations prove that specific separated bioactivity-guided fractions demonstrate better efficacies against specific target diseases compared to pure phytocompounds and crude extracts [9,10].

Considering the foregoing scenario, this chapter will address why specific bioactive fractions provide better and promising bioactivity against target diseases in comparison to crude extracts and pure isolated compounds. This chapter will also emphasize the strategy developed for characterizing and standardizing target-specific bioactivity-guided fractions separated from crude extracts as an effective, safe, and potential natural drug discovery tool.

2. WHY DO BIOACTIVITY-GUIDED PHYTOFRACTIONS EXHIBIT PROMISING ALLEVIATION AGAINST DISEASE MODELS

Herbs and other medicinal plants contain mixtures of numerous types of phytoconstituents, which have versatile biological activity, from disease mitigation to other pharmaceutical uses. Plants' organ parts such as leaf-, stem-, twig-, root-, flower-, fruit-, and bark-derived phytocompounds are obtained through different extraction methods.

Extraction and chromatographic isolation provide three forms of bioactive agents. These are crude extracts, bioactivity-guided phytofractions, and isolated pure compounds. From microbial natural-product library data, 79.9% of biologically protective activities against molecular in vitro disease models have been reported on bioactivity-guided fractions, whereas 12.5% of activities have been found in crude extracts [11]. Crude extracts are a combination of a maximum number of different categories of secondary and primary bioactive metabolites such as amino acids, macro- and micronutrients, vitamins, steroids, terpenoids, alkaloids, flavonoids, coumarins, glycosides, saponins, etc. [12]. Each category of metabolite includes many derivatives of biomolecules with versatile structural modifications in their pharmacophore skeleton. It is well recognized that not all these primary and secondary metabolites as a whole are responsible for eliciting pharmacological activity against a specific target disease [13]. Different types of phytometabolites exhibit numerous biological activities against many diseases for management, mitigation, and cure. Therefore plant extracts have flexible pharmacological activities against several diseases. It is interesting to note that crude extracts derived from plant organs do not contain a uniform ratio of concentrations of all phytomolecules [14]. This is why the percentage of yield or extractive value of each phytolead is not identical in the crude extract. Therefore pharmacologically relevant doses of lead phytomolecules differ from each other and are formulated as crude extracts for therapeutic purposes. As is well known, a particular bioactive molecule shows its pharmacological action in a dose-dependent manner toward the target disease [15]. Moreover, when the combination of all phytomolecules is present in the crude extract, then target disease-specific bioactive compounds may be present in low concentrations in comparison to whole extract concentrations. If completely crude extracts were in a high concentration for therapeutic purposes, then the pharmacological dose of phytocompound fractions would not be sufficient to elicit a therapeutic action. Sometimes other fractions present in the crude extract may interfere with the therapeutic potency of active fractions functioning as pharmacological enzyme inhibitors or receptor antagonists [16,17]. This phenomenon may diminish the overall attenuating effect of crude extract against particular target diseases. Furthermore, as an example, unwanted interactions may be caused by the presence of tannins in a herbal drug, which may hold back the absorption of proteins and alkaloids, or enzymes such as cytochrome P450 may be induced, which may hasten drug metabolism resulting in the blood levels of actives being too low for therapeutic effect [18].

However, separation of target disease-specific bioactive fractions of crude extracts containing two or three active biomarker phytocompounds along with their numerous derivatives may offer prospective pharmacological action toward the mitigation or management of desired diseases. This amelioration conferred by the bioactivity-guided active fraction may be due to the synergistic interaction among the mixtures of multiple components [19]. Bioactivity-guided active fractions that mediate these kinds of protective activity are achievable if only proper isolation, characterization, and standardization of the active fractions from crude extract are undertaken and updated in the phytofraction screening library databases for future reference. Hence the bioactive fractions demonstrate better results compared to crude extracts against the target disease for the mitigation and management of pathological syndromes.

The active standardized fraction not only exhibits better results than crude extract but also could be a better potential agent than isolated pure phytomolecules individually present in the active fraction. Picerno et al. [20] reported that *n*-butanol fraction derived from the crude methanol extract of *Paeonia rockii* ssp. *rockii* is a better antifungal agent than whole crude methanol extract. The bioactive *n*-butanol fraction has also contributed prospective free radical scavenging and antifungal activity in comparison to isolated pure compounds such as paeoniflorin and gallic acid, respectively. The concurrent presence of various bioactive compounds such as paeoniflorin in combination with gallic acid and methyl and ethyl gallate could augment the activity of the *n*-butanol active fraction, suggesting a probable synergy in the action or improvement of bioavailability of the antioxidant and antimicrobial molecules.

To understand the foregoing theory in a convincing manner, a rational mathematical expression behind the better potency of bioactive fractions over crude extracts and isolated pure compounds is presented as follows:

Say,
Crude extract = CE
Bioactive fractions = BF
Isolated pure compounds = PC
The onset scheme of extraction is as follows:

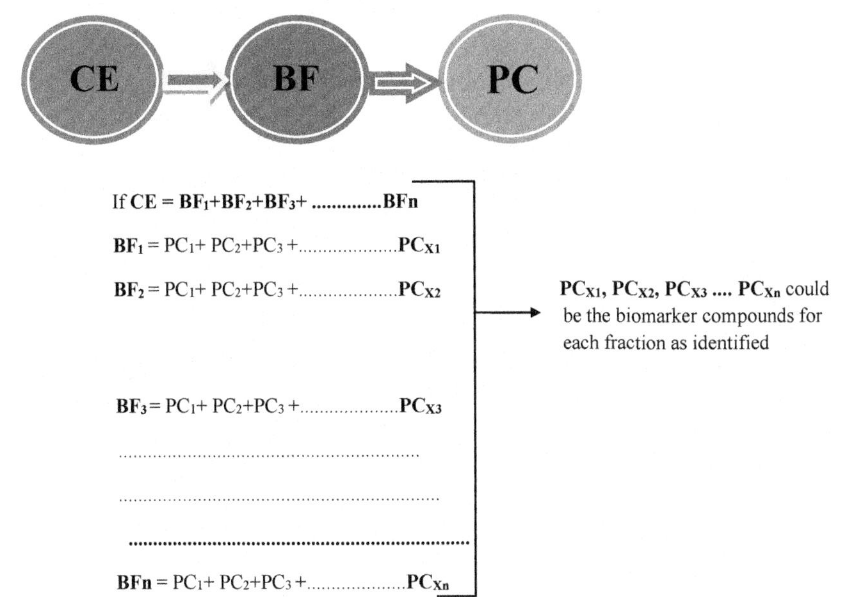

Suppose one crude extract contains **n** number of bioactive fractions and each fraction from one to **n** number contains **X1**, **X2**, **X3**,…, **Xn** number of isolated pure compounds, respectively, then the total number of (**X1** + **X2** + **X3** + **X4**, …, **Xn**) compounds (including primary and secondary metabolites) are present in the whole crude extract.

As per earlier scientific reports and hypothesis, the order of biopotentiality of three agents is as follows.

In respect to one specific target disease:

$$\mathbf{BF_n > CE < PC_{Xn} < BF_n}$$

Out of different bioactive fractions such as ($\mathbf{BF_1}$, $\mathbf{BF_2}$, $\mathbf{BF_3}$, …, and $\mathbf{BF_n}$) if $\mathbf{BF_n}$ is optimized as the lead fraction against one particular disease, then it should be characterized with a biomarker compound such as $\mathbf{PC_{Xn}}$. In that case, it elicits a potential action against the specific disease due to the presence of synergistic compounds such as PC_1, PC_2, PC_3, …, and $\mathbf{PC_{Xn}}$. This is why $\mathbf{BF_n}$ shows a better result than isolated pure compound $\mathbf{PC_{Xn}}$ and **CE**.

We know that pharmacological action is drug dose or concentration dependent.

If $\mathbf{PC_{Xn}}$, PC_2, and PC_3, available in the bioactive fraction $\mathbf{BF_n}$, are responsible for eliciting disease-alleviating action against one particular disease, then the optimum concentrations or doses (say **a**, **b**, and **c** mg/mL) of those compounds are necessary in the crude extract or fraction for showing the desired bioactivity.

The total weight of crude extract is **y** mg after completing extraction, and denoted as extractive matter, which is constant according to the standardized principle of extraction from its mother medicinal herb/plant.

To obtain the desired concentration of responsible compounds in the crude extract, it should be taken in higher concentration, which could be a toxic dose due to the presence of other unwanted phytoconstituent-enriched fractions in high concentration. Separation of disease-specific bioactive fraction ($\mathbf{BF_n}$), the undesirable compound mixture, including its fraction ($\mathbf{BF_1}$ to $\mathbf{BF_{n-1}}$), could be avoided and toxicity might be reduced. Moreover, synergism between the disease-specific target compounds ($\mathbf{PC_{Xn}}$, PC_2, and PC_3) and their derivatives may show powerful activity.

Sometimes it has been found that any undesirable fraction among $\mathbf{BF_1}$ to $\mathbf{BF_{n-1}}$ may act as an antagonist of the main target fraction ($\mathbf{BF_n}$), containing active compounds such as ($\mathbf{PC_{Xn}}$, PC_2, and PC_3). After separation and identification of a disease-specific bioactive fraction it may exhibit better results compared to original crude extract as well as pure isolated compounds in similar doses, bypassing the previously cited problems. This may demonstrate the combinatorial effect of a combination therapy elicited by three or more compounds.

3. SEPARATION, CHARACTERIZATION, AND STANDARDIZATION OF TARGET DISEASE-SPECIFIC BIOACTIVITY-GUIDED FRACTIONS

To develop proper treatment of target diseases in a pharmacologically relevant manner, bioassay-guided active herbal fractions are characterized and standardized followed by separation from crude extract. Here the bioactive herbal fraction is extracted, characterized, and standardized according to suitable methodologies for disease management purposes. During separation of bioassay-guided fractions from crude extract, different techniques such as solvent fractionation and preparative thin layer chromatography (TLC) are used. Column chromatographies are adopted to optimize the lead fraction as described in Figs. 3.1 and 3.2. Different solvents are used in an increasing order of polarity index to obtain the different fractions to undertake bioactivity for optimization of lead bioactive fractions (Fig. 3.1).

For a better understanding, the schematic outline of separation and standardization of target disease-specific bioactive fractions is presented in Figs. 3.1 and 3.2. Optimized protocol for the separation of target disease-specific bioactive fractions from completely crude extract should be recorded in natural drug databases for future reference to obtain reproducible results from the same fraction. Each standardized herbal bioactive lead fraction contains many peaks or bands along with marker compounds; these are assessed through high-performance liquid chromatography (HPLC) or TLC, respectively, and addressed in the pharmacopeia for future guidance. The peaks and bands of bioactive fractions developed in HPLC or TLC are expressed in terms of R_t (retention time) and R_f (retention factor) values, correspondingly. This approach may facilitate the relevant corelation between the fraction–bioactivity relationship and the identification process of well-characterized lead fractions for drug-like properties [21–23]. The identification of one or two of the most potential compounds (known phytomarker compounds) is required to characterize the target fraction. Structural elucidation followed by preparative TLC/HPLC, liquid chromatography-mass spectrometry, ^{13}C-nuclear magnetic resonance (NMR), ^{1}H-NMR, and Fourier transform infrared spectroscopy are followed to standardize the phytomarkers. After phytomarker identification, the optimized bioactive fractions are then evaluated for their druggability followed by in vivo bioassay (Fig. 3.1).

4. SIGNIFICANCE AND UTILITY OF TARGET-ORIENTED, DISEASE-SPECIFIC, BIOACTIVITY-GUIDED PHYTOFRACTIONS

Bioactivity-guided target disease-specific active fractions possess significant advantages over isolated pure compounds and their mother crude extracts

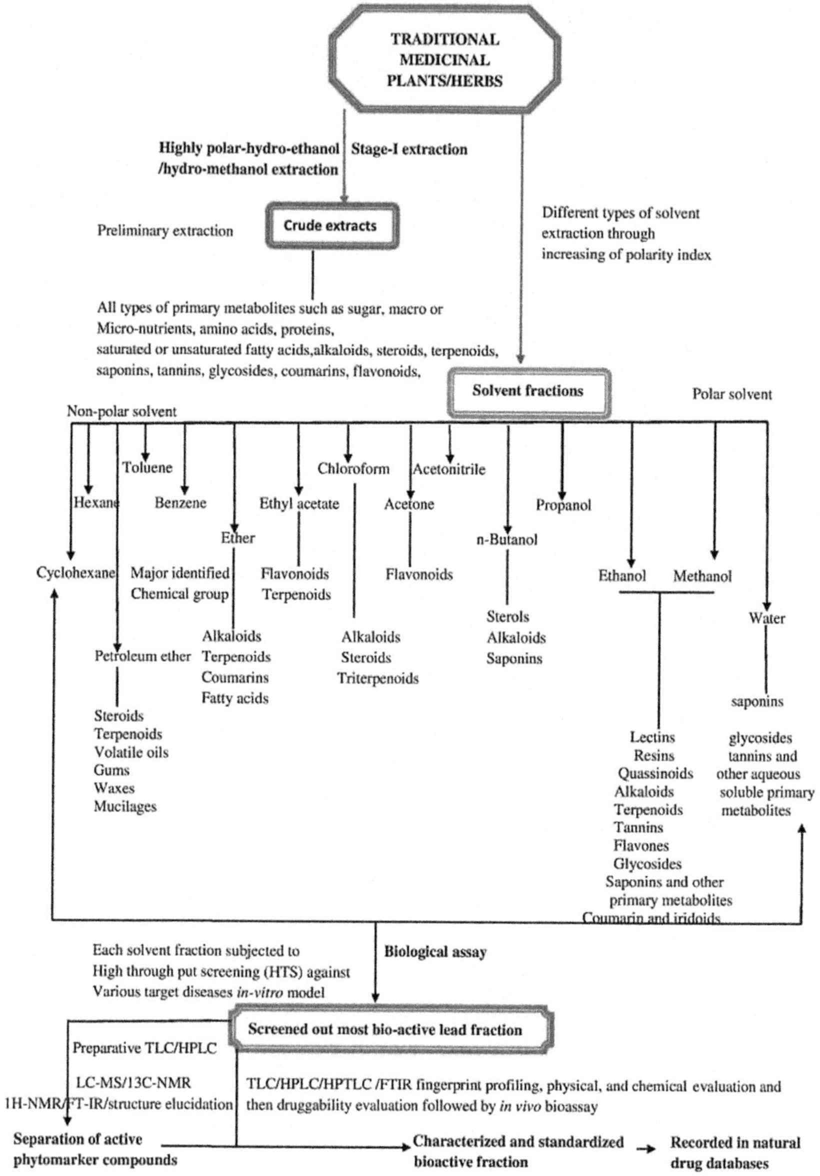

TRADITIONAL MEDICINAL PLANTS/HERBS

Highly polar-hydro-ethanol /hydro-methanol extraction

Stage-I extraction

Preliminary extraction

Crude extracts

Different types of solvent extraction through increasing of polarity index

All types of primary metabolites such as sugar, macro or Micro-nutrients, amino acids, proteins, saturated or unsaturated fatty acids,alkaloids, steroids, terpenoids, saponins, tannins, glycosides, coumarins, flavonoids,

Solvent fractions

Non-polar solvent

Polar solvent

Toluene

Chloroform

Acetonitrile

Hexane

Benzene

Ethyl acetate

Acetone

Propanol

Ether

n-Butanol

Cyclohexane

Major identified Chemical group

Flavonoids Terpenoids

Flavonoids

Ethanol

Methanol

Sterols Alkaloids Saponins

Water

Alkaloids Terpenoids Coumarins Fatty acids

Petroleum ether

Alkaloids Steroids Triterpenoids

saponins

Steroids Terpenoids Volatile oils Gums Waxes Mucilages

Lectins Resins Quassinoids Alkaloids Terpenoids Tannins Flavones Glycosides Saponins and other primary metabolites Coumarin and iridoids

glycosides tannins and other aqueous soluble primary metabolites

Each solvent fraction subjected to High through put screening (HTS) against Various target diseases *in-vitro* model

Biological assay

Screened out most bio-active lead fraction

Preparative TLC/HPLC

LC-MS/13C-NMR 1H-NMR/FT-IR/structure elucidation

TLC/HPLC/HPTLC /FTIR fingerprint profiling, physical, and chemical evaluation and then druggability evaluation followed by *in vivo* bioassay

Separation of active phytomarker compounds

Characterized and standardized bioactive fraction

Recorded in natural drug databases

FIGURE 3.1 The depicted flow chart summarizes the generalized methodology of obtaining bioassay-guided plant fractions by extracting using different organic solvents in the order of increasing polarity index one by one, followed by standardization and characterization for the development of potent and safe drug candidates. *FTIR*, Fourier transform infrared; *HPLC*, high-performance liquid chromatography; *HPTLC*, high-performance thin layer chromatography; *LC-MS*, liquid chromatography-mass spectrometry; *NMR*, nuclear magnetic resonance; *TLC*, thin layer chromatography.

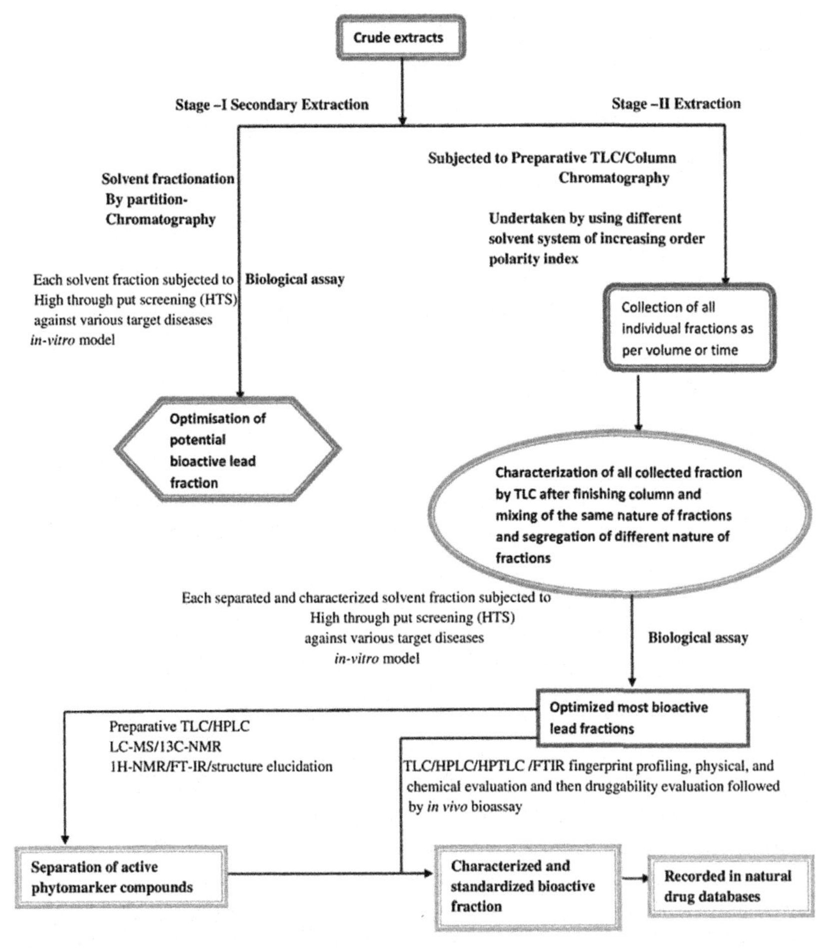

FIGURE 3.2 The schematic chart depicts the method of extraction of lead bioactive fractions from crude phytoextracts by solvent fractionation and chromatographic techniques followed by identification and standardization for designing novel natural drug candidates.

subjected to ideal and proper separation, characterization, and subsequent standardization. This is only possible if the specific bioactive fractions are separated from crude extract with proper and reproducible methods, including characterization of well-defined two or three potent bioactive compounds.

The bioactive fraction is highly advantageous, because it confers a combinatorial effect on the locus of action for mitigating the target disease syndrome due to the presence of many pharmacophore scaffolds or therapeutic lead metabolites. Pharmacologically active and lead phytomarkers in combination with other supporting derivatives offer a better effect than with only one of them in the corresponding bioactive fraction [12]. The premise suggests that a combination of potential leads such as antioxidants is healthy, while high doses of single supplemental antioxidants may cause harm. An antioxidant-rich food, which includes small doses of a huge number of antioxidants, is therefore supposed to alleviate oxidative stress and related other human diseases more efficiently than large doses of single antioxidants, as supported by earlier reports [24,25]. What beneficial aspect we can expect from the bioassay-guided active fractions through their combinatorial therapy is that chances of drug resistance toward the target microorganism might be lessened. Because the separated and well-standardized bioactive fractions contain numerous disease-specific synergistic phytoconstituents along with their derivatives, they may remain in different concentrations. Therefore if a few therapeutic phytocompounds become resistant toward the target microorganism, then structurally unrelated derivatives of resistant pharmacophore scaffold present in the active fraction may exhibit prospective microbicidal or static activity due to the variation of their doses and molecular structures [26]. Moreover, the number of phytoconstituents present in the bioactive fraction may function as multidrug resistance inhibitors [27]. Furthermore, multidrug target-mediated ligand components present in the bioactive fraction may also reduce the chances of drug resistance. This is why synergism of herbal bioactive fractions can minimize the chance of treatment failure and the instance of drug withdrawal prior to a disease remedial approach. The reduced risk of side effects may be observed in the case of bioassay-guided active herbal fraction-mediated treatment strategies compared to isolated pure molecules [28,29]. Since a number of protective phytocompounds available in the target fraction may attenuate the side effects exerted by other compounds in the bioactive fraction, such as monoterpenoid-like compounds of ginger, they may reduce the symptoms of untoward effects such as nausea, flatulence, and loss of appetite produced by the other phytoconstituents. One phytocompound may act as an absorption enhancer for the overall bioassay-guided target disease-specific active fraction to demonstrate their significant optimal activity by increasing bioavailability. Such beneficial pharmacokinetic and pharmacodynamic interactions between the active phytoconstituents among the bioactive fraction may hold the promise for maximizing drug action [30].

5. EXPERIMENTAL EVIDENCE IN FAVOR OF BIOASSAY-GUIDED PHYTOFRACTION AS A THERAPEUTIC TOOL

Bioactive plant fractions of *A. annua* contain various flavonoids (artemetin, casticin, and chrysosplenetin) along with artemisinin. The improved antimalarial activity of artemisinin and flavonoid-enriched fraction over pure artemisinin is due to enhanced bioavailability of artemisinin. This is because of the inhibitory effect of flavonoids on the hepatic and intestinal cytochrome P450 that metabolize artemisinin. Thus flavonoids present in the fraction may help to potentiate and improve the pharmacological activity of artemisinin against malaria parasites [31,32]. Stermitz et al. [33] claimed that the antibacterial agent berberine fraction containing flavonolignan (5′-methoxyhydnocarpin) displayed everlasting and promising antimicrobial action compared to berberine on its own. MDR efflux pumps protect the *Staphylococcus aureus* bacteria from the bactericidal effect of berberine. Flavonolignan (5′-methoxyhydnocarpin) acts as an efflux pump inhibitor and helps to accumulate sufficient concentration inside the bacterial cells to elicit a strong antibacterial action. Thus bioactivity-guided fractions containing multicompounds effectively disabled the bacterial resistance mechanism against the berberine antimicrobial. The earlier report suggested that significantly improved pharmacodynamic activity was observed against a *Plasmodium falciparum* strain as well as a quinine-resistant strain, while three well-known alkaloids, quinine with its D-isomer, quinidine, and cinchonine, were used in combination in equal parts rather than as single compounds. It was a well-documented report of synergism shown by cinchona bark-derived quinoline alkaloid-based bioactive fractions that displayed a more consistent effect than any of the alkaloids used singly [34]. Cannabis-derived bioactive fraction is a better antispastic than tetrahydrocannabinol at an equivalent dose [35]. On the other hand, our lab postulated that a steroid- and terpenoid-rich bioactive fraction from *Euphorbia neriifolia* Linn. exhibited better protection against a nociceptive pain, inflammation, and in vitro arthritis model compared to the equivalent dose of crude extract from the same plant [36]. The dichloromethane subfractions of *Parastrephia lucida* have shown significant inhibition against the phospholipase enzymatic pathway compared to crude dicholoromethane extract and pure single standard drugs such as acetylsalicylic acid, naproxen, and indomethacin for the management of inflammation and pain [37]. The anxiolytic and anticonvulsant activity elicited by Kava Kava (*Piper methysticum*) bioactive fraction consisting of kava lactones, particularly kavain, dihydrokavain, yangonin, dimethoxyyangonin, methysticin, and dihydromethysticin, are superior to when they are applied separately. One constituent, namely, **dihydromethysticin**, seems to be particularly important for synergy. In some experiments, it was observed that the oral bioavailability of kavain was greater if it was administered in a fraction of bioactive compounds when compared to an equivalent quantity of the pure constituent in mice and dogs [38].

6. DISCUSSION AND CONCLUSION

These studies on bioactive fractions highlight the separation, characterization, and methodology of bioactive lead fractions and justification behind showing their superior protective activity against target diseases in comparison to crude extracts and isolated marker compounds. The addressed information and general concepts of this review may conclude that lead bioactive fraction is derived from mother crude extract of medicinal herbs/shrubs or plants. This crude extract is composed of several bioactive fractions consisting of many primary and secondary metabolites that may confer biological action. However, which type of secondary and primary metabolites is present in each fraction needs to be investigated and characterized thoroughly before undertaking in vitro high-throughput screening and in vivo bioevaluation studies. The separation of bioactivity-guided fractions through preparative HPLC/TLC or column chromatography is mandatory to characterize and optimize the lead fractions, which confer excellent and improved bioactivity over crude extracts and isolated pure compounds (phytomarkers) due to synergistic combinatorial action. We may forecast from our review studies that few bioactive fractions present in the crude extract could interfere with desired bioactive fractions' pharmacological action against target diseases as a pharmacokinetic/dynamic antagonist. Moreover, the responsible phytocompounds for eliciting pharmacological action against specific target diseases are present in the lead bioactive fraction in optimized and higher concentrations compared to the same amount of crude extracts, as explained earlier. This could be the rationality behind the exploration of bioactivity-guided fractions over crude extracts and isolated pure compounds (phytomarkers) for efficient natural drug discovery approaches. However, collection time, collection place, climatic condition, and soil quality must be mentioned and considered against the each herb and medicinal plant during botanical drug development from bioactivity-guided fractions. The presence of phytoconstituents and marker compounds in the lead bioactive fraction from differently formulated batches may vary due to the previously cited influencing factors. Therefore standardization of these bioactive fractions and reproducibility of the bioactivity results are becoming much more complicated.

Still, what we gain from the bioassay-guided fraction-mediated therapeutic strategies are synergistic effects by the same categories of biometabolite derivatives and sufficient concentrations of target-specific compound mixtures in the locus of action. The chances of drug resistance and treatment failure are extremely unlikely in the case of bioassay-guided fractions, as they will be designed as dosage form due to the inhibition of multiple pathways. Side effects are fewer for bioactive fractions due to the presence of toxicity-healing phytocompounds. Therefore the combination therapy exhibited by bioassay-guided plant fractions is expected to reduce untoward effects but at the same time exhibit augmented therapeutic action.

Other compounds are present in the fractions as assisting agents and may act as adjuvants, excipients, and stabilizers for the main therapeutic active ingredients to increase the biological action by modifying their pharmacokinetic parameters. Moreover, flavonoid-containing bioactive fractions may stabilize the active ingredient of bioactive fractions from oxidation-mediated decomposition [18].

We may therefore conclude that bioactive fractions could be the emerging tool for an efficient drug discovery approach in traditional medicine subject to their proper phytochemical characterization, fingerprint profiling, marker compound identification for future study, and proper disease management.

ACKNOWLEDGMENTS

We are thankful to DST (SERB), Government of India, for funding (Ref. No. **SB/FT/LS-269/2012**) to undertake thiswork.

REFERENCES

[1] S. Frantz, Pharma faces major challenges after a year of failures and heated battles, Nat. Rev. Drug Discov. 6 (2007) 5–7.
[2] A.M. Thayer, Blockbuster model breaking down, Mod. Drug Discov. 7 (2004) 23–24.
[3] G.R. Zimmermann, J. Lehar, C.T. Keith, Multi-target therapeutics: when the whole is greater than the sum of the parts, Drug Discov. Today 12 (2007) 34–42.
[4] T.W. Corson, C.M. Crews, Molecular understanding and modern application of traditional medicines: triumphs and trials, Cell 130 (2007) 769–774.
[5] B. Patwardhan, A.D. Vaidya, M. Chorghade, S.P. Joshi, Reverse pharmacology and systems approaches for drug discovery and development, Curr. Bioactive Compounds 4 (2008) 201–212.
[6] P. Agarwal, A. Shashi, A. Fatima, A. Verma, Current scenario of Herbal Technology worldwide: an overview, Int. J. Pharm. Sci. Res. 4 (2013) 4105–4117, https://doi.org/10.13040/IJPSR.0975-8232.4(11).4105-4117.
[7] K.G. Ramawat, S. Goyal, The indian herbal drugs scenario in global perspectives, Bioactive Mol. Med. Plants 9 (2008) 325–347.
[8] I. Ahmad, F. Aqil, M. Owais (Eds.), Modern Phytomedicine: Turning Medicinal Plants into Drugs, John Wiley & Sons, 2006.
[9] X. Xu, F. Li, X. Zhang, P. Li, X. Zhang, Z. Wu, D. Li, In vitro synergistic antioxidant activity and identification of antioxidant components from *Astragalus membranaceus* and *Paeonia lactiflora*, PLoS One 9 (5) (2014) e96780.
[10] G. Uddin, Ismail, A. Rauf, M. Raza, H. Khan, Nasruddin, M. Khan, U. Farooq, A. Khan, Arifullah, Urease inhibitory profile of extracts and chemical constituents of *Pistacia atlantica* ssp. cabulica Stocks, Nat. Prod. Res. 20 (2015) 1–6.
[11] A.L. Harvey, R. Edrada-Ebel, R.J. Quinn, The re-emergence of natural products for drug discovery in the genomics era, Nat. Rev. Drug Discov. 14 (2015) 111–129.
[12] A. Bernhoft, Bioactive compounds in plants – benefits and risks for man and animals, in: Proceedings from a Symposium Held at the Nor Academy of Science and Letters, Oslo, Norway, 13–14 November, 2008.
[13] A.G. Atanasov, B. Waltenberger, E.M. Pferschy-Wenzig, T. Linder, C. Wawrosch, P. Uhrin, et al., Discovery and resupply of pharmacologically active plant-derived natural products: a review, Biotechnol. Adv. 33 (2015) 1582–1614.
[14] P.S. Patil, R. Shettigar, An advancement of analytical techniques in herbal research, J. Adv. Sci. Res. 1 (2010) 08–14.
[15] M. Qiang, Y.H.L. Anthony, Pharmacogenetics, pharmacogenomics, and individualized medicine, Pharmacol. Rev. 63 (2011) 437–459.
[16] T.K. Milugo, L.K. Omosa, J.O. Ochanda, B.O. Owuor, F.A. Wamunyokoli, J.O. Oyugi, J.W. Ochieng, Antagonistic effect of alkaloids and saponins on bioactivity in the quinine tree (*Rauvolfia caffra* sond.): further evidence to support biotechnology in traditional medicinal plants, BMC Complement. Altern. Med. 13 (1) (2013) 285.

[17] P.E.D. Resende, S. Kaiser, V. Pittol, A.L. Hoefel, R.D. Silva, C.V. Marques, et al., Influence of crude extract and bioactive fractions of *Ilex paraguariensis* A. St. Hil. (yerba mate) on the Wistar rat lipid metabolism, J. Func. Foods 15 (2015) 440–451.

[18] E.M. Williamson, Synergy and other interactions in phytomedicines, Phytomedicine 8 (5) (2001) 401–409. www.worldwidescience.org/topicpages/h/herbal+plant+extracts.html.

[19] J.C. Yeh, I.J. Garrard, C.W. Cho, S.W. Annie Bligh, G.H. Lu, T.P. Fan, D. Fisher, Bioactivity-guided fractionation of the volatile oil of Angelica sinensis radix designed to preserve the synergistic effects of the mixture followed by identification of the active principles, J. Chromatogr. A 1236 (2012) 132–138.

[20] P. Picerno, T. Mencherini, F. Sansone, P. Del Gaudio, I. Granata, A. Porta, R.P. Aquino, Screening of polar extract of *Paeonia rockii*: composition and antioxidant and antifungal activities, J. Ethnopharmacol. 138 (2011) 705–712.

[21] D.G. Corley, R.C. Durley, Strategies for database dereplication of natural products, J. Nat. Prod. 57 (11) (1994) 1484–1490.

[22] L.K. Pannell, N. Shigematsu, Increased speed and accuracy of structural determination of biologically active natural products using LC-MS, Am. Lab. 30 (7) (1998) 28–30.

[23] G. Saxena, S. Farmer, G.H.N. Towers, R.E.W. Hancock, Use of specific dyes in the detection of antimicrobial compounds from crude plant extracts using a thin layer chromatography agar overlay technique, Phytochem. Anal. 6 (3) (1995) 125–129.

[24] R. Blomhoff, Dietary antioxidants and cardiovascular disease, Curr. Opin. Lipidol. 16 (2005) 47–54.

[25] A.D. Crawford, S. Liekens, A.R. Kamuhabwa, J. Maes, S. Munck, R. Busson, J. Rozenski, C.V. Esguerra, P.A. de Witte, Zebrafish bioassay-guided natural product discovery:isolation of angiogenesis inhibitors from East African medicinal plants, PLoS One 6 (2011) e14694, https://doi.org/10.1371/journal.pone.0014694.

[26] P.R. Marinho, R.S. Guilherme, M. Mara, F.L. Silva, M.G.D. Marval, M.S. Laport, Antibiotic-resistant bacteria inhibited by extracts and fractions from Brazilian marine sponges, Rev. Bras. Farmacogn. 20 (2010). S0102-695X2010000200022.

[27] H. Hirt, B. M'Pia, Natural Medicine in the Tropics I: Foundation Text, Anamed, Winnenden, 2008.

[28] S. Khatri, M. Kumar, N. Phougat, R. Chaudhary, A.K. Chhillar, Perspectives on phytochemicals as antibacterial agents: an outstanding contribution to modern therapeutics, Mini Rev. Med. Chem. 16 (2016) 290–308.

[29] L. Mundy, B. Pendry, M.M. Rahman, Antimicrobial resistance and synergy in herbal medicine, J. Herb. Med. 6 (2) (2016) 53–58.

[30] P. Rasoanaivo, C.W. Wright, M.L. Willcox, B. Gilbert, Whole plant extracts versus single compounds for the treatment of malaria:synergy and positive interactions, Malar. J. 10 (Suppl. 1) (2011) S4, https://doi.org/10.1186/1475-2875-10-S1-S4.

[31] P.J. Weathers, M.A. Elfawal, M.J. Towler, G.K. Acquaah-Mensah, S.M. Rich, Pharmacokinetics of artemisinin delivered by oral consumption of *Artemisia annua* dried leaves in healthy vs. *Plasmodium chabaudi*-infected mice, J. Ethnopharmacol. 153 (3) (2014) 732–736.

[32] P.J. Weathers, M.J. Towler, The flavonoids casticin and artemetin are poorly extracted and are unstable in an *Artemisia annua* tea infusion, Planta Med. 78 (10) (2012) 1024–1026.

[33] F.R. Stermitz, P. Lorenz, J.N. Tawara, L.A. Zenewicz, K. Lewis, Synergy in a medicinal plant: antimicrobial action of berberine potentiated by 5′-methoxyhydnocarpin, a multidrug pump inhibitor, Proc. Natl. Acad. Sci. 97 (4) (2000) 1433–1437.

[34] P. Druilhe, O. Brandicourt, T. Chongsuphajaisiddhi, J. Berthe, Activity of a combination of three cinchona bark alkaloids against *Plasmodium falciparum* in vitro, Antimicrob. Agents Chemother. 32 (2) (1988) 250–254.

[35] H. Wagner, G. Ulrich-Merzenich, Synergy research: approaching a new generation of phytopharmaceuticals, Phytomedicine 16 (2) (2009) 97–110.

[36] P. Palit, S.C. Mandal, B. Bhunia, Total steroid and terpenoid enriched fraction from *Euphorbia neriifolia* Linn offers protection against nociceptive-pain, inflammation, and in vitro arthritis model: an insight of mechanistic study, Int. Immunopharmacol. 41 (2016) 106–115.

[37] R.E. D'Almeida, M.I. Isla, E.D.L. Vildoza, C. Quispe, G. Schmeda-Hirschmann, M.R. Alberto, Inhibition of arachidonic acid metabolism by the Andean crude drug *Parastrephia lucida* (Meyen) Cabrera, J. Ethnopharmacol. 150 (3) (2013) 1080–1086.

[38] M.H. Pittler, E. Ernst, Efficacy of kava extract for treating anxiety: systematic review and meta-analysis, J. Clin. Psychopharmacol. 20 (2000) 84–89.

Chapter 4

Development of Chinese Herbal Health Products for the Prevention of Aging-Associated Diseases

Pou K. Leong, Jihang Chen, Kam M. Ko
Hong Kong University of Science & Technology, Hong Kong SAR, China

1. MITOCHONDRIAL DYSFUNCTION IN AGING-ASSOCIATED DISEASES

The "Mitochondrial Free Radical Theory of Aging" posits that mitochondrial reactive oxygen species (mtROS) play a crucial role in aging and age-associated diseases [1]. During aerobic respiration, mtROS are unavoidably produced from the electron transport chain located in the inner mitochondrial membrane. Due to their proximity to the sites of ROS generation, mitochondrial membrane structural components are prone to oxidative damage, with the resultant disruption of mitochondrial structural and functional integrity. A mutation in mitochondrial DNA, for example, can cause an impairment in mitochondrial energy metabolism [2], and the resulting oxidative damage to mitochondrial membranes may further increase the production of mtROS from the electron transport chain, leading to a destructive "vicious cycle." In addition to the primordial role of the mitochondrion in energy metabolism, it also serves as a central coordinator of cell survival and death, as evidenced by its role in regulating apoptosis [3]. In support of this, accumulating experimental and clinical findings have demonstrated the involvement of mitochondrial dysfunction in the pathogenesis of certain age-associated diseases, including cardiovascular diseases, neurodegenerative diseases, osteoporosis, and compromised immune function. Hence an important approach in preventive health would be to focus on preserving mitochondrial structural and functional integrity.

Natural Products and Drug Discovery. https://doi.org/10.1016/B978-0-08-102081-4.00004-6

1.1 Cardiovascular Diseases

Cardiovascular diseases, including atherosclerosis, myocardial infarction, cardiohypertrophy, and cardiomyopathy, are the leading causes of morbidity and mortality in industrialized countries [4]. Age-associated changes in the vascular system, such as arterial thickening and stiffening, are predisposing factors for the pathogenesis of cardiovascular diseases [5]. Among the pathological changes in the vascular system, atherosclerosis, which refers to the accumulation of oxidized low-density lipoprotein (LDL) in arterial walls, is a major risk factor for myocardial infarction. Under normal physiological conditions, endothelial cells regulate vascular tone by the release of nitric oxide (NO), a vasodilator. However, under pathological conditions, such as chronic inflammation, superoxide radicals arising from a reduced nicotinamide adenine dinucleotide phosphate (NADPH) oxidase-catalyzed reaction in activated macrophages can react with NO to form peroxynitrite, which is a strong oxidizing species [6]. ROS have been shown to be involved in the development of atherosclerosis by causing the oxidation of LDL, vascular inflammation, and endothelial damage [7]. The blockage of coronary arteries can lead to myocardial ischemic injury and eventual necrotic cell death if nutrients and oxygen are not restored in a timely manner. Ironically, the resumption of oxygen to previously ischemic cardiomyocytes can result in a burst of ROS production, with resultant oxidative injury to the myocardium, which is termed ischemia/reperfusion injury. Studies have shown that reperfusion-induced ROS production is, at least in part, responsible for the increased electron leakage in Ca^{2+}-overloaded mitochondria, which is an event secondary to a stimulatory effect on the Krebs cycle and oxidative phosphorylation [8]. The excessive production of ROS can impair the optimal metabolic functioning of mitochondria in cardiac tissue and result in the reduction of mitochondrial capacity in oxidative phosphorylation, leading to cardiac dysfunction. Consistent with this postulation, recent findings have shown that the production of ROS is significantly increased in the failing myocardium [9]. During the aging process, bioenergetics in myocardial mitochondria decreases in functional capacity, presumably due to the accumulation of oxidant-induced mitochondrial damage, with a consequent insufficient contractile force generated by the myocardium. To compensate for the decline in the strength of cardiac contraction, cardiomyocytes can undergo a remodeling process of hypertrophy, during which new sarcomeres are produced, with a resultant increase in the thickness of the ventricular wall, as well as the strength of cardiac contraction. However, if the capacities of fatty acid oxidation and oxidative phosphorylation decline further with aging, cardiac hypertrophy can develop pathologically into myocardial infarction [10,11]. To cope with cardiac hypertrophy and myocardial infarction, interventions targeting mitochondria are likely to produce beneficial effects in cardiovascular diseases [12].

1.2 Neurodegenerative Diseases

Neurodegenerative diseases refer to neurological disorders that lead to progressive loss of function in the central nervous system. Because of the high energy demand for neuronal survival and excitability, neurons are highly susceptible to damage caused by mitochondrial dysfunction [13]. Many lines of evidence have shown that a common manifestation of neurodegenerative diseases, such as Alzheimer's disease, Parkinson's disease, Huntington's disease, and Friedreich's ataxia, is the accumulation of misfolded or unfolded proteins in neurons. Interestingly, the mis/unfolded protein aggregates are associated with an impairment of mitochondrial bioenergetics and excessive production of ROS, all of which contribute to the pathogenesis of neurodegenerative disorders [14–16].

Alzheimer's disease is characterized by a progressive decline in cognitive function and memory. A growing body of evidence has shown that the extent of mitochondrial amyloid β-peptide (Aβ) accumulation in the brain correlates with the degree of mitochondrial dysfunction as well as the severity of cognitive functional impairment in patients with Alzheimer's disease. Studies have shown that a high concentration of Aβ can induce ROS production, with a subsequent oxidation of mitochondrial proteins such as voltage-dependent anion channels, aconitase, glyceraldehyde phosphate dehydrogenase, and lactate dehydrogenase [17]. The activities of mitochondrial α-ketoglutarate dehydrogenase [18] and complex IV [19] were also directly inhibited by Aβ [20], with a resultant impairment in mitochondrial energy production leading to neuronal cell death. In this regard, peroxisome proliferator-activated receptor gamma coactivator 1-α, a transcription coactivator regulating mitochondrial biogenesis, can decrease the generation of Aβ via a proliferator/activated receptor-γ-dependent mechanism in cultured N2a neuroblastoma cells [21]. In support of this, the degree of peroxisome proliferator-activated receptor gamma coactivator 1-α expression was found to be decreased in the brains of patients with Alzheimer's disease [22]. In addition to Aβ, an accumulation of intraneuronal neurofibrillary fibers, which consist of hyperphosphorylated tau protein, is also a hallmark of Alzheimer's disease. Studies have revealed that hyperphosphorylated tau can selectively impair the function of mitochondrial complex I, resulting in an increased production of ROS and a reduced generation of adenosine triphosphate (ATP) [23]. These observations strongly suggest a causative role of mitochondrial dysfunction in the pathogenesis of Alzheimer's disease.

Parkinson's disease is a neurological locomotor disorder characterized by bradykinesia, tremor, gait difficulty, postural instability, and rigidity [24]. The hallmark of Parkinson's disease is the formation of Lewy bodies and the loss of dopaminergic neurons in the substantia nigra of the brain. A study has indicated that mitochondrial complex I activity is decreased in the substantia nigra of patients with Parkinson's disease [25], presumably due to an increase in oxidative modification (protein carbonylation) of complex I, with the resultant misassembly and dysfunction of the protein complex [15]. The oxidative metabolism of dopamine and the relatively high concentration of ferrous ions in dopaminergic neurons increase their susceptibility to oxidative stress-induced functional impairment [16], which in turn leads to impairment in mitochondrial bioenergetics and hence neuronal cell death. It has been reported that the inhibition of complex I can facilitate the formation of α-synuclein cytosolic inclusion in vitro and in vivo. In this connection, Lewy bodies, which are a hallmark of Parkinson's disease and an aggregation of α-synuclein, may be causally related to the impairment of complex I. The aggregation of α-synuclein (as α-synuclein protofibrils) was found to interact with various mitochondrial components. As such, α-synuclein, by virtue of its cryptic mitochondrial targeting signals, was preferentially imported into mitochondria, and then incorporated into their inner membrane. The increased importing of α-synuclein into mitochondria was found to correlate with the reduced activity of mitochondrial complex I [26]. Taken together, these observations support a role of α-synuclein in the inhibition of complex I activity. The functionally impaired mitochondrial complex I and the permeabilization of mitochondrial inner membrane [27] lead to the loss of mitochondrial membrane potential, cytochrome *c* release, and apoptosis [28], with a resultant destruction of dopaminergic neurons.

1.3 Osteoporosis

Osteoporosis, which is an age-associated pathological condition involving bone structure, is characterized by a reduction in bone mass density and the deterioration of bone tissue [29]. Despite the fact that postmenopausal osteoporosis (also known as primary type 1 osteoporosis) is predominantly found in postmenopausal women (who possess reduced blood levels of estrogen), senile osteoporosis (also called primary type 2 osteoporosis) predominantly occurs in the elderly resulting from a reduced activity of osteoblasts (which are a type of bone cell). Bone is an active metabolic tissue undergoing the process of remodeling in response to physiological or pathological conditions. Bone remodeling, which is responsible for maintaining the structural and functional integrity of bone tissue, involves bone formation and bone resorption [30]. The major types of cells involved in bone remodeling are osteoblasts, osteocytes, and osteoclasts [31]. In brief, osteoblasts play an important role in bone formation as well as bone matrix production. After depositing proteins and minerals in the bone matrix, osteoblasts subsequently differentiate into osteocytes. Osteoclasts, which are descended from the monocyte/macrophage lineage, dock onto the bone surface and trigger bone resorption via the release of lysosomal enzymes. In response to bone remodeling stimuli, such as microdamage, disuse, ovariectomy, or aging, osteocytes undergo an apoptotic cell death program, which has been found to be an initiating step in bone remodeling. Frikha-Benayed et al. showed that the regional differences in mitochondrial activity among cortical bone osteocytes likely determine the susceptibility of osteocytes to undergo apoptosis and associated bone remodeling [32,33]. By virtue of the high metabolic activity in bone tissues, bone cells (i.e., osteoblasts, osteocytes, and osteoclasts) are highly dependent on mitochondrial bioenergetics. In this regard, age-associated mitochondrial dysfunction can impair the normal physiological functions of bone cells, resulting in the pathogenesis of osteoporosis. Studies have shed light on the role of mitochondrial lesions/mitochondrial oxidative stress in the development of osteoporosis. As such, the number of mitochondria in periosteum osteoblasts was found to be negatively correlated with age in rats [34], implicating the possible reduction of bone formation in the aging process. ROS, presumably arising from damaged mitochondria, were shown to be involved in the differentiation of osteoclasts and hence bone resorption [35]. Furthermore, osteoclasts isolated from aged rats were reported to exhibit a reduced content of mitochondrial DNA as well as intracellular ATP levels and a higher activity of bone resorption, all of which implicate a role of mitochondrial dysfunction in osteoporosis [36]. The involvement of mitochondrial ROS in the pathogenesis of osteoporosis has recently been investigated using superoxide dismutase knockdown mice. As expected, the ablation of superoxide dismutase in osteocytes was found to increase the level of superoxide radicals, which was paralleled with the disorganization of the canalicular network in osteocytes and a reduction in the number of viable osteocytes in these superoxide dismutase-deficient mice [33]. Osteoporosis in superoxide dismutase knockdown mice was also associated with the suppression of bone formation and activation of bone resorption. The molecular mechanism underlying mitochondrial dysfunction in relation to osteoporosis requires further investigation.

1.4 Dysregulation of Immune Function

The immune system, which consists of innate immunity and adaptive immunity, works to eliminate exogenous pathogens as well as host cells with lesions (such as mutated cells and apoptotic cells). Innate immunity is a nonspecific immune response, encompassing the cytotoxic action of natural killer cells, phagocytosis afforded by macrophages, and inflammatory response. Adaptive immunity is a specific immune response comprising a cell-mediated immune response, which is regulated by effector T cells and T helper cells, as well as a humoral immune response, which is regulated by B cells. Mitochondria in the immune system are considered as the major source of energy for supporting the high energy demand in eliciting an immune response. A growing body of evidence has demonstrated differences in the energy metabolism of different immune cells. For instance, quiescent T cells rely on glycolysis and mitochondrial oxidative metabolism, whereas activated T helper cells (presumably due to the activation of T cell receptors) prefer aerobic glycolysis with a concomitant downregulation of mitochondrial oxidative metabolism [37]. Similarly, proinflammatory macrophages (also called M1 macrophages) also preferentially stimulate glycolysis as well as the pentose phosphate pathway, with a concomitant suppression of mitochondrial oxidative metabolism to achieve a balanced production of ATP and NADPH [38]. The metabolic changes in immune cells are hypothesized to play a role in their immunological function. As such, NADPH is consumed in the biosynthetic pathways of fatty acids (which can serve as the precursor of prostaglandins, a family of proinflammatory molecules) as well as in the NADPH oxidase-catalyzed generation of superoxide radicals (which is also known as the "oxidative burst" involved in the bactericidal action of neutrophils). Studies have shown that two reactions catalyzed by isocitrate dehydrogenase and succinate dehydrogenase in the tricarboxylic acid cycle are blocked in M1 macrophages, with resultant increases in the levels of citrate and succinate, respectively [39,40]. Citrate is a substrate for the synthesis of itaconic acid (an antibacterial compound) as well as for the production of fatty acids (the precursors of prostaglandins), whereas succinate can stabilize hypoxia-inducible factor-1α that can lead to the expression of interleukin 1β (IL-1β) [41]. These findings suggest a novel role of mitochondrial energy metabolism in influencing the function of immune cells. Whether or not mitochondrial dysfunction in the aging process can negatively influence the relationship between the function and energy metabolism of M1 macrophages/T helper cells and hence cause the impairment of immunological function is deserving of further investigation.

Immunosenescence refers to the age-associated decline in immunological competence, with the expansion of highly differentiated CD28^{-} T memory cell (which has a reduced proliferative capacity as well as a decreased T cell receptor repertoire), reduction in the number of naive T/B cells, and decrease in the T cell repertoire. In contrast, various functions of innate immunity seem to be enhanced during the aging process, as evidenced by age-associated increases in plasma levels of proinflammatory cytokines and the level of expression of Toll-like receptors in innate immune cells [42]. Franceschi et al. proposed a theory of "inflamm-aging," which emerged from an understanding of immunosenescence, in an effort to explain the phenomenon of chronic low-grade inflammation in aging/age-associated diseases [43]. Macrophages, which are involved in both inflammatory and stress responses, are likely to be coordinators and/or mediators of "inflamm-aging." Macrophages can recognize pathogen-associated molecular patterns (PAMPs) and damage-associated molecular patterns (DAMPs) with the subsequent induction of an inflammatory response. PAMPs are bacterial surface molecules, such as lipopolysaccharide and peptidoglycan, while DAMPs are biomolecules, such as ATP, uridine triphosphate, nucleic/mitochondrial DNA, monosodium urate, and cholesterol crystal, that exist intracellularly under physiological conditions [44–46]. During the aging process, various misfolded/pathological proteins, such as lipofuscin, tau protein aggregates, α-synuclein fibrils, and Aβ, can accumulate in nonproliferating cells (e.g., neurons) [44]. A high level of misfolded proteins can be recognized by receptors on macrophages (or residential macrophages) as "nonself" molecules, with the resultant induction of an inflammatory response. In this connection, neuroinflammation has been shown to accompany various neurodegenerative diseases [44]. In addition, the misplaced biomolecules, presumably arising from the membrane rupture of senescent cells, can be recognized as DAMPs by macrophages, causing chronic inflammation. Studies have further revealed novel roles for mitochondria in the "inflamm-aging." First, various mitochondrial components (such as mitochondrial DNA, *N*-formyl peptides, and cardiolipin) can be recognized as DAMPs by macrophages [47]. In this regard, the release of mitochondrial DAMPs as a result of age-associated mitochondrial dysfunction can induce inflammation. Second, the age-associated impairment of mitophagy (i.e., the autophagy of mitochondria) can indirectly increase the number of damaged mitochondria and hence elevate the production of ROS. Mitochondrial ROS likely activate the NACHT, LRR, and PYD domains-containing protein 3 (NLRP3) inflammasome complex [48,49], which is comprised of NLRP3, apoptosis-associated speck-like protein containing a CARD (ASC, the caspase 1 recruiting protein), and caspase 1 [50]. The activated NLRP3 in inflammasomes can then cause the cleavage of pro-IL-1β into IL-1β, which is a proinflammatory cytokine detectable in patients with age-associated diseases/metabolic disorders [51,52]. Accumulating evidence has also supported the involvement of inflammasomes in chronic inflammation in age-associated metabolic diseases [52]. Understanding the role of inflammasomes in the development of "inflamm-aging" may provide a new approach for developing interventions for age-associated metabolic diseases.

2. CONCEPTUAL BASIS OF PREVENTIVE HEALTH IN CHINESE MEDICINE

2.1 Yin-Yang Theory

Chinese medicine (CM) views the human body as an organic entity, which consists of major organs ("Liver," "Heart," "Spleen," "Lung," and "Kidney") that function in a mutually interdependent manner [53]. Yin-Yang theory, which is rooted in ancient Chinese philosophy, is the conceptual framework of CM. According to Yin-Yang theory, the universe is formed by a combination/interaction of the two complementary but opposing forces, namely, Yin and Yang. The dynamic equilibrium between Yin and Yang determines the physiological status/phase of a given matter [54]. Any disturbance in the balance of Yin and Yang in a human subject can lead to the development of various subhealthy or diseased conditions. For example, an absolute excess of Yin can cause an overconsumption of Yang; an absolute excess of Yang can lead to an overconsumption of Yin; a relative excess of Yin is referred to as a deficiency of Yang; and a relative excess of Yang is viewed as a deficiency of Yin. With this health conceptual framework, CM classifies body structures, explains clinical symptoms, and guides the treatment of diseases based on Yin-Yang theory [55]. Vital substances (namely, essence, Qi, blood, and body fluids) are fundamental to life and constitute the material and functional basis of the human body [53]. According to Yin-Yang theory, functional activities of the body (such as Qi) are classified as Yang, while material elements (such as essence, blood, and body fluids) of vital functions belong to Yin [56]. In addition, any internal or external causative agent of diseases (i.e., pathogenic factors) can be classified as belonging to either Yin or Yang categories according to their pathological characteristics. Therapeutic interventions in CM include the use of Chinese herbs, Chinese herbal formulations, and acupuncture, all of which are capable of modulating Yin-Yang activities and hence can be of use in the treatment of diseases with Yin-Yang deficiency [55].

2.2 Qi and Body Function

In the realm of CM theory, the interaction between Yin and Yang generates Qi that possesses dual properties; it can be referred to as: (1) the refined and nutritive substances flowing in the body and (2) the functional manifestation of organs. Qi can flow through meridians and nourishes organs to provide vital energy for supporting normal physiological activities. Within the conceptual framework of CM, the consequence of the complete deprivation of Qi is death [57]. With regard to the role of Qi in regulating physiological functions, Qi can be categorized into three functionally related types, namely, primordial Qi, pectoral Qi, and normal Qi, with the latter being subdivided into nutritive Qi and defensive Qi [57–59] (Fig. 4.1). In brief, primordial Qi, which is referred to as the congenital essence of the "kidney," is inherited from parents. Primordial Qi is the primary motive force for stimulating growth and development, as well as invigorating vital activities of organs in the body, and it is Yang in nature. In this regard, the deprivation of primordial Qi due to congenital defects or the aging process can lead to organ dysfunction and/or degeneration. Despite the fact that primordial Qi is determined by congenital essence, the acquired essence derived from food essence/Chinese tonifying herbs can supplement the primordial Qi to a certain extent, resulting in restoration of optimal physiological function in the organs. Pectoral Qi is formed by combining "natural air" inhaled by the "Lung" and the "grain Qi" (i.e., the food essence) from the transformation of food and water by the "Spleen" and "Stomach," and these are Yin in nature. Pectoral Qi circulates in the chest of the human body. The principal actions of pectoral Qi are to facilitate air exchange in the lungs and regulate blood circulation propelled by the heart as well as the rate of heart beating. Therefore pectoral Qi is viewed as a manifestation of the coordination between physiological activities of the

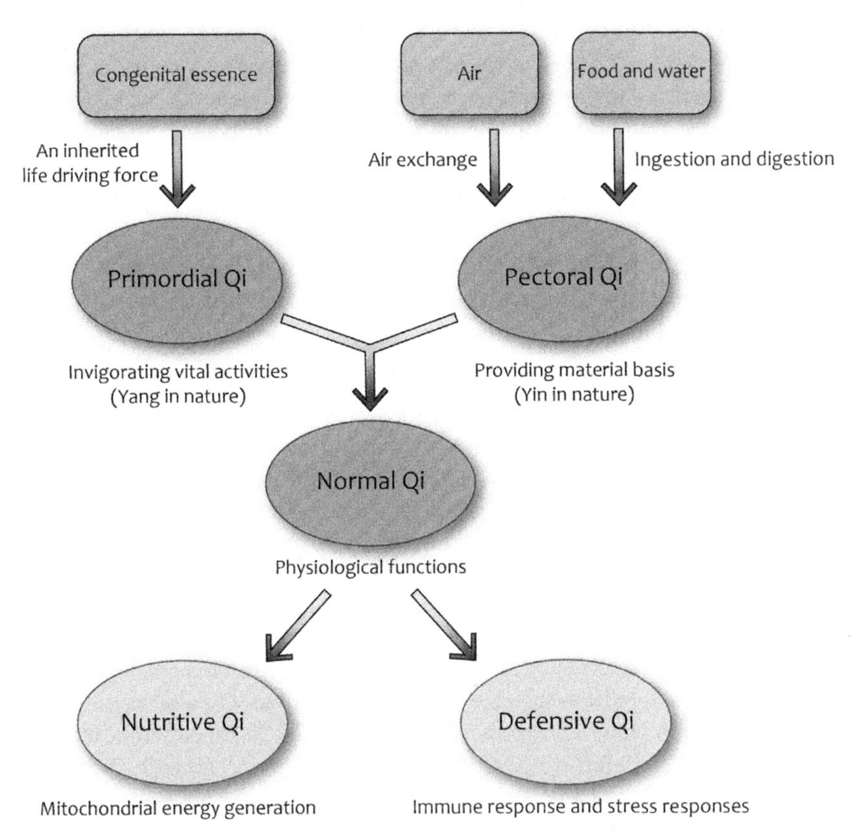

FIGURE 4.1 The physiological basis of Qi formation. *Modified from P.K. Leong, H.S. Wong, J. Chen, K.M. Ko, Yang/Qi invigoration: an herbal therapy for chronic fatigue syndrome with yang deficiency? Evid. Based Complement. Altern. Med. 2015 (2015) 945901.*

lungs and the heart. Primordial Qi combines with pectoral Qi to form normal Qi (also called Zheng Qi in Chinese) that circulates in the body to support normal physiological functions. The interrelationship between primordial Qi and pectoral Qi is consistent with the notion that normal Qi (or simply Qi) arises from an interaction between Yin and Yang. Normal Qi manifests in two physiological functions, namely, nutritive Qi and defensive Qi. Nutritive Qi is responsible for the generation of "Blood" and hence nourishes the internal organs to sustain physiological functions of the body. Nutritive Qi, when compared with defensive Qi, is Yin in nature by virtue of its role in generating vital substances. Defensive Qi protects the body against disease-causing internal (such as inflammation, cancer, internal stress, and mental activity) and external (such as bacteria, viruses, and environmental stress) factors. Defensive Qi also warms and nourishes the internal organs to prevent the invasion of pathogenic factors. Interestingly, in the context of modern medicine, the physiological role of defensive Qi in CM is not limited to the functioning of the immune system. Defensive Qi can also be functionally viewed as stress responsive (such as the antioxidant response, the unfolded protein response, the heat shock response, and the antiinflammatory response) in the body against any internal and external pathogenic challenges. Therefore in view of its protective action against pathogenic factors, defensive Qi is Yang in nature.

2.3 Restoring the Dynamic Balance Between Yin and Yang and Hence the Generation of Normal Qi Using Chinese Tonifying Herbs

In the framework of CM, the etiology of a disease is described in terms of an imbalance between Yin and Yang in the body, and therefore factors that can interrupt the Yin-Yang status in the body are disease causative factors. As previously mentioned, normal Qi is generated from the interaction between Yin (pectoral Qi) and Yang (primordial Qi); the manifestation of disease can also be considered as a consequence of a deficiency of normal Qi as well as the presence of pathogenic factors. According to CM theory, pathogenic factors can be internal (such as the dysfunction of internal organs, a disturbance of vital substances, and emotion), external (such as environment and climate), and other factors (such as lifestyle, physical activity, and mental activity) [55,58]. The antagonistic interaction between normal Qi and pathogenic factor(s) results in either an "excessive" syndrome (i.e., a collection of related symptoms), which refers to a situation where a strong normal Qi fights offensively against the pathogenic factor(s) with a manifestation of an acute and rapid syndrome, or a "deficiency" syndrome, which refers to a situation where weak normal Qi fights defensively against pathogenic factor(s) with a resultant chronic and protracted syndrome [58]. In an effort to diagnose a disease/syndrome, CM practitioners collect and analyze data obtained from patients by inspection, auscultation and olfaction, questioning, and pulse sensing and palpation. Then, the differentiation of the syndrome can be achieved by following the Eight Guiding Principles, which systemically categorize the syndrome into Yin/Yang, cold/heat, interior/exterior, and excessive/deficiency, based on the nature and location of the disease and the relative strength of normal Qi and pathogenic factors [58]. Following the diagnosis, the ultimate goal of the treatment is to reestablish the balance of Yin and Yang, restore the physiological function of internal organs, and revitalize the normal Qi in the body. In doing so, the prescription of Chinese herbal formulations and acupuncture are commonly used interventions for diseases.

The practice of CM always emphasizes disease prevention. According to CM theory, a "Deficiency" in body function (referred to as a subhealthy status in modern medicine) can result from a genetic predisposition, acquired environmental and lifestyle factors, and aging (Fig. 4.2). To avoid the worsening of an unhealthy or "Deficient" body, it is important to correct the condition by

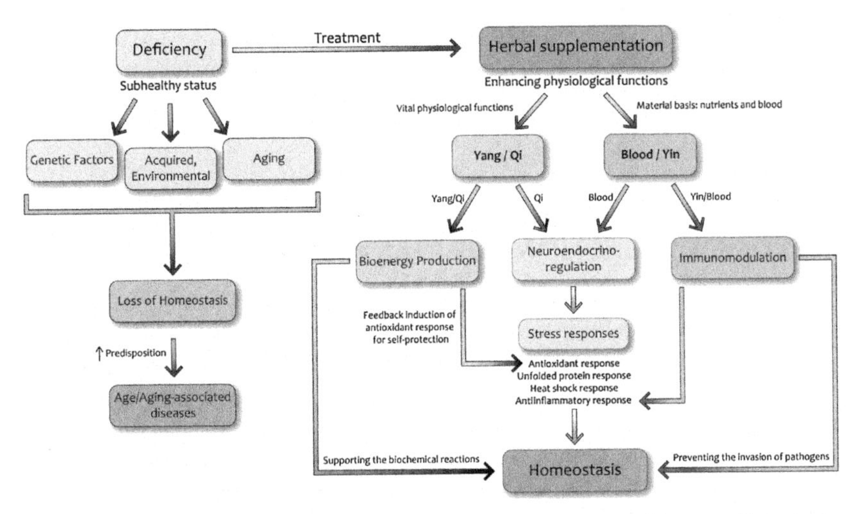

FIGURE 4.2 Causes of body "Deficiency" and their treatment using Chinese tonifying herbs with corresponding health-promoting functions.

restoring Yang, Qi, Yin, and Blood to normal levels. This is the CM approach used in preventing diseases. The practice of CM adopts various means to correct a body that is functionally deficient. One of these involves the use of Chinese tonifying herbs that act within the body, and another makes use of changes in lifestyle to upregulate the deficient bodily status from the exterior. Generally speaking, one can use Chinese tonifying herbs in Yang, Qi, Yin, and Blood to restore the balance of body function.

Generally, Chinese herbs are divided into two categories: tonifying herbs and therapeutic herbs. Tonifying herbs have been subcategorized into four functional groups: Yang-invigorating, Qi-invigorating, Yin-nourishing, and Blood-enriching [60]. The conceptual approach to health in modern medicine has emerged from half a century or more of scientific investigation in preventive health (i.e., prevention of diseases, particularly those that are age related). Recent biomedical research advocates the adoption of three approaches in preventive health: (1) antioxidation, (2) immunomodulation, and (3) neuroendocrine regulation. For example, by using vitamins or naturally occurring herbs, one can enhance antioxidant (and energy metabolism), immune, and neuroendocrinological regulatory functions in the body, in the hope of preventing or delaying the development of diseases, particularly those associated with aging. Research findings indicate that while most Yang-invigorating and Qi-invigorating Chinese tonifying herbs enhance mitochondrial ATP generation and antioxidant capacity in the body, many Yin-nourishing Chinese tonifying herbs also produce modulatory actions on the body's immunity. In addition, some Qi-invigorating and Blood-enriching Chinese tonifying herbs can facilitate the regulation of neuroendocrine function, with the maintenance of homeostasis by virtue of a host of antistress responses. Examples of each category of Chinese tonifying herbs/herbal formulations and their pharmacological actions will be discussed in the following sections.

3. PHARMACOLOGICAL BASIS OF THE HEALTH-PROMOTING ACTIONS OF CHINESE TONIFYING HERBS

According to CM theory, Yang is a manifestation of various body functions, all of which are dependent on cellular activities. In this regard, ATP, which is generated from mitochondria (the "power house" in the cell) energizes a wide range of cellular activities. With this in mind, the Yang/Qi-invigorating action of Chinese tonifying herbs may be related to mitochondrial ATP generation. As mentioned earlier, Yin represents the material basis for manifesting body function. Therefore Yin-nourishment and Blood-enrichment likely involve the regulation of nutrient supply and the blood circulatory system in the body. In the context of preventive health, Yin-nourishing herbs also produce effects on the immune system. Our laboratory has investigated this aspect of the pharmacological actions of Yang and Yin tonifying herbs. While Yang tonifying herbs were consistently found to increase ATPgeneration capacity in cultured H9c2 cardiomyocytes in vitro as well as rat heart mitochondria ex vivo [61], Yin tonifying herbs were shown to produce a comitogenic effect on concanavalin A-stimulated mouse splenocytes in vitro and ex vivo, indicative of immunostimulatory properties [62]. While Qi tonifying herbs also increased mitochondrial ATP generation [63], blood tonifying herbs stimulated the production of peripheral blood cells and premature blood cells in the bone marrow [64]. The pharmacological actions of the active compounds isolated from each category of Chinese tonifying herbs will be discussed in the following sections.

3.1 Yang/Qi-Invigorating Action: Effects on Cellular Energy Metabolism and Mitochondrial Function

3.1.1 β-Sitosterol (Yang-Invigorating)

β-sitosterol (BSS) is an active ingredient found in *Cistanche deserticola* Y. C. Ma (Cistanches Herba), one of the most commonly used Yang-invigorating tonifying herbs in CM [65] (Fig. 4.3). Our previous studies have demonstrated that BSS, which consists of a structural skeleton of 3-hydroxy cyclopentanoperhydrophenanthrene, can perturb mitochondrial membranes by fluidization [66]. The ability of BSS to increase mitochondrial membrane fluidity stimulates mitochondrial electron transport, as indicated by a significant increase in ATP generation capacity and mitochondrial state 3 respiration observed in both H9c2 cells and C2C12 myotubes as well as in mouse skeletal

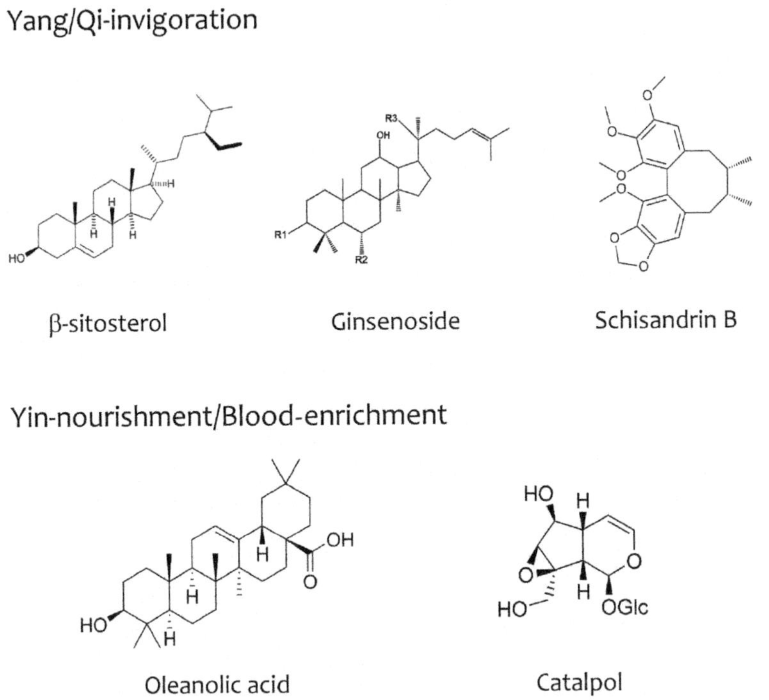

FIGURE 4.3 Chemical structures of active components of Yang/Qi-invigorating, Yin-nourishing, and Blood-enriching herbs.

muscle and rat heart mitochondria ex vivo. In addition, incubation with BSS caused a transient increase in mitochondrial membrane potential, which was paralleled by an increase in mitochondrial ROS production. ROS production induced by BSS was found to trigger glutathione reductase-catalyzed glutathione redox cycling, with resultant protection against oxidant injury, as assessed by hypoxia/reoxygenation-induced apoptosis in H9c2 cells in vitro and ischemia/reperfusion injury in rat hearts ex vivo [67].

BSS was also shown to cause a redox-sensitive induction of mitochondrial uncoupling in C2C12 myotubes through the activation of uncoupling protein 3 [68]. The induction of mitochondrial uncoupling could in turn reactivate mitochondrial electron transport, thereby maintaining a sustained low level of mitochondrial ROS production. ROS production further induced mitochondrial biogenesis, via the activation of the adenosine monophosphate-dependent protein kinase/peroxisome proliferator-activated receptor γ coactivator-1 pathway. BSS can produce an increase in mitochondrial number, together with an augmented mitochondrial responsiveness to energy demand, allowing sufficient energy generation to maintain physical and mental activities and thereby produce a beneficial effect in patients with Yang deficiency [66].

3.1.2 Ginsenosides (Qi-Invigorating)

Ginseng Radix (the root of *Panax ginseng* CA Meyer) is one of the most popular Qi-invigorating Chinese tonifying herbs for energizing the body [69]. According to CM theory, Ginseng Radix acts through the "Spleen," "Heart," "Lung," and "Kidney" meridians. Qi-invigoration by ginseng implies energy and life force. A growing body of evidence has demonstrated that the Qi-invigorating action of Ginseng Radix is closely related to the enhancement in mitochondrial function [63,70]. Ginseng Radix extract was found to increase mitochondrial ATP generation capacity in H9c2 cardiomyocytes in situ and rat hearts ex vivo [69]. Long-term administration of a Ginseng Radix extract maintained mitochondrial function and reduced oxidative stress in skeletal muscle following exhaustive exercise in rats, as evidenced by an increase in mitochondrial glutathione redox status and decreased levels of lipid peroxidation [70].

Ginsenosides (the active components of Ginseng Radix) play a vital role in producing their pharmacological effects (Fig. 4.3). There are approximately 40 types of ginsenosides, among which Rb1, Rg1, Rg3, Rh1, Re, and Rd are the most widely studied [71]. Ginsenosides exhibit a broad spectrum of pharmacological activities relevant to the prevention and/or treatment of cardiovascular diseases, which include the inhibition of mitochondrial ROS production, stimulation of NO production, improvement in blood circulation, enhancement of vasomotor tone, and improvement of blood lipid profiles [72]. Yang et al. showed that ginsenoside Rg5 protected against ischemic injury in cardiomyocytes [73]. Ginsenoside Rg5 promoted protein kinase B (Akt) translocation to mitochondria, which prevented dynamin-related protein 1 recruitment and mitochondrial hexokinase-II dissociation, thereby suppressing mitochondrial fission in cardiomyocytes. Rg5 also inhibited the opening of mitochondrial permeability transition pores and increased ATP production, resulting in an increased resistance to hypoxia/reoxygenation injury in cardiomyocytes. Meanwhile, Rg5 suppressed cell apoptosis with increased mitochondrial hexokinase-II binding and reduced dynamin-related protein 1 recruitment in mitochondria in mouse hearts subjected to isoproterenol-induced ischemic injury [74]. In addition, ginsenoside Rd treatment afforded significant protection against focal cerebral ischemia in rats, as indicated by a decrease in infarct size. The neuroprotective effect of Rd was associated with an improvement in neurological function, as evidenced by improved activities of respiratory chain complexes and aconitase, lowered mitochondrial hydrogen peroxide production, and hyperpolarized mitochondrial membrane potential.

3.1.3 Schisandrin B (Qi-Invigorating)

Schisandrin B (Sch B) is the most abundant active dibenzocyclooctadiene derivative isolated from the fruit of *Schisandra chinensis* (Turcz.) Baill. (Schisandrae Fructus [SF]), which is traditionally used for its astringent and sedative effects as well as in tonifying herbal formulations for Qi-invigoration [75] (Fig. 4.3). Sch B was found to consistently induce a glutathione antioxidant response in cultured H9c2 cardiomyocytes [76], AML12 hepatocytes [77], PC12 neuronal cells [78], and rodent brain, heart, and liver tissues [79–81]. The Sch B-induced glutathione antioxidant response was found to be associated with a cyto/tissue protective effect against oxidant-induced injury in brain, heart, and liver cells/tissues [75–81]. More importantly, the ability of Sch B to enhance mitochondrial glutathione redox status (i.e., the ratio of reduced glutathione to oxidized glutathione) was found to be paralleled by the suppression of mitochondrial permeability transition pore formation, suggesting a role of Sch B in preserving mitochondrial structural integrity under conditions of oxidative stress [82,83]. In view of the "Mitochondrial Free Radical Theory of Aging," a longevity study has been undertaken, and demonstrated that Sch B can increase the mean lifespan of mice, presumably by sustaining mitochondrial glutathione redox status and functional integrity of various tissues of aging mice [84]. Studies have shown that the mechanism underlying this Sch B-elicited glutathione antioxidant response involves the redox-sensitive extracellular signal-regulated kinase/nuclear factor (erythroid-derived 2)-like-2 factor (Nrf2)/antioxidant response element pathway [76,77]. The ability of Sch B to activate Nrf2 is related to its biotransformation via cytochrome P_{450}-catalyzed reactions [85]. The resultant metabolite of Sch B (which is a catechol) can then be oxidized into the corresponding quinone that may give rise to the generation of ROS. ROS can then activate the redox-sensitive extracellular signal-regulated kinase or modify the redox-sensitive cysteine residue of Kelch-like ECH-associated protein 1 (a repressor of Nrf2), with the resultant activation of Nrf2. Nrf2 is the master transcription factor regulating the expression of an array of antioxidant proteins/enzymes, such as superoxide dismutase, glutathione peroxidase, NAD(P)H quinone dehydrogenase 1, hemeoxygenase 1, and enzymes involved in glutathione synthesis and redox cycling, by binding to an antioxidant response element [86]. The ability of Sch B to reduce oxidative stress may also be mediated by the induction of a "heat shock" response, as indicated by the finding that Sch B afforded protection against mercury chloride-induced hepatotoxicity [87] as well as Aβ-induced neurotoxicity [88] in rodents. In addition, Sch B induced the Nrf2-mediated expression of thioredoxin, which can indirectly suppress the activation of NLRP3 inflammasomes in cultured mouse peritoneal macrophages, presumably via binding to thioredoxin-interacting protein [89]. Sch B also suppressed the proinflammatory Toll-like receptor/c-Jun kinase/nuclear factor-κB signal transduction pathway and its associated expression of proinflammatory effectors in cultured lipopolysaccharide-activated RAW264.7 macrophages and lipopolysaccharide-activated microglia-neuron cocultured cells [90,91]. Furthermore, Sch B ameliorated endoplasmic reticulum stress caused by the accumulation of unfolded proteins and the extent of hepatic steatosis in cultured tunicamycin-incubated HepG2 cells and tunicamycininjected mice [92]. Taken together, these results strongly suggest that Sch B can protect against a number of stress challenges (such as oxidative stress, endoplasmic reticulum stress, and inflammation) via the induction of an adaptive response (including an antioxidant response, a heat shock response, an unfolded protein response, and an antiinflammatory response) in vitro and in vivo. Interestingly, the ability of Sch B to protect against various stresses is consistent with the CM concept of defensive Qi, which can counteract disturbances in the homeostatic status in the body.

3.2 Yin-Nourishing/Blood-Enriching Action: Effects on Immune and Blood/Circulatory Functions

3.2.1 Oleanolic Acid (Yin-Nourishing)

Oleanolic acid is an active pentacyclic triterpenoid compound isolated from many medicinal plants [93], such as the fruit of *Ligustrum lucidum* Ait. (Ligustri Lucidi Fructus [LF]), and which possesses hepatoprotective, antiinflammatory, antitumor, and antihyperlipidemic properties, as reviewed in [93,94] (Fig. 4.3). LF is a Yin tonifying herb that is traditionally used for nourishing Yin in the "Kidney" and "Liver," improving the eyesight as well as promoting the growth of black hair [95]. Previous studies using activity-guided fractionation of LF have demonstrated that an oleanolic acid-enriched fraction of LF can produce an immune-stimulatory effect on concanavalin A-activated splenocytes in vitro and ex vivo as well as protect against CCl_4 hepatotoxicity in mice [96,97]. Consistent with our working hypothesis that Yin-nourishing herbs can modulate immune and cardiovascular function, oleanolic acid has been found to produce an immunomodulatory effect and also protect against cardiovascular diseases. Oleanolic acid can produce an antibacterial effect in *Mycobacterium tuberculosis*-infected mice via an increased expression of interferon γ (IFN-γ) and tumor necrosis factor α (TNF-α) in the lungs (i.e., an immune-stimulatory effect) [98]. Oleanolic acid can also reduce the extent of parasitemia in *Trypanosoma cruzi*-challenged mice (which is an experimental model of Chagas' disease) via the reduction of IFN-γ and the increase of interleukin 10 (IL-10) in the plasma of infected mice (i.e., an immune-suppressive effect) [99]. In addition, oleanolic acid can improve the survival and function of pancreatic islet allografts in mice with pancreatic islet transplants through suppression of the release of proinflammatory cytokines (such as IL-10 and interleukin 4) [100]. The ability of oleanolic acid to elicit an antiinflammatory response was also shown in myelin oligodendrocyte protein-immunized mice (which is an experimental model of autoimmune encephalomyelitis), in which plasma levels of proinflammatory cytokines (such as interleukin 17A, IFN-γ, interleukin 6 [IL-6], and TNF-α) were suppressed [101]. Oleanolic acid was also shown to reduce the extent of myocarditis and hence protect against cardiac injury in cardiac α-myosin (MyHc-α614–629)-immunized mice [102]. The ability of oleanolic acid to produce immune-stimulatory or immune-suppressive actions seems to be dependent on the pathological condition in question. However, the aforementioned findings suggest that oleanolic acid can consistently attenuate the pathogenesis of diseases and restore homeostasis via its immunomodulatory action. Furthermore, oleanolic acid was found to reduce hyperglycemia in rats with streptozotocin-induced diabetes [103,104], presumably via the reduction of postprandial hyperglycemia, the improvement of pancreatic β-cell function, insulin sensitivity, and the attenuation of proinflammatory factors [105,106]. In Western medicine, type 2 diabetes mellitus is viewed as a metabolic disorder, which is associated with insulin resistance, chronic inflammation, and vascular impairment. In the context of CM theory, type 2 diabetes mellitus is a pathological condition caused by a Yin deficiency in the "Kidney" and therefore Yin-nourishment should attenuate its pathogenesis. In this connection, the ability of oleanolic acid to elicit an antidiabetic effect supports its use in diseases associated with Yin deficiency. In summary, these findings strongly support our hypothesis that Yin-nourishing herbs are able to modulate the immune and cardiovascular systems.

3.2.2 Catalpol (Blood-Enriching)

The root of *Rehmannia glutinosa* Libosch. (Rehmanniae Radix [RR]) is used in three types of Chinese herbal preparations, namely, fresh RR with the properties of removing "heat" from the blood, dried RR with the properties of Yin-nourishing and promoting the production of body fluid, and steam-prepared RR with the properties of Yin-nourishing and Blood-enriching. The differential pharmacological activities of the different RRs are due to the changes in chemical profile with various processing procedures [107]. Studies have demonstrated the ability of RR to induce blood production, as evidenced by an enhanced proliferation of cultured bone marrow hematopoietic cells and an increase in the number of red blood cells in hemorrhagic anemic mice [108]. A furfural disaccharide (or 5-hydroxymethyl-2-furfural), which is only found in the stream-prepared RR preparation, was also investigated in a clinical trial for sickle cell anemia [109]. RR is also commonly used in various Chinese tonifying herbal formulations for producing hemopoietic effects [110,111]. For example, an herbal formulation, namely, Hemomine, which is composed of Angelica Radix, Cnidii Fructus, Paeoniae Radix, RR, and Glycyrrhizae Radix, was found to increase the proliferation of hematopoietic stem cells in hemorrhagic anemia in rats [110]. Si-Wu-Tang (SWT), another Blood-enriching decoction consisting of RR, Angelica Radix, Chuanxiong Rhizoma, and Paeoniae Radix, increased serum levels of erythropoietin (EPO) in mice with phenylhydrazine-induced anemia [111].

Catalpol is the major active iridoid compound found in RR (Fig. 4.3). Although the effects of catalpol in hemorheology and hemopoiesis of bone marrow are yet to be defined, it has been shown to protect against ischemia/reperfusion injury in the brain [112–114], heart [115], and kidney [116]. Thus catalpol improved cerebral angiogenesis and ameliorated the extent of cerebral ischemia/reperfusion injury in rats via the activation of the Janus kinase 2/signal transducer and activator of transcription 3 pathway and the associated production of vascular endothelial growth factor and EPO [112,114]. The ability of catalpol to protect against cerebral ischemia/reperfusion injury is likely related to the induction of an antioxidant response in the brain [113]. Similarly, catalpol suppressed the generation of reactive oxygen/nitrogen species and induced the production of NO (a vasodilator), with a resultant protection against myocardial ischemia/reperfusion injury in rats [115]. In addition, catalpol inhibited the phosphoinositide 3-kinase/Akt pathway, which in turn reduced the release of proinflammatory cytokines (TNF-α, IL-1β, IL-6, and IL-10) in the ischemic-reperfused kidneys in mice, with subsequent protection against renal ischemia/reperfusion injury [116]. Generalized tissue protection against ischemia/reperfusion injury afforded by catalpol suggests the involvement of a generalized mechanism underlying its protective effect, i.e., the induction of an antioxidant response as well as an improvement in angiogenesis. Catalpol-induced angiogenesis in relation to hemorheology and hemopoiesis of bone marrow remains to be determined.

4. CHINESE HERBAL TONIFYING FORMULAS

4.1 Wu-Zi-Yan-Zong-Wan (Yang-Invigorating)

Wu-Zi-Yan-Zhong-Wan (WZ), also known as Five Seeds Combo, is a renowned Chinese Yang-invigorating herbal formulation composed of five herbs, namely, Lycii Fructus, Cuscutae Semen, Rubi Fructus, Plantaginis Semen, and SF [117]. In the practice of CM, WZ is considered as "the first and fundamental antiaging recipe." Studies have demonstrated that the potential antiaging effect of WZ may be associated with its ability to increase mitochondrial antioxidant status and the maintenance of mitochondrial function [117,118].

WZ extracts have been shown to concentration dependently protect against ethanol-induced toxicity in CYP2E1 cDNA-transfected human HepG2 (E47) cells [117]. The cytoprotective effect afforded by WZ was likely due to its potent antioxidant activity, as evidenced by decreases in ROS production and the extent of lipid peroxidation, as well as the increase in cellular/mitochondrial glutathione levels and mitochondrial membrane potential. In addition, preincubation with WZ extracts significantly suppressed ethanol-induced DNA fragmentation in E47 cells. To determine the chemical basis underlying cytoprotection by WZ, ultraperformance liquid chromatography-tandem mass spectrometer analysis was utilized to compare the extracted contents of marker compounds in the WZ formulation and its component herbs. The results demonstrated the presence of betaine (5.38 μg/g) (from Lycii Fructus), quercetin (19.06 μg/g) (Cuscutae Semen, Rubi Fructus, and Plantaginis Semen), hyperin (30.32 μg/g) (Cuscutae Semen and SF), Sch B (11.10 μg/g), schisandrin (17.08 μg/g), and deoxyschisandrin (7.14 μ/g) (SF) in the WZ formulation, all of which were of higher concentrations than those in single herb extracts. The higher concentrations of active components detectable in the WZ formulation suggest a potentially higher efficacy of WZ when compared with its individual component herbs, not to mention the possibility of synergistic interactions among them.

The ability of WZ to enhance mitochondrial antioxidant status and functional capacity has also been demonstrated in a rat model of long-term ethanol intoxication [118]. The long-term consumption of ethanol induces oxidative damage in various organs and accelerates the process of aging. Long-term treatment (42 days) with a WZ extract significantly ameliorated ethanol-induced mortality as well as reversed the decrease in body weight in alcohol-treated rats. WZ treatment also decreased the hepatic index (the ratio of liver weight to body weight) when compared with the untreated ethanol group. In addition, the extent of hepatic damage was dramatically decreased in WZ-treated rats, as assessed by decreases in serum alanine aminotransferase and aspartate aminotransferase activities. Histopathological analysis of liver tissues showed that WZ extract treatment attenuated ethanol-induced macrovesicular steatosis, vacuole formation, and inflammation. The hepatoprotective effect of the WZ extract was associated with decreases in the extents of ROS production, lipid peroxidation, and oxidative modification of proteins. The antioxidant effect afforded by WZ was found to be mainly attributable to the enhancement in mitochondrial glutathione levels in rat livers. The active ingredients present in the WZ extract, such as Sch B, verbascoside, astragalin, kaempferol, and hyperoside, are all capable of enhancing hepatic glutathione redox status, and they may work synergistically to maintain cellular antioxidant defense systems and thereby contribute to the hepatoprotective effect of the WZ extract.

In addition to hepatoprotection, the WZ extract also increased viability and suppressed early- and middle/late-stage apoptosis in $CoCl_2$-induced neurotoxicity in PC12 cells [119]. $CoCl_2$-induced neurotoxicity is a commonly used model for investigating neuronal damage and apoptosis resulting from hypoxic/ischemic insult. The neuroprotective effect of the WZ extract was associated with the suppression of intracellular ROS production and the reversal of the collapsed mitochondrial membrane potential, which were indicative of inhibition of the mitochondrial apoptotic signaling pathway. In this regard, incubation with WZ increased antiapoptotic proteins, such as B-cell lymphoma 2 protein and B-cell lymphoma-XL protein, and suppressed cytochrome *c* release from mitochondria into the cytosol. This further prevented the activation of caspase-3 and poly(ADP-ribose) polymerase. Moreover, WZ was also found to inhibit extracellular signal-regulated kinases, c-Jun kinase, and p38 mitogen-activated protein kinase phosphorylation. Taken together, WZ extract acts as neuroprotective and antiapoptosis agents in PC12 cells subjected to hypoxic/ischemic challenge, possibly through a mitochondrial protective pathway.

4.2 Er-Zhi-Wan (Yin-Nourishing)

Er-Zhi-Wan (EZW) is a Chinese herbal formulation consisting of two herbs: LF and Ecliptae Herba. According to different versions of Chinese Materia Medica, various processing methods for producing LF have been documented for the preparation of EZW, such as soaking LF with wine, steaming LF with wine, or steaming LF over water. Conventionally, EZW is prepared using an equal weight ratio of LF and Eclipta Herba. Based on CM theory, EZW is used for the treatment of diseases with "Kidney" Yin deficiency by virtue of its Yin tonifying action in nourishing the "Kidney" and strengthening tendons and bone. A study has also shown that EZW can produce hepatoprotection against CCl_4 toxicity [120].

Based on syndrome differentiation in the practice of CM, postmenopausal osteoporosis can be related to "Kidney" Yin deficiency [121,122]. With this notion in mind, pharmacological studies have revealed that EZW can ameliorate postmenopausal osteoporosis, as indicated by a reduction in the extent of ovariectomy-induced osteoporosis in rodents [123−126]. As such, EZW, which was prepared using a mixture of 95% ethanol extract of LF and 95% Ecliptae Herba in a ratio of 1:1, induced the estrogen-driven expression of luciferase in cultured MCF-7 cells without causing any increased proliferation, suggesting that EZW is an effective and safe herbal formulation with estrogenic activity [123]. In addition, incubation of cultured osteoclasts with serum isolated from EZW-treated rats inhibited osteoclast proliferation via the suppression of release of nuclear factor B ligand receptor activator and macrophage-colony stimulating factor [124]. Cheng et al. demonstrated that EZW treatment can reduce the extent of osteoporosis in ovariectomized rats, as evidenced by increases in serum levels of bone-building minerals (Ca^{2+} and phosphorus), biochemical markers of bone formation, decreases in biochemical markers of bone resorption, and an associated increase in bone mass density of the femur, fourth lumbar vertebra, and the tibia [125]. Similarly, Sun et al. also showed that EZW can suppress the pathogenesis of alveolar bone osteoporosis in ovariectomized rats, as indicated by increases in serum levels of estradiol and bone alkaline phosphatase, and a decrease in tartrate-resistant acid phosphatase via the activation of the wnt/β-catenin pathway [126].

4.3 Shengmai San (Qi-Invigorating)

Sheng-Mai-San (SMS), a Chinese herbal formulation containing Ginseng Radix, Ophiopogonis Radix, and SF, has been clinically prescribed in China for more than 800 years for Qi-invigoration and body fluid retention, particularly in the treatment of coronary heart disease [127]. Pharmacological studies have indicated that SMS can protect against oxidant-induced injury in brain, heart, kidney, and liver in vitro and/or in vivo [128−137], presumably via the elicitation of antioxidant and antiinflammatory responses.

An experimental study has revealed that SMS can induce an antioxidant response and thereby protect against cerebral ischemia/reperfusion injury in rats, as indicated by an increase in the activity of glutathione peroxidase and a decrease in the level of thiobarbituric acid reactive substances. The mechanism underlying the cerebral protection by SMS may be related to the suppression of apoptotic cell death in the brain [128]. In this regard, four active components isolated from SMS were found to inhibit the proapoptotic activation of caspase 3 in H_2O_2-intoxicated PC12 neuronal cells [129,130]. In addition, acute treatment with SMS can ameliorate ex vivo myocardial ischemia/reperfusion injury in rats, as shown by the suppression of lactate dehydrogenase leakage from isolated-perfused hearts, presumably via the induction of mitochondrial K_{ATP} channel opening [131]. Long-term treatment with SMS can induce the expression of antioxidant enzymes (superoxide dismutase and hemeoxygenase-1) [132] and suppress proapoptotic factors (Bcl-2-associated X protein, cytochrome *c*, and caspase 3) [133], with a resultant cardiac protection against chronic intermittent hypoxia in mice. Similarly, You et al. demonstrated that long-term treatment with SMS can induce an antioxidant response and hence protect against adriamycin-induced cardiomyopathy in rats [134]. Regarding the traditional use of SMS for treating retention of body fluid associated with heat stroke, a recent study has indicated that SMS can protect against heat stroke-induced renal injury in rats, as evidenced by a decrease in plasma levels of NO, creatinine, and blood urea nitrogen. The ability of SMS to induce an antioxidant response seems to be beneficial to the metabolic activity of the liver. In this regard, SMS can increase GSH synthesis in the liver and reduce levels of triglyceride and cholesterol in high-fat diet-induced obesity in rats [135]. Interestingly, a manufactured herbal product based on an SMS formulation (namely, Wei-Kang-Su) was found to elicit an antioxidant response in various organs in rats (brain, heart, kidneys, and liver) and protect against CCl_4 hepatotoxicity and long-term ethanol-induced liver injury [136,137]. Taken together, SMS may be an effective health-promoting agent for enhancing the resistance of mitochondria to oxidative stress and offering the prospect of disease prevention.

4.4 Si-Wu-Tang (Blood-Enriching)

SWT (Si Wu decoction) is a Chinese herbal formulation comprised of four herbs, namely, RR, Angelica Radix, Chuanxiong Rhizoma, and Paeoniae Radix. According to CM theory, SWT with its Blood-enriching action has been widely used for the treatment of diseases characterized as "Blood deficiency"

[138]. Several experimental studies have indicated that SWT possesses potent antiinflammatory and antioxidant activities, and may therefore produce beneficial effects in the prevention or treatment of aging-related diseases [139,140].

SWT has been shown to prevent the triphasic response of Lewis on skin and scratching behavior, which is probably an itch-associated behavior after a challenge, in a passively sensitized Balb/c mice model [141]. In this model, passive sensitization with a murine monoclonal IgE antibody specific for the dinitrophenol group (anti-DNP IgE mAb) followed by challenge of the mouse ear with dinitrofluorobenzene (DNFB) induces a triphasic skin reaction with an immediate phase response (IPR) and a late phase response (LPR) at 1 and 24 h after challenge. It is then followed by a third inflammatory phase response, termed the "very late phase response (vLPR)" at 8 days after challenge. SWT dose dependently inhibited ear swelling in the LPR and vLPR after DNFB challenge and slightly reduced the scratching behavior. The inhibitory effects on the LPR and vLPR have been attributed, in part, to the action of the Chuanxiong Rhizoma in the SWT formulation. On the other hand, although Angelica Radix did not inhibit ear swelling in IPR, it (but not Chuanxiong Rhizoma) suppressed the scratching behavior in SWT. These results indicated that the combination of the four herbs in the SWT formulation works synergistically to produce the inhibition of cutaneous inflammatory responses.

Radiation therapy can be an effective treatment for cancer. However, serious gastrointestinal side effects, such as vomiting and diarrhea, limit the clinical use of this important interventional therapy. Experimental studies have demonstrated that high doses of ionizing radiation cause frank ulceration or areas of complete epithelial denudation, which results in inflammation of the intestinal mucosa (mucositis). SWT has been shown to protect against radiation damage by increasing the numbers of various populations of circulating white blood cells, especially lymphocytes, and prevents side effects on the intestinal lining in γ-ray irradiated female mice [140]. The results of hematoxylin and eosin staining showed that histopathological scores, which quantify the severity of morphological changes, were much lower in SWT-treated mice when compared with the control group. Immunohistochemical analysis of fibrinogen also indicated that SWT treatment reduced perivascular staining of fibrinogen (which is a marker of γ-ray-induced inflammation), indicative of a protection against intestinal inflammation in the γ-ray irradiated mice.

A randomized, double-blind, placebo-controlled clinical trial has shown that administration of SWT significantly increases the antioxidant index and antioxidant enzyme activities in the skin and liver of healthy adults [139]. Treatment with SWT was found to substantially increase plasma glutathione levels, which is associated with an enhancement of antioxidant defense systems consisting of the endogenous antioxidants superoxide dismutase, catalase, glutathione peroxidase, and glutathione reductase. The ability of

SWT to enhance antioxidant status is associated with an improvement in hepatic function, resulting in improved lipid profile parameters and decreased hepatic damage as well as normalizing abnormal conditions such as fatty liver, gallbladder stones, and splenomegaly. Similarly, SWT treatment can improve skin elasticity and texture, presumably through the intermediacy of free radical scavenging activities.

5. CONCLUSIONS

The "Mitochondrial Free Radical Theory of Aging" proposes that the unavoidable production of mitochondrial ROS during aerobic respiration can lead to the disruption of mitochondrial structural and functional integrity, which in turn can favor the development of several age-associated diseases. In the realm of CM, Qi arising from the interaction of Yin and Yang is the vital energy that drives physiological functions in the body. An imbalance between Yin and Yang (such as Yin and/or Yang deficiency) in the body can disrupt the generation of Qi and the complete depletion of Qi ultimately results in death. Given that the mitochondrion serves as a coordinator of cell survival (energy production) and cell death (apoptosis), the concept of Qi in CM is likely ascribable to mitochondrial function under both physiological and pathological conditions. To restore different types (Yang, Yin, Qi, Blood, or their combinations) of deficiency in the body, the practice of CM uses tonifying herbs (or herbal formulations) that are divided into functional categories of Yang-invigoration, Yin-nourishment, Qi-invigoration, and Blood-enrichment, or their combination. Therefore tonifying herbs/herbal formulations offer a promising prospect in the search for preventive interventions in promoting health. As mentioned earlier, recent pharmacological studies have demonstrated the effectiveness of various Chinese herbal formulations in preserving mitochondrial structural and functional integrity, producing immunomodulatory effects as well as eliciting adaptive responses (such as antioxidant, anti-inflammatory, and heat shock responses) and the associated protection against pathological processes involved in age-associated diseases in vitro and in vivo. As an herbal formulation is comprised of various herbs with a host of active components, the elucidation of the biochemical/molecular mechanism(s) underlying the pharmacological action of the herbal formulation can identify the relevant active compound(s) isolated from the component herb(s). While beneficial effects of these active components have been demonstrated by pharmacological studies, the long-term administration of a single compound for preventive health is not optimal in the practice of CM that emphasizes a holistic regulation of body function in achieving a healthy status. To optimize efficacy and ensure their safety, herbal formulations, which are made of a combination of tonifying herbs under the guidance of CM principles, are much preferred for the development of herbal health products, particularly those aimed at preventing age-associated diseases.

REFERENCES

[1] D.J. Herman, Aging: a theory based on free radical and radiation chemistry, Gerontology 11 (1956) 298–300.

[2] G.C. Kujoth, P.C. Bradshaw, S. Haroon, T.A. Prolla, The role of mitochondrial DNA mutations in mammalian aging, PLoS Genet. 3 (2007) e24.

[3] L. Galluzzi, O. Kepp, C. Trojel-Hansen, G. Kroemer, Mitochondrial control of cellular life, stress, and death, Circ. Res. 111 (2012) 1198–1207.

[4] H.M. Honda, P. Korge, J.N. Weiss, Mitochondria and ischemia/reperfusion injury, Ann. N.Y. Acad. Sci. 1047 (2005) 248–258.

[5] S.V. Raju, L.A. Barouch, J.M. Hare, Nitric oxide and oxidative stress in cardiovascular aging, Sci. Aging Knowl. Environ. 2005 (2005) re4.

[6] P. Pacher, J.S. Beckman, L. Liaudet, Nitric oxide and peroxynitrite in health and disease, Physiol. Rev. 87 (2007) 315–424.

[7] J. Chang, S. Kou, W. Lin, C. Liu, Regulatory role of mitochondria in oxidative stress and atherosclerosis, World J. Cardiol. 2 (2010) 150–159.

[8] P.S. Brookes, Y. Yoon, L. Robotham, M.W. Anders, S. Sheu, Calcium, ATP, and ROS: a mitochondrial love-hate triangle, Am. J. Physiol. Cell. Physiol. 287 (2004) 817–833.

[9] J.S. Bhatti, G.K. Bhatti, P.H. Reddy, Mitochondrial dysfunction and oxidative stress in metabolic disorders – a step towards mitochondria based therapeutic strategies, Biochim. Biophys. Acta 1863 (2017) 1066–1077.

[10] M.G. Rosca, B. Tandler, C.L. Hoppel, Mitochondria in cardiac hypertrophy and heart failure, J. Mol. Cell. Cardiol. 55 (2013) 31–41.

[11] H. Tsutsui, S. Kinugawa, S. Matsushima, Mitochondrial oxidative stress and dysfunction in myocardial remodeling, Cardiovasc. Res. 81 (2009) 449–456.

[12] D.A. Brown, J.B. Perry, M.E. Allen, H.N. Sabbah, B.L. Stauffer, S.R. Shaikh, et al., Expert consensus document: mitochondrial function as a therapeutic target in heart failure, Nat. Rev. Cardiol. 14 (2017) 238–250.

[13] M. Golpich, E. Amini, Z. Mohamed, R. Azman Ali, N. Mohamed Ibrahim, A. Ahmadiani, Mitochondrial dysfunction and biogenesis in neurodegenerative diseases: pathogenesis and treatment, CNS Neurosci. Ther. 23 (2017) 5–22.

[14] M. Recuero, T. Munoz, J. Aldudo, M. Subias, M.J. Bullido, F. Valdivieso, A free radical-generating system regulates APP metabolism/processing, Febs. Lett. 584 (2010) 4611–4618.

[15] P.M. Keeney, J. Xie, R.A. Capaldi, J.P. Bennett Jr., Parkinson's disease brain mitochondrial complex I has oxidatively damaged subunits and is functionally impaired and mis-assembled, J. Neurosci. 26 (2006) 5256–5264.

[16] S.J. Chinta, J.K. Andersen, Redox imbalance in Parkinson's disease, Biochim. Biophys. Acta 1780 (2008) 1362–1367.

[17] R. Sultana, D.A. Butterfield, Oxidatively modified, mitochondria-relevant brain proteins in subjects with Alzheimer disease and mild cognitive impairment, J. Bioenerg. Biomembr. 41 (2009) 441–446.

[18] S. Sorbi, E.D. Bird, J.P. Blass, Decreased pyruvate dehydrogenase complex activity in Huntington and Alzheimer brain, Ann. Neurol. 13 (1983) 72–78.

[19] S.J. Kish, C. Bergeron, A. Rajput, S. Dozic, F. Mastrogiacomo, L.J. Chang, Brain cytochrome oxidase in Alzheimer's disease, J. Neurochem. 59 (1992) 776–779.

[20] X. Wang, B. Su, G. Perry, M.A. Smith, X. Zhu, Insights into amyloid-beta-induced mitochondrial dysfunction in Alzheimer disease, Free Radic. Biol. Med. 43 (2007) 1569–1573.

[21] L. Katsouri, C. Parr, N. Bogdanovic, M. Willem, M. Sastre, PPARγ co-activator-1α (PGC-1α) reduces amyloid-β generation through a PPARγ-dependent mechanism, J. Alzheimers Dis. 25 (2011) 151–162.

[22] W. Qin, V. Haroutunian, P. Katsel, C.P. Cardozo, L. Ho, J.D. Buxbaum, G.M. Pasinetti, PGC-1α expression decreases in the Alzheimer disease brain as a function of dementia, Arch. Neurol. 66 (2009) 352–361.

[23] A. Eckert, R. Nisbet, A. Grimm, J. Götz, March separate, strike together–role of phosphorylated TAU in mitochondrial dysfunction in Alzheimer's disease, Biochim. Biophys. Acta 1842 (2014) 1258–1266.

[24] J. Hardy, H. Cai, M.R. Cookson, K. Gwinn-Hardy, A. Singleton, Genetics of Parkinson's disease and parkinsonism, Ann. Neurol. 60 (2006) 389–398.

[25] A.H. Schapira, J.M. Cooper, D. Dexter, J.B. Clark, P. Jenner, C.D. Marsden, Mitochondrial complex I deficiency in Parkinson's disease, J. Neurochem. 54 (1990) 823–827.

[26] L. Devi, V. Raghavendran, B.M. Prabhu, N.G. Avadhani, H.K. Anadatheerthavarada, Mitochondrial import and accumulation of alpha-synuclein impair complex I in human dopamineric neuronal cultures and Parkinson disease brain, J. Biol. Chem. 283 (2008) 9089–9100.

[27] H.A. Lashuel, B.M. Petre, J. Wall, M. Simon, R.J. Nowak, T. WalT, et al., Alpha-synuclein, especially the Parkinson's disease-associated mutants, forms pore-like annular and tubular protofibrils, J. Mol. Biol. 322 (2002) 1089–1102.

[28] W.W. Smith, H. Jiang, Z. Pei, Y. Tanaka, H. Morita, A. Sawa, et al., Endoplasmic reticulum stress and mitochondrial cell death pathways mediate A53T mutant alpha-synuclein-induced toxicity, Hum. Mol. Genet. 14 (2005) 3801–3811.

[29] M.T. Drake, B.L. Clarke, E.M. Lewiecki, The pathophysiology and treatment of osteoporosis, Clin. Ther. 37 (2015) 1837–1850.

[30] M. Shahi, A. Peymani, M. Sahmani, Regulation of bone metabolism, Rep. Biochem. Mol. Biol. 5 (2017) 73–82.

[31] W. Xiao, Y. Wang, S. Pacios, S. Li, D.T. Graves, Cellular and molecular aspects of bone remodeling, Front. Oral Biol. 18 (2016) 9–16.

[32] D. Frikha-Benayed, J. Basta-Pljakic, R.J. Majeska, M.B. Schaffler, Regional differences in oxidative metabolism and mitochondrial activity among cortical bone osteocytes, Bone 90 (2016) 15–22.

[33] K. Kobayashi, H. Nojiri, Y. Saita, D. Morikawa, Y. Ozawa, K. Watanabe, et al., Mitochondrial superoxide in osteocytes perturbs canalicular networks in the setting of age-related osteoporosis, Sci. Rep. 5 (2015) 9148.

[34] A. Kalani, P.K. Kamat, M.J. Voor, S.C. Tyagi, N. Tyagi, Mitochondrial epigenetics in bone remodeling during hyperhomocysteinemia, Mol. Cell. Biochem. 395 (2014) 89–98.

[35] D.A. Callaway, J.X. Jiang, Reactive oxygen species and oxidative stress in osteoclastogenesis, skeletal aging and bone diseases, J. Bone Min. Metab. 33 (2015) 359–370.

[36] T. Miyazaki, M. Iwasawa, T. Nakashima, S. Mori, K. Shigemoto, H. Nakamura, et al., Intracellular and extracellular ATP coordinately regulate the inverse correlation between osteoclast survival and bone resorption, J. Biol. Chem. 287 (2012) 37808–37823.

[37] V.A. Gerriets, R.J. Kishton, A.G. Nichols, A.N. Macintyre, M. Inoue, O. Ilkayeva, et al., Metabolic programming and PDHK1 control CD4+ T cell subsets and inflammation, J. Clin. Invest. 125 (2015) 194–207.

[38] A.S. Yazdi, S.K. Drexler, J. Tschopp, The role of the inflammasome in nonmyeloid cells, J. Clin. Immunol. 30 (2010) 623–627.

[39] A.K. Jha, S.C. Huang, A. Sergushichev, V. Lampropoulou, Y. Ivanova, E. Loginicheva, et al., Network integration of parallel metabolic and transcriptional data reveals metabolic modules that regulate macrophage polarization, Immunity 42 (2015) 419–430.
[40] L.A. O'Neill, A broken Krebs cycle in macrophages, Immunity 42 (2015) 393–394.
[41] V. Infantino, P. Convertini, L. Cucci, M.A. Panaro, M.A. Di Noia, R. Calvello, et al., The mitochondrial citrate carrier: a new player in inflammation, Biochem. J. 438 (2011) 433–436.
[42] V. Morrisette-Thomas, A.A. Cohen, T. Fülöp, É. Riesco, V. Legault, Q. Li, et al., Inflamm-aging does not simply reflect increases in pro-inflammatory markers, Mech. Ageing Dev. 139 (2014) 49–57.
[43] C. Franceschi, M. Bonafè, S. Valensin, F. Olivieri, M. De Luca, E. Ottaviani, et al., Inflamm-aging. An evolutionary perspective on immunosenescence, Ann. N.Y. Acad. Sci. 908 (2000) 244–254.
[44] L. Lin, S. Park, E.G. Lakatta, RAGE signaling in inflammation and arterial aging, Front. Biosci. 14 (2009) 1403–1413.
[45] B.K. Davis, H. Wen, J.P. Ting, The inflammasome NLRs in immunity, inflammation, and associated diseases, Annu. Rev. Immunol. 29 (2011) 707–735.
[46] O.A. Anderson, A. Finkelstein, D.T. Shima, A2E induces IL-1ß production in retinal pigment epithelial cells via the NLRP3 inflammasome, PLoS One 8 (2013) e67263.
[47] C. Franceschi, P. Garagnani, G. Vitale, M. Capri, S. Salvioli, Inflammaging and 'Garb-aging', Trends Endocrinol. Metab. 28 (2017) 199–212.
[48] S.R. Kim, D.I. Kim, S.H. Kim, H. Lee, K.S. Lee, S.H. Cho, et al., NLRP3 inflammasome activation by mitochondrial ROS in bronchial epithelial cells is required for allergic inflammation, Cell Death Dis. 5 (2014) e1498.
[49] M.E. Heid, P.A. Keyel, C. Kamga, S. Shiva, S.C. Watkins, R.D. Salter, Mitochondrial ROS induces NLRP3-dependent lysosomal damage and inflammasome activation, J. Immunol. 191 (2013) 5230–5238.
[50] A. Lu, H. Wu, Structural mechanisms of inflammasome assembly, Febs. J. 282 (2015) 435–444.
[51] A. Salminen, J. Ojala, K. Kaarniranta, A. Kauppinen, Mitochondrial dysfunction and oxidative stress activate inflammasomes: impact on the aging process and age-related diseases, Cell. Mol. Life Sci. 69 (2012) 2999–3013.
[52] H.B. Li, C. Jin, Y. Chen, R.A. Flavell, Inflammasome activation and metabolic disease progression, Cytokine Growth Factor Rev. 25 (2014) 699–706.
[53] Z. Zhen, Advanced Textbook on Traditional Chinese Medicine and Pharmacology, vol. 1, New World Press, Beijing, China, 1995.
[54] H. Yin, X. Shuai, Fundamentals of Traditional Chinese Medicine, Foreign Languages Press, Beijing, China, 1992.
[55] K.A. O'Brien, C.C. Xue, The theoretical framework of Chinese medicine, in: P.C. Leung, C.C. Xue, Y.C. Chen (Eds.), A Comprehensive Guide to Chinese Medicine, World Scientific Publishing Co. Pvt. Ltd., Singapore, 2003, pp. 47–84.
[56] Z. Liu, L. Liu, Essentials of Chinese Medicine, Springer, London, UK, 2009.
[57] D. Zhang, X. Wu, Qi, blood, body fluid, essence of life and spirit, in: Y. Liu (Ed.), The Basic Knowledge of Traditional Chinese Medicine, Hai Feng Publishing Co., Hong Kong, 1991, pp. 49–53 (Chapter 5).
[58] Y. Lu, C. Liu, The basic substances constituting the human body: Qi, Blood and Body Fluids, in: P. Chen (Ed.), Advanced TCM Series, Concepts and Theories of Traditional Chinese Medicine, vol. 2, Science Press, Beijing, 1997, pp. 98–119 (Chapter 4).

[59] P.K. Leong, H.S. Wong, J. Chen, K.M. Ko, Yang/Qi invigoration: an herbal therapy for chronic fatigue syndrome with yang deficiency? Evid. Based Complement. Altern. Med. 2015 (2015) 945901.

[60] J. Geng, W. Huang, T. Ren, X, Practical Traditional Chinese Medicine & Pharmacology, New World Press, Beijing, China, 1991, pp. 208–234.

[61] K.M. Ko, T.Y. Leon, D.H. Mak, P.Y. Chiu, Y. Du, M.K. Poon, A characteristic pharmacological action of 'Yang-invigorating' Chinese tonifying herbs: enhancement of myocardial ATP-generation capacity, Phytomedicine 13 (2006) 636–642.

[62] K.M. Ko, H.Y. Leung, Enhancement of ATP generation capacity, antioxidant activity and immunomodulatory activities by Chinese Yang and Yin tonifying herbs, Chin. Med. 2 (2007) 3.

[63] H.S. Wong, W.F. Cheung, W.L. Tang, K.M. Ko, "Qi-Invigorating" Chinese tonic herbs (Shens) stimulate mitochondrial ATP generation capacity in H9c2 cardiomyocytes in situ and rat hearts ex vivo, Chin. Med. 3 (2012) 101–105.

[64] P. Guo, Q.D. Liang, J.J. Hu, J.F. Wang, S.Q. Wang, The effect of Siwu Tang on EPO and G-CSF gene expression in bone marrow of irradiated blood deficiency mice, Zhongguo Zhong Yao Za Zhi 30 (2005) 1173–1176 (in Chinese).

[65] H.S. Wong, K.M. Ko, Herba Cistanches stimulates cellular glutathione redox cycling by reactive oxygen species generated from mitochondrial respiration in H9c2 cardiomyocytes, Pharm. Biol. 51 (2013) 64–73.

[66] H.S. Wong, P.K. Leong, J.H. Chen, H.Y. Leung, W.M. Chan, K.M. Ko, beta-Sitosterol increases mitochondrial electron transport by fluidizing mitochondrial membranes and enhances mitochondrial responsiveness to increasing energy demand by the induction of uncoupling in C2C12 myotubes, J. Funct. Foods 23 (2016) 253–260.

[67] H.S. Wong, N. Chen, P.K. Leong, K.M. Ko, beta-Sitosterol enhances cellular glutathione redox cycling by reactive oxygen species generated from mitochondrial respiration: protection against oxidant injury in H9c2 cells and rat hearts, Phytother. Res. 28 (2014) 999–1006.

[68] H.S. Wong, J. Chen, P.K. Leong, H.Y. Leung, W.M. Chan, K.M. Ko, A cistanches herba fraction/beta -sitosterol causes a redox-sensitive induction of mitochondrial uncoupling and activation of adenosine monophosphate-dependent protein kinase/peroxisome proliferator-activated receptor gamma Coactivator-1 in C2C12 myotubes: a possible mechanism underlying the weight reduction effect, Evid. Based Complement. Altern. Med. 2015 (2015) 142059.

[69] Y.S. Lee, H.S. Park, D.K. Lee, M. Jayakodi, N.H. Kim, S.C. Lee, et al., Comparative analysis of the transcriptomes and primary metabolite profiles of adventitious roots of five Panax ginseng cultivars, J. Ginseng Res. 41 (2017) 60–68.

[70] J. Voces, A.C. Cabral de Oliveira, J.G. Prieto, L. Vila, A.C. Perez, I.D. Duarte, et al., Ginseng administration protects skeletal muscle from oxidative stress induced by acute exercise in rats, Braz. J. Med. Biol. Res. 37 (2014) 1863–1871.

[71] C.Y. Li, D.T. Lau, T.T. Dong, J. Zhang, R. Choi, H.Q. Wu, et al., Dual-index evaluation of character changes in Panax ginseng C. A. Mey stored in different conditions, J. Agric. Food Chem. 61 (2013) 6568–6573.

[72] J.H. Kim, Cardiovascular diseases and Panax ginseng: a review on molecular mechanisms and medical applications, J. Ginseng Res. 36 (2012) 16–26.

[73] Y.L. Yang, J. Li, K. Liu, L. Zhang, Q. Liu, B. Liu, et al., Ginsenoside Rg5 increases cardiomyocyte resistance to ischemic injury through regulation of mitochondrial hexokinase-II and dynamin-related protein 1, Cell Death Dis. 8 (2017) e2625.

[74] R. Ye, X. Zhang, X. Kong, J. Han, Q. Yang, Y. Zhang, et al., Ginsenoside Rd attenuates mitochondrial dysfunction and sequential apoptosis after transient focal ischemia, Neuroscience 178 (2011) 169–180.

[75] G. Tu (Ed.), Pharmacopoeia of the People's Republic of China, The Peoples Medical Publishing House Beijing, China, 1988.

[76] P.Y. Chiu, N. Chen, P.K. Leong, H.Y. Leung, K.M. Ko, Schisandrin B elicits a glutathione antioxidant response and protects against apoptosis via the redox-sensitive ERK/Nrf2 pathway in H9c2 cells, Mol. Cell. Biochem 350 (2011) 237–250.

[77] P.K. Leong, P.Y. Chiu, N. Chen, H.Y. Leung, K.M. Ko, Schisandrin B elicits a glutathione antioxidant response and protects against apoptosis via the redox-sensitive ERK/Nrf2 pathway in AML12 hepatocytes, Free Radic. Res. 45 (2011) 483–495.

[78] P.Y. Lam, K.M. Ko, (-)Schisandrin B ameliorates paraquat-induced oxidative stress by suppressing glutathione depletion and enhancing glutathione recovery in differentiated PC12 cells, Biofactors 37 (2001) 51–57.

[79] N. Chen, P.Y. Chiu, K.M. Ko, Schisandrin B enhances cerebral mitochondrial antioxidant status and structural integrity, and protects against cerebral ischemia/reperfusion injury in rats, Biol. Pharm. Bull. 31 (2008) 1387–1391.

[80] N. Chen, K.M. Ko, Schisandrin B-Induced glutathione antioxidant response and cardioprotection are mediated by reactive oxidant species production in rat hearts, Biol. Pharm. Bull. 33 (2010) 825–829.

[81] P.Y. Chiu, M.H. Tang, D.H. Mak, M.K. Poon, K.M. Ko, Hepatoprotective mechanism of schisandrin B: role of mitochondrial glutathione antioxidant status and heat shock proteins, Free Radic. Biol. Med. 35 (2003) 368–380.

[82] P.Y. Chiu, H.Y. Leung, A.H. Siu, M.K. Poon, K.M. Ko, Schisandrin B decreases the sensitivity of mitochondria to calcium ion-induced permeability transition and protects against carbon tetrachloride toxicity in mouse livers, Biol. Pharm. Bull. 30 (2007) 1108–1112.

[83] P.Y. Chiu, H.Y. Leung, A.H. Siu, M.K. Poon, K.M. Ko, Schisandrin B decreases the sensitivity of mitochondria to calcium ion-induced permeability transition and protects against ischemia-reperfusion injury in rat hearts, Acta. Pharmacol. Sin. 28 (2007) 1559–1565.

[84] K.M. Ko, N. Chen, H.Y. Leung, E.P. Leong, M.K. Poon, P.Y. Chiu, Long-term schisandrin B treatment mitigates age-related impairments in mitochondrial antioxidant status and functional ability in various tissues, and improves the survival of aging C57BL/6J mice, Biofactors 34 (2008) 331–342.

[85] T. Qian, P.K. Leong, K.M. Ko, W. Chan, Investigation of in vitro and in vivo metabolism of schisandrin B from schisandrae fructus by liquid chromatography coupled electrospray ionization tandem mass spectrometry, Pharmacol. Pharm. 6 (2015) 363–373.

[86] Y. Huang, W. Li, Z.Y. Su, A.N. Kong, The complexity of the Nrf2 pathway: beyond the antioxidant response, J. Nutr. Biochem 26 (2015) 1401–1413.

[87] A. Stacchiotti, G. Li Volti, A. Lavazza, R. Rezzani, L.F. Rodella, Schisandrin B stimulates a cytoprotective response in rat liver exposed to mercuric chloride, Food Chem. Toxicol. 47 (2009) 2834–2840.

[88] V.V. Giridharan, R.A. Thandavarayan, S. Arumugam, M. Mizuno, H. Nawa, K. Suzuki, Schisandrin B ameliorates ICV-infused amyloid β induced oxidative stress and neuronal dysfunction through inhibiting RAGE/NF-κB/MAPK and up-regulating HSP/beclin expression, PLoS One 10 (2015) e0142483.

[89] P.K. Leong, K.M. Ko, Schisandrin B induces an Nrf2-mediated thioredoxin expression and suppresses the activation of inflammasome in vitro and in vivo, Biofactors 41 (2015) 314–323.

[90] P.K. Leong, H.S. Wong, J. Chen, W.M. Chan, H.Y. Leung, K.M. Ko, Differential action between schisandrin a and schisandrin B in eliciting an anti-inflammatory action: the depletion of reduced glutathione and the induction of an antioxidant response, PLoS One 11 (2016) e0155879.

[91] K.W. Zeng, T. Zhang, H. Fu, G.X. Liu, X.M. Wang, Schisandrin B exerts anti-neuroinflammatory activity by inhibiting the Toll-like receptor 4-dependent MyD88/IKK/NF-κB signaling pathway in lipopolysaccharide-induced microglia, Eur. J. Pharmacol. 692 (2012) 29–37.

[92] M. Jang, Y. Yun, S.H. Kim, J.H. Kim, M.H. Jung, Protective effect of gomisin N against endoplasmic reticulum stress-induced hepatic steatosis, Biol. Pharm. Bull. 39 (2016) 832–838.

[93] J. Pollier, A. Goossens, Oleanolic acid, Phytochemistry 77 (2012) 10–15.

[94] J. Liu, Pharmacology of oleanolic acid and ursolic acid, J. Ethnopharmacol. 49 (1995) 57–68.

[95] Z. Pang, Z. Zhi-yan, W. Wang, Y. Ma, N. Feng-ju, X. Zhang, et al., The advances in research on the pharmacological effects of Fructus Ligustri Lucidi, Biomed. Res. Int. 2015 (2015) 281873.

[96] T.K. Yim, W.K. Wu, W.F. Pak, K.M. Ko, Hepatoprotective action of an oleanolic acid-enriched extract of *Ligustrum lucidum* fruits is mediated through an enhancement on hepatic glutathione regeneration capacity in mice, Phytother. Res. 15 (2001) 589–592.

[97] T.K. Yim, K.M. Ko, Effects of Fructus Ligustri Lucidi extracts on concanavalin a-stimulated proliferation of isolated murine splenocytes, Pharm. Biol. 39 (2001) 146–151.

[98] A. Jiménez-Arellanes, J. Luna-Herrera, J. Cornejo-Garrido, S. López-García, M.E. Castro-Mussot, M. Meckes-Fischer, et al., Ursolic and oleanolic acids as antimicrobial and immunomodulatory compounds for tuberculosis treatment, Bmc. Complement. Altern. Med. 13 (2013) 258.

[99] D. da Silva Ferreira, V.R. Esperandim, M.P. Toldo, C.C. Kuehn, J.C. do Prado Júnior, W.R. Cunha, et al., In vivo activity of ursolic and oleanolic acids during the acute phase of *Trypanosoma cruzi* infection, Exp. Parasitol. 134 (2013) 455–459.

[100] A. Nataraju, D. Saini, S. Ramachandran, N. Benshoff, W. Liu, W. Chapman, et al., Oleanolic Acid, a plant triterpenoid, significantly improves survival and function of islet allograft, Transplantation 88 (2009) 987–994.

[101] R. Martín, M. Hernández, C. Córdova, M.L. Nieto, Natural triterpenes modulate immune-inflammatory markers of experimental autoimmune encephalomyelitis: therapeutic implications for multiple sclerosis, Br. J. Pharmacol. 166 (2012) 1708–1723.

[102] R. Martín, C. Cordova, J.A. San Román, B. Gutierrez, V. Cachofeiro, M.L. Nieto, Oleanolic acid modulates the immune-inflammatory response in mice with experimental autoimmune myocarditis and protects from cardiac injury. Therapeutic implications for the human disease, J. Mol. Cell. Cardiol. 72 (2014) 250–262.

[103] A. Mukundwa, S. Mukaratirwa, B. Masola, Effects of oleanolic acid on the insulin signaling pathway in skeletal muscle of streptozotocin-induced diabetic male Sprague-Dawley rats, J. Diabetes 8 (2016) 98–108.

[104] C.T. Musabayane, M.A. Tufts, R.F. Mapanga, Synergistic antihyperglycemic effects between plant-derived oleanolic acid and insulin in streptozotocin-induced diabetic rats, Ren. Fail. 32 (2010) 832–839.

[105] D. Camer, Y. Yu, A. Szabo, X.F. Huang, The molecular mechanisms underpinning the therapeutic properties of oleanolic acid, its isomer and derivatives for type 2 diabetes and associated complications, Mol. Nutr. Food Res. 58 (2014) 1750–1759.

[106] J.M. Castellano, A. Guinda, T. Delgado, M. Rada, J.A. Cayuela, Biochemical basis of the antidiabetic activity of oleanolic acid and related pentacyclic triterpenes, Diabetes 62 (2013) 1791–1799.

[107] R.X. Zhang, M.X. Li, Z.P. Jia, Rehmannia glutinosa: review of botany, chemistry and pharmacology, J. Ethnopharmacol. 117 (2008) 199–214.

[108] Y. Yuan, S. Hou, T. Lian, Y. Han, Studies of Rehmannia glutinosa Libosch. f. hueichingensis as a blood tonic, Zhongguo Zhong Yao Za Zhi 17 (1992) 366–368 (in Chinese).

[109] A.S. Lin, K. Qian, Y. Usami, L. Lin, H. Itokawa, C. Hsu, et al., 5-Hydroxymethyl-2-furfural, a clinical trials agent for sickle cell anemia, and its mono/di-glucosides from classically processed steamed Rehmanniae Radix, J. Nat. Med. 62 (2008) 164–167.

[110] S.K. Ku, H. Kim, J.W. Kim, K.S. Kang, H.J. Lee, Ameliorating effects of herbal formula hemomine on experimental subacute hemorrhagic anemia in rats, J. Ethnopharmacol. 198 (2017) 205–213.

[111] H.W. Lee, H. Kim, J.A. Ryuk, K.J. Kil, B.S. Ko, Hemopoietic effect of extracts from constituent herbal medicines of Samul-tang on phenylhydrazine-induced hemolytic anemia in rats, Int. J. Clin. Exp. Pathol. 7 (2014) 6179–6185.

[112] W. Dong, Y. Xian, W. Yuan, Z. Huifeng, W. Tao, L. Zhiqiang, Catalpol stimulates VEGF production via the JAK2/STAT3 pathway to improve angiogenesis in rats' stroke model, J. Ethnopharmacol. 191 (2016) 169–179.

[113] Y.R. Liu, P.W. Li, J.J. Suo, Y. Sun, B.A. Zhang, H. Lu, et al., Catalpol provides protective effects against cerebral ischaemia/reperfusion injury in gerbils, J. Pharm. Pharmacol. 66 (2014) 1265–1270.

[114] H.F. Zhu, D. Wan, Y. Luo, J.L. Zhou, L. Chen, X.Y. Xu, Catalpol increases brain angiogenesis and up-regulates VEGF and EPO in the rat after permanent middle cerebral artery occlusion, Int. J. Biol. Sci. 6 (2010) 443–453.

[115] C. Huang, Y. Cui, L. Ji, W. Zhang, R. Li, L. Ma, et al., Catalpol decreases peroxynitrite formation and consequently exerts cardioprotective effects against ischemia/reperfusion insult, Pharm. Biol. 51 (4) (April 2013) 463–473.

[116] J. Zhu, X. Chen, H. Wang, Q. Yan, Catalpol protects mice against renal ischemia/reperfusion injury via suppressing PI3K/Akt-eNOS signaling and inflammation, Int. J. Clin. Exp. Med. 8 (2015) 2038–2044.

[117] M.L. Chen, S.P. Ip, S.H. Tsai, K.M. Ko, C.T. Che, Biochemical mechanism of Wu-Zi-Yan-Zong-Wan, a traditional Chinese herbal formula, against alcohol-induced oxidative damage in CYP2E1 cDNA-transfected HepG2 (E47) cells, J. Ethnopharmacol. 128 (2010) 116–122.

[118] M.L. Chen, S.H. Tsai, S.P. Ip, K.M. Ko, C.T. Che, Long-term treatment with a "Yang-invigorating" Chinese herbal formula, Wu-Zi-Yan-Zong-Wan, reduces mortality and liver oxidative damage in chronic alcohol-intoxicated rats, Rejuvenation Res. 13 (2010) 459–467.

[119] K.W. Zeng, X.M. Wang, H. Ko, H.O. Yang, Neuroprotective effect of modified Wu-Zi-Yan-Zong granule, a traditional Chinese herbal medicine, on CoCl2-induced PC12 cells, J. Ethnopharmacol. 130 (2010) (2010) 13–18.

[120] W. Yao, H. Gu, J. Zhu, G. Barding, H. Cheng, B. Bao, et al., Integrated plasma and urine metabolomics coupled with HPLC/QTOF-MS and chemometric analysis on potential biomarkers in liver injury and hepatoprotective effects of Er-Zhi-Wan, Anal. Bioanal. Chem. 406 (2014) 7367–7378.

[121] J. Ge, L. Xie, J. Chen, et al., Chin. J. Integr. Med. (2006). https://doi.org/10.1007/s11655-016-2744-2.

[122] Y.Y. Chen, Y.T. Hsue, H.H. Chang, M.J. Gee, The association between postmenopausal osteoporosis and kidney-vacuity syndrome in traditional Chinese medicine, Am. J. Chin. Med. 27 (1999) 25–35.

[123] H. Xu, Z.R. Su, W. Huang, R.C. Choi, Y.Z. Zheng, D.T. Lau, et al., Er Zhi Wan, an ancient herbal decoction for woman menopausal syndrome, activates the estrogenic response in cultured MCF-7 cells: an evaluation of compatibility in defining the optimized preparation method, J. Ethnopharmacol. 143 (2012) 109–115.

[124] H. Zhang, W.W. Xing, Y.S. Li, Z. Zhu, J.Z. Wu, Q.Y. Zhang, et al., Effects of a traditional Chinese herbal preparation on osteoblasts and osteoclasts, Maturitas 61 (2008) 334–339.

[125] M. Cheng, Q. Wang, Y. Fan, X. Liu, L. Wang, R. Xie, et al., A traditional Chinese herbal preparation, Er-Zhi-Wan, prevents ovariectomy-induced osteoporosis in rats, J. Ethnopharmacol. 38 (2011) 279–285.

[126] W. Sun, Y.Q. Wang, Q. Yan, R. Lu, B. Shi, Effects of Er-Zhi-Wan on microarchitecture and regulation of Wnt/β-catenin signaling pathway in alveolar bone of ovariectomized rats, J. Huazhong Univ. Sci. Technol. Med. Sci. 34 (2014) 114–119.

[127] K.M. Ko (Ed.), Traditional Herbal Medicines for Modern Times – Shengmai San, Taylor Francis Ltd., United Kingdom, 2002, pp. 1–15.

[128] R.H. Ichikawa, X. Wang, T. Konishi, Role of component herbs in antioxidant activity of shengmai san–a traditional Chinese medicine formula preventing cerebral oxidative damage in rat, Am. J. Chin. Med. 31 (2013) 509–521.

[129] G.S. Cao, S.X. Li, Y. Wang, Y.Q. Xu, Y.N. Lv, J.P. Kou, et al., A combination of four effective components derived from Sheng-mai san attenuates hydrogen peroxide-induced injury in PC12 cells through inhibiting Akt and MAPK signaling pathways, Chin. J. Nat. Med. 14 (2016) 508–517.

[130] K. Shen, Y. Wang, Y. Zhang, H. Zhou, Y. Song, Z. Cao, et al., Cocktail of four active components derived from Sheng Mai San inhibits hydrogen peroxide-induced PC12 cell apoptosis linked with the caspase-3/ROCK1/MLC pathway, Rejuvenation Res. 18 (2015) 517–527.

[131] N. Wang, S. Minatoguchi, M. Arai, Y. Uno, Y. Nishida, K. Hashimoto, et al., Sheng-Mai-San is protective against post-ischemic myocardial dysfunction in rats through its opening of the mitochondrial KATP channels, Circ. J. 66 (2002) 763–768.

[132] W.L. Mo, C.Z. Chai, J.P. Kou, Y.Q. Yan, B.Y. Yu, Sheng-Mai-San attenuates contractile dysfunction and structural damage induced by chronic intermittent hypoxia in mice, Chin. J. Nat. Med. 13 (2015) 743–750.

[133] C.Z. Chai, W.L. Mo, X.F. Zhuang, J.P. Kou, Y.Q. Yan, B.Y. Yu, Protective effects of Sheng-Mai-San on right ventricular dysfunction during chronic intermittent hypoxia in mice, Evid. Based Complement. Altern. Med. 2016 (2016) 4682786.

[134] X.F. Ding, G. Chen, Y.L. Liu, Effects of shengmai for injection on cardiogenic shock, Zhongguo Zhong Yao Za Zhi 32 (2007) 2298–2305 (in Chinese).

[135] H.T. Yao, Y.W. Chang, C.T. Chen, M.T. Chiang, L. Chang, T.K. Yeh, Shengmai San reduces hepatic lipids and lipid peroxidation in rats fed on a high-cholesterol diet, J. Ethnopharmacol. 116 (2008) 49–57.

[136] P.K. Leong, N. Chen, P.Y. Chiu, H.Y. Leung, C.W. Ma, Q.T. Tang, et al., Long-term treatment with shengmai san-derived herbal supplement (Wei Kang Su) enhances antioxidant response in various tissues of rats with protection against carbon tetrachloride hepatotoxicity, J. Med. Food 13 (2010) 427–438.

[137] P.Y. Chiu, P.Y. Lam, H.Y. Leung, P.K. Leong, C.W. Ma, Q.T. Tang, et al., Co-treatment with Shengmai San-derived herbal product ameliorates chronic ethanol-induced liver damage in rats, Rejuvenation Res. 14 (2011) 17–23.

[138] J.Y. Zhan, K.Y. Zheng, K.Y. Zhu, W.L. Zhang, C.W. Bi, J.P. Chen, et al., Importance of wine-treated Angelica Sinensis Radix in Si Wu Tang, a traditional herbal formula for treating women's ailments, Planta Med. 70 (2013) 533–537.

[139] H.F. Chiu, Y.H. Wu, Y.C. Shen, S.J. Wang, K. Venkatakrishnan, C.K. Wang, Antioxidant and physiological effects of Si-Wu-Tang on skin and liver: a randomized, double-blind, placebo-controlled clinical trial, Chin. Med. 11 (2016) 30.

[140] J. Ni, A.L. Romero-Weaver, A.R. Kennedy, Potential beneficial effects of Si-Wu-Tang on white blood cell numbers and the gastrointestinal tract of gamma-ray irradiated mice, Int. J. Biomed. Sci. 10 (2014) 182–190.

[141] E. Tahara, T. Satoh, K. Toriizuka, J. Nagai, S. Nunome, Y. Shimada, et al., Effect of Shimotsu-to (a Kampo medicine, Si-Wu-Tang) and its constituents on triphasic skin reaction in passively sensitized mice, J. Ethnopharmacol. 68 (1999) 219–228.

Chapter 5

Ethnobotany/Ethnopharmacology, and Bioprospecting: Issues on Knowledge and Uses of Medicinal Plants by Moroccan People

Mostafa Elachouri
Mohammed first University, Oujda, Morocco

1. INTRODUCTION

Cultural knowledge in Morocco is as old as the existence of the Moroccan people themselves. Biocultural diversity is another feature that characterizes these ancestral practices. Spirituality and religious beliefs are the bedrock of this knowledge system, which makes it so remarkable. This knowledge base has provided sustenance for Moroccan people in a diverse, complex, and risk-prone environment. The use of plants as medicine is an integral part of the variety of cultures in Morocco, resulting in the traditional medical system pluralism [1]. Herbal drugs have been used by Moroccan society since ancient times as medicines for the treatment of a range of diseases. The acquisition of this traditional knowledge, which is passed from one generation to another by oral communication, is due to historical, sociological, and cultural reasons. This empirical health care system was deeply rooted in classical Arab medicine, which was further developed and enriched by other knowledge brought in by various ethnic groups that migrated to Morocco from many areas, including Arabs from the Middle East, the Andalusians, Jews from Europe, and the Blacks from Sudan, Senegal, and Niger. This knowledge is becoming popular and widespread in Morocco and has a rich and vibrant history, enriched with various documented uses of plants, which are often unique to geographical features of the region and traditional knowledge related to medicine practices [2].

Natural Products and Drug Discovery. https://doi.org/10.1016/B978-0-08-102081-4.00005-8

This ancestral medical approach is still in use by Moroccan society to meet their primary health care needs. In fact, many studies indicate that 70% of the Moroccan population use plants as medicine [3–6]. Such an observation indicates that Indigenous people in this country have a vast knowledge of this wealthy ancestral heritage. The rush toward these herbal drugs used to treat their ailments, instead of conventional medicine, is due to different parameters: lack of health facilities, poor access to basic services, availability of medicinal plants, vulnerability of sociocultural status in Morocco, and the historic beliefs of the people. In Moroccan society, medicinal plants are considered to be a vital part of biological diversity and play a central role, not only as medicine but also as trade commodities and an essential resource for the well-being of poor people, especially in rural areas [7–9]. The plants used were considered as a "stockroom" of natural medicinal herbs with potential use as medicines. The most recent surveys indicate that the region has been distinguished throughout generations with a tremendous inventory of medicinal plants and very rich tradition in the use of these herbs for treating various diseases [6]. Despite the use of these herbs being common practice in the region, traditional knowledge is not integrated into the Moroccan health system. Furthermore, anthropogenic activities lead to the erosion and loss of these traditional medical practices. This is why urgent measures must be undertaken to safeguard these valuable herbs as well as the traditional knowledge related to medicinal plants from ongoing destruction forever. In this framework, the main aim of this chapter is to summarize the information regarding the whole set of plants used by Moroccan society for the treatment of different ailments as well as to better understand and highlight the evolution and state of the phytotherapeutic traditions that are deeply rooted in ancient and Arabic civilizations.

Three essential components constitute this chapter:

1. A brief history of medical sciences in Muslim–Arab civilization.
2. The evolution of medical ethnobiology in Morocco.
3. Constraints and challenges facing the medicinal plant's sector.

2. BRIEF HISTORY OF MEDICAL SCIENCES IN MUSLIM-ARAB CIVILIZATION

2.1 Sciences at a Glance in the Golden Age

By expanding upon the wisdom of the Greeks over the centuries, the indigenous Arab civilization, during the period from the 9th to the 12th centuries has contributed greatly to the development of sciences.

The scientific glory of the Arabic nation was born in the 7th century CE in the Arabian Peninsula, the beginning of the Islamic state, with the preaching of the prophet Mohammed (Peace be upon him) in CE 622, a date related to the migration of Mohammed and his disciples from Mecca to Medina, the unification of the Arab tribes, and the inauguration of the Muslim religion.

Within a century after his death (CE 632) a large part of the globe, from southern Europe throughout North Africa to Central Asia and on to India, was controlled and influenced by the new Arabic-Muslim Empire with a flowering of knowledge and intellect that later spread throughout Europe and greatly influenced both medical practice and education. At that period, Islam covered nearly two-thirds of the known world and contacts with the West were already established through Byzantium and Sicily crossing Spain. Between the 7th and 13th centuries the Arabic world enjoyed great splendor. The Greco-Latin texts (including the Hebrew Bible) were translated into Arabic. In the 9th century, Baghdad was the center of Muslim culture.

During that period, called "the Golden Age," science expanded under the Abbasid caliphs of Baghdad around AD 750 and gradually spread its influence throughout Spain and eastward into Central Asia for more than 600 years. The origins of these developments date back to the Greeks and Indians as well as to the empiric knowledge of the Indigenous population. Indexing many subcultures and languages into its circle, Arab-Islamic medicine was a field of medical writing that was influenced by several medical systems, including traditional Arabian medicine and ancient Hellenistic medicine. Islamic medicine was built on tradition; chiefly the theoretical and practical knowledge developed in Greece and Rome. For Islamic scholars, Galen (AD 210) and Hippocrates (5th century BC) were preeminent authorities, followed by Hellenic scholars in Alexandria. The early Muslim scholars accumulated the greatest body of scientific knowledge in the world and built on it through their own discoveries. The tool serving as an international communication of sciences was the common Arab language. The works of ancient Greek and Roman physicians Hippocrates, Dioscorides, Soranus, Celsus, and Galen also had lasting impact on this ancestral medicine [10,11].

2.1.1 Brief History of Medical Sciences

Chronologically, the evolution of the history of Arab medicine can be divided into three periods:

1. **The first period** started in the 7th century when the Arab-Islamic scholars were invited by the caliphs of Baghdad, where scientists and researchers studied the past and created the future. A variety of texts taken by the ancient Egyptians, Hebrews, Persians, Greeks, and Romans such as Hippocrates, Rufus of Ephesus, Dioscorides, and Galen were analyzed and translated into the Arabic language. In fact, in the reign of Abbasid, the Khalifs Al-Ma'mun (reigned 813–833 in Baghdad), very interested by the richness of Greek sciences and made aware by translating this wealthy knowledge, ordered the foundation of "The House of Wisdom," the first institution destined for this purpose. Hence ever since that period hospitals

and medical schools have flourished, first in Baghdad and later in the main provincial cities. The most famous scientist of all translators was Hunayn Ibn-Is'haq, court physician to the caliph al-Mutawakkil, who translated a large number of medical works of Hippocrates and Galen, philosophical works by Plato and Aristotle, and mathematical works by Euclid and Archimedes.

2. **During the second period** the Arab scholars synthesized and further elaborated the knowledge they had gathered from ancient manuscripts, adding their own experience. They translated classical medical texts into the Arab language, not only from Greece, but also from Persia, India, and China. By this, the principal works of their predecessors (Galen, Hippocrates, etc.) were translated into Arabic and became available. Then, Muslim researchers controlled the monopoly of the medicine of their predecessors and reached a stature in medical science roughly on a level with the greatest of the Greeks. There were some great physicians too that shaped the history of Islamic medicine. Most of them were Muslim. Mention has to be made of the physicians in Al Andalus or Islamic Spain; these scholars contributed to the biomedical sciences of the ArabicIslamic world during the 8th to 13th centuries CE.

Numerous Arab pioneers are mentioned in medical history, among the most famous scholars of outstanding repute, who shaped the history of Islamic medicine, we can cite only two, well known to Medieval Europe:

Abu-Bakr Mohammed Ibn-Zakarya Al-Razi (846–930): Known to the Western world as Razes, was a scientist who played an important role in keeping the medical heritage that developed over thousands of years. His outstanding book *AlHawy*, which means "the complete text," is composed of 22 volumes on subjects of medical knowledge with the ninth volume devoted to pharmacology. The manuscript represents a unique document containing medical knowledge accumulated through many civilizations until his period.

Ibn Sina (AD 980–1037), known in the Western world as Avicenna: Considered a great scientist in his time; his writings influenced many who appeared after him. He was a unique phenomenon, not only because of his encyclopedic accomplishments in medicine, but also because of the versatility of his genius. Ibn Sina wrote 100 treaties, most of them in medicine. Among these documents is "Al-Qunun Fi al-Tibb" or "Canon of Medicine," an encyclopedia containing more than 1 million words. It was considered the greatest book of all time and contained all medical knowledge up to the 10th century. The document was translated into many languages and was used by Western civilizations for several centuries as a standard work of medical reference. At this period, intense effort to translate and analyze the works of Hippocrates, Rufus of Ephesus, Dioscorides, and Galen took place [12]. So, Arab scholars synthesized and further elaborated the knowledge they had gathered from ancient manuscripts, adding their own experience.

3. **The third phase** of Arab medicine is characterized by the reversal of scientific knowledge from Arab to Latin language. It started in the 12th century when European scholars interested in the sciences from Islamo-Arab civilizations, drew attention to the necessity to translate the chief works related to Arab and Muslim scientists into the Latin language.

   It was only later in the 12th and 13th centuries, when most of the Arabic works began to be translated into Latin, that such knowledge passed to the West. These valuable works, which are considered as an encyclopedic accomplishment in medicine, were rightly acclaimed to be the peak of Arabo-Muslim sciences, and were translated into different languages to become the standard works and medical references of medicine in Europe up to the 17th century.

2.1.2 Evolution of Herbal Medical Sciences

Plants had been used for medicinal purposes long before recorded history. Ancient Chinese and Egyptian papyrus writings describe medicinal uses for plants as early as 3000 BC. Indigenous cultures (such as African and Native American) used herbs in their healing rituals, while others developed traditional medical systems (such as Siddha, Ayurveda, Unani, and traditional Chinese medicine [TCM]) in which herbal therapies were used.

During the domination of the Arab Empire in AD 632–1150, physicians combined different sciences such as chemistry, pharmacology, medicine, and plant sciences to develop new aspects and promote knowledge of herbs and their potential medical efficacy and safety in their dealings with the treatment of different diseases. Broadly speaking, the history of medicine, pharmacy development, and discovery of drugs can be divided into two great periods.

Greek and Roman period: Greek medicine, which was greatly influenced by Egyptian medicine, was the oldest and most famous medical school of classic Greece, under the leadership of Hippocrates. In this school, all medical knowledge was employed to overcome the disease. The writings or medical treatises of this school were known as the *Hippocratic Corpus*, (around 280 BC). The philosopher Aristotle (384–322 BC) was succeeded by the botanist Theophastrus (about 370 to 287 BC), who wrote the book *Historia Plantarum* in which he classified and described over 500 plants [12]. Theophastrus classified plants into trees, shrubs, subshrubs, and herbs—a very simple classification, but the most rational up to Linnaeus.

Homer (9th or 8th century BC), in the *Odyssey*, used the word "pharmakon" to refer to a drug, and the derived words "pharmacy" and "pharmacology" deal with the sciences of drugs or medicines.

In Rome the knowledge of medicinal plants reached an early peak with Dioscorides (1st century AD) and Galen (2nd century AD).

Pedanias Dioscorides' work *De Materia Medica* (the compendium of medicine) was written in Greek between AD 60 and 78, and is divided into five books. Dioscorides gives a careful description of over 900 drugs from plant, animal, and mineral origin.

Dioscorides gives comprehensive details of the drugs deriving from the three realms (animal, plants, and minerals), the preparations obtained from them, and even recommendations for special diseases. He also describes some rudimentary chemical procedures such as distillation, sublimation, mercury extraction from cinnabar, and others. *De Materia Medica* was considered to be an indisputable reference work on drugs until the 19th century.

Claudius Galen or Galenus, a physician-pharmacist who performed at that time a great deal of work, wrote about 400 works, among them the outstanding work called *Opera Omnia* in which he mentions numerous drugs and methods that use natural products, covering all aspects of medicine. Galenus is the first scholar who prescribed medicines in simple or complex mixtures, called "Galenicals," specifically devised for each therapy.

In the 4th century, after the decline and fall of the Roman Empire, the cultural center transferred to the East to the Byzantine Empire, and then to the Arab world; this gave rise to the exchange of knowledge between the West and East (Persia, China, and India), allowing the flourishing of all sciences, especially medicine and pharmacology [13,14].

Great Arabian Period: At the period when the great Arab-Islamic culture reached its summit, Islamo-Arab scholars combined different sciences such as chemistry, medicine, pharmacology, and plant sciences to develop new aspects and upgrade the knowledge of herbs and their potential medical efficacy and safety in their dealings with the treatment of different ailments. Arab chemists such as Ibn Hayyan and others were able to extract different anesthetic compounds from local herbs for local or general anesthetization. Among the herbs used for extractions were *Hyoscyamus aureus*, opium, and *Cannabis sativa*. Great scientific figures wrote several comprehensive treatises where herbs, or mixtures of them, represented the "corpus therapeuticum" to alleviate and treat disease. One of the main niches of Arab medicine was Abulcasis (Abul-Qasim al-Zahrawi, 936–1013), an Arab physician who published a book on medicinal herbs called "*al-Tasrif*," an encyclopedia devoted to the therapy and practice of pharmacy, in which he described rudimentary physicochemical procedures such as distillation or sublimation and pharmaceutical formulations. Another scientist, Ibn Al-Baytar, was able to introduce around 350 new plant species as medicinal herbs for treating human diseases [15]. Mention also needs to be made of the famous Arabic scientist Avicenna (Ibn Sina) who wrote one of the greatest encyclopedic medical texts ever written, the *Canon Medicinae*, a compendium which holds most of the Greco-Roman knowledge of medicine, where he describes the use and efficacy of about 750 drugs arranged alphabetically. Mention should also be made of the physician Razes who wrote about 200 books, 50 of which were related to collecting medical knowledge; some of these are compilations from Hindi, Syrian, Persian, Greek, and Latin sources [10,16,17].

3. TRADITIONAL MEDICINE IN MOROCCO

Morocco is one of the most megadiverse countries in the Mediterranean region. Morocco has a distinguished geographical position, situated in North Africa between 21° and 36°N and 1°−17°W, and limited in the north and west by very large coastlines (about 3400 km) corresponding, respectively, to the Mediterranean Sea and Atlantic Ocean. The eventful geological history of Morocco has resulted in the dissected topography of the country in four areas by the chain of mountains, the Rif, the Middle Atlas, the High Atlas, and the Anti-Atlas, with the highest peaks (4200 m at Tobkal) in North Africa (Fig. 5.1).

The altitudinal variation in Morocco creates a large number of isolated areas, several biotopes with a special ecology, and a Mediterranean climate grouping a series of bioclimates (from humid to Saharan). These heterogeneous ecological conditions have bestowed Morocco with an enormous wealth of phytogenetic resources estimated to number 3913 native vascular plant species [18], of which around 879 are endemic [19]. Thus Morocco hosts enormous biodiversity resources, including plant species with a potential use in medical ethnobotany. Furthermore, the country exhibits a high level of cultural diversity characterized by a huge sociocultural background rooted in an extensive ethnohistory, including the indigenous Berber and Amazighi

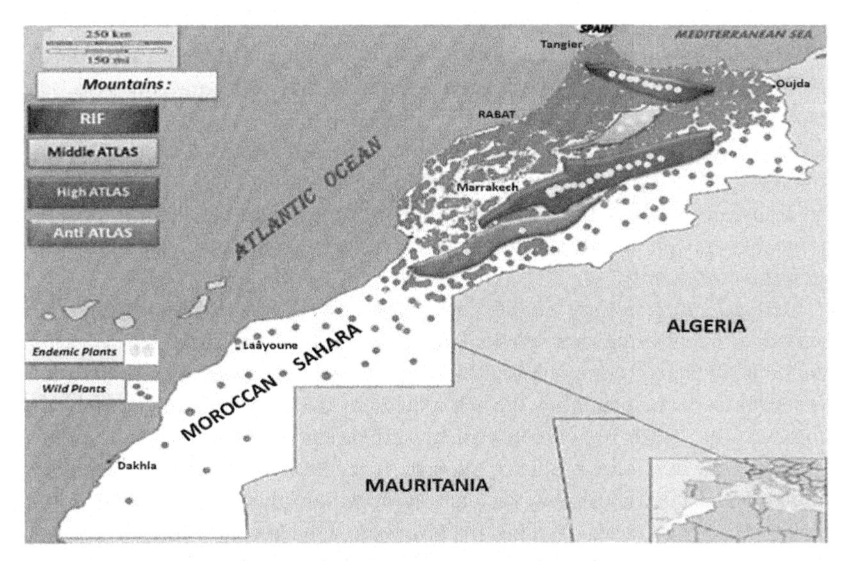

FIGURE 5.1 Map of Morocco.

societies, Arabo-Muslim influence during the Golden Age (7th to 15th centuries), and colonial conquests such as Roman, French, Spanish, etc., ... [20]. Moroccan people have for centuries maintained until now that their ancestral medical practices, especially phytotherapy traditions, are based on wild plants.

Thus the currently widespread folk medicine practices in Moroccan society are the result of an accumulation of knowledge from various sources and different ethnic traditions, coupled with the long exposure to and experience of this people with natural resources. In fact, several ethnobotanical and ethnopharmacological surveys carried out in different regions of Morocco indicate that traditional cultures and uses of plants by Moroccan society are better preserved and remain active even today [3,6].

3.1 Current Medical Ethnobiological Studies in Morocco

Most cultures possess a huge reserve of undocumented traditional knowledge of applying herbal remedies in the treatment of ailments. For centuries this knowledge has been widely used as a starting point for drug discovery, and considered to be of relevance for drug development. It has now become more important than ever to record and conserve the traditional medical system to aid the discovery of new drugs and possibly to find improved applications of traditional medicine [21]. Medical ethnobiological sciences, including ethnobotany, ethnopharmacology, and ethnomedicine, are known to serve as significant bridges between conservation scientists and local communities; it allows interaction between researchers and the local people who have knowledge of the use of plants. Ethnobotany can also provide useful information on drug development, thus saving time and money [22,23]. In the broadest sense, medical ethnobiology, ethnobotany, and ethnopharmacology attempt to make sense and to understand traditional knowledge of medicinal plants. The historical perspective of this discipline has focused on the medicinal use of natural products that has preceded recorded human history probably by thousands of years. Documenting traditional medical knowledge and transforming it into a scientific language may help the conservation of indigenous people's cultural heritage for future generations, including the scientific community.

Medical ethnobiology studies are considered to be the most effective method of identifying new medicinal plants and the possible extraction of beneficial bioactive compounds. Many developing countries have intensified their efforts in documenting the ethnomedical data on medicinal plants. In Morocco the field of ethnobotanical, ethnopharmacological, and ethnomedicinal studies is in progress; in fact, from the first study conducted by Bellakhdar in 1991 until now, we can regroup more than 60 published works carried out in different regions of Morocco. Our results indicate that, in Morocco, there has been increasing ethnobotanical and ethnopharmacological investigations during the last few decades (Fig. 5.1). In the following, we discuss in depth all data gathered from the studies selected.

3.2 Knowledge of Ethnobotany, Ethnomedicine, and Medicinal Plant Uses

The consumption of plant-based medicines and other botanicals in the West has increased manifold in recent years. About two centuries ago, our medicinal practices were largely dominated by plant-based medicines. However, the medicinal use of herbs went into rapid decline in the West when more predictable synthetic drugs were made commonly available. In contrast, many developing nations continued to benefit from the rich knowledge of medical herbalism. For example, Siddha and Ayurveda medicines in India, Kampo medicine in Japan, TCM, and Unani medicine in the Middle East and South Asia are still used by a large majority of people.

To guarantee the sustainable use of those resources and the conservation of local biodiversity, as well as equitable sharing of medical traditional knowledge, medical ethnobiology remained the suitable approach to categorize and document plant species and folk practices of Indigenous people. In this respect, an endeavor has been made recently by our researcher team to evaluate the richness of medicinal plants and the knowledge held by the Moroccan people. We undertook a synthetic work based on the information obtained from reviewing several ethnobotanical surveys carried out in different regions of Morocco during the period between 1991 and 2015, with a focus on the use of botanicals in folk medical practices. Through an exhaustive review of articles published during the period cited, we assessed altogether 60 papers. We have performed this research by using as search engines the databases PubMed, Science Direct, and Google Scholar.

Our results indicated an increasing number of investigation in the field of ethnobiology research have been documented. In fact, we note that the number of articles published in this subject increases progressively from two papers published during the period 1991−95 to 41 articles during the period 2011−15. Furthermore, the number of plants recorded increased similarly, from 207 to 688 species.

After reviewing the 60 selected articles, tremendous information was regrouped and analyzed. The results indicate that 25,468 informants have been interviewed and 841 medicinal plant species, belonging to 117 families, were listed. The dominant families were Lamiaceae and Asteraceae followed by Fabaceae and Apiaceae. The plants recorded have been used as medicines to treat 26 ailment categories. All these ailments were treated with the highest diversity of medicinal plant species. Different categories of ailments have been recorded. Focusing our attention only on the most cited, we note, respectively, diabetes with 1238 citations followed by digestive diseases with 1630 citations and dermal diseases with 1083 citations). Equally, we report most commonly plants used specifically for treating these ailments. In the same table, we mention their corresponding Informant Consensus, an efficient method for identifying potentially effective medicinal plants. This parameter allowed us to assess consent within a community and between cultural groups, indicating which plants are widely used for the treatments of different ailments.

With the aim of knowing the distribution of knowledge in a community and to have novel values from analysis of the results gathered in our work, we used a quantitative ethnobotany calculation proposed by Soussa Arojo in 2012 [24]: the knowledge richness index (KRI) and knowledge sharing index (KSI). These were calculated as follows: a binary matrix was constructed that contained the record of citation for each species (Si) known to each study (Ri). KRI is calculated by the following method:

$$\mathrm{KRI} = 1\Big/\sum J_{\mathrm{i}}^{2}, \text{ where } J_{\mathrm{i}} = \mathrm{RS_i}\Big/\sum \mathrm{RS_i}.$$

where $\mathrm{RS_i}$ is the record of species (Si) cited for study (Ri) and $\sum \mathrm{RS_i}$ is the total record of species (Si) cited for the region or in the total literature (60 articles). The KSI expresses the degree of sharing among a determined region and other regions of Morocco. KSI = KRI/KRI max, where KSI is the KRI of each region and KRI max is the maximum richness of all regions.

The results confirm that informants are distributed according to the richness and uniqueness of their knowledge of a particular cultural field. It means that the sharing of information is important between the individuals of the Moroccan community, which share more than 97% of the information.

The information compiled in this chapter reveals the existence of a rich wealth of indigenous knowledge related to medicinal plant uses. Indeed, traditional medicine in Morocco is still playing a significant role in meeting the basic health care needs of the population. This interest in the use of medicinal plants is due to the inaccessibility of effective health care facilities, the high cost of conventional medicine, and easy access to and availability of medicinal plants.

3.3 The Trading of Medicinal Plants

Plants based on medicine have been known for millennia and are highly esteemed all over the world as a rich source of health products for the prevention of diseases and ailments. The rise of the demand for these phytogenetic resources, in both developing and developed countries, is due to the growing recognition that the natural products are nontoxic, have fewer side effects, and are easily available at affordable prices. Furthermore, these herbal drugs are important for pharmacologically active compounds and pharmacological research. A great number of modern pharmaceutical drugs originate from medicinal plants, thus the medicinal plant-based industry is a promising sector and has enormous economic growth potential. These valuable herbs play a central role, not only as phytopharmacy or a source of income, but also in trade commodities, which meet the demand of the markets at local, regional, national, and international levels. To know about medicinal and aromatic plant commodities, in particular on aspects of trade streams, structure, volumes, and values, is essential for assessing the economic importance of these activities [7,9,25].

At the international level, Morocco holds a significant place on the AMPs trade, with a production marked by the wealth and diversity of these herbs. Most AMPs national production is destined entirely for export as essential oils or in dried form. The main export commodities are France and the United States but export quantities are increasing due to the opening of other markets such as Spain, Japan, Germany, Switzerland, and Canada. The values of AMP exported by Morocco play a very important socioeconomic role. Despite extreme difficulty in estimating with accuracy sales data regarding commercial AMPs the calculations are likely underestimated. Nonetheless, several published reports indicate that the sector generated average annual earnings of 56 million Moroccan Dirhams in 2000, 112.4 million Moroccan Dirhams in 2003, and 550 million Moroccan Dirhams in 2015, at an exchange rate of about US$1 = 9.8 Moroccan Dirhams [26,27]. Among the plants exported by Morocco are argan kernel (*Argania spinosa*), atlas cedar wood (*Cedrus atlantica*), carob fruit (*Ceratonia siliqua*), khella fruit (*Ammi visnaga*), lavender flower (*Lavandula* spp.), mastic resin (*Pistacia lentiscus*), oregano herb (*Origanum compactum*), thyme herb (*Thymus satureioides*), pennyroyal flowering herb (*Mentha pulegium*), pomegranate fruit (*Punica granatum*), thyme herb (*Thymus vulgaris*), etc.

It is worth noting that the Moroccan AMPs play a very important socioeconomic role. The sector provides much employment to local and rural communities, generating an estimated 500,000 work days per year [27].

4. CONSTRAINTS AND CHALLENGES FACING THE MEDICINAL PLANTS SECTOR

While many benefits can be derived from the use of AMPs, and the demand for plant-derived products has increased worldwide, potential negative outcomes cannot be ignored. Two major constraints have disabled the AMPs sector: the harmful effect on humans and adverse consequences on natural habitats. In fact, recent work conducted in Morocco has revealed examples of incorrect use of numerous herbs in traditional pharmacopeia leading to toxic effects [4,28,29]. There are also shortages arising from overharvesting of wild species, detrimental climatic and environmental changes, and natural disasters leading to the disappearance of living species. This is why it is important to have an awareness of these problems. Urgent measures must be undertaken to protect medicinal plants from the ongoing destruction of their natural habitats and to avoid the damage affecting the user of medical herbs. Regarding the conservation and protection of AMPs, the Convention on Biological Diversity adopted in Rio de Janeiro (in June 1992) has implemented three major goals: "*the conservation of biological diversity, the sustainable use of its components, and the fair and equitable sharing of the benefits from the use of genetic resources*" [30]. The principal objectives of this Convention are the conservation and sustainable use of biological diversity, as well as the fair and equitable sharing of benefits arising from utilization of natural resources.

Concerning the control and standardization of medicines, the shortage of safe and good quality Moroccan medicinal plants is further compromised by the fact that currently there is no efficient phytovigilance and a lack of regulation. Although the law of Dahir of February 27, 1923 prohibits the retail of toxic plants by herbalists [31], its application remains largely anecdotal. In fact, a recent claim issued by the Moroccan Poison Center stated that medicinal herbs were responsible for about 3%–5% of all reported intoxications, of which 17% were associated with fatal events [32].

Thus we can note that medicines made from plants must be assessed with the aim of checking the safety, efficacy, quality, and toxicity of any herbal products, taking into account the quality of the material used. In addition, standards for minimum acceptable quality are necessary in Moroccan pharmacopoeia [33]. The challenge for Moroccan society is to develop policies aimed at safeguarding natural resources and improve public health to avoid unnecessary harm.

5. CONCLUSION

This chapter attempted to highlight the importance and potential of medicinal plants from Moroccan biodiversity, which have short- and long-term potential to be developed as future phytopharmaceuticals to treat and/or manage a panoply of problem diseases. It was designed to provide background knowledge related to medicinal plants, which deserves more attention for both policy makers and researchers.

REFERENCES

[1] T. Martinson, Medical Pluralism in Morocco: The Cultural, Religious, Historical and Political-Economic Determinants of Health and Choice, Independent Study Project (ISP) Collection. Paper 1001, 2011.

[2] J. Bellakhdar, La Pharmacopée Marocaine Traditionnelle, Ibis Press, Paris, 1997, p. 764.

[3] M. Eddouks, M. Ajebli, M. Hebi, Ethnopharmacological survey of medicinal plants used in Daraa-Tafilalet region (Province of Errachidia), J. Ethnopharmacol. (2017). https://doi.org/10.1016/j.jep.2016.12.017.

[4] A. Yamani, V. Bunel, M.-H. Antoine, C. Husson, C. Stévigny, P. Duez, M. Elachouri, J. Nortier, Substitution between Aristolochia and Bryoniagenusin North-Eastern Morocco: toxicological implications, J. Ethnopharmacol. 166 (2015) 250–260.

[5] N. Elkhouri, A. Baali, H. Amour, Maternal morbidity and the use of medicinal herbs in the city of Marakech, Indian J. Tradit. Knowl. 15 (1) (2016) 79–85.
[6] J. Fakchich, M. Elachouri, Ethnobotanical survey of medicinal plants used by people in Oriental Morocco to manage various ailments, J. Ethnopharmacol. 154 (2014) 76–87.
[7] L. El Mansouri, A. Ennabili, D. Bousta, Socio-economic interest and valorization of medicinal plants from the Rissani oasis (SE of Morocco), Bol. Latinoam. del Caribe Plantas Med. Aromát. 10 (2011) 30–45.
[8] J. El-Hilaly, M. Hmammouchi, B. Lyoussi, Ethnobotanical studies and economic evaluation of medicinal plants in Taounate province (Northern Morocco), J. Ethnopharmacol. 86 (2003) 149–158.
[9] A. Ennabili, N. Gharnit, E. El Hamdaouni, Inventory and social interest of medicinal, aromatic and honey-plants from Mokrisset region (NW of Morocco), Stud. Bot. 19 (2000) 57–74.
[10] B. Saad, Greco-Arab and Islamic herbal medicine: a review, Eur. J. Med. Plants 4 (3) (2014) 249–258. www.sciencedomain.org.
[11] D.L. Cowen, V.H. Helfand, Pharmacy: An Illustrated History, Harry N. Abrams, New York, 1990.
[12] O. Lafont, D'Aristote à Lavoisier.Les Etapes de la Naissance d'une Science, Ellipses, Paris, 1994.
[13] E. Ravina, H. Kubinyi, The Evolution of Drug Discovery: From Traditional Medicines to Modern Drugs, WILEY-VCH Verlag GmbH & Co. KGaA, Weinheim, 2011, ISBN 978-3-527-32669-3.
[14] E.R. Habermann, Rudolf Buchheim and the beginning of pharmacology as a science, Annu. Rev. Pharmacol. 14 (1974) 1–8.
[15] A. Ibn Al Baytar, Traité des simples wFrench translation by L. Leclerc, 1, 2, 3x, vol. 1883, Imprimerie Nationale, Paris, 1877, p. 1448.
[16] B. Saad, H. Azaizeh, O. Said, Tradition and perspectives of Arab herbal medicine: a Review, eCAM 2 (2005) 475–479.
[17] F.S. Haddad, Pioneers of Arabian medicine, Bull. Soc. Liban. Hist. Med. 3 (1993) 74–83.
[18] A. Fennane, M. Ibn Tattou, J. Mathez, A. Ouyahya, J. El Oualidi, Flore pratique du Maroc, in: Manuel de détermination des plantes vasculaires, vols. 1–3, Institut Scientifique, Université Mohammed V, Rabat, 1999, 2007, 2015.
[19] H. Rankou, A. Culham, S.L. Jury, M.J.M. Christenhusz, The endemic flora of Morocco, Phytotaxa 78 (2013) 1–69.
[20] J. Bellakhdar, La Pharmacopée marocaine traditionnelle: Médecine arabe ancienne et savoirs populaires, Mazars Guy Rev. d'histoire la Pharm. 30 (1998) 84–86.
[21] W. Balée, Footprints of the Forest: Ka'apor Ethnobotany. The Historical Ecology of Plant Utilization by an Amazonian People, Columbia University Press, New York, 1994.
[22] M.C.M. Amorozo, A abordagem etnobotânica na pesquisa de plantas medicinais, in: L.C. Di Stasi (Ed.), Plantas medicinais brasileiras: Arte e Ciência: Um guia de estudo inter-disciplinas, São Paulo, UNESP, 1995.
[23] Brito ARMS, A Farmacologia de plantas medicinais na pesquisa de plantas, in: L.C. Di Stasi (Ed.), Plantas medicinais brasileiras: Arte e Ciência: Um guia de estudo inter-disciplinas, São Paulo, UNESP, 1995.
[24] T.A.S. Araújo, A.L.S. Almeida, J.G. Melo, M.F.T. Medeiros, M.A. Ramos, R.R.V. Silva, C.F.C.B.R. Almeida, U.P. Albuquerque, A new technique for testing distribution of knowledge and to estimate sampling sufficiency in ethnobiology studies, J. Ethnobiol. Ethnomed. 8 (2012) 11.

[25] M.R. González-Tejero, M. Casares-Porcel, C.P. Sánchez-Rojas, J.M. Ramiro-Gutiérrez, J. Molero-Mesa, A. Pieroni, M.E. Giusti, E. Censorii, C. De Pasquale, A. Della, D. Paraskeva-Hadijchambi, A. Hadjichambis, Z. Houmani, M. El-Demerdash, M. El-Zayat, M. Hmamouchi, S. ElJohrig, Medicinal plants in the Mediterranean area: synthesis of the results of the project Rubia, J. Ethnopharmacol. 116 (2008) 341–357.

[26] USAID, United States Agency for International Development, National Development Strategy for the Aromatic and Medicinal Plants Sector, Morocco Integrated Agriculture and Agribusiness Program, July 2008.

[27] HCEFLCD, Haut-Commissariat aux Eaux et Foret et à la lutte Contre la Désertification, Rapport, 2015.

[28] S. Skalli, A. Chebat, N. Badrane, R.S. Bencheikh, Side effects of cade oil in Morocco: an analysis of reports in the Moroccan herbal products database from 2004 to 2012, Food Chem. Toxicol. 64 (2014) 81–85.

[29] A. Ouarghidi, B. Powell, G.J. Martin, H.J. De Boer, A. Abbad, Species substitution in medicinal roots and possible implications for toxicity in Morocco, Econ. Bot. 66 (2012) 370–382.

[30] CBD, Handbook was Updated by the Secretariat of the Convention on Biological Diversity, 2005, ISBN 92-9225-011-6. Available on line: https://www.cbd.int/doc/handbook/cbd-hb-all-en.pdf.

[31] Bulletin Official, Résidence Générale de France à Rabat, 1923.

[32] CAPM – Centre Anti-Poison du Maroc, Toxicologie Maroc, in: Ministère de la Santé (Ed.), Annual Report, Société Empreintes Edition ed, 2010.

[33] M.E. Mostafa, The stake of the phytotherapy in Morocco, in: H. Greche, A. Ennabili (Eds.), Recherches sur les plantes aromatiques et médicinales, 2009, pp. 218–225, 978-9954-1-4960-0. Maroc.

Chapter 6

Chemotaxonomy of Medicinal Plants: Possibilities and Limitations

Ram Singh[1], Geetanjali[2]
[1]Department of Applied Chemistry, Delhi Technological University, Delhi, India; [2]Department of Chemistry, Kirori Mal College, University of Delhi, Delhi, India

1. INTRODUCTION

The flora and fauna represent one of the greatest assets of ecosystems on this earth, with a poorly explored richness. Biodiversity is the most important natural resource available, a resource whose potential for sustainable development is a challenge to be addressed at national and international levels. Medicinal plants constitute a large part of the world's flora. They have always played a pivotal role as sources for drug lead (leading) compounds [1–6]. Since early times, humans have been using medicinal plants for the treatment of various diseases based on their instinct, taste, and experience. People have been using whole plants or parts such as leaves, stem, bark, root, flower, fruits, etc. or in the form of a decoction to treat and prevent diseases. The term natural product refers to the primary and secondary metabolites produced by any living organisms naturally. The natural products are molecules obtained from plants, insects, fungi, algae, prokaryotes and other living organisms either in pure form or in combination with other molecules. All of these living organisms coexist in ecosystems and interact with each other in various ways in which chemistry plays a major role [7,8].

Plants are the major contributors of natural products, and their classification based on the chemicals they produce is always the center of research. Many approaches have evolved toward the taxonomy of plants, including morphologic classification, anatomic classification, and chemotaxonomic classification. The first two methods can be grouped under traditional classifications, whereas the third method is the modern approach to classify the plants. The determination of the taxonomic classification of a group of organisms using their chemical constituents, especially their secondary

Natural Products and Drug Discovery. https://doi.org/10.1016/B978-0-08-102081-4.00006-X

metabolites, is known as the *chemotaxonomic classification*. This is based on the fact that all the living organisms produce secondary metabolites that are derived from primary metabolites. The presence of these chemical constituents is the basis of their chemotaxonomic classifications. The chemical structure of the secondary metabolites and their biosynthetic pathways is often specific and restricted to taxonomically related organisms and, therefore, useful in classification. This method of classification has advantages over the traditional method due to the ease of working methodology. In this method of classification, the materials to be analyzed can be dried or fresh [9]. The concept of chemotaxonomy was elaborated on in 1816 by De Candolle [10]:

1. Plant taxonomy will be the most useful guide to man in his search for new industrial and medicinal plants; and
2. Chemical characteristics of plants will be most valuable to plant taxonomy in the future."

Both of these statements are still valid in the present-day study of natural product chemistry and chemotaxonomic classification. However, there exist several possibilities and limitations of this classification so far. The rise of chemotaxonomic classification is mainly due to the advancement in the analytical techniques for chemical purification and characterization that can detect even trace amount of present chemical compounds [11]. In plants, the more popular families that have been studied through chemotaxonomy are Malvaceae, Ranunculaceae, Magnoliaceae, Polygonaceae, and Solanaceae [12,13]. The findings of chemotaxonomic studies help in the determination of plant classification, phylogeny, and evolution through correlation between phytochemical compounds and morphologic data [14–17]. They are useful to taxonomists, phytochemists, and pharmacologists to solve various taxonomic problems.

2. SECONDARY METABOLITES AS GUIDE FOR CLASSIFICATIONS

Secondary metabolites are the chemical compounds produced by plants. These compounds have no function in the growth, metabolism, and other important primary functions of plant. Their roles are not defined in most plants; however, these compounds play key roles in human lives. The alkaloids, phenolics, and terpenoids represent the major class of secondary metabolites. They play important roles in chemotaxonomy.

2.1 Alkaloid in Chemotaxonomy

Alkaloids are heterocyclic nitrogen-containing basic compounds [18,19]. Chemotaxonomic analysis based on alkaloids depends on the type of parent base compound such as pyridine, piperidine, pyrazole, indole, etc. present in the alkaloids. The indole alkaloids contain indole as the parent base. More than 2500 indole alkaloids were isolated mainly from three plant families: Rubiaceae, Loganiaceae, and Apocynaceae [20]. These are formed from two building blocks, secologanin (**1**) and tryptamine (**2**) or tryptophane (**3**), through a single precursor, strictosidine (**4**), which suggests a close relationship between these families [21,22]. Other plant species that contain indole alkaloids include *Physostigma venenosum* (family Leguminosae), which contains physostigmine (**5**) [23]; *Rauwolfia serpentina* (family Apocyanaceae) [24] and *Corynanthe yohimbe* (family Rubiaceae) [25], which contain yohimbine (**6**); and *Vinca rosea* (family Apocyanaeae), which contains vinblastine (**7**) [26]. The subfamilies such as Ixoroideae, Cinchonoideae, and Rubioideae show good correlation between the biosynthetic pathways and morphologic aspects in the evaluation of their chemical data and other parameters given by Robbrecht [27]. These subfamilies showed correlation in their chemical constituents possessing the typical profile of anthraquinones, indole alkaloids, and iridoids that are considered Rubiaceae chemotaxonomic markers [28]. The Rubiaceae contains another predominant class, which consists of derivatives of tryptamine (**2**) and monoterpene (iridoid) secologanin (**1**) as monoterpene indole alkaloids [29]. Other types of alkaloids like quinoline alkaloids are also characteristic of Rubiaceae; however, the presence of other alkaloids does not support any chemotaxonomic correlation [29].

The presence of pyridine and piperidine alkaloids in important in chemotaxonomic analyses. The alkaloids like lobeline (**8**) obtained from *Lobelia inflata* (family Lobeliaceae) [30], nicotine (**9**) obtained from *Nicotiana tobaccum* (family Solanaceae) [31], and anabasine (**10**) obtained from *Nicotiana gluaca* (family Chenopodiaceae) [32] show their importance in chemotaxonomic analysis. Anabasine (**10**) occurs in tobacco, where it is formed from lysine and nicotinic acid (**11**); as in the legume and chenopod species, this can be synthesized from two molecules of lysine [32]. Similarly, the alkaloids like isoquinoline alkaloids, tropane alkaloids, indole alkaloids, etc. have been useful tools for chemotaxonomic classification of plants [33].

The aporphine alkaloids were isolated from the leaves of *Annona salzmannii* and *Annona vepretorum* (family Annonaceae) and used for the chemotaxonomic analysis [34]. The alkaloids such as anonaine (**12**), asimilobine (**13**), and norcorydine (**14**) were identified from the leaves of *A. salzmannii,* whereas oxonantenine (**15**), lanuginosine (**16**), and lysicamine (**17**) were found from the leaves of *A. vepretorum* [34]. Another aporphine alkaloid, liriodenine (**18**), was found in both the plants [34]. These phytochemical characteristics were helpful in establishing the chemotaxonomic relationship of *A. salzmannii* and *A. vepretorum* with other *Annona* species [34]. A close relationship between the genera *Stemona* and *Stichoneuron* was studied by DNA analysis, and the result favored keeping them in the same Stemonaceae family [35]. The *Stemona* genus possesses a very large chemical diversity and hence gave a significant chemotaxonomic clue [36].

The phytochemical investigation showed stenine- and stemoninine-type alkaloids identified in the *Stemona* plant. This further supported the DNA analysis data [35].

The herb Sceletium has been used as complementary medicine and sold without information regarding the phytochemical constituents [37]. Phytochemical investigations showed that *Sceletium emarcidum*, *Sceletium exalatum*, and *Sceletium rigidum* are the emarcidum type whereas *Sceletium tortuosum*, *Sceletium expansum*, and *Sceletium strictum* are of the tortuosum type [38]. The phytochemical analysis was done with the use of high-performance liquid chromatography with UV. The isolated alkaloids were identified with the use of online mass spectroscopy. The alkaloids, mesembrine (**19**), mesembrenone (**20**), mesembranol (**21**), and epimesembranol (**22**), were present in *S. tortuosum* and *S. expansum*. *S. strictum* also contained mesembrine (**19**), mesembrenone (**20**), and either 4′-*O*-demethylmesembrenone or 4′-*O*-demethylmesembrenol [38]. All of these species are associated with the tortuosum type. The emarcidum-type specimens showed a complete absence of the major alkaloid mesembrine (**19**). This phytochemical study clearly showed the difference between the two types from the same species. For medicinal applications, it is important to identify the correct *Sceletium* species [38].

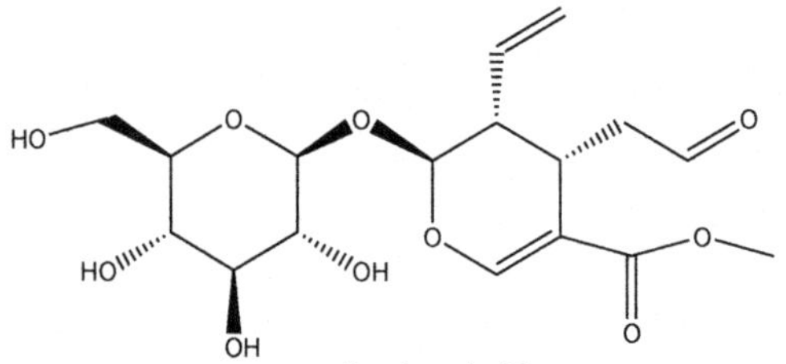

Secologanin (**1**)

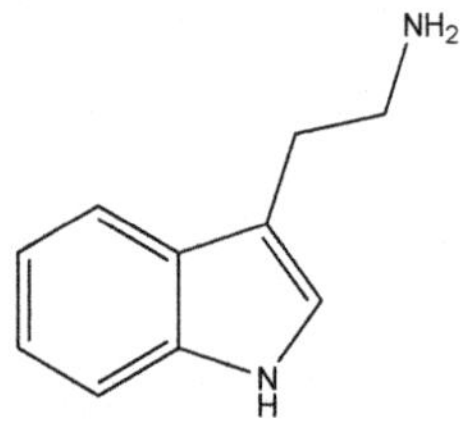

Tryptamine (**2**)

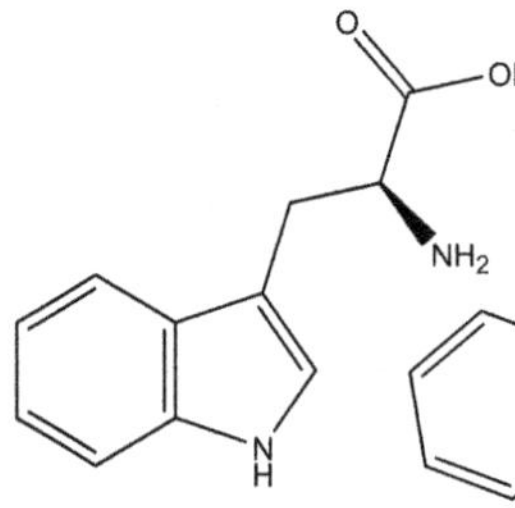

Tryptophane (**3**)

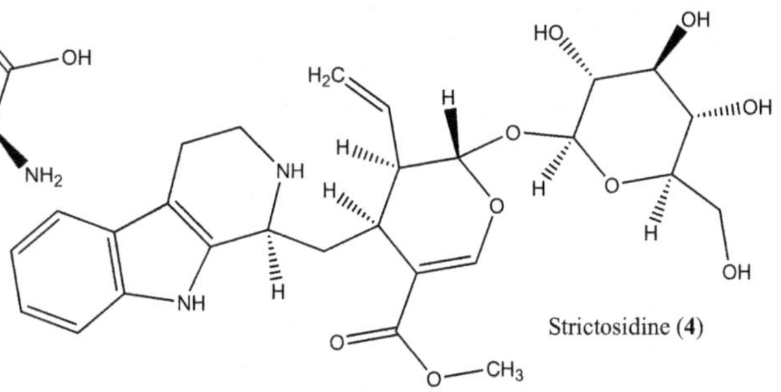

Strictosidine (**4**)

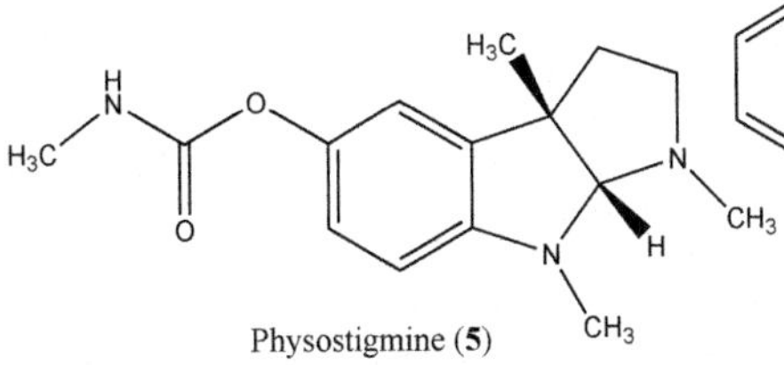

Physostigmine (**5**)

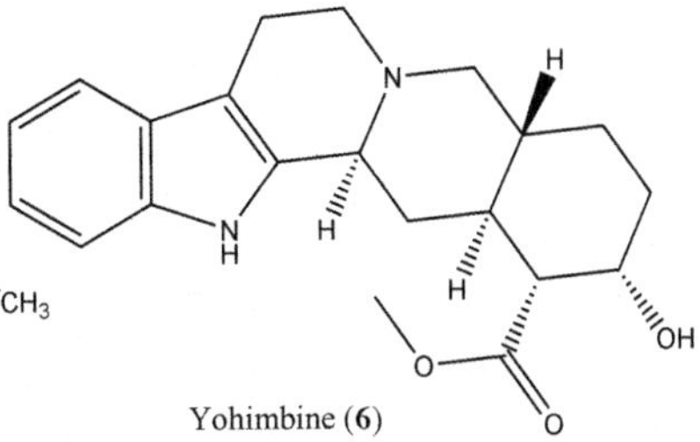

Yohimbine (**6**)

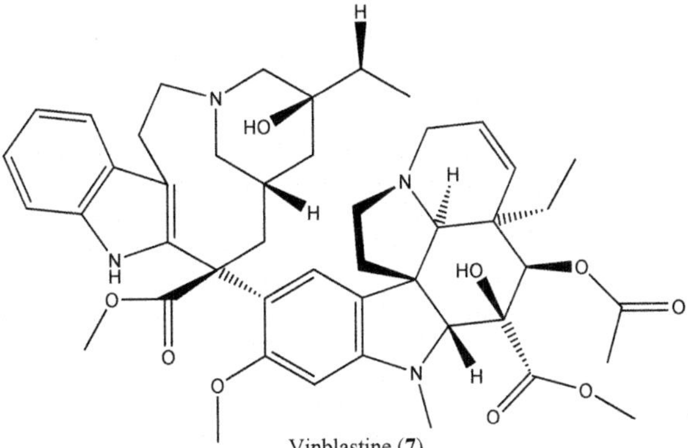

Vinblastine (**7**)

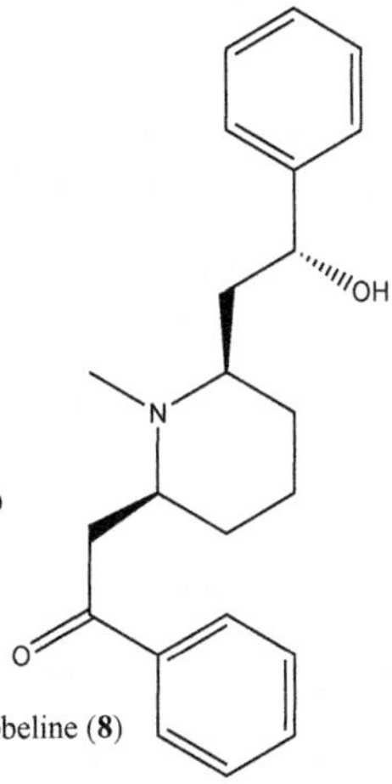

Lobeline (**8**)

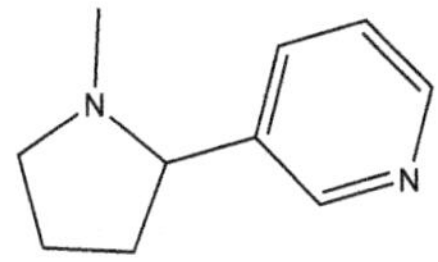

Nicotine (**9**)

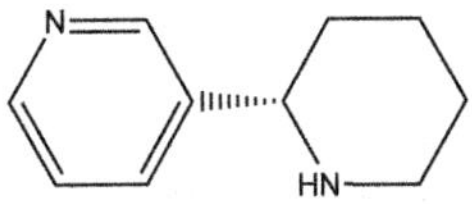

Anabasin (**10**)

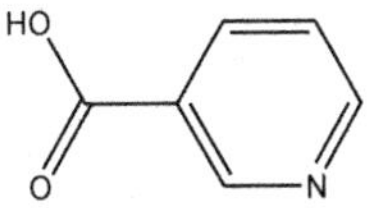

Nicotinic acid (**11**)

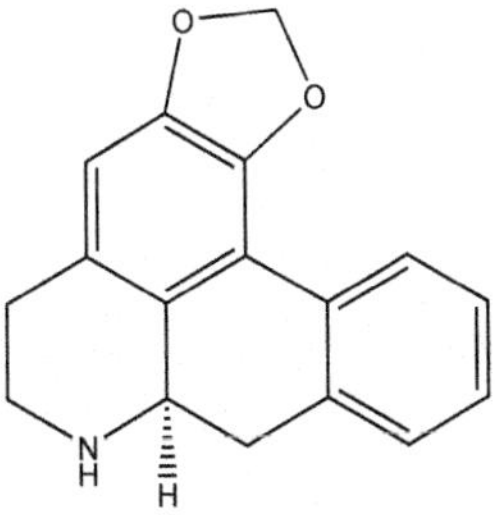

Anonaine (**12**)

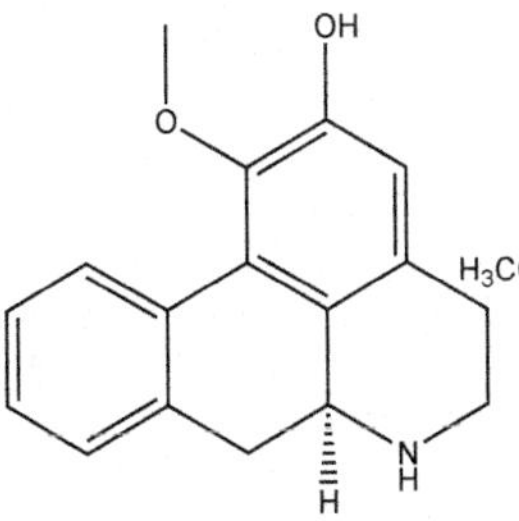

Asimilobine (**13**)

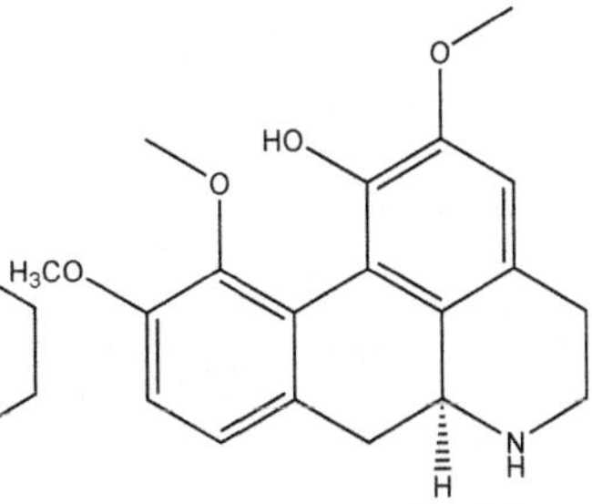

Norcorydine (**14**)

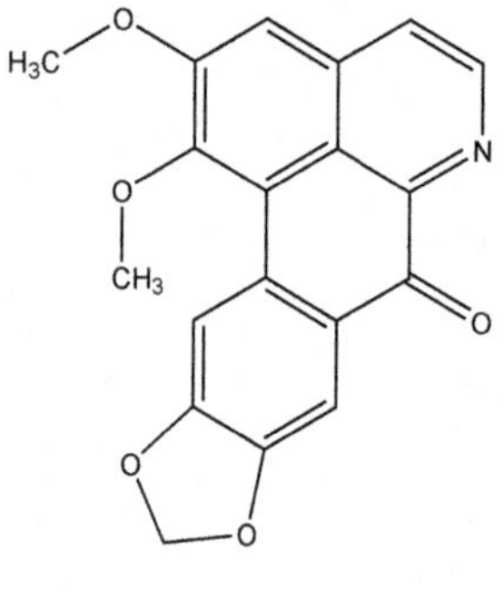

Oxonantenine (**15**) 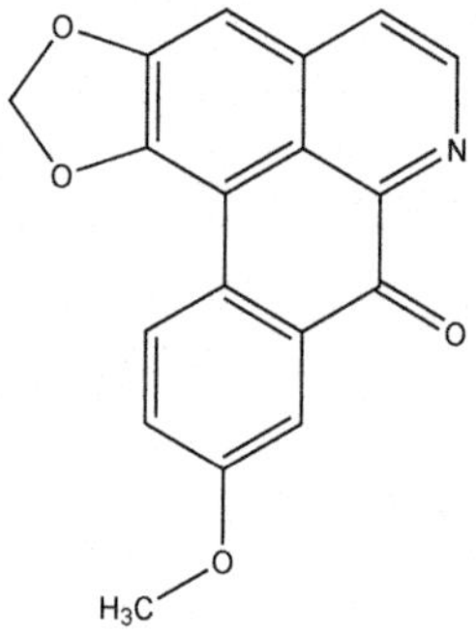 Lanuginosine (**16**) 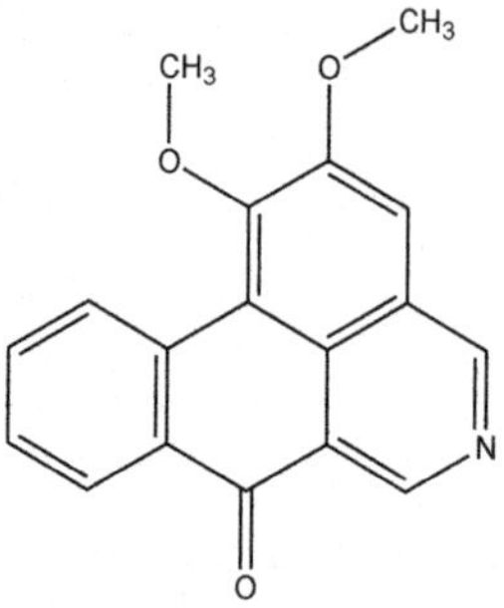 Lysicamine (**17**)

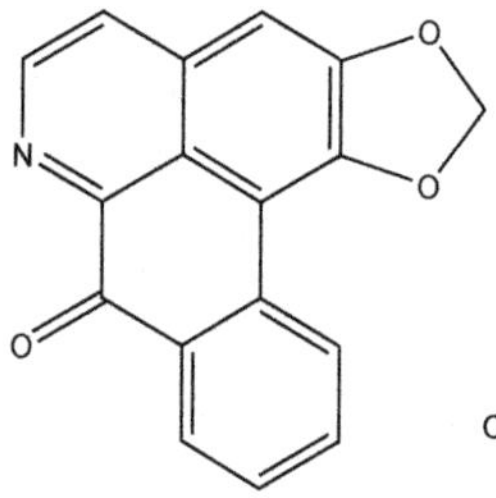

Liriodenine (**18**) 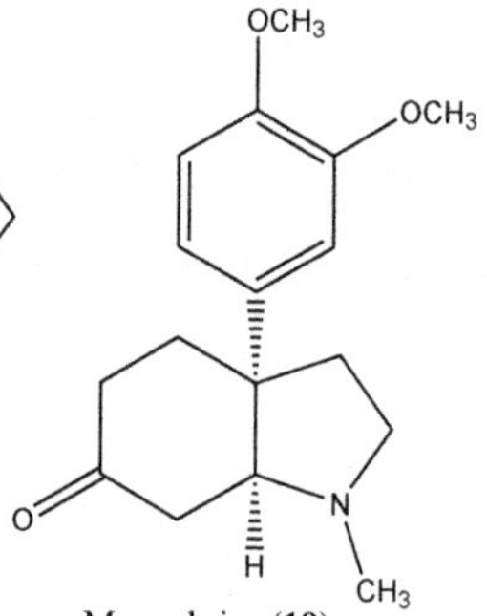 Mesembrine (**19**) 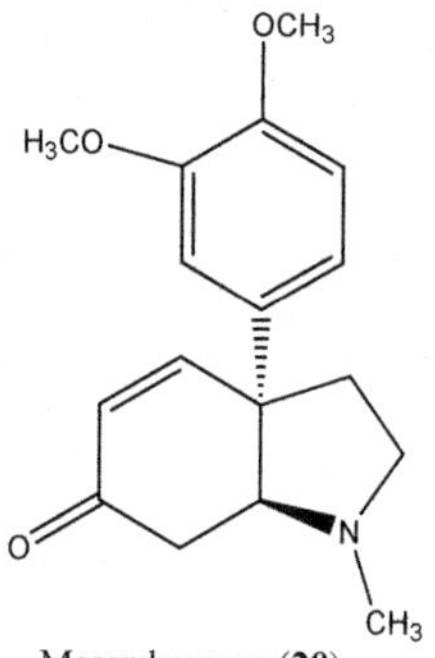 Mesembrenone (**20**)

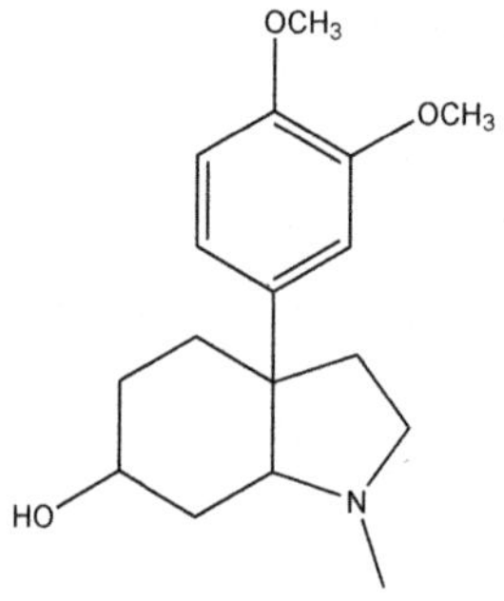

Mesembranol (**21**) 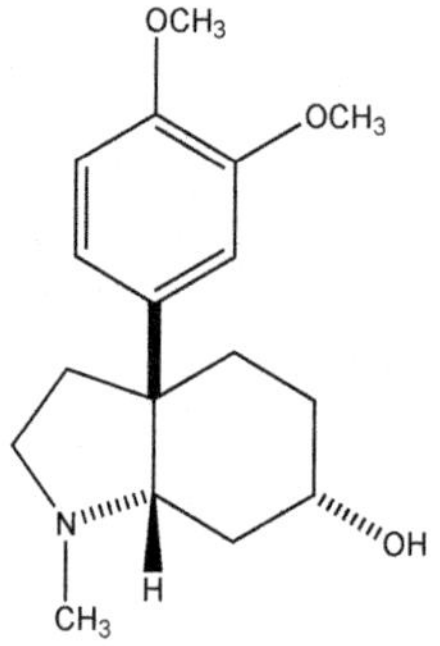 Epimesembranol (**22**)

2.2 Plant Phenol in Chemotaxonomy

Along with alkaloids, plant phenols are also widespread class of metabolites in nature, and their distribution is almost ubiquitous. The majority of the natural-occurring phenolics followed the phenylpropanoid pathway and use some 20% of the carbon fixed by photosynthesis [39,40]. This is the most popular class of secondary plant metabolites due to their potential benefits to human health [41]. Different classes of plant phenols include flavones, flavanones, isoflavanones, anthocyanidins, and chalcones. Flavonoids are the largest group of phenolic compounds and are mostly found in higher plant. These compounds have been proved to be of chemotaxonomic importance [42–45]. All flavonoids have a common biosynthetic origin and therefore possess the same basic structural elements, like the 2-phenylchromone skeleton. They may be present in many classes depending on the degree of oxidation of the pyran ring, which may be open and cyclize into a furan ring like 2-phenyl benzo pyrilium (anthocyanin) and 2-phenyl chromone (flavone, flavanol, isoflavone) [46]. Middleton et al. has well summarized the pharmaceutical view of the flavonoid effects on mammalian cells [47]. Flavonoids have found their applications as antiviral, antitoxic, cytoprotective, and antioxidant compounds. They have played their role in inflammatory processes, from coronary vascular diseases to cancer [47]. Hence, the plant species that possess this class of compounds are of high medicinal value.

A chemotaxonomic study of practically all the species of the genus *Aloe* showed that the flavonoids occur as major compounds in 31 of a total of 380 species investigated [48]. The chemotaxonomic survey showed that the infrageneric level of classification can be done with the help of leaf exudate compounds [49]. The chemotaxonomic value of 5-hydroxyaloin A (**23**) in the genus *Aloe* was discussed by Rauwald and Beil [50]. The six species, *Aloe alooides*, *Aloe castanea*, *Aloe dolomitica*, *Aloe spicata*, *Aloe tauri*, and *Aloe vryheidensis* ,are characterized by 6-*O*-coumaroylaloesin (**24**) [51].

The order Ericales members have been chemotaxonomically analyzed with the help of flavonoid chemistry [52]. This order consists of around 25 families and 347 genera distributed in four clades (Balsaminoids, Polemonioids, Primuloids, and Ericoids). Ericales showed the higher production of flavonols in comparison to the flavones [52]. In Ericlaes, all the clades showed a similar related ratio of flavone to flavonol and pattern of hydroxyl protection [52].

The flavonoid analysis of 31 species of *Carmichaelia* showed the presence of isoflavones flavonol 3-*O*-glycosides, flavonol 7-*O*-glycosides, flavonol-3,7-di-*O*-glycosides, and flavone C-glycosides. Acid hydrolysis of 3-*O*-glycosides has yielded three flavonol aglycones—quercetin (**25**), isorhamnetin (**26**), and kaempferol (**27**)—which are common to nearly all species, with one aglycone being dominant in each of them [53].

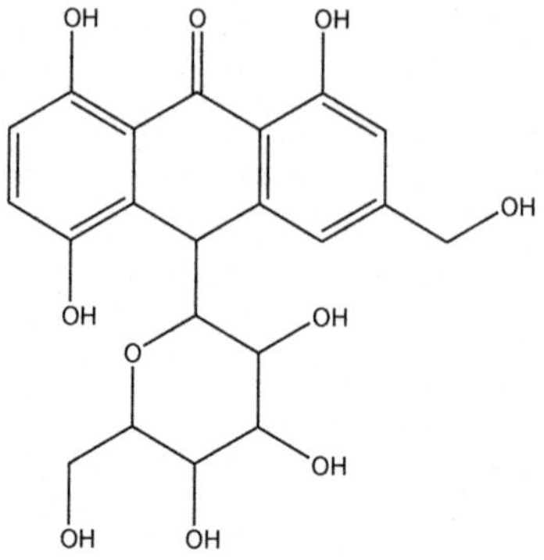

5-Hydroxyaloin A (**23**)

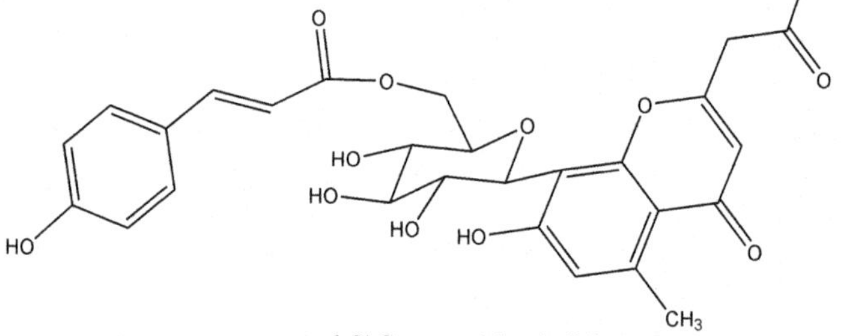

6-O'-Coumaroylaloesin (**24**)

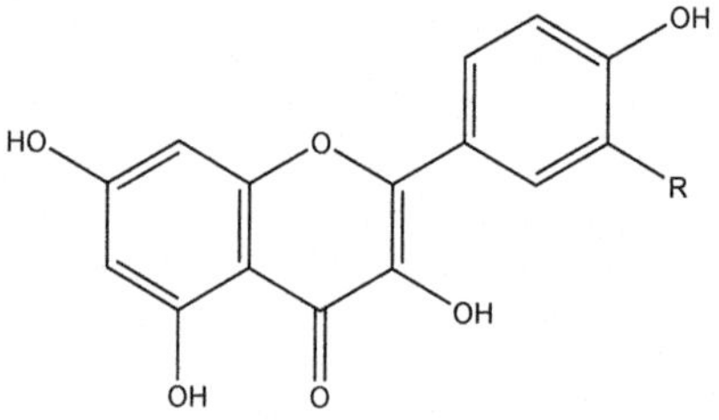

Quercetin (**25**) R = OH

Isorhamnetin (**26**) R = OCH3

Kaempferol (**27**) R = H

2.3 Quinones in Chemotaxonomy

Anthraquinones, naphthoquinones, and their glycosides are another group of secondary metabolites useful in chemotaxonomy [54,55]. The plant families such as Verbenaceae [56], Bignoniaceae [57], Rhamnaceae [58], Polygonaceae [59], Leguminosae [60], and Rubiaceae [61] contain anthraquinones, naphthoquinones, and their glycosides and hence can be used for chemotaxonomic classification. In plants, anthraquinones originate from different precursors and pathways [62,63]. The two main biosynthetic pathways present in higher plants include the polyketide pathway [64] and the chorismate/*o*-succinylbenzoic acid pathway [65], the latter occurring in Rubiaceae.

The family Rubiaceae consists of 450 genera and 6500 species and includes trees, shrubs, and, infrequently, herbs. The genus *Rubia* belongs to this family and has more than 60 species distributed widely, including Africa, temperate Asia, and North and South America. About 15 species are reported from India [54,66]. This species have high commercial, economical, and medicinal importance. Plants belonging to this species are known to contain substantial amounts of anthraquinones, especially in the roots [67,68]. The Indian Ayurvedic system of medicine listed this species as a reputed species. In India and neighboring countries, *Rubia* has long history in skin care and treatment and has been used for disorders of the urinary tract [69,70]. The phytochemical analysis of *Rubia* species led to the isolation of a number of physiologically active compounds, viz., anthraquinones, naphthoquinones, terpenes, bicyclic hexapeptides, iridoids, and carbohydrates. More than 65 anthraquinones and their glycosides were obtained from the genus *Rubia* that differ in the nature of their substituents and the substitution patterns [54].

2.4 Glycosides in Chemotaxonomy

The utilization of glycosides in chemotaxonomy is mostly based on the nonsugar molecules. Glycosides can be defined as the compounds in which one or more sugars are combined with nonsugar molecules through glycosidic linkage. They are classified on the basis of their linkage with nonsugar molecules as O-glycoside, C-glycoside, N-glycoside, and S-glycoside. Some glycosides are very common and hence are of little chemotaxonomic value. However, some flavonol glycosides are useful in the chemotaxonomic analysis. The *Rosa* species uses the flavonol glycosides as chemotaxonomic markers for the classification [71]. The petals of *Rosa* species belonging to the section Gallicanae produced 15 flavonol glycosides. These flavonol glycosides in the petals of 31 *Rosa* species belonging to sections Gallicanae, Cinnamomeae, Caninae, and Synstylae indicated that the species belonging to these sections could be classified into four types based on the pattern of flavonol glycoside contents [71].

The C-glycosides are the compounds that possess a direct carbon linkage between the sugar and nonsugar parts. They are not prevalent in nature. They are found in some plants as the derivative of anthraquinones [20,54], such as aloin (**28**) present in the species *Aloe* (family liliaceae) [72] and cascaroside (**29**) present in *Cascara* (family rhamnaceae) [73].

The S-glycosides are the compounds that possess a direct sulfur linkage between the sugar and nonsugar parts and produce isothiocyanate on hydrolysis. The families of plants like cruciferae, moringaceae, and capparaceae possess these types of compounds, so these families have a phylogenetic relationship [74]. The sinigrin (**30**), an S-glycoside, is present in the plants of the Brassicaceae family and can act as chemotaxonomic marker.

Other studies based on chemotaxonomic data obtained via gas chromatography coupled to mass spectrometry show that the iridoid glycosides are present in several different species belonging to the Rubiaceae subfamilies [75,76]. Literature is available on the utility of glycosides in chemotaxonomy [77–81].

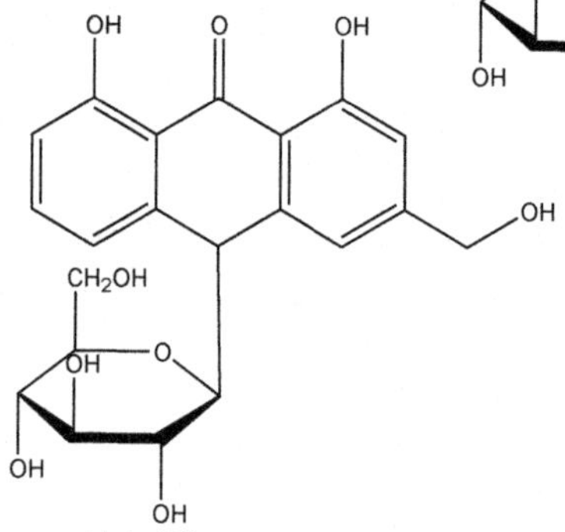

Aloin (**28**)

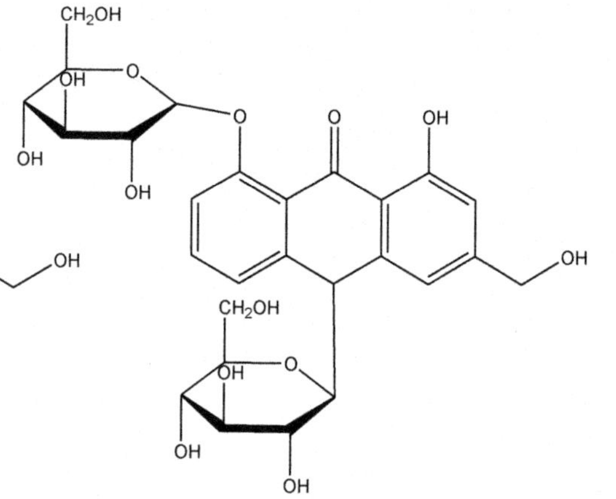

Cascaroside (**29**)

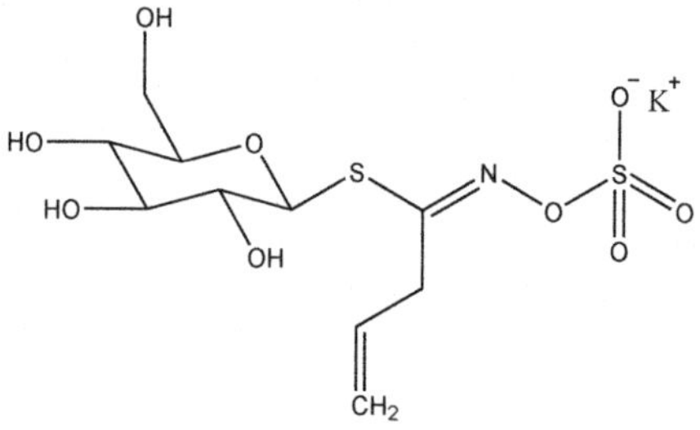

Sinigrin (**30**)

3. LIMITATIONS OF CHEMOTAXONOMIC CLASSIFICATION

The plant secondary metabolites play a basic role in the chemotaxonomic classification of plants, with some support from primary metabolites. The content and sometimes presence of secondary metabolites fluctuate in plants due to various factors that play a vital role in their biosynthetic pathways and further accumulation [82]. This fluctuation in plant secondary metabolites has limitations in chemotaxonomic classification. Environment is a major factor responsible for fluctuation in plant secondary metabolites along with other contributing factors, such as genetic, ontogenic, and mesphogenetic factors [82].

The biosynthesis of plant secondary metabolites is influenced by both biotic and abiotic factors [83]. The contents of secondary metabolites for the same plant species differ with respect to the environment in which they are grown [84]. To combat environmental stress, plants produce specific secondary metabolites. Same plants present in different environments come in contact with different types and qualities of abiotic components like soil, water, light, temperature etc. These abiotic components are necessary for the developmental growth and survival of plant species. Hence, their variations lead to variations in the production of their secondary metabolites, which causes limitations in their chemotaxonomic classification.

Water is an essential component for the growth and development of plants. Water is responsible for causing three basic types of stress—water stress, drought stress, and salinity stress—in plants and affects the production and accumulation of plant secondary metabolites, which in turn affects the chemotaxonomic classification. Insufficient water causes drought stress that affects the normal physiologic activities [85,86] and the biochemical properties of plants [87–89]. During this difficult time of stress, the secondary metabolites of the plant help them to survive. For example, the medicinal plants like *Artemisia annua, Hypericum perforatum,* and *Catharanthus roseus* showed an increase in their secondary metabolite production [82].

It was observed that the quantity of secondary metabolites like artemisinin (**31**) in *Artemisia* and betulinic acid (**32**), quercetin (**25**), and rutin (**33**) in *Hypericum brasiliense* increased. There is no trend observed that the water stress only increases the secondary metabolite production. Lowering in the concentration of secondary metabolites like hypericin (**34**) and pseudohypericin (**35**) while increasing hyperforin (**36**) has been observed in water-deficient conditions [90].

Saline water also affects the phytochemical constituents in plants. Due to salt stress, the concentration of reserpine (**37**) in *Rauvolfia tetraphylla* and vincristine alkaloids in *C. roseus* increases. The salt stress also affects the presence of the same secondary metabolites in different parts of plants. For example, in *Ricinus communis*, the concentration of ricinine (**38**) alkaloid increases in shoots while it decreases in roots [91]. The alkaloid content in plants like *Achillea fragrantissima, Solanum nigrum,* and *C. Roseus* increased due to salt stress. The variation in the content of phenolic acids like protocatechuic acid (**39**), chlorogenic acid (**40**), and caffeic acid (**41**) was observed in *Matricaria chamomilla* with the increase of salinity [92,93].

The variation in the production and accumulation of plant secondary metabolites has also been observed with temperature change. *Panax quinquefolius* showed an increase in the concentration of ginsenosides (**42**) with the rise in temperature [94]. The light intensity and photoperiod also cause variation in the biosynthesis of plant secondary metabolites. The formation of coumarins in *Mikania glomerata* favored in sunlight [95].

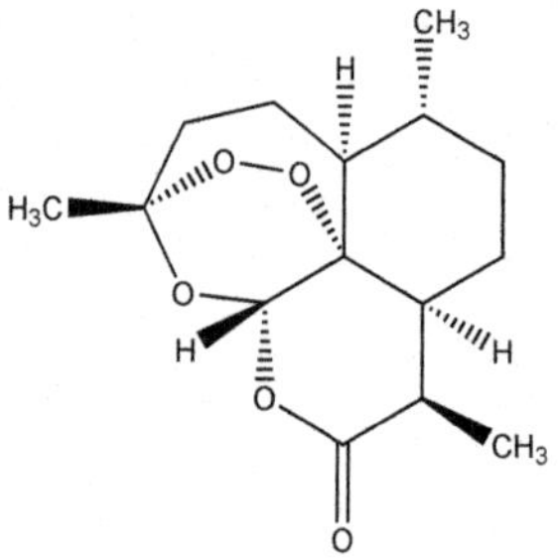

Artemisinin (**31**)

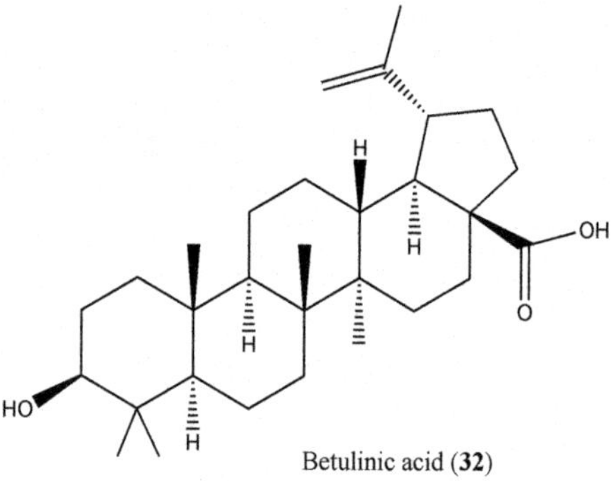

Betulinic acid (**32**)

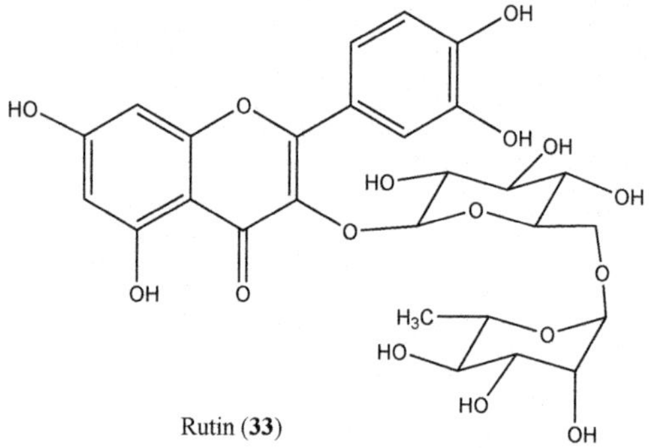

Rutin (**33**)

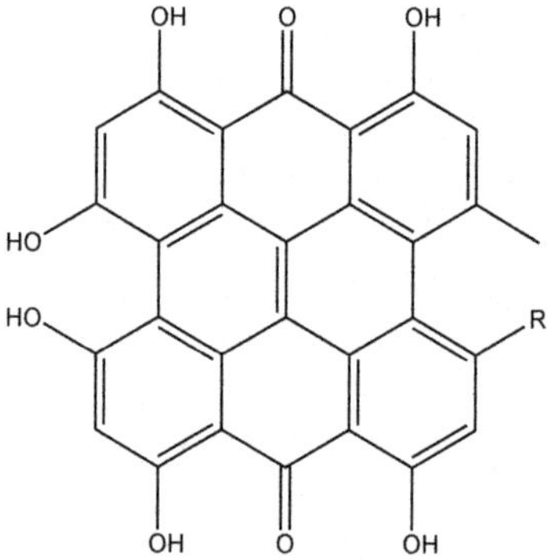

Hypericin (**34**) R = CH_3

Pseudohypericin (**35**) R = CH_2OH

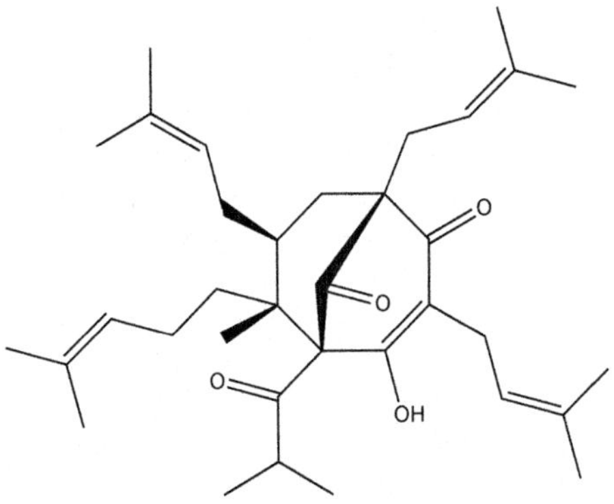

Hyperforin (**36**)

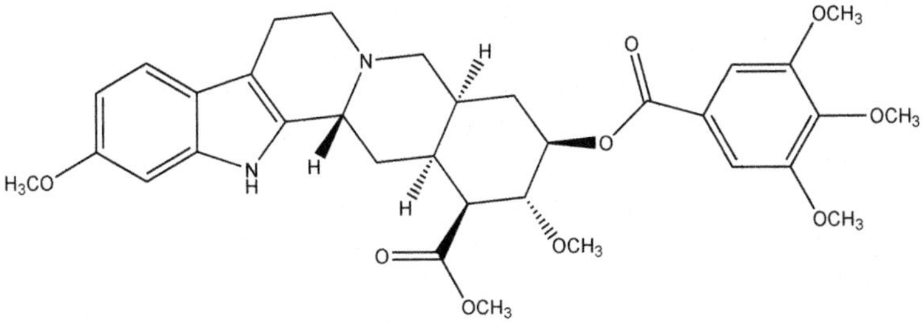

Reserpine (**37**)

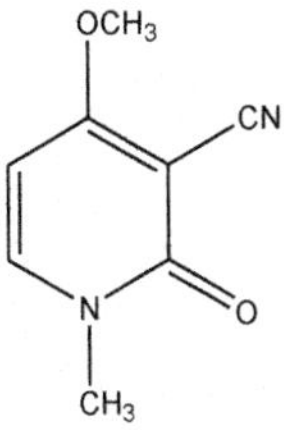

Ricinine (**38**)

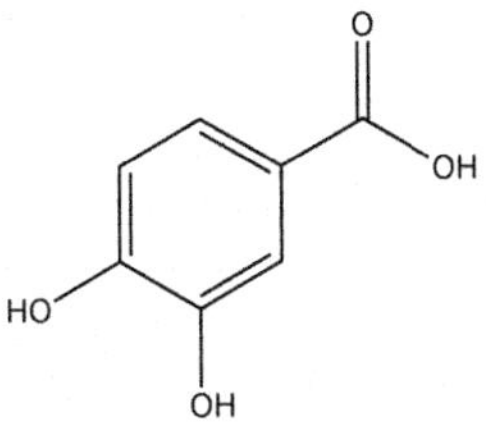

Protocatechuic acid (**39**)

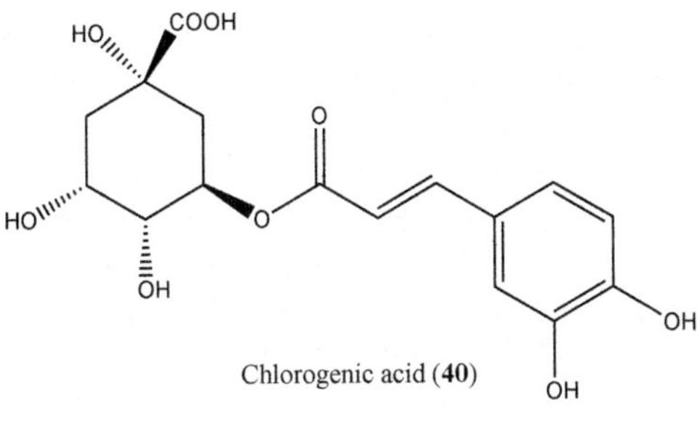

Chlorogenic acid (**40**)

4. SUMMARY AND FUTURE PROSPECTS

The role of medicinal plants in human life is immense. One challenge that the evolution of medicinal plants faces is their classification. There have been controversies about the taxonomic classification for various plant families and species. The concept of classification of medicinal plants based on chemical character, which is known as chemotaxonomic classification, has certain limitations. The chemotaxonomic classification is based on the chemical compounds that are produced by the plants. These are mainly secondary metabolites. The alkaloids, phenolics, and terpenoids are the major class of secondary metabolites. They play important roles in chemotaxonomy.

Due to the advancement of analytical techniques, the characterization of secondary metabolites and their biosynthetic pathways contributed to extensive taxonomic improvements. The presence or absence of a particular phytochemical in a plant can be used to assign its taxonomic position. There is a lot in the future for the chemotaxonomic classification because of the estimated 400,000–500,000 plant species around the globe, only a small percentage have been investigated phytochemically.

REFERENCES

[1] D.J. Newman, G.M. Cragg, K.M. Snader, The influence of natural products upon drug discovery, Nat. Prod. Rep. 17 (2000) 215–234.

[2] D.J. Newman, G.M. Cragg, K.M. Snader, Natural products as sources of new drugs over the period 1981–2002, J. Nat. Prod. 66 (2003) 1022–1037.

[3] F.E. Koehn, G.T. Carter, The evolving role of natural products in drug discovery, Nat. Rev. Drug Discov. 4 (2005) 206–220.

[4] I. Paterson, E.A. Anderson, The renaissance of natural products as drug candidates, Science 310 (2005) 451–453.

[5] M.J. Balunas, A.D. Kinghorn, Drug discovery from medicinal plants, Life Sci. 78 (2005) 431–441.

[6] W.P. Jones, Y.W. Chin, A.D. Kinghorn, The role of pharmacognosy in modern medicine and pharmacy, Curr. Drug Targets 7 (2006) 247–264.

[7] T. Reynolds, The evolution of chemosystematics, Phytochemisty 68 (2007) 2887–2895.

[8] T.O. Larsen, J. Smedsgaard, K.F. Nielsen, M.E. Hansen, J.C. Frisvad, Phenotypic taxonomy and metabolite profiling in microbial drug discovery, Nat. Prod. Rep. 22 (2007) 672–695.

[9] S. Ankanna, D. Suhrulatha, N. Savithramma, Chemotaxonomical studies of some important monocotyledons, Bot. Res. Int. 5 (2012) 90–96.

[10] A.P. De Candolle, Essaisur les propridtdsmédicales des plantes, cornparees avec leurs forms extérieuresetleur classification naturelle, second ed., 1816. Paris.

[11] V.V. Bhargava, S.C. Patel, K.S. Desai, Importance of terpenoids and essential oils in chemotaxonomic approach, Int. J. Herb. Med. 1 (2013) 14–21.

[12] V.V. Sivarajan, Introduction to the Principles of Plant Taxonomy, Cambridge University Press, 1991.

[13] B. Bremer, Combined and separate analyses of morphological and molecular data in the plant family Rubiaceae, Cladistics 12 (1996) 21–40.

[14] A. Otto, V. Wilde, Sesqui-, di-, and triterpenoids as chemosystematic markers in extant conifers – a review, Bot. Rev. 67 (2001) 141–238.

[15] C.A. Carbonezi, L. Hamerski, O.A. Flausino Jr., M. Furlan, V.D.S. Bolzani, M.C.M. Young, Determinação por RMN das configurações relativas e conformações de alcalóides oxindólicos isolados de *Uncaria guianensis*, Quim. Nova 27 (2004) 878–881.

[16] R. Dahlgren, A revised system of classification of the angiosperms, Bot. J. Linn. Soc. 80 (1980) 91–124.

[17] N. Kharazian, Chemotaxonomy and flavonoid diversity of *Salvia L.* (Lamiaceae) in Iran, Acta Bot. Bras. 28 (2014) 281–292.

[18] M. Saxena, J. Saxena, R. Nema, D. Singh, A. Gupta, Phytochemistry of medicinal plants, J. Pharmacog. Phytochem. 1 (2013) 168–182.

[19] R. Singh, Geetanjali, V. Singh, Exploring alkaloids as inhibitors of selected enzyme, Asian J. Chem. 23 (2011) 483–490.

[20] R. Singh, Chemotaxonomy: a tool for plant classification, J. Med. Plants Stud. 4 (2) (2016) 90–93.

[21] L.F. Szabó, Molecular evolutionary lines in the formation of indole alkaloids derived from secologanin, ARKIVOC 3 (2008) 167–181.

[22] L.F. Szabó, Rigorous biogenetic network for a group of indole alkaloids derived from strictosidine, Molecules 13 (2008) 1875–1896.

[23] P.K. Mukherjee, V. Kumar, M. Mal, P.J. Houghton, Acetylcholinesterase inhibitors from plant, Phytomedicine 14 (2007) 289–300.

[24] F.E. Bader, D.F. Dickel, E. Schlittler, Rauwolfia alkaloids. IX.[1] isolation of yohimbine from *Rauwolfia serpentina* Benth, J. Am. Chem. Soc. 76 (1954) 1695–1696.

[25] A.P. Singh, R. Singh, Potent natural aphrodisiacs for the management of erectile dysfunction and male sexual debilities, Front. Biosci. 1 (2012) 167–180.

[26] M.R. Kramers, H. Stebbings, The insensitivity of *Vinca rosea* to vinblastine, Chromosoma 61 (1977) 277–287.

[27] E. Robbrecht, Tropical woody Rubiaceae, Oper. Bot. Belg. 1 (1988) 599–602.

[28] H. Inouye, Y. Takeda, H. Nishimura, A. Kanomi, T. Okuda, C. Puff, Chemotaxonomic studies of rubiaceous plants containing iridoid glycosides, Phytochemisty 27 (1988) 2591–2598.

[29] O.R. Gottlieb, Micromolecular Evolution, Systematics and Ecology: An Essay into a Novel Botanical Discipline, vol. 19, Springer Science & Business Media, Berlin, Germany, 1982, p. 94.

[30] H. Yonemitsu, K. Shimomura, M. Satake, S. Mochida, M. Tanaka, T. Endo, A. Kaji, Lobeline production by hairy root culture of *Lobelia inflata* L. Plant Cell Rep. 9 (1990) 307–310.

[31] Q. Shi, C. Li, F. Zhang, Nicotine synthesis in Nicotiana tabacum L. induced by mechanical wounding is regulated by auxin, J. Exp. Bot. 57 (11) (2006) 2899–2907.

[32] P.A. Steenkamp, F.R. van Heerden, B.E. van Wyk, Accidental fatal poisoning by Nicotiana glauca: identification of anabasine by high performance liquid chromatography/photodiode array/mass spectrometry, Forensic. Sci. Int. 127 (3) (2002) 208–217.

[33] K.W. Bentley, β-Phenylethylamines and the isoquinoline alkaloids, Nat. Prod. Rep. 9 (4) (1992) 365–391.

[34] M.N.O. Teles, L.M. Dutra, A. Barison, E.V. Costa, Alkaloids from leaves of *Annona salzmannii* and *Annona vepretorum* (Annonaceae), Biochem. Syst. Ecol. 61 (2015) 465–469.

[35] R.R. Zhang, H.Y. Tian, Y. Wu, X.H. Sun, J.L. Zhang, Z.G. Ma, R.W. Jiang, Isolation and chemotaxonomic significance of stenine- and stemoninine-type alkaloids from the roots of *Stemona tuberose*, Chin. Chem. Let. 25 (9) (2014) 1252–1255.

[36] S. Kongkiatpaiboon, J. Schinnerl, S. Felsinger, V. Keeratiniakal, S. Vajrodaya, W. Gritsanapan, L. Brecker, H. Greger, Structural relationships of *Stemona alkaloids*: assessment of species-specific accumulation trends for exploiting their biological activities, J. Nat. Prod. 74 (9) (2011) 1931–1938.

[37] M.T. Smith, C.R. Field, N.R. Crouch, M. Hirst, The distribution of mesembrine alkaloids in selected taxa of the Mesembryanthemaceae and their Modification in the Sceletium Derived 'Kougoed', Pharm. Biol. 36 (3) (1998) 173–179.

[38] S. Patnala, I. Kanfer, Chemotaxonomic studies of mesembrine-type alkaloids in Sceletium plant species, South Afr. J. Sci. 109 (3/4) (2013) 1–5.

[39] R.L. Metcalf, Plant volatiles as insect attractants, CRC Crit. Rev. Plant Sci. 5 (1987) 251–301.

[40] L. Ralston, S. Subramanian, M. Matsuno, O. Yu, Partial reconstruction of flavonoid and isoflavonoid biosynthesis in yeast using soyabean type I and type II chalcone isomerases, Plant Physiol. 137 (2005) 1375–1388.

[41] D. Treutter, Managing phenol contents in crop plants by phytochemical farming and breeding-visions and constraints, Int. J. Mol. Sci. 11 (3) (2010) 807–857.

[42] M. Nakiboglu, The classification of the *Salvia* L. (Labiatae) species distributed in west Anatolia according to phenolic compounds, Turk. J. Bot. 26 (2) (2002) 103–108.

[43] K.M. Valant-Vestachera, J.N. Roitman, E. Wollenweber, Chemodiversity of exudate flavonoids in some members of the Lamiaceae, Biochem. Syst. Ecol. 31 (11) (2003) 1279–1289.

[44] N. Kharazian, M.R. Rahiminejad, Chemotaxonomy of wild diploid *Triticum* L. (Poaceae) species in Iran, Int. J. Bot. 4 (3) (2008) 260–268.

[45] N. Kharazian, Taxonomy and morphology of *Salvia spinosa* in Iran, Taxonomy Biosyst. J. 1 (2009) 9–20.

[46] M. Wink, P.G. Waterman, Chemotaxonomy in relation to molecular phyologeny of plants, in: M. Wink (Ed.), ., Biochemistry of Plant Secondary Metabolites Ann. Plants Rev, vol. 2, CRC Press, Boca Raton, 1999, pp. 300–341.

[47] E. Middleton, C. Kandaswami, T.C. Theoharides, The effects of plant flavonoids on mammalian cells: implications for inflammation, heart disease, and cancer, Pharmacol. Rev. 52 (4) (2000) 673–751.

[48] A.M. Viljoen, B.E. van Wyk, F.R. van Heerden, Distribution and chemotaxonomic significance of flavonoids in *Aloe* (Asphodelaceae), Plant Syst. Evol. 211 (1) (1998) 31–42.

[49] A.M. Viljoen, B.E. van Wyk, The chemotaxonomic value of the phenyl pyrone, aloenin in the genus *Aloe*, Biochem. Syst. Ecol. 28 (10) (2000) 1009–1017.

[50] H.W. Rauwald, A. Beil, 5-Hydroxyaloin A in the genus *Aloe*: thin layer chromatographic screening and high performance liquid chromatographic determination, Z. Naturforsch. 48c (1993) 1–4.

[51] A.M. Viljoen, B.E. van Wyk, A chemotaxonomic and morphological appraisal of *Aloe* series Purpurascentes, *Aloe* section Anguialoe and their hybrid, *Aloe broomii*, Biochem. Syst. Ecol. 29 (6) (2001) 621–631.

[52] M.E.N. Rocha, M.R. Figueiredo, M.A.C. Kaplan, T. Durst, J.T. Arnason, Chemotaxonomy of the *Ericales*, Biochem. Syst. Ecol. 61 (2015) 441–449.

[53] A.W. Purdie, Some flavonoid components of *Carmichaelia* (Papilionaceae)– a chemotaxonomic survey, N. Z. J. Bot. 22 (1) (1984) 7–14.

[54] R. Singh, Geetanjali, S.M. Chauhan, 9,10-Anthraquinones and other biologically active compounds from the genus *Rubia*, Chem. Biodiv. 1 (9) (2004) 1241–1264.

[55] R. Singh, Geetanjali, Isolation and synthesis of anthraquinones and related compounds of *Rubia cordifolia*, J. Ser. Chem. Soc. 70 (7) (2005) 937–942.

[56] R.A. Muzychina, Natural Anthraquinones: Biological and Physicochemical Properties, 1998. Phasis, Moscow.

[57] A.R. Burnett, R.H. Thomson, Naturally occurring quinines. Part X. The quinonoid constituents of *Tabebuia avellanedae* (Bignoniaceae), J. Chem. Soc. C Org. (1967) 2100–2104.

[58] J.E. Richardson, M.F. Fay, Q.C.B. Cronk, D. Bowman, M.W. Chase, A phylogenetic analysis of Rhamnaceae using rbcL AND trnL-F plastid DNA sequences, Amer. J. Bot. 87 (9) (2000) 1309–1324.

[59] Y. Kimura, M. Kozawa, K. Baba, K. Hata, New constituents of roots of polygonum cuspidatum, Planta Med. 48 (7) (1983) 164–168.

[60] S.K. Bhattacharjee, Handbook of Aromatic Plants, Pointer Publications, Jaipur, India, 2000.

[61] A.R. Burnett, R.H. Thomson, Naturally occurring quinines. Part XIII. Anthraquinones and related naphthalenic compounds in *Galium* spp. and in *Asperula odorata*, J. Chem. Soc. C Org. (1968) 854–857.

[62] E. Leistner, Biosynthesis of Plant Quinones: The Biochemistry of Plants, vol. 7, Academic Press, London, 1981, pp. 403–423.

[63] H. Inouye, E. Leistner, Biosynthesis of Quinones: The Chemistry of Quinonoid Compounds, vol. 2, John Wiley & Sons, New York, 1988, pp. 1293–1349.

[64] A.J.J. Van den Berg, R.P. Labadie, Quinones: Methods in Plant Biochemistry, vol. 1, Academic Press, London, 1989, pp. 451–491.

[65] E. Leistner, Biosynthesis of Chorismate-Derived Quinones in Plant Cell Cultures: Primary and Secondary Metabolism of Plant Cell Cultures, Springer, Berlin, 1985, pp. 215–224.

[66] Anonymous, The wealth of India: raw materials, PID, CSIR, New Delhi, 1985.

[67] R.H. Thomson, Naturally Occurring Quinones IV, Chapman & Hall, London, 1996.

[68] R. Weinsma, R. Verpoorte, Progress in the Chemistry of Organic Natural Products, vol. 49, Springer, 1986, p. 79.

[69] E.M. Williams, Major Herbs of Ayurveda, Churchill Livingstone, Elsevier Science Ltd., 2002, pp. 257–260.

[70] R.N. Chopra, S.L. Nayar, I.C. Chopra, Glossary of Indian Medicinal Plants, CSIR, New Delhi, 1957, p. 215.

[71] O. Sarangowa, T. Kanazawa, M. Nishizawa, T. Myoda, C. Bai, T. Yamagishi, Flavonol glycosides in the petal of *Rosa* species as chemotaxonomic markers, Phytochemisty 107 (2014) 61–68.

[72] H.M. Chiang, Y.T. Lin, P.L. Hsiao, Y.H. Su, H.T. Tsao, K.C. Wen, Determination of marked components – aloin and aloe-emodin – in *Aloe vera* before and after hydrolysis, J. Food Drug Anal. 20 (2012) 646–652.

[73] J.W. Fairbairn, S. Simic, Vegetable purgatives containing anthracene derivatives: part XI, J. Pharm. Pharmacol. 12 (S1) (1960), 45T–51T.

[74] S.J. Kim, S. Kawaguchi, Y. Watanabe, Glucosinolates in vegetative tissues and seeds of twelve cultivars of vegetable turnip rape (*Brassica rapa* L.), Soil Sci. Plant Nutr. 49 (2003) 337–346.

[75] K.M. Valant-Vetschera, E. Wollenweber, Exudate flavonoid aglycones in the alpine species of *Achillea* sect. Ptarmica: chemosystematics of *A. moschata* and related species (Compositae–Anthemideae), Biochem. Syst. Ecol. 29 (2001) 149–159.

[76] D.S. Rycroft, Chemosystematics and the liverwort genus *Plagiochila*, J. Hattori Bot. Lab. 93 (2003) 331–342.

[77] E.E. Conn, The metabolism of a natural product: lessons learned from cyanogenic glycosides, Planta Med. 57 (1991) S1–S9.

[78] R.E. Miller, R. Jensen, I.E. Woodrow, Frequency of Cyanogenesis in tropical rainforests of far North Queensland, Australia, Ann. Bot. 97 (2006) 1017–1044.

[79] I.R. Redovnikovi, T. Gliveti, K. Delonga, J. Vorkapi-Fura, Glucosinolates and their potential role in the plant, Period. Biol. 110 (2008) 297–309.

[80] J. Barillari, R. Cervellati, M. Paolini, A. Tatibouët, P. Rollin, R. Iori, Isolation of 4-methylthio-3-butenyl glucosinolate from Raphanus sativus sprouts (kaiware daikon) and its redox properties, J. Agric. Food Chem. 53 (26) (2005) 9890–9896.

[81] N. Frank, M. Dubois, T. Goldmann, A. Tarres, E. Schuster, F. Robert, Semiquantitative analysis of 3-butenyl isothiocyanate to monitor an off-flavor in mustard seeds and glycosinolates screening for origin identification, J. Agric. Food Chem. 58 (6) (2010) 3700–3707.

[82] N. Verma, S. Shukla, Impact of various factors responsible for fluctuation in plant secondary metabolites, J. Appl. Res. Med. Aromat. Plants 49 (2015) 1–9.

[83] Y. Zhi-lin, D. Chuan-chao, C. Lian-qing, Regulation and accumulation of secondary metabolites in plant-fungus symbiotic system, Afr. J. Biotechnol. 6 (2007) 1266–1271.

[84] J. Radusiene, B. Karpaviciene, Z. Stanius, Effect of external and internal factors on secondary metabolites accumulation in St. John's wort, Bot. Lith. 18 (2012) 101–108.

[85] H.F. Tippmann, U. Schluter, D.B. Collinge, Common themes in biotic and abiotic stress signalling in plants, Floric. Ornam. Plant Biotechnol. 3 (2006) 52–67.

[86] Y.S.S. Lisar, R. Motafakkerazad, M.M. Hossain, I.M.M. Rahman, Water Stress Inplants: Causes, Effects and Responses, In Tech Publishers, Crotia, 2012, pp. 1–14.

[87] D. Aimar, M. Calafat, A.M. Andrade, L. Carassay, G.I. Abdala, M.L. Molas, Vasanthaiah, Drought Tolerance and Stress Hormones: From Model Organisms to Foragecrops, Plants and Environment, In Tech Publishers, Croatia, 2011, pp. 137–164.

[88] N. Azhar, B. Hussain, M.Y. Ashraf, K. Abbasi, Water stress mediated changes ingrowth: physiology and secondary metabolites of desi ajwain (*Trachyspermum ammi* L.), Pak. J. Bot. 43 (2011) 15–19.

[89] P. Valentovic, M. Luxova, L. Kolarovic, O. Gasparikova, Effect of osmotic stress on compatible solutes content, membrane stability and water relations in two maize cultivars, Plant Soil Environ. 52 (2006) 186–191.

[90] S.M.A. Zobayed, F. Afreen, T. Kozai, Phytochemical and physiological changes in the leaves of St. John's wort plants under a water stress condition, Environ. Exp. Bot. 59 (2007) 109–116.

[91] H.A.H. Said-Al Ahl, E.A. Omer, Medicinal and aromatic plants production under salt stress: a review, Herba. Pol. 57 (2011) 72–87.

[92] W.M. Abd EL-Azim, S.T.H. Ahmed, Effect of salinity and cutting date on growth and chemical constituents of *Achillea fragrantissima* Forssk, under Ras Sudr conditions, Res. J. Agric. Biol. Sci. 5 (2009) 1121–1129.

[93] J.K. Cik, B. Klejdus, J. Hedbavny, M. Backor, Salicylic acid alleviates NaCl-induced changes in the metabolism of *Matricaria chamomilla* plants, Ecotoxicol 18 (2009) 544–554.

[94] G.M. Jochum, K.W. Mudge, R.B. Thomas, Elevated temperatures increase leaf senescence and root secondary metabolite concentrations in the understory herb *Panax quinquefolius* (Araliaceae), Am. J. Bot. 94 (2007) 819–826.

[95] E.M. Castro, J.E.B.P. Pinto, S.K.V. Bertolucci, M.R. Malta, M.D.G. Cardoso, F.A.M. Silva, Coumarin contents in young *Mikania glomerata* plants (Guaco) under different radiation levels and photoperiod, Acta Farm. Bonaer. 25 (2006) 387–392.

Section II

Leads From Natural Products

Chapter 7

The Role of Natural Products From Plants in the Development of Anticancer Agents

Danielle Twilley, Namrita Lall
Department of Plant and Soil Sciences, University of Pretoria, Pretoria, South Africa

1. INTRODUCTION

Plants have been used for many years as a source of medicine for the treatment of various diseases. Early medicines from plants were used in the forms of powders, tinctures, teas, poultices, and various other herbal formulations. Plants are important sources for the discovery of new drugs due to their abundance of natural products. Natural products are small molecule secondary metabolites, such as terpenoids, flavonoids, and alkaloids, which are used for the survival of the plant [1]. These secondary metabolites can also be found in various microorganisms such as yeast and bacteria, as well as marine organisms [2,3].

One of the earliest records for using natural products as a source of medicine was found on clay tablets in cuneiform, a system of writing developed by Sumerians in Mesopotamia (2600 BC). The clay tablet depicted the use of Cypress and myrrh oils and more than 1000 plant-derived medicines. The ancient Egyptians were also well known for using plants as a source of medicine. Their pharmaceutical uses of plants were recorded in the Ebers Papyrus (2900 BC), in which more than 700 plant-based medicines were described by using them as ointments, infusions, mouthwashes, and pills [4,5].

It is interesting to note that of the available anticancer drugs currently on the market, 70% have been sourced from natural products or have been derived from plants [3]. It has been reported that approximately 65%–80% of the world's population in developing countries depends primarily on plants for their health care needs [6]. It is further estimated that 35,000–70,000 plant species around the world have been used for their medicinal value, which equates to 14%–28% of the 250,000 plant species found around the world [7].

Natural Products and Drug Discovery. https://doi.org/10.1016/B978-0-08-102081-4.00007-1

2. NATURAL PRODUCTS AND THEIR ANTICANCER ACTIVITY

In this chapter, the role of different natural products such as terpenoids, flavonoids, and alkaloids for their potential as anticancer agents is discussed. According to a prediction [8], the number of cancer cases is expected to increase by 70% from 14 million cancer cases in 2012 to 23.6 million cancer cases by 2030, so it is important to discover new anticancer drugs and plants provide a great source.

2.1 Terpenoids

Terpenoids make up the largest class of natural products, consisting of approximately 25,000 various chemical structures [9]. Terpenoids can also be referred to as terepenes or isoprenoid, as terpenoids are synthesized from two five-carbon building blocks knows as isoprenoids. The terpenoids further consist of several subclasses: monoterpenoids (C_{10}), diterpenoid (C_{20}), triterpenoids (C_{30}), tetraterpenoids (C_{40}), polytepenoids $(C_5)_n$, and sesquiterpenoids (C_{15}) [2,3].

2.1.1 Mechanism of Action of Terpenoids

Terpenoids act through a wide network by modulating various regulator factors, signaling pathways, enzymes, and proteins to inhibit the growth, proliferation, invasion, metastasis, and angiogenesis of different types of cancer cells (Table 7.1). Huang et al. [51] and Sobral et al. [52] provide a detailed review of various terpenoids, their anticancer activity, and their mechanism of action. Some of these mechanisms are summarized next.

- Apoptosis: Apoptosis can be induced through the extrinsic/death receptor pathway or through the intrinsic/mitochondrial pathway: upregulation of Fas expression; cleavage of caspase-3, -8, and -9 and poly(ADP) ribose polymerase (PARP); stimulation of p38 mitogen-activated protein kinase (MAPK); release of cytochrome *c*; downregulation of signal transduction proteins such as the antiapoptotic members B-cell lymphoma 2 (Bcl-2), Bcl-xL, Bcl-w, and Mcl-1; upregulation of proapoptotic members Bax, Bak, and surviving; activation of p53; and regulation of murine double minute 2 (*mdm2*).
- Cell cycle: Cell cycle arrest in G0/G1, G2/M, and S phases and regulation of cyclins A, B, D (1, 2, and 3), and E as well as cyclin-dependent kinases (cdk) 1, 2, and 4 and inhibition of small G protein p21 and p27 and E2F transcription factor 1 (*E2F1*).
- Generation of reactive oxygen species (ROS) such as H_2O_2.
- Decrease in mitochondrial membrane potential.
- Regulation of stress-induced transcription factor, nuclear factor (NK)-κB.

TABLE 7.1 Anticancer Activity of Various Classes of Terpenoids and Their Mechanism of Action

Terpenoid Compounds	Cell Lines[a]/ Mechanism	Anticancer Activity	References
Monoterpenoids			
Carvone	UACC-62	TGI[b] > 1664 μM	[10,11]
	MCF-7	TGI > 1664 μM/ IC_{50}[c] = 166.2 μM/ IC_{50} = 0.63 μM	
	NCI-ADR/RES	TGI > 1664 μM	
	786-O	TGI > 1664 μM	
	NCI-H460	TGI > 1664 μM	
	PC-3	TGI > 1664 μM	
	NIH:004FVCAR-3	TGI > 1664 μM	
	HT-29	TGI > 1664 μM/ IC_{50} = 454.1 μM	
	K-562	TGI > 1664 μM	
	LNCap	IC_{50} > 100 μM	[12]
	HL-60 (induces apoptosis; increased levels of H_2O_2; reduced MMP levels and activated caspase-8)	IC_{50} = 115.6 μM	[11]
	Hep-2	IC_{50} = 0.62 mM	[13]
	P-815	IC_{50} = 0.16 μM	[14]
	K-562	IC_{50} = 0.11 μM	
	CEM	IC_{50} = 0.17 μM	
	MCF-7/g	IC_{50} = 0.91 μM	
	CT26.WT	IC_{50} > 100 μg/mL	[15]
	A549	IC_{50} = 47.80 μg/mL	
	MDA MB-231	IC_{50} > 100 μg/mL	
	CaCo2	IC_{50} > 100 μg/mL	

Continued

TABLE 7.1 Anticancer Activity of Various Classes of Terpenoids and Their Mechanism of Action—cont'd

Terpenoid Compounds	Cell Lines[a]/ Mechanism	Anticancer Activity	References
Carveol	LNCaP	$IC_{50} = >100\ \mu M$	[12]
	P-815	$IC_{50} = 0.11\ \mu M$	[14]
	K-562 (cell cycle arrest in the S phase)	$IC_{50} = 0.11\ \mu M$	
	CEM	$IC_{50} = 0.13\ \mu M$	
	MCF-7	$IC_{50} = 0.26\ \mu M$	
	MCF-7/g	$IC_{50} = 0.45\ \mu M$	
	HCT-116	9% growth inhibition at 25 μg/mL	[16]
	OVCAR-8	3.61% growth inhibition at 25 μg/mL	
	SF-295	21.16% growth inhibition at 25 μg/mL	
Citral	HeLa	$IC_{50} = <0.1\ \mu g/mL$	[17]
	Ishikawa and ECC-1 (induces apoptosis by activating p53)	$IC_{50} = 15-25\ \mu M$ (2.3 −3.8 μg/mL)	[18]
	NB4 (induces apoptosis by activating caspase-3; downregulation of Bcl-2 and NF-κB; upregulation of Bax; reduction of mitochondrial membrane potential)	$IC_{50} = 3.995\ \mu g/mL$	[19]
	MCF-7 (induces apoptosis at low concentrations and necrosis at high concentrations; cell cycle arrest at G2/M phase; inhibition of cyclooxygenase 2)	$IC_{50} = 18 \times 10^{-5}$ M	[20]

TABLE 7.1 Anticancer Activity of Various Classes of Terpenoids and Their Mechanism of Action—cont'd

Terpenoid Compounds	Cell Lines[a]/ Mechanism	Anticancer Activity	References
	Jurkat, U937, and BS 24-1 (induces apoptosis by activating pro-caspase-3)	58%–90% cell death	[21]
Geraniol	HeLa	$IC_{50} = 131$ μg/mL	[17]
	Inhibition of MCF-7 cell growth seen after 7 days' incubation: induced apoptosis by arresting G0/G1 phase in cell cycle. Reduction in protein levels of cyclin D1, cdk4, cyclin E, and cyclin A.		[22]
	A549	$IC_{50} = 727.2$ μM/ $IC_{50} > 100$ μg/mL	[15,23]
	A549 bearing mice: reduction of membrane-bound Ras; elevated caspase-3 activity	Reduction in tumor weight at 25, 50, and 75 mmol/kg	
	CaCo2 (cell cycle arrest at the S phase; reduction of ornithine decarboxylase activity)	70% inhibition at 400 μM/ $IC_{50} > 100$ μg/mL	[15,24]
	PC-3 (induces apoptosis by increasing G1 phase arrest and caspase-3 activity; reduction in cyclins A, B, D, and E and cdk1, cdk4 and Bcl-2, Bcl-w; increased expression of p21, p27, and Bax)	$IC_{50} < 0.25$ mM	[25]
	Reduction in PC-3 nude mice tumor models at 60 and 300 mg/kg: induced apoptosis by increasing caspase-3 and reducing Ki-67; reduction in cyclins A, B, D, and E and cdk1, cdk4 and Bcl-2, Bcl-w; increased expression of p21, p27, and Bax.		
	CT26.WT	$IC_{50} > 100$ μg/mL	[15]
	MDA MB-231	$IC_{50} > 100$ μg/mL	

Continued

TABLE 7.1 Anticancer Activity of Various Classes of Terpenoids and Their Mechanism of Action—cont'd

Terpenoid Compounds	Cell Lines[a]/ Mechanism	Anticancer Activity	References
B-ionone	Inhibition of MCF-7 cell growth seen after 7 days incubation: Induced apoptosis by arresting G0/G1 phase in cell cycle.		[22]
	DU145 (induces apoptosis by arresting the cell cycle in the G1 and G2 phases; downregulation of cdk4 and cyclin D1)	$IC_{50} = 210\ \mu M$	[26]
	LNCaP (induces apoptosis)	$IC_{50} = 130\ \mu M$	
	PC-3 (induces apoptosis by inhibiting cell cycle at G1 phase)	$IC_{50} = 130\ \mu M$	
	MDA-MB-435 (regulates MAPK pathway)	$IC_{50} = 42\ \mu M$	[27]
	SGC-7901 (induces apoptosis)	$IC_{50} = 89\ \mu M$	[28]
Eugenol	HeLa	$IC_{50} = 86.6\ \mu g/mL$	[17]
	MDA-MB-231 (induces apoptosis; increases cleavage of PARP-1 and caspase-3; increased Bax levels; increased levels of cleaved caspase-9 and cytochrome *c*; downregulation of *E2F1*)	$IC_{50} = 1.7\ \mu M$	[29]
	T47-D	$IC_{50} = 0.9\ \mu M$	
	P815	$IC_{50} = 0.10\ \mu M$	[14]
	CEM	$IC_{50} = 0.09\ \mu M$	
	K-562	$IC_{50} = 0.24\ \mu M$	
	MCF-7 (induces apoptosis)	$IC_{50} = 0.41\ \mu M/1.5\ \mu M$	
	MCF-7/g	$IC_{50} = 0.87\ \mu M$	

TABLE 7.1 Anticancer Activity of Various Classes of Terpenoids and Their Mechanism of Action—cont'd

Terpenoid Compounds	Cell Lines[a]/ Mechanism	Anticancer Activity	References
Carvacrol	P815	$IC_{50} = 0.067\ \mu M$	[14,20]
	CEM	$IC_{50} = 0.042\ \mu M$	
	K-562 (cell cycle arrest in the S phase)	$IC_{50} = 0.067\ \mu M$	
	MCF-7	$IC_{50} = 0.125\ \mu M/23.6 \times 10^{-5}\ M$	
	MCF-7/g	$IC_{50} = 0.067\ \mu M$	
	Hep-2	$IC_{50} = 0.32\ mM$	[13]
	CaCo2 (induces apoptosis)	$IC_{50} = 343\ \mu M$	[30]
Menthol	MCF-7	$IC_{50} > 50 \times 10^{-5}\ M$	[20]
	T24 (decrease in mitochondrial membrane potential; decreases cell viability through the TRPM8 channel)	Cell viabilities of 87.9%, 76.64%, 67.35%, 56.89%, and 44.56% at concentrations of 10, 100, 500, 1000, and 2000 μM menthol, respectively	[31]
	SNU-5 (decrease in topoisomerase I, IIα, IIβ, and IIIβ levels and increase in NF-κB levels)	$IC_{50} = 1.62\ mg/mL$	[32]
Sesquiterpenoids			
Artemisinin	HepG2 (cell cycle arrest in G1 phase; increased expression of Rb; reduction in levels of Bcl-2 and increased levels of Bax, caspase-3, and PARP)	$IC_{50} = 13.98\ \mu M$	[33]
	Hep3B (cell cycle arrest in G1 phase)	$IC_{50} = 10.4\ \mu M$	
	Huh7	$IC_{50} = 8.9\ \mu M$	
	Bel-7404	$IC_{50} = 9.9\ \mu M$	

Continued

TABLE 7.1 Anticancer Activity of Various Classes of Terpenoids and Their Mechanism of Action—cont'd

Terpenoid Compounds	Cell Lines[a]/ Mechanism	Anticancer Activity	References
Zerumbone	MCF-7	$IC_{50} = 10\ \mu M$	[34]
	HT-29 (cell cycle arrest in G2 phase)	$IC_{50} = 10\ \mu M$	
	MKN-1	$IC_{50} = 10\ \mu M$	[35]
	MKN-28	$IC_{50} > 100\ nM$	
	MKN-45	$IC_{50} > 1\ \mu M$	
	MKN-74	$IC_{50} > 25\ \mu M$	
	NUGC-4	$IC_{50} > 1\ \mu M$	
	AGS (inhibition of vascular endothelial growth factor/B2M ratio and NF-κβ expression)	$IC_{50} > 10\ \mu M$	
	HeLa (induces apoptosis by increased expression of caspase-3)	$IC_{50} = 11.3\ \mu M$	[36]
	KB	$IC_{50} = 33.5\ \mu M$	[37]
	BC	$IC_{50} = 20.2\ \mu M$	
	NC1–H187	$IC_{50} = 11.1\ \mu M$	
Diterpenoids			
Carnosic acid	CaCo-2 (inhibited MMP-9 and MMP-2; downregulated COX2 expression)	$IC_{50} = 92.1\ \mu M$	[38,39]
	HT-29	$IC_{50} = 48.5\ \mu M/30.89\ \mu M$	
	LoVo	$IC_{50} = 26.4\ \mu M$	
	HL-60 (increase in caspase-3)	$IC_{50} = 5.7\ \mu M$	[40]
	HT-1080	$IC_{50} = 9.0\ \mu M$	
	HCT-116 (induced cleavage of caspase-9 and -3 and PARP; increased expression of p53 and Bax;	$IC_{50} = 28.95\ \mu M$	[39]

TABLE 7.1 Anticancer Activity of Various Classes of Terpenoids and Their Mechanism of Action—cont'd

Terpenoid Compounds	Cell Lines[a]/ Mechanism	Anticancer Activity	References
	inhibition of Bcl-2, Bcl-xL, and *mdm2* expression; increased ROS; inhibited phosphorylation of STAT3; attenuation of cyclins D1, D2, and D3 and survivin)		
	SW-480	$IC_{50} = 26.54\ \mu M$	
Carnosol	HL-60 (increase in caspase-3)	$IC_{50} = 5.3\ \mu M$	[40]
	HT-1080	$IC_{50} = 6.6\ \mu M$	
	MDA-MB-231 (cell cycle arrest in G2 phase; activation of caspase-8 and -9; depolarization of mitochondrial potential; decreased expression of p27 and Bcl-2; increased expression of p21, Bax, and γH2AX; accumulation of ROS)	$IC_{50} = 25\ \mu M$	[41]
Triterpenoids			
Lupeol	HeLa	$IC_{50} = 13.09\ \mu g/mL$	[42,43]
	MCF-7 (induced apoptosis by reducing the expression of Bcl-2 and Bcl-xL)	$IC_{50} = 25\ \mu g/mL/80\ \mu M$	
	HT-29	$IC_{50} > 30\ \mu g/mL$	
	WM-35	$IC_{50} = 32\ \mu M$	[44]
	451-Lu (induces apoptosis by increases caspase-3 and PARP cleavage; decreased	$IC_{50} = 38\ \mu M$	

Continued

TABLE 7.1 Anticancer Activity of Various Classes of Terpenoids and Their Mechanism of Action—cont'd

Terpenoid Compounds	Cell Lines[a]/ Mechanism	Anticancer Activity	References
	expression of Bcl-2 and increased expression of Bax protein; cell cycle arrest in G1 phase; decreased in cyclin D1 and D2 and cdk2)		
Ursolic acid	HepG2	$IC_{50} = 68.82\ \mu M$	[45]
	Bac-823	$IC_{50} = 66.38\ \mu M$	
	SH-SY5Y	$IC_{50} = 54.62\ \mu M$	
	HeLa	$IC_{50} = 33.12\ \mu M$	
	HCT-15 (cell cycle arrest in G0/G1 phase)	$IC_{50} = 30\ \mu M$	[46]
	HONE-1	$IC_{50} = 8.8\ \mu M$	[47]
	KB	$IC_{50} = 8.2\ \mu M$	
	HT-29	$IC_{50} = 4.7\ \mu M$	
	Jurkat (cell cycle arrest in the sub-G1 phase)	$IC_{50} = 10\ \mu M$	[48]
Tetraterpenoids			
Carotene	PC-3 (cell cycle arrest in sub-G1 phase; release of cytochrome *c*; activation of caspase-3 and -8)	$IC_{50} = 30\ \mu M$	[49]
	MCF-7	$IC_{50} = 14.58\ \mu g/mL$	[50]
	HepG2	$IC_{50} = 7.44\ \mu g/mL$	

[a]*Description of cell line abbreviations are given in Table 7.5.*
[b]*Total growth inhibition.*
[c]*50% inhibitory concentration.*

- In some cases, terpenoids can also increase autophagy.
- Regulation of various factors associated with inflammation such as tumor necrosis factor α and cyclooxygenase 2 (COX2).
- Inhibition of angiogenesis (growth of new blood vessels) and metastasis by regulating various factors such as urokinase plasminogen activator (u-PA), vascular endothelial growth factor (VEGF), and matrix metalloproteinase-9 (MMP-9).
- Inhibition of tumor metastasis, motility, and invasion by regulating various factors and pathways such as phosphatidylinositol 3-kinase/protein kinase B (PI3K/AKT) intracellular signaling pathway and Janus kinase/signal transducers and activators of transcription (JAK/STAT) pathway MMP-1, -2, and -7.
- Inhibition of enzymes resulting in the overwinding or underwinding of DNA: topoisomerases I, II, and III.
- Disruption of the actin cytoskeleton of cancer cells mainly through F-actin.
- Inhibition of AKT-associated gene products such as forkhead box class O proteins, glycogen synthase kinase 3 (GSK3), and I-κB kinase α (iKKα).
- Inhibition of tyrosine kinase signaling pathways involved in cell proliferation, differentiation, migration, metabolism, and programmed cell death.
- Inhibition of heat shock transcription factor 1, which plays an important role in cell proliferation, metastasis, invasion, and migration.
- Regulation of tissue inhibitor of matrix metalloproteinase proteins (TIMP1 and TIMP2).
- Decrease in expression of transient receptor potential melastatin subfamily member 8 involved in proliferation, survival, and invasion.

2.2 Flavonoids

Flavonoids consist of six main subclasses: flavonols, flavanones, anthocyanins, flavones, isoflavones, and flavan-3-ols (catechins). The basic chemical structure of a flavonoid (C6−C3−C6) consists of two phenolic rings that are attached through an oxygen-containing heterocyle. The type of subclass in which a flavonoid falls under depends on the saturation and opening of the central pyran ring. There are currently over 4000 various types of favonoids that have been identified [2,53,54].

2.2.1 Mechanism of Action of Flavonoids

There is a wide range of mechanisms by which flavonoids act to inhibit processes involved in carcinogenesis such as proliferation, inflammation, invasion, metastasis, and activation of apoptosis. Many of these flavonoids have been tested on various cancer cell lines and evaluated for activity against various molecular targets (Table 7.2). A detailed explanation of these

TABLE 7.2 Activity of Various Flavonoids Against Cancer Cell Lines and Their Mechanism of Action

Flavonoid Compounds	Cell Line[a]/Mechanism	Anticancer Activity	References
Flavonols			
Quercetin	LoVo (induces apoptosis by arresting the cell cycle in the S phase)	IC_{50}^{b} = 40.2 μM	[42,54a,55,56]
	MCF-7 (induces apoptosis by arresting the cell cycle in the S phase)	IC_{50} = 30.8 μM/31.04 μg/mL/19 μg/mL/25.04 μg/mL	
	HL-60 (induces apoptosis, decreases COX-2 and Bcl-2; increases Bax and caspase-3 cleavage)	IC_{50} = 61.11 μM	[56a]
	HeLa	IC_{50} = 5.78 μg/mL/11.93 μg/mL	[42,54a,56b]
	NCI-H-460	IC_{50} = 12.57 μg/mL	
	HepG2 (decreased expression of cyclin D1; induces G1 phase arrest)	IC_{50} = 20.58 μg/mL	
	CEM	IC_{50} = 55 μM	[57,57a]
	K562	IC_{50} = 40 μM/6 μM	
	Nalm6 (induced cell cycle arrest in the S phase followed by apoptosis; increase in p53 and Bax; reduction in Bcl-2 and Bcl-xL; release of cytochrome *c*; increase in activated caspase-3, cleaved caspase-9 and PARP1)	IC_{50} = 20 μM	
	T47D	IC_{50} = 160 μM	
	EAC	IC_{50} = 50 μM	

	K562/R	$IC_{50} = 7.8\ \mu M$	[57]
	GLC4	$IC_{50} = 6.0\ \mu M$	
	GLC4/R	$IC_{50} = 11.5\ \mu M$	
	BxPC-3	$IC_{50} = 43.4\ \mu M$	[58]
	PANC-1	$IC_{50} = 114.5\ \mu M$	
	Mgc-803	31.2% inhibition at 20 μM	[59]
	Bcap-37	16.1% inhibition at 20 μM	
	HT-29	$IC_{50} = 3.82\ \mu g/mL$	[42]
Kaempferol	Miapaca-2 (inhibited cell migration and invasion; decreased phosphorylation of EGFR, Src, AKT, and ERK1/2)	40% inhibition at 50 μM	[59a]
	Panc-1 (inhibited cell migration and invasion; decreased phosphorylation of EGFR, Src, AKT, and ERK1/2)	15% inhibition at 50 μM	
	SNU-123 (inhibited cell migration and invasion)	10% inhibition at 50 μM	
	HCT-15	$IC_{50} = 120\ \mu g/mL$	[59b]
	KB (induced cell cycle arrest in G2/M phase)	$IC_{50} = 50\ \mu M$	[59c]

Continued

TABLE 7.2 Activity of Various Flavonoids Against Cancer Cell Lines and Their Mechanism of Action—cont'd

Flavonoid Compounds	Cell Line[a]/Mechanism	Anticancer Activity	References
	GLC4	$IC_{50} = 6.5\ \mu M$	[57]
	GLC4/R	$IC_{50} = 16.8\ \mu M$	
	K562	$IC_{50} = 8.5\ \mu M$	
	K562/R	$IC_{50} = 7.0\ \mu M$	
	BxPC-3	$IC_{50} = 76.4\ \mu M$	[58]
	PANC-1	$IC_{50} = 193.1\ \mu M$	
Flavones			
Luteolin	BxPC-3	$IC_{50} = 14.3\ \mu M$	[58]
	PANC-1	$IC_{50} = 27.3\ \mu M$	
	HeLa	$IC_{50} = 15.41\ \mu M$	[60]
	U2OS	$IC_{50} = 36.35\ \mu M$	
	NCI-H460 (cell cycle arrest in S phase; increased Bax/Bcl-2 ratio; decreased expression of Sirt1)	53.48% inhibition at 40 μM	[61]
	Mgc-803	50.7% inhibition at 20 μM	[59]
	Bcap-37	28.8% inhibition at 20 μM	
	MCF-7	$IC_{50} = 14\ \mu M$	[56]

Apigenin	GLC4	$IC_{50} = 4.0\ \mu M$	[57]
	GLC4/R	$IC_{50} = 6.0\ \mu M$	
	K562	$IC_{50} = 9.0\ \mu M$	
	K562/R	$IC_{50} = 9.0\ \mu M$	
	T24 (induces G2/M phase cell cycle arrest; induces apoptosis by decreasing p-Akt, inhibits PI3K/PDK pathway, inhibits cleaved PARP and caspase-3; increase in Bax and Bad; decrease in Bcl-2 and Bcl-xL)	$IC_{50} = 43.8\ \mu M$	[62]
	HL60 (cell cycle arrest in G2/M phase; inhibits P13K/PKB pathway; induces caspase-dependent pathway)	$IC_{50} = 30\ \mu M$	[63]
	BxPC-3 (cell cycle arrest in G2/M phase, inhibition of GSK-3β/NF-κβ signaling pathway; decrease in cyclin B1, activation of mitochondrial pathway of apoptosis; upregulation of IL17F, IL17C, IL17A, and IFNB1)	$IC_{50} = 12.4\ \mu M$	[58]
	PANC-1 (cell cycle arrest in G2/M phase, inhibition of GSK-3β/NF-κβ signaling pathway; decrease in cyclin B1, activation of mitochondrial pathway of apoptosis)	$IC_{50} = 41.2\ \mu M$	

Continued

TABLE 7.2 Activity of Various Flavonoids Against Cancer Cell Lines and Their Mechanism of Action—cont'd

Flavonoid Compounds	Cell Line[a]/Mechanism	Anticancer Activity	References
Flavanones			
Naringenin	BxPC-3	IC_{50} = 292.9 μM	[58]
	PANC-1	IC_{50} = >300 μM	
	Mgc-803	4.9% inhibition at 20 μM	[59]
	Bcap-37	12% inhibition at 20 μM	
	A431 (increased reactive oxygen species generation in A431 cells; change in mitochondrial membrane potential; induces DNA fragmentation; cell cycle arrest in G0/G1 phase; increase in caspase-3 activity)	46.35% cell viability at 300 μM	[64]
Hesperidin	BxPC-3	IC_{50} = 258.2 μM	[58]
	PANC-1	IC_{50} = 268.7 μM	
	MSTO-211H (increase in sub-G1 population, suppression of specificity protein 1 [Sp1], p27, p21, cyclin D1, Mcl-1, and survivin; induced cleavage of Bid, caspase-3, and PARP; upregulation of Bax and downregulation of Bcl-xL).	IC_{50} = 152.3 μM	[65]
	HepG2 (increase in caspase-9, -8, and 3; reduction in mitochondrial transmembrane potential; downregulation of Bcl-xL; upregulation of Bax, Bak, and Bid)	IC_{50} = 150.43 μM	[66]
	HeLa	IC_{50} = 5.275 μM (3.22 μg/mL)	[67]

Flavanols			
Catechin	GLC4	$IC_{50} = 8.0\ \mu M$	[57]
	GLC4/R	$IC_{50} = 10\ \mu M$	
	K562	$IC_{50} = >100\ \mu M$	
	K562/R	$IC_{50} = 4.6\ \mu M$	
	T47D	$IC_{50} = 72.97\ \mu M$	[68]
	MCF-7	$IC_{50} = 128.86\ \mu M$	
Epicatechin	T47D	$IC_{50} = 946.6\ \mu M$	
	MCF-7	$IC_{50} = 135.97\ \mu M$	
Epigallocatechin	T47D	$IC_{50} = 103.14\ \mu M$	
	MCF-7	$IC_{50} = 1408.27\ \mu M$	
Isoflavones			
Genistein	BxPC-3	$IC_{50} = 187.7\ \mu M$	[58]
	PANC-1	$IC_{50} = 174.6\ \mu M$	
	HL-60	$IC_{50} = 60\ \mu M$	[69]
	MCF-7	$IC_{50} = 78\ \mu M$	
	Colo-205	40% inhibition at 100 μM	
	PC-3	28% inhibition at 100 μM	
	H-446 (cell cycle arrest in G2/M phase; inhibition of FoxM1, cyclin B1 and survivin expression)	$IC_{50} = 25\ \mu M$	[70]
	SK-BR-3	$IC_{50} = 138.13\ \mu M$	[71]

Continued

TABLE 7.2 Activity of Various Flavonoids Against Cancer Cell Lines and Their Mechanism of Action—cont'd

Flavonoid Compounds	Cell Line[a]/Mechanism	Anticancer Activity	References
Daidzein	SK-BR-3	$IC_{50} = 211.70\ \mu M$	[71]
	MCF-7 (decrease in mitochondrial membrane potential; increase in production of ROS; activation of caspase-9 and -7)	$IC_{50} = 100\ \mu M$	[72]
	BEL-7402 (increases DNA fragmentation and ROS production; decrease in mitochondrial membrane potential; cell cycle arrest in G2/M phase; decreased expression of Bcl-2, Bcl-x, and Bid; upregulation of caspase-7 and Bim)	$IC_{50} = 59.7\ \mu M$	[73]
	HeLa	$IC_{50} = 97.9\ \mu M$	
	A549	$IC_{50} > 100\ \mu M$	
	HepG2	$IC_{50} > 100\ \mu M$	
	MG-63	$IC_{50} > 100\ \mu M$	
Anthocyanidins			
Cyanidin	MCF-7	$IC_{50} = 47.18\ \mu M$	[74]
	LoVo	$IC_{50} = 46.9\ \mu M$	[75]
	LoVo/ADR	$IC_{50} = 26.2\ \mu M$	
	U937 (cell cycle arrest in G2/M phase)	$IC_{50} = 60\ \mu g/mL$	[76]

Delphinidin	MCF-7	$IC_{50} = 120\ \mu M$	[74]
	LoVo	$IC_{50} = 37.6\ \mu M$	[75]
	LoVo/ADR	$IC_{50} = 16.4\ \mu M$	
	A549	$IC_{50} = 55\ \mu M$	[77]
	NCI-H441 (reduced expression of VEGFR2, p85, and P110α of PI3K; reduced phosphorylation of AKT, ERK1/2, JNK1/2, p38; reduced expression of cyclin D1, PCNA protein expression, Bcl-2, Bcl-xL and Mcl-1; increased expression of Bak and Bax; activation of caspase-9 and -3; cleavage of PARP)	$IC_{50} = 58\ \mu M$	
	SK-MES-1 (Reduced expression of p85 and P110α of PI3K; reduced phosphorylation of EGFR, AKT, ERK1/2, JNK1/2, p38; reduced expression of cyclin D1, PCNA protein expression, Bcl-2, Bcl-xL, and Mcl-1; increased expression of Bak and Bax; activation of caspase-9 and -3; cleavage of PARP)	$IC_{50} = 44\ \mu M$	

[a] *Description of cell line abbreviations are given in Table 7.5.*
[b] *50% inhibitory concentration.*

mechanisms of action is described by Chahar et al. [78] and Ren et al. [79]. A summary of the mechanism of action of flavonoids is given next:

- Decrease phosphorylation of epidermal growth factor receptor, which is responsible for cell growth and proliferation.
- Inhibition of phase I metabolizing enzymes such as cytochrome P450 (CYP), which is responsible for activating procarcinogens.
- Inhibition of P450 isoenzymes CYP1A1 and CYP1A2, thereby having a protective role against DNA damage.
- Increase in DNA fragmentation.
- Increase in ROS generation.
- Inhibition of angiogenesis in cancer cells through factors such as VEGF.
- Regulation of mitochondrial membrane potential and NF-κβ.
- Inducing phase II metabolizing enzymes that detoxifie carcinogens. Examples include glutathione-*S*-transferase, quinine reductase, and UDP-glucuronyl transferase.
- Inhibition of cancer cell proliferation by inhibiting enzymes such as xanthine oxidase, COX2, lipo-oxygenases, and ornithine decarboxylase.
- Cell cycle arrest at G2/M and G1/S phases by regulating various cyclin-dependent kinases, cyclins, Ras dependent extracellular signal-regulated kinase (ERK) $^1/_2$, and small G proteins p21 and p27.
- Inducing apoptosis in cancer cells and not in normal cells. This is achieved through a number of ways such as inhibition of topoisomerase I/II and Mcl-1 protein, decrease in ROS, regulating heat shock proteins, down-regulation of NF-κβ, Bcl-2, Bcl-xL, survivin, and Mcl-1; activation of endonucleases, caspase-9, -3, and -7, cleavage of PARP, release of cytochrome *c*, and increase of Bax, Bak, Bid, and p53.
- Inhibition of signal transduction enzymes such as protein tyrosine kinase, protein kinase C, and phosphoinositide 3-kinase (PIP_3), which all play a role in the regulation of cell proliferation.
- Inhibition of tumor metastasis, motility, and invasion by regulating various factors and pathways such as PI3K and AKT and associated genes such as GSK3.
- Inhibition of regulator factors associated with cell proliferation such as FoxM1, which is part of the forkhead box (Fox) transcription factor class.
- Decrease phosphorylation of Src proto-oncogene associated with the development, progression, growth, and metastasis of various cancers.
- Decrease in the expression of silent information regulator 1, which is involved in progression, differentiation, and apoptosis.
- Regulation of various cytokines involved in cancer such as IL17F, IL17C, IL17A, and interferon β1.

2.3 Alkaloids

Alkaloids are some of the most important secondary metabolites from plants that are used in the treatment of cancer, especially steroidal alkaloids. The basic structure of an alkaloid consists of a ring structure that contains a nitrogen atom. The nitrogen atom can be located either in the ring structure or in the side chain of the molecule. The steroidal alkaloids, which play one of the biggest roles in the development of anticancer agents, can be divided into three subclasses, namely the pregnane and cyclopreganane alkaloids, which consist of a 21- and 24-carbon heterocyclic skeleton, respectively. The C-27 steroidal alkaloids are further separated into two groups known as the cholestane and C-nor-D-homosteroidal alkaloids [80,81]. Apart from the steroidal alkaloids, there are many other classes of alkaloids such as; isoquinoline, quinolone, pyrrolidine, indolizidine, piperidine and pyridine, organic amine, tropane, indole, diterpene, and cyclopeptide alkaloids.

2.3.1 Mechanism of Action of Alkaloids

Alkaloids act in many different ways to inhibit the proliferation of cancer cells (Table 7.3). One of the main mechanisms of vinca alkaloids (an important source of anticancer drugs) is their ability to interact with tubulin and to disrupt the function of the microtubule [93]. The mechanistic action of alkaloids as described by Jiang et al. [80] and Lu et al. [81] are as follows:

- Induces apoptosis by activating cell cycle arrest at G1 or G2/M phases, and various cell cycle–related factors such as p21, p27, phosphorylation of p53, death receptor 5 (DR-5), and ERK1/2.
- Inhibition of various enzymes such as *N*-acetyltransferase, COX2, and telomerase.
- Regulation of various cyclin-dependent kinase proteins, B-cell lymphoma 2 (Bcl-2) proteins such as survivin, Bax, Bcl-2, and Bcl-xL, and caspase-3, -8, and -9.
- Inducing the production of ROS in cancer cells.
- Ability to inhibit various regulatory factors of metastasis and angiogenesis such as Nf-κβ focal adhesion kinase, u-PA, MMP-2, -3, and -9, VEGF, and HIF-1.
- Inhibition of topoisomerase I, c-Fos, cAMP response element-binding, activated transcription factor 2, P-glycoprotein (P-gp), CYP3A4, mitogen-activated protein kinase phosphatase 1, Wnt/β-catenin signaling, and STAT-3.
- Decrease in mitochondrial membrane potential.
- Upregulation of tumor necrosis factor receptors (TNFRs) such as TNFR1 and 2, TNFR-1 associated death domain, Fas-associated death domain, Fas-receptor, and TIMP-1.

TABLE 7.3 Activity of Steroidal Alkaloids Against Cancer Cell Lines and Their Mechanism of Action

Alkaloid Compounds	Cell Line[a]/ Mechanism	Anticancer Activity	References
Pregnane Steroidal Alkaloids			
Sarcovagine D	HL-60	IC_{50}[b] = 2.87 μM/ 2.96 μM	[82,83]
	K562	IC_{50} = 3.53 μM	
	SGC7901	IC_{50} = 4.87 μM	
	SK-BR-3	IC_{50} = 2.25 μM/4.17 μM	
	PANC-1	IC_{50} = 2.70 μM/ 10.76 μM	
	SMMC-7721	IC_{50} = 16.69 μM	[83,84]
	A549	IC_{50} = 11.17 μM/ 5.35 μM	
	H1299	IC_{50} = 20.23 μM	[84]
	HCT-116	IC_{50} = 2.89 μM	
	HT-29	IC_{50} = 1.35 μM	
	U87	IC_{50} = 6.13 μM	
	U251	IC_{50} = 10.08 μM	
	MCF-7	IC_{50} = 16.24 μM	
	BT549	IC_{50} = 3.37 μM	
Sarcorucinine A1	HL-60	IC_{50} = 6.15 μM	[82]
	K562	IC_{50} = 5.00 μM	
	SGC7901	IC_{50} = 13.34 μM	
	SK-BR-3	IC_{50} = 3.59 μM	
	PANC-1	IC_{50} = 4.82 μM	
Cholestane Steroidal Alkaloids			
Tomatidine	MCF-7	IC_{50} = 7.17 μM	[85]
	A549 (inhibits expression of MMP-2 and MMP-3; reduces phosphorylation of Akt and ERK1/2;	IC_{50} > 50 μM	[86]

TABLE 7.3 Activity of Steroidal Alkaloids Against Cancer Cell Lines and Their Mechanism of Action—cont'd

Alkaloid Compounds	Cell Line[a]/ Mechanism	Anticancer Activity	References
	downregulation of NF-κβ)		
	HL-60 (loss of mitochondrial membrane potential; inhibition of survivin; induces release of apoptosis-inducing factor [AIF])	$IC_{50} = 1.92\ \mu M$	[87]
	K562 (loss of mitochondrial membrane potential; inhibition of survivin; induces release of AIF)	$IC_{50} = 1.51\ \mu M$	
Solamargine	Hep3B (cell cycle arrest in G2/M phase; increased expression of TNFR-I and –II)	$IC_{50} = 5\ \mu g/mL$	[88]
	WM115 (induces cellular necrosis by inducing lysosomal membrane permeabilization; release of cytochrome *c*; upregulation of TNFR1, Bcl-xL and Bcl-2; downregulation of hILP/XIAP, Apaf-1 and Baxl; cleavage of caspase-3)	$IC_{50} = 6\ \mu M$	[89]
	WM239-A (induces cellular necrosis by inducing lysosomal membrane permeabilization; release of	$IC_{50} = 6\ \mu M$	

Continued

TABLE 7.3 Activity of Steroidal Alkaloids Against Cancer Cell Lines and Their Mechanism of Action—cont'd

Alkaloid Compounds	Cell Line[a]/ Mechanism	Anticancer Activity	References
	cytochrome *c*; upregulation of TNFR1, Bcl-xL and Bcl-2; downregulation of hILP/XIAP, Apaf-1 and Baxl; cleavage of caspase-3)		
	B16F10	$IC_{50} = 10.15$ μg/mL	[90,91]
	HT-29	$IC_{50} = 9.88$ μg/mL	
	MCF-7	$IC_{50} = 18.23$ μg/mL/ 3.13 μg/mL	
	HeLa	$IC_{50} = 7.48$ μg/mL/ 2.97 μg/mL	
	HepG2	$IC_{50} = 4.58$ μg/mL/ 2.67 μg/mL	
	MO59 J	$IC_{50} = 9.59$ μg/mL	
	U343	$IC_{50} = 16.30$ μg/mL	
	U251	$IC_{50} = 8.09$ μg/mL	
	HCT116	$IC_{50} = 2.97$ μg/mL	[91]
	Hep2	$IC_{50} = 3.13$ μg/mL	
Solasonine	B16F10	$IC_{50} = 24.19$ μg/mL	[90,91]
	HT29	$IC_{50} = 22.67$ μg/mL	
	MCF7	$IC_{50} = 22.25$ μg/mL/ 8.46 μg/mL	
	HeLa	$IC_{50} = 16.04$ μg/mL/ 9.07 μg/mL	
	HepG2	$IC_{50} = 6.01$ μg/mL/ 4.34 μg/mL	
	MO59 J	$IC_{50} = 21.72$ μg/mL	
	U343	$IC_{50} = 23.09$ μg/mL	
	U251	$IC_{50} = 26.21$ μg/mL	
	HCT-116	$IC_{50} = 6.79$ μg/mL	[91]
	Hep2	$IC_{50} = 6.94$ μg/mL	

TABLE 7.3 Activity of Steroidal Alkaloids Against Cancer Cell Lines and Their Mechanism of Action—cont'd

Alkaloid Compounds	Cell Line[a]/ Mechanism	Anticancer Activity	References
C-nor-D Homosteroidal Alkaloids			
Verticine	LLC	$IC_{50} = 16.53$ μg/mL	[92]
	A2780	$IC_{50} = 48.41$ μg/mL	
	HepG2	$IC_{50} = 53.38$ μg/mL	
	A549	$IC_{50} = 97.58$ μg/mL	
Vericinone	LLC	$IC_{50} = 3.03$ μg/mL	[92]
	A2780	$IC_{50} = 17.03$ μg/mL	
	HepG2	$IC_{50} = 14.64$ μg/mL	
	A549	$IC_{50} = 47.39$ μg/mL	

[a]*Description of cell line abbreviations are given in Table 7.5.*
[b]*50% inhibitory concentration.*

- Release of cytochrome *c* and apoptosis inducing factor and regulation of apoptotic peptidase activating factor 1 (Apaf-1) and X-linked inhibitor of apoptosis protein, which are associated with apoptosis.
- Inhibition of tumor metastasis, motility, and invasion by regulating various factors and pathways such as AKT, GSK3, and p38 MAPK.

3. PLANT-DERIVED ANTICANCER DRUGS CURRENTLY IN USE AND IN CLINICAL TRIALS

Many plant-derived drugs that have been developed for the treatment of cancer by biotechnology and pharmaceutical companies [1] (Fig. 7.1). The U.S. National Cancer Institute (NCI) has also greatly contributed to the discovery of potential plant-derived anticancer agents by screening a large amount of plant extracts (approximately 114,000) for cytotoxicity against various types of cancer cell lines. This plant collection program was active from 1960 to 1982 and was reinitiated in 1986 [98]. A new project funded by the NCI titled "Novel Strategies for Plant-derived Anticancer Agents" was initiated, where from 2003, a total of 5886 plant accessions were collected from 288 different families and tested on a panel of human tumor cells [99]. There are also many plant-derived agents that are undergoing clinical trials for the treatment of various types of cancer (Table 7.4; Fig. 7.2).

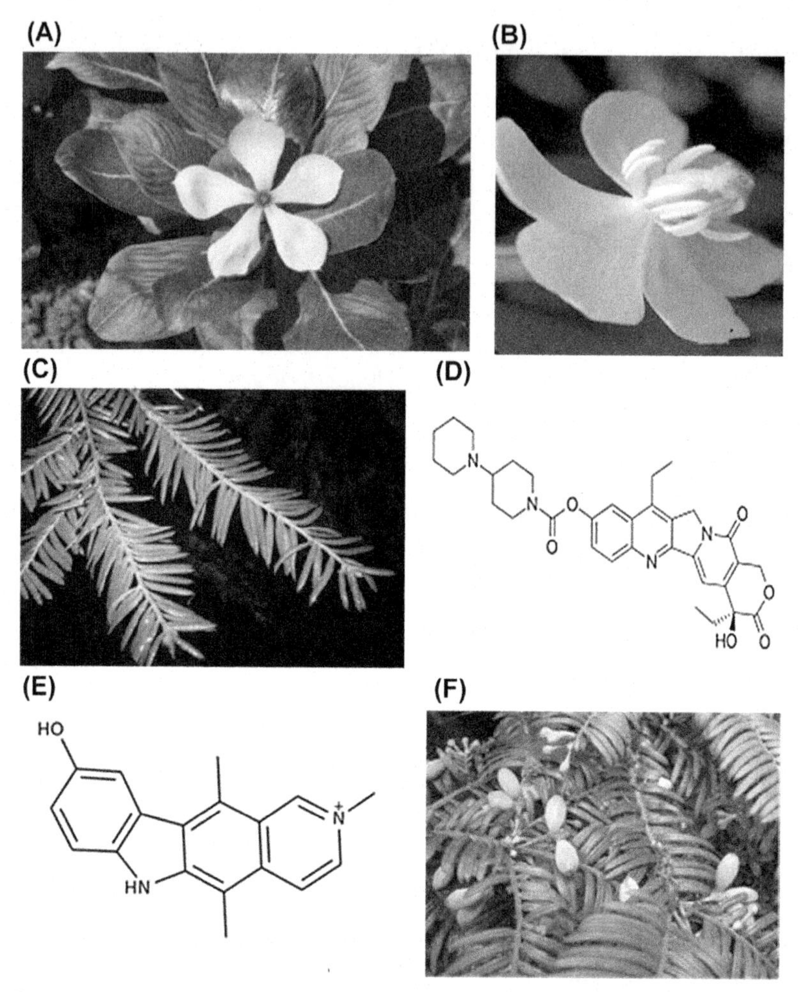

FIGURE 7.1 Plant-derived anticancer drugs currently in use. (A) *Catharanthus roseus* G. Don, commonly known as the Madagascar periwinkle, the source of vincristine and vinblastine [94]. (B) *Podophyllum peltatum* is one of the known sources of podophyllotoxin [95]. (C) Taxol was first isolated from *Taxus brevifolia* Nutt, commonly known as the Pacific Yew [96]. (D) Irinotecan isolated from *Camptotheca acuminata* is used in the treatment of colorectal cancer. (E) Elliptinium derived from ellipticine is used for the treatment of breast cancer. (F) *Cephalotaxus harringtonia* var. *drupacea* is the source of homoharringtonine, used for the treatment of leukemia [97].

3.1 Vincristine and Vinblastine

These two well-known vinca alkaloids were isolated from *Catharanthus roseus* G. Don (Apocynaceae), commonly known as the Madagascar periwinkle (Fig. 7.1A). During an in vivo experiment, it was noted that an extract of the Madagascar periwinkle was able to inhibit lymphocytic leukemia in

TABLE 7.4 Plant-Derived Natural Products Currently in Clinical Trials for Cancer Treatment

Compound	Class of Compound	Clinical Trial Phase	Use	References
Betulinic acid	Pentacyclic triterpenoid	I/II	An ointment for the treatment of dysplastic nevi	[100]
Combretastin A1 phosphate	stilbene	II	Anaplastic thyroid cancer/blood flow inhibition to tumor/head and neck cancer	[101]
Curcumin	Phenolic diarylheptanoid	II	Colorectal and pancreatic cancer	[102]
Flavopiridol	Alkaloid derivative	I/II	Gastric carcinoma	[103]
		II	Acute myelogenous leukemia	[104]
Ingenol mebutate	Diterpene ester	I/II	Basal cell carcinoma, squamous cell carcinoma and intraepidermal carcinoma	[105]
Meisoindigo	*Bis*-indole alkaloid derivative	II	Chronic myelogenous leukemia	[106]
Lycopene	Tetraterpenoid	II	Prevention and treatment of prostate cancer	[107]
Perillyl alcohol	Monoterpenoid	I/II	Breast and ovarian cancer and glioblastoma	[108]
Protopanaxadiol	Triterpenoid	I	Treatment of lung, gastric, breast, and pancreatic cancer alone and in combination with paclitaxel	[109]
Resveratrol	Phenolic	I	Prevention and treatment of colon cancer	[110]

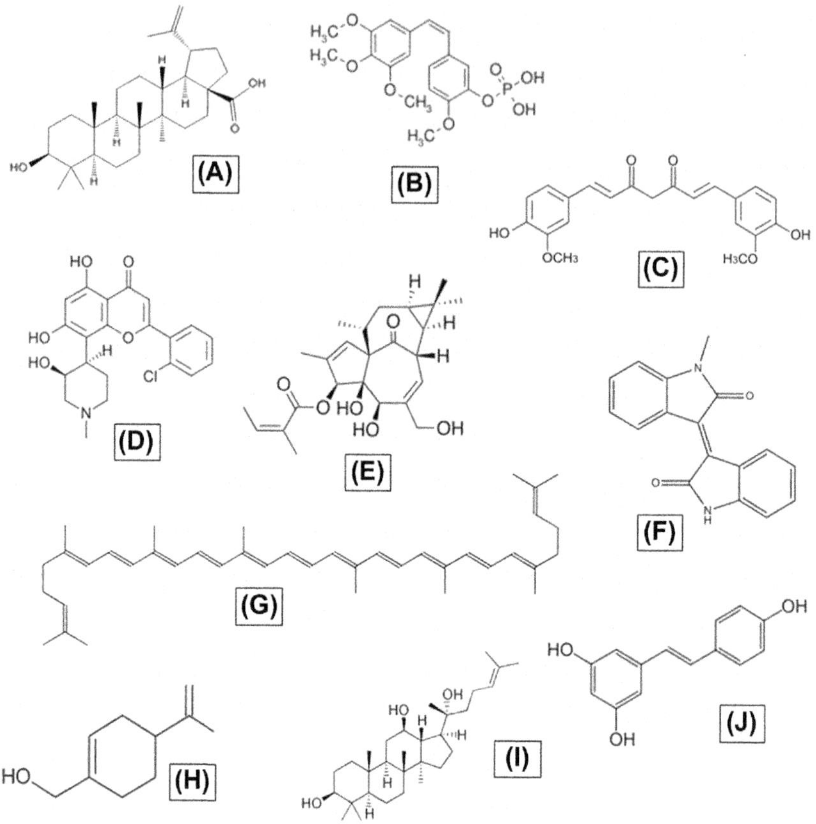

FIGURE 7.2 Plant-derived anticancer agents currently undergoing clinical trials: (A) betulinic acid, (B) combrestatin A4 phosphate, (C) curcumin, (D) flavopiridol, (E) ingenol mebutate, (F) meisoindigo, (G) lycopene, (H) perillyl alcohol, (i) protopanaxadiol, and (J) resveratrol.

mice, which led to the isolation and identification of vincristine and vinblastine. Analogs of vincristine and vinblastine were also synthesized and are known as vinorelbine and vindesine. These are used in combination with other cancer drugs for the treatment of leukemia, lymphoma, late-stage testicular cancer, breast and lung cancer, and Kaposi' sarcoma [111]. These vinca alkaloids act by binding to tubulin and therefore disrupt the assembly of microtubules in mitosis leading to tubulin depolymerization [112].

3.2 Etoposide and Teniposide

These anticancer agents have been derived from epipodophyllotoxin, which is an isomer of podophyllotoxin. Podophyllotoxin was isolated from the roots of *Podophyllum peltatum* Linnaeus and *Podophyllum emodi* Wallich, both from Podophyllaceae (Fig. 7.1B). It acts by binding to tubulin and therefore interferes with spindle formation during mitosis [113]. The semisynthetic derivatives, etoposide and teniposide, are used for the treatment of bronchial and testicular cancer as well as lymphomas [111]. These two derivatives act by inhibiting topoisomerase II, in turn arresting the cell cycle in the metaphase by stabilizing the covalent DNA—enzyme cleavable complex, and by inducing topoisomerase II—mediated DNA breakage [113].

3.3 Paclitaxel and Docetaxel

Paclitaxel is better known by its trade name Taxol, which was first isolated from the bark of *Taxus brevifolia* Nutt, also known by its common name as the Pacific Yew from the Taxaceae (Fig. 7.1C). This anticancer drug forms part of the taxane diterpenoids, which include the paclitaxel analog docetaxel, also known by its trade name as Taxotere. These two anticancer drugs have widespread use against cancer. Paclitaxel is used for the treatment of breast, ovarian, and non—small cell lung cancer and, in some cases, Kaposi sarcoma. It acts as a microtubule stabilizer and, therefore, interferes with cell division through disruption of the normal breakdown of microtubules. Docetaxel, on the other hand, is used for breast and non—small cell lung cancer [111,112,114].

3.4 Topotecan and Irinotecan

These two anticancer drugs, which are used for the treatment of ovarian and small cell lung cancers (Topotecan) and of colorectal cancer (irinotecan), were derived from camptothecin and form part of the camptothecin quinoline alkaloid derivatives (Fig. 7.1D). Camptothecin was first isolated from the Chinese ornamental tree *Camptotheca acuminata* Decne, from the Nyssaceae family. Camptothecin itself was not considered as a drug for cancer treatment as it led to severe bladder toxicity in clinical trials [111]. These anticancer agents inhibit DNA replication and transcription by arresting the cell cycle at the S phase and inhibiting topoisomerase I [115,116].

3.5 Elliptinium

Elliptinium is a derivative of ellipticine, which was isolated from species within the Apocynaceae family including *Bleekeria vitensis* (Fig. 7.1E). It is used in France for the treatment of breast cancer. It acts as a topoisomerase II inhibitor and intercalating agent. It causes DNA breakage by stabilizing topoisomerase II, causing DNA replication, RNA synthesis, and protein synthesis inhibition [111,117].

3.6 Homoharringtonine

Homoharringtonine, a cephalotaxine alkaloid ester, was isolated from the Chinese tree *Cephalotaxus harringtonia* var. *drupacea* (Sieb and Zucc) from the Cephalotaxaceae family (Fig. 7.1F). Homoharringtonine is used in combination with harringtonine for the treatment of acute myelogenous leukemia and chronic myelogenous leukemia in China [111]. These two structures differ only in that homoharringtonine has an additional methylene group in the ester side-chain. These compounds exhibit anticancer activity in inducing apoptosis and inhibiting protein synthesis at the ribosomal level [1].

4. CELL LINE ABBREVIATIONS

Various cancer cell lines, both human and murine, are mentioned in this chapter due to the cytotoxicity of various flavonoids, terpenoids, and alkaloids against these cell lines (Tables 7.1–7.3). Table 7.5 gives the abbreviations of the mentioned cell lines as well as a description of the type of cancer and its origin.

TABLE 7.5 List of Cell Lines Used in Tables 7.1–7.3

Cell Line Code	Cell Line Description	Cell Line Code	Cell Line Description
451-Lu	Human metastatic melanoma	MCF-7	Human breast cancer
786-O	Human kidney carcinoma	MCF-7/g	Human gemcitabine resistant breast adenocarcinoma
A2780	Human ovarian carcinoma	MDA-MB-231	Human mammary gland carcinoma
A431	Human epidermoid carcinoma	MDA-MB-435	Human breast carcinoma
A549	Non–small cell human pulmonary carcinoma	MG-63	Human osteosarcoma
AGS	Human gastric carcinoma	Mgc-803	Human gastric carcinoma
B16F10	Murine melanoma	Miapaca-2	Human pancreatic carcinoma
Bac 823	Human gastric carcinoma	MKN-1	Human gastric carcinoma
BC	Human breast carcinoma	MKN-28	Human gastric carcinoma
Bcap-37	Human breast carcinoma	MKN-45	Human gastric carcinoma

TABLE 7.5 List of Cell Lines Used in Tables 7.1–7.3—cont'd

Cell Line Code	Cell Line Description	Cell Line Code	Cell Line Description
Bel-7402	Human hepatocellular carcinoma	MKN-74	Human gastric carcinoma
Bel-7404	Human hepatocellular carcinoma	MO59 J	Human glioblastoma
BS 24-1	Human leukemia	MSTO-211H	Human mesothelioma carcinoma
BT549	Human breast carcinoma	Nalm-6	Human leukemia
BxPC-3	Human pancreatic carcinoma	NB-4	Human acute promyelocytic leukemia
BxPC-3	Human pancreatic carcinoma	NCI-ADR/RES	Human breast adenocarcinoma, multidrug-resistant phenotype
CaCo2	Human colorectal carcinoma	NCI-H-187	Human small cell lung carcinoma
CEM	Human acute T lymphoblastoid leukemia	NCI-H-441	Human non–small cell lung carcinoma
Colo-205	Human colorectal	NCI-H-460	Non–small cell lung carcinoma
CT26.WT	Mouse colon carcinoma	NIH-004-FVCAR-3	Human ovarian carcinoma
DU-145	Human prostate carcinoma	NUGC-4	Human gastric carcinoma
EAC	Human breast carcinoma	OVCAR-8	Human ovarian adenocarcinoma
ECC-1	Human endometrial carcinoma	P-815	Murine masocytoma
GCL4/R	Drug resistant human small cell lung cancer	PANC-1	Human pancreatic carcinoma
GLC-4	Human small cell lung cancer	PC-3	Human prostate carcinoma
H-1299	Human lung carcinoma	SF-295	Human glioblastoma
H-446	Human small cell lung carcinoma	SGC-7901	Metastatic human gastric carcinoma
HCT-116	Human colon carcinoma	SH-SY5Y	Human neuroblastoma

Continued

TABLE 7.5 List of Cell Lines Used in Tables 7.1–7.3—cont'd

Cell Line Code	Cell Line Description	Cell Line Code	Cell Line Description
HCT-15	Human colon carcinoma	SK-BR-3	Human breast carcinoma
HeLa	Human cervical epithelial adenocarcinoma	SK-MES-1	Human non–small cell lung carcinoma
Hep-2	Human larynx carcinoma	SMMC-7721	Human pancreatic carcinoma
Hep3B	Human hepatocellular carcinoma	SNU-123	Human pancreatic carcinoma
HepG2	Human hepatocellular carcinoma	SNU-5	Human gastric carcinoma
HL-60	Human leukemia	SW-480	Human colon carcinoma
HONE-1	Human nasopharyngeal carcinoma	T24	Human bladder carcinoma
HT-1080	Human fibrosarcoma	T47-D	Human breast carcinoma
HT-29	Human colorectal adenocarcinoma	U251	Human neuroblastoma
Huh-7	Human hepatocellular carcinoma	U251	Human glioblastoma
Ishikawa	Human endometrial carcinoma	U2OS	Human osteosarcoma
Jurkat	Human leukemia	U343	Human glioblastoma
K-562	Human chronic myeloid leukemia	U87	Human neuroblastoma
K-562/R	Human drug-resistant leukemia	U937	Human leukemia
KB	Human oral epidermoid carcinoma	UACC-62	Human melanoma
LLC	Lewis lung carcinoma	WM115	Human melanoma, vertical growth phase
LNCap	Human prostate carcinoma	WM239-A	Human metastatic melanoma
LoVo	Human metastatic colorectal carcinoma	WM35	Human primary melanoma
LoVo/ADR	Doxorubicin-resistant human metastatic colorectal cancer		

5. CONCLUSION

There has been a great amount of research done to determine the anticancer activity of various phytochemicals on cancer cell lines, and it is clear that natural products are the lead role players for the development of new anti-cancer drugs. In this chapter, many phytochemicals have been highlighted that have been tested for their anti-cancer activity and mechanism of action to inhibit carcinogenesis which include factors relating to proliferation, growth, metastasis, angiogenesis, and invasion. There is, however, still a need to prove the efficacy of these promising phytochemical in preclinical and clinical trials to determine their safety and efficacy and pharmacokinetic profiles.

REFERENCES

[1] L. Pan, H.B. Chai, A.D. Kinghorn, Discovery of new anticancer agents from higher plants, Front. Biosci. 4 (2013) 142–156.

[2] T.N. Chinembiri, L.H. du Plessis, M. Gerber, J.H. Hamman, J. du Plessis, Review of natural compounds for potential skin cancer treatment, Molecules 19 (2014) 11679–11721.

[3] M.B.J. Heinrich, S. Gibbons, E.M. Williamson, Fundamentals of Pharmacognosy and Phytotherapy, first ed., Churchill Livingstone, Edinburgh, UK, 2004.

[4] A.G. Atanasov, B. Waltenberger, E.M. Pferschy-Wenzig, T. Linder, C. Wawrosch, P. Uhrin, V. Temml, L. Wang, S. Schwaiger, E.H. Heiss, J.M. Rollinger, D. Schuster, J.M. Breuss, V. Bochkov, M.D. Mihovilovic, B. Kopp, R. Bauer, V.M. Dirsch, H. Stuppner, Discovery and resupply of pharmacologically active plant-derived natural products: a review, Biotechnol. Adv. 33 (2015) 1582–1614.

[5] D.A. Dias, S. Urban, U. Roessner, A historical overview of natural products in drug discovery, Metabolites 2 (2) (2012) 303–336.

[6] H. Tag, P. Kalita, P. Dwivedi, A.K. Das, N.D. Namsa, Herbal medicines used in the treatment of diabetes mellitus in Arunachal Himalaya, northeast, India, J. Ethnopharmacol. 141 (2012) 786–795.

[7] S. Padulosi, D. Leaman, P. Quek, Challenges and opportunities in enhancing the conservation and use of medicinal and aromatic plants, J Herbs Spices Med. Plants 9 (4) (2002) 243–267.

[8] WHO, Skin Cancers- How Common Is Skin Cancer?, 2015. Available at: http://www.who.int/uv/faq/skincancer/en/index1.html.

[9] J. Gershenzon, N. Dudareva, The function of terpene natural products in the natural worlds, Nat. Chem. Biol. 3 (2007) 408–414.

[10] J.L. Bicas, I.A. Neri-Numa, A.L.T.G. Ruiz, J.E. De Carvalho, G.M. Pastore, Evaluation of the antioxidant and antiproliferative potential of bioflavors, Food Chem. Toxicol. 49 (7) (2011) 1610–1615.

[11] Z. Yu, W. Wang, L. Xu, J. Dong, Y. Jing, D-Limonene and D-Carvone induce apoptosis in HL-60 cells through activation of caspase-8, Asian J. Traditional Med. 3 (4) (2008) 134–143.

[12] J. Chen, M. Le, Y. Jing, J. Dong, The synthesis of L-Carvone and limonene derivatives with increased antiproliferative effect and activation of ERK pathway in prostate cancer cells, Bioorg. Med. Chem. 14 (19) (2006) 6539–6547.

[13] A. Stammati, P. Bonsi, F. Zucco, R. Moezelaar, H.L. Alakomi, A. von Wright, Toxicity of selected plant volatiles in microbial and mammalian short-term assays, Food Chem. Toxicol. 37 (8) (1999) 813–823.

[14] A. Jaafari, M. Tilaoui, H.A. Mouse, L.A. M'bark, R. Aboufatima, A. Chait, M. Lepoivre, A. Zyad, Comparative study of the antitumor effect of natural monoterpenes: relationship to cell cycle analysis, Rev. Farmacogn. 22 (3) (2012).

[15] M. Gomide, F. Lemos, M.T.P. Lopes, T.M. Alves, L.F. Viccini, C.M. Coelho, The effect of the essential oils from five different *Lippia* species in the viability of tumor cell lines, Rev. Farmacogn. 23 (6) (2016) 895–902.

[16] L.N. Andrade, T.C. Lima, R.G. Amaral, C. Pessoa, M.O. Filho, B.M. Soares, L.G. do Nascimento, A.A. Carvalho, D.P. de Sousa, Evaluation of the cytotoxicity of structurally correlated p-methane derivatives, Molecules 20 (2015) 13264–13280.

[17] A.C. Mesa-Arango, J. Montiel-Ramos, B. Zapata, C. Durán, L. Betancur-Galvis, E. Stashenko, Citral and carvonechemotypes from the essential oils of Colombian *Lippiaalba* (Mill.) N.E. Brown: composition, cytotoxicity and antifungal activity, Memórias do Instituto Oswaldo Cruz 104 (6) (2009) 878–884.

[18] Y. Liu, R.J. Whelan, B.R. Pattnaik, K. Ludwig, E. Subudhi, H. Rowland, N. Claussen, N. Zucker, S. Uppal, D.M. Kushner, M. Felder, M.S. Patankar, A. Kapur, Terpenoids from *Zingiberofficinale* (Ginger) induce apoptosis in endometrial cancer cells through the activation of p53, PLoS One 7 (12) (2012) e53178.

[19] H. Xia, W. Liang, Q. Song, X. Chen, X. Chen, J. Hong, The *in vitro* study of apoptosis in NB4 cell induced by citral, Cytotechnology 65 (10) (2013) 49–57.

[20] W. Chaouki, D.Y. Leger, B. Liagre, J.L. Beneytout, M. Hmamouchi, Citral inhibits cell proliferation and induces apoptosis and cell cycle arrest in MCF-7 cells, Fundam. Clinic. Pharmacol. 23 (5) (2009) 549–556.

[21] N. Dudai, Y. Weinstein, M. Krup, T. Rabinski, R. Ofir, Citral is a new inducer of caspase-3 in tumor cell lines, Planta Medica 71 (5) (2005) 484–488.

[22] R.E. Duncan, D. Lau, A. El-Sohemy, M.C. Archer, Geraniol and beta-ionone inhibit proliferation, cell cycle progression. And cyclin-dependent kinase 2 activity in MCF-7 breast cancer cells independent of effects on HMG-CoA reductase activity, Biochem. Pharmacol. 68 (9) (2004) 1739–1747.

[23] M. Galle, R. Crespo, B.R. Kladniew, S.M. Villegas, M. Polo, M.G. de Bravo, Supression by geraniol of the growth of A549 human lung adenocarcinoma cells and inhibition of the mevalonate pathway in culture and *in vivo*: potential use in cancer chemotherapy, Nutr. Cancer 66 (5) (2014) 888–895.

[24] S. Carnesecchi, Y. Schneider, J. Caraline, B. Duranton, F. Gosse, N. Seiler, F. Raul, Geraniol. A component of plant essential oils, inhibits growth and polyamine biosynthesis in human colon cancer cells, J. Pharmacol. Exp. Ther. 298 (1) (2001) 197–200.

[25] S.H. Kim, H.C. Bae, E.J. Park, C.R. Lee, B.J. Kim, S. Lee, H.H. Park, S.J. Kim, I. So, T.W. Kim, J.H. Jeon, Geraniol inhibits prostate cancer growth by targeting cell cycle and apoptosis pathways, Biochem. Biophys. Res. Commun. 407 (1) (2011) 129–134.

[26] S. Jones, N.V. Fernandes, H. Yeganehjoo, R. Katuru, H. Qu, Z. Yu, H. Mo, B-ionone induces cell cycle arrest and apoptosis in human prostate tumor cells, Nutr. Cancer 65 (4) (2013) 600–610.

[27] J.R. Liu, Y.M. Yang, H.W. Dong, X.R. Sun, Effect of beta-ionone in human mammary cancer cells (Er-) by MAPK pathway, J. Hyg. Res. 34 (6) (2005) 706–709.

[28] J.R. Liu, B.Q. Chen, B.F. Yang, H.W. Dong, C.H. Sun, Q. Wang, G. Song, Y.Q. Sing, Apoptosis of human gastric adenocarcinoma cells induced by beta-ionone, World J. Gastroenterol. 10 (3) (2004) 348–351.

[29] I. Al-Sharif, A. Remmal, A. Aboussekhra, Eugenol triggers apoptosis in breast cancer cells through E2F1/survival down-regulation, BMC Cancer 13 (2013) 600.

[30] M.L.R. Cabello, D.G. Praena, S. Pichardo, F.J. Moreno, J.M. Bermúdez, S. Auceji, A.M. Cameán, Cytotoxicity and morphological effects induced by carvacrol and thymol on the human cell line caco-2, Food Chem. Toxicol. 64 (2014) 281–290.

[31] Q. Li, X. Wang, Z. Yang, B. Wang, S. Li, Menthol induces cell death via the TRPM8 channel in the human bladder cancer cell line T24, Oncology 77 (6) (2009) 335–341.

[32] J.P. Lin, H.F. Lu, J.H. Lee, J.G. Lin, T.C. Hsia, L.T. Wu, J.G. Chung, (-)-Menthol inhibits DNA topoisomerases I,II α and β, and promotes NF-îB expression in human gastric cancer SNU-5 cells, Anticancer Res. 25 (2005) 2069–2074.

[33] J. Hou, D. Wang, R. Zhang, H. Wang, Experimental therapy of hepatoma with artemisinin and its derivatives: *in vitro* and *in vivo* activity, chemosensitization, and mechanism of action, Cancer Ther. Preclin. 14 (17) (2008) 5519–5530.

[34] C. Kirana, G.H. McIntosh, I.R. Record, G.P. Jones, Antitumor activity of extract of *Zingiber aromaticum* and its bioactive sesquiterpenoid Zerumbone, Nutr. Cancer 45 (2) (2009) 210–225.

[35] K. Tsuboi, Y. Matsuo, T. Shamoto, T. Shibata, S. Koide, M. Morimoto, S. Guha, B. Sung, B.B. Aggarwal, H. Takahashi, H. Takeyama, Zerumbone inhibits tumor angiogenesis via NF-κβ in gastric cancer, Oncol. Rep. 31 (1) (2013) 57–64.

[36] A.B.H. Abdul, A.S. Al-Zubairi, N.D. Tailan, S.I.A. Wahab, Z.N.M. Zain, S. Ruslay, M.M. Syam, Anticancer activity of natural compound (zerumbone) extracted from *Zingiber zerumbet* in human HeLa cervical cancer cells, Int. J. Parmacol. 4 (3) (2008) 160–168.

[37] U. Sriphana, S. Pitchuanchom, P. Kongsaeree, C. Yenjai, Antimalarial activity and cytotoxicity of zerumbone derivatives, Sci. Asia 39 (2013) 95–99.

[38] M.V. Barni, M.J. Carlini, E.G. Cafferata, L. Puricelli, S. Moreno, Carnosic acid inhibits the proliferation and migration capacity of human colorectal cancer cells, Oncol. Rep. 27 (4) (2012) 1041–1048.

[39] D.H. Kim, K.W. Park, I.G. Chae, J. Kundu, E.H. Kim, J.K. Kundu, K.S. Chun, Carnosic acid inhibits STAT3 signaling and induces apoptosis through generation of ROS in human colon cancer HCT116 cells, Mol. Carcinog. 55 (2016) 1096–1110.

[40] A. López-Jiménez, M. García-Caballero, M.A. Medina, A.R. Quesada, Anti-angiogenic properties of carnosol and carnosic acid, two major dietry compounds from rosemary, Eur. J. Nutr. 52 (1) (2013) 85–95.

[41] Y.A. Dhaheri, S. Attoub, G. Ramadan, K. Arafat, K. Bajbouj, N. Karuvantevida, S. AbuQamar, A. Eid, R. Iratni, Carnosol induces ROS-mediated Veclin1-independent autophagy and apoptosis in triple negative breast cancer, PLoS One 9 (10) (2014) e109630.

[42] S. Ahmad, M.A. Sukari, N. Ismail, I.S. Ismail, A.B. Abdul, M.F.A. Bakar, N. Kifli, G.C.L. Ee, Phytochemicals from *Mangiferapajang* Kostern and their biological activities, BMC Complement. Altern. Med. 15 (2015) 83–91.

[43] D. Pitchai, A. Roy, C. Ignatius, *In vitro* evaluation of anticancer potentials of lupeol isolated from *Elephantopus scaber* L. on MCF-7 cell line, J. Adv. Pharm. Technol. Res. 5 (4) (2014) 179–184.

[44] M. Saleem, N. Maddodi, M.A. Zaid, N. Khan, B. bin Hafeez, M. Asim, Y. Suh, J.M. Yun, V. Setaluri, Mukhtar, Lupeol inhibits growth of highly aggressive human metastatic melanoma cells *in vitro* and *in vivo* by inducing apoptosis, Clin. Cancer Res. 14 (7) (2008) 2119–2127.

[45] J.W. Shao, Y.C. Dai, J.P. Xue, J.C. Wang, F.P. Lin, Y.H. Guo, *In vitro* and *in vivo* anticancer activity evaluation of ursolic acid derivatives, Eur. J. Med. Chem. 46 (7) (2011) 2652–2661.

[46] J. Li, W.J. Guo, Q.Y. Yang, Effects of ursolic acid and oleanolic acid on human colon carcinoma cell line HCT15, World J. Gastroenterol. 8 (3) (2002) 493–495.

[47] Y.M. Chjang, J.Y. Chang, C.C. Kuo, C.Y. Cang, Y.H. Kuo, Cytotoxic triterpenes from the aerial roots of *Ficus microcarpa*, Phytochemistry 66 (4) (2005) 495–501.

[48] R.K.I. Panucci, A. Mellitto, C.R. Oliveira, W. de Mello Marin, C. Bincoletto, *In vitro* study of anti-leukemic potential of ursolic acid in jurkat cell line, J. Clin. Exp. Oncol. 5 (3) (2016). https://doi.org/10.4172/2324-9110.1000161.

[49] K.R. Jayappriyan, R. Raikumar, V. Venkatakrishnan, S. Nagaraj, R. Rengasamy, *In vitro* anticancer activity of natural β-carotene from *Dunaliella salina* EU5891199 in PC-3 cells, Biomed. Prev. Nutr. 3 (2) (2013) 99–105.

[50] A.M. Badr, E.F. Shabana, H.H. Senousy, H.Y. Mohammad, Anti-inflammatory and anticancer effects of β-carotene, extracted from *Dunaliella bardawil* by milking, J Food Agric. Environ. 12 (3&4) (2014) 24–31.

[51] M. Huang, J.J. Lu, M.Q. Huang, J.L. Bao, X.P. Chen, Y.T. Wang, Teroenoids: natural products for cancer therapy, Expert Opin. Invest. Drugs 21 (12) (2012) 1801–1818.

[52] M.V. Sobral, A.L. Xavier, T.C. Lima, D.P. de Sousa, Antitumor activity of monoterpenes found in essential oils, Sci. World J. 2014 (2014), 953451.

[53] C. Busch, M. Burkard, C. Leischner, U.M. Lauer, J. Frank, S. Venturelli, Epigenetic activities of flavonoids in the prevention and treatment of cancer, Clin. Epigenet. 7 (2015) 64–82.

[54] P.C. Hollman, M.B. Katan, Dietary flavonoids: intake, health effects and bioavailability, Food Chem. Toxicol. 37 (9–10) (1999) 937–942;
[54a] H. Son, N. Anh, Phytochemical composition, in vitro antioxidant and anticancer activities of quercetin from methanol extract of Asparagus cochinchinensis (LOUR.) Merr. tuber, J. Med. Plants Res. 7 (46) (2013) 3360–3366.

[55] Z.R. Zhang, M.A. Zaharna, M.M.K. Wong, S.K. Chiu, H.Y. Cheung, Taxifolin enhances andrographolide-induced mitotic arrest and apoptosis in human prostate cancer cells via spindle assembly checkpoint activation, PLoS One 8 (1) (2013) e54577.

[56] A. Vijayalakshmi, K. Masilamani, E. Nagarajan, V. Ravichandiran, *In vitro* antioxidant and anticancer activity of flavonoids from *Cassia tora* linn. leaves against human breast carcinoma cell lines, Der. Pharma. Chem. 7 (9) (2015) 122–129;
[56a] G. Niu, S. Yin, S. Xie, Y. Li, D. Nie, L. Ma, X. Wang, Y. Wu, Quercetin induces apoptosis by activating caspase-3 and regulating Bcl-2 and cyclooxygenase-2 pathway in human HL-60 cells, Acta Biochim. Biophys. Sin 43 (1) (2011) 30–37;
[56b] J. Zhou, L. Li, L. Fang, H. Xie, W. Yao, X. Zhou, Z. Xiong, L. Wang, Z. Li, F. Luo, Quercetin reduces cyclin D1 activity and induces G1 phase arrest in HepG2 cells, Oncol. Lett. 12 (1) (2016) 516–522.

[57] M. Tungjai, W. Poompimon, C. Loetchutinat, S. Kothan, N. Dechsupa, S. Mankhetkorn, Spectrophotometric characterization of behavior and the predominant species of flavonoids in physiological buffer: determination of solubility, lipophilicity and anticancer efficacy, Open Drug Deliv. J. 2 (2008) 10–19;

[57a] S. Srivastava, R.R. Somasagara, M. Hegde, M. Nishana, S.K. Tadi, M. Srivastava, B. Choudhary, S.C. Raghavan, Quercetin, a natural flavonoid interacts with DNA, arrests cell cycle and causes tumor regression by activating mitochondrial pathway of apoptosis, Sci. Rep. 6 (2016) 24049.

[58] J.L. Johnson, E.G. de Mejia, Flavonoid apigenin modified gene expression associated with inflammation and cancer and induced apoptosis in human pancreatic cancer cells through inhibition of GSK-3/NF-B signaling cascade, Mol. Nutr. Food Res. 57 (2013) 2112–2127.

[59] J. Wu, W. Yi, L. Jin, D. Hu, B. Song, Antiproliferative and cell apoptosis-inducing activities of compounds from *Buddleja davidii* in Mgc-803 cells, Cell Div. 7 (2012) 20–31;

[59a] J. Lee, J.H. Kim, Kaempferol inhibits pancreatic cancer cell growth and migration through the blockade of EGFR-related pathway in vitro, PLoS One 11 (5) (2016) e0155264;

[59b] C. Kalyani, M.L. Narasu, Y. Devi, Synergistic growth inhibitory effect of flavonol-kaempferol and conventional chemotherapeutic drugs on cancer cells, Int. J. Pharm. Pharm. Sci. 9 (2) (2016) 123–127;

[59c] S. Sivapriyadharshini, P.R. Padma, Kaempferol exerts a differential effect on KB oral carcinoma cells and normal human buccal cells, Int. J. Pharm. Bio. Sci. 7 (3) (2016) 1244–1252.

[60] Y.S. Kim, S.H. Kim, J. Shin, A. Harikishore, J.K. Lim, Y. Jung, H.N. Lyu, N.I. Baek, K.Y. Choi, H.S. Yoon, K.T. Kim, Luteolin suppresses cancer cell proliferation by targeting vaccinia-related kinase 1, PLoS One 9 (10) (2014) e109655.

[61] L. Ma, H. Peng, K. Li, R. Zhao, L. Li, Y. Yu, X. Wang, Z. Han, Luteolin exerts an anticancer effect on NCI-H460 human non-small cell lung cancer cells through the induction of Sirt1-mediated apoptosis, Mol. Med. Rep. 12 (3) (2015) 4196–4202.

[62] Y. Zhu, Y. Mao, H. Chen, Y. Lin, Z. Hu, J. Wu, X. Xu, X. Xu, J. Qin, L. Xie, Apigenin promotes apoptosis, inhibits invasion and induces cell cycle arrest of T24 human bladder cancer cells, Cancer Cell Int. 13 (2013) 54–61.

[63] R.R. Ruela de Sousa, G.M. Fuhler, N. Blom, C.V. Ferreira, H. Aoyama, M.P. Pepperlenbosch, Cytotoxicity of apigenin on leukemia cell lines: implications for prevention and therapy, Cell Death Dis. 1 (2010) e19.

[64] M.S. Ahamad, S. Siddiqui, A. Jafri, S. Ahmad, M. Afzal, M. Arshad, Induction of apoptosis and antiproliferative activity of naringenin in human epidermoid carcinoma cell through ROS generation and cell cycle arrest, PLoS One 9 (10) (2014) e110003.

[65] K.A. Lee, S.H. Lee, Y.J. Lee, S.M. Baeg, J.H. Shim, Hesperidin induces apoptosis by inhibiting Sp1 and its regulators protein in MSTO-211H cells, Biomol. Ther. 20 (3) (2012) 273–279.

[66] R. Banjerdpongchai, B. Wudtiwai, P. Khaw-on, W. Rachakhom, N. Duangnil, Kongtawelert, Hesperidin from *Citrus* seed induces human hepatocellular carcinoma HepG2 cell apoptosis via both mitochondrial and death receptor pathways, Tumor Biol. 37 (1) (2016) 227–237.

[67] R. Bartoszewski, A. Hering, M. Marszall, J.S. Hajduk, S. Bartoszewska, N. Kapoor, K. Kochan, R. Ochocka, Mangiferin has an additive effect on the apoptotic properties of hesperidin in *Cyclopia* sp. tea extracts, PLoS One 9 (3) (2014) e92128.

[68] E. Evacuasiany, H. Ratnawati, L.K. Liana, W. Widowati, M. Maesaroh, T. Mozef, C. Risdian, Cytotoxic and antioxidant activities of catechin in inhibiting the malignancy of breast cancer, Oxid. Antioxidants Med. Sci. 3 (2) (2014) 141–146.

[69] K. Polkowski, J. Popiołkiewicz, P. Krzeczyński, J. Ramza, W. Pucko, O. Zegrocka-Stendel, J. Boryski, J.S. Skierski, A.P. Mazurek, G. Grynkiewicz, Cytostatic and cytotoxic activity of synthetic genistein glycosides against human cancer cell lines, Cancer Lett. 203 (1) (2004) 59–69.

[70] T. Tian, J. Li, B. Li, Y. Wang, M. Li, D. Ma, X. Wang, Genistein exhibits anti-cancer effects via down-regulating FoxM1 in H446 small-cell lung cancer cells, Tumor Biol. 35 (5) (2014) 4137–4145.

[71] E.J. Choi, G.H. Kim, Antiproliferative activity of daidzein and genistein may be related to ERα/c-erbB-2 expression in human breast cancer cells, Mol. Med. Rep. 7 (3) (2013) 781–784.

[72] S. Jin, Q.Y. Zhang, X.M. Kang, J.X. Wang, W.H. Zhao, Daidzein induces MCF-7 breast cancer cell apoptosis via the mitochondrial pathway, Ann. Oncol. 21 (2) (2010) 263–268.

[73] B.J. Han, W. Li, G.B. Jiang, S.H. Lai, C. Zhang, C.C. Zeng, Y.J. Liu, Effects of daidzein in regards to cytotoxicity *in vitro*, apoptosis, reactive oxygen species level, cell cycle arrest and the expression of caspase and Bcl-2 family proteins, Oncol. Rep. 34 (3) (2015) 1115–1120.

[74] J. Tang, E. Oroudjev, L. Wilson, G. Ayoub, Delphinidin and cyaniding exhibit anti-proliferative and apoptotic effects in MCF7 human breast cancer cells, Integr. Cancer Sci. Ther. 2 (1) (2015) 82–86.

[75] J. Cyorovic, F. Tramer, M. Granzotto, L. Candussio, G. Decorti, S. Passamonti, Oxidative stress-based cytotoxicity of delphinidin and cyaniding in colon cancer cells, Arch. Biochem. Biophys. 501 (1) (2010) 151–157.

[76] J.W. Hyun, H.S. Chung, Cyanidin and Malvidin from *Ozyra sativa* cv. Heugjinjubyeo mediate cytotoxicity against human monocytic leukemia cells by arrest of G2/M phase and induction of apoptosis, J. Agric. Food Chem. 52 (8) (2004) 2213–2217.

[77] H.C. Pal, S. Sharma, L.R. Strickland, J. Agarwal, M. Athar, C.A. Elmets, F. Afaq, Delphinidin reduces cell proliferation and induces apoptosis of non-small-cell lung cancer cells by targeting EGFR/VEGFR2 signaling pathways, PLoS One 8 (10) (2013) e77270.

[78] M.K. Chahar, N. Sharma, M.P. Dobhal, Y.C. Joshi, Flavonoids: a versatile source of anticancer drugs, Pharmacogn. Rev. 5 (9) (2011) 1–12.

[79] W. Ren, Z. Qiao, H. Wang, L. Zhu, L. Zhang, Flavonoids: promising anticancer agents, Med. Res. Rev. 23 (4) (2003) 519–534.

[80] Q.W. Jiang, M.W. Chen, K.J. Cheng, P.Z. Yu, X. Wei, Z. Shi, Therapeutic potential of steroidal alkaloids in cancer and other diseases, Med. Res. Rev. 36 (1) (2016) 119–143.

[81] J.J. Lu, J.L. Bao, X.P. Chen, M. Huang, Y.T. Wang, Alkaloids isolated from natural herbs as the anticancer agents, Evidence Based Complement. Altern. Med. 2012 (2012), 485042.

[82] Y.X. Yan, Y. Sun, J.C. Chen, Y.Y. Wang, Y. Li, M.H. Qiu, Cytotxic steroids from *Sarcococca saligna*, Planta Medica 77 (15) (2011) 1725–1729.

[83] Y. Sun, Y.X. Yan, J.C. Chen, L. Lu, X.M. Zhang, Y. Li, M.H. Qiu, Pregnane alkaloids from *Pachysandra axillaris*, Steroids 75 (12) (2010) 818–824.

[84] P. Zhang, L. Shao, Z. Shi, Y. Zhang, J. Du, K. Cheng, P. Yu, Pregnane alkaloids from *Sarcococca ruscifolia* and their cytotoxic activity, Phytochem. Lett. 14 (2015) 31–34.

[85] L. Sucha, M. Hroch, M. Rezacova, E. Rudolf, R. Havelek, L. Sispera, J. Cmielova, R. Kohlerova, A. Bezrouk, P. Tomsik, The cytotoxic effects of α-tomatine in MCF-7 human adenocarcinoma breast cancer cells depends on its interaction with cholesterol in incubation media and does not involve apoptosis induction, Oncol. Rep. 30 (6) (2013) 2593–2602.

[86] K.H. Yan, L.M. Lee, S.H. Yan, H.C. Huang, C.C. Li, H.T. Lin, P.S. Chen, Tomatidine inhibits invasion of human lung adenocarcinoma cell A549 by reducing matrix metalloproteinases expression, Chem. Biol. Interact. 203 (3) (2013) 580–587.

[87] M.W. Chao, C.H. Chen, Y.L. Chang, C.M. Teng, S.L. Pan, α-Tomatine-mediated anticancer activity *in vitro* and *in vivo* through cell cycle and caspase-independent pathways, PLoS One 7 (9) (2012) e44093.

[88] K.W. Kuo, S.H. Hsu, Y.P. Li, W.L. Lin, L.F. Liu, L.C. Chang, C.C. Lin, C.N. Lin, H.M. Sheu, Anticancer activity evaluation of the *Solanum* glcoalkaloid solamargine: triggering apoptosis in human hepatoma cells, Biochem. Pharmacol. 60 (12) (2000) 1865–1873.

[89] S.S.A. Sinani, E.A. Eltayeb, B.L. Coomber, S.A. Adham, Solamargine triggers cellular necrosis selectively in different types of human melanoma cancer cells through extrinsic lysosomal mitochondrial death pathway, Cancer Cell Int. 16 (2016) 11–23.

[90] C.C. Munari, P.F. de Oliveira, J.C.L. Campos, S. de Paula Lima Martina, J.C. da Costa, J.K. Bastos, D.C. Tavares, Antiproliferative activity of *Solanum lycocarpum* alkaloidic extract and their consituents, solamargine and solasonine, in tumor cell lines, J. Nat. Med. 68 (1) (2014) 236–241.

[91] M.M. Shabana, M.M. Salama, S.M. Ezzat, L.R. Ismail, *In vitro* and *in vivo* anticancer activity of the fruit peels of *Solanum melongena* L. against hepatocellular carcinoma, J. Carcinog. Mutagen. 4 (2013) 3–9.

[92] D. Wang, Y. Jiang, K. Wu, S. Wang, Y. Wang, Evaluation of antitumor property of extracts and steroidal alkaloids from the cultivated Bulbus *Fritillariae ussuriensis* and preliminary investigation of its mechanism of action, BMC Complement. Altern. Med. 15 (2015) 29–40.

[93] M. Moudi, R. Go, C.Y.S. Yien, M. Nazre, Vinca alkaloids, Int. J. Prev. Med. 4 (11) (2013) 1231–1235.

[94] B. Gangulyb, *Catharanthus roseus* 6576, 2010. Available at: https://commons.m.wikimedia.org/wiki/File:Catharanthus_roseus_6576.JPG.

[95] T. Barnes, White Flowered Mayapple Plant Flower *Podophyllum peltatum*, 2013. Available at: https://commons.m.wikimedia.org/wiki/File:White_flowered_mayapple_plant_flower_podophyllum_peltatum.jpg.

[96] W. Siegmund, Taxus brevifolia, 2008. Available at: https://commons.m.wikimedia.org/wiki/File:PacificYew_8538.jpg.

[97] A. Barra, *Cephalotaxus harringtonia* Var. Drupacea, 2008. Available at: https://commons.m.wikimedia.org/wiki/File:Cephalotaxus_harringtonia_var_drupacea.JPG.

[98] M. Shoeb, Anticancer agents from medicinal plants, Bangladesh J. Pharmacol. 1 (2007) 35–41.

[99] M.J. Balunas, A.D. Kinghorn, Drug discovery from medicinal plants, Life Sci. 78 (2005) 431–441.

[100] US National Health Institute, Evaluation of 20% Betulinic Acid Ointment for Treatment of Dysplastic Nevi (Moderate to Severe Dysplasia), 2006. Available at: https://clinicaltrials.gov/ct2/show/NCT00346502.

[101] US National Health Institute, Combretastatin A4 Phosphate in Treating Patients with Advanced Anaplastic Thyroid Cancer, 2003. Available at: https://clinicaltrials.gov/ct/show/NCT00060242?order=2.

[102] N. Dhillon, B.B. Aggarwal, R.A. Newman, R.A. Wolff, A.B. Kunnumakkara, J.L. Abbruzzese, C.S. Ng, V. Badmaey, R. Kurzrock, Phase II clinical trials of curcumin in patients with advanced pancreatic cancer, Cancer Ther. Clin. 14 (14) (2008) 4491–4499.

[103] G.K. Schwartz, D. Ilson, L. Saltz, E. O'Reilly, W. Tong, P. Maslak, J. Werner, P. Perkins, M. Stoltz, D. Kelsen, Phase II study of the cyclin-dependent kinase inhibitor flavopiridol administered to patients with advanced gastric carcinoma, J. Clin. Oncol. 19 (7) (2001) 1985–1992.

[104] J.E. Karp, A. Blackford, D. Douglas Smith, K. Alino, A.H. Seung, J. Bolanos-Meade, J.M. Greer, H.E. Carraway, S.D. Gore, R.J. Jones, M.J. Levis, M.A. McDevitt, L. Austin Doyle, J.J. Wright, Clinical activity of sequential flavopiridol, cytosine arabinoside, and mitoxantrone for adults with newly diagnosed, poor-risk acute myelogenous leukemia, Leukemia Res. 34 (2010) 877–882.

[105] J.R. Ramsay, A. Suhrbier, J.H. Aylward, S. Ogbourne, S.J. Cozzi, M.G. Poulsen, K.C. Baumann, P. Welburn, G.L. Redlich, P.G. Parsons, The sap from *Euphorbia peplus* is effective against human nonmelanoma skin cancers, Br. J. Dermatol. 164 (3) (2011) 633–636.

[106] Cooperative Study Group of Phase III Clinical Trials on Meisoindigo, Phase II clinical trial on meisoindigo in the treatment of chronic myelogenous leukemia, Zhonghua Xueyexue Zazhi 18 (1997) 69–72.

[107] US National Health Institute, Lycopene in Treating Pateints with Metastatic Prostate Cancer, 2003. Available at: https://clinicaltrials.gov/ct2/show/record/NCT00068731.

[108] J. da Gama Fischer, P.C. Carvalho, A.G. da Costa Neves-Ferreira, C.O. da Fonseca, J. Perales, M. da Costa Carvalho, G.B. Domont, Anti-thrombin as a prognostic biomarker candidate for patients with recurrent glioblastoma multiform under treatment with perillyl alcohol, J. Exp. Ther. Oncol 7 (2009) 285–290.

[109] X. Ouyang, Z. Yu, Z. Chen, F. Xie, W. Fang, Y. Peng, X. Chen, W. Chen, W. Wang, P. Qi, W. Jia, A pilot study of safety and efficacy of pandimex with or without paclitaxel in the treatment of advances solid tumors, J. Clin. Oncol. 23 (2005) 3188.

[110] US National Health Institute, Resveratrol for Patients with Colon Cancer, 2005. Available at: https://clinicaltrials.gov/ct2/show/NCT00256334.

[111] G.M. Cragg, D.J. Newman, Plants as a source of anti-cancer agents, J. Ethnopharmacol. 100 (2005) 72–79.

[112] M. Jordan, L. Wilson, Microtubules as a target for anticancer drugs, Nat. Rev. 4 (2004) 253–265.

[113] K.R. Hande, Etoposide: four decades of development of a topoisomerase II inhibitor, Eur. J. Cancer 34 (1998) 1514–1521.

[114] D.G.I. Kingston, Tubulin-interactive natural products as anticancer agents, J. Nat. Prod. 72 (2009) 507–515.

[115] G.M. Cragg, D.J. Newman, A tale of two tumor targets: topoisomerase I and tubulin. The Wall and Wani contribution to cancer chemotherapy, J. Nat. Prod. 67 (2004) 232–244.

[116] Y.H. Hsiang, M.G. Lihou, L.F. Liu, Arrest of replication forks by drug-stabilized topoisomerase I-DNA cleavable complex as a mechanism of killing by camptothechin, Cancer Res. 49 (1989) 5077–5082.

[117] Y. Zhu, Y. Mao, H. Chen, Y. Lin, Z. Hu, J. Wu, X. Xu, X. Xu, J. Qin, L. Xie, Apigenin promotes apoptosis, inhibits invasion and induces cell cycle arrest of T24 human bladder cancer cells, Cancer Cell Int. 13 (2013) 54–61.

Chapter 8

Plant Drugs in the Treatment of Osteoporosis

Sudhir Kumar, Rakesh Maurya
Medicinal and Process Chemistry Division, Central Drug Research Institute, Lucknow, India

1. INTRODUCTION

Bone is a dynamic connective tissue that is continuously undergoing formation and resorption by opposing actions of osteoblasts that form bone and osteoclasts that resorb bone to maintain calcium homeostasis and a healthy bone state. The imbalance between bone formation and bone resorption leads to the development of fragile bones characterized by loss of bone mass and bone microarchitecture, with associated risk of bone fracture [1]. Symptoms of the disease remain silent unless a fracture occurs during a minor fall, hence the disease is also known as the "silent disease" [2]. Severe loss of bone mass, deteriorated bone microarchitectural parameters, and subsequent progression of the disease increase the risk of complications associated with osteoporosis. Primary osteoporosis mainly affects women after the menopause, hence it is called postmenopausal osteoporosis, when the amount of estrogen in the body greatly decreases. Postmenopausal decrease in estrogen levels leads to a higher rate of bone resorption than bone formation and results in a rapid depletion of calcium from the skeleton. Secondary osteoporosis is associated with aging, calcium deficiency, and diminished bone formation due to increased parathormone activity, endocrine disease (hyperparathyroidism, hyperthyroidism, hypogonadism, hyperprolactinemia, and diabetes mellitus), use of certain drugs, and age-related reduction in vitamin D synthesis or resistance to vitamin D activity. Secondary osteoporosis occurs with age in everyone to some degree, when the process of resorption and formation of bone are no longer coordinated and bone breakdown overcomes bone building. The prevalence of osteoporosis in women is twice as frequent as in men [3].

Since the disease is asymptomatic, prevention of postmenopausal reabsorption and promotion of bone formation are critical aspects for management of the disease. Vitamin D and calcium are integral parts of bone health and are common preventive measures. Use of calcitonin to modulate serum levels of calcium and

Natural Products and Drug Discovery. https://doi.org/10.1016/B978-0-08-102081-4.00008-3

phosphorus, hormone reposition, bisphosphonates, estrogen replacement therapies, and selective estrogen receptor (ER) modulators are current strategies used for both the prophylaxis and treatment of osteoporosis [4–6].

Plant-based therapies are safe alternatives and the need of the hour due to side effects and cost factors associated with the current treatment of osteoporosis; in this context, several plant extracts have been evaluated for their possible role in the treatment of osteoporosis. Some of them have resulted in isolation of interesting bioactive molecules, estrogenic enough to induce bone formation along with promising prevention of osteoclast formation and/or differentiation. Some of the dietary plants have been investigated to be effective nutraceuticals for optimal bone health. This chapter provides an account of plant formulations and plant-based natural compounds for the prevention and treatment of osteoporosis.

2. GLOBAL BURDEN OF OSTEOPOROSIS

Osteoporosis is the leading cause of mortality and morbidity in both sexes and all races. The prevalence of osteoporosis is expected to significantly increase with increased loss of bone mass associated with worldwide population growth, an aging population, and longer life expectancy [7]. Twenty-five to fifty percent of older adults aged more than 85 years are estimated to have weak bones susceptible to fracture [8]. The annual incidence of osteoporotic hip fracture, forearm fracture, vertebral fracture, and fracture at other sites is more than the combined annual incidence of heart attacks, stroke, and new breast cancer. The National Osteoporosis Foundation estimated that the prevalence of osteopenia and osteoporosis will increase by 61 million by 2020 in the United States, of which almost 14 million individuals older than 50 years are likely to develop osteoporosis and another 47 million are expected to have osteopenia [9]. Osteoporosis is the most common underlying worldwide cause of more than 8.9 million fractures annually [10]. Hip fracture is the leading cause of morbidity and mortality. The annual incidence of osteoporotic hip fracture is estimated to be as large as 6.26 million by 2050 [9]. Globally, twice the number of women are affected than men [11]. The incidence of osteoporotic fracture is more probable in older age. Worldwide, one in three women over age 50 and one in five men over age 50 are susceptible to osteoporotic fracture [12].

3. MARKERS OF BONE METABOLISM

Bone tissue is continuously maintained during the lifespan of an individual. The dynamicity of bone tissue is maintained by opposing but balanced involvement of osteoblasts and osteoclasts. They are always involved in the dedicated business of bone formation and bone resorption. These remodeling processes maintain the healthy state of bone unless there is a balance between the activities of two opposing actions. The process is no longer balanced in aged people, especially in postmenopausal women. Several factors are responsible for the inclination toward excessive bone resorption. Among them aging and the menopause are leading factors for the development of osteopenia over time, which lead to weak bones prone to fracture. Understanding the process of remodeling and factors responsible for bone health is central toward preventive and curative measures against osteoporosis [13].

Counteracting processes of bone formation and bone resorption are replicated by markers of resorption and formation. Under normal conditions the balance between bone resorption and formation is regulated through various systemic hormones and local mediators so that bone removed is always equal to bone newly formed. Various biochemical assays are now available that are helpful tools to assess the dynamics of bone formation and bone resorption, which involve measurement of enzymes and proteins released from osteoblasts and osteoclasts during bone degradation and formation. Although the currently available bone turnover markers are not disease specific, they provide useful information about the imbalance in the metabolism of the skeleton. Although bone mineral density (BMD) is still a standard criterion for evaluation and diagnosis of the disease, such bone markers are not always used for diagnosis of osteoporosis, but may provide a useful assessment of imbalances in bone turnover and fracture risk [14]. Additionally, because bone turnover markers include enzymes and nonenzymatic products of bone remodeling, they are usually classified to reflect catabolic and anabolic processes.

3.1 Markers of Bone Formation

Markers specific to bone formation are products of active osteoblasts, which include serum total alkaline phosphatase (ALP), bone-specific ALP, osteocalcin (OC), and by-products of collagen neosynthesis (procollagen type I propeptides). Serum total ALP or ALP is a membrane-bound enzyme located on the outer cell surface and produced by the liver, bone, intestine, and placenta. In normal adults with no liver disease, almost half of the serum total ALP is produced by bone and the rest is produced mainly by the liver [15]. In bone, ALP is produced by osteoblasts and plays an important role in osteoid formation and mineralization [16]. In individuals with normal liver function, serum total ALP serves as a nonspecific marker of bone formation and osteoblast activity. Although the measurement may give a false-positive result, particularly in subjects with high liver ALP, serum total ALP is still a widely used nonspecific marker of bone metabolism. Since patients with chronic hepatic failure, hyperparathyroidism, rickets and osteomalacia, malignant tumors, chronic obstructive pulmonary disease, chronic renal failure, and Paget's disease also have an increased level of ALP, detection of the bone-specific ALP isoenzyme has higher specificity [17]. Several methods have been developed to detect bone-specific ALP, which include electrophoresis, precipitation, heat denaturation, selective inhibition, and immunoassays [18].

OC, a noncollagen protein produced by mature osteoblasts during bone formation, is most abundant in bone matrix and regulates mineralization [19]. OC is found in relatively high quantities in bone and represents approximately 15% of the noncollagenous protein fraction in bone. Osteoblasts release OC in the extracellular matrix, a part of which is released into blood circulation, thus serum OC level is directly related to osteoblastic activity. Since OC is expressed mainly during the phase of bone anabolic activities, serum OC level is considered a specific marker of osteoblastic activity [20]. However, OC is also released into the blood circulation during bone resorption, even though serum levels of immunoreactive OC have been shown to reflect well the bone formation rate (BFR) [21].

Procollagen type I propeptides are a precursor of collagen type I, the profuse form of collagen present in bone [22]. The amino (N-)terminal propeptide (PINP) and the carboxy (C-)terminal propeptide (PICP) are short terminal extension peptides attached to precursor molecules of collagen-I (procollagen type I). These short terminal peptides are enzymatically cleaved by specific proteases during the formation of collagen-I in bone matrix, which are released into circulation [23]. Since both PICP and PINP production is associated with collagen-I synthesis in a stoichiometric way, their detection in circulation provides an indication of proliferating osteoblasts, fibroblasts, and bone formation. Both propeptides may be considered as markers of bone formation; however, PINP is recommended by the International Osteoporosis Foundation as having greater diagnostic validity, lower individual variability, and higher stability than PICP [24].

3.2 Markers of Bone Resorption

Markers specific to bone resorption are products of collagen breakdown such as hydroxyproline (OHP) or the various collagen cross-links and telopeptides. Other markers include receptor activator of nuclear factor κB ligand (RANKL) and osteoprotegerin (OPG), osteoclast-specific enzymes such as tartrate-resistant acid phosphatase (TRAP) and cathepsin K. TRAP released during bone resorption from osteoclasts may be an appropriate biochemical marker for osteoclast function. TRAP is a metalloprotein enzyme expressed in bone-resorbing osteoclasts as well as macrophages, dendritic cells, and other cells from the liver, spleen, kidneys, thymus, placenta skin, lung, and heart. Other than osteoporosis and metabolic bone diseases, TRAP is also expressed in certain pathological conditions such as leukemic Gaucher's disease [25]. TRAP serves as a nonspecific marker of bone-resorbing activity. However, TRAP-5b, an isoform among two isoforms TRAP-5a and TRAP-5b, is characteristic of osteoclasts [26]. Although both isoforms are present in osteoclasts, TRAP-5a has not responded to alendronate treatment, while a significant decrease in TRAP-5b activity was observed in postmenopausal women, suggesting TRAP-5b is specific to osteoclast activity. Colorimetric methods of measurement of TRAP in blood detected total TRAP without differentiating TRAP-5a and 5b. Immunoassays specific to TRAP-5b have been developed to assess activity specific to osteoclasts [27].

RANKL is produced by osteoblasts that coordinate between osteoblasts and osteoclasts. RANKL binds to the RANK on the surface of osteoclast precursor cells, which leads to the activation of mitogen-activated protein kinases (MAPKs), Nuclear factor kappa-light-chain-enhancer of activated B cells (NF-κB), and the Nuclear factor of activated T-cells, cytoplasmic 1 (NFATc1) signaling pathway. Activation of RANK leads to the development of fully differentiated osteoclasts, thus RANKL is involved in the stimulation of osteoclast differentiation [28].

Cathepsin K is an enzyme of the cysteine protease family expressed in osteoclasts, essential for normal bone resorption. Therefore cathepsin K is a striking target for therapeutic intervention and a marker of osteoclastic activity. Cathepsin K knockout mice exhibited impaired matrix digestion [29]. Serum cathepsin K level is found to be elevated in postmenopausal women with osteoporosis. Immunoassay of circulating cathepsin K levels is an inevitable marker of bone turnover [30].

Osteopontin (OPN), also known as bone sialoprotein (BSP), is an important constituent of the noncollagenous matrix of bone, which plays an important role in the osteoclast–matrix–adhesion process [31]. Serum OPN levels were found to be significantly higher directly correlated with other biochemical markers of bone turnover [32]. OPG is secreted by osteoblasts and osteogenic stromal cells and interfaces with the interaction between RANK and RANKL by binding to RANKL. Thus osteoclastogenic activity is counteracted by OPG. The RANKL/OPG ratio is an important indicator of bone mass and skeletal health [33].

OHP is liberated during degradation of bone collagen and subsequently excreted into urine. Detection of urinary OHP is a good approximation of bone degradation. Although detection of OHP cannot be considered as the sole indicator of bone resorption, OHP is also present in other tissues such as skin, and is liberated also by the degradation of elastin, Clq, and newly synthesized collagen [21]. Two types of hydroxylysine glycosides, namely, glycosyl-galactosyl-hydroxylysine (GGHL) and galactosyl-hydroxylysine (GHL), are liberated into blood and subsequently into urine as a result of the catabolic degradation of collagen. In addition, GGHL is present in skin and C1q, whereas GHL, which is more specific to bone, may provide a good indication of bone resorption [34]. Although no immunoassay is available for the detection of hydroxylysines, high-performance liquid chromatography (HPLC) detection of hydroxylysines after suitable derivatization is a potential marker of bone resorption [35]. Pyridinoline (PYD) and deoxypyridinoline (DPD) are trifunctional cross-links between several collagen peptides to stabilize the collagen protein. During collagen breakdown these cross-link components are released into circulation and urine. Since cross-linking is absent in newly formed collagen, detection of these cross-link components is a good indicator of degradation of mature collagen. Both are independent from food sources. Although PYD is also found in cartilage, vessels, and ligaments, DPD is more specific to bone. HPLC detection and immunoassay of these cross-links are widely used osteocatabolic markers. The specific antibody-based assay of carboxy-terminal cross-linked telopeptide of type I collagen (CTX) and amino-terminal cross-linked telopeptide of type I collagen (NTX) is an indicator of bone resorption. CTX assay is more specific to bone and recommended by the International Osteoporosis Foundation [24].

4. SCREENING ASSAYS FOR OSTEOPOROSIS

In vitro bone cell cultures are optimal tools for analysis of the direct effect of compounds and formulations. Finally, the compounds that have confirmed positive effects on bone metabolism are further tested on in vivo models. Osteoblast cultures, osteoclast cultures, and ovariectomized (OVx) rat models for postmenopausal osteoporosis are important screening assays for the discovery as well as validation of active compounds.

4.1 Osteoblast Cultures

Bone formation and remodeling is the coordinated action of various cell types involved in this process, including osteoblasts and osteoclasts. Due to inadequate availability of primary human osteoblast cells, various in vitro culture models of different origins are important tools in developing new therapies. Other than primary human osteoblast cells, isolated osteoblasts from murine calvaria, mesenchymal stem cells from bone marrow, and rat calvarial osteoblasts are employed in cell culture models. In addition to these, stabilized cell lines have been developed, which include ROS 17/2.8 or UMR 106, human MG-63 or SaOS-2 cells derived from bone tumors, and MC3T3-E1, UMR 201, and RCJ cell lines derived from primary bone cell cultures. The cells are grown in basic culture environments and allowed to differentiate into mature osteoblasts in the presence of ascorbic acid and β-glycerophosphate, except SaOs-2 cell lines, which are cultured in the presence of ascorbic acid 2-phosphate [36].

The extent of formation of osteoblasts can be visualized and quantified under a microscope. Currently, the traditional microscope method has been replaced by staining for ALP activity, allowing the most convenient determination and quantification of osteoblast activity. PINP measurement provides a reasonable marking of bone matrix formation, which can be measured using commercially available immunoassay. Similarly, the mineralization of osteoblasts can be quantified by counting mineralized nodule formation under a microscope or by Alizarin Red staining or von Kossa staining [37].

4.2 Osteoclast Cultures

Osteoclasts are multinucleated cells responsible for bone resorption and are therapeutic targets for antiresorptive drugs. In vitro cell culture of osteoclasts is difficult as they are fragile cells adhered to bones. Osteoclasts are either obtained directly from bone as the primary source or derived from hematopoietic stem cells that differentiate under the influence of RANKL, macrophage colony-stimulating factor, and tumor necrosis factor (TNF)-α. In addition, giant cell tumors of bone are a useful source of human osteoclasts [38]. In vitro cell culture assays are important tools for investigating the effects of compounds or extracts on bone cells. In vitro osteoclast culture could be conveniently performed using CD14* and CD34 osteoclast precursor cells, derived from human bone marrow. CD14* and CD34 cells have osteoclastogenic potential that can be stimulated on bone slice to convert into osteoclast. The differentiation of osteoclasts can be determined by assessment of TRAP-5b isoforms. It has been proved that TRAP-5b activity is directly proportional to the number of mature osteoclasts formed in culture. Bone catabolic activity can also be performed by calculating the formation of resorption pits under a microscope on bone slices cultured with osteoclasts. Although time-consuming and laborious, it provides a more natural profile of bone resorption. Osteoclast activity, measured in terms of CTX, correlates strongly with absorption pit formation. CTX measurements provide a convenient approximation of volume of resolved bone in vitro.

4.3 The Ovariectomy Rat Model

The ovariectomy model is most important and commonly used as an in vivo model for mimicking postmenopausal osteoporosis. The extent of bone loss followed by ovariectomy is dependent on the time since ovariectomy and site of the skeleton. Significant bone loss is induced soon after ovariectomy in female rats after 14 days in proximal tibial metaphysis and after 30 days in femoral neck, and reaches a steady state after 90 days in proximal tibial metaphysis and after 270 days in femoral neck [39]. Ovariectomy is a Food and Drug Administration-approved preclinical model [40] to study the effect of potential involvements that can preserve bone metabolism in post-menopausal conditions. Sexually and skeletally mature female Sprague–Dawley rats (12 weeks of age) are most commonly used for modeling bone loss due to estrogen deficiency. Although longitudinal bone growth is still continued at the age of 12 weeks, peak bone mass and sexual maturity had been achieved at the age of 12 months [39]. Extent of bone loss is reflected by enlargement of bone cavity in cortical bone because of endosteal bone resorption and simultaneous periosteal bone formation. Enlargement of the bone marrow cavity is a sensitive index of bone loss because of increased endosteal bone resorption and simultaneous bone formation at the periosteum [41].

Further investigation of bone turnover is determined by measurement of calcium, phosphorus, and magnesium in urine as well as in serum, all biochemical markers of bone turnover. Dual-energy X-ray absorptiometry is often used to determine BMD. Peripheral quantitative computerized tomography (*p*-QCT) and microcomputerized tomography (μ-CT) are powerful imaging techniques that capture the image of an area of bone. *p*-QCT can analyze cancellous and cortical bone but is unable to evaluate histomorphological characteristics at individual trabecular levels. However, μ-CT is superior over *p*-QCT because μ-CT is capable of capturing high-resolution images at an area of bone as well as at individual trabecular levels. Additionally, μ-CT is capable of evaluating histomorphometric parameters such as trabecular thickness and trabecular separation. Mechanical strength evaluation is a rational measurement of actual gain or loss of bone strength in experimental rat models, including three-point bending, four-point bending, and torsion testing to assess bone mechanical strength in diaphyses of long bones. Compression testing and cantilever testing have also been developed to evaluate mechanical strength at the vertebral and femoral head.

5. OSTEOPROTECTIVE PLANT FORMULATIONS

Potential efficacy and fewer side effects of traditional medicines have motivated researchers to dig for better treatment for chronic diseases. Medicinal plants that are described in traditional systems of medicine have been evaluated for their potential role in the management of osteoporosis. Some of the screened plants have shown promising activity in the form of extracts, formulations, and polyherbal formulations. This section will cover plant-based formulations that have shown preventive as well as curative activity toward osteoporosis.

Achyranthes bidentata root extract (100, 300, or 500 mg/kg/day) given to OVx female Sprague–Dawley rats at 16 weeks significantly prevented deterioration of trabecular microarchitecture induced by ovariectomy. The antibone resorption effect of a *A. bidentata* was explored by He et al. *n*-Butanol fractions prepared from methanol extract of the roots of *A. bidentata* have the highest protective effect against parathyroid hormone (PTH)-induced bone resorption. Furthermore, oral administration at doses of 25, 50, and 100 mg/kg/day to OVx rats significantly prevented loss of BMD without affecting the uterus [42,43].

Butea monosperma Lam. is a traditional medicine of India used for the promotion of fracture healing. The medicinal plant is effective in preventing osteoporotic fractures as it stimulates new bone formation. In vivo study demonstrates that daily oral administration of an ethanolic extract of stem bark and its acetone-soluble fraction exhibited a considerable increase in ALP activity and increased mineralization of osteoblast cultures [44].

Cissus quadrangularis is a medicinal plant from India, described in Ayurveda for syphilis, venereal disease, piles, and leucorrhea. It is also known as bone setter in English. Bone fracture healing and antiosteoporotic activity were evaluated in ovariectomized rat models at dose levels of 500 and 750 mg/kg/day of ethanol extract of *C. quadrangularis*. OVx rats treated at these dose levels showed increased biomechanical, biochemical, and histopathological parameters that proved to have definite antiosteoporotic effects [45]. *C. quadrangularis* extract-treated patients with mandible fractures showed better healing of fractures with significant levels of expression of OPN protein [46].

The total coumarin fraction obtained from the fruits of *Cnidium monnieri* has antiosteoporotic effects in OVx rats. BMD of the femur metaphysis, uterus weight, as well as estradiol, OC, and calcitonin levels in serum were increased. Coumarin-rich fraction has antiosteoporotic effects by increasing the production of estradiol and calcitonin [47].

Dalbergia sissoo Roxb. (family Fabaceae) is an erect timber tree that can grow up to 35 m. Traditionally, water decoction of its leaves has been used for the treatment of gonorrhea and wood for leprosy. Animal studies have shown that leaves have a protective effect against postmenopausal bone loss. An experimental drill-hole injury model revealed fracture-healing properties of ethanolic extract, which induced bone regeneration at the fracture site. An experimental fracture model administered at a dose of 250 mg/kg/day significantly improved fracture healing much earlier (day 15) than the normal healing process [48]. Butanol-soluble standardized fraction (BSSF) made from ethanolic extract of the leaves of *D. sissoo* exhibited the prevention of pathological bone loss under estrogen deficiency. Standardized fraction treatment in ovariectomized rats resulted in an increased new bone formation, improved trabecular microarchitecture of the long bones, decreased bone turnover markers (OC and type I collagen) and expression of osteogenic genes. On the whole, the osteoprotective effects of BSSF were comparable to those of 17β-estradiol and no uterine estrogenicity was observed [49].

Eucommia ulmoides Oliv. is a traditional Chinese medicine used for the treatment of kidney disease, bone fractures, and joint diseases in China. The dried cortex of *E. ulmoides* Oliv. was extracted with 60% boiling ethanol. Treatment with extract at a dose of 300 or 500 mg/kg/day was able to prevent estrogen deficiency-induced osteoporosis in female Sprague–Dawley rats [50].

Glycine max is a common dietary supplement, and contains plenty of relevant dietary flavonoids such as genistein, daidzein, and biochanin A. These soy flavonoids have strong effects on bone metabolism, bone mechanical strength, and bone turnover markers in postmenopausal women. Animal studies have signified suppressed bone loss induced by ovariectomy without uterotrophic side effects. Ethanolic extract of *G. max* increased survival and DNA synthesis in osteoblasts of MC3T3-E1 cells at the concentration range of 0.01–0.0.1 g/L. The osteogenic effect of soy extract on MC3T3-E1 cells on proliferation, ALP activity, and collagen synthesis is eliminated by antiestrogen tamoxifen, indicating estrogen-like effects of soy extract [51].

Fathilah et al. examined the potential of *Labisia pumila* var. Alata extract for prevention of postmenopausal osteoporosis. Oral administration to the OVx female rats at the daily dose of 17.5 and 64.5 mg/kg for 2 months prevented bone histomorphological changes associated with osteoporosis. This study confirmed that supplementation of *L. pumila* var. Alata is as effective as estrogen replacement therapy for prevention of postmenopausal osteoporosis [52].

Passiflora foetida butanolic fraction showed a positive skeletal effect in an estrogen-deficient OVx female Balb/c mice model. Oral treatment with doses of 50 and 100 mg/kg/day for 8 weeks presented better microarchitectural parameters, bone biomechanical strength, and more osteoprogenitor cells in the bone marrow. The antiosteoporotic effect of the extract was exerted through stimulation of osteoblast function and inhibition of osteoclast function without uterine estrogenicity [53].

Peperomia pellucida (L.) (family Piperaceae) is used in Cameroon to treat fracture healing. The fracture-healing efficacy of this plant was investigated by F.T. Ngueguim et al. Ethanol extract of the whole plant of *P. pellucida*, administered at 100 and 200 mg/kg to adult female Sprague–Dawley rats having a drill-hole injury, induced bone regeneration at the fracture site. The most effective dose (200 mg/kg extract) significantly increased mineral deposition and bone microarchitecture. In vitro studies on bone marrow stromal cells (BMSCs) from the left femora of Sprague–Dawley female rats showed increased expression of osteogenic genes, including type I collagen, OC, and bone morphogenetic protein-2 (BMP-2). In addition, *P. pellucida* extract also showed an increase in mineralized nodule formation [54].

K. Porwal et al. studied the effect of guava fruit (*Psidium guajava*) extract for its osteoanabolic effect. A dose of 250 mg/kg increased BMD, biomechanical strength, peak bone mass, and femur length in growing female rats. The ethanolic extract is able to restore osteopenia associated with estrogen deficiency. Strong osteoanabolic effects suggest that guava fruits are an effective dietary supplement for achieving high peak bone mass and length during growth periods. Daily supplementation of guava fruit extract (GE) at doses of 250 and 500 mg/kg to recently weaned female rats for 9 weeks resulted in an 11.3% and 6.8% increase in femur length compared to the vehicle-treated group. In an in vitro study of osteogenic differentiation by ALP activity, GE significantly increased ALP activity at 1.25, 1.5, and 1.75 mg/mL concentrations. In addition, osteogenic genes, including Runx2, BMP-2, ColI, and Wnt3a, were significantly upregulated [55].

Salvia miltiorrhiza, commonly known as red sage, is a traditional Chinese medicine to treat skeletal diseases. Several clinical trials have been performed to establish the efficacy of the herbal medicine cotreated with calcium and vitamin D. The daily application of 1.6 g of the extract as tablets for 20 days, in addition to 600 mg calcium and vitamin D 125 IU per day to menopausal women, showed a result comparable to the administration of 2 mg nylestriol for 2 weeks [56,57].

The ethyl acetate fraction of the methanolic extract of stems of *Sambucus sieboldiana* inhibited the PTH-stimulated bone resorption of neonatal mouse bones in organ culture. In vivo administration of ethyl acetate fraction at doses of 50 and 100 mg/kg/day to OVx rats protected the BMD of the lumbar (L_{2-4}) vertebra in estrogen-deficient condition [58].

Spinacia oleracea L. is a familiar dietary plant of the family Amaranthaceae, commonly known as spinach. *S. oleracea* extract administered orally to osteopenic adult OVx Sprague–Dawley rats prevented ovariectomy-induced bone loss as evidenced by a 122% increase in bone volume/tissue volume (BV/TV) and a 29% decline in trabecular separation. This effect was further supported by a more than twofold increase in expression in osteogenic runt-related transcription factor 2 (Runx2), bone morphogenetic protein 2, collagen-I, osterix, and OC. Free-flowing granule-formulated *S. oleracea* extract showed better activity in fracture-healing models. Formulated granules provided a 39% increase in BV/TV at the fracture site compared to a 25% increase in BV/TV by nonformulated *S. oleracea* extract [59].

Ulmus wallichiana Planch. is a plant of the family Ulmaceae, found growing throughout the Himalayas, particularly in the Kumaon and Garhwal regions of India. Traditionally, it has been used for fracture healing by established healers around the Kumaon region of India. OVx rats are the most widely used animal model for mimicking bone loss postmenopausal conditions in humans. OVx rats treated with ethanolic extract at 750 mg/kg/day and its butanolic fraction at 50 mg/kg/day from the stem bark once daily for 12 weeks prevented bone loss in both cortical and trabecular regions and exhibited increased BMD [60].

The traditional medicine system of India describes the plant *Tinospora cordifolia* as a remedy for rheumatoid arthritis, inflammation, and other -diseases of the musculoskeletal system. The antiosteoporotic potential of *T. cordifolia* was established by G. Abiramasundari et al. A cell growth and viability study on human osteoblast-like cells MG-63 and primary osteoblast cells isolated from the femur of rats indicated that alcoholic extract at a dosage of 25 μg/mL is effective stimulation of growth of osteoblasts as evaluated by colorimetric MTT assay. In addition, significant differentiation and increased mineralization of these two cell types have been achieved at the aforementioned concentration [61]. The in vivo effect of *T. cordifolia* extract was evaluated by P. Kapur et al. Ethanolic stem extract given to female OVx Sprague–Dawley rats at a dose of 10 mg/kg bw showed an osteoprotective effect. Additionally, significant reduction in serum OC and increases in ALP activity were observed followed by treatment of *T. cordifolia* extract [62].

The Chinese polyherbal medicine "Hoehu ekki to" is composed of 10 herbal medicines consisting of *Astragalus* roots, *Atractylodes lanceae* rhizome, *Panax ginseng* roots, *Angelica* roots, *Bupleuri* roots, *Zyzyphus* fruits, *Aurantis nobilis* pericarp, *Glycyrrhiza* roots, *Cimifugae* rhizome, and *Zingiberus* rhizome in a ratio of 8:8:8:6:4:4:4:3:2:1. Hot water extracted from this medicine has been used for the treatment of bone loss associated with a gonadotropin-releasing hormone (GnRH) agonist in rats. GnRH agonist administration reduced the BMD in the whole femur by 91.0% compared to the control group. However, administration of "Hoehu ekki to" increases the serum estradiol levels by 5.3-fold, compared with the GnRH agonist-treated group, and causes 106.2% augmentation of the BMD [63].

6. ACTIVE CONSTITUENTS FROM PLANTS

Plant-based natural compounds have an imperative function in the improvement of bone-related diseases such as osteoporosis. Natural products that exert their antiosteoporotic action by binding with ERs are called phytoestrogens and are a better way to treat fragile bones [64]. The three main types of phytoestrogens include isoflavones, coumestans, and lignans. This section will cover the role of selected natural products in an integrated drug discovery approach toward osteoporosis.

6.1 Flavonoids

Flavonoids are among the most important natural molecules that possess antiosteoporotic activity. They have anabolic effects on osteoblasts and promote proliferation, differentiation, and mineralization of osteoblasts. They also affect bone formation by prevention of bone loss by inhibiting osteoclast activity, forming osteoclasts, or inducing apoptosis in osteoclasts. A number of flavonoids of plant origin have been reported to have significant potent activity (Figs. 8.1A and B). Kaempferol (**1**), an important natural flavonol found in many plants, exerts a potent inhibitory effect on the bone resorption of osteoclast cells with an IC_{50} value of 1.6 μM [65]. Kaempferol showed estrogenic properties via transcriptional activity of pERE-Luc (3.98-fold at 50 μM), including its effects on ERα phosphorylation and mineralization processes (4.02 ± 0.41-fold at 50 μM) in cultured rat osteoblasts [66]. Quercetin (**2**), another structurally similar flavonol isolated from several plants, exhibited inhibitory concentration IC_{50} 5.3 μM on osteoclasts cells obtained from rabbit long bones [65]. Baicalein (**3**) and wogonin (**4**) were isolated from *Scutellaria baicalensis* and induced cell proliferation and calcified nodule formation on human periodontal ligament cells at concentrations ranging from 0.15 to 0.6 mg/L [67]. Apigenin (**5**) is significant

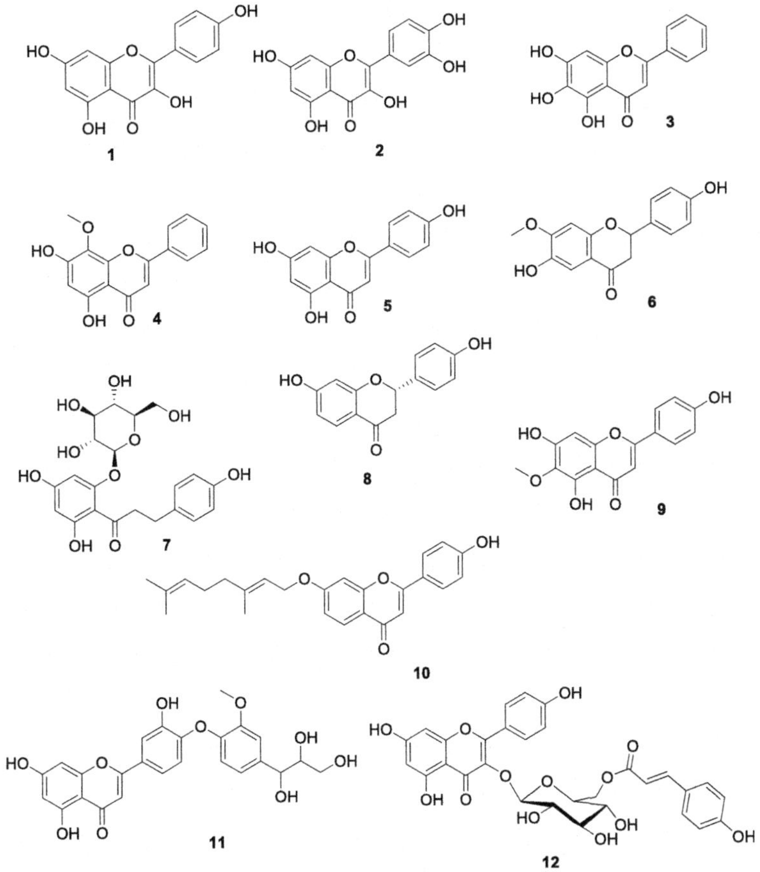

FIGURE 8.1A Chemical structure of natural antiosteoporotic flavonoids.

among flavonoids for antiosteoporotic activity. Significant osteoprotection was observed in the presence of apigenin because it inhibited osteoclast differentiation from the RAW 264.7 cell line by reducing RANKL-induced TRAP expression and calcitonin receptor resulting in reduced multinucleated osteoclast formation. Additionally, apigenin inhibited osteoclast differentiation indicated by downregulated osteoclast differentiation markers such as TRAP, RANK, and c-Fms in osteoclast precursor cells obtained from mouse bone marrow. Furthermore, apigenin-induced apoptosis of mature osteoclasts was obtained from rabbit long bone and inhibited bone resorption. Apigenin inhibits osteoclastogenic cytokine secretion induced by TNF-α and interferon gamma in the MC3T3-E1 mouse calvarial osteoblast cell line [68].

A flavone, 6,4′-dihydroxy-7-methoxyflavanone (**6**), derived from the heart wood of *Dalbergia odorifera*, inhibited RANKL-induced osteoclastogenesis [69]. Phloridzin (**7**), a polyphenol found in apple, *Malus domestica* (Rosaceae),

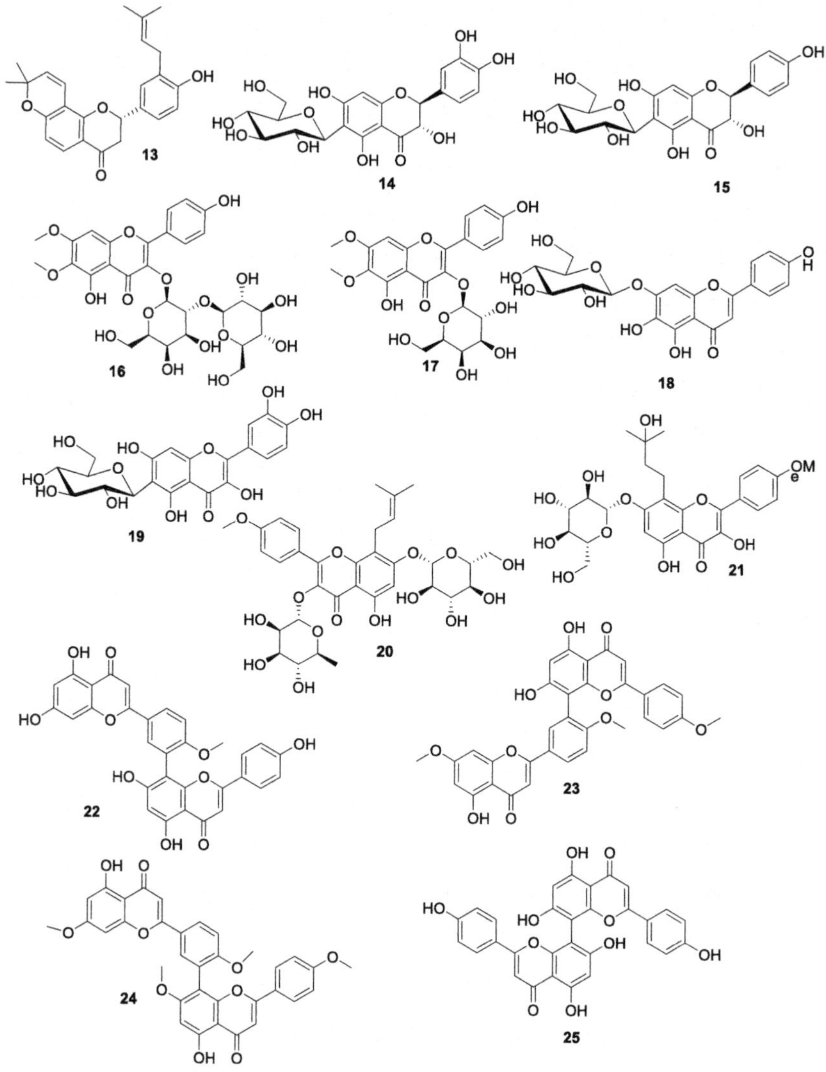

FIGURE 8.1B Chemical structure of natural antiosteoporotic flavonoids.

may exert a protective effect in the pathogenesis of osteoporosis on aging and sex hormone-related changes. Phloridzin prevented ovariectomy-induced bone loss as indicated by improved BMD at the diaphyseal site in OVx rats [70]. Liquiritigenin (**8**), isolated from *Glycyrrhizae radix*, significantly elevated ALP activity, collagen synthesis, glutathione content, osteoblast differentiation, and mineralization in osteoblastic MC3T3-E1 cells. In addition, liquiritigenin significantly decreased osteoclast differentiation by upregulating interleukin-6

(IL-6), TNF-α, and RANKL in the presence of antimycin A [71]. Furthermore, in an in vitro culture, liquiritigenin also inhibited osteoclast differentiation from bone marrow-derived macrophages (BMMs) and RAW-D cells into mature osteoclasts. Thus the compound has beneficial effects on both bone resorption activity and osteoblastic activity [72]. Hispidulin (**9**), isolated from different plants such as *Millingtonia hortensis* Linn., *Salvia plebeian* R. Br. *Salvia officinalis*, *Scoparia dulcis* Linn., and *Artimisia* species [73], showed anti-osteoclastogenic effects by reducing RANKL-induced NF-κB, c-Jun N-terminal kinase, and p38 and NFATc1 in osteoclast precursors. Hispidulin (**9**) showed osteoprotective effects in RANKL-stimulated RAW 264.7 cells and BMM cells [74]. Root and stem bark of *Millettia griffoniana* (Leguminosae) contains griffonianone E (**10**), which shows ALP activity at 10^{-5} M in MVLN cells and Ishikawa cells. Additionally, griffonianone E showed a significant induction of β-galactosidase activity in a yeast-based ERα assay [75]. Pterospermin A (**11**) and *trans*-tiliroside (**12**) are significantly active osteogenic constituents isolated from the chloroform fraction and *n*-butanol fraction of the ethanol extract of *Cupressus sempervirens* flowers, respectively. Pterospermin A and *trans*-tiliroside showed a significant increase in ALP activity at 10 and 100 pM; however, they failed to increase osteoblast viability assessed by MTT. These data suggest that the compounds increase osteoblast differentiation without osteoblast proliferation. Osteogenic gene expression assay of these compounds showed upregulation of mRNA levels of BMP-2, Runx2, and ColI in osteoblasts. Additionally, these compounds significantly increased mineralization of bone marrow cells extracted from the femur of 21-day-old rats [76].

Shinflavanone (**13**) is a promising candidate for osteoporosis therapy and has been isolated from *Glycyrrhiza glabra* L. (Leguminosae). Shinflavanone (**13**) shows strong inhibition of 1α,25-dihydroxyvitamin D3-induced bone resorption pit formation by osteoclast-like cells with $IC_{50} = 0.70$ μg/mL [77].

Ulmoside A (**14**) and ulmoside B (**15**) are two flavonoid 6-*C*-glycosides isolated from the ethanolic extract of the stem bark of *U. wallichiana*. Compounds, namely, (2*S*,3*S*)-(+)-3,4,5,7-tetrahydroxydihydroflavonol-6-*C*-β-D-glucopyranoside (**14**, ulmoside A), (2*S*,3*S*)-(+)-4,5,7-trihydroxydihydro flavonol-6-*C*- β-D-glucopyranoside (**15**, ulmoside B), and quercetin-6-*C*-β-D-glucopyranoside significantly stimulated osteoblast ALP activity, a phenotypic marker for the mature osteoblast. These compounds stimulated osteoblast differentiation comparable to BMP-2 [78]. Runx2, BMP-2, ColI, and OC are osteogenic gene markers. Compound **14** elevated all these osteogenic gene expression markers in the preosteoblasts. In addition, compound **14** induced differentiation of preosteoblasts and osteoclastogenesis [79]. In an in vivo study, extract given at a dose of 5.0 mg/kg/day orally to OVx rats improved bone biomechanical quality through positive modifications of BMD in the trabecular (distal femur, proximal tibia, and vertebrae) and cortical (femur shaft) regions and trabecular microarchitecture without uterotropic effect [80].

Phytochemical investigation of the methanolic extract of *Boerhavia repens* (Nyctaginaceae) leads to isolation of two flavonoid glycosides, namely, eupalitin 3-*O*-β-D-galactopyranosyl-(1 → 2)-β-D-glucopyranoside (**16**) and eupalitin 3-*O*-β-D-galactopyranoside (**17**), which showed a protective effect on PTH-stimulated bone resorption at 200 μM concentration [81]. Scutellarin 7-*O*-β-D-glucopyranoside, or plantaginin (**18**), isolated from an *n*-butanol fraction of ethanolic extract of *Kigelia pinnata* leaves, showed osteogenic activity in primary calvarial osteoblast cultures. Scutellarin 7-*O*-β-D-glucopyranoside (**18**) showed increased ALP activity at 10 nM concentration, whereas significant stimulation of mineralization was observed at 1 μM [82]. 3,3′,4′,5,7-Pentahydroxyflavone-6-*C*-β-D-glucopyranoside (**19**) is a *C*-glucoside abundant in *U. wallichiana*. Compound (**19**) showed bone anabolic effects by osteoblast stimulation, and much more effectively increased bone mineral osteoprogenitors, BMD, and BFR than did quercetin [83]. Compound (**19**) at 1.0 nM showed osteoprotective effects as well. Compound (**19**) considerably inhibited expression of osteoclastogenic genes and differentiation of multinucleated osteoclasts from bone marrow cells. Comparatively, (**19**) is 1000-fold more effective than quercetin in terms of these effects [84].

Icariin (**20**), a flavonoid glycoside, is isolated from hot water extract of aerial parts of the plant *Epimedium koreanum* (Berberidaceae). These compounds are useful for the prevention and treatment of osteoporosis in estrogen-deficient ovariectomized rats and showed upgraded bone P and Ca contents, BMD, and femur strength in animal models [85]. Maohuoside A was first time isolated from *Epimedium platyetalum* [86]. Icariin (**20**) was also isolated from the water extract of Er-xian decoction, a traditional Chinese formula for the treatment of osteoporosis and menopausal syndrome. Icariin not only increased bone formation by promoting osteoblast proliferation and ALP activity, but also inhibited bone resorption and decreased the TRAP activity of osteoclasts [87]. Maohuoside A (**21**) promotes osteogenesis by 15.8% on day 11 in rat bone marrow-derived mesenchymal stem cells. The mechanism of osteogenesis involves inhibition of ERK1/2 and p38 MAPK pathways and stimulation of BMP-2 [88,89].

Biflavonoids isolated from the leaves and twigs of *Cephalotaxus koreana* significantly increased ALP activity at a concentration range of 1.0–20.0 μM. Among these bioflavonoids, three compounds, namely, bilobetin (**22**), sciadopitysin (**23**), and 7,4′,7″,4‴-*O*-methyl-amentoflavone (**24**), significantly increased collagen synthesis and showed significant mineralization, which was determined by Alizarin Red staining. Compound sciadopitysin was found to be most active, which increased collagen syntheses by up to 180% compared to that of control cells at a concentration of 10.0 μM [90]. 8,8″-Biapigeninyl or cupressuflavone (BA) (**25**) is a biflavonoid of two apigenin molecules found in *C. sempervirens*. Cupressuflavone at 10^{-10} and 10^{-8} M concentration dependently inhibited osteoclastogenesis of bone marrow cells and promoted proliferation, differentiation, and mineralization at 10^{-10} M. Comparatively, cupressuflavone is 10,000-fold more active than apigenin. In vivo studies showed dose-dependent inhibition of expression of osteoclastic genes at 1.0, 5.0, and 10.0 mg/kg/day doses, in addition to significant stimulation of osteogenic genes core binding factor alpha-1, CollI, and BMP-2[91].

6.2 Isoflavonoids

Natural isoflavones represent the most potent class of phytoestrogens and are shown to promote osteoblast formation, differentiation, and mineralization. Isoflavones display their effects on BMD, bone mechanical strength, and bone turnover markers in postmenopausal women by suppressing bone resorption and stimulating bone formation. Diets high in isoflavones have positive effects on the risk of developing osteoporosis. Isoflavonoids have been the promising center of attention for research into natural therapeutic phytoestrogens.

Daidzein (**26**), glycitin (**27**), and genistein (**28**) along with glycosides daidzin (**29**) and genistin (**30**) (Fig. 8.2) are effective antiosteoporotic constituents of soybean *G. max* (Leguminosae). Daidzein is among the main isoflavones in soybeans, which showed a reduction of osteoclastogenesis induced by 1,25-dihydroxyvitamin D3. The effect of daidzein on the development and activity of osteoclasts was compared with that of 17β-estradiol on nonadherent porcine bone marrow cells. The presence of daidzein (10^{-8} M) with 1,25-dihydroxyvitamin D3 (10^{-8} M) reduced the number of osteoclasts formed in response to vitamin D3 by $58 \pm 8\%$. However, the presence of 17β-estradiol (10^{-8} M) with 1,25-dihydroxyvitamin D3 (10^{-8} M) showed a $52 \pm 5\%$ reduction in the number of osteoclasts formed in response to vitamin D3 [92]. Results showed that daidzein regulates bone formation through ER. Daidzein (**26**) promoted osteoblast proliferation at a lower dose via ERs, whereas at higher dose it mainly acted on peroxisome proliferator-activated receptor-λ to inhibit the proliferation of osteoblasts [93]. The mechanism for the antiosteoclastogenic effect of daidzein (**26**) was investigated by Tyagi et al. Estrogen-deficient conditions induce osteoclastogenesis by upregulation of TNF-α producing T cells, TNF-α secreting $CD4^{+}CD28^{null}$ T cells, and B-lymphopoiesis. OVx mice treated with compound (**26**) showed a reduction in the expansion of $CD4^{+}$ T cells, whereas it increased the percentage of $CD4^{+}CD28^{+}$ T cells. Treatment with daidzein reduced TRAP expression in $CD4^{+}CD28^{null}$ T cells and bone marrow cell coculture. Daidzein (**26**) also regulated B-lymphopoiesis and decreased expression of RANKL in $B220^{+}$ cells. Thus compound (**26**) produced osteoprotective effects by reversing the damaging immune changes arising under estrogen-deficient conditions [94]. Daidzein put forth osteoprotective effects by stimulated thyroid C cells, which produce calcitonin, a hormone that inhibits bone resorption. Subcutaneous injections of daidzein (30 mg/kg/day) for 3 weeks significantly increased the volume of C cells in thyroid and increased cancellous bone area, trabecular thickness, and trabecular number in

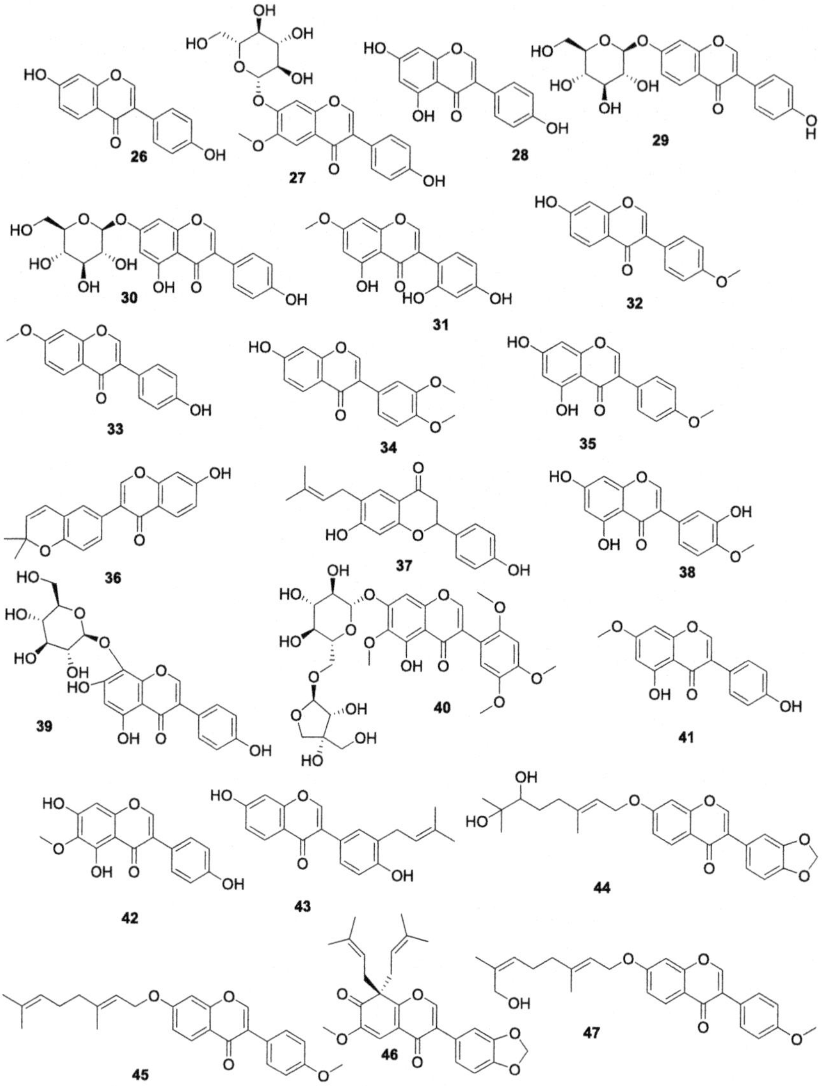

FIGURE 8.2 Chemical structure of osteoactive natural isoflavonoids.

the metaphyseal region of the proximal tibia. Additionally, daidzein also significantly reduced serum OC and urinary calcium secretion [95]. Glycitin (**27**) promotes BMSC proliferation and osteoblast formation from BMSCs, and increases ColI mRNA expression and ALP activity. Administration of 1 and 5 μM glycitin (**27**) significantly promoted ColI mRNA expression, TGF-β protein expression, and increased *p*-AKT levels in BMSCs and ALP activity [96]. Genistein (**28**) has nonuterotropic and osteoprotective effects by regularizing the accretion of pre-B lymphocytes in bone marrow and inhibiting bone resorption. Subcutaneous injection of genistein (**28**) at a dose of 0.4 mg/day to OVx rats shows a noteworthy increase in BMD [97]. Genistein (**28**) treatment at a dosage of 4.5 mg/kg or 9 mg/kg to ovariectomy operated female Sprague–Dawley rats appreciably recovered the loss of bone density after 4 weeks. These doses of 4.5 mg/kg/day and 9 mg/kg/day showed 47% and 43% increment in bone density, which was 24% and 40% more superior than the increment of bone density at a higher dose of genistein (18 mg/kg/day) and nylestriol (1.5 mg/kg/week), respectively [98]. In a comparative study of glycitin (**27**), daidzin (**29**), and genistin (**30**) on bone loss and lipid metabolism abnormality in OVx rats, genistin showed bone loss prevention at a dose of 50 mg/kg/day, whereas glycitin and daidzin are effective in the restoration of abnormality of lipid metabolism due to suppression of bone turnover [99]. Cajanin (**31**), formonentin (**32**), isoformonentin (**33**), and cladrin (**34**) are active constituents isolated from *B. monosperma*. These compounds showed promising osteoblast proliferation, differentiation, and mineralization in neonatal rat calvarial primary osteoblast cultures. These compounds induced a marked increase in ALP expression, Alizarin Red S staining of nascent calcium deposition in osteoblasts cultured for 48 h, and von Kossa silver-stained nodule formation by the osteoblasts 15 days after culture with these compounds [44]. Cajanin (**31**) is also found in *Cajanus cajan* (Fabaceae), *Canavalia ensiformis* (jack bean), coffee, and coffee products. In vitro studies of cajanin (**31**) on osteoblast proliferation, differentiation, and mineralization of bone marrow osteoprogenitor cells led to increased osteoblast ALP production at 10^{-11} M. Osteoblast differentiation–promotion involves activation of MEK-Erk and Akt pathways. A once daily oral treatment of cajanin at a dose of 10.0 mg/kg/day to recently weaned female Sprague–Dawley rats for 30 consecutive days increased BMD, BFR, bone biomechanical strength, and mineral apposition rate [100]. Isoformonentin (**33**), a methoxydaidzein present in several medicinal plants, reverses bone loss in osteopenic rats and exerts bone anabolic action by increasing osteoblast survival [101], differentiation, and mineralization. Promotion of osteoblast differentiation involves simultaneous activation of MEK-Erk and Akt pathways. Treatment of calvarial osteoblast cells with isoformononetin (**33**) from 10^{-9} to 10^{-7} M resulted in enhanced mRNA levels of ALP in osteoblasts. Furthermore, in addition to calvarial osteoblasts, isoformononetin (**33**) also promoted differentiation of BMSCs at 10 nM concentration. In contrast to genistein or daidzein, the mechanism of action of cajanin (**31**) and isoformononetin (**33**) does not act via ERs in osteoblasts. ER antagonist ICI-182780 has no effect on cajanin- (**31**) or isoformononetin (**33**)-induced osteoblast proliferation and differentiation [100]. Cladrin (**34**) is a 3′,4′-dimethoxy analog of daidzein. Increased lipophilicity due to *O*-methylation enhances its oral absorption and systemic bioavailability. Subsequently, cladrin is more potent than daidzein because of superior oral bioavailability of cladrin over daidzein. Cladrin, at 10 nM, stimulated both osteoblast proliferation and differentiation. Cladrin showed increased ALP production and hence differentiation in rat neonatal calvarial osteoblasts. Cladrin showed increased BrdU incorporation and hence proliferation in rat neonatal calvarial osteoblasts. Osteoblast proliferation and differentiation were abashed by MEK1/2 inhibitors suggesting that the compound has an osteogenerative effect by activating the MEK-Erk pathway. Cladrin significantly induced formation of mineralized nodules at 10 nM concentration in rat BMSCs. In vivo studies showed increased mineral apposition and BFRs. Daily oral gavages of cladrin at a dose of 10.0 mg/kg/day to recently weaned female Sprague–Dawley rats for 30 consecutive days exhibited increased BMD compared with control [102,103]. Cladrin also has a protective effect against high-fat diet-induced obesity and associated bone loss as indicated by suppressed bone marrow adipogenesis on cladrin treatment at doses of 5 and 10 mg/kg/day in obese male mice. Thus cladrin showed osteoprotective effects. Both in vivo and in vitro conditions increased osteogenic markers [104]. Formononetin (**32**) is a soy isoflavonoid found abundantly in traditional Chinese medicine from *Astragalus mongholicus* and *Trifolium pretense* L. [105], and in an Indian medicinal plant, *B. monosperma* [44]. Formononetin has a strong effect on osteoblast differentiation as evaluated by increased osteoblast ALP activity, which was abolished by p38 MAPK inhibitor SB203580 (10 μM) suggesting that the compound works by activating the p38 MAPK pathway. Results show no effect of formononetin on osteoblast proliferation. However, 100 nM formononetin in rat BMSCs increased formation of mineralized nodules. Thus formononetin stimulated the formation of mature osteoblasts from osteoprogenitor cells in bone marrow. In addition, formononetin has no effect on mineral apposition and bone formation; neither effected osteoclast differentiation nor activated ER in osteoblasts [102]. Biochanin A (**35**), a natural isoflavone identified in various diets and plant species, caused a significant increase in cell growth, collagen content, ALP activity, and OC secretion in osteoblastic MC3T3-E1 cells at 1–50 μM [106,107]. Corylin (**36**) and bavachin (**37**) were isolated from ethyl acetate fraction of ethanol extract of *Psoralea coryfolia* (Leguminosae) fruits. Corylin and bavachin exhibited strong stimulation of osteoblastic differentiation on UMR 106 cells. A cell differentiation assay of corylin (**36**) and bavachin (**37**) on UMR 106 cells showed an 11.3% and 20.5% increase in ALP activity at 0.3 μM concentrations, respectively [108]. Genistein, biochanin A, pratensein (**38**), biochanin 7-*O*-glucoside (**39**), and Caviunin 7-*O*-[β-D-apiofuranosyl-(1-6)-β-D-glucopyranoside] or dalsissooside (**40**) are significantly active isoflavonoids isolated from the *n*-butanol fraction of *D. sissoo* Roxb. leaves [109]. These compounds showed significant ALP activity in calvarial osteoblasts at 10 nM, whereas dalsissooside showed significant ALP activity at 100 pM. Studies showed osteoblast mineralization at these concentrations. Dalsissooside is a very active compound and showed robust osteoblast ALP production and mineralization at

1 pM concentration. Compound **40** significantly upregulates the expression of BMP-2 mRNA by almost 15-fold. Dalsissooside showed overall protective and osteogenic effects in an ER-independent way. ALP activity is mediated by p38 MAPK activation [110].

Prunetin (**41**) is an isoflavonoid present in red clover and the fruit of *Prunus avium* (red cherry). In vitro culture studies on primary osteoblasts, osteoblastic cell lines, and HEK293 T cells as well as in vivo studies have established osteoprotective effects of the compound. Prunetin at 10 nM significantly increased proliferation and formation of mineralized nodules in an in vitro culture of bone marrow stromal/osteoprogenitor cells. Osteoblast proliferation and differentiation are promoted via a G-protein-coupled receptor, GPR30/GPER1, as evidenced by elimination of these effects by G15, a selective GPR30 antagonist [111]. Tectorigenin (**42**) is an *O*-methylated isoflavone isolated from the rhizome *Belamcanda chinensis* (Iridaceae). This compound has osteogenic effects on bone and positive effects on BMD without any effect on the uterus. The compound has selective ER modulator activities via binding to both ERα and ERβ. Intravenous treatment of compound **42** (7 mg/animal) to OVx rat models inhibits pulsatile luteinizing hormone (LH) secretion from the pituitary. Suppression of LH surge may be used in menopausal women suffering from hot flushes [112]. Neobavaisoflavone (**43**) is a natural isoflavone of *Psoralea corylifolia* L. and demonstrated promoted osteogenesis and bone matrix protein, including (ColI), OC, and BSP in MC3T3-E1cells. mRNA levels of Runx2 and Osx are upregulated by neobavaisoflavone (**43**), which was abolished by p38 inhibitor SB203580, suggesting that it probably acted through a p38-dependent signaling pathway [113].

4′-*O*-Geranylisoquiritigenin (**44**), 7-*O*-geranylformononetin (**45**), griffonianone C (**46**), and 4′-methoxy-7-*O*-[(*E*)-3-methyl-7-hydroxymethyl-2,6-octadi enyl]isoflavone (**47**) are potential estrogenic isoflavonoids isolated from the root bark of *M. griffoniana* [114,115]. ALP activity of these phytochemicals in Ishikawa cells indicated an effective concentration of 10^{-6} M for 4′-*O*-geranylisoquiritigenin, 7-*O*-geranylformononetin, griffonianone C, and 4′-methoxy-7-*O*-[(*E*)-3-methyl-7-hydroxymethyl-2,6-octadienyl]isoflavone. These compounds showed moderate estrogenic activity, which was suppressed by cotreatment of estrogen antagonist ICI 182,780 suggesting that these compounds show activity by binding to ER [75].

6.3 Lignans

Lignans are important antiosteoporotic agents among phytoestrogens distributed extensively in the plant kingdom. Lignans are dimers of two C6-C3 moieties biogenetically originated by the shikimic acid pathway. Some very interesting molecules with promising osteoprotective and curative activity (Fig. 8.3) have been isolated from different plants. Vitexdoin F (**48**) is a

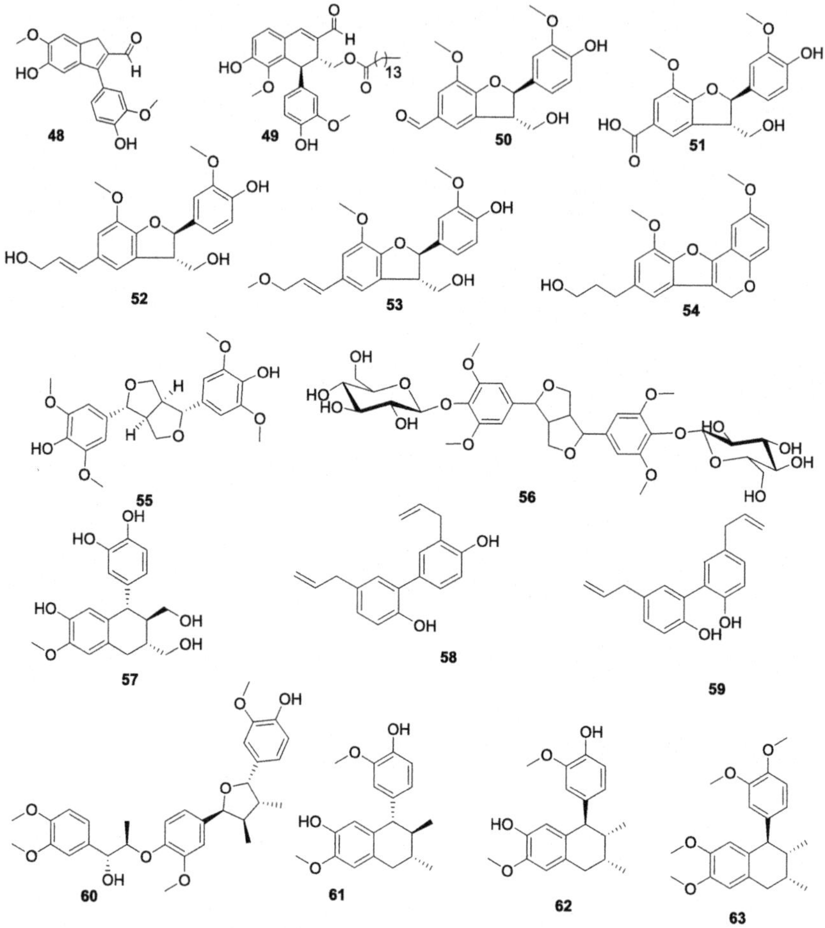

FIGURE 8.3 Chemical structure of natural lignans having antiosteoporotic activity.

phenylindene-type lignan isolated from the ethanolic (80% v/v) extract of *Vitex negundo* seeds [116]. Antiosteoporotic capacity of the isolated compounds was examined on osteoblast-like UMR 106 and osteoclastic cells. Vitexdoin F (**48**) exhibited substantial stimulation of proliferation and ALP activity of osteoblast-like UMR 106 cells at 10^{-7} M concentration with a significant increase (nearly twofold) in the OPG/RANKL ratio. However, the compound was devoid of preventive activity of osteoclastic cells. Another lignan isolated from the same plant, vitexdoin H (**49**), showed 8.0% inhibition of osteoclastic TRAP activity at 10^{-7} M concentration [117]. (7*R*,8*S*)-Ficusal (**50**), (7*R*,8*S*)-ceplignan (**51**), (7*R*,8*S*)-dehydrodiconiferyl alcohol (**52**), (7*R*,8*S*)-dehydrodiconiferyl alcohol-γ′-methyl ether (**53**), and 2,10-dimethoxy-8-hydroxypropyl-benzofuran [3,2-c][1]benzopyran (**54**) (Samwinol) have been isolated from *Sambucus williamsii*. These natural lignans exerted significant promoting effects on cell proliferation of osteoblastic-like UMR 106 cells. (7*R*,8*S*)-Ficusal, (7*R*,8*S*)-ceplignan, (7*R*,8*S*)-dehydrodiconiferyl alcohol, (7*R*,8*S*)-dehydrodiconiferyl alcohol-γ′-methyl ether, and Samwinol exerted a 31.3%, 28.3%, 25.6%, 25.1%, and 26.0% increase is cell numbers at 10^{-10}, 10^{-12}, 10^{-7}, 10^{-10}, and 10^{-10} M concentrations, respectively [118]. Phytochemical investigation of 90% ethanol fraction of the rhizomes of *Dioscorea spongiosa* led to the isolation of (+)-syringaresinol (**55**). Compound **55** increased osteocyte proliferation by 120% at a concentration of 10 μM higher than that of 17β-estradiol, with positive control and moderate mineralizing activity at 2 μM [119]. 3,3′,5,5′-Tetramethoxy-7,9′,7′,9-diepoxylignin-4,4′-di-*O*-β-D-glucopyranoside (**56**) is a main lignan glycoside isolated from *Curculigo orchioides* (Hypoxidaceae). The compound has an osteogenic effect by increasing osteoblast proliferation, production of ALP, and osteoprotective effects through decreasing the area of the bone resorption pit, osteoclastic formation, and TRAP activity [120]. Isotaxiresinol (**57**) was isolated from the aqueous extract of the wood of *Taxus yunnanensis*. Oral administration of isotaxiresinol (50 and 100 mg/kg/day) for 6 weeks to OVx rats increased bone mineral content and BMD in total and cortical bones. Serum biochemical markers for bone remodeling indicated significant inhibited bone resorption without any uterotropic effect [121]. Honokiol (**58**) and magnolol (**59**) are important lignan compounds isolated from *Magnolia officinalis*. These lignans have osteogenic as well as osteoprotective effects. Osteogenic effects are supported by significant elevation of cell growth, ALP activity, collagen synthesis, mineralization, and glutathione content in osteoblasts, whereas decreased osteoclast differentiation inducing RANKL, TNF-α, and IL-6 supported osteoprotective effects [122–124]. (−)-Saucerneol (**60**) is an important antiresorptive lignan isolated from *Saururus chinensis* (Saururaceae). (−)-Saucerneol showed significant inhibition of osteoclast formation as indicated by inhibition of RANKL-induced activity of TRAP [125].

Arylnaphthalene-type lignans such as guaiacin (**61**), isoguaiacin (**62**), and isoguaiacin dimethylether (**63**) isolated from the bark of *Machilus thunbergii* significantly increased ALP activity. Isoguaiacin dimethylether was found to be most active, followed by isoguaiacin and guaiacin. Isoguaiacin dimethylether increased ALP activity by 125.7 ± 7.4, 147.8 ± 4.0, and $149.0 \pm 9.4\%$, whereas guaiacin increased ALP activity by 111.4 ± 5.1, 122.0 ± 6.8, and $128.2 \pm 8.7\%$ compared to control at a concentration of 1, 10, and 25 μM, respectively. Additionally, isoguaiacin and isoguaiacin dimethylether increased osteoblast mineralization by about 126% and 138% as compared to control, respectively [126].

6.4 Coumarins

Coumarins are phytoestrogenic constituents that modulate bone metabolism through ER. Coumestrol (**64**) inhibited RANKL-induced TRAP activity, the

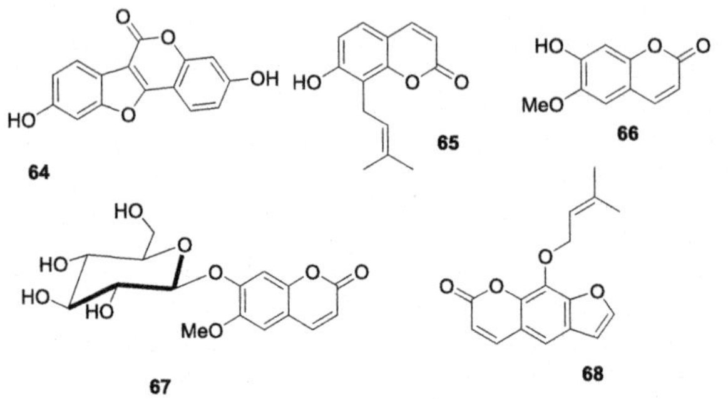

FIGURE 8.4 Chemical structure of coumarins.

formation of TRAP-positive multinucleated cells, and the formation of resorption pits, suggesting its role in the inhibition of osteoclast differentiation [127]. Coumestrol (**64**) (Fig. 8.4) enhanced differentiation and mineralization of the MC3T3-E1 cell line as well as enhanced cellular Ca and P contents [128]. In primary osteoblast cultures, coumestrol increased proliferation and differentiation by a likely estrogen-independent mechanism [129].

Osthole (**65**) is a constituent of the fruits of *C. monnieri* Cusson. (Apiaceae). Compound **65**, tested for the proliferation of osteoblast-like UMR 106 cells in vitro, significantly promoted the cells' activity. Osthole stimulated osteoblast proliferation and differentiation through β-catenin/BMP signaling [130]. Scopoletin (**66**) and scopolin (**67**) (scopoletin-7-*O*-β-D-glucopyranoside) were isolated from the ethanolic extract of the roots of *Artemisia iwayomogi* [131]; the former was also isolated from the ethanolic extract of the roots of *Morinda officinalis*. Ethanolic extract of *M. officinalis* and its ethyl acetate fraction showed osteogenic and osteoprotective activity. Ethyl acetate fraction at a concentration of 20 μg/mL increased osteoblast proliferation and ALP activity by 90.1% and 33.9%, respectively, and inhibited osteoclastic TRAP activity by 55.0%. Scopoletin (**66**) and scopolin (**67**) suppressed the differentiation of preosteoclastic RAW 264.7 cells [132] and stimulated differentiation of osteoblastic MC3T3-E1 cells [133].

Their antiosteoporotic activity was established by investigation of RANKL-induced osteoclast differentiation and TRAP activity of murine macrophage RAW 264.7 cells, which showed dose-dependent decrease in osteoclast differentiation and TRAP activity. Scopoletin was found to be more active than scopolin, probably due to lower membrane permeability associated with glycosides [131]. Osthole (**65**) and imperatorin (**68**) showed proliferative effects in MCF-7 cells and Saos-2 cells. Osthole and imperatorin showed maximum stimulation of MCF-7 cell proliferation at concentrations of 1 and 10 μM/L, respectively, which is comparable to the proliferative effect of 1 nmol/L estradiol. The proliferative effects are completely dependent on ER as suggested by complete block of proliferative effects by ER antagonist ICI182,780 [134].

6.5 Alkaloids

Cathepsin K, matrix metalloproteinase-9 (MMP-9), and TRAP activity inhibitors are the effective antiresorption drug agents for prevention and treatment in osteoporosis therapy. Berberine (**69**) (Fig. 8.5) is an osteoprotective active constituent of a traditional Chinese formula for bone resorption [135]. Berberine inhibited TRAP activity of osteoclasts and has an inhibitory effect on bone loss without any osteogenic effects. In vivo studies on OVx rats showed decreased osteoclast number, increased bone density, and trabecular thickness at the dose of 30 mg/kg of berberine [87]. Daily oral administration of berberine at a dose of 30 or 50 mg/kg to ovariectomized rats inhibited the formation and bone-resorbing activity of osteoclasts without estrogenic effects [136]. 8,8″-Biskoenigine (**70**), a carbazole alkaloid isolated from the ethanolic extract of aerial parts of *Murraya koenigii* (Rutaceae), showed inhibition of cathepsin K with IC_{50} 1.18 ± 0.04 μg/mL, and activity in the CAT-B model with IC_{50} of 1.3 μg/mL [137]. Rutaccarpine (**71**), an alkaloid isolated from *Winchia calophylla*, showed MMP-9 inhibition with IC_{50} 7.95 ± 0.43 μg/mL [138]. Coptisine (**72**) is an isoquinoline alkaloid that was isolated from *Coptidis* rhizome, which inhibited bone resorption significantly by suppressing formation and differentiation of osteoclasts through regulation of RANKL and OPG gene and RANKL-induced expression of NFATc1 [139]. Harmine (**73**), a β-carboline alkaloid present in a number of different plants, significantly

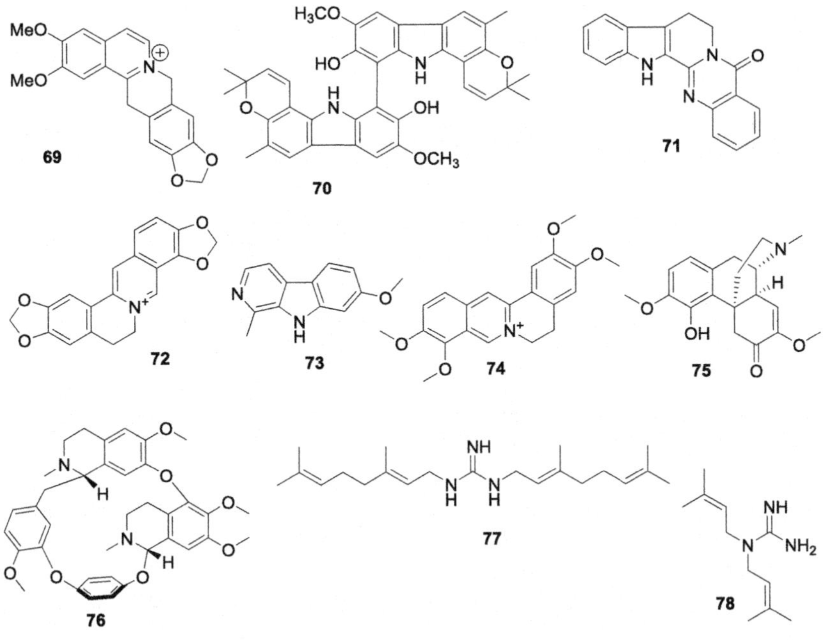

FIGURE 8.5 Chemical structure of plant-derived alkaloids.

inhibits osteoclast formation by inhibiting RANKL-induced TRAP activity in RAW 264.7 with IC_{50} 3.3 μM [140]. Palmatine (**74**) is an isoquinoline alkaloid found in many plants such as *Enantia chlorantha*, *Rhizoma coptidis*, *Radix tinosporae*, and *Cortex phellodendri*. Palmatine inhibited osteoclast formation and reduced the viability of mature osteoclasts by disrupting actin ring formation [141]. Sinomenine (**75**) is a quinoline alkaloid isolated from the roots and stems of *Sinomenium acutum*. The alkaloid shows osteoprotective effects as the compound is found to attenuate the viability of mature osteoclasts. The resorbing ability of osteoclasts is also hampered by the inhibition of actin ring formation, a marker colligated with actively resorbing osteoclasts, which induces the apoptosis of mature osteoclasts at 0.5 mmol/L concentrations [142]. Tetrandrine (**76**) is a bisbenzylisoquinoline alkaloid found in the root of *Stephania tetrandra*. Compound tetrandrine shows osteoprotective effects in sciatic-neurectomized osteoporosis model mice. Tetrandrine (**76**) inhibits osteoclast differentiation from osteoclast precursors and suppresses RANKL-induced amplification of NFATc1 [143]. Nitensidine A (**77**) and pterogynine (**78**) are guanidine alkaloids extracted from the South American plant *Pterogyne nitens*. Both compounds exhibited inhibition of osteoclastogenesis with IC_{50} 0.93 ± 0.25 and IC_{50} 2.7 ± 0.40 μM, respectively [144].

6.6 Conclusions

Osteoporosis is an emerging issue for the elderly society, especially women, all over the world, therefore discovery and development of plant-based formulations for the treatment of osteoporosis is essential to promote healthy long life. The present scenario of antiosteoporotic drugs is mainly occupied by synthetic drugs, which although successful in the treatment of osteoporosis are associated with side effects. Certain plants are described in traditional medicine systems for the achievement of healthy bones. Further investigation of plants leads to identification of several plants with better preventive and curative properties. Enormous studies have explored the role of natural products in osteoporosis. A wide diversity of chemical structures has been reported to exhibit antiosteoporotic activity. Secondary metabolites such as flavonoids, isoflavonoids, lignins, coumarins, and a number of alkaloids from plants are the prime class of plant secondary metabolites to develop medicines or dietary supplements for the prevention and/or cure of osteoporosis. These chemical skeletons could provide a scaffold for the development of new antiosteoporotic drugs. The diversity of chemical structures with potential biological activity presented in this chapter provides a glimpse at plant-based drugs and will provide a guide map to the development of new antiosteoporotic drugs.

REFERENCES

[1] K. Matsuo, N. Irie, Osteoclast-osteoblast communication, Arch. Biochem. Biophys. 473 (2) (2008) 201–209.

[2] M. Mafi Golchin, et al., Osteoporosis: a silent disease with complex genetic contribution, J. Genet. Genomics 43 (2) (2016) 49–61.

[3] P.M. Cawthon, Gender differences in osteoporosis and fractures, Clin. Orthopaedics Relat. Res. 469 (7) (2011) 1900–1905.

[4] R. Civitelli, et al., Bone turnover in postmenopausal osteoporosis. Effect of calcitonin treatment, J. Clin. Invest. 82 (4) (1988) 1268–1274.

[5] Y. Shen, D.L. Gray, D.S. Martinez, Combined pharmacologic therapy in postmenopausal osteoporosis, Endocrinol. Metab. Clin. North Am. 46 (1) (2017) 193–206.

[6] L. Bandeira, J.P. Bilezikian, Novel therapies for postmenopausal osteoporosis, Endocrinol. Metab. Clin. North Am. 46 (1) (2017) 207–219.

[7] S. Boonen, et al., Osteoporosis and osteoporotic fracture occurrence and prevention in the elderly: a geriatric perspective, Best Pract. Res. Clin. Endocrinol. Metab. 22 (5) (2008) 765–785.

[8] A. Clegg, et al., Frailty in elderly people, Lancet 381 (9868) (2013) 752–762.

[9] F.D. Shuler, et al., Understanding the burden of osteoporosis and use of the World Health Organization FRAX, Orthopedics 35 (9) (2012) 798–805.

[10] O. Johnell, J.A. Kanis, An estimate of the worldwide prevalence and disability associated with osteoporotic fractures, Osteoporos. Int. 17 (12) (2006) 1726–1733.

[11] J.A. Cauley, et al., Geographic and ethnic disparities in osteoporotic fractures, Nat. Rev. Endocrinol. 10 (6) (2014) 338–351.

[12] L.J. Melton 3rd, et al., Perspective. How many women have osteoporosis? J. Bone Miner Res. 7 (9) (1992) 1005–1010.

[13] G.A. Rodan, Bone homeostasis, Proc. Natl. Acad. Sci. USA 95 (23) (1998) 13361–13362.

[14] P. Garnero, P.D. Delmas, Biochemical markers of bone turnover: clinical usefulness in osteoporosis, Ann. Biol. Clin. (Paris) 57 (2) (1999) 137–148.

[15] S. Green, C.L. Anstiss, W.H. Fishman, Automated differential isoenzyme analysis. II. The fractionation of serum alkaline phosphatases into "liver", "intestinal" and "other" components, Enzymologia 41 (1) (1971) 9–26.

[16] H. Harris, The human alkaline phosphatases: what we know and what we don't know, Clin. Chim. Acta 186 (2) (1990) 133–150.

[17] L. Gorman, B.E. Statland, Clinical usefulness of alkaline phosphatase isoenzyme determinations, Clin. Biochem. 10 (5) (1977) 171–174.

[18] C.S. Hill, R.L. Wolfert, The preparation of monoclonal antibodies which react preferentially with human bone alkaline phosphatase and not liver alkaline phosphatase, Clin. Chim. Acta 186 (2) (1990) 315–320.

[19] G. Karsenty, M. Ferron, The contribution of bone to whole-organism physiology, Nature 481 (7381) (2012) 314–320.

[20] J.P. Brown, et al., Serum bone Gla-protein: a specific marker for bone formation in postmenopausal osteoporosis, Lancet 1 (8386) (1984) 1091–1093.

[21] J. Shukla, et al., Assessment of bone loss in postmenopausal women by evaluation of urinary hydroxyproline and serum status of osteocalcin, Int. Res. J. Biol. Sci. 2 (9) (2013) 11–14.

[22] S.H. Liu, et al., Collagen in tendon, ligament, and bone healing. A current review, Clin. Orthop. Relat. Res. (318) (1995) 265–278.

[23] A.H. Merry, et al., Identification and partial characterisation of the non-collagenous amino- and carboxyl-terminal extension peptides of cartilage procollagen, Biochem. Biophys. Res. Commun. 71 (1) (1976) 83–90.

[24] D. Bauer, et al., National Bone Health Alliance Bone Turnover Marker Project: current practices and the need for US harmonization, standardization, and common reference ranges, Osteoporos. Int. 23 (10) (2012) 2425–2433.

[25] G.W. Oddie, et al., Structure, function, and regulation of tartrate-resistant acid phosphatase, Bone 27 (5) (2000) 575–584.

[26] C. Minkin, Bone acid phosphatase: tartrate-resistant acid phosphatase as a marker of osteoclast function, Calcif Tissue Int. 34 (3) (1982) 285–290.

[27] J.M. Halleen, et al., Tartrate-resistant acid phosphatase 5b: a novel serum marker of bone resorption, J. Bone Miner Res. 15 (7) (2000) 1337–1345.

[28] S. Khosla, Minireview: the OPG/RANKL/RANK system, Endocrinology 142 (12) (2001) 5050–5055.

[29] P. Saftig, et al., Impaired osteoclastic bone resorption leads to osteopetrosis in cathepsin-K-deficient mice, Proc. Natl. Acad. Sci. USA 95 (23) (1998) 13453–13458.

[30] M. Muñoz-Torres, et al., Serum cathepsin K as a marker of bone metabolism in postmenopausal women treated with alendronate, Maturitas 64 (3) (2009) 188–192.

[31] F.P. Reinholt, et al., Osteopontin–a possible anchor of osteoclasts to bone, Proc. Natl. Acad. Sci. 87 (12) (1990) 4473–4475.

[32] D. Fodor, et al., The value of osteopontin in the assessment of bone mineral density status in postmenopausal women, J. Investig. Med. 61 (1) (2013) 15–21.

[33] B.F. Boyce, L. Xing, The RANKL/RANK/OPG pathway, Curr. Osteoporos. Rep. 5 (3) (2007) 98–104.

[34] L. Moro, et al., The glycosides of hydroxylysine are final products of collagen degradation in humans, Biochim. Biophys. Acta 1156 (3) (1993) 288–290.

[35] L. Moro, et al., High-performance liquid chromatographic analysis of urinary hydroxylysyl glycosides as indicators of collagen turnover, Analyst 109 (12) (1984) 1621–1622.

[36] E.M. Czekanska, et al., In search of an osteoblast cell model for in vitro research, Eur. Cell Mater. 24 (2012) 1–17.

[37] V. Kartsogiannis, K.W. Ng, Cell lines and primary cell cultures in the study of bone cell biology, Mol. Cell Endocrinol. 228 (1–2) (2004) 79–102.

[38] I.E. James, et al., Purification and characterization of fully functional human osteoclast precursors, J. Bone Mineral Res. 11 (11) (1996) 1608–1618.

[39] P.P. Lelovas, et al., The laboratory rat as an animal model for osteoporosis research, Comp. Med. 58 (5) (2008) 424–430.

[40] D.D. Thompson, et al., FDA Guidelines and animal models for osteoporosis, Bone 17 (Suppl. 4) (1995) 125s–133s.

[41] C.C. Danielsen, L. Mosekilde, B. Svenstrup, Cortical bone mass, composition, and mechanical properties in female rats in relation to age, long-term ovariectomy, and estrogen substitution, Calcif Tissue Int. 52 (1) (1993) 26–33.

[42] C.C. He, et al., Osteoprotective effect of extract from *Achyranthes bidentata* in ovariectomized rats, J. Ethnopharmacol. 127 (2) (2010) 229–234.

[43] R. Zhang, et al., *Achyranthes bidentata* root extract prevent OVX-induced osteoporosis in rats, J. Ethnopharmacol. 139 (1) (2012) 12–18.

[44] R. Maurya, et al., Osteogenic activity of constituents from *Butea monosperma*, Bioorg. Med. Chem. Lett. 19 (3) (2009) 610–613.

[45] A. Shirwaikar, S. Khan, S. Malini, Antiosteoporotic effect of ethanol extract of *Cissus quadrangularis* Linn. on ovariectomized rat, J. Ethnopharmacol. 89 (2–3) (2003) 245–250.

[46] N. Singh, et al., Osteogenic potential of *Cissus qudrangularis* assessed with osteopontin expression, Natl. J. Maxillofac. Surg. 4 (1) (2013) 52–56.

[47] Y.-M. Li, et al., *Cnidium monnieri*: a review of traditional uses, phytochemical and ethnopharmacological properties, Am. J. Chin. Med. 43 (5) (2015) 835–877.

[48] V. Khedgikar, et al., Ethanolic extract of *Dalbergia sissoo* promotes rapid regeneration of cortical bone in drill-hole defect model of rat, Biomed. Pharmacother. 86 (2017) 16–22.

[49] V. Khedgikar, et al., A standardized phytopreparation from an Indian medicinal plant (*Dalbergia sissoo*) has antiresorptive and bone-forming effects on a postmenopausal osteoporosis model of rat, Menopause 19 (12) (2012) 1336–1346.

[50] R. Zhang, et al., Du-Zhong (*Eucommia ulmoides* Oliv.) cortex extract prevent OVX-induced osteoporosis in rats, Bone 45 (3) (2009) 553–559.

[51] E.M. Choi, et al., Soybean ethanol extract increases the function of osteoblastic MC3T3-E1 cells, Phytochemistry 56 (7) (2001) 733–739.

[52] S.N. Fathilah, et al., *Labisia pumila* protects the bone of estrogen-deficient rat model: a histomorphometric study, J. Ethnopharmacol 142 (1) (2012) 294–299.

[53] N. Ahmad, et al., Evaluation of anti-osteoporotic activity of butanolic fraction from Passiflora foetida in ovariectomy-induced bone loss in mice, Biomed. Pharmacother. 88 (2017) 804–813.

[54] F.T. Ngueguim, et al., Ethanol extract of *Peperomia pellucida* (Piperaceae) promotes fracture healing by an anabolic effect on osteoblasts, J. Ethnopharmacol. 148 (1) (2013) 62–68.

[55] K. Porwal, et al., Guava fruit extract and its triterpene constituents have osteoanabolic effect: stimulation of osteoblast differentiation by activation of mitochondrial respiration via the Wnt/β-catenin signaling, J. Nutr. Biochem. 44 (2017) 22–34.

[56] Y. Guo, et al., *Salvia miltiorrhiza*: an ancient Chinese herbal medicine as a source for anti-osteoporotic drugs, J. Ethnopharmacol. 155 (3) (2014) 1401–1416.

[57] Y. Cui, et al., Characterization of *Salvia miltiorrhiza* ethanol extract as an anti-osteoporotic agent, BMC Complement Altern. Med. 11 (2011) 120.

[58] H. Li, et al., Antiosteoporotic activity of the stems of *Sambucus sieboldiana*, Biol. Pharm. Bull. 21 (6) (1998) 594–598.

[59] S. Adhikary, et al., Dried and free flowing granules of *Spinacia oleracea* accelerate bone regeneration and alleviate postmenopausal osteoporosis, Menopause (2017). Published Ahead-of-Print).

[60] K. Sharan, et al., Extract and fraction from *Ulmus wallichiana* Planchon promote peak bone achievement and have a nonestrogenic osteoprotective effect, Menopause 17 (2) (2010) 393–402.

[61] G. Abiramasundari, K.R. Sumalatha, M. Sreepriya, Effects of Tinospora cordifolia (Menispermaceae) on the proliferation, osteogenic differentiation and mineralization of osteoblast model systems in vitro, J. Ethnopharmacol. 141 (1) (2012) 474–480.

[62] P. Kapur, et al., Evaluation of the antiosteoporotic potential of Tinospora cordifolia in female rats, Maturitas 59 (4) (2008) 329–338.

[63] S. Sakamoto, et al., Preventive effects of a herbal medicine on bone loss in rats treated with a GnRH agonist, Eur. J. Endocrinol. 143 (1) (2000) 139–142.

[64] N. Chattopadhyay, D.K. Sharma, Postmenopausal Osteoporosis and its Therapies, in Reference Module in Biomedical Sciences, Elsevier, 2016.

[65] A. Wattel, et al., Potent inhibitory effect of naturally occurring flavonoids quercetin and kaempferol on in vitro osteoclastic bone resorption, Biochem. Pharmacol. 65 (1) (2003) 35–42.

[66] A.J. Guo, et al., Kaempferol as a flavonoid induces osteoblastic differentiation via estrogen receptor signaling, Chin. Med. (London, UK) 7 (2012) 10.

[67] J. Liu, et al., Screening of osteoanagenesis-active compounds from Scutellaria baicalensis Georgi by hPDLC/CMC-online-HPLC/MS, Fitoterapia 93 (2014) 105–114.

[68] S. Bandyopadhyay, et al., Attenuation of osteoclastogenesis and osteoclast function by apigenin, Biochem. Pharmacol. 72 (2) (2006) 184–197.

[69] N.K. Im, et al., 6,4′-Dihydroxy-7-methoxyflavanone inhibits osteoclast differentiation and function, Biol. Pharm. Bull. 36 (5) (2013) 796–801.

[70] C. Puel, et al., Prevention of bone loss by phloridzin, an apple polyphenol, in ovariectomized rats under inflammation conditions, Calcif Tissue Int. 77 (5) (2005) 311–318.

[71] E.M. Choi, Liquiritigenin isolated from Glycyrrhiza uralensis stimulates osteoblast function in osteoblastic MC3T3-E1 cells, Int. Immunopharmacol. 12 (1) (2012) 139–143.

[72] K. Uchino, et al., Dual effects of Liquiritigenin on the proliferation of bone cells: promotion of osteoblast differentiation and inhibition of osteoclast differentiation, Phytotherapy Res. 29 (11) (2015) 1714–1721.

[73] M. Atif, et al., Pharmacological assessment of hispidulin – a natural bioactive flavone, Acta Pol. Pharm. 73 (3) (2016) 565–578.

[74] M. Nepal, et al., Hispidulin attenuates bone resorption and osteoclastogenesis via the RANKL-induced NF-kappaB and NFATc1 pathways, Eur. J. Pharmacol. 715 (1–3) (2013) 96–104.

[75] G.J. Wanda, et al., Estrogenic properties of isoflavones derived from *Millettia griffoniana*, Phytomedicine 13 (3) (2006) 139–145.

[76] P. Dixit, et al., Osteogenic constituents from *Pterospermum acerifolium* Willd. flowers, Bioorg. Med. Chem. Lett. 21 (15) (2011) 4617–4621.

[77] H. Suh, et al., Syntheses of (±)-shinflavanone and its structural analogues as potent inhibitors of bone resorption pits formation, Bioorg. Med. Chem. Lett. 9 (10) (1999) 1433–1436.

[78] P. Rawat, et al., Ulmosides A and B: flavonoid 6-C-glycosides from *Ulmus wallichiana*, stimulating osteoblast differentiation assessed by alkaline phosphatase, Bioorg. Med. Chem. Lett. 19 (16) (2009) 4684–4687.

[79] G. Swarnkar, et al., A novel flavonoid isolated from the steam-bark of *Ulmus wallichiana* Planchon stimulates osteoblast function and inhibits osteoclast and adipocyte differentiation, Eur. J. Pharmacol. 658 (2–3) (2011) 65–73.

[80] K. Sharan, et al., A novel flavonoid, 6-C-beta-d-glucopyranosyl-(2S,3S)-(+)-3′,4′,5,7-tetrahydroxyflavanone, isolated from *Ulmus wallichiana* Planchon mitigates ovariectomy-induced osteoporosis in rats, Menopause 17 (3) (2010) 577–586.

[81] J. Li, et al., Effects on cultured neonatal mouse calvaria of the flavonoids isolated from *Boerhaavia repens*, J. Nat. Prod. 59 (11) (1996) 1015–1018.

[82] E. Ramakrishna, et al., Phytochemical investigation of *Kigelia pinnata* leaves and identification of osteogenic agents, Med. Chem. Res. (2017). Ahead of Print.

[83] J.A. Siddiqui, et al., A naturally occurring rare analog of quercetin promotes peak bone mass achievement and exerts anabolic effect on osteoporotic bone, Osteoporos. Int. 22 (12) (2011) 3013–3027.

[84] J.A. Siddiqui, et al., Quercetin-6-C-β-D-glucopyranoside isolated from *Ulmus wallichiana* planchon is more potent than quercetin in inhibiting osteoclastogenesis and mitigating ovariectomy-induced bone loss in rats, Menopause 18 (2) (2011) 198–207.

[85] Y.J. Chen, P.C. Kung, Preparations Containing Licariin WO1999047137 A1, 1999.

[86] J. Wang, et al., Chemical constituents of *Epimedium platyetalum*, Acta Bot. Sin. 44 (10) (2002) 1258–1260.

[87] L. Qin, et al., Antiosteoporotic chemical constituents from Er-Xian Decoction, a traditional Chinese herbal formula, J. Ethnopharmacol. 118 (2) (2008) 271–279.

[88] L. Yang, N.L. Wang, G.P. Cai, Maohuoside A promotes osteogenesis of rat mesenchymal stem cells via BMP and MAPK signaling pathways, Mol. Cell Biochem. 358 (1–2) (2011) 37–44.

[89] M. Cai, et al., Maohuoside A acts in a BMP-dependent manner during osteogenesis, Phytother Res. 27 (8) (2013) 1179–1184.

[90] M.K. Lee, et al., Osteoblast differentiation stimulating activity of biflavonoids from *Cephalotaxus koreana*, Bioorg. Med. Chem. Lett. 16 (11) (2006) 2850–2854.

[91] J.A. Siddiqui, et al., 8,8′-Biapigeninyl stimulates osteoblast functions and inhibits osteoclast and adipocyte functions: osteoprotective action of 8,8′-biapigeninyl in ovariectomized mice, Mol. Cell. Endocrinol. 323 (2010) 256–267.

[92] C.M. Rassi, et al., Down-regulation of osteoclast differentiation by daidzein via caspase 3, J. Bone Miner Res. 17 (4) (2002) 630–638.

[93] L. Bao, S.E. Zou, S.F. Zhang, Dose-dependent effects of daidzein in regulating bone formation through estrogen receptors and peroxisome proliferator-activated receptor gamma, Zhong Xi Yi Jie He Xue Bao 9 (2) (2011) 165–172.

[94] A.M. Tyagi, et al., Daidzein prevents the increase in CD4+CD28 null T cells and B lymphopoesis in ovariectomized mice: a key mechanism for anti-osteoclastogenic effect, PLoS One 6 (6) (2011) e21216.

[95] B. Filipovic, et al., Daidzein administration positively affects thyroid C cells and bone structure in orchidectomized middle-aged rats, Osteoporos. Int. 21 (9) (2010) 1609–1616.

[96] L. Zhang, et al., Glycitin regulates osteoblasts through TGF-beta or AKT signaling pathways in bone marrow stem cells, Exp. Ther. Med. 12 (5) (2016) 3063–3067.

[97] J. Wu, et al., Cooperative effects of exercise training and genistein administration on bone mass in ovariectomized mice, J. Bone Miner Res. 16 (10) (2001) 1829–1836.

[98] Z.L. Wang, et al., Pharmacological studies of the large-scaled purified genistein from Huaijiao (*Sophora japonica*-Leguminosae) on anti-osteoporosis, Phytomedicine 13 (9–10) (2006) 718–723.

[99] T. Uesugi, et al., Comparative study on reduction of bone loss and lipid metabolism abnormality in ovariectomized rats by soy isoflavones, daidzin, genistin, and glycitin, Biol. Pharm. Bull. 24 (4) (2001) 368–372.

[100] B. Bhargavan, et al., Methoxylated isoflavones, cajanin and isoformononetin, have non-estrogenic bone forming effect via differential mitogen activated protein kinase (MAPK) signaling, J. Cell. Biochem. 108 (2) (2009) 388–399.

[101] K. Srivastava, et al., Isoformononetin, a methoxydaidzein present in medicinal plants, reverses bone loss in osteopenic rats and exerts bone anabolic action by preventing osteoblast apoptosis, Phytomedicine 20 (6) (2013) 470–480.

[102] A.K. Gautam, et al., Differential effects of formononetin and cladrin on osteoblast function, peak bone mass achievement and bioavailability in rats, J. Nutr. Biochem. 22 (4) (2011) 318–327.

[103] K. Khan, et al., Positive skeletal effects of cladrin, a naturally occurring dimethoxydaidzein, in osteopenic rats that were maintained after treatment discontinuation, Osteoporos. Int. 24 (4) (2013) 1455–1470.

[104] J. Gautam, et al., An isoflavone cladrin prevents high-fat diet-induced bone loss and inhibits the expression of adipogenic gene regulators in 3T3-L1 adipocyte, J. Pharm. Pharmacol. 68 (8) (2016) 1051–1063.

[105] J.H. Wu, et al., Formononetin, an isoflavone, relaxes rat isolated aorta through endothelium-dependent and endothelium-independent pathways, J. Nutr. Biochem. 21 (7) (2010) 613–620.

[106] K.H. Lee, E.M. Choi, Biochanin A stimulates osteoblastic differentiation and inhibits hydrogen peroxide-induced production of inflammatory mediators in MC3T3-E1 cells, Biol. Pharm. Bull. 28 (10) (2005) 1948–1953.

[107] K. Sharan, et al., Role of phytochemicals in the prevention of menopausal bone loss: evidence from in vitro and in vivo, human interventional and pharmacokinetic studies, Curr. Med. Chem. 16 (9) (2009) 1138–1157.

[108] Z. Xiong, et al., Osteoblastic differentiation bloassay and its application to investigating the activity of fractions and compounds from *Psoralea corylifolia* L. Pharmazie 58 (12) (2003) 925–928.

[109] P. Dixit, et al., Constituents of *Dalbergia sissoo* Roxb. leaves with osteogenic activity, Bioorg. Med. Chem. Lett. 22 (2) (2012) 890–897.

[110] P. Kushwaha, et al., A novel therapeutic approach with Caviunin-based isoflavonoid that en routes bone marrow cells to bone formation via BMP2/Wnt-β-catenin signaling, Cell Death Dis. 5 (9) (2014) e1422.

[111] K. Khan, et al., Prunetin signals via G-protein-coupled receptor, GPR30(GPER1): stimulation of adenylyl cyclase and cAMP-mediated activation of MAPK signaling induces Runx2 expression in osteoblasts to promote bone regeneration, J. Nutr. Biochem. 26 (12) (2015) 1491–1501.

[112] D. Seidlova-Wuttke, et al., *Belamcanda chinensis* and the thereof purified tectorigenin have selective estrogen receptor modulator activities, Phytomedicine 11 (5) (2004) 392–403.

[113] M.-J. Don, L.-C. Lin, W.-F. Chiou, Neobavaisoflavone stimulates osteogenesis via p38-mediated up-regulation of transcription factors and osteoid genes expression in MC3T3-E1 cells, Phytomedicine 19 (6) (2012) 551–561.

[114] E. Yankep, Z.T. Fomum, E. Dagne, An O-geranylated isoflavone from *Millettia griffoniana*, Phytochemistry 46 (3) (1997) 591–593.

[115] E. Yankep, et al., O-Geranylated isoflavones and a3-phenylcoumarin from *Millettia griffoniana*, Phytochemistry 49 (8) (1998) 2521–2523.

[116] M. Kumar, et al., Anti-osteoporotic constituents from Indian medicinal plants, Phytomedicine 17 (13) (2010) 993–999.

[117] C.-J. Zheng, et al., Anti-inflammatory and anti-osteoporotic lignans from *Vitex negundo* seeds, Fitoterapia 93 (2014) 31–38.

[118] H.H. Xiao, et al., New lignans from the bioactive fraction of *Sambucus williamsii* Hance and proliferation activities on osteoblastic-like UMR106 cells, Fitoterapia 94 (2014) 29–35.

[119] J. Yin, et al., The in vitro anti-osteoporotic activity of some diarylheptanoids and lignans from the rhizomes of *Dioscorea spongiosa*, Planta Med. 74 (12) (2008) 1451–1453.

[120] L. Jiao, et al., Antiosteoporotic activity of phenolic compounds from *Curculigo orchioides*, Phytomedicine 16 (9) (2009) 874–881.

[121] J. Yin, et al., In vivo anti-osteoporotic activity of isotaxiresinol, a lignan from wood of Taxus yunnanensis, Phytomedicine 13 (1–2) (2006) 37–42.

[122] E.M. Choi, Honokiol isolated from *Magnolia officinalis* stimulates osteoblast function and inhibits the release of bone-resorbing mediators, Int. Immunopharmacol. 11 (10) (2011) 1541–1545.

[123] E.M. Choi, Honokiol protects osteoblastic MC3T3-E1 cells against antimycin A-induced cytotoxicity, Inflamm. Res. 60 (11) (2011) 1005–1012.

[124] E.M. Choi, Magnolol protects osteoblastic MC3T3-E1 cells against antimycin A-induced cytotoxicity through activation of mitochondrial function, Inflammation 35 (3) (2012) 1204–1212.

[125] S.N. Kim, et al., Inhibitory effect of (-)-saucerneol on osteoclast differentiation and bone pit formation, Phytother Res. 23 (2) (2009) 185–191.

[126] M.K. Lee, et al., Stimulatory activity of Lignans from *Machilus thunbergii* on osteoblast differentiation, Biol. Pharm. Bull. 30 (4) (2007) 814–817.

[127] S. Kanno, S. Hirano, F. Kayama, Effects of the phytoestrogen coumestrol on RANK-ligand-induced differentiation of osteoclasts, Toxicology 203 (1–3) (2004) 211–220.

[128] S. Kanno, S. Hirano, F. Kayama, Effects of phytoestrogens and environmental estrogens on osteoblastic differentiation in MC3T3-E1 cells, Toxicology 196 (1–2) (2004) 137–145.

[129] J.A. Lee, et al., 2,3,7,8-Tetrachlorodibenzo-p-dioxin modulates functional differentiation of mouse bone marrow-derived dendritic cells Downregulation of RelB by 2,3,7,8-tetrachlorodibenzo-p-dioxin, Toxicol. Lett. 173 (1) (2007) 31–40.

[130] Q. Zhang, et al., Coumarins from *Cnidium monnieri* and their antiosteoporotic activity, Planta Med. 73 (1) (2007) 13–19.

[131] S.H. Lee, et al., Scopoletin and scopolin isolated from *Artemisia iwayomogi* suppress differentiation of osteoclastic macrophage RAW 264.7 cells by scavenging reactive oxygen species, J. Nat. Prod. 76 (4) (2013) 615–620.

[132] Y.B. Wu, et al., Antiosteoporotic activity of anthraquinones from *Morinda officinalis* on osteoblasts and osteoclasts, Molecules 14 (1) (2009) 573–583.

[133] Y. Ding, et al., Phenolic compounds from *Artemisia iwayomogi* and their effects on osteoblastic MC3T3-E1 cells, Biol. Pharm. Bull. 33 (8) (2010) 1448–1453.

[134] M. Jia, et al., Estrogenic activity of osthole and imperatorin in MCF-7 cells and their osteoblastic effects in Saos-2 cells, Chin J. Nat. Med. 14 (6) (2016) 413–420.

[135] L. Xue, et al., Effects and interaction of icariin, curculigoside, and berberine in er-xian decoction, a traditional Chinese medicinal formula, on osteoclastic bone resorption, Evid. Based Complement. Alternat. Med. 2012 (2012) 490843.

[136] H. Li, et al., The effect of kampo formulae on bone resorption in vitro and in vivo. II. Detailed study of berberine, Biol. Pharm. Bull. 22 (4) (1999) 391–396.

[137] Y.S. Wang, et al., Two new carbazole alkaloids from *Murraya koenigii*, J. Nat. Prod. 66 (3) (2003) 416–418.

[138] G.Z. Zeng, et al., Natural inhibitors targeting osteoclast-mediated bone resorption, Bioorg. Med. Chem. Lett. 16 (24) (2006) 6178–6180.

[139] J.W. Lee, et al., Coptisine inhibits RANKL-induced NF-kappaB phosphorylation in osteoclast precursors and suppresses function through the regulation of RANKL and OPG gene expression in osteoblastic cells, J. Nat. Med. 66 (1) (2012) 8–16.

[140] T. Yonezawa, et al., Harmine, a beta-carboline alkaloid, inhibits osteoclast differentiation and bone resorption in vitro and in vivo, Eur. J. Pharmacol. 650 (2–3) (2011) 511–518.
[141] J.W. Lee, et al., Palmatine attenuates osteoclast differentiation and function through inhibition of receptor activator of nuclear factor-kappab ligand expression in osteoblast cells, Biol. Pharm. Bull. 33 (10) (2010) 1733–1739.
[142] L.G. He, et al., Sinomenine induces apoptosis in RAW 264.7 cell-derived osteoclasts in vitro via caspase-3 activation, Acta Pharmacol. Sin. 35 (2) (2014) 203–210.
[143] T. Takahashi, et al., Tetrandrine prevents bone loss in sciatic-neurectomized mice and inhibits receptor activator of nuclear factor kappaB ligand-induced osteoclast differentiation, Biol. Pharm. Bull. 35 (10) (2012) 1765–1774.
[144] Y. Tajima, et al., Nitensidine A, a guanidine alkaloid from *Pterogyne nitens*, induces osteoclastic cell death, Cytotechnology 67 (4) (2015) 585–592.

Chapter 9

Phytodrugs and Immunomodulators for the Therapy of Leishmaniasis

C. Benjamin Naman[1,*], Ciro M. Gomes[2,*], Gaurav Gupta[3,*]
[1]*Center for Marine Biotechnology and Biomedicine, University of California, San Diego, California, United States;* [2]*Hospital Universitário de Brasília, Brasília, Brazil;* [3]*Department of Immunology, University of Manitoba, Winnipeg, MB, Canada*

1. LEISHMANIASIS

1.1 Overview of the Disease

Leishmaniasis represents a set of infectious diseases that mainly affect those living in the tropical and subtropical regions of the world and in underdeveloped countries [1,2]. It is considered by the World Health Organization (WHO) to be one of the world's most neglected diseases that significantly reduces the quality of life in affected areas [2]. This disease is caused by protozoan parasites of the genus *Leishmania,* and is almost always transmitted by insects. The clinical manifestation of the disease is widely variable and depends both on genetic characteristics of the infecting parasite and on the host immune response [3]. This heterogeneity of clinical manifestation justifies the necessity of a detailed study of host–parasite interactions, diagnostic procedures, and treatment of leishmaniasis.

Control measures for leishmaniasis have been widely debated by the scientific community [2]. Recently, there has been an alarming growth of this disease, and epidemic outbreaks of anthroponotic leishmaniasis have been reported [4]. Uncontrolled urban expansion, climate change, and poor local health assistance are some of the factors that influence the disease expansion, especially in urban and periurban areas [3]. The treatment of leishmaniasis is still largely based on drugs that were developed more than half a century ago and have serious adverse side effects and contraindications to usage [5]. In addition, there has been an increased number of incidences of drug resistance in

* Equal contribution.

Natural Products and Drug Discovery. https://doi.org/10.1016/B978-0-08-102081-4.00009-5

endemic areas of leishmaniasis [6]. Moreover, the use of immunosuppressant drugs or patients with HIV coinfection are other major concerns that must also be addressed, as these have a direct influence on drug resistance in *Leishmania* [7].

The current situation provides an urgent need for new strategies of treatment and control of leishmaniasis. In this chapter, various clinical aspects of the disease leishmaniasis are discussed, including the current available drug modalities. Also covered is the importance of new emerging and investigational phytodrugs and immunomodulators for the therapy of leishmaniasis, as well as specific biochemical pathways to target in the parasite or the host for drug discovery and development.

1.2 Epidemiology

The epidemiology of leishmaniasis is complex, due to the diversity of causal species and vectors. It is estimated that about 2 million cases occur yearly in the world and that 350 million people are potentially at risk of acquiring leishmaniasis [2]. The disease is endemic in most tropical and subtropical countries and also affects southern Europe [2]. Leishmaniasis can be caused by more than 20 types of protozoan parasites of the genus *Leishmania* [1], and several of these are listed in Table 9.1. The parasites are transmitted by female bloodsucking insect vectors from more than 30 species of the genera *Phlebotomus* and *Lutzomyia* [1]. Leishmaniasis is classically considered to be a zoonosis of wild mammals such as rodents, marsupials, and primates. Humans usually acquire the infection on occasions of contact with forest areas during work or leisure activities. This epidemiological profile has been changing with the emergence of periurban or urban leishmaniasis, in which domestic animals and even humans can act as disease reservoirs [8,9]. Recently, multifactorial genetic analyses showed that the *Leishmania* genotype is strongly correlated to its continental origin [10]. Different geographical occurrence of leishmaniasis is also related to different species of vectors and parasites, ultimately resulting in disease peculiarities that depend on the place of infection. These attributes are correlated with different clinical manifestations and responsiveness to treatment.

Countries such as Brazil, Ethiopia, India, Somalia, South Sudan, and Sudan are associated with the detection of about 90% of all cases of visceral leishmaniasis (VL). In contrast, most cases of cutaneous and mucocutaneous leishmaniasis occur in Afghanistan, Algeria, Brazil, Colombia, the Islamic Republic of Iran, Pakistan, Peru, Saudi Arabia, and the Syrian Arab Republic [2]. According to the WHO, recurrent epidemics of VL are reported in East Africa. Also, major epidemics of cutaneous leishmaniasis have affected Middle Eastern countries, such as Afghanistan [2]. Large disease outbreaks have been reported in densely populated cities, especially in war and conflict zones, refugee camps, and other settings of large-scale population migration [2].

TABLE 9.1 Involved Species and Correspondent Clinical Manifestation of Leishmaniasis

Involved Species	Clinical Manifestation
New World Leishmaniasis	
Leishmania (Viannia) braziliensis	CL, MCL
Leishmania (Viannia) panamensis	CL, MCL
Leishmania (Viannia) peruviana	CL
Leishmania (Viannia) guyanensis	CL
Leishmania (Viannia) colombiensis	CL
L. (Leishmania) amazonensis	CL, VL, DCL
L. (Leishmania) mexicana	CL, DCL
L. (Leishmania) pifanoi	CL
L. (Leishmania) venezuelensis	CL
L. (Leishmania) garnhami	CL
L. (Leishmania) infantum (= L. chagasi)	VL
Old World Leishmaniasis	
L. (Leishmania) aethiopica	CL, DCL
L. (Leishmania) major	CL
L. (Leishmania) tropica	CL
L. (Leishmania) donovani	VL, CL
L. (Leishmania) infantum	VL, CL

CL, cutaneous leishmaniasis; *DCL*: diffuse cutaneous leishmaniasis; *MCL*, mucocutaneous leishmaniasis; *VL*, visceral leishmanias.

1.3 Life Cycle

The life cycle of *Leishmania* starts in parasite reservoirs. The classic reservoirs are animals from forests, most commonly rodents. In periurban leishmaniasis cases, domestic dogs and other animals must also be considered [1]. Humans in densely populated areas can also act as reservoirs and contribute to the cycle of anthroponotic leishmaniasis. The disease reservoir hosts have amastigote forms of *Leishmania* inside infected tissues. Bloodsucking insect vectors feed on an infected reservoir and will be contaminated with the macrophages infested with *Leishmania*. The amastigote forms transform into promastigotes in the midgut of the insect and then migrate to the proboscis [11]. Transmission to a new host occurs when infected vectors bite humans and other mammals, because they regurgitate some protozoan promastigotes during their blood meal. In the infected tissue, promastigotes will transform into amastigotes and, depending on host immunity multiply in macrophages and in other mononuclear phagocytic cells [2].

1.4 Disease Manifestation

There are several methods for classifying the clinical manifestations of leishmaniasis, but the high heterogeneity of clinical forms makes it difficult. The most used classifications of the disease consider the geographical occurrence of the disease and the main affected organs (Tables 9.1 and 9.2).

1.4.1 Geographical Classification

The geographical classification of leishmaniasis differentiates it into Old World leishmaniasis and New World leishmaniasis. Old World leishmaniasis occurs in the Eastern hemisphere, mostly in Middle East, Mediterranean

TABLE 9.2 Remarkable Clinical Differences Between New World and Old World Leishmaniasis

	New World Leishmaniasis	Old World Leishmaniasis
Visceral leishmaniasis	• Post–kala-azar dermal leishmaniasis is rare. Potentially fatal.	• Darkening of skin and mucosa is more common, typically found in India. Potentially fatal.
Cutaneous leishmaniasis	• Disease caused by *Leishmania* of the *Viannia* subgenus can be complicated by mucosal dissemination. • Systemic therapies are encouraged to avoid parasite dissemination.	• Disease caused by *L. infantum* and *L. tropica* tends to heal spontaneously. • This allows the use of local therapy.
Diffuse leishmaniasis	Clinically similar in the new world and in the old world	
Disseminated leishmaniasis	• Described in the Americas and related to local species. The disease coursed with several acneiform disseminated lesions.	–
Mucocutaneous or mucosal leishmaniasis	• Mucosal involvement is more common in the New World where the term "mucocutaneous leishmaniasis" is properly used.	Less frequently seen in the Old World.

littoral, the Arabian Peninsula, Africa, Asia, and the Indian Subcontinent of Asia [2]. The most important etiologic agents are *Leishmania tropica, Leishmania major, Leishmania aethiopica, Leishmania infantum*, and *Leishmania donovani* [2]. New World leishmaniasis occurs in the Western hemisphere, in the Americas. The most typically involved species are the *Leishmania mexicana* species complex (*Leishmania mexicana, Leishmania amazonensis,* and *Leishmania venezuelensis*) or the *Leishmania* (*Viannia*) *braziliensis* species complex [3]. Table 9.1 lists the relationship between the main causative agents of leishmaniasis, clinical manifestations, and the geographical occurrence of the parasite, and Table 9.2 lists the main differences between the Old World and the New World forms of leishmaniasis, as well as the most frequent clinical manifestations.

1.4.2 Clinical Classification

The clinical manifestations of leishmaniasis are broadly classified into visceral leishmaniasis (VL), cutaneous leishmaniasis (CL), and mucocutaneous leishmaniasis (MCL). However, other alternative and specific clinical forms are also described. Each is discussed here in more detail.

1.4.2.1 Visceral Leishmaniasis

VL is also known as kala-azar and is considered the most severe form of leishmaniasis [12]. In the Old World, the disease is caused by parasites of the *L. donovani–L. infantum* complex. In the New World, *L. infantum* (sometimes referred to as *L. chagasi*) is the causal agent resulting in a similar clinical presentation with that of VL caused by *L. infantum* in the Old World. Patients tend to take years to develop severe symptoms, and many infected patients remain asymptomatic. Previously infected patients who present a relative or severe reduction of cellular immunity are at risk of developing classic signs of VL. Some of the most important risk factors for disease progression are age (children and elderly patients are most susceptible), malnutrition, genetic factors, and immunosuppression [13]. This disease is characterized by lymphadenopathy, continuous and chronic fever, shivering, and a reticuloendothelial hyperplasia resulting in enlargement of the spleen (Fig. 9.1). In some cases, enlargement of the liver is also observed [13]. The patient can present with severe cases of liver and renal failure and ultimately death [2]. The fatality rate for VL is estimated to be 10%–20%, with the understanding that most deaths due to this disease likely go causally unrecognized [2,14]. Furthermore, the WHO states that "if the disease is not treated, the fatality rate in developing countries can be as high as 100% within 2 years."

1.4.2.2 Post–Kala-Azar Dermal leishmaniasis

Post–kala-azar dermal leishmaniasis is a frequent manifestation associated with VL caused by *L. donovani* in Africa and Asia [15]. It frequently occurs

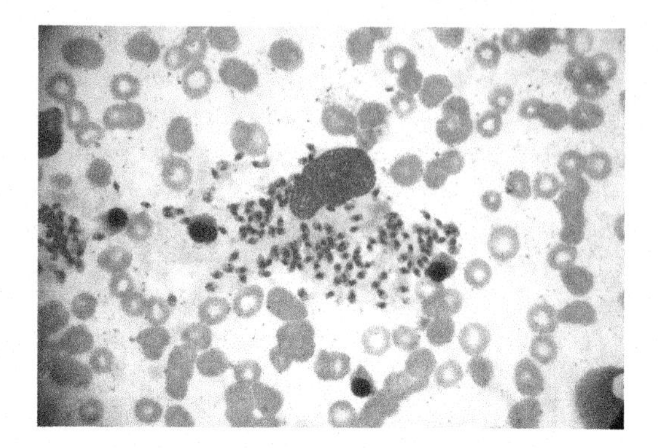

FIGURE 9.1 Smear from a bone marrow aspiration in a patient with visceral leishmaniasis showing great quantity of amastigote forms of the parasite.

after a successful treatment of VL and represents an intermediate state between disease and cure. Less frequently, post–kala-azar dermal leishmaniasis can occur during the evolution of an active VL or without any previous signs or symptoms of VL. The clinical presentation consists mainly in a macular or papular skin rash, representing an immune response localized to places near infecting parasites [15].

1.4.2.3 Cutaneous Leishmaniasis

CL is the most common form of leishmaniasis, and it initiates with an erythematous papule at the site of an insect bite. The lesion progressively ulcerates after an incubation period ranging from 2 weeks to 6 months [3]. The classic lesion of CL consists of a round painless ulcer with an elevated border and a button shape without any secretion [3], as shown in Fig. 9.2. In the Old

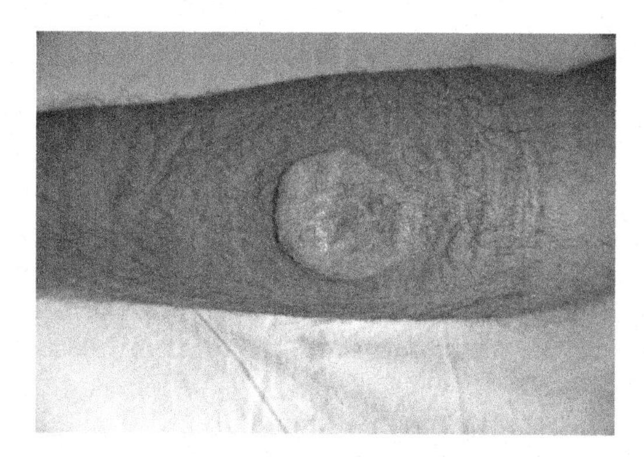

FIGURE 9.2 Classical ulcer of American cutaneous leishmaniasis caused by *L.* (*Viannia*) *braziliensis*.

World, the clinical presentation of CL strongly depends on the causal species of *Leishmania.* In some cases, the disease caused by *L. infantum* can heal spontaneously and confer some immunity to reinfection. In the New World, the term "mucocutaneous leishmaniasis" is frequently used instead of CL because the parasites of the subgenus *L.* (*Viannia*) metastasize to mucosa resulting in disfiguring lesions (described in more detail later). Exceptionally, other types of clinical presentation, such as plaques, verrucous, and lupoid lesions of localized CL, are also found [16]. These are considered chronic and resistant forms of leishmaniasis that result from unsatisfactory treatment [16]. The relapsing form of CL acquires a clinical aspect similar to that of the chronic lupoid form of CL and is represented by a central scar with active nodules on the edges of the lesion [17]. This is considered by some to be a specific form of CL and is called "leishmaniasis *recidiva cutis*," a difficult-to-treat and chronic disease form.

The disseminated form of leishmaniasis is a rare clinical presentation and is characterized by the appearance several cutaneous lesions as a result of hematogenic or lymphatic dissemination of the parasites [18]. This disease form occurs in the New World and is related to local species, especially in northeastern Brazil. Affected patients typically present more than 10 disseminated pleomorphic lesions spanning two or more noncontiguous body areas. Clinical lesions can appear as papules, acneiform lesions, or ulcerated. The mucosa can also be involved, and systemic symptoms such as fever may be present. Disseminated leishmaniasis is considered as an emerging form of the disease. Recent data showed a threefold increase in the number of patients affected, and this form represents 2.4% of the total number of tegumentary leishmaniasis cases in America [19].

Diffuse CL (DCL) is a rare manifestation of CL. It affects patients with an impaired cellular immune response [3]. The clinical presentation of this disease is represented by the appearance of diffuse nodular or plaque lesions, sometimes ulcerated. These lesions are rich in infected macrophages, due to an insufficient immune response against the parasite. The treatment of DCL is difficult as it tends to be relapse and several courses of treatments are necessary with poor response [18].

1.4.2.4 Mucosal and Mucocutaneous Leishmaniasis

The mucosa-infected leishmaniasis, or mucocutaneous leishmaniasis (MCL), is considered as an evolution of existing CL since it can occur in patients previously diagnosed as having CL and improperly treated [20]. This phenomenon is especially true in cases of New World CL where *L.* (*V.*) *braziliensis* and *Leishmania* (*Viannia*) *panamensis* are present. MCL is infrequent in the Old World, but virtually any species of *Leishmania* can cause clinical lesions in mucosa. It is believed that after hematogenic dissemination, the parasites start to affect the mucosa. Other factors favoring the development of

FIGURE 9.3 Initial phases of mucocutaneous leishmaniasis in the anterior nasal septum caused by *L.* (*Viannia*) *braziliensis*.

MCL include improper host immune response, severe cases of CL, or malnutrition, and males are more susceptible than females [21]. The clinical presentation is characterized by mucosal infiltration, most commonly initiated in the nasal septum, that evolves to ulceration and ultimately results in the destruction of facial structures (Figs. 9.3 and 9.4).

1.4.2.5 HIV and Leishmaniasis Coinfection

The occurrence of HIV infection and AIDS in endemic regions of leishmaniasis has changed the clinical scenario of both diseases. HIV infection and leishmaniasis occur with specific clinical manifestations and require specific and specialized care. Both infections can have a synergistic immunosuppressive effect on patients [22,23]. At the same time, prior HIV infection facilitates the infectivity and pathogenicity of *Leishmania*, and leishmaniasis can also

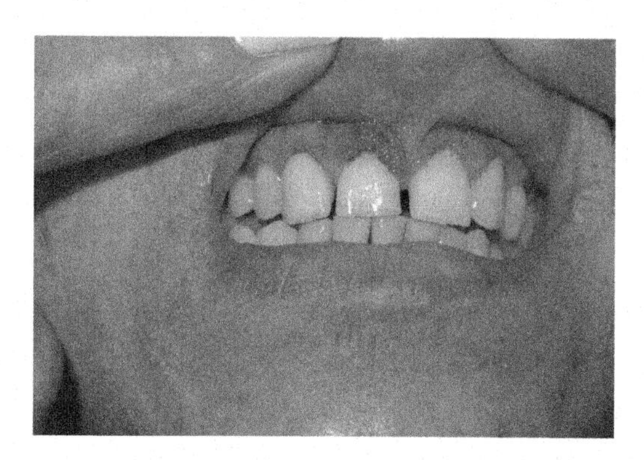

FIGURE 9.4 Oral mucocutaneous leishmaniasis caused by *L.* (*Viannia*) *braziliensis*.

reduce efficacy of antiretroviral therapy in patients [2,24]. In the early stages of HIV infection, patients present with similar clinical manifestations of CL and MCL, and their chances of acquiring these diseases are higher and accompanied by poor response to treatment [2]. In severely immunosuppressed HIV patients, otherwise atypical manifestations such as diffuse and disseminated leishmaniasis become more common. Another atypical symptom of leishmaniasis observed in HIV patients is the immune reconstitution inflammatory syndrome [25].

1.5 Immunological Alterations in Host

On infection by *Leishmania*, the interaction of the parasites with the host and the host immune response define the risk of disease development, its clinical presentation, as well as morbidity and mortality [26]. T lymphocytes are extremely important determinants for disease pathogenicity [27]. Although many new pathways for immunological response against *Leishmania* have been recently described, the knowledge of T helper (Th)1/Th2 balance is crucial for the understanding the disease manifestations. The understood goal of cell activation and cytokine production is to stimulate phagocytes to produce nitric oxide and reactive oxygen species, which are important mediators for killing parasites [26,28]. After the initial insect vector bite, the presence of an adequate cellular Th1 response will protect patients from developing leishmaniasis. Tumor necrosis factor (TNF)-α and interleukin (IL)-1β are produced soon after the recognition of invasive pathogens by macrophages [29]. In endemic areas, most people having contact with *Leishmania* parasites will still not develop leishmaniasis. This fact can be proved by the frequent positive result of the Montenegro skin test in inhabitants of endemic areas. This test consists of intradermic injection of *Leishmania* antigens [30], and a positive outcome means that the patient already had contact with the parasite and elicited a cellular immune response against *Leishmania* [3].

Disease development of leishmaniasis begins in patients who do not present a competent Th1 response. To initiate disease pathogenicity, *Leishmania* parasites must escape the initial cellular response of the host. The promastigotes transform into amastigotes inside of blood-borne macrophages and then spread to the spleen and liver in VL or skin and mucosa in CL and MCL [11]. Virtually all patients who develop VL are considered to have an insufficient Th1 response against the parasite. However, in cases of CL and MCL, the Th1/Th2 balance will determine the clinical form of the disease [1]. In New World tegumentary leishmaniasis, the importance of this immune balance can be easily observed. In American MCL, there are formations of granulomas due to an ineffective host response in resolving the infection. At the same time, this ineffective host response results in extensive tissue destruction, and the disease remains limited to certain parts of the skin and mucosa. The most important cytokines involved are interferon (IFN)-γ, TNF-α, and IL-12 [2,26].

In patients having a Th2 immune response, the infection will not be effectively controlled [31]. They frequently present a strong and ineffective humoral response and a negative Montenegro skin test. The parasite easily disseminates because of cytokines like IL-4, IL-10, and transforming growth factor (TGF)-β. This immunological response ultimately results in nitric oxide in tissues at insufficient levels to effect parasite killing, thus leading to the development of anergic forms of leishmaniasis such as the diffused form. Heavily infected macrophages can be found in disease lesions along with low levels of IFN-γ [1].

Cytokines like IL-12 and IL-4 are critical for T-cell polarization during *Leishmania* infection and pathogenesis [32]. IL-12 influences the differentiation toward Th1, resulting in production of nitric oxide from macrophages residing in infected tissues. IL-4 promotes Th2 polarization along with production of IL-4, IL-5, and IL-13, thereby suppressing nitric oxide production in tissues [29,32]. The role of an innate immune response has been recently explored in leishmaniasis. Early host responses drive the generation of an effective Th1 immunity [33]. The inflammasome pathway and its dependent cytokines, IL-1β and IL-18, contribute to the development of disease [34,35]. IL-18 was proved to have a central role in Th1/Th2 differentiation during infection by *L. major* [33]. This complex interaction of cytokines showed a new direction in research for clear understanding of disease pathogenicity. Moreover, there is the implication of novel drug targets for the development of immunomodulators, drugs, and vaccines for leishmaniasis.

2. THERAPEUTIC AGENTS FOR LEISHMANIASIS

2.1 Current Antileishmanial Drugs

Although recent advances in molecular mechanisms of leishmaniasis have been discovered, the therapeutic interventions of the disease are still based on several old, expensive, and/or toxic medications. The currently used drugs for the treatment of leishmaniasis are presented here in delineated subsections, along with observed side effects and emerging drug resistances that are all important to consider, as well as a discussion of new perspectives on chemotherapy and immunomodulation (Table 9.3).

2.1.1 Pentavalent Antimonials

Pentavalent antimonials form a group of drugs that has been used for years for the treatment of leishmaniasis. The exact chemical structure of the drugs and mechanism of action are still under debate [36]. Nowadays, two pentavalent antimonials are commercialized: meglumine antimoniate and sodium stibogluconate [2]. Antimonials are usually administered intravenously or through intramuscular injection but can also be administered by intralesional injections to treat localized cutaneous leishmaniasis. Pentavalent antimonials were

TABLE 9.3 Overview of the Current Drugs Used for the Treatment of Leishmaniasis. Most Relevant Indications, Side Effects, and Resistance Profile

	Most Relevant Indications	Most Important Side Effects	Drug Resistance
Pentavalent antimonials	• New World cutaneous or mucocutaneous leishmaniasis caused by *L. braziliensis, L. amazonensis, L. peruviana,* and *L. venezuelensis.* • Post–kala-azar dermal leishmaniasis in East Africa.	• Diffuse myalgia, malaise and arthralgia. • Cardiotoxicity, hepatotoxicity nephrotoxicity and pancreatitis.	• Great number of reported cases in the Old World especially in Bangladesh, Bhutan, India, and Nepal.
Amphotericin B deoxycholate	• Relapse cases of new world cutaneous or mucocutaneous leishmaniasis.	• Extremely nephrotoxic. Infusion reactions such as high fever, chills, and phlebitis are common.	• The exact mechanism of resistance is not well described nor its clinical impact. • Resistance reports are growing due to its large-scale use in areas with a high occurrence of pentavalent antimonials resistance.
Liposomal amphotericin B	• Anthroponotic visceral leishmaniasis caused by *L. donovani* in Bangladesh, Bhutan, India, and Nepal. • Visceral leishmaniasis caused by *L. infantum* in the Mediterranean Basin, Middle East, Central Asia, and South America.	• Less nephrotoxic and more tolerable in comparison to the deoxycholate form.	
Pentamidine isethionate	• New World cutaneous leishmaniasis caused by *L. guyanensis* and *L. panamensis.*	• Secondary infection of the intramuscular injection site. • Hypotension, hypoglycemia, and nephrotoxicity.	• Unknown clinical impact.
Miltefosine	• An option for anthroponotic visceral leishmaniasis caused by *L. donovani* and for post–kala-azar dermal leishmaniasis.	• Teratogenicity and gastrointestinal effects.	• Resistance reports are rapidly growing because of its large-scale use and because of its long half-life period.
Paromomycin	• Topical treatment for Old World cutaneous leishmaniasis. • In association to pentavalent antimonials for VL caused by *L. donovani* in East Africa.	• Reversible ototoxicity.	• Unknown clinical impact.

considered for a long time, in most parts of the world, as the first-line therapy for leishmaniasis. In some places, though, the drug has been discontinued due to reports of developed drug resistance, in particular for the treatment of Old World VL. However, antimonials remain the first line of chemotherapy against MCL in the New World [37].

2.1.2 Amphotericin B

Amphotericin B (AmpB) is a polyene antibiotic that is largely used to treat fungal infections. In its classic form, AmpB deoxycholate, this drug presents a broad range of adverse reactions, making treatment and patient compliance difficult. Different new formulations for the delivery of AmpB have been developed in order to reduce the adverse events related to treatment with this drug. For example, liposomal AmpB, AmpB lipid complex, and AmpB colloidal dispersions tend to have lower levels of toxic side effects than AmpB alone [38].

2.1.3 Pentamidine Isethionate

Pentamidine is an aromatic diamidine with antiprotozoal activity, effective for the treatment of leishmaniasis and also used for treating diseases such as African trypanosomiasis and pneumocystis pneumonia. It has been used with success in some parts of the globe, especially on the north and northeastern parts of South America [39]. This drug is administered intramuscularly or by intravenous infusion.

2.1.4 Miltefosine

Miltefosine is the only oral drug that when used alone proved to be effective for the treatment of leishmaniasis. The medication was originally developed for cancer treatment, but it has a proven antileishmanial effect [40].

2.1.5 Paromomycin

Paromomycin is an antimicrobial agent from the aminoglycoside classically used to treat intestinal amebiasis. For Old World VL it is usually used intramuscularly isolated or in combination with other effective drugs [2,41,42]. Topical medications that contain 15% paromomycin are used to treat cutaneous leishmaniasis in the Old World [2]. The cost of this drug is considered relatively low in comparison to the above-cited medications [2]. In the New World, the mode of action and effectiveness of this drug is still not well established, especially for the treatment of American tegumentary leishmaniasis [37].

2.1.6 Other Medications

Several other drugs, such as ketoconazole, fluconazole, and itraconazole (ITZ), were also found to be effective against *Leishmania* [2,37]. However, the lack of reliable studies with these drugs precludes the introduction of these medications as an effective therapy against leishmaniasis. Antimicrobial drugs such as rifampicin and azithromycin have already been studied for their antileishmanial activity, but none of these drugs have been approved for the treatment of leishmaniasis [2,43,44].

2.2 Drug Resistance

Drug resistances in leishmaniasis have increased in the past two decades, and mostly this has been related to the use of pentavalent antimonials in the Old World [6]. In most states of Bangladesh, Bhutan, India, and Nepal, VL patients are unresponsive to pentavalent antimonials, and the WHO recommends liposomal AmpB as the first-line choice for therapy [2].

Some studies have suggested that the growing incidence of antimonial resistance in VL is due to irresponsible use of this drug [6]. Initially, in some parts of the Old World, a daily 10 mg/kg dose of pentavalent antimonial for 6–10 days was sufficient to cure most cases [45]. Overly short course treatments, frequent interruptions of those treatments, and the occurrence of anthroponotic transmission have resulted in the spreading of antimonial resistant *Leishmania* strains [6]. Initially, local health authorities tried to introduce higher dosages and prolonged durations of pentavalent antimonial treatment, but no significant change was observed [46]. It has also been discussed that due to the severe side effects of antimonial drug therapy, patient compliance is poor and treatment is ceased prematurely, allowing the more rapid development of drug resistance than would be otherwise expected.

The parasite harbors inhibitory factors that reduce the intracellular drug concentration or inactivate the drug and thereby contribute to drug resistance [45,47]. One such parasite drug transporter, aquaglyceroporin, is downregulated in resistant parasites and is a key determinant of trivalent antimonial accumulation and clinical drug resistance in VL and post kalazar dermal leishmaniasis (PKDL) patients [48]. The ATP-binding cassette (ABC) transporter multidrug-resistance protein A (MRPA), a thiol transporter, was also found to confer antimony resistance in the amastigote stage of the parasite and in clinical isolates [49,50], thereby showing the involvement of thiol pathway in antimonial resistance [6]. Immunosuppressive conditions jeopardize clinical response to treatment. The emergence of HIV infection or diseases such as cancer or the use of immunosuppressive drugs is intimately related to the emergence of resistant cases of leishmaniasis [6]. Although reports of resistance are growing in the New World, pentavalent antimonials remain the first-line therapy for New World CL and MCL.

Resistance against AmpB has not been extensively reported, but this is basically inevitable due to of the large-scale use of this drug in some parts of the world. AmpB resistance in some clinical strains of *Leishmania* is associated with alterations in the thiol metabolic pathway, ABC transporter MDR1, and altered sterol profile of the membrane in the parasite [51]. Moreover, increased levels of silent information regulator 2 (Sir2) regulated MDR1, reactive oxygen species (ROS) concentration, and apoptosis-like process, which led to AmpB resistance in clinical isolates of *L. donovani* [51] However, the clinical impacts on the treatment of leishmaniasis are unknown [51,52].

Miltefosine resistance is rapidly developing in the Old World as the drug is considered an effective option for VL patients with infection resistant to pentavalent antimonials. Accordingly, it has been largely used in countries such as India [53]. This large-scale use and the oral availability enhance the preoccupation with the improper use of this drug, because the drug need not be administered directly by a healthcare professional [54]. Additionally, the long half-life period of this drug facilitates resistance generation especially in areas where the anthroponotic transmission is endemic [6]. A recent study has shown cross-resistance to miltefosine and liposomal AmpB in *Leishmania* parasites due to different mutations in P-type ATPase transporter that affected the lipid species [55].

2.3 Drug Toxicity and Side Effects

Most effective existing drugs against leishmaniasis are considered to have considerable toxicity and detrimental side effects associated with their use. These are outlined here in accordance with the particular drug in question.

2.3.1 Pentavalent Antimonials

Pentavalent antimonials present a wide variety of side effects [36]. The most common effects are diffuse mild myalgia, malaise, and arthralgia. Electrographic changes show a prolonged Q-T interval that can degenerate to malignant arrhythmia depending on the cumulative dose and on the previous cardiac condition of the patient treated [1]. Other severe manifestations can be also found, including pancreatitis, hepatotoxicity, renal failure, leukopenia, anemia, and thrombocytopenia.

2.3.2 AmpB Deoxycholate

Nephrotoxicity is the most common adverse reaction to AmpB deoxycholate treatment, but other manifestations such as high fever, chills, and phlebitis are common during infusion [2]. Those frequent reactions jeopardize treatment. Hypokalemia and myocarditis are rarely reported effects, but physicians must be aware of this possibility.

2.3.3 Liposomal Formulation of AmpB

The lipid formulation of AmpB was developed as a less nephrotoxic drug in comparison to the deoxycholate form. However, nephrotoxicity with the use of this type of drug must be considered and treatment must be carefully prescribed. Furthermore, the cost of this drug is increased in comparison to AmpB, which may preclude many leishmaniasis patients from using it in chemotherapy.

2.3.4 Pentamidine Isethionate

During the administration of pentamidine, the risk of hypotension, hypoglycemia, and nephrotoxicity must be considered [39]. A high-dose intramuscular injection of this drug can result in secondary infection and necrosis, but those last reactions are rare if basic precautions are taken.

2.3.5 Miltefosine

Miltefosine is considered to have milder effects in comparison to pentavalent antimonials and AmpB [53]. The undesirable effects of this drug are mostly related to gastrointestinal side reactions such as anorexia, vomiting, and diarrhea. Other rarely observed effects are skin rash hepatotoxicity and renal failure. The drug is considerably teratogenic and is contraindicated for use in pregnant women and administered together with contraceptives to women of child-bearing age [2,53,56].

2.3.6 Paromomycin

Paromomycin is considered a cost-effective drug in some regions like East Africa [41] due to the rarity of severe adverse effects. The most common reactions are pain or secondary infection at the injection site and, rarely, nephrotoxicity, hepatotoxicity, reversible ototoxicity, and pyrexia [57]. The topical formulation can cause local irritation and allergy [2]. The drug can cause ototoxicity in the fetus if used during pregnancy [2].

2.4 Polychemotherapy for Leishmaniasis

Polychemotherapy for leishmaniasis is based on the association of antileishmanial drugs with topical or systemic immunomodulators. Machado et al. showed that the association of the vasodilator pentoxifylline with pentavalent antimonials improved the effectiveness and adverse effects of antimonials for the treatment of mucocutaneous leishmaniasis [58]. Pentoxifylline modulates TNF-α levels, but the exact mechanism of synergism when associated with antimonials in unknown [59]. In Brazil, the first-line treatment for mucocutaneous leishmaniasis is the association of pentavalent antimonials and pentoxifylline.

The granulocyte macrophage colony-stimulating factor (GM-CSF) stimulates the production and function of granulocytes and macrophages. Its association with pentavalent antimonials accelerated healing time in patients with cutaneous leishmaniasis infected by *L.* (*V.*) *braziliensis* [60,61]. The association of GM-CSF with AmpB also seems to be beneficial. In patients with VL and AIDS who suffer from severe pancytopenia, this combined treatment can mobilize and activate monocytes and macrophages [62].

Imiquimod is a topical agent that stimulates the Th1 response resulting in the production of TNF-α, IFN-γ, and IL-12. It is classically used to treat skin cancer and warts [63]. In vitro studies showed nitric oxide production and an antileishmanial action [64]. Although the topical application of 5% imiquimod can result in local irritation, its association with pentavalent antimonials were beneficial for the treatment of New World CL [61].

2.5 Herbal Remedies

Plants are an original source of new and more effective drugs and have been exploited for the development of new antileishmanial drugs. For thousands of years, plants and their extracts have traditionally been used against different pathologies throughout the world, and still, in some poor and underdeveloped regions, medicinal plants are a primary and vital source of treatment [65]. According to a report of the WHO, approximately 70% of the world population uses medicinal plants to cure diseases through their traditional practitioners. These complex mixtures of natural products have also played a prominent role in ancient traditional medicine systems, such as Chinese, Ayurveda, and Egyptian, which are still in common use today.

However, to date, no herbal medicines have been validated for the treatment of leishmaniasis in a large scale. However, anecdotally reported successes of native plants in the healing of leishmaniasis ulcers are frequent. In fact, there are several reports of therapeutic success by the use of plants and herbal extracts. Several plant-derived phytochemicals showed antiprotozoal and anti-inflammatory activities, making them possible leads for the development of new drugs against leishmaniasis [29,66]. In fact, some plant extracts showed antileishmanial activity along with strong Th1 response [29].

It is important to note that most herbal remedies are used orally or in topical applications. Effectiveness of those plants against leishmaniasis has never been proven in large controlled clinical trials, and the frequent self-healing of cutaneous lesions has put in doubt the real effects of this type of treatment.

An interesting additional action of some herbal remedies is the capacity that some plant extracts have to act as photosensitizing agents. This property can lead to the destruction of skin lesions, especially when exposed to a controlled source of light such as lasers and photodynamic therapy devices, which are under investigation but will not be described in this chapter. In the following sections, a detailed revision of possible research strategies and of promising natural products for use in the development of new effective drugs for the treatment of leishmaniasis are presented.

3. DRUG TARGETS IN *LEISHMANIA*

3.1 Polyamine Pathway

Activated macrophages produce antileishmanial molecules such as free radicals and other reactive oxygen species, and kill parasites effectively, whereas "suboptimally" and "alternatively" activated macrophages turn on the arginase (ARG) pathway to produce polyamines (PAs) and enhance parasite replication and persistence [67]. These PAs such as putrescine, spermidine (Spd), and spermine occur widely in nature [68], are essential metabolites in eukaryotes that participate in a variety of proliferative processes [69], and, in trypanosomatid protozoa, play an additional role in the synthesis of the critical thiol trypanothione [70].

The PA biosynthetic pathway commences with the synthesis of the polyamine precursor L-ornithine, catalyzed by ARG. L-Ornithine is then decarboxylated by ornithine decarboxylase (ODC), which produces the PA, putrescine. Putrescine is used as the substrate for the constitutive spermidine synthase (SpdS) that adds the aminopropyl group donated from decarboxylated *S*-adenosylmethionine (dAdoMet) provided by the *S*-adenosylmethionine decarboxylase (AdoMetDC) [71]. Finally, spermine synthase performs a similar function on Spd, producing spermine by the addition of another aminopropyl group. All these enzymes are in principal good drug targets in *Leishmania* parasites [72].

The enzyme ODC is essential for parasite growth, as was confirmed from studies with $odc^{-/-}$ *L. donovani* promastigotes and amastigotes, which showed that these require putrescine or Spd supplementation for survival. *Leishmania* require ODC activity, as they are incapable of scavenging PAs from the host environment in amounts sufficient to sustain an infection [73,74]. In addition, ODC from *Leishmania* was much more stable than that of the mammalian host, which is a therapeutic advantage when attempting to achieve a long-lasting inhibition of the enzyme [71,75]. The ODC enzyme-activated irreversible inhibitor α-difluoromethylornithine (DFMO) and several other fluorinated ornithine analogs were growth inhibitory and cytotoxic to *L. donovani* [76] and *L. infantum* [77]. Another ODC inhibitor, 3-aminooxy-1-aminopropane (APA) that binds to the catalytic site of ODC, is more potent than DFMO against *Leishmania* [78,79].

On the other hand, studies with $adometdc^{-/-}$ *Leishmania* showed that they were incapable of growth in media without PAs. This auxotrophy was rescued by Spd but not by putrescine, spermine, or methylthioadenosine [71]. Inhibitors of AdoMetDC such as MDL73811 (structural analogue of dAdoMet) [80] and CGP 40,215A (competitive inhibitor of AdoMetDC) had strong antileishmanial activity [81].

SpdS is an essential enzyme for *L. donovani* amastigotes, as $spds^{-/-}$ parasites were unable to convert putrescine to Spd and were auxotrophic for PAs, required spermidine for growth in its insect vector form, and adversely impacted in ability to infect mice. Thus, SpdS was essential for maintaining a robust infection in mammals and indicate that the pharmacologic inhibition of it is a valid therapeutic strategy for the treatment of leishmaniasis [72]. However, SpdS was also targeted by pentamidine in drug-resistant *Leishmania*, resulting in increased affinity of SpdS toward putrescine along with decreased affinity toward pentamidine [82,83].

3.2 Thiol Metabolism

Leishmania and other related trypanosomatids have evolved a unique and unusual system to maintain their intracellular redox balance by a mechanism that is completely different from that of their insect vectors and mammalian hosts. They lack the ubiquitous glutathione reductase (GR) [84,85] and have thereby replaced the glutathione (GSH) as the main intracellular reductant with a system using N^1,N^8-bis(glutathionyl)spermidine (trypanothione) and its metabolic precursor, N^1-glutathionylspermidine, as their main thiols. These GSH conjugates are returned to the reduced state in redox cycling by the action of trypanothione reductase (TR) [84,86,87]. TRs belong to the NADPH-dependent flavoprotein oxidoreductase family and are structurally and mechanistically related to GRs. Both TR and GR have mutually exclusive substrate specificities [86,88] that provide an ideal opportunity to design selective inhibitors against the parasite thiol system. Gene deletion studies have shown the functional role of TRs in protecting the parasites from oxidative stress and its absence results in attenuated virulence [85,89,90]. Trypanothione neutralizes nitric oxide and labile iron by forming a dinitrosyl—iron complex with at least 600 times higher affinity than GSH [91], thereby providing protection against oxidative/nitrosative stress in vivo [85,92].

Tryparedoxin-dependent peroxidases (TPs) belong to the 2-cysteine peroxiredoxin family and are categorized according to their compartmentalization in the cytosol or mitochondria. TP protects the parasites against hydroperoxides but not nitric oxide produced by host defense mechanisms, by detoxifying them to water and alcohol [93—95]. This enzyme has also been shown to be important for metastasis of *Leishmania guyanensis* [96]. Overexpression of cytosolic tryparedoxin peroxidase (cTP) in *L. donovani* promastigotes and amastigotes has shown that these transfected parasites had enhanced infectivity and were less responsive to antimony drugs [95].

Studies have also shown thiols to be important in drug resistance. The pentavalent antimony compound Sb(V), a pro drug, is converted to its active form, Sb(III) using parasite specific thiol dependent reductase 1 (TDR1) and ACR2 enzymes [97—99]. Other thiols, including parasite-specific thiols such as trypanothione as well as macrophage-specific thiols such as glycylcysteine, reduces Sb(V) to Sb(III) nonenzymatically [84,100]. These events result in the formation of an Sb(III)-thiol conjugate that inhibits TR and leads to increased redox potential. This Sb(III)-thiol conjugate gets sequestered by ABC transporter MRPA [101] into intracellular organelles or is directly pumped out [102].

3.3 Sterol Pathway

The sterol biosynthetic pathway is an important drug target in *Leishmania*, as the parasite's main sterol component, ergosterol, is a C24 alkylated sterol that is absent in the mammalian host. Moreover, *Leishmania* is incapable of synthesizing its own cholesterol and imports it from the mammalian host cell and the environment [103]. The antileishmanial drug AmpB has high affinity for ergosterol and acts on the parasite by forming pores in the plasma membrane resulting in its lysis [104]. Dynamic changes in the sterol composition of the parasite were associated with promastigote metacyclogenesis and thereby affected its virulence [105].

The enzyme *S*-adenosyl-L-methionine:C24-Δ-sterol-methyltransferase (SMT) catalyzes the sterol alkylation in the C24 position during the last step of sterol biosynthesis in trypanosomatids. This enzyme is a potential drug target because its inhibition led to parasite death due to changes in its morphology and depletion of endogenous sterol [106]. However, SMT is also responsible for the development of drug resistance, as evidenced by AmpB drug-resistant *Leishmania* lacking functional SMT and yielding an absence of ergosterol, which considerably reduced AmpB insertion within the parasite plasma membrane [107]. Moreover, SMT's expression in AmpB resistant parasites made them sensitive to AmpB again [107]. Treatment with imipramine, an SMT inhibitor, resulted in accumulation of cholesta-5,7,22-trien-3β-ol and cholesta-7-24-dien-3β-ol sterols [108] and was effective in killing antimony resistant *Leishmania* [109].

The C14α-sterol demethylase (14DM) is positioned upstream to SMT in the sterol biosynthesis pathway and performs sterol demethylation in C14 position of C4-monomethylated or C4-dimethylated sterols [73]. 14DM is also a good drug target and the primary target of azole drugs, such as ITZ [104]. Studies with $14dm^{-/-}$ *Leishmania* promastigotes has shown that these parasites exhibited morphological and cytokinetic defects but can undergo replication despite of having a slower growth rate. These $14dm^{-/-}$ promastigotes show a loss of C24 alkylated sterols with an accumulation of 14-methyl sterols. Moreover, they also develop a resistance to ITZ as well as AmpB with a 5 and 10–100 times higher IC_{50}, respectively [110]. Studies have also shown that ITZ in combination with aryl-quinuclidines exhibits potent antiproliferative synergism against *L. amazonensis* [111].

3.4 Glucose Metabolism

Glucose provides a major source of carbon and energy to *Leishmania* [112]. Interestingly, glucose catabolism is 10- to 20-fold higher in the promastigotes compared with amastigotes [113]. Promastigotes grow in a sugar rich microenvironment due to ingestion of plant sap by the sandfly [114]; whereas amastigotes experience sugar-limited environment in the macrophage phagolysosome [115], which explains the higher rates of glucose uptake and utilization by promastigotes compared with amastigotes. Furthermore, amastigotes transport glucose at a much lower rate than promastigotes and derive metabolic energy primarily from fatty acid oxidation [113,116]. Thus, it is likely that a major role of sugar uptake is to provide glucose and other hexoses for biosynthesis of glycoconjugates and β-mannan. The genetic ablation of the glucose transporters in the parasite could affect the biosynthesis of these molecules and thus might affect parasite development and survival [117]. An *L. mexicana* promastigote glucose transporter "knock-out" (Δ*lmgt*) exhibited no detectable glucose transport activity and was able to grow, although at a reduced rate [118]. In contrast, Δ*lmgt* cells exhibited dramatically reduced viability inside macrophages and were unable to grow as axenic amastigotes, demonstrating that glucose transporters are essential for amastigote viability [118].

There is a marked shift in metabolic flux of the parasites from glycolysis toward the pentose phosphate pathway under oxidative stress that might be triggered by changes in enzyme activities of the central glucose metabolism. *Leishmania* parasites overexpressing glucose-6-phosphate dehydrogenase (G6PDH) and transaldolase were more resistant to lethal doses reactive oxygen species and to antileishmanial drugs such as sodium antimony gluconate (SAG), compared to their wild-type (WT) counterparts. These changes might be due to metabolic reconfiguration for replenishment of the NADPH pool [119]. Steroids such as dehydroepiandrosterone and epiandrosterone have been shown to inhibit trypanosomes expressing *L. mexicana* G6PDH [120] that could be exploited for drug targeting. Zinc sulfate has a direct antileishmanial effect by inhibiting *L major* and *L. tropica* enzymes such as fructophosphokinase, glucose phosphate isomerase, hexokinase, G6PDH, ribose-5-phosphate isomerase, and others that are involved in glucose metabolism [121]. Another important drug target is phosphomannomutase (PMM), which catalyzes the conversion of mannose-6-phosphate to mannose-1-phosphate, an essential step in the biosynthesis of glycoconjugates. Deletion of PMM from *L. mexicana* results in loss of virulence, suggesting that PMM is a promising drug target for the development of antileishmanial inhibitors [122].

A recent study showed the contribution of key enzymes such as glycerol kinase (GK), phosphoenolpyruvate carboxykinase (PEPCK), and pyruvate phosphate dikinase (PPDK) using null mutants in gluconeogenesis, an important metabolic pathway for de novo biosynthesis of glucose in *Leishmania*. Interestingly, GK participated in the entry of glycerol in promastigotes and amastigotes; PEPCK participated in the entry of aspartate in promastigotes, and PPDK was involved in the entry of alanine in amastigotes [123].

3.5 Proteasome Pathway

The proteasome is a nonlysosomal, ubiquitin-dependent protein degradation pathway, where substrates are first covalently linked to multiple ubiquitin molecules. Ubiquitin-conjugated proteins are then rapidly degraded by the 26S proteasome, a 2000-kDa ATP-dependent proteolytic complex [124]. The 26S proteasome comprises a barrel-shaped 20S catalytic core complex capped at one or both ends by a 19S complex. Another regulator complex termed 11S (or PA28) can replace 19S and activate the proteolysis of short peptides. Proteasomes have several distinct peptidase activities. The main roles of the proteasome are proteolysis of abnormal, misfolded, or improperly assembled proteins and control of cell cycle by selective degradation of regulatory proteins such as transcription factors and cyclins.

In protozoan parasites, the proteasome is involved in cell differentiation and replication and could therefore be a promising therapeutic target [125]. Proteasomes from *L. mexicana* showed 10 distinct spots with molecular masses of 22–32 kDa, indicating subunit complexity similar to that of eukaryotes [126]. Lactacystin, a proteasome-specific inhibitor, was effective on *L. mexicana* replication only when used at a high concentration, whereas MG132, a peptide aldehyde inhibitor, was effective at the usual level of concentration [126]. A novel *Leishmania* antigen (LePa), similar to the human 20S proteasome α-type subunit, was immunogenic in humans [127]. Similarly, another *Leishmania* proteasome antigen (antigen 24) was also found to elicit strong immunogenicity [128]. A recent study by Khare et al. has led to the identification of a selective inhibitor of the kinetoplastid proteasome (GNF6702), which inhibits the kinetoplastid proteasome through a noncompetitive mechanism, does not inhibit the mammalian proteasome or growth of mammalian cells and is well tolerated in mice [129]. Capitalizing on key structural differences between the parasitic and mammalian proteasomes may allow for the development of potent and selective antileishmanial agents acting on this drug target.

4. APPROACHES FOR DRUG SCREENING OF NATURAL PRODUCT LIBRARIES AGAINST *LEISHMANIA*

Several approaches have been used for the testing of crude extracts, purified phytochemicals, and natural product libraries against *Leishmania*. The medium and high-throughput screening assays prepared for drug screens against *Leishmania*, described later, can use different stages of the parasite, such as the flagellated promastigotes, aflagellated axenic amastigotes (without host macrophages), and intracellular amastigotes (inside host macrophages). Among them, the promastigotes and axenic amastigote stages are facile for primary or pilot evaluations of natural products for their direct killing activity. However, an ideal drug must be functional against the intracellular amastigote stage that is present in the mammalian host. Some screening assays, such as high-content microscopy assays, Flowsight, and others can identify the natural products that are active against the stringent intracellular amastigotes. Furthermore, intracellular amastigotes clearance activity of these natural products may also highlight their immunomodulatory activity on host cells separately from antileishmanial activity against amastigotes.

4.1 Fluorescence Activated Cell Sorter–Based Assays

The incorporation of a fluorescence activated cell sorter (FACS) has been used in biological assays to identify and evaluate potential antileishmanial drugs. The main advantages of using it are rapid, highly sensitive, and accurate analysis of a large number of cells compared to those obtained by the conventional microscopic techniques, especially in determining IC_{50} values of drugs against the intracellular amastigotes or extracellular promastigotes [130]. Fluorescent dyes such as CFSE, PKH-26, or SYTO-17 were used for labeling the promastigotes or, in some cases, allowed to infect the macrophages, followed by drug treatment and FACS analysis [131–133]. This technique permitted accurate study of cellular infection but could not detect intracellular amastigotes after prolonged incubation periods. Increased accuracy in detection of intracellular amastigotes could be obtained, however, after labeling with monoclonal antibodies [134].

The recent development of genetically modified *Leishmania* species transfected with the reporter genes such as GFP [135], DsRed [136], or mCherry [137] has proved to be a powerful tool in *Leishmania* research and requires no additional antibody staining or permeabilization. However, *Leishmania* strains, expressing the GFP reporter gene as an episomal transgene, were useful for screening promastigotes but resulted in background noise when used for evaluating intracellular amastigotes [135,138,139]. By using flow cytometry analysis, it was possible to see that expression of GFP was not homogeneous in the parasites because the number of copies of the GFP gene was not the same in all the parasites [135]. Moreover, *Leishmania* having episomal expression of GFP require constant drug pressure for its continuous expression that might limit the sensitivity of the drug assay [140]. On the contrary, stable and constitutive expression of fluorescent protein like GFP in *Leishmania* species has been shown by targeted integration of exogenous genes in the transcriptionally active 18S ribosomal locus through homologous recombination [141,142]. Introduction of the GFP gene in the *Leishmania* genome did not alter their sensitivity profiles to the reference drugs and supports the use of such parasites for drug screening using FACS [143,144]. Thus, the transgenic fluorescent protein (GFP)-expressing *Leishmania* appears to be suitable for flow cytometry–based drug assays.

Recently, a new flow cytometer (Flowsight), which combines flow cytometry and imaging, has been shown to estimate parasitic loads using DsRed transgenic *Leishmania* in an automated manner and could also distinguish between an activated infected and bystander macrophages [145]. This and other future developments will have a dramatic impact on large-scale screening for antileishmanial agents.

4.2 Plate Reader–Based Assays

Plate reader–based assays are inexpensive, effective, simple, and rapid, thus forming an integral part of antileishmanial drug discovery [146]. Classic colorimetric plate reader–based assays, which use MTT or Alamar blue, are useful for screening the efficacy of drugs against *Leishmania* promastigotes and axenic amastigotes. However, these classic assays have apparent drawbacks such as unreliable readouts, high background noise and inconclusive observation following longer incubation periods with parasites [140,147,148]. Thus, there is a strong need to evolve an appropriate screening technology, combined with combinatorial or medicinal chemistry and computational biology, for increasing the efficiency of target-based drug discovery against protozoan parasites [140].

Plate-based reporter gene assays have proved to be extremely helpful in filling up the lacuna in classic colorimetric drug assays. Typically, a reporter gene would encode a product that has a readily measurable phenotype and is easily distinguishable over endogenous cellular background. Intracellular reporter gene products are expressed and retained inside the expressing cells, for example, chloramphenicol acetyltransferase (CAT), β-galactosidase, GFP, firefly and bacterial luciferase, and glucuronidase [140]. *Leishmania* cell lines expressing firefly luciferase reporter genes have been developed using an episomal vector, for in vitro screening of antileishmanial agents [149]. This *Leishmania*-expressing luciferase system is used to evaluate potential antileishmanial compounds in a 96-well microplate format [150,151]. The luciferase-expressing *Leishmania* were shown to be useful in studying the high-throughput screening of drugs acting on intracellular macrophages with an adequate level of sensitivity, detecting one parasite per 10 macrophages in a 96-well plate [152]. However, this system has certain drawbacks including the transient luminescence readout time, proper timing of the mixing of the sample and the reagents, and high cost of the reagents and plate luminometers [153].

Leishmania parasites have also been engineered to express β-galactosidase [154], but high background activity from host macrophages prevents the use of this reporter system in amastigote-macrophage drug screening assays. Alternatively, the use of a β-lactamase reporter gene for quantifying *Leishmania* amastigotes in macrophages grown in microtiter plates has shown promising results. The β-lactamase gene [155,156] was integrated into the rRNA region of the genome, thereby allowing for high-level stable expression of the enzyme. A colorimetric readout for quantifying the growth of *Leishmania* inside the macrophages was developed by expressing the β-lactamase gene in *Leishmania*. The assay is simple to perform, and the reagents are relatively inexpensive [153].

Another high-throughput plate-based drug assay uses the previously described TR component of the kinetoplast-thiol-redox metabolism [157,158] for measuring drug activity against intracellular *Leishmania* amastigotes by monitoring its 5,5′-dithiobis(2-nitrobenzoic acid) (DTNB)-coupled reducing activity. This reaction combines the TR-catalyzed reduction of trypanothione disulfide with its in situ regeneration through DTNB, resulting in a colorimetric assay with improved sensitivity and efficiency [159]. No cross-reactions with the host cells were observed, as a result of the poor affinity of substrate for the mammalian homologue. The assay was designed for microtiter plate format and colorimetric detection, enabling automation and high throughput measurements with basic equipment [160].

4.3 Microscopy-Based High Content Assay

There are many advantages of microscopy-based assays over plate reader–based assays. Typically, a microscopy-based assay mimics the traditional direct-counting assay and provides cell-by-cell analysis, which helps to determine parameters such as the number of infected cells, the number of amastigotes per individual cell, and information about cell cytotoxicity. This information provides a more direct understanding of the assay results and reduces the occurrence of artifacts [161].

Intracellular *Leishmania* amastigotes reside in the parasitophorous vacuole inside of macrophages, and high-content screening using automated microscopy provides a suitable platform for assessing such intracellular parasites; recently, both a 96-well and a 384-well high-content assay have been reported [162,163]. Moreover, studies have also shown that the high-content plate assay using promastigotes or axenic amastigotes were less sensitive and accurate compared to the intracellular amastigote high-throughput microscope assay [162]. A new and improved 384-well microscopy-based intracellular *Leishmania* assay was recently developed, with sufficient capacity to allow primary screening of medium-sized small molecule libraries. In this assay, the authors have used axenic amastigotes in place of metacyclic promastigotes, as they do not elicit an oxidative burst when entering the macrophage [133]. Thus, the high-content microscopy assay provides a powerful and accurate tool for screening of drug candidates against intracellular *Leishmania.*

4.4 Animal Models Suitable for Drug Discovery and Development

A suitable animal model that is desired for leishmaniasis drug discovery research needs to closely mimic the pathological and immunological changes observed in humans. However, none of the experimental models that have been developed so far for leishmaniasis closely depict the human disease. Thus, the most developed in vivo systems developed to date will be described in this section.

Small rodents such as mice, rats, hamsters, mastomys, squirrel, gerbil, and others have been used for establishing *Leishmania* infection [164]. Among these models, hamsters and mouse models are the most common ones that are used for drug studies against leishmaniasis. Mouse models provide the flexibility of using small amount of a drug for evaluating its therapeutic and immunoprophylactic activity [165]. These murine models can also be exploited for setting up self-healing or nonhealing models of leishmaniasis using inbred mouse strains such as C57BL6 and BALB/c [166,167]. Recently, humanized mouse models of leishmaniasis have been used for drug studies that induce systemic infections and thereby provoke a nonprotective human immune response [168]. The hamster model of leishmaniasis using golden Syrian hamsters provides a nonhealing model of VL, as depicted by a synchronous infection in the liver and spleen accompanied by severe anemia as observed in human VL [169,170]. Systemic infection of the hamster with *L. donovani* results in a progressive increase in visceral parasite burden, progressive cachexia, hepatosplenomegaly, pancytopenia, hypergamma-globulinemia, and, ultimately, death [170]. The major advantage of this hamster model is that through biopsy, it is possible to monitor pre- and post-treatment efficacy of antileishmanial drugs. The golden hamster is considered to be the best experimental model to study VL, because it most closely reproduces the clinical pathogenesis of the disease as seen in humans and dogs. However, the wide use of hamsters is still limited due to lack of available reagents such as antibodies to cell markers and cytokines of these animals [171].

The infection of dogs with *L. infantum* or *L. chagasi* is an important laboratory model because it closely mimics the human infections [172]. Experimental models of leishmaniasis have also been achieved with *L. donovani*, for which the dog is not a natural reservoir [173]. Interestingly, German shepherd dogs are better experimental models than beagles [174], but some workers claim highly successful infection rate with a mixed breed [175].

Nonhuman primate models of leishmaniasis are seen as important, given the fact that they provide an opportunity to conduct studies on different aspects of the disease that would not be ethically achievable with humans. However, for still ethical considerations, as well as practical financial reasons, the use of primates in biomedical research is limited. Nonetheless, studies with New and Old World monkeys showed that owl monkeys [176] and squirrel monkeys [177] develop an acute and fulminating, but short-lived, infection. An antileishmanial screening was also performed in owl and squirrel monkeys. However, Old World monkeys such as *Macaca* spp. and African vervet monkeys developed low or inconsistent infections [164]. Studies on the Indian langur, *Presbytis entellus,* showed that this species was highly susceptible to *L. donovani* and that infection resulted in a progressive acute disease, leading to fatality after 110–150 days [178,179].

5. PHYTOCHEMICALS WITH ANTILEISHMANIAL AND IMMUNOMODULATORY ACTIVITIES

5.1 Natural Products Drug Discovery

Because there is an overwhelming need for novel agents for the therapy of leishmaniasis, many different research directions have been pursued. Due to the expectedly low return on investment for research costs for new antileishmanial drugs, these efforts have mainly taken place in academic, government, and other nonprofit laboratories. However, large pharmaceutical companies acting as corporate partners or in philanthropic capacities should be expected to help bring any major discoveries to wide production and market. One particularly productive strategy for drug discovery and development in general (not only for leishmaniasis) has been the biochemical investigation of natural products, the so-called secondary metabolites of natural organisms, as well as chemical derivatives designed to specifically improve the drug-like properties of these molecules [180]. Recent advances in the field have allowed for natural product discovery to greatly improve and take advantage of automated fraction generation and high-throughput screening, use genomic information and metabolomics, and capitalize on incredibly sensitive modern analytical instrumentation [181,182]. Still, many current drugs were discovered by traditional natural product research methods, such as bioactivity-guided isolation and structural characterization [180]. For antiparasitic studies not directly related to leishmaniasis research, this concept was already well proven before the 2015 Nobel Prize in Physiology or Medicine was awarded to three scientists—Youyou Tu, Satoshi Omura, and William C. Campbell—for the discovery and development of the natural product antiparasitic agents artemisinin (for malaria) and avermectin (for onchocerciasis, or "river blindness") [183]. The structures and biological activities of these molecules have also inspired medicinal chemistry efforts to produce semisynthetic analogues with improved properties, such as the clinically crucial drugs artemisinin and ivermectin [183].

The phytochemical investigation of natural materials can be seen as improving on the long-standing foundation of research on herbal medicine and natural toxins, as this field was initially focused on the discovery of plant natural products, or phytochemicals [184]. Improved access to isolated ecosystems as well as the development of modern technologies, such as the self-contained underwater breathing apparatus (SCUBA), has also allowed for sampling of organisms in remote and sometimes extreme environments. Thus, the chemical investigation of all sorts of naturally occurring organisms has been broadly undertaken, including medicinal and non-medicinal plants, marine and terrestrial microorganisms, as well as several "lower animal" life forms such as amphibians, insects, and sea sponges [185,186]. Research outcomes from international efforts for antileishmanial natural product drug discovery will be shown in the following text, which is divided into sections including plant-derived purified natural products (phytochemicals) with demonstrated in vivo antileishmanial activities, crude extracts with in vitro or in vivo antileishmanial activities, and categorized structural classes of phytochemicals so far only determined to have in vitro antileishmanial activities. Additionally, one section has been included for the description of natural products from other sources, such as marine, microbial, and fungal organisms.

Although the natural products here presented are described as being antileishmanial at various stages of development, none has yet been demonstrated as being safe or efficacious for the therapy of leishmaniasis in humans.

5.2 In Vivo Antileishmanial Phytochemicals

In recent decades, many natural product research investigations have been conducted for the discovery of new antileishmanial agents, especially those purified from plants. The outcome of these experiments has included observation of in vitro antileishmanial activity for many member compounds of many diverse structural classes of natural products [187–189]. Relatively few of the natural products with in vitro activity against *Leishmania* species have been demonstrated as effective in vivo, however. Some of the natural products and natural product derivatives that are not currently approved for use as drugs but that have demonstrated antileishmanial activity in animal models are shown in Fig. 9.5 and will be discussed here. This set of molecules is not meant to be comprehensive, and several of these compounds were only shown to be active in vivo when delivered in liposomal, niosomal, or polymer-coated nanoparticle encapsulations [190]. Indeed, future natural product research efforts that include in vivo drug delivery techniques such as encapsulation may improve the discovery rate of new antileishmanial agents with enriched safety and efficacious properties compared to the molecule(s) alone.

Berberine is a benzyltetrahydroisoquinoline alkaloid type of phytochemical with wide distribution in nature's plant families and is an especially prevalent molecule in the genus *Berberis* [191]. Berberine has been isolated from many herbal medicines that have been used as remedies for leishmaniasis [191]. Some natural and semisynthetic derivatives of this molecule, including tetrahydroberberine and *N*-methyl-tetrahydroberberine, reduced parasite burden to the livers of *L. donovani* infected hamsters by 50%–56% when administered in the relatively large quantity of 416 mg/kg over 4 days by intramuscular injection [192]. Several other derivatives of tetrahydroberberine have since been developed that demonstrated potent in vitro antiprotozoal activities, but none of these demonstrated *Leishmania* parasite burden reduction $\geq$50% in murine livers at any test concentrations that did not accompany overt toxicity [193,194]. However, the synthetic analogue 5,6-didehydro-8,8-diethyl-13-oxodihydroberberine chloride reduced the liver parasite burden in an animal model of VL by 47% when delivered at 1 mg/kg/day for 5 days, and further developmental research for this class of molecules is ongoing [193,194].

Another phytochemical with widespread distribution is (−)-α-bisabolol, which is a fragrant monocyclic sesquiterpene that is present in the essential oil of many plants, including those used in traditional medicines [195]. Since this molecule had been extensively studied previously for other biological activities, as well as for toxicity to animals and drug absorption and distribution

tetrahydroberberine | *N*-methyl-tetrahydroberberine | 5,6-didehydro-8,8-diethyl-13-oxodihydroberberine chloride | (−)-α-bisabolol

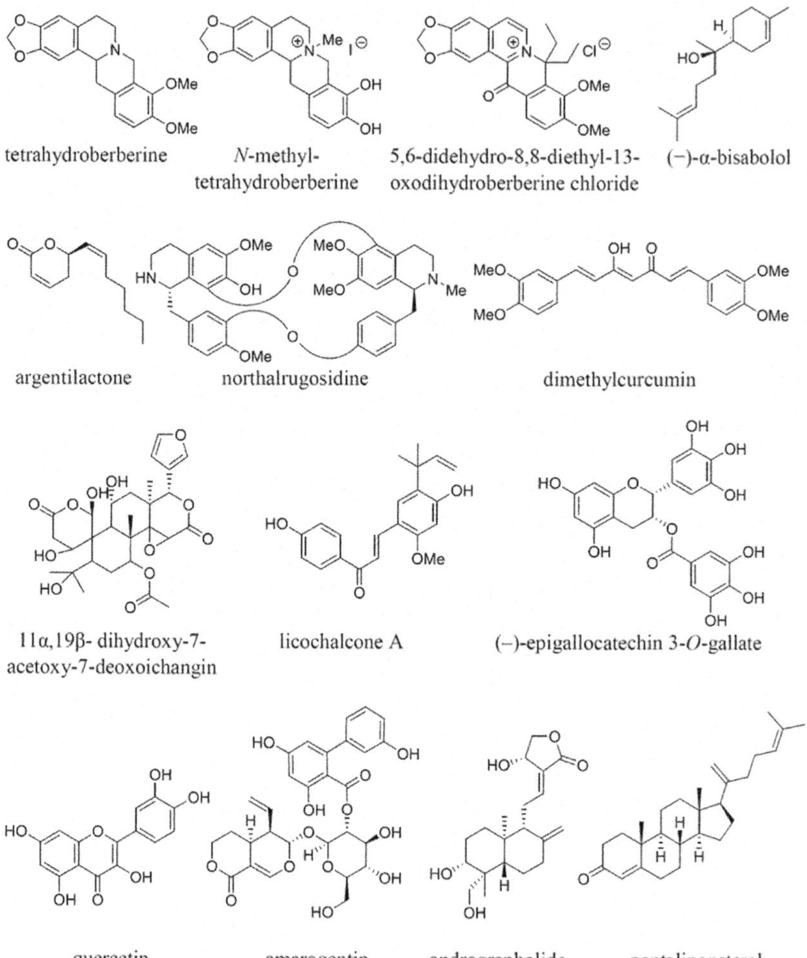

argentilactone | northalrugosidine | dimethylcurcumin

11α,19β- dihydroxy-7-acetoxy-7-deoxoichangin | licochalcone A | (−)-epigallocatechin 3-*O*-gallate

quercetin | amarogentin | andrographolide | pentalinonsterol

FIGURE 9.5 Structures of some natural products with in vivo antileishmanial activity.

properties, this molecule was selected for antileishmanial studies [195]. Although it demonstrated weak antileishmanial activities in vitro against *L. infantum* and *L. donovani* intracellular amastigotes ($IC_{50} = 56.9$ and 39.4 μM, respectively), (−)-α-bisabolol was selective and not cytotoxic at up to 1 mM in concentration [195]. Thus an in vivo trial was conducted using *L. infantum*–infected mice, and animals treated orally with (−)-α-bisabolol at 200 mg/kg/day for 14 days were shown to have a significantly reduced parasite burden of the liver and spleen when compared to the control group (89.2% and 71.6% reductions, respectively) [195]. It was also suggested that the mode of action for this phytochemical could include the disruption of the parasite plasma membrane [195]. Later, the same group was able to show that the treatment of *L. tropica*–infected hamsters, an in vivo model for cutaneous leishmaniasis, by topical application of (−)-α-bisabolol at 1.0%, 2.5%, and 5% in ointment led to 49%, 56%, and 55% reductions in lesion thickness, respectively [195]. The oral and topical efficacy of this molecule, coupled with its apparently low toxicity ($LD_{50} = 14$ g/kg in rats) and with widespread ability to be produced may make (−)-α-bisabolol a promising drug lead for the therapy of leishmaniasis if these properties are able to be translated for the human population [195,196].

Argentilactone is a relatively simple δ-lactone–containing compound that was isolated from the essential oil of the flowering plant *Annona haematantha* L. (Annonaceae), and is also present in some other plant varieties [197]. Although this phytochemical was not investigated due to its occurrence in a traditional medicine used for the therapy of leishmaniasis, it was observed to have in vitro antileishmanial activity against *L. donovani*, *L. major*, and *L. amazonensis* [197]. When argentilactone was tested on mice infected with *L. amazonensis* by subcutaneous injection at 25 mg/kg/day for 14 days, a 97% reduction in the parasite burden of the cutaneous leishmaniasis lesions was observed [197]. Furthermore, a 75% reduction in the parasite burden of the cutaneous leishmaniasis lesions was observed when argentilactone was tested in the same way but with oral delivery of the same dosage [197]. Due to the simple structure and apparent orally delivered antileishmanial activity of this molecule, it may hold some promise in the future as a drug candidate. However, no further studies appear to have ever been conducted to advance the development of this phytochemical toward human trials.

The naturally occurring bisbenzylisoquinoline alkaloid, northalrugosidine, was selected for in vivo testing after the in vitro screening of a plant-derived natural product library led to the observation of in vitro antileishmanial activity by three bisbenzyltetrahydroisoquinoline alkaloids purified from *Thalictrum alpinum* L. (Ranunculaceae) [198]. Northalrugosidine was the most active and selective among those tested for in vitro antileishmanial activity against *L. donovani* promastigotes ($IC_{50} = 0.28$ μM) and cytotoxicity to robust HT-29 human colon adenocarcinoma cells (29.3-fold selectivity) [198]. Thus, this phytochemical was tested in vivo using a murine model of VL, which resulted in the observation of a dose-dependent reduction to the parasitic burden in the murine livers and spleens without overt toxicity effects at 2.8, 5.6, and 11.1 mg/kg per animal when administered intravenously [198]. Several other bisbenzyltetrahydroisoquinoline alkaloids have been previously shown to have in vitro antileishmanial activity, and are suspected to act on trypanothione reductase, but none of these have yet been shown to be efficacious against VL in vivo [199].

Among diarylheptanoid natural products, curcumin is a well-known molecule derived from turmeric root, or *Curcuma longa* L. (Zingiberaceae). Several analogues have been isolated from natural sources, and many more have been synthetically generated. Dimethylcurcumin is one such molecule that was tested on mice infected with *L. amazonensis* [200]. This compound was delivered once subcutaneously at 20 mg/kg, and reduced the cutaneous leishmaniasis lesion size by 65.5% when measured 45 days later [200]. The observed reduction in lesion size continued over time, as it was measured again after 60 and 75 cumulative days, but there was a diminished statistical significance compared to the untreated controls by these time points [200]. It should be noted that curcumin and its derivatives have been excessively reported on in the literature and have been shown to possess a number of biological activities in vitro, in vivo, and even through clinical trials [201]. However, the bioavailability of these molecules due to absorption and metabolism is markedly low, and so their formulation and delivery are of great importance [201].

The seco-limonoid compound, 11α,19β-dihydroxy-7-acetoxy-7-deoxoichangin, is a natural product, isolated from the bark of *Raputia heptaphylla* Pittier (Rutaceae), which has been shown to exert an immunomodulatory effect on infected murine and human phagocytic cells and to impart on them a renewed microbicidal function [202]. Thus, although the in vitro IC_{50} value of this compound against *L. panamensis* was 59 μM, it was selected for further studies using a hamster model of cutaneous leishmaniasis. When diseased animals were treated intralesionally with this phytochemical at 8.7 mg/kg/day, every second day for 20 day, the result was promising. Of the animals that were infected with *L. panamensis* and treated with 11α,19β-dihydroxy-7-acetoxy-7-deoxoichangin, 40% were shown to have had cutaneous lesions reduced in size by at least 50% in comparison with control groups [202]. Furthermore, of the animals that were infected with *L. amazonensis* and treated intralesionally with this molecule at 8.7 mg/kg/4 days for 16 days, one-third resulted in a complete cure of the lesion and one-half had cutaneous lesions reduced in size by at least 50% in comparison with control groups [202]. Furthermore, the difference in parasite burden of the lesion tissue between treated and untreated groups was statistically significant, but there was no significant difference between the animals treated compared to those given meglumine antimoniate at 120 mg/kg/day for 10 days was not statistically significant [202].

Licochalcone A is a flavonoid-type phytochemical that has been isolated from licorice root, *Glycyrrhiza glabra* L. (Fabaceae), which is used as a Chinese traditional medicinal plant [203]. This molecule was shown to have potent in vitro antileishmanial activity against *L. major* [203]. The antileishmanial activity of licochalcone A was tested on mice infected with *L. major* by intraperitoneal administration of 2.5 and 5 mg/kg/day, which "completely prevented lesion development" [204]. When tested on hamsters infected with *L. donovani* by intraperitoneal delivery of licochalcone A at 20 mg/kg/day × 6, parasite burden to the livers was reduced by more than 96% [204]. Many analogues of licochalcone A have been produced, and several also showed in vivo antileishmanial activity [205]. Another flavonoid, (−)-epigallocatechin-3-*O*-gallate (EGCG) is a well-known chemical constituent of green tea, *Camellia sinensis* (L.) Kuntze (Theaceae) [206]. This phytochemical is also present in the FDA-approved botanical drug, sinecatechins, which is a topically applied cream for treating perianal and genital warts [206]. It has been demonstrated that EGCG has a wide range of biological activities, at least in vitro. Furthermore, when EGCG was tested on mice infected with *L. amazonensis* at 30 mg/kg/day × 5 day/week by oral administration, the result was a statistically significant reduction in cutaneous leishmaniasis lesion size by about 50% after about 4 weeks [206].

Another widely distributed phytochemical is quercetin, a natural product with in vitro antileishmanial activity that induces reactive oxygen species generation *L. amazonensis*–infected macrophages [207]. This effect was observed to be selective for the *L. amazonensis*–infected macrophages compared with uninfected macrophages, implying a direct or indirect mechanism of action for this biological activity [207]. Of more than 50 flavonoid and flavonoid glycoside compounds tested together, quercetin was selected as one of the most attractive leads because it demonstrated potent in vitro activity ($IC_{50} = 1.0$ μg/mL) against *L. donovani* axenic amastigotes and 37.1-fold selectivity over L6 rat skeletal myoblast cell toxicity [208]. Next, mice infected with *L. donovani* were administered with quercetin at 30 mg/kg/day × 5 intraperitoneally, but the parasite burden to their livers was reduced by significantly less than 50% compared to the control group [208]. Although it was also reported that quercetin was not efficacious at reducing the parasite burden to the spleens of *L. donovani*-infected golden hamsters as a "free drug" when administered 3 mg/kg/3 days × 6 subcutaneously, an equivalent dose administration was found to be efficacious in reducing the parasite burden of hamster spleens if encapsulated and delivered in liposomes (51% reduction), niosomes (68% reduction), or polylactide polymer-coated nanoparticles (87% reduction) [209].

Amarogentin is a *seco*-iridoid glycoside phytochemical that was originally isolated as the bitterness principal of the Ayurvedic medicinal plant, *Swertia chirayita* (Roxb.) Buch.-Ham. ex C.BClarke [basionym *Gentiana chirayita* Roxb. (Gentianaceae)] [210]. This compound has been since reported to inhibit in vitro the topoisomerase I of *L. donovani* [211]. It was also demonstrated that amarogentin was not particularly efficacious at reducing the parasite burden to the spleens of *L. donovani*–infected golden hamsters (34% reduction) as a "free drug" when administered 2.5 mg/kg/3 days × 6 subcutaneously, but similarly to the case with quercetin, efficacious reduction to the parasite burden of the hamster spleens was observed when amarogentin was encapsulated in liposomes (69% reduction) or niosomes (90% reduction) [212]. Along the same lines, the labdane diterpenoid phytochemical andrographolide was originally isolated as the principal bitterness component of another Ayurvedic medicinal plant, *Andrographis paniculata* (Burm. f.) Wall.

ex Nees (Acanthaceae) [213]. This plant is also found in use as a traditional Chinese medicine, and many of the bioactivities attributed to the herbal remedy can be attributed to a high concentration of the active principal [192,214]. The first report of andrographolide as an antileishmanial agent appears to regard an in vivo test, where it was also not particularly efficacious at reducing the parasite burden to the spleens of *L. donovani*–infected golden hamsters (37% reduction) as a "free drug" when administered 2.5 mg/kg/3 days × 6 subcutaneously [215]. It was demonstrated, however, that using an equivalent dose administration of andrographolide with the same animal model led to efficacious reduction to the parasite burden of the hamster spleens when it was encapsulated in liposomes (67% reduction), or mannosylated liposomes (86% reduction) [215]. A natural derivative of andrographolide was later demonstrated to possess similar in vivo antileishmanial activity when tested in the same way [216].

Pentalinonsterol is a plant sterol that was discovered and first described as a phytochemical and in vitro antileishmanial agent by bioactivity-guided fractionation from the roots of *Pentalinon andrieuxii* Muell.-Arg. (Apocynaceae) [217]. The roots of this plant are used for the topical treatment of cutaneous leishmaniasis in traditional Mexican medicine, and especially in the Yucatan peninsula [218]. A patent application including a method for synthesizing pentalinonsterol reported the in vivo efficacy of pentalinonsterol for treating *L. donovani*–infected mice [97]. In this study, the parasite burden in the murine liver and spleen were both reduced by more than 50% after intravenous treatment with 2.5 mg/kg of pentalinonsterol in a 100 μL liposomal formulation [97]. Furthermore, the mode of action for pentalinonsterol appears to include the induction of a host-protective mediation of a strong proinflammatory cytokine response, indicative of its potential for immunotherapy in the treatment of VL [219].

5.3 Antileishmanial Crude Plant Extracts

The U.S. FDA recognized an unmet need in the pharmacopeia for herbal medicines and complex mixtures of phytochemicals and implemented in 2004 documented guidance for a new class of drugs called "botanical drug products" [220]. These botanical drug products must have demonstrated safety and efficacy profiles, established quality controls for consistent production, and are permitted to contain undetermined chemical constituents and/or act by unspecified mechanisms of action [220]. Only two botanical drug products have passed FDA approval through the end of 2016, and neither of these were approved for therapy of leishmaniasis [221–223]. By contrast, though, quite a few medicinal plants and plant extracts appear to be in use for the therapy of leishmaniasis in endemic areas where previously described treatment options are unavailable for various reasons that often includes lack of access to modern medicine and healthcare [224,225]. Accordingly, the worldwide investigation of some of these medicinal plants and also an herbal medicine repurposing effort have been under way for the potential herbal therapy of leishmaniasis [226]. However, no registered clinical trials have been recorded to investigate the safety or efficacy of crude natural product extracts or herbal medicines for the therapy of leishmaniasis, most probably due to the significant cost associated with such research efforts, so these should not be taken as medicines until such time that their safety and effectiveness are understood and proven.

This has, however, invigorated the investigation of mixtures of natural products that can exist as either crude extracts or as chemically refined fractions. Over the years, a wealth of information has been deposited into the scientific literature about the in vitro antileishmanial activities of natural extracts, especially from plants that grow in regions of endemic leishmaniasis. Some thorough reviews of the literature were published covering articles about in vitro antileishmanial crude phytochemical mixtures that were printed between the years 2000 and 2010, or earlier, which should be consulted [65,226–228]. However, some authors have chosen to use abnormally high concentration thresholds for considering test samples as being bioactive, such as in vitro IC_{50} values of 100 or 250 μg/mL. For the purpose of this section, crude or refined mixtures of natural products were considered to be active only if they demonstrated IC_{50} values $\leq$10 μg/mL in vitro or a statistically significant reduction in parasite burden to infected animals in vivo. Thus, several plant extracts shown more recently to have this strictly defined antileishmanial activity are tabulated here in Table 9.4. Some of these have been investigated for the determination of purified antileishmanial natural products, while others may represent good directions for future research efforts.

5.4 Plant Natural Product In Vitro Antileishmanial Agents and Immunomodulators

Although there are relatively few phytochemicals with demonstrated in vivo antileishmanial activity, there has been a significant amount of research conducted on in vitro bioactive natural products. Typically, the mechanism of action for in vitro active antileishmanial agents will go undetermined until the molecule shows promise as a drug candidate or lead molecule, perhaps due to the observation of in vivo efficacy in an animal model of leishmaniasis. Even then, the determination of mechanisms of action is not a trivial research question, as described in more detail earlier in this chapter. Many review articles have cataloged research outcomes from antileishmanial natural product drug discovery efforts, and several stand out in recent history that should be consulted, including research up to the years 2001 [229], 2002 [188], 2003 [230], 2006 [187], 2008 [231], and 2014 [189]. The content of these articles will not be duplicated in this chapter. However, it is worth highlighting that in vitro antileishmanial activity of many naturally occurring chemicals has been observed, and these belong to effectively all molecular families,

TABLE 9.4 Selected Plant Extracts With Demonstrated In Vitro Anti-leishmanial Activities

Plant Name	Plant Part, Extraction Solvent, Observed Activity	References
Artemesia afra	Leaf, DCM, in vitro against *L. donovani* axenic amastigotes	[258]
Catha edulis	Root, DCM, in vitro against *L. donovani* axenic amastigotes	[258]
Conyza scabrida	Leaf, DCM/MeOH (1:1), in vitro against *L. donovani* axenic amastigotes	[258]
Cupressus sempervirens	Cones, EtOH, in vitro against *L. donovani* promastigotes	[242,259]
Croton caudatus	Leaf, hexane, in vitro and in vivo against *L. donovani*	[260]
Ekebergia capensis	Fruit, DCM/MeOH (1:1), in vitro against *L. donovani* axenic amastigotes	[258]
Eucomis autumnalis	Flowers/buds, DCM, in vitro against *L. donovani* axenic amastigotes	[258]
Harpagophytum procumbens	EtOH, in vitro against *L. donovani* axenic amastigotes	[261]
Hypericum aethiopicum	Leaf, DCM/MeOH (1:1), in vitro against *L. donovani* axenic amastigotes	[258]
Hypericum perforatum	EtOH, in vitro against *L. donovani* axenic amastigotes	[261]
Leonotis leonurus	Leaf, DCM/MeOH (1:1), in vitro against *L. donovani* axenic amastigotes	[258]
Maytenus undata	Root, DCM, in vitro against *L. donovani* axenic amastigotes	[258]
Pentalinon andrieuxii	Root, hexanes, in vitro and in vivo against *L. mexicana*	[218,262]
Salvia repens	Whole plant, DCM/MeOH (1:1), in vitro against *L. donovani* axenic amastigotes	[258]
Schefflera umbellifera	Root, DCM, in vitro against *L. donovani* axenic amastigotes	[258]
Serenoa repens	Fruit, EtOH in vitro against *L. donovani* axenic amastigotes	[261]
Silybum marianum	EtOAc, in vitro against *L. donovani* axenic amastigotes	[261]
Tarchonanthus camphorates	Whole plant, DCM/MeOH (1:1), in vitro against *L. donovani* axenic amastigotes	[258]

TABLE 9.4 Selected Plant Extracts With Demonstrated In Vitro Antileishmanial Activities—cont'd

Plant Name	Plant Part, Extraction Solvent, Observed Activity	References
Taxodium distichum	Cones, MeOH, in vitro and in vivo against *L. donovani*	[244]
Tetradenia riparia	Leaves, steam, in vitro against *L. amazonensis* intracellular amastigotes and immunomodulatory	[263]
Valeriana officinalis L.	EtOH, in vitro against *L. donovani* axenic amastigotes	[261]

including many types of alkaloids, coumarins, curcuminoids, chalcones and flavonoids, iridoids, lactones, quinones, tannins, terpenoids, and more. The active phytochemicals associated with scientific reports published between the years 2014 and 2016, which would not appear in the mentioned review articles, are outlined in the following molecular class–delineated subsections.

5.4.1 Alkaloids

Among the classes of natural products that exist, the alkaloids are the most structurally diverse and divergent in biosynthetic pathways [232]. Unsurprisingly, quite a few of these have been shown to have antileishmanial activities in vitro. Indeed, alkaloids may be particularly well suited for the targeting of intracellular parasites due to their ability to exist in charged protonated forms under acidic conditions and neutral deprotonated forms under neutral conditions, with the obvious exception of alkaloids that contain permanently charged quaternary amines. This is important, because it has been established that small molecules must pass three membranes to reach any biochemical molecular target in the parasite even in vitro: that of the macrophage, the intracellular parasitophorous vacuole, and the parasite itself [161]. A set of thorough reviews specifically covering the alkaloids with antiprotozoal activities that were reported through 2008 is available [191,233], and these should be consulted as their contents will not be duplicated here.

The indole alkaloid demethylaspidospermine was isolated from the leaves and bark of *Geissospermum reticulatum* A. Genttry (Apocynaceae), and was shown to be an in vitro growth inhibitor of *L. infantum* promastigotes with $GI_{50} = 7.7\ \mu M$ [234]. Similarly, The tetrahydroprotoberberine alkaloid govaniadine was isolated from *Corydalis govaniana* Wall. (Papaveraceae) and shown to be potently antileishmanial in vitro to promastigotes of *L. amazonensis* [235]. However, although this molecule was nearly 284-fold

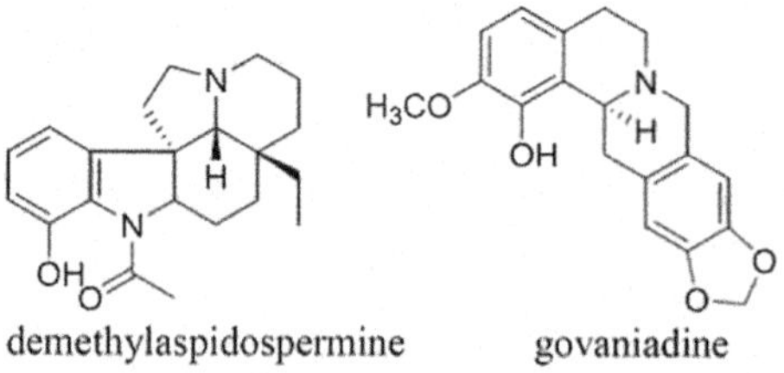

FIGURE 9.6 Structures of some alkaloids with in vitro antileishmanial activity.

less toxic to J774 macrophages, its antileishmanial activity to intracellular amastigotes was not reported with its isolation [235]. The structures of these described alkaloidal phytochemicals with in vitro antileishmanial activities are shown in Fig. 9.6.

5.4.2 Alkanes

The alkane aldehydes (*E*)-2-decenal, (*E*)-2-undecenal, and (*E*)-2-dodecenal were reported to be the in vitro antileishmanial components of the essential oil of coriander or cilantro, *Coriandrum sativum* L. (Apiaceae) [236]. Although these molecules appeared to be relatively bioactive as in vitro antileishmanial agents against axenic amastigotes of *L. donovani* ($IC_{50} = 1-5$ μg/mL), their strikingly low molecular weight means that a proportionately higher molar concentration of the molecules is necessary for antileishmanial activity [236]. What was quite interesting, however, was that the related saturated aldehydes decenal, undecenal, and dodecanal were not found to be bioactive in this test system [236]. Furthermore, several other *cis*-unsaturated alkane aldehydes were isolated and also did not exhibit antileishmanial activities in this study [236]. The structures of these described alkyl aldehyde phytochemicals with in vitro antileishmanial activities are shown in Fig. 9.7.

5.4.3 Flavonoids

The flavonoids are widely spread phytochemicals that have characteristic structures resulting from the combination of shikimate and acetate derived aromatic building blocks. However, special tailoring enzymes that decorate the core skeleton of various types of flavonoids can impart some interesting and specific biological functions. The phytochemical apigenin is a widely

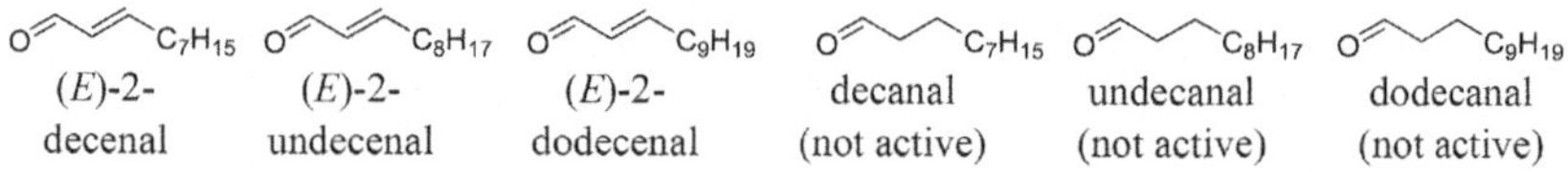

FIGURE 9.7 Structures of some alkanes with in vitro antileishmanial activity.

distributed flavone with a wide range of reported in vitro biological activities [237]. As was the case with the in vivo antileishmanial flavonoid described earlier, quercetin, apigenin was found to induce the production of reactive oxygen species when exposed to *L. amazonensis* [237]. The in vitro antileishmanial activity of apigenin was also reduced or eliminated when infected promastigotes were pretreated with reduced GSH or the antioxidant molecule, *N*-acetyl-L-cysteine, indicating that its likely mechanism of action involves the generation of ROS [237]. Furthermore, severe damage to the mitochondrial membrane along with a reduced greatly reduced mitochondrial membrane potential of the parasite was observed after treatment with this molecule, indicating the mode of cell death [237].

Other possible mechanisms of action for the antileishmanial activity of flavonoids have also been proposed. For example, the flavonoid dimers agathisflavone and tetrahydrorobustaflavone were reported to be in vitro inhibitors of cathepsin-like cysteine protease recombinant protein rCPB2.8 from *L. mexicana* [238]. Based on competition profiles against the protein's reporter substrate Z-Phe-Arg-7-amino-4-methylcoumarin, it was determined that the mode of enzyme inhibition was partially noncompetitive for agathisflavone, and uncompetitive for tetrahydrorobustaflavone [238]. Although these molecules were potent inhibitors (IC_{50} = 0.4 and 2.2 μM, respectively) of the rCPB2.8 enzyme, neither was specific for this protein compared with inhibitory activities against human cathepsins L and B [238]. Additionally, several flavonoids that were isolated from the leaves and stems of *Byrsonima coccolobifolia* Kunth. (Malpighiaceae) were demonstrated to be in vitro inhibitors of *L. amazonensis* ARG, a protein essential to parasite survival [239]. For example, the naturally occurring flavonoids isoquercitrin and epicatechin, as well as several semi-synthetic derivatives such as 3,5,7-triacetylepicatechin, inhibited ARG in vitro with IC_{50} = 0.9–2.0 μM [239]. These phytochemicals were shown to be noncompetitive ARG inhibitors in vitro, and furthermore, other analogues had in vitro IC_{50} values of up to 223 μM, implying that the effects observed were not unspecific [239]. The authors note the broad array of enzymes that flavonoids like these have been shown to inhibit, and suggest that perhaps a combination of mechanisms is responsible for observed antileishmanial activities [238,239]. The structures of these described flavonoid phytochemicals with in vitro antileishmanial activities are shown in Fig. 9.8.

5.4.4 Phenylpropanoids

The phenylpropanoids are a family of phytochemicals derived from the shikimic acid biosynthetic pathway. These have easily identifiable substructures of distinct phenylpropanoid units, but can be modified and connected in a variety of ways that impart different biological functions on the molecule. The phenylpropanoid dimers dehydrodieugenol B and 1-(8-propenyl)-3-[3′-methoxy-1′-(8-propenyl)phenoxy]-4,5-dimethoxybenzene were isolated from

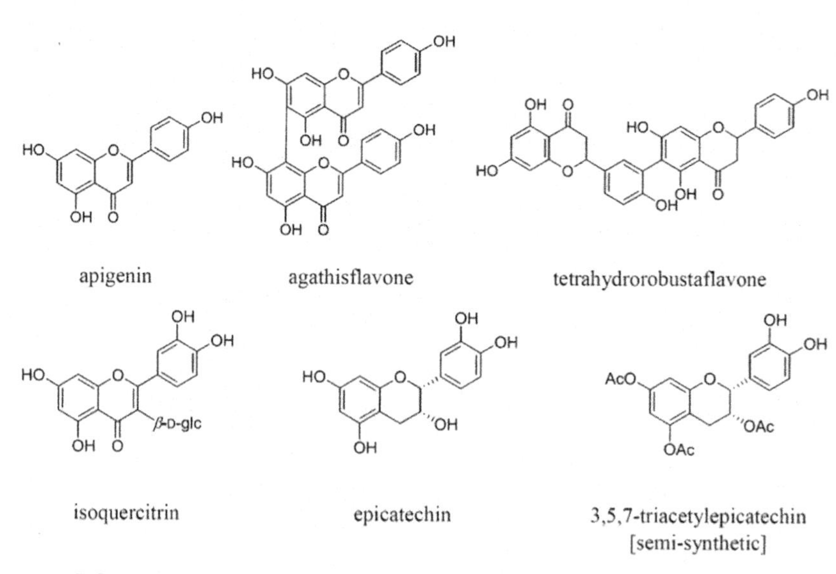

FIGURE 9.8 Structures of some flavonoids with in vitro antileishmanial activity.

the twigs of *Nectandra leucantha* Nees and Mart (Lauraceae) and shown to be potentially immunomodulating antileishmanial agents in vitro [240]. In this study, these molecules were shown to have moderate direct bioactivity on *L. donovani* amastigotes ($IC_{50} \approx 20\ \mu M$), but were significantly less toxic to the promastigote form of the parasite and to murine peritoneal macrophages [240]. In addition, the compounds had little were shown to not increase nitric oxide production but still suppressed cytokines IL-6 and IL-10, indicating a possible immunomodulatory action by an NO-independent mechanism that may be useful for future drug development and discovery [240]. The structures of these described phenylpropanoid phytochemicals with in vitro anti-leishmanial activities are shown in Fig. 9.9.

5.4.5 Quinones

Quinones are not necessarily a structurally distinct family of phytochemicals from a biosynthetic or structural point of view, but it is convenient to consider

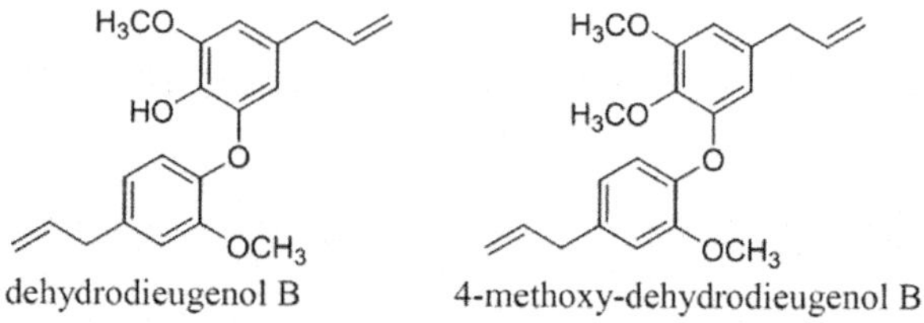

dehydrodieugenol B

4-methoxy-dehydrodieugenol B

FIGURE 9.9 Structures of some phenylpropanoids with in vitro antileishmanial activity.

them as being related for the purpose of this chapter. Quinones are commonly regarded as nuisance molecules in medicinal chemistry screening efforts due to their pan–assay activity and frequent toxicity profiles. This is not due to an induction of false positive test results, however, rather the reactivity of these molecules makes them bioactive in a widespread set of biological systems. A cheminformatics analysis of the extracts of 36 antiprotozoal extracts from *Juglans* spp. (Juglandaceae) led to the isolation of an active antileishmanial agent, hydrojuglone glucoside [241]. Several quinones related to this quinol phytochemical, including juglone, plumbagin, lapachol, shikonin, and 1,4-napthoquinone, are known in vitro antileishmanial agents against axenic amastigotes of *L. dononvani* (IC_{50} = 2.0, 0.9, 3.3, 0.1, 2.4 μM, respectively) [241]. Although hydrojuglone glucoside was less active in this bioassay (IC_{50} = 16.7 μM), it should be anticipated that β-glucosidases acting on it in other in vitro systems or in vivo would release the bioactive and cytotoxic aglycone, juglone [241]. The structures of these described quinone phytochemicals with in vitro antileishmanial activities are shown in Fig. 9.10.

5.4.6 Terpenoids

The terpenoids are an incredibly large family of phytochemicals. Since isoprenyl building blocks can often be easily observed in the final structure of these natural products, they are occasionally regarded as isoprenoids in the literature. Based on the number of five-carbon isoprenoid subunits, the terpenoid family is further subdivided into classes, such as hemiterpenoids (C_5), monoterpenoids (C_{10}), sesquiterpenoids (C_{15}), diterpenoids (C_{20}), sesterterpenoids (C_{25}), triterpenoids (C_{30}), and so on. Furthermore, the various ways that the isoprenoid monomer units are attached together with ring junctions and configurations are used to further delineate the subclasses of this molecular family.

Two abietane diterpenoids with quinone methide functionality, taxodione and deoxotaxodione, were isolated from the cones of *Cupressus sempervirens* L. (Cupressaceae), and shown to have potent in vitro antileishmanial activity against *L. donovani* promastigotes (IC_{50} = 0.08 and 0.26 μM, respectively) [242]. These phytochemicals were markedly less potent when also tested against *L. donovani* intracellular amastigotes (IC_{50} = 9.0 and 17.2 μM, respectively), and neither compound was reported to be cytotoxic to THP1

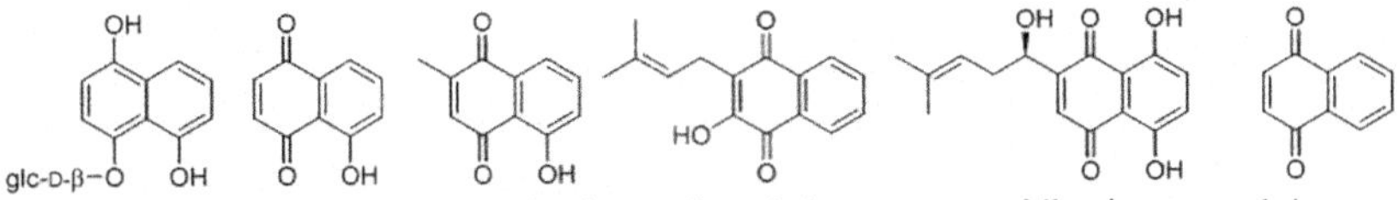

hydrojuglone glucoside juglone plumbagin lapachol shikonin 1,4-napthoquinone

FIGURE 9.10 Structures of some quinones with in vitro antileishmanial activity.

human acute monocyclic leukemia cells [242]. This result is striking, because taxodione itself was initially discovered and reported as a potent cytotoxic and antitumor agent [243]. Indeed, when taxodione was reisolated from a methanolic extract of the cones of *Taxodium distichum* L. Rich. (Cupressaceae), which demonstrated in vivo antileishmanial activity against *L. donovani* in mice, this compound was shown to be both active in vitro against *L. donovani* promastigotes as well as being cytotoxic to mammalian cells [244]. A structurally related *seco*-abietane diterpenoid, taxotrione, was further isolated from *T. distichum*, and was also toxic to both *L. donovani* promastigotes and mammalian cells in vitro ($IC_{50} = 6.9$ and 4.4 μM, respectively) [244]. Taxotrione was further shown to be even more toxic in vitro to amastigotes of *L. amazonensis* ($IC_{50} = 0.52$ μM), showing some slight selectivity over mammalian cells [244].

In a semisynthetic investigation of modifications to the diterpenoid dehydroabietylamine for antileishmanial drug discovery, several amide derivatives have been reported with significantly enhanced activity in vitro against axenic amastigotes of *L. donovani* [245]. Furthermore, the most active of these molecules, *N*-(dehydroabietyl)prop-2-enamide ($IC_{50} = 0.37$ μM), was shown to have 63-fold selectivity as compared to L6 rat skeletal myoblasts [245]. Although much work still needs to be done to advance this lead molecule in testing, the fact that it can be synthesized from affordable starting materials means that it may hold promise in drug development in the future.

The nor-clerodane diterpenoid, *trans*-dehydrocrotonin, is an antileishmanial phytochemical isolated from the Amazonian folk medicinal plant *Croton cajucara* Benth. (Euphorbiaceae) [246]. Interestingly, this molecule was a much more potent in vitro antileishmanial agent when tested using *L. amazonensis* intracellular amastigotes ($IC_{50} = 1.5$ μM), as it was markedly less active against promastigotes and axenic amastigotes of the same species [246]. Furthermore, the difference in potency could not be attributed to macrophage cytotoxicity, since this was not observed at any tested concentration [246]. The mode of action for *trans*-dehydrocrotonin likely includes the inhibition of trypanothione reductase, as it was demonstrated to do this in vitro [246]. The structures of all diterpenoid phytochemicals with in vitro antileishmanial activities described to this point in the chapter are shown in Fig. 9.11.

From the roots of *P. andrieuxii* Mueller-Arg. (Apocynaceae), several steroids and sterols were isolated and shown to have antileishmanial activity against *L. mexicana* amastigotes in vitro [217]. These include the phytochemical pentalinonsterol, which was shown to have in vivo antileishmanial activity as was described in section 5.3. What was most striking about the report of the antileishmanial activities of pentalinonsterol, 24-methylcholesta-4,24(28)-dien-3-one, cholest-4-en-3-one, neridienone, and 6,7-dihydroneridienone was that although the IC_{50} values for these against *L. mexicana* amastigotes were moderate to potent (0.03–3.5 μM), their

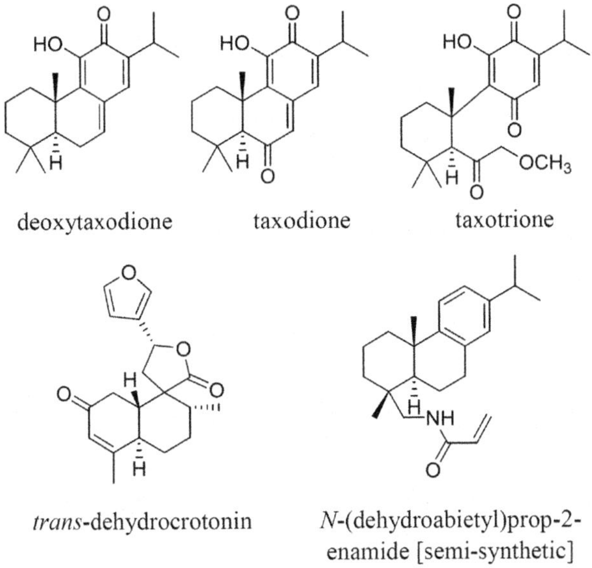

deoxytaxodione taxodione taxotrione

trans-dehydrocrotonin *N*-(dehydroabietyl)prop-2-enamide [semi-synthetic]

FIGURE 9.11 Structures of some terpenoids with in vitro antileishmanial activity.

selectivity for killing amastigotes over promastigotes ranged from about sevenfold to a striking 2700-fold [217].

A widely distributed plant sterol, fucosterol, was isolated in mass from the brown alga *Lessonia vadosa* Searles (Lessoniaceae), and tested for its antileishmanial activity [247]. This compound effectively killed intracellular amastigotes of *L. infantum* ($IC_{50} = 10.3\ \mu M$) and *L. amazonensis* ($IC_{50} = 7.9\ \mu M$), but was about fourfold and sevenfold less potent against the promastigote form of each species, respectively [247]. Since fucosterol is widely distributed in nature, this molecule may serve as an advanced starting material for future antileishmanial medicinal chemistry optimization studies.

In addition to several flavonoids mentioned before, the pentacyclic triterpenoid, ursonic acid, was reported to be an in vitro inhibitor of cathepsin-like cysteine protease recombinant protein rCPB2.8 from *L. mexicana* [238]. Based on competition profiles against the protein's reporter substrate Z-Phe-Arg-7-amino-4-methylcoumarin, it was determined that the mode of enzyme inhibition was partially noncompetitive for this phytochemical [238]. As was the case with the flavonoids described, the activity of this rCPB2.8 enzyme inhibitor ($IC_{50} = 3.8\ \mu M$) was not specific for this protein compared with inhibitory activities against human cathepsins L and B [238]. The structures of all triterpenoid phytochemicals with in vitro antileishmanial activities described in this section are shown in Fig. 9.12.

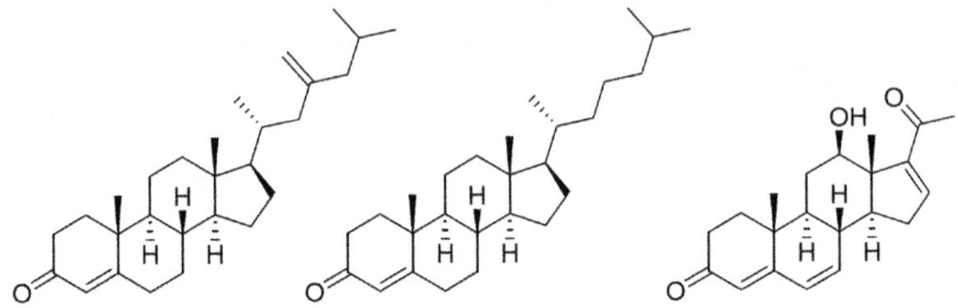

24-methylcholesta-4,24(28)-dien-3-one cholest-4-en-3-one neridienone

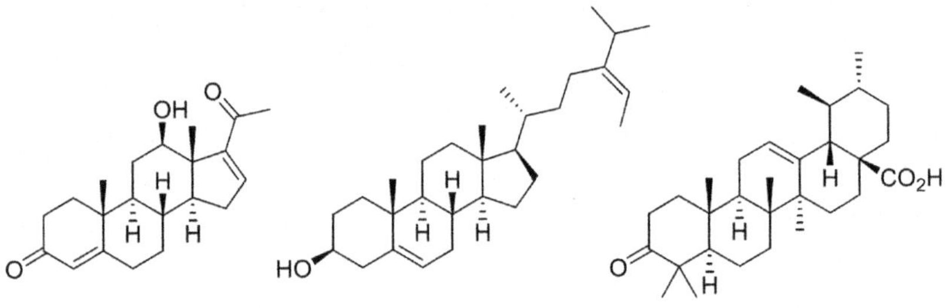

6,7-dihydroneridienone fucosterol ursonic acid

FIGURE 9.12 Structures of some steroids and sterols with in vitro antileishmanial activity.

5.5 Marine and Fungal Natural Products With In Vitro Antileishmanial Activity

As with the review articles described, which have cataloged many in vitro antileishmanial natural products from plants and other sources, one recent document should be consulted as it pertains specifically to marine natural products in antileishmanial drug discovery [248]. The field of marine natural products research has been especially active in recent years, as rediscovery rates for known compounds from plant and microbial sources have been seen as a nuisance to many investigators [249]. However, it is expected that natural products from all sources can be fruitful lead molecules for antileishmanial drug discovery.

After having been isolated from the marine sponge, *Monanchora arbuscular* Duchassaing and Michelotti (Crambeidae), the pyrimidine alkaloid monalidine A, as well as guanidine alkaloids batzelladines D, F, and L and norbatzelladine L were shown have in vitro antileishmanial activity against *L. infantum* promastigotes, with IC_{50} values ranging from 2 to 4 μM, although they were also inactive against amastigotes of the same species [250]. These molecules were also demonstrated in vitro to induce mitochondrial membrane depolarization as well as a profound generation of ROS, indicating potential mechanisms of antileishmanial action [250]. Furthermore, a method for the large-scale synthetic production of monalidine A from affordable starting materials has been reported, and research efforts using this as a biochemical probe appear to be ongoing [250,251].

A potently antileishmanial xenicane diterpenoid, cristaxenicin A, was isolated from the deep sea gorgonian *Acanthoprimnoa cristata* Kükenthal and Gorzawsky (Primnoidae) [252]. This molecule demonstrated potent in vitro antileishmanial activity against promastigotes of *L. amazonensis* ($IC_{50} = 0.088$ μM), and 23–53-fold selectivity over in vitro mammalian cell cytotoxicity [252]. Later, the core scaffold analogue of cristaxenicin A was synthesized, and this racemic material was active in vitro against promastigotes of *L.amazonensis* ($IC_{50} = 2.4$ μM), but much less potent than the natural product [252]. However, further studies appear to be ongoing to utilize this synthetic material for medicinal chemical optimization and mode of action studies for the cristaxenicin class of antileishmanial agents.

A traditional bioassay-guided fractionation of the filamentous fungus *Geosmithia langdonii* M. Kolarík, Kubátová and Pazoutová (Ascomycota: Hypocreales) yielded a series of simple and related antileishmanial natural products [253]. Specifically, (+)-epiepoformin, gentisaldehyde, gentisin alcohol, and *m*-cresol were all shown to inhibit *L. donovani* promastigotes in vitro ($IC_{50} = 6.9$, 3.3, 8.5, and 9.2 μM, respectively [253]. Although these are not particularly attractive lead molecules, the natural products research investigation of organisms that can be cultured in laboratory settings may be a promising avenue for future antileishmanial drug discovery. The structurers of all marine and fungal natural products with in vitro antileishmanial activities described in this section are shown in Fig. 9.13.

5.6 Combination Therapy

Phytomedicines have shown promise against leishmaniasis when used alone for monotherapy or used in combination with known antileishmanial drugs. Combination therapy provides a less-expensive alternative therapeutic route, which has an increased treatment efficacy and tolerance and reduced treatment duration and limits the emergence of drug resistance [254]. These factors may also contribute to an enhancement of patient compliance with suggested treatment regimens in the future.

Studies with combination with glucantime with garlic showed that garlic alone induced host protective Th1 immune responses compared to glucantime alone. Combination therapy of garlic with glucantime induced a prominent Th1 response with the highest level of protection against *Leishmania* [255]. This highlights that the combined immunomodulatory effect of garlic together with parasite killing activity of glucantime [255]. In another study, combination therapy with triterpenoid glycyrrhizic acid (GA) with SAG against in vivo SAG resistant *Leishmania* showed that GA could regulate the levels of drug

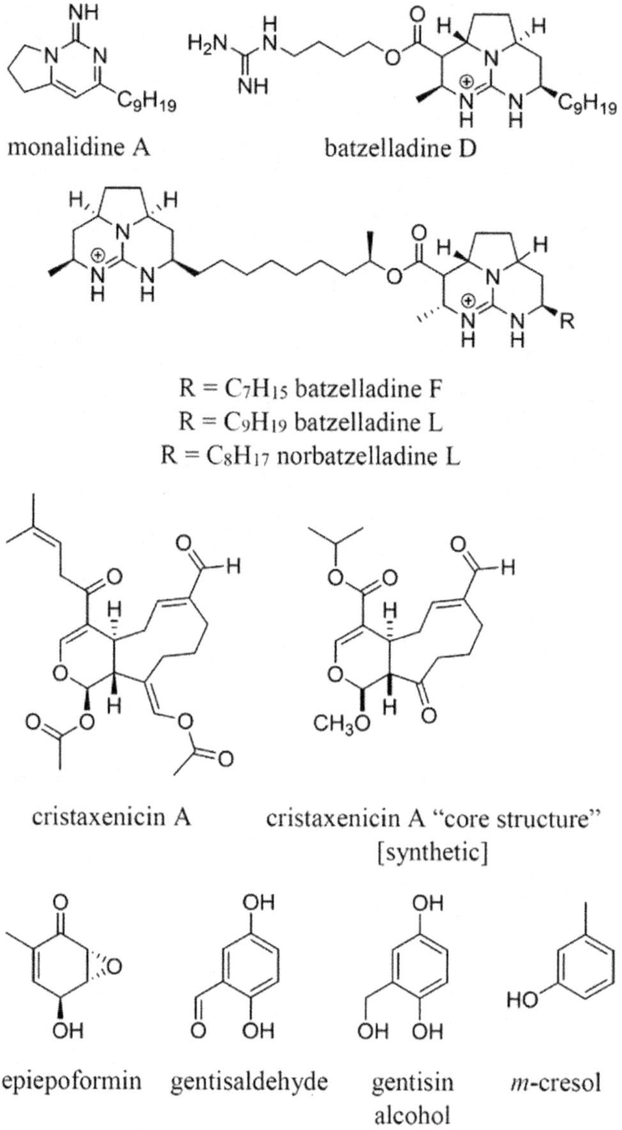

monalidine A

batzelladine D

$R = C_7H_{15}$ batzelladine F
$R = C_9H_{19}$ batzelladine L
$R = C_8H_{17}$ norbatzelladine L

cristaxenicin A

cristaxenicin A "core structure" [synthetic]

epiepoformin gentisaldehyde gentisin alcohol *m*-cresol

FIGURE 9.13 Structures of some marine and fungal natural products with in vitro anti-leishmanial activity.

transporter pumps and also induce a Th1 immune response, whereas SAG mediated parasite killing activity [256]. Saponins isolated from ivy, including α-hederin, β-hederin, and hederacolchiside A_1, have also been shown to combine with antileishmanial drugs such as pentamidine and AmpB in enhancing the cumulative antiparasitic activity of these drugs [257]. Combined therapy in humans using oral azithromycin and allopurinol, a drug with antileishmanial activity, for two months and intramuscular injection of glucantime for 20 days showed slightly better outcome after a 2-month follow-up in 86 patients with no severe adverse effect [43]. More studies are expected to continue to determine opportunities for synergistic chemotherapy and immunomodulation for treating leishmaniasis.

6. FUTURE DIRECTIONS AND CONCLUSIONS

The control of leishmaniasis, a WHO-designated neglected tropical disease, is a highly desirable achievement for public health. This disease is uncontrolled in many parts of the world where social problems and poverty are also present. Lack of serious investments in healthcare and research, global warming and large population movements, as well as the emergence of drug resistance are among the major obstacles that must be confronted by the medical community. Drug toxicity for most of the existing treatments for leishmaniasis is also a great cause of morbidity and mortality among patients affected with leishmaniasis, and new drugs are urgently needed.

Natural products remain a vast source of materials for the development of new lead molecules for medicinal chemistry programs, as well as drugs, for the prevention and treatment of infectious diseases. Plant-derived materials can be effective, low in cost, and abundant in nature. Previous studies have discovered many herbal medicines with demonstrated experimental activity against leishmaniasis, however, the commercialization of most of these as drugs is far away from happening. To date, no herbal medicine or experimental phytochemical has been shown to be safe or effective for use in human leishmaniasis through large controlled clinical trials. The expansion of antileishmanial research into the natural products of marine and terrestrial microbial organisms may be a particularly fruitful endeavor, since the discovery of useful lead molecules and drug candidates will be able to be produced in a scaled fashion more easily after the initial discovery.

Simultaneously, screening of natural products for their immunomodulatory activity on host immune cells such as macrophages, dendritic cells and T lymphocytes, should be done for evaluating their effect on host cytokine response. These natural products having immunomodulatory activity can be combined with known antileishmanial drugs for developing potential therapy against leishmaniasis. In addition, advanced methods for the delivery of drug(s) to affected target areas by means including liposomal, neosomal, or polymer-coated nanoparticle encapsulations need to be considered for the enhancement of safety and efficacy of emerging therapies for leishmaniasis.

To this end, significant investments must necessarily be made in order to favor more basic and translational research for the treatment of leishmaniasis. The sequential performance of increasingly advanced research (e.g., in vitro studies, in vivo studies, pharmacokinetic and pharmacodynamic studies) must be conducted by the scientific community prior to the essential controlled clinical trials for the therapy of leishmaniasis. Ongoing research is aimed at advancing more and more candidates through the drug discovery pipeline. Academic, government, and nonprofit organizations must continue to work toward new antileishmanial drugs, and collaborative cooperation with the philanthropic arms of large pharmaceutical companies may prove to be invaluable in the eventual development and production of new safe and effective antileishmanial drugs.

REFERENCES

[1] M. da Saúde BR Secretaria de Vigilância, Manual de vigilância da leishmaniose tegumentar americana, 2007.

[2] World Health Organization, Control of the Leishmaniases: Report of a Meeting of the WHO Expert Committee on the Control of Leishmaniases, 2010.

[3] C.M. Gomes, N.A. de Paula, O.O. de Morais, K.A. Soares, A.M. Roselino, R.N.R. Sampaio, Complementary exams in the diagnosis of American tegumentary leishmaniasis, Bras. Dermatol. 89 (2014) 701–709, https://doi.org/10.1590/abd1806-4841.20142389.

[4] R. Reithinger, M. Mohsen, K. Aadil, M. Sidiqi, P. Erasmus, P.G. Coleman, Anthroponotic Cutaneous Leishmaniasis, Kabul, Afghanistan, 2003.

[5] C.M. Gomes, M.V. Cesetti, N.A. de Paula, S. Vernal, G. Gupta, R.N.R. Sampaio, et al., Field validation of SYBR® Green- and TaqMan®-based real-time PCR using biopsy and swab samples to diagnose American tegumentary leishmaniasis in a *Leishmania (V.) braziliensis*-endemic area, J. Clin. Microbiol. (2016) JCM.01954–16, https://doi.org/10.1128/JCM.01954-16.

[6] S. Mohapatra, Drug resistance in leishmaniasis: newer developments, Trop. Parasitol. 4 (2014) 4–9, https://doi.org/10.4103/2229-5070.129142.

[7] F.L. Carvalho, D.L.S. Aires, Z.F. Segunda, C.M.P.E.S. de Azevedo, R.D.G.C.F. Corrêa, D.M.C. de Aquino, et al., The epidemiological profile of HIV-positive individuals and HIV-Leishmaniasis co-infection in a referral center in São Luis, Maranhão, Brazil, Cien Saude Colet 18 (2013) 1305–1312.

[8] I.N.G. Rosário, A.J. Andrade, R. Ligeiro, R. Ishak, I.M. Silva, Evaluating the adaptation process of sandfly fauna to anthropized environments in a leishmaniasis transmission area in the Brazilian amazon, J. Med. Entomol. 54 (2) (2016), https://doi.org/10.1093/jme/tjw182.

[9] F. Abedi-Astaneh, H. Hajjaran, M.R. Yaghoobi-Ershadi, A.A. Hanafi-Bojd, M. Mohebali, M.R. Shirzadi, et al., Risk mapping and situational analysis of cutaneous leishmaniasis in an endemic area of Central Iran: a GIS-based survey, PLoS One 11 (2016) e0161317, https://doi.org/10.1371/journal.pone.0161317.

[10] Evolutionary and geographical history of the *Leishmania donovani* complex with a revision of current taxonomy, Proc. Natl. Acad. Sci. USA 104 (2007) 9375–9380. https://doi.org/10.1073/pnas.0703678104.

[11] R. Reithinger, J.-C. Dujardin, H. Louzir, C. Pirmez, B. Alexander, S. Brooker, Cutaneous leishmaniasis, Lancet Infect. Dis. 7 (2007) 581–596. https://doi.org/10.1016/S1473-3099(07)70209-8.

[12] E.E. Zijlstra, M.S. Ali, A.M. El-Hassan, I.A. El-Toum, M. Satti, H.W. Ghalib, et al., Kala-azar: a comparative study of parasitological methods and the direct agglutination test in diagnosis, Trans. R. Soc. Trop. Med. Hyg. 86 (1992) 505–507. https://doi.org/10.1016/0035-9203(92)90086-R.

[13] J.H. Kolaczinski, R. Reithinger, D.T. Worku, A. Ocheng, J. Kasimiro, N. Kabatereine, et al., Risk factors of visceral leishmaniasis in East Africa: a case-control study in Pokot territory of Kenya and Uganda, Int. J. Epidemiol. 37 (2008) 344–352. https://doi.org/10.1093/ije/dym275.

[14] J. Alvar, I.D. Vélez, C. Bern, M. Herrero, P. Desjeux, J. Cano, et al., Leishmaniasis worldwide and global estimates of its incidence, PLoS One 7 (2012) e35671. https://doi.org/10.1371/journal.pone.0035671.

[15] E.E. Zijlstra, The immunology of post-kala-azar dermal leishmaniasis (PKDL), Parasit. Vectors 9 (2016) 464. https://doi.org/10.1186/s13071-016-1721-0.

[16] C.M. Gomes, M.V. Cesetti, O.O. Morais, M.S.T. Mendes, A.M. Roselino, R.N.R. Sampaio, The influence of treatment on the development of leishmaniasis recidiva cutis: a 17-year case–control study in Midwestern Brazil, J. Eur. Acad. Dermatol. Venereol. 29 (2015) 109–114. https://doi.org/10.1111/jdv.12473.

[17] C.M. Gomes, C.D.R. de Paula, R.N.R. Sampaio, F.D.S. Damasco, O.O. de Morais, C.D.R. de Paula, Recurrent cutaneous leishmaniasis, Bras. Dermatol. 88 (2013) 462–464. https://doi.org/10.1590/abd1806-4841.20131885.

[18] S. Vernal, N.A. de Paula, C.M. Gomes, A.M. Roselino, Disseminated leishmaniasis by *Leishmania Viannia* subgenus: a series of 18 cases in southeastern Brazil, Open Forum Infect. Dis. 3 (2016) ofv184. https://doi.org/10.1093/ofid/ofv184.

[19] L. Jirmanus, M.J. Glesby, L.H. Guimarães, E. Lago, M.E. Rosa, P.R. Machado, et al., Epidemiological and clinical changes in American tegumentary leishmaniasis in an area of *Leishmania (Viannia) braziliensis* transmission over a 20-year period, Am. J. Trop. Med. Hyg. 86 (2012) 426–433. https://doi.org/10.4269/ajtmh.2012.11-0378.

[20] C.M. Gomes, N.A. de Paula, M.V. Cesetti, A.M. Roselino, R.N.R. Sampaio, Mucocutaneous leishmaniasis: accuracy and molecular validation of noninvasive procedures in a L. (V.) braziliensis-endemic area, Diagn. Microbiol. Infect. Dis. 79 (2014) 413–418. https://doi.org/10.1016/j.diagmicrobio.2014.05.002.

[21] G.L.L. Machado-Coelho, W.T. Caiaffa, O. Genaro, P.A. Magalhães, W. Mayrink, Risk factors for mucosal manifestation of American cutaneous leishmaniasis, Trans. R. Soc. Trop. Med. Hyg. 99 (2005) 55–61. https://doi.org/10.1016/j.trstmh.2003.08.001.

[22] J. Alvar, Leishmaniasis and AIDS co-infection: the Spanish example, Parasitol. Today (Regul. Ed.) 10 (1994) 160–163.

[23] A.M. Da-Cruz, A.C.S.A. Rodrigues, M. Mattos, M.P. Oliveira-Neto, V. Sabbaga-Amato, M.P. Posada, et al., Immunopathologic changes in HIV-Leishmania co-infection, Rev. Soc. Bras. Med. Trop. 39 (Suppl. 3) (2006) 75–79.

[24] J. Alvar, P. Aparicio, A. Aseffa, M. den Boer, C. Cañavate, J.-P. Dedet, et al., The relationship between leishmaniasis and AIDS: the second 10 years, Clin. Microbiol. Rev. 21 (2008) 334–359. https://doi.org/10.1128/CMR.00061-07 table of contents.

[25] M.P. Posada-Vergara, J.A.L. Lindoso, J.E. Tolezano, V.L. Pereira-Chioccola, M.V. Silva, H. Goto, Tegumentary leishmaniasis as a manifestation of immune reconstitution inflammatory syndrome in 2 patients with AIDS, J. Infect. Dis. 192 (2005) 1819–1822. https://doi.org/10.1086/497338.

[26] G. Gupta, S. Oghumu, A.R. Satoskar, Mechanisms of immune evasion in leishmaniasis, Adv. Appl. Microbiol. 82 (2013) 155–184. https://doi.org/10.1016/B978-0-12-407679-2.00005-3.

[27] M.C.A. Brelaz-de-Castro, A.F. de Almeida, A.P. de Oliveira, M. de Assis-Souza, L.F. da Rocha, V.R.A. Pereira, Cellular immune response evaluation of cutaneous leishmaniasis patients cells stimulated with *Leishmania (Viannia) braziliensis* antigenic fractions before and after clinical cure, Cell. Immunol. 279 (2012) 180–186. https://doi.org/10.1016/j.cellimm.2012.11.006.

[28] G. Gupta, S. Bhattacharjee, S. Bhattacharyya, P. Bhattacharya, A. Adhikari, A. Mukherjee, et al., CXC chemokine-mediated protection against visceral leishmaniasis: involvement of the proinflammatory response, J. Infect. Dis. 200 (2009) 1300–1310. https://doi.org/10.1086/605895.

[29] I.A. Rodrigues, A.M. Mazotto, V. Cardoso, R.L. Alves, A.C. Amaral, J.R. Silva, et al., Natural products: insights into leishmaniasis inflammatory response, Mediat. Inflamm. 2015 (2015) 835910–835912. https://doi.org/10.1155/2015/835910.

[30] L. Antônio, A. Schubach, K.B.F. Marzochi, A. Fagundes, Comparison between in vivo measurement of the Montenegro skin test and paper recording, Int. J. Dermatol. 51 (2012) 618–619. https://doi.org/10.1111/j.1365-4632.2010.04530.x.

[31] K.A. Soares, A.A.A. Urdapilleta, G.M.D. Santos, A.L. Carneiro, C.M. Gomes, A.M. Roselino, et al., Field validation of a *Leishmania (Leishmania) mexicana* exo-antigens ELISA for diagnosing tegumentary leishmaniasis in regions of *Leishmania (Viannia)* predominance, Braz. J. Infect. Dis. 19 (2015) 302–307. https://doi.org/10.1016/j.bjid.2015.03.010.

[32] F.Y. Liew, T(H)1 and T(H)2 cells: a historical perspective, Nat. Rev. Immunol. 2 (2002) 55–60. https://doi.org/10.1038/nri705.

[33] P. Gurung, R. Karki, P. Vogel, M. Watanabe, M. Bix, M. Lamkanfi, et al., An NLRP3 inflammasome-triggered Th2-biased adaptive immune response promotes leishmaniasis, J. Clin. Invest. 125 (2015) 1329–1338. https://doi.org/10.1172/JCI79526.

[34] D.S. Zamboni, D.S. Lima-Junior, Inflammasomes in host response to protozoan parasites, Immunol. Rev. 265 (2015) 156–171. https://doi.org/10.1111/imr.12291.

[35] G. Michel, B. Ferrua, P. Munro, L. Boyer, N. Mathal, D. Gillet, et al., Immunoadjuvant properties of the Rho activating factor CNF1 in prophylactic and curative vaccination against *Leishmania infantum*, PLoS One 11 (2016) e0156363. https://doi.org/10.1371/journal.pone.0156363.

[36] M.I.S. Lima, V.O. Arruda, E.V.C. Alves, A.P.S. de Azevedo, S.G. Monteiro, S.R.F. Pereira, Genotoxic effects of the antileishmanial drug Glucantime, Arch. Toxicol. 84 (2010) 227–232. https://doi.org/10.1007/s00204-009-0485-0.

[37] L. Reveiz, A.N.S. Maia-Elkhoury, R.S. Nicholls, G.A.S. Romero, Z.E. Yadon, Interventions for American cutaneous and mucocutaneous leishmaniasis: a systematic review update, PLoS One 8 (2013) e61843. https://doi.org/10.1371/journal.pone.0061843.

[38] J.P. Adler-Moore, J.-P. Gangneux, P.G. Pappas, Comparison between liposomal formulations of amphotericin B, Med. Mycol. 54 (2016) 223–231. https://doi.org/10.1093/mmy/myv111.

[39] M. Nacher, B. Carme, D. Sainte Marie, P. Couppié, E. Clyti, P. Guibert, et al., Influence of clinical presentation on the efficacy of a short course of pentamidine in the treatment of cutaneous leishmaniasis in French Guiana, Ann. Trop. Med. Parasitol. 95 (2001) 331–336. https://doi.org/10.1080/00034980120064355.

[40] A.K. Mukherjee, G. Gupta, A. Adhikari, S. Majumder, S. Kar Mahapatra, S. Bhattacharyya Majumdar, et al., Miltefosine triggers a strong proinflammatory cytokine response during visceral leishmaniasis: role of TLR4 and TLR9, Int. Immunopharmacol 12 (2012) 565–572. https://doi.org/10.1016/j.intimp.2012.02.002.

[41] A.M. Musa, B. Younis, A. Fadlalla, C. Royce, M. Balasegaram, M. Wasunna, et al., Paromomycin for the treatment of visceral leishmaniasis in Sudan: a randomized, open-label, dose-finding study, PLoS Negl. Trop. Dis. 4 (2010) e855. https://doi.org/10.1371/journal.pntd.0000855.

[42] S. Sundar, J. Chakravarty, Paromomycin in the treatment of leishmaniasis, Expert Opin. Investig. Drugs 17 (2008) 787−794. https://doi.org/10.1517/13543784.17.5.787.

[43] L. Dastgheib, M. Naseri, Z. Mirashe, Both combined oral azithromycin plus allopurinol and intramuscular Glucantime yield low efficacy in the treatment of Old World cutaneous leishmaniasis: a randomized controlled clinical trial, Int. J. Dermatol. 51 (2012) 1508−1511. https://doi.org/10.1111/j.1365-4632.2012.05610.x.

[44] D.K. Kochar, G. Saini, S.K. Kochar, R.A. Bumb, A double blind, randomised placebo controlled trial of rifampicin with omeprazole in the treatment of human cutaneous leishmaniasis, J. Vector Borne Dis. 43 (2006).

[45] N. Singh, Drug resistance mechanisms in clinical isolates of *Leishmania donovani*, Indian J. Med. Res. 123 (2006) 411−422.

[46] M. Mishra, U.K. Biswas, D.N. Jha, A.B. Khan, Amphotericin versus pentamidine in antimony-unresponsive kala-azar, Lancet 340 (1992) 1256−1257.

[47] A.K. Haldar, P. Sen, S. Roy, Use of antimony in the treatment of leishmaniasis: current status and future directions, Mol. Biol. Int. 2011 (2011) 571242−571323. https://doi.org/10.4061/2011/571242.

[48] S. Mandal, M. Maharjan, S. Singh, M. Chatterjee, R. Madhubala, Assessing aquaglyceroporin gene status and expression profile in antimony-susceptible and -resistant clinical isolates of *Leishmania donovani* from India, J. Antimicrob. Chemother. 65 (2010) 496−507. https://doi.org/10.1093/jac/dkp468.

[49] K. El Fadili, N. Messier, P. Leprohon, G. Roy, C. Guimond, N. Trudel, et al., Role of the ABC transporter MRPA (PGPA) in antimony resistance in *Leishmania infantum* axenic and intracellular amastigotes, Antimicrob. Agents Chemother. 49 (2005) 1988−1993. https://doi.org/10.1128/AAC.49.5.1988-1993.2005.

[50] A. Mukherjee, P.K. Padmanabhan, S. Singh, G. Roy, I. Girard, M. Chatterjee, et al., Role of ABC transporter MRPA, gamma-glutamylcysteine synthetase and ornithine decarboxylase in natural antimony-resistant isolates of *Leishmania donovani*, J. Antimicrob. Chemother. 59 (2007) 204−211. https://doi.org/10.1093/jac/dkl494.

[51] B. Purkait, R. Singh, K. Wasnik, S. Das, A. Kumar, M. Paine, et al., Up-regulation of silent information regulator 2 (Sir2) is associated with amphotericin B resistance in clinical isolates of *Leishmania donovani*, J. Antimicrob. Chemother. 70 (2015) 1343−1356. https://doi.org/10.1093/jac/dku534.

[52] B. Purkait, A. Kumar, N. Nandi, A.H. Sardar, S. Das, S. Kumar, et al., Mechanism of amphotericin B resistance in clinical isolates of *Leishmania donovani*, Antimicrob. Agents Chemother. 56 (2012) 1031−1041. https://doi.org/10.1128/AAC.00030-11.

[53] V. Ramesh, G.K. Katara, S. Verma, P. Salotra, Miltefosine as an effective choice in the treatment of post-kala-azar dermal leishmaniasis, Br. J. Dermatol. 165 (2011) 411−414. https://doi.org/10.1111/j.1365-2133.2011.10402.x.

[54] V. Bhandari, A. Kulshrestha, D.K. Deep, O. Stark, V.K. Prajapati, V. Ramesh, et al., Drug susceptibility in Leishmania isolates following miltefosine treatment in cases of visceral leishmaniasis and post kala-azar dermal leishmaniasis, PLoS Negl. Trop. Dis. 6 (2012) e1657. https://doi.org/10.1371/journal.pntd.0001657.

[55] C. Fernandez-Prada, I.M. Vincent, M.-C. Brotherton, M. Roberts, G. Roy, L. Rivas, et al., Different mutations in a P-type ATPase transporter in Leishmania parasites are associated with cross-resistance to two leading drugs by distinct mechanisms, PLoS Negl. Trop. Dis. 10 (2016) e0005171. https://doi.org/10.1371/journal.pntd.0005171.

[56] L. Monzote, Current treatment of leishmaniasis: a review, Open Antimicrob. Agents J. 1 (2009).

[57] S. Sundar, T.K. Jha, C.P. Thakur, P.K. Sinha, S.K. Bhattacharya, Injectable Paromomycin for Visceral Leishmaniasis in India, vol. 356, 2009, pp. 2571–2581. https://doi.org/10.1056/NEJMoa066536.

[58] P.R.L. Machado, H. Lessa, M. Lessa, L.H. Guimarães, H. Bang, J.L. Ho, et al., Oral pentoxifylline combined with pentavalent antimony: a randomized trial for mucosal leishmaniasis, Clin. Infect. Dis. 44 (2007) 788–793. https://doi.org/10.1086/511643.

[59] B.K. Das, S. Mishra, P.K. Padhi, R. Manish, R. Tripathy, P.K. Sahoo, et al., Pentoxifylline adjunct improves prognosis of human cerebral malaria in adults, Trop. Med. Int. Health 8 (2003) 680–684. https://doi.org/10.1046/j.1365-3156.2003.01087.x.

[60] J.B. Santos, A.R. de Jesus, P.R. Machado, A. Magalhães, K. Salgado, E.M. Carvalho, et al., Antimony plus recombinant human granulocyte-macrophage colony-stimulating factor applied topically in low doses enhances healing of cutaneous Leishmaniasis ulcers: a randomized, double-blind, placebo-controlled study, J. Infect. Dis. 190 (2004) 1793–1796. https://doi.org/10.1086/424848.

[61] R. Almeida, A. D'Oliveira, P. Machado, O. Bacellar, A.I. Ko, A.R. de Jesus, et al., Randomized, double-blind study of stibogluconate plus human granulocyte macrophage colony-stimulating factor versus stibogluconate alone in the treatment of cutaneous Leishmaniasis, J. Infect. Dis. 180 (1999) 1735–1737. https://doi.org/10.1086/315082.

[62] A. Mastroianni, Liposomal amphotericin B and rHuGM-CSF for treatment of visceral leishmaniasis in AIDS, Infez. Med. 12 (2004) 197–204.

[63] O.L.S. Almeida, J.B. Santos, Advances in the treatment of cutaneous leishmaniasis in the new world in the last ten years: a systematic literature review, Bras. Dermatol. 86 (2011) 497–506. https://doi.org/10.1590/S0365-05962011000300012.

[64] S. Buates, G. Matlashewski, Treatment of experimental leishmaniasis with the immunomodulators imiquimod and S-28463: efficacy and mode of action, J. Infect. Dis. 179 (1999) 1485–1494. https://doi.org/10.1086/314782.

[65] L.F.D. Passero, M.D. Laurenti, G. Santos-Gomes, B.L. Soares Campos, P. Sartorelli, J.H.G. Lago, Plants used in traditional medicine: extracts and secondary metabolites exhibiting antileishmanial activity, Curr. Clin. Pharmacol. 9 (2014) 187–204.

[66] M. Salem, K. Werbovetz, Natural products from plants as drug candidates and lead compounds against leishmaniasis and trypanosomiasis, Curr. Med. Chem. 13 (2006) 2571–2598. https://doi.org/10.2174/092986706778201611.

[67] F.O. Martinez, L. Helming, S. Gordon, Alternative activation of macrophages: an immunologic functional perspective, Annu. Rev. Immunol. 27 (2009) 451–483. https://doi.org/10.1146/annurev.immunol.021908.132532.

[68] L.F. Schnur, U. Bachrach, C.L. Greenblatt, M. Ben Joseph, Polyamine synthesis and levels during the growth and replication of *Leishmania tropica* minor and *Leishmania aethiopica*, FEBS Lett. 106 (1979) 202–206.

[69] N. Seiler, F. Raul, Polyamines and apoptosis, J. Cell. Mol. Med. 9 (2005) 623–642.

[70] M.L. Cunningham, A.H. Fairlamb, Trypanothione reductase from *Leishmania donovani*. Purification, characterisation and inhibition by trivalent antimonials, Eur. J. Biochem. 230 (1995) 460–468.

[71] S.C. Roberts, J. Scott, J.E. Gasteier, Y. Jiang, B. Brooks, A. Jardim, et al., S-adenosylmethionine decarboxylase from *Leishmania donovani*. Molecular, genetic, and biochemical characterization of null mutants and overproducers, J. Biol. Chem. 277 (2002) 5902–5909. https://doi.org/10.1074/jbc.M110118200.

[72] C. Gilroy, T. Olenyik, S.C. Roberts, B. Ullman, Spermidine synthase is required for virulence of *Leishmania donovani*, Infect. Immun. 79 (2011) 2764–2769. https://doi.org/10.1128/IAI.00073-11.

[73] Y. Jiang, S.C. Roberts, A. Jardim, N.S. Carter, S. Shih, M. Ariyanayagam, et al., Ornithine decarboxylase gene deletion mutants of *Leishmania donovani*, J. Biol. Chem. 274 (1999) 3781–3788.

[74] J.M. Boitz, P.A. Yates, C. Kline, U. Gaur, M.E. Wilson, B. Ullman, et al., *Leishmania donovani* ornithine decarboxylase is indispensable for parasite survival in the mammalian host, Infect. Immun. 77 (2009) 756–763. https://doi.org/10.1128/IAI.01236-08.

[75] S. Hanson, J. Adelman, B. Ullman, Amplification and molecular cloning of the ornithine decarboxylase gene of *Leishmania donovani*, J. Biol. Chem. 267 (1992) 2350–2359.

[76] K. Kaur, K. Emmett, P.P. McCann, A. Sjoerdsma, B. Ullman, Effects of DL-alpha-difluoromethylornithine on *Leishmania donovani* promastigotes, J. Protozool. 33 (1986) 518–521.

[77] R.M. Reguera, R.B. Fouce, J.C. Cubría, M.L. Bujidos, D. Ordóñez, Fluorinated analogues of L-ornithine are powerful inhibitors of ornithine decarboxylase and cell growth of *Leishmania infantum* promastigotes, Life Sci. 56 (1995) 223–230.

[78] A.R. Khomutov, Inhibition of enzymes of polyamine biosynthesis by substrate-like O-substituted hydroxylamines, Biochem. Mosc. 67 (2002) 1159–1167.

[79] S. Singh, A. Mukherjee, A.R. Khomutov, L. Persson, O. Heby, M. Chatterjee, et al., Antileishmanial effect of 3-aminooxy-1-aminopropane is due to polyamine depletion, Antimicrob. Agents Chemother. 51 (2007) 528–534. https://doi.org/10.1128/AAC.01055-06.

[80] S.C. Roberts, Y. Jiang, J. Gasteier, B. Frydman, L.J. Marton, O. Heby, et al., *Leishmania donovani* polyamine biosynthetic enzyme overproducers as tools to investigate the mode of action of cytotoxic polyamine analogs, Antimicrob. Agents Chemother. 51 (2007) 438–445. https://doi.org/10.1128/AAC.01193-06.

[81] R. Mukhopadhyay, P. Kapoor, R. Madhubala, Antileishmanial effect of a potent S-adenosylmethionine decarboxylase inhibitor: CGP 40215A, Pharmacol. Res. 33 (1996) 67–70. https://doi.org/10.1006/phrs.1996.0011.

[82] U. Bachrach, S. Brem, S.B. Wertman, L.F. Schnur, C.L. Greenblatt, *Leishmania* spp.: effect of inhibitors on growth and on polyamine and macromolecular syntheses, Exp. Parasitol. 48 (1979) 464–470.

[83] M. Basselin, M.A. Badet-Denisot, F. Lawrence, M. Robert-Gero, Effects of pentamidine on polyamine level and biosynthesis in wild-type, pentamidine-treated, and pentamidine-resistant Leishmania, Exp. Parasitol. 85 (1997) 274–282. https://doi.org/10.1006/expr.1996.4131.

[84] A.H. Fairlamb, P. Blackburn, P. Ulrich, B.T. Chait, A. Cerami, Trypanothione: a novel bis(glutathionyl)spermidine cofactor for glutathione reductase in trypanosomatids, Science 227 (1985) 1485–1487.

[85] C. Dumas, M. Ouellette, J. Tovar, M.L. Cunningham, A.H. Fairlamb, S. Tamar, et al., Disruption of the trypanothione reductase gene of Leishmania decreases its ability to survive oxidative stress in macrophages, Embo J. 16 (1997) 2590–2598. https://doi.org/10.1093/emboj/16.10.2590.

[86] S.L. Shames, A.H. Fairlamb, A. Cerami, C.T. Walsh, Purification and characterization of trypanothione reductase from *Crithidia fasciculata*, a newly discovered member of the family of disulfide-containing flavoprotein reductases, Biochemistry 25 (1986) 3519–3526.

[87] A.H. Fairlamb, A. Cerami, Metabolism and functions of trypanothione in the Kinetoplastida, Annu. Rev. Microbiol. 46 (1992) 695–729. https://doi.org/10.1146/annurev.mi.46.100192.003403.

[88] R.L. Krauth-Siegel, B. Enders, G.B. Henderson, A.H. Fairlamb, R.H. Schirmer, Trypanothione reductase from *Trypanosoma cruzi*. Purification and characterization of the crystalline enzyme, Eur. J. Biochem. 164 (1987) 123–128.

[89] D.J. Steenkamp, Trypanosomal antioxidants and emerging aspects of redox regulation in the trypanosomatids, Antioxid. Redox Signal. 4 (2002) 105–121. https://doi.org/10.1089/152308602753625906.

[90] J. Tovar, M.L. Cunningham, A.C. Smith, S.L. Croft, A.H. Fairlamb, Down-regulation of *Leishmania donovani* trypanothione reductase by heterologous expression of a trans-dominant mutant homologue: effect on parasite intracellular survival, Proc. Natl. Acad. Sci. USA 95 (1998) 5311–5316.

[91] A. Bocedi, K.F. Dawood, R. Fabrini, G. Federici, L. Gradoni, J.Z. Pedersen, et al., Trypanothione efficiently intercepts nitric oxide as a harmless iron complex in trypanosomatid parasites, Faseb J. 24 (2010) 1035–1042. https://doi.org/10.1096/fj.09-146407.

[92] A. Mukherjee, G. Roy, C. Guimond, M. Ouellette, The gamma-glutamylcysteine synthetase gene of Leishmania is essential and involved in response to oxidants, Mol. Microbiol. 74 (2009) 914–927. https://doi.org/10.1111/j.1365-2958.2009.06907.x.

[93] H. Castro, A.M. Tomás, Peroxidases of trypanosomatids, Antioxid. Redox Signal. 10 (2008) 1593–1606. https://doi.org/10.1089/ars.2008.2050.

[94] J.F. Turrens, Oxidative stress and antioxidant defenses: a target for the treatment of diseases caused by parasitic protozoa, Mol. Aspects Med. 25 (2004) 211–220. https://doi.org/10.1016/j.mam.2004.02.021.

[95] J.P. Iyer, A. Kaprakkaden, M.L. Choudhary, C. Shaha, Crucial role of cytosolic tryparedoxin peroxidase in *Leishmania donovani* survival, drug response and virulence, Mol. Microbiol. 68 (2008) 372–391. https://doi.org/10.1111/j.1365-2958.2008.06154.x.

[96] J. Walker, N. Acestor, R. Gongora, M. Quadroni, I. Segura, N. Fasel, et al., Comparative protein profiling identifies elongation factor-1beta and tryparedoxin peroxidase as factors associated with metastasis in *Leishmania guyanensis*, Mol. Biochem. Parasitol. 145 (2006) 254–264. https://doi.org/10.1016/j.molbiopara.2005.10.008.

[97] A.R. Satoskar, J.R. Fuchs, A.D. Kinghorn, L. Pan, Antileishmanial Compositions and Methods of Use, 2014.

[98] H. Denton, J.C. McGregor, G.H. Coombs, Reduction of anti-leishmanial pentavalent antimonial drugs by a parasite-specific thiol-dependent reductase, TDR1, Biochem. J. 381 (2004) 405–412. https://doi.org/10.1042/BJ20040283.

[99] Y. Zhou, N. Messier, M. Ouellette, B.P. Rosen, R. Mukhopadhyay, *Leishmania major* LmACR2 is a pentavalent antimony reductase that confers sensitivity to the drug pentostam, J. Biol. Chem. 279 (2004) 37445–37451. https://doi.org/10.1074/jbc.M404383200.

[100] C.D.S. Ferreira, P.S. Martins, C. Demicheli, C. Brochu, M. Ouellette, F. Frézard, Thiol-induced reduction of antimony(V) into antimony(III): a comparative study with trypanothione, cysteinyl-glycine, cysteine and glutathione, Biometals 16 (2003) 441–446.

[101] D. Légaré, D. Richard, R. Mukhopadhyay, Y.D. Stierhof, B.P. Rosen, A. Haimeur, et al., The Leishmania ATP-binding cassette protein PGPA is an intracellular metal-thiol transporter ATPase, J. Biol. Chem. 276 (2001) 26301–26307. https://doi.org/10.1074/jbc.M102351200.

[102] S. Dey, M. Ouellette, J. Lightbody, B. Papadopoulou, B.P. Rosen, An ATP-dependent As(III)-glutathione transport system in membrane vesicles of *Leishmania tarentolae*, Proc. Natl. Acad. Sci. USA 93 (1996) 2192–2197.

[103] L.J. Goad, G.G. Holz, D.H. Beach, Sterols of *Leishmania* species. Implications for biosynthesis, Mol. Biochem. Parasitol. 10 (1984) 161–170.

[104] S. Pomel, S. Cojean, P.M. Loiseau, Targeting sterol metabolism for the development of antileishmanials, Trends Parasitol. 31 (2015) 5–7. https://doi.org/10.1016/j.pt.2014.11.007.

[105] C. Yao, M.E. Wilson, Dynamics of sterol synthesis during development of *Leishmania* spp. parasites to their virulent form, Parasit. Vectors 9 (2016) 200. https://doi.org/10.1186/s13071-016-1470-0.

[106] W. de Souza, J.C.F. Rodrigues, Sterol biosynthesis pathway as target for anti-trypanosomatid drugs, Interdiscip. Perspect. Infect. Dis. 2009 (2009) 642502–642519. https://doi.org/10.1155/2009/642502.

[107] M. Pourshafie, S. Morand, A. Virion, M. Rakotomanga, C. Dupuy, P.M. Loiseau, Cloning of S-adenosyl-L-methionine:C-24-Delta-sterol-methyltransferase (ERG6) from *Leishmania donovani* and characterization of mRNAs in wild-type and amphotericin B-Resistant promastigotes, Antimicrob. Agents Chemother. 48 (2004) 2409–2414. https://doi.org/10.1128/AAC.48.7.2409-2414.2004.

[108] V.V. Andrade-Neto, T.M. Pereira, M.D. Canto-Cavalheiro, E.C. Torres-Santos, Imipramine alters the sterol profile in *Leishmania amazonensis* and increases its sensitivity to miconazole, Parasit. Vectors 9 (2016) 183. https://doi.org/10.1186/s13071-016-1467-8.

[109] S. Mukherjee, B. Mukherjee, R. Mukhopadhyay, K. Naskar, S. Sundar, J.-C. Dujardin, et al., Imipramine exploits histone deacetylase 11 to increase the IL-12/IL-10 ratio in macrophages infected with antimony-resistant *Leishmania donovani* and clears organ parasites in experimental infection, J. Immunol. 193 (2014) 4083–4094. https://doi.org/10.4049/jimmunol.1400710.

[110] W. Xu, F.-F. Hsu, E. Baykal, J. Huang, K. Zhang, Sterol biosynthesis is required for heat resistance but not extracellular survival in Leishmania, PLoS Pathog. 10 (2014) e1004427. https://doi.org/10.1371/journal.ppat.1004427.

[111] S.T. de Macedo-Silva, G. Visbal, J.A. Urbina, W. de Souza, J.C.F. Rodrigues, Potent in vitro antiproliferative synergism of combinations of ergosterol biosynthesis inhibitors against *Leishmania amazonensis*, Antimicrob. Agents Chemother. 59 (2015) 6402–6418. https://doi.org/10.1128/AAC.01150-15.

[112] J.J. Cazzulo, Aerobic fermentation of glucose by trypanosomatids, Faseb J. 6 (1992) 3153–3161.

[113] D.T. Hart, G.H. Coombs, *Leishmania mexicana*: energy metabolism of amastigotes and promastigotes, Exp. Parasitol. 54 (1982) 397–409.

[114] Y. Schlein, Sandfly diet and Leishmania, Parasitol. Today (Regul. Ed.) 2 (1986) 175–177.

[115] R.J. Burchmore, M.P. Barrett, Life in vacuoles–nutrient acquisition by *Leishmania* amastigotes, Int. J. Parasitol. 31 (2001) 1311–1320.

[116] R.J. Burchmore, D.T. Hart, Glucose transport in amastigotes and promastigotes of *Leishmania mexicana mexicana*, Mol. Biochem. Parasitol. 74 (1995) 77–86.

[117] D. Rodríguez-Contreras, S.M. Landfear, Metabolic changes in glucose transporter-deficient *Leishmania mexicana* and parasite virulence, J. Biol. Chem. 281 (2006) 20068–20076. https://doi.org/10.1074/jbc.M603265200.

[118] R.J.S. Burchmore, D. Rodríguez-Contreras, K. McBride, P. Merkel, M.P. Barrett, G. Modi, et al., Genetic characterization of glucose transporter function in *Leishmania mexicana*, Proc. Natl. Acad. Sci. USA 100 (2003) 3901–3906. https://doi.org/10.1073/pnas.0630165100.

[119] A.K. Ghosh, A.H. Sardar, A. Mandal, S. Saini, K. Abhishek, A. Kumar, et al., Metabolic reconfiguration of the central glucose metabolism: a crucial strategy of *Leishmania donovani* for its survival during oxidative stress, Faseb J. 29 (2015) 2081–2098. https://doi.org/10.1096/fj.14-258624.

[120] S. Gupta, A.T. Cordeiro, P.A.M. Michels, Glucose-6-phosphate dehydrogenase is the target for the trypanocidal action of human steroids, Mol. Biochem. Parasitol. 176 (2011) 112–115. https://doi.org/10.1016/j.molbiopara.2010.12.006.

[121] Y.M. Al-Mulla Hummadi, N.M. Al-Bashir, R.A. Najim, The mechanism behind the antileishmanial effect of zinc sulphate. II. Effects on the enzymes of the parasites, Ann. Trop. Med. Parasitol. 99 (2005) 131–139. https://doi.org/10.1179/136485905X19937.

[122] L. Kedzierski, R.L. Malby, B.J. Smith, M.A. Perugini, A.N. Hodder, T. Ilg, et al., Structure of *Leishmania mexicana* phosphomannomutase highlights similarities with human isoforms, J. Mol. Biol. 363 (2006) 215–227. https://doi.org/10.1016/j.jmb.2006.08.023.

[123] D. Rodríguez-Contreras, N. Hamilton, Gluconeogenesis in *Leishmania mexicana*: contribution of glycerol kinase, phosphoenolpyruvate carboxykinase, and pyruvate phosphate dikinase, J. Biol. Chem. 289 (2014) 32989–33000. https://doi.org/10.1074/jbc.M114.569434.

[124] K. Ferrell, C.R. Wilkinson, W. Dubiel, C. Gordon, Regulatory subunit interactions of the 26S proteasome, a complex problem, Trends Biochem. Sci. 25 (2000) 83–88.

[125] A. Paugam, A.L. Bulteau, J. Dupouy-Camet, C. Creuzet, B. Friguet, Characterization and role of protozoan parasite proteasomes, Trends Parasitol. 19 (2003) 55–59.

[126] C.D. Robertson, The *Leishmania mexicana* proteasome, Mol. Biochem. Parasitol. 103 (1999) 49–60.

[127] C.B. Christensen, L. Jørgensen, A.T. Jensen, S. Gasim, M. Chen, A. Kharazmi, et al., Molecular characterization of a *Leishmania donovanii* cDNA clone with similarity to human 20S proteasome a-type subunit, Biochim. Biophys. Acta 1500 (2000) 77–87.

[128] B. Couvreur, A. Bollen, J.C. Dujardin, More panantigens in Leishmania, Trends Parasitol. (2001).

[129] S. Khare, A.S. Nagle, A. Biggart, Y.H. Lai, F. Liang, L.C. Davis, et al., Proteasome inhibition for treatment of leishmaniasis, Chagas disease and sleeping sickness, Nature 537 (2016) 229–233. https://doi.org/10.1038/nature19339.

[130] A. Bolhassani, T. Taheri, Y. Taslimi, S. Zamanilui, F. Zahedifard, N. Seyed, et al., Fluorescent Leishmania species: development of stable GFP expression and its application for in vitro and in vivo studies, Exp. Parasitol. 127 (2011) 637–645. https://doi.org/10.1016/j.exppara.2010.12.006.

[131] D. Liu, C. Kebaier, N. Pakpour, A.A. Capul, S.M. Beverley, P. Scott, et al., Leishmania major phosphoglycans influence the host early immune response by modulating dendritic cell functions, Infect. Immun. 77 (2009) 3272–3283. https://doi.org/10.1128/IAI.01447-08.

[132] S.M. Abdullah, B. Flath, W. Presber, Mixed infection of human U-937 cells by two different species of Leishmania, Am. J. Trop. Med. Hyg. 59 (1998) 182–188.

[133] S.M. Abdullah, B. Flath, H.W. Presber, Comparison of different staining procedures for the flow cytometric analysis of U-937 cells infected with different Leishmania-species, J. Microbiol. Methods 37 (1999) 123–138.

[134] F. Guinet, A. Louise, H. Jouin, J.C. Antoine, C.W. Roth, Accurate quantitation of Leishmania infection in cultured cells by flow cytometry, Cytometry 39 (2000) 235–240.

[135] S.A. Pulido, D.L. Muñoz, A.M. Restrepo, C.V. Mesa, J.F. Alzate, I.D. Vélez, et al., Improvement of the green fluorescent protein reporter system in Leishmania spp. for the in vitro and in vivo screening of antileishmanial drugs, Acta Trop. 122 (2012) 36–45. https://doi.org/10.1016/j.actatropica.2011.11.015.

[136] B.K. Kolli, J. Kostal, O. Zaborina, A.M. Chakrabarty, K.-P. Chang, Leishmania-released nucleoside diphosphate kinase prevents ATP-mediated cytolysis of macrophages, Mol. Biochem. Parasitol. 158 (2008) 163–175. https://doi.org/10.1016/j.molbiopara.2007.12.010.

[137] E. Calvo-Álvarez, N.A. Guerrero, R. Alvarez-Velilla, C.F. Prada, J.M. Requena, C. Punzón, et al., Appraisal of a Leishmania major strain stably expressing mCherry fluorescent protein for both in vitro and in vivo studies of potential drugs and vaccine against cutaneous leishmaniasis, PLoS Negl. Trop. Dis. 6 (2012) e1927. https://doi.org/10.1371/journal.pntd.0001927.

[138] M.R.E. Varela, D.L. Muñoz, S.M. Robledo, B.K. Kolli, S. Dutta, K.-P. Chang, et al., *Leishmania (Viannia) panamensis*: an in vitro assay using the expression of GFP for screening of antileishmanial drug, Exp. Parasitol. 122 (2009) 134–139. https://doi.org/10.1016/j.exppara.2009.02.012.

[139] A. Dube, N. Singh, S. Sundar, N. Singh, Refractoriness to the treatment of sodium stibogluconate in Indian kala-azar field isolates persist in in vitro and in vivo experimental models, Parasitol. Res. 96 (2005) 216–223. https://doi.org/10.1007/s00436-005-1339-1.

[140] A. Dube, R. Gupta, N. Singh, Reporter genes facilitating discovery of drugs targeting protozoan parasites, Trends Parasitol. 25 (2009) 432–439. https://doi.org/10.1016/j.pt.2009.06.006.

[141] A. Misslitz, J.C. Mottram, P. Overath, T. Aebischer, Targeted integration into a rRNA locus results in uniform and high level expression of transgenes in Leishmania amastigotes, Mol. Biochem. Parasitol. 107 (2000) 251–261.

[142] N. Singh, R. Gupta, A.K. Jaiswal, S. Sundar, A. Dube, Transgenic *Leishmania donovani* clinical isolates expressing green fluorescent protein constitutively for rapid and reliable ex vivo drug screening, J. Antimicrob. Chemother. 64 (2009) 370–374. https://doi.org/10.1093/jac/dkp206.

[143] A.P. Patel, A. Deacon, G. Getti, Development and validation of four Leishmania species constitutively expressing GFP protein. A model for drug discovery and disease pathogenesis studies,, Parasitology 141 (2014) 501–510. https://doi.org/10.1017/S0031182013001777.

[144] C. Di Giorgio, O. Ridoux, F. Delmas, N. Azas, M. Gasquet, P. Timon-David, Flow cytometric detection of Leishmania parasites in human monocyte-derived macrophages: application to antileishmanial-drug testing, Antimicrob. Agents Chemother. 44 (2000) 3074–3078.

[145] C. Terrazas, S. Oghumu, S. Varikuti, D. Martinez-Saucedo, S.M. Beverley, A.R. Satoskar, Uncovering Leishmania-macrophage interplay using imaging flow cytometry, J. Immunol. Methods 423 (2015) 93–98. https://doi.org/10.1016/j.jim.2015.04.022.

[146] M. Ginouves, B. Carme, P. Couppie, G. Prevot, Comparison of tetrazolium salt assays for evaluation of drug activity against *Leishmania* spp, J. Clin. Microbiol. 52 (2014) 2131–2138. https://doi.org/10.1128/JCM.00201-14.

[147] A. Dutta, S. Bandyopadhyay, C. Mandal, M. Chatterjee, Development of a modified MTT assay for screening antimonial resistant field isolates of Indian visceral leishmaniasis, Parasitol. Int. 54 (2005) 119−122. https://doi.org/10.1016/j.parint.2005.01.001.

[148] J. Mikus, D. Steverding, A simple colorimetric method to screen drug cytotoxicity against Leishmania using the dye Alamar Blue, Parasitol. Int. 48 (2000) 265−269.

[149] S. Ashutosh, S. Gupta, Ramesh, S. Sundar, N. Goyal, Use of *Leishmania donovani* field isolates expressing the luciferase reporter gene in in vitro drug screening, Antimicrob. Agents Chemother. 49 (2005) 3776−3783. https://doi.org/10.1128/AAC.49.9.3776-3783.2005.

[150] S. Pandey, S.N. Suryawanshi, Nishi, N. Goyal, S. Gupta, Chemotherapy of leishmaniasis. Part V: synthesis and in vitro bioevaluation of novel pyridinone derivatives, Eur. J. Med. Chem. 42 (2007) 669−674. https://doi.org/10.1016/j.ejmech.2006.11.011.

[151] L. Gupta, A. Talwar, S. Nishi, S. Palne, P.M.S. Gupta, Chauhan, Synthesis of marine alkaloid: 8,9-dihydrocoscinamide B and its analogues as Novel class of antileishmanial agents, Bioorg. Med. Chem. Lett. 17 (2007) 4075−4079. https://doi.org/10.1016/j.bmcl.2007.04.035.

[152] T. Lang, S. Goyard, M. Lebastard, G. Milon, Bioluminescent Leishmania expressing luciferase for rapid and high throughput screening of drugs acting on amastigote-harbouring macrophages and for quantitative real-time monitoring of parasitism features in living mice, Cell. Microbiol. 7 (2005) 383−392. https://doi.org/10.1111/j.1462-5822.2004.00468.x.

[153] F.S. Buckner, A.J. Wilson, Colorimetric assay for screening compounds against Leishmania amastigotes grown in macrophages, Am. J. Trop. Med. Hyg. 72 (2005) 600−605.

[154] J.H. LeBowitz, C.M. Coburn, D. McMahon-Pratt, S.M. Beverley, Development of a stable Leishmania expression vector and application to the study of parasite surface antigen genes, Proc. Natl. Acad. Sci. USA 87 (1990) 9736−9740.

[155] J.G. Sutcliffe, Nucleotide sequence of the ampicillin resistance gene of *Escherichia coli* plasmid pBR322, Proc. Natl. Acad. Sci. USA 75 (1978) 3737−3741.

[156] J.T. Moore, S.T. Davis, I.K. Dev, The development of beta-lactamase as a highly versatile genetic reporter for eukaryotic cells, Anal. Biochem. 247 (1997) 203−209. https://doi.org/10.1006/abio.1997.2092.

[157] J. Tovar, S. Wilkinson, J.C. Mottram, A.H. Fairlamb, Evidence that trypanothione reductase is an essential enzyme in Leishmania by targeted replacement of the tryA gene locus, Mol. Microbiol. 29 (1998) 653−660.

[158] R.L. Krauth-Siegel, G.H. Coombs, Enzymes of parasite thiol metabolism as drug targets, Parasitol. Today (Regul. Ed.) 15 (1999) 404−409.

[159] C.J. Hamilton, A. Saravanamuthu, I.M. Eggleston, A.H. Fairlamb, Ellman's-reagent-mediated regeneration of trypanothione in situ: substrate-economical microplate and time-dependent inhibition assays for trypanothione reductase, Biochem. J. 369 (2003) 529−537. https://doi.org/10.1042/BJ20021298.

[160] E. van den Bogaart, G.J. Schoone, P. England, D. Faber, K.M. Orrling, J.-C. Dujardin, et al., Simple colorimetric trypanothione reductase-based assay for high-throughput screening of drugs against Leishmania intracellular amastigotes, Antimicrob. Agents Chemother. 58 (2014) 527−535. https://doi.org/10.1128/AAC.00751-13.

[161] M. De Rycker, I. Hallyburton, J. Thomas, L. Campbell, S. Wyllie, D. Joshi, et al., Comparison of a high-throughput high-content intracellular *Leishmania donovani* assay with an axenic amastigote assay, Antimicrob. Agents Chemother. 57 (2013) 2913−2922. https://doi.org/10.1128/AAC.02398-12.

[162] G. De Muylder, K.K.H. Ang, S. Chen, M.R. Arkin, J.C. Engel, J.H. McKerrow, A screen against Leishmania intracellular amastigotes: comparison to a promastigote screen and identification of a host cell-specific hit, PLoS Negl. Trop. Dis. 5 (2011) e1253. https://doi.org/10.1371/journal.pntd.0001253.

[163] J.L. Siqueira-Neto, S. Moon, J. Jang, G. Yang, C. Lee, H.K. Moon, et al., An image-based high-content screening assay for compounds targeting intracellular *Leishmania donovani* amastigotes in human macrophages, PLoS Negl. Trop. Dis. 6 (2012) e1671. https://doi.org/10.1371/journal.pntd.0001671.

[164] M. Hommel, C.L. Jaffe, B. Travi, G. Milon, Experimental models for leishmaniasis and for testing anti-leishmanial vaccines, Ann. Trop. Med. Parasitol. 89 (Suppl. 1) (1995) 55–73.

[165] H.W. Murray, E.B. Brooks, J.L. DeVecchio, F.P. Heinzel, Immunoenhancement combined with amphotericin B as treatment for experimental visceral leishmaniasis, Antimicrob. Agents Chemother. 47 (2003) 2513–2517. https://doi.org/10.1128/AAC.47.8.2513-2517.2003.

[166] N. Courret, T. Lang, G. Milon, J.C. Antoine, Intradermal inoculations of low doses of *Leishmania major* and *Leishmania amazonensis* metacyclic promastigotes induce different immunoparasitic processes and status of protection in BALB/c mice, Int. J. Parasitol. 33 (2003) 1373–1383.

[167] J. Louis, A. Gumy, H. Voigt, M. Röcken, P. Launois, Experimental cutaneous Leishmaniasis: a powerful model to study in vivo the mechanisms underlying genetic differences in Th subset differentiation, Eur. J. Dermatol. 12 (2002) 316–318.

[168] A.K. Wege, C. Florian, W. Ernst, N. Zimara, U. Schleicher, F. Hanses, et al., Leishmania major infection in humanized mice induces systemic infection and provokes a nonprotective human immune response, PLoS Negl. Trop. Dis. 6 (2012) e1741. https://doi.org/10.1371/journal.pntd.0001741.

[169] W.P. Lafuse, R. Story, J. Mahylis, G. Gupta, S. Varikuti, H. Steinkamp, et al., *Leishmania donovani* infection induces anemia in hamsters by differentially altering erythropoiesis in bone marrow and spleen, PLoS One 8 (2013) e59509. https://doi.org/10.1371/journal.pone.0059509.

[170] J.P. Farrell, *Leishmania donovani*: acquired resistance to visceral leishmaniasis in the golden hamster, Exp. Parasitol. 40 (1976) 89–94.

[171] S. Gupta, Nishi, Visceral leishmaniasis: experimental models for drug discovery, Indian J. Med. Res. 133 (2011) 27–39.

[172] J.A. Rioux, Y.J. Golvan, H. Croset, R. Houin, Leishmanioses in the Mediterranean "Midi": results of an ecologic survey, Bull Soc. Pathol. Exot. Filiales 62 (1969) 332–333.

[173] W.L. Chapman, W.L. Hanson, V.B. Waits, K.E. Kinnamon, Antileishmanial activity of selected compounds in dogs experimentally infected with *Leishmania donovani*, Rev. Inst. Med. Trop. Sao Paulo 21 (1979) 189–193.

[174] C.M. Keenan, L.D. Hendricks, L. Lightner, H.K. Webster, A.J. Johnson, Visceral leishmaniasis in the German shepherd dog. I. Infection, clinical disease, and clinical pathology, Vet. Pathol. (2016). https://doi.org/10.1177/030098588402100113.

[175] P. Abranches, G. Santos-Gomes, N. Rachamim, L. Campino, L.F. Schnur, C.L. Jaffe, An experimental model for canine visceral leishmaniasis, Parasite Immunol. 13 (1991) 537–550. https://doi.org/10.1111/j.1365-3024.1991.tb00550.x.

[176] W.L.J. Chapman, W.L. Hanson, L.D. Hendricks, Toxicity and efficacy of the antileishmanial drug meglumine antimoniate in the owl monkey (*Aotus trivirgatus*), J. Parasitol. (1983).

[177] W.L. Chapman, W.L. Hanson, Visceral leishmaniasis in the squirrel monkey (*Saimiri sciurea*), J. Parasitol. 67 (1981) 740–741.

[178] A. Dube, J.K. Srivastava, P. Sharma, A. Chaturvedi, J.C. Katiyar, S. Naik, *Leishmania donovani*: cellular and humoral immune responses in Indian langur monkeys, *Presbytis entellus*, Acta Trop. 73 (1999) 37–48.

[179] R. Anuradha, K. Pal, J.C. Zehra, N. Katiyar, G. Sethi Bhatia, et al., The Indian langur: preliminary report of a new nonhuman primate host for visceral leishmaniasis,, Bull. World Health Organ. 70 (1992) 63–72.

[180] D.J. Newman, G.M. Cragg, Natural products as sources of new drugs from 1981 to 2014, J. Nat. Prod. 79 (2016) 629–661. https://doi.org/10.1021/acs.jnatprod.5b01055.

[181] A.L. Harvey, R. Edrada-Ebel, R.J. Quinn, The re-emergence of natural products for drug discovery in the genomics era, Nat. Rev. Drug Discov. 14 (2015) 111–129. https://doi.org/10.1038/nrd4510.

[182] Y. Luo, R.E. Cobb, H. Zhao, Recent advances in natural product discovery, Curr. Opin. Biotechnol. 30 (2014) 230–237.

[183] C. Hertweck, Natural products as source of therapeutics against parasitic diseases, Angew. Chem. Int. Ed. 54 (2015) 14622–14624. https://doi.org/10.1002/anie.201509828.

[184] M.J. Balunas, A.D. Kinghorn, Drug discovery from medicinal plants, Life Sci. 78 (2005) 431–441.

[185] W.P. Jones, Y.-W. Chin, A.D. Kinghorn, The role of pharmacognosy in modern medicine and pharmacy, Cdt 7 (2006) 247–264. https://doi.org/10.2174/138945006776054915.

[186] A.D. Kinghorn, Y.-W. Chin, S.M. Swanson, Discovery of natural product anticancer agents from biodiverse organisms, Curr. Opin. Drug Discov. Dev. 12 (2009) 189–196.

[187] M.M. Salem, K.A. Werbovetz, Natural products from plants as drug candidates and lead compounds against leishmaniasis and trypanosomiasis, Curr. Med. Chem. 13 (2006) 2571–2598.

[188] A. Fournet, V. Muñoz, Natural products as trypanocidal, antileishmanial and antimalarial drugs, Curr. Top. Med. Chem. 2 (2002) 1215–1237.

[189] N. Singh, B.B. Mishra, S. Bajpai, R.K. Singh, V.K. Tiwari, Natural product based leads to fight against leishmaniasis, Bioorg. Med. Chem. 22 (2014) 18–45. https://doi.org/10.1016/j.bmc.2013.11.048.

[190] A.B. Vermelho, C.T. Supuran, V. Cardoso, Leishmaniasis: Possible New Strategies for Treatment, 2014.

[191] E.J. Osorio, S.M. Robledo, J. Bastida, Alkaloids with Antiprotozoal Activity, Elsevier, 2008, pp. 113–190 (Chapter 2).

[192] J.L. Vennerstrom, J.K. Lovelace, V.B. Waits, W.L. Hanson, D.L. Klayman, Berberine derivatives as antileishmanial drugs, Antimicrob. Agents Chemother. 34 (1990) 918–921. https://doi.org/10.1128/AAC.34.5.918.

[193] M. Bahar, Y. Deng, X. Zhu, S. He, T. Pandharkar, M.E. Drew, et al., Potent antiprotozoal activity of a novel semi-synthetic berberine derivative, Bioorg. Med. Chem. Lett. 21 (2011) 2606–2610. https://doi.org/10.1016/j.bmcl.2011.01.101.

[194] M. Endeshaw, X. Zhu, S. He, T. Pandharkar, E. Cason, K.V. Mahasenan, et al., 8,8-dialkyldihydroberberines with potent antiprotozoal activity, J. Nat. Prod. 76 (2013) 311–315. https://doi.org/10.1021/np300638f.

[195] V. Corpas-López, F. Morillas-Márquez, M.C. Navarro-Moll, G. Merino-Espinosa, V. Díaz-Sáez, J. Martín-Sánchez, (-)-α-Bisabolol, a promising oral compound for the treatment of visceral leishmaniasis, J. Nat. Prod. 78 (2015) 1202–1207. https://doi.org/10.1021/np5008697.

[196] V. Corpas-López, G. Merino-Espinosa, M. López-Viota, P. Gijón-Robles, M.J. Morillas-Mancilla, J. López-Viota, et al., Topical treatment of *Leishmania tropica* infection using (−)-α-Bisabolol ointment in a hamster model: effectiveness and safety assessment, J. Nat. Prod. 79 (2016) 2403–2407. https://doi.org/10.1021/acs.jnatprod.6b00740.

[197] A.I. Waechter, M.E. Ferreira, A. Fournet, A.R. de Arias, H. Nakayama, S. Torres, et al., Experimental treatment of cutaneous leishmaniasis with argentilactone isolated from *Annona haematantha*, Planta Med. 63 (2007) 433–435. https://doi.org/10.1055/s-2006-957728.

[198] C.B. Naman, G. Gupta, S. Varikuti, H. Chai, R.W. Doskotch, A.R. Satoskar, et al., Northalrugosidine is a bisbenzyltetrahydroisoquinoline alkaloid from *Thalictrum alpinum* with in vivo antileishmanial activity, J. Nat. Prod. 78 (2015) 552–556. https://doi.org/10.1021/np501028u.

[199] A. Fournet, A. Inchausti, G. Yaluff, A.R. de Arias, H. Guinaudeau, J. Bruneton, et al., Trypanocidal bisbenzylisoquinoline alkaloids are inhibitors of trypanothione reductase, J. Enzym. Inhib. 13 (2009) 1–9. https://doi.org/10.3109/14756369809035823.

[200] C.A. Araujo, L.V. Alegrio, D.C. Gomes, M.E.F. Lima, L. Gomes-Cardoso, L.L. Leon, Studies on the effectiveness of diarylheptanoids derivatives against *Leishmania amazonensis*, Mem. Inst. Oswaldo Cruz 94 (1999) 791–794. https://doi.org/10.1590/S0074-02761999000600015.

[201] S. Prasad, A.K. Tyagi, B.B. Aggarwal, Recent developments in delivery, bioavailability, absorption and metabolism of curcumin: the golden pigment from golden spice, Cancer Res. Treat. 46 (2014) 2–18. https://doi.org/10.4143/crt.2014.46.1.2.

[202] D. Granados-Falla, A. Gomez-Galindo, A. Daza, S. Robledo, C. Coy-Barrera, L. Cuca, et al., Seco-limonoid derived from *Raputia heptaphylla* promotes the control of cutaneous leishmaniasis in hamsters (*Mesocricetus auratus*), Parasitology 143 (2016) 289–299. https://doi.org/10.1017/S0031182015001717.

[203] M. Chen, S.B. Christensen, J. Blom, E. Lemmich, L. Nadelmann, K. Fich, et al., Licochalcone A, a novel antiparasitic agent with potent activity against human pathogenic protozoan species of Leishmania, Antimicrob, Agents Chemother. 37 (1993) 2550–2556. https://doi.org/10.1128/AAC.37.12.2550.

[204] M. Chen, S.B. Christensen, T.G. Theander, A. Kharazmi, Antileishmanial activity of licochalcone A in mice infected with *Leishmania major* and in hamsters infected with *Leishmania donovani*, Antimicrob. Agents Chemother. 38 (1994) 1339–1344. https://doi.org/10.1128/AAC.38.6.1339.

[205] L. Zhai, M. Chen, J. Blom, T.G. Theander, S.B. Christensen, A. Kharazmi, The antileishmanial activity of novel oxygenated chalcones and their mechanism of action, J. Antimicrob. Chemother. 43 (1999) 793–803. https://doi.org/10.1093/jac/43.6.793.

[206] J.D.F. Inacio, M.M. Canto-Cavalheiro, E.E. Almeida-Amaral, Vitro and in vivo effects of (−)-Epigallocatechin 3-O-gallate on *Leishmania amazonensis*, J. Nat. Prod. 76 (2013) 1993–1996. https://doi.org/10.1021/np400624d.

[207] F. Fonseca-Silva, J.D.F. Inacio, M.M. Canto-Cavalheiro, E.E. Almeida-Amaral, Reactive oxygen species production by quercetin causes the death of *Leishmania amazonensis* intracellular amastigotes, J. Nat. Prod. 76 (2013) 1505–1508. https://doi.org/10.1021/np400193m.

[208] D. Tasdemir, M. Kaiser, R. Brun, V. Yardley, T.J. Schmidt, F. Tosun, et al., Antitrypanosomal and antileishmanial activities of flavonoids and their analogues: in vitro, in vivo, structure-activity relationship, and quantitative structure-activity relationship studies, Antimicrob. Agents Chemother. 50 (2006) 1352–1364. https://doi.org/10.1128/AAC.50.4.1352-1364.2006.

[209] S. Sarkar, S. Mandal, J. Sinha, S. Mukhopadhyay, N. Das, M.K. Basu, Quercetin: critical evaluation as an antileishmanial agent in vivo in hamsters using different vesicular delivery modes, J. Drug Target 10 (2002) 573–578. https://doi.org/10.1080/106118021000072681.

[210] F. Korte, Amarogentin, ein neuer Bitterstoff aus Gentianaceen. Charakteristische Pflanzeninhaltsstoffe, IX. Mitteil, Eur. J. Inorg. Chem. 88 (1955) 704–707. https://doi.org/10.1002/cber.19550880518.

[211] S. Ray, H.K. Majumder, A.K. Chakravarty, S. Mukhopadhyay, R.R. Gil, G.A. Cordell, Amarogentin, a naturally occurring secoiridoid glycoside and a newly recognized inhibitor of topoisomerase I from *Leishmania donovani*, J. Nat. Prod. 59 (1996) 27–29. https://doi.org/10.1021/np960018g.

[212] S. Medda, S. Mukhopadhyay, M.K. Basu, Evaluation of the in-vivo activity and toxicity of amarogentin, an antileishmanial agent, in both liposomal and niosomal forms, J. Antimicrob. Chemother. 44 (1999) 791–794.

[213] K. Gorter, Sur le principe amer de l'*Andrographis paniculata* N, Recl. Trav. Chim. Pays-Bas 30 (1911) 151–160. https://doi.org/10.1002/recl.19110300404.

[214] A. Varma, H. Padh, N. Shrivastava, Andrographolide: a new plant-derived antineoplastic entity on horizon, Evid. Based Complement Alternat Med. 2011 (2011) 815390–815399. https://doi.org/10.1093/ecam/nep135.

[215] J. Sinha, S. Mukhopadhyay, N. Das, M.K. Basu, Targeting of liposomal andrographolide to *L. donovani*-infected macrophages in vivo, Drug Deliv. 7 (2000) 209–213. https://doi.org/10.1080/107175400455137.

[216] S. Lala, A.K. Nandy, S.B. Mahato, M.K. Basu, Delivery in vivo of 14-deoxy-11-oxoandrographolide, an antileishmanial agent, by different drug carriers, Indian J. Biochem. Biophys. 40 (2003) 169–174.

[217] L. Pan, C.M. Lezama-Davila, A.P. Isaac-Marquez, E.P. Calomeni, J.R. Fuchs, A.R. Satoskar, et al., Sterols with antileishmanial activity isolated from the roots of *Pentalinon andrieuxii*, Phytochemistry 82 (2012) 128–135. https://doi.org/10.1016/j.phytochem.2012.06.012.

[218] C.M. Lezama-Davila, L. Pan, A.P. Isaac-Marquez, C. Terrazas, S. Oghumu, R. Isaac-Márquez, et al., *Pentalinon andrieuxii* root extract is effective in the topical treatment of cutaneous leishmaniasis caused by *Leishmania mexicana*, Phytother Res. 28 (2014) 909–916. https://doi.org/10.1002/ptr.5079.

[219] G. Gupta, K.J. Peine, D. Abdelhamid, H. Snider, A.B. Shelton, L. Rao, et al., A novel sterol isolated from a plant used by mayan traditional healers is effective in treatment of visceral leishmaniasis caused by *Leishmania donovani*, ACS Infect. Dis. 1 (2015) 497–506. https://doi.org/10.1021/acsinfecdis.5b00081.

[220] U.F.A.D. Administration, FDA Guidance for Industry-Botanical Drug Products (Draft Guidance), US Food and Drug Administration, 2000.

[221] J.E. Frampton, Crofelemer: a review of its use in the management of non-infectious diarrhoea in adult patients with HIV/AIDS on antiretroviral therapy, Drugs 73 (2013) 1121–1129. https://doi.org/10.1007/s40265-013-0083-6.

[222] S.T. Chen, J. Dou, R. Temple, R. Agarwal, K.-M. Wu, S. Walker, New therapies from old medicines, Nat. Biotechnol. 26 (2008) 1077–1083. https://doi.org/10.1038/nbt1008-1077.

[223] S.L. Lee, J. Dou, R. Agarwal, R. Temple, J. Beitz, C. Wu, A. Mulberg, L.X. Yu, J. Woodcock, Evolution of traditional medicines to botanical drugs, Science 347 (2015) S32–S34.

[224] M.M. Iwu, J.E. Jackson, B.G. Schuster, Medicinal plants in the fight against leishmaniasis, Parasitol. Today (Regul. Ed.) 10 (1994) 65–68.

[225] G.A. Cordell, M.D. Colvard, Natural products and traditional medicine: turning on a paradigm, J. Nat. Prod. 75 (2012) 514–525. https://doi.org/10.1021/np200803m.

[226] A.M.G. Brito, D. Dos Santos, S.A. Rodrigues, R.G. Brito, L. Xavier-Filho, Plants with anti-Leishmania activity: integrative review from 2000 to 2011, Pharmacogn. Rev. 7 (2013) 34–41. https://doi.org/10.4103/0973-7847.112840.

[227] R. Sen, M. Chatterjee, Plant derived therapeutics for the treatment of Leishmaniasis, Phytomedicine 18 (2011) 1056–1069. https://doi.org/10.1016/j.phymed.2011.03.004.

[228] L.G. Rocha, J.R.G.S. Almeida, R.O. Macêdo, J.M. Barbosa-Filho, A review of natural products with antileishmanial activity,, Phytomedicine 12 (2005) 514–535. https://doi.org/10.1016/j.phymed.2003.10.006.

[229] M.J. Chan-Bacab, L.M. Peña-Rodríguez, Plant natural products with leishmanicidal activity, Nat. Prod. Rep. 18 (2001) 674–688.

[230] O. Kayser, A.F. Kiderlen, S.L. Croft, Natural products as antiparasitic drugs, Parasitol. Res. 90 (Suppl. 2) (2003) S55–S62. https://doi.org/10.1007/s00436-002-0768-3.

[231] T. Polonio, T. Efferth, Leishmaniasis: drug resistance and natural products (review), Int. J. Mol. Med. 22 (2008) 277–286.

[232] P.M. Dewick, Alkaloids, John Wiley & Sons, Ltd, Chichester, UK, 2001. https://doi.org/10.1002/0470846275.ch6.

[233] B.B. Mishra, R.K. Singh, A. Srivastava, V.J. Tripathi, V.K. Tiwari, Fighting against Leishmaniasis: search of alkaloids as future true potential anti-Leishmanial agents, Mini Rev. Med. Chem. 9 (2009) 107–123.

[234] M. Reina, W. Ruiz-Mesia, M. López-Rodríguez, L. Ruiz-Mesia, A. González-Coloma, R. Martínez-Díaz, Indole alkaloids from *Geissospermum reticulatum*, J. Nat. Prod. 75 (2012) 928–934. https://doi.org/10.1021/np300067m.

[235] D.R. Callejon, T.B. Riul, L.G.P. Feitosa, T. Guaratini, D.B. Silva, A. Adhikari, et al., Leishmanicidal evaluation of tetrahydroprotoberberine and spirocyclic erythrina-alkaloids, Molecules 19 (2014) 5692–5703. https://doi.org/10.3390/molecules19055692.

[236] M.A. Donega, S.C. Mello, R.M. Moraes, S.K. Jain, B.L. Tekwani, C.L. Cantrell, Pharmacological activities of cilantro's aliphatic aldehydes against *Leishmania donovani*, Planta Med. 80 (2014) 1706–1711. https://doi.org/10.1055/s-0034-1383183.

[237] F. Fonseca-Silva, M.M. Canto-Cavalheiro, R.F.S. Menna-Barreto, E.E. Almeida-Amaral, Effect of apigenin on *Leishmania amazonensis* is associated with reactive oxygen species production followed by mitochondrial dysfunction, J. Nat. Prod. 78 (2015) 880–884. https://doi.org/10.1021/acs.jnatprod.5b00011.

[238] L.R.F. de Sousa, H. Wu, L. Nebo, J.B. Fernandes, M.F.D.G.F. da Silva, W. Kiefer, et al., Natural products as inhibitors of recombinant cathepsin L of *Leishmania mexicana*, Exp. Parasitol. 156 (2015) 42–48. https://doi.org/10.1016/j.exppara.2015.05.016.

[239] L.R.F. de Sousa, S.D. Ramalho, M.C. Burger, L. Nebo, J.B. Fernandes, M.F.D.G.F. da Silva, et al., Isolation of arginase inhibitors from the bioactivity-guided fractionation of *Byrsonima coccolobifolia* leaves and stems, J. Nat. Prod. 77 (2014) 392–396. https://doi.org/10.1021/np400717m.

[240] T.A. da Costa-Silva, S.S. Grecco, F.S. de Sousa, J.H.G. Lago, E.G.A. Martins, C.A. Terrazas, et al., Immunomodulatory and antileishmanial activity of phenylpropanoid dimers isolated from *Nectandra leucantha*, J. Nat. Prod. 78 (2015) 653–657. https://doi.org/10.1021/np500809a.

[241] T. Ellendorff, R. Brun, M. Kaiser, J. Sendker, T.J. Schmidt, Pls-prediction and confirmation of hydrojuglone glucoside as the antitrypanosomal constituent of *Juglans* spp, Molecules 20 (2015) 10082–10094. https://doi.org/10.3390/molecules200610082.

[242] J. Zhang, A.A. Rahman, S. Jain, M.R. Jacob, Antimicrobial and antiparasitic abietane diterpenoids from *Cupressus sempervirens*, Res. Rep. Med. Chem. (2012).

[243] S.M. Kupchan, A. Karim, C. Marcks, Tumor inhibitors. XLVIII. Taxodione and taxodone, two novel diterpenoid quinone methide tumor inhibitors from *Taxodium distichum*, J. Org. Chem. 34 (2002) 3912–3918. https://doi.org/10.1021/jo01264a036.

[244] C.B. Naman, A.D. Gromovsky, C.M. Vela, J.N. Fletcher, G. Gupta, S. Varikuti, et al., Antileishmanial and cytotoxic activity of some highly oxidized abietane diterpenoids from the Bald Cypress, *Taxodium distichum*, J. Nat. Prod. 79 (2016) 598–606. https://doi.org/10.1021/acs.jnatprod.5b01131.

[245] M. Pirttimaa, A. Nasereddin, D. Kopelyanskiy, M. Kaiser, J. Yli-Kauhaluoma, K.-M. Oksman-Caldentey, et al., Abietane-type diterpenoid amides with highly potent and selective activity against *Leishmania donovani* and *Trypanosoma cruzi*, J. Nat. Prod. 79 (2016) 362–368. https://doi.org/10.1021/acs.jnatprod.5b00990.

[246] G.S. Lima, D.B. Castro-Pinto, G.C. Machado, M.A.M. Maciel, A. Echevarria, Antileishmanial activity and trypanothione reductase effects of terpenes from the Amazonian species *Croton cajucara* Benth (Euphorbiaceae), Phytomedicine 22 (2015) 1133–1137. https://doi.org/10.1016/j.phymed.2015.08.012.

[247] M. Becerra, S. Boutefnouchet, O. Córdoba, G.P. Vitorino, L. Brehu, I. Lamour, et al., Antileishmanial activity of fucosterol recovered from *Lessonia vadosa* Searles (Lessoniaceae) by SFE, PSE and CPC, Phytochem. Lett. 11 (2015) 418–423.

[248] A. Tempone, C. Martins de Oliveira, R. Berlinck, Current approaches to discover marine antileishmanial natural products, Planta Med. 77 (2011) 572–585. https://doi.org/10.1055/s-0030-1250663.

[249] M. Tulp, L. Bohlin, Rediscovery of known natural compounds: nuisance or goldmine? Bioorg. Med. Chem. 13 (2005) 5274–5282. https://doi.org/10.1016/j.bmc.2005.05.067.

[250] M.F.C. Santos, P.M. Harper, D.E. Williams, J.T. Mesquita, G.É. Pinto, T.A. da Costa-Silva, et al., Anti-parasitic guanidine and pyrimidine alkaloids from the marine sponge *Monanchora arbuscula*, J. Nat. Prod. 78 (2015) 1101–1112. https://doi.org/10.1021/acs.jnatprod.5b00070.

[251] L.F. Martins, J.T. Mesquita, G.É. Pinto, T.A. Costa-Silva, S.E.T. Borborema, A.J. Galisteo Junior, et al., Analogues of marine guanidine alkaloids are in vitro effective against trypanosoma cruzi and selectively eliminate *Leishmania (L.) infantum* intracellular amastigotes, J. Nat. Prod. 79 (2016) 2202–2210. https://doi.org/10.1021/acs.jnatprod.6b00256.

[252] S.-T. Ishigami, Y. Goto, N. Inoue, S.-I. Kawazu, Y. Matsumoto, Y. Imahara, et al., Cristaxenicin A, an antiprotozoal xenicane diterpenoid from the deep sea gorgonian *Acanthoprimnoa cristata*, J. Org. Chem. 77 (2012) 10962–10966. https://doi.org/10.1021/jo302109g.

[253] L.G. Malak, M.A. Ibrahim, D.W. Bishay, A.M. Abdel-baky, A.M. Moharram, B. Tekwani, et al., Antileishmanial metabolites from *Geosmithia langdonii*, J. Nat. Prod. 77 (2014) 1987–1991. https://doi.org/10.1021/np5000473.

[254] J. van Griensven, M. Balasegaram, F. Meheus, J. Alvar, L. Lynen, M. Boelaert, Combination therapy for visceral leishmaniasis, Lancet Infect Dis. 10 (2010) 184–194.

[255] T. Ghazanfari, Z.M. Hassan, M. Ebtekar, A. Ahmadiani, G. Naderi, A. Azar, Garlic induces a shift in cytokine pattern in *Leishmania major*-infected BALB/c mice, Scand. J. Immunol. 52 (2000) 491–495.

[256] A. Bhattacharjee, S. Majumder, S.B. Majumdar, S.K. Choudhuri, S. Roy, S. Majumdar, Co-administration of glycyrrhizic acid with the antileishmanial drug sodium antimony gluconate (SAG) cures SAG-resistant visceral leishmaniasis, Int. J. Antimicrob. Agents 45 (2015) 268–277. https://doi.org/10.1016/j.ijantimicag.2014.10.023.

[257] O. Ridoux, C. Di Giorgio, F. Delmas, R. Elias, V. Mshvildadze, G. Dekanosidze, et al., In vitro antileishmanial activity of three saponins isolated from ivy, alpha-hederin, beta-hederin and hederacolchiside A(1), in association with pentamidine and amphotericin B, Phytother. Res. 15 (2001) 298–301.

[258] T.A. Mokoka, S. Zimmermann, T. Julianti, Y. Hata, N. Moodley, M. Cal, et al., In vitro screening of traditional South African malaria remedies against *Trypanosoma brucei* rhodesiense, *Trypanosoma cruzi*, *Leishmania donovani*, and *Plasmodium falciparum*, Planta Med. 77 (2011) 1663–1667. https://doi.org/10.1055/s-0030-1270932.

[259] N.M. Al-Musayeib, R.A. Mothana, A. Matheeussen, P. Cos, L. Maes, In vitro antiplasmodial, antileishmanial and antitrypanosomal activities of selected medicinal plants used in the traditional Arabian Peninsular region, BMC Complement Altern. Med. 12 (2012) 49. https://doi.org/10.1186/1472-6882-12-49.

[260] S. Dey, D. Mukherjee, S. Chakraborty, S. Mallick, A. Dutta, J. Ghosh, et al., Protective effect of *Croton caudatus* Geisel leaf extract against experimental visceral leishmaniasis induces proinflammatory cytokines in vitro and in vivo, Exp. Parasitol. 151–152 (2015) 84–95. https://doi.org/10.1016/j.exppara.2015.01.012.

[261] N. Llurba Montesino, M. Kaiser, R. Brun, T.J. Schmidt, Search for antiprotozoal activity in herbal medicinal preparations; new natural leads against neglected tropical diseases, Molecules 20 (2015) 14118–14138. https://doi.org/10.3390/molecules200814118.

[262] C.M. Lezama-Davila, A.P. Isaac-Marquez, P. Zamora-Crescencio, M.D.R. Uc-Encalada, S.Y. Justiniano-Apolinar, L. del Angel-Robles, et al., Leishmanicidal activity of *Pentalinon andrieuxii*, Fitoterapia 78 (2007) 255–257. https://doi.org/10.1016/j.fitote.2006.12.005.

[263] I.G. Demarchi, S. MdeTerron, M.V. Thomazella, C.A. Mota, Z.C. Gazim, D.A.G. Cortez, et al., Antileishmanial and immunomodulatory effects of the essential oil from *Tetradenia riparia* (Hochstetter) Codd, Parasite Immunol. 38 (2016) 64–77. https://doi.org/10.1111/pim.12297.

Chapter 10

Natural Products Targeting Inflammation Processes and Multiple Mediators

G. David Lin[1], Rachel W. Li[2]
[1]*Research School of Chemistry, The Australian National University, Canberra, ACT, Australia;*
[2]*ANU Medical School, The Australian National University, Canberra, ACT, Australia*

1. INFLAMMATION RESPONSES AND PATHWAYS

The word "inflammation" is derived from the Latin *inflammare*, meaning "to set on fire." In pathophysiology, inflammation is an adaptive response triggered by noxious stimuli and conditions, such as infection and tissue injury [1].

A precise definition of the inflammation is difficult, but a useful description is that it is a process that enables the body's defensive and regenerative resources to be channeled into tissues that have suffered damage or are contaminated with abnormal material, such as invading microorganisms. It is a process that aims to limit the damaging effect of any contaminating material, to cleanse and remove any foreign particles and damaged tissue debris, and to allow healing processes to restore the tissues to a level of normality.

Inflammation is a phenomenon well known to us since ancient civilizations. The classic description of inflammation was given by Aulus Cornelius Celsus in the 1st century AD. In his *De Medicina*, Celsus stated that "the signs of inflammation are four: redness, and swelling with heat and pain." To these four signs was added a fifth, loss of function, by Virchow, the founder of modern cellular pathology in his classic work *Cellular Pathology* (*Die Cellularpathologie in ihrer Begründung auf physiologische und pathologische Gewebelehre*) published in 1858.

Loss of function is a consequence of the swelling and pain associated with inflammation and is clearly an indication that affected tissues require rest for rapid rehabilitation. The loss of function is also equivalent to tissue stress or malfunction, which induces an adaptive response, a so-called para-inflammation that is probably responsible for the chronic inflammatory conditions associated with modern human diseases [1].

Natural Products and Drug Discovery. https://doi.org/10.1016/B978-0-08-102081-4.00010-1

The sequence of events in inflammation was summarized by Hansen [2] as follows:

- cellular injury;
- local vasodilation, increased capillary permeability, cellular adhesion, and exudation to bring inflammatory cells and chemical mediators to the injured area;
- destruction of injurious agents by phagocytes;
- clearing away of cellular debris; and
- deposition of a protein framework for tissue healing.

During the course of inflammation, cellular mediators are released, and in the plasma a chain reaction (cascade) response activates other protein mediators. These cellular and molecular interactions result in the formation of substances that augment the inflammatory response and produce systemic signs and symptoms. While there are many and varied kinds of inflammation, they can be generally understood from changes at:

- cellular levels (tissue damage or foreign material–invaded infection), and
- molecular levels (signaling inducers and mediators).

1.1 Cellular Changes of Inflammatory Responses

At the cellular level, the basic components of an inflammatory reaction are due to the combined effects of changes of the microcirculation, alteration of permeability of the blood vessel walls, and migration of leukocytes and phagocytes [3].

The immediate response to cellular injury is a brief spasm of arterioles in the injury area. This vasoconstriction is mediated by endothelial factors and is believed to be a protective reflex designed to minimize bleeding or to limit the extent of the injury [4]. The dominant vascular response in inflammation is not vasoconstriction but vasodilation with increased blood flow (hyperemia). The redness on the skin is due to an increased volume of blood flowing through the inflamed area. Consequently, the temperature of the inflamed skin rises and approaches that of deep body temperature. This effect on microcirculation is seen clearly if a firm line is drawn with a blunt point over the surface of the forearm. A red line appears in exactly the position that stimulus is applied.

Detailed studies of vascular changes occurring during inflammation have been made [5]. It has been demonstrated that the whole capillary bed at the damaged site becomes suffused with blood at an increased pressure. Capillaries dilate and many closed ones open up. The venules dilate, and there is an increased flow of blood in the draining veins. This rapid flow gradually slows in the central capillaries and venules even though the vessels are still dilated. This slowing may gradually spread to the peripheral areas of the lesion and the flow may even stop completely. Despite this stasis the capillary pressure remains high, probably due to a resistance to outflow. The pressure in the small veins may be increased by a rise in the pressure of interstitial fluid due to edema. Two of the cardinal signs of inflammation, heat and redness, are caused by this increase in blood flow to the affected area.

The dilated vessels in the injured area become more permeable because increased pressure within them widens the spaces between the endothelial cells forming the vessel walls. These changes result in exudation of plasma into the interstitial space. Swelling (or edema) is the consequence of this change. Under the activating stimulus of several cellular mediators, circulating neutrophils are attracted by chemotaxis to the area of injury where their motility is slowed within postcapillary venules. The neutrophils adhere to the endothelium and eventually transmigrate into the tissues to engulf and digest bacteria or other injurious agents and dead cells [6]. The plasma protein fibrinogen also exudes into the interstitium. The consequence of edema formation is the development of pain. Pain is, at least in part, due to the increased pressure on sensory nerves caused by the accumulation of the edematous fluid. Pain may also be due to the release of pain-inducing chemicals at the site of the reaction.

The cardinal signs of inflammation can thus be accounted for by the various changes that occur to the vasculature of the affected area. Vasodilation results in increased redness and temperature of the affected area, while the change to the protein permeability of blood vessels results in the development of swelling and pain.

The process of inflammation may, on occasions, be capable of causing more harm to the organism than the initiating noxious stimulus itself. For example, the necrotic lesions produced on dogs at the feeding site of a tick are brought about by the dog's defense system; they are not caused directly by the tick. The necrosis can be virtually eliminated by the prior destruction of polymorphonuclear leukocytes. Similarly, some immunologically induced reactions to seemingly harmless agents result in tissue damage out of all proportion to the threat from the sensitizing agent [7]. There is no obvious reason why the inflammatory reactions that occur in diseases such as rheumatic fever or rheumatoid arthritis could benefit the host. Clearly not all inflammation is useful. The combined effect of the components of the body's defense mechanism may often be excessive.

Inflammation covers a host of pathophysiological events and means different things to different people—acute or chronic, organ specific such as asthma, reversible or irreversible—but each inflammatory reaction is unique. Inflammation in autoimmune conditions such as rheumatoid arthritis is fundamentally the same as that of simple infections or wound; however, the trigger of the reaction is very different.

1.2 Signaling Mediators of Inflammatory Responses

We know that very similar reactions can be induced by widely differing stimuli. This led to the idea that some intermediary control system exists to link the stimulus and the effect. The most popular idea is that the inflammatory insult causes the release of chemicals within the body, which then trigger the inflammatory reactions. This mediator concept was exemplified by Lewis [8] in 1927, who proposed that the local vasodilation and increased vascular permeability observed in the triple response could be mediated by a substance, liberated by the tissue, which he termed "H-substance." Production of numerous inflammatory mediators, in turn, alters the functionality of many tissues and organs and the downstream effectors of the inflammatory pathway.

The search for mediators and effectors of inflammation has been particularly directed toward finding substances interacting with local vasodilation, increased capillary permeability, cellular adhesion, and exudation. According to their biochemical properties, seven inflammatory mediator groups have been established [1]:

- vasoactive amines (e.g., histamine and serotonin);
- vasoactive peptides (e.g., substance P and bradykinin);
- complement fragments C3a, C4a, and C5a (anaphylatoxins);
- lipid mediators (e.g., eicosanoids and platelet-activating factor [PAF]);
- cytokines (e.g., tumor necrosis factor [TNF]-α, interleukin [IL]-1, and IL-6);
- chemokines (e.g., IL-8 and eotaxin) in response to inflammatory inducers; and
- proteolytic enzymes (e.g., cathepsins and matrix metalloproteinases).

It is unclear how these mediators are activated in sequence or in hierarchy in responses to inflammatory inducers. But it is clear that the inflammatory mediators specifically affect the functional states of the cells and tissues, which are the effectors of an inflammatory response. For example, the most obvious effect of the mediators acting on the vasculature and on leukocyte migration is to induce the formation of an exudate. Many inflammatory mediators also affect neuroendocrine and metabolic functions.

Here, we focus on inflammatory mediators involved in the following pathways:

- Arachidonic acid (AA) pathways,
- Nuclear transcription factor κB (NF-κB) pathways,
- Nitric oxide (NO) pathway, and
- Proinflammatory cytokines.

Common inflammatory mediators involved in these inflammatory response pathways are discussed in the next section.

2. COMMON MEDIATORS OF INFLAMMATION PATHWAYS

The most strongly implicated mediators of inflammatory pathways are:

- eicosanoids (prostaglandins, thromboxanes, leukotrienes [LTs], and PAF), catalyzed by phospholipases, cyclooxygenases (COXs), and lipoxygenases (LOs) in their formation;

- inhibitors of κB (I*k*B) and IκB kinases (IKK);
- inducible nitric oxide (iNO); and
- IL-1 and TNFα.

These common mediators are discussed in detail next.

2.1 Eicosanoids

Eicosanoids are a blanket term to describe a group of bioactive lipids derived from phospholipids, for example phosphatidylcholine, which produces AA and lysophosphatidic acid catalyzed by cytosolic phospholipase A_2 ($cPLA_2$).

AA is metabolized to produce more bioactive lipids either by COXs, which generate prostaglandins and thromboxanes, or by LOs, which generate LTs and lipoxins (see Fig. 10.1).

Main eicosanoid groups as shown in Fig. 10.1 are:

- prostaglandins (mediators of inflammatory and anaphylactic reactions),
- prostacyclins (mediators of vasodilation, antiinflammation),
- thromboxanes (mediators of vasoconstriction),
- LTs (mediators of asthmatic and allergic reactions), and
- lipoxins (antiinflammatory mediators, dampening inflammation).

One productive area in searching for chemical mediators of inflammation has been the finding of prostaglandins and prostacyclins, which are capable of proinflammation and antiinflammation, respectively. Sir John Robert Vane (1927–2004) was a pioneer in the field through his seminal research published in *Nature* [9,10] and other journals [11–13].

Indeed, eicosanoids are the most studied chemical mediations of inflammatory reactions. Every cell in the body has the capacity to generate some eicosanoids from AA. AA is usually found only at very low levels in the free form, but an abundant supply is normally available in a bound form, principally as cell membrane phospholipid, which constitutes the major source of AA. Following perturbation of the cell, the phospholipids are split by the action of enzymes, such as phospholipase A_2 (PLA_2), releasing AA. This metabolism is generally rapid, since it is the availability of AA that is the rate-limiting factor [14].

Following the release of AA, two enzyme systems are available for metabolizing AA—COXs and LOs—each giving rise to a distinct array of products (Fig. 10.1).

2.1.1 PLA_2 and the Production of AA and PAF

PLA_2 is an upstream enzyme involved in the production of many important chemical mediators of inflammation. It catalyzes a hydrolysis of the sn-2 ester of a phospholipid to AA and lysophospholipid [15].

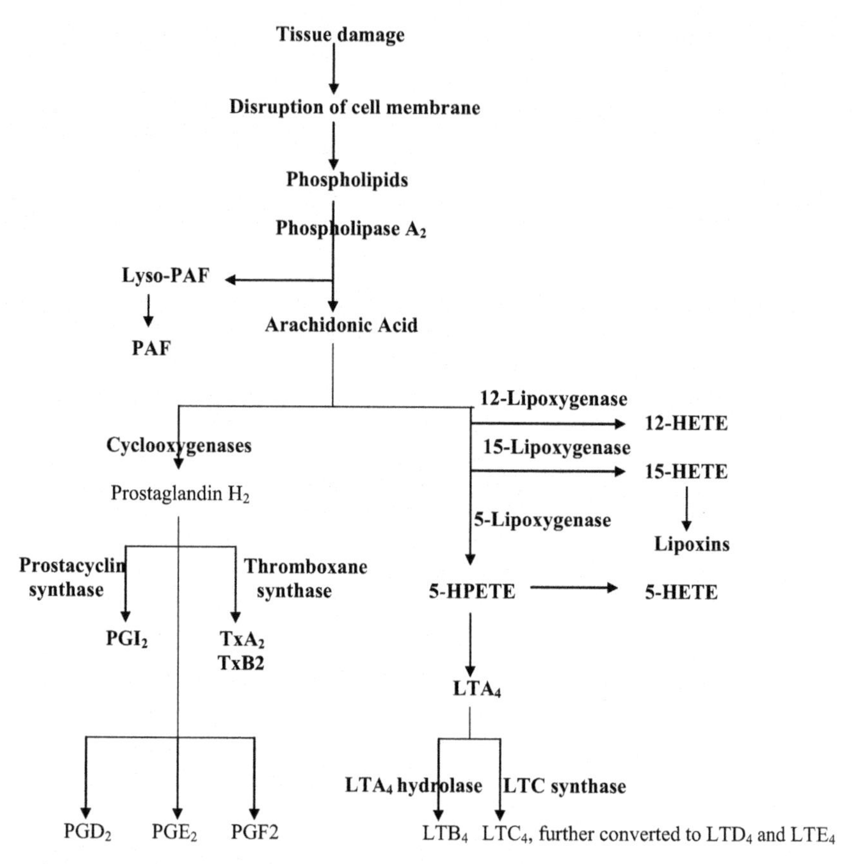

FIGURE 10.1 Key mediators generated from arachidonic acid metabolic pathway. Arachidonic acid, liberated by PLA_2, is further metabolized via the cyclooxygenase pathways and the lipoxygenase pathways to form eicosanoids, including platelet-activating factor (PAF), prostacyclin (PGI_2), thromboxanes (Tx) A_2 and B_2, prostaglandins (PGD_2, PGE_2, and PGF_2), hydroxyeicosatetraenoic acid (HETE), lipoxins, hydro(pero)xyeicosatetraenoic acid (HPETE) and leukotrienes (LT) A_4 to E_4.

Lysophospholipids are converted into PAF, which has an important role in pathological conditions like asthma, anaphylaxis, and inflammation of the kidney.

PAF induces inflammatory reaction in various animal species and in humans. At the cellular level, PAF causes contraction of smooth muscle, an increase in vascular permeability and platelet aggregation, as well as exocytosis in both platelets and leukocytes. PAF also mimics the main clinical features of asthma and is particularly effective in producing hyperreactivity and accumulation of eosinophils in lung tissue. Asthmatic patients have high levels of circulating PAF and their eosinophils make more PAF than those of normal controls [16]. For the development of natural PAF antagonists as therapeutic agents, see a review by Braquet and Hosford [17].

AA, liberated by PLA_2, is further metabolized via the COX pathways and the LO pathways (Fig. 10.1).

2.1.2 COXs and the Production of Prostaglandins, Thromboxanes, and Prostacyclin

Apart from non-nucleated erythrocytes, all cells are capable of synthesizing prostaglandins, which are released in response to many kinds of trauma or any disturbance of the cell membrane. Prostaglandins are formed by the action of the COX (prostaglandin endoperoxide synthase or prostaglandin synthase) on AA. COX catalyzes the conversion of AA to prostaglandin endoperoxide H_2, which is the precursor to a series of mediators that include prostaglandin I_2 (i.e., prostacyclin), thromboxanes (e.g., A_2 and B_2) and several important prostaglandins (e.g., PGD_2, PGE_2, and PGE_{2F}); see Fig. 10.1.

In 1975, Samuelsson discovered that AA in the platelet was metabolized to the proaggregatory thromboxane A_2 [18]. This work and the identification of LTs [19] earned Samuelsson a share of the Nobel Prize in Physiology or Medicine in 1982.

Soon after the discovery of thromboxane A_2, another prostaglandin (PGI_2, also called prostacyclin) was discovered by John Vane's group and showed to have the opposite activity to that of thromboxane A_2 [20]. PGI_2 relaxes blood vessels and inhibits platelet aggregation. Its synthesis in endothelial cells of blood vessel walls is of special importance.

These inflammatory mediators are completely or partially responsible for increased blood vessel permeability, dilation of blood vessels, chemotaxis of granulocytes, degradation of cartilage, and proliferation of B- and T-lymphocytes [21]. LTs and prostaglandins play a critical role in activation of inflammatory cells (neutrophils and macrophages) and secretion of other mediators (e.g., histamine secretion in mast cells and basophils; serotonin secretion in platelets).

Two different isoforms of cyclooxygenase have been identified: COX-1 and COX-2. The regulations of these two isozymes and their respective genes are quite different [22]. COX-1 is constitutively expressed, whereas COX-2 is highly sensitive to extracellular stimuli. COX-2 messenger RNA (mRNA) is rapidly induced in cells treated with serum, growth factors, forskolin, or phorbol ester. Moreover, the induction of the COX-2 gene in response to stimulation is inhibited by the concomitant treatment of cells with glucocorticoids, whereas the COX-1 gene is not inhibited by glucocorticoids [23]. Evidence suggests that COX-1 has a housekeeping function, producing prostaglandins necessary for normal cellular process, where *COX-2* is responsible for inflammatory and mitogenic responses, producing prostaglandins involved in inflammation and growth regulation [24].

2.1.3 5-LOs and the Production of LTs

An alternate pathway of AA oxidation is catalyzed by LOs. The three main groups of LOs (5-, 12-, and 15-LOs) specifically introduce molecular oxygen into a specific carbon of AA and form hydroperoxy acids [25]. The physiological function of individual LO is uncertain aside from 5-LO. The products of 5-LO include LTs, which constitute an important class of inflammatory mediators.

The enzyme 5-LO is the "gatekeeper" of the LTs pathway, catalyzing the first committed steps in the metabolic reactions leading to the synthesis of all the LTs [26]. Because of its central role and potential to block all LTs production with one specific inhibitor, 5-LO has been the subject of intense investigation. 5-LO is a cytosolic enzyme that requires Ca^{2+} and adenosine triphosphate (ATP) for maximal activity. The human enzyme has been purified to homogeneity from peripheral blood leukocytes and is a 78-kDa protein. When neutrophils are activated by calcium ionophore A23187, 5-LO, normally cytosolic, is translocated to a membrane fraction, where it gains access to its substrate, AA.

The biological pathway leading to LT formation has been well characterized [19,27]. 5-LO catalyzes the formation of LTA_4 and 5-(*S*)-hydroxy-6,8,11,14-eicosatetraenoic acid (5-HETE) from AA through the intermediary molecule of 5-(*S*)-hydroperoxy-6,8,11,14-eicosatetraenoic acid (5-HPETE). LTA_4 can be further metabolized by specific enzymes to produce biologically active compounds, for example, LTB_4 by LTA_4 hydrolase, and LTC_4 by LTC_4 synthase (Fig. 10.1).

Historically, slow-reacting substance of anaphylaxis (SRS-A) was identified as a product of the 5-LO pathway of AA metabolism, and Samuelsson [19] renamed the constituents of SRS-A as LTs. There is some evidence that LO products contribute to vascular changes in inflammation.

The biological effects of LTs are numerous. Although all LTs are primarily products of leukocytes, including neutrophils, mast cells, and macrophages, LTB_4 alone acts as a calcium ionophore, leading to further stimulation of 5-LO activity. In addition, LTB_4 is a potent chemotactic substance for leukocytes and has been shown to influence lymphocyte activity. In contrast, the peptidyl derivatives of LTs (LTC_4, LTD_4, and LTE_4) primarily affect smooth muscle contractility and other cells with contractile capacity. For example, on a molar basis, LTC_4 and LTD_4 are 100 times more potent in inducing constriction of the airway than histamine.

Zileuton is currently available on the market as a specific inhibitor of 5-LO and is indicated for the treatment of asthma. Antagonists of LT receptors have received more attention now following the market launch of montelukast and zafirlukast, widely used as antiasthmatic agents [28].

2.2 Protein Kinase–Mediated IκB Degradation in the NF-κB Pathway

We know that NF-κB plays important roles in the pathogenesis of various diseases including cancer, atherosclerosis, myocardial infarction, allergy, asthma, diabetes, arthritis, Crohn's disease, multiple sclerosis, osteoporosis, and septic shock. Among many functions, NF-κB controls the expression of genes encoding the proinflammatory cytokines, chemokines, adhesion molecules, and inducible COX-2 and NO [29].

NF-κB is maintained in an inactive state in the cytoplasm by binding to the inhibitor of Kappa B (IκB). Phosphorylation of IκB by IκB kinases (IKKα and IKKβ), via the canonical NF-κB pathway, results in ubiquitination of IκB and its proteasomal degradation. The degradation of IκB liberates NF-κB, which moves to the nucleus and activates the transcription of its target genes.

Many natural products act like IKK inhibitors, blocking IκB phosphorylation degradation and thus preventing NF-κB activation [30]. Examples include curcumin, resveratrol, epigallocatechin gallate, quercetin, capsaicin, and silybin.

2.3 Inducible NO

NO, a simple gas, was discovered as a mediator of macrophage-mediated tumor cytotoxicity, vessel dilation, and neurotransmission in the late 1980s and since then has been demonstrated to possess many important physiological activities, including smooth muscle relaxation, tumor cell lysis, and destruction of microorganisms [31,32].

NO is synthesized from the guanidino nitrogen of L-arginine by three isoforms of NO synthases (NOS): a neural form (nNOS, or NOS-1) and an endothelial form (eNOS, or NOS-3), both being constitutive and differentiated from an inducible form (iNOS, or NOS-2) in response to the stimulation of cytokines or endotoxins.

In response to inflammatory stimuli, both NOS-2 and COX-2 are induced and the production of NO further stimulates the formation of COX-2. This leads to the production of both NO and PGE_2 and significant cross-talks between PGE_2/NO and COX-2/NOS-2 [33]. It is obvious that inhibiting both NOS-2 and COX-2 would provide a better antiinflammatory action than a single inhibition.

2.4 Proinflammatory and Antiinflammatory Cytokines

Cytokines are a collective name of lymphokines (cytokines secreted by lymphocytes), monokines (cytokines secreted by monocytes), interleukins (cytokines secreted by one leukocyte and acting on other leukocytes), chemokines (cytokines with chemotactic activities), etc. Cytokines are regulators of host responses to infection, trauma, inflammation and immune responses and can be broadly fallen into proinflammatory or antiinflammatory group.

Proinflammatory cytokines, predominantly produced by activated macrophages, tend to make a disease or a condition worse. They are involved in the upregulation of inflammatory reactions and include IL-1β, IL-6, and TNFα [34].

Antiinflammatory cytokines act to reduce inflammation and promote healing by controlling the proinflammatory cytokine response and include IL-1 receptor antagonist, IL-4, IL-10, IL-11, and IL-13 [35]. In particular, IL-10 is a potent antiinflammatory cytokine, which downregulates proinflammatory cytokine receptors.

Cytokines form a cytokine network, which provides for functional redundancy within the inflammatory and immune systems. In many instances, individual cytokines have multiple biological activities; different cytokines can also have the same activity, thus requiring more than a single defect in the network to alter drastically the outcome of the cytokine-mediated inflammatory processes [36]. It is widely known that cytokines often act by activating transcriptional factors such as NF-κB and protein kinases such as protein kinase C (PKC) and, in turn, regulating target gene expression to maintain the inflammatory state—for example, the induction of iNOS and COX-2.

2.5 Antiinflammatory Drugs Targeting Common Mediators

Antiinflammatory drugs targeting mediators can be broadly divided into three groups: glucocorticoids, nonsteroidal antiinflammatory drugs (NSAIDs), and biologics [37].

2.5.1 Glucocorticoids

Glucocorticoids are synthetic steroids and are based on natural steroids that occur in the body. They include hydrocortisone, cortisone, beclomethasone, betamethasone, budesonide, dexamethasone, and methylprednisolone.

Glucocorticoids bind and activate the glucocorticoid receptor, a type of nuclear receptor, leading to antiinflammatory action, by producing lipocortin (annexin-1), which inhibits PLA_2, COX-1/COX-2, and leukocyte inflammatory events, including epithelial adhesion, emigration, chemotaxis, phagocytosis, and respiratory burst.

Glucocorticoids are potent antiinflammatories, but prolonged use causes systemic side effects including immunosuppression.

2.5.2 NSAIDs

NSAIDs are also known as antiinflammatories, which are a group of related pain medicines including aspirin, ibuprofen, indomethacin, naproxen, piroxicam, diclofenac, and methyl salicylate. These drugs block eicosanoid production, particularly prostaglandins.

Prostaglandins play an important role in the induction of pain and fever, but they also have important roles in the maintenance of normal physiological functions such as gastric cytoprotection and supporting platelets and blood clotting. NSAIDs can cause ulcers in the stomach and promote bleeding. These adverse effects observed with most of the current clinically available NSAIDs are believed to stem from an inhibition of COX-1 activity.

The discovery of two isoforms of constitutive COX-1 and inducible COX-2 renewed the interest in developing new NSAIDs for the therapy of inflammation with the hope of obtaining a better therapeutic index (efficacy vs. side effects) than that of aspirin [38]. COX-2 is cytokine inducible and is expressed in inflammatory cells [39]. COX-2 is responsible for the biosynthesis of prostaglandins under acute inflammatory conditions and was a focus of attention for the development of NSAIDs in 1990s [40] when two new highly selective COX-2 inhibitors, celecoxib and rofecoxib, were marketed, but halted in 2004 when rofecoxib was withdrawn due to serious cardiovascular events. The serious events are caused by an imbalance of antithrombotic and vasodilatory PGI_2 and prothrombotic thromboxane A_2 [41].

2.5.3 Biologics

The increasing understanding of the role of cytokines in many inflammatory disorders that are intimately associated with an imbalance of the cytokine network has enable the pharmaceutical industry to make tremendous effort to develop cytokine modulators not only for the treatment of the inflammatory disorders but also for the control of atherosclerosis and Alzheimer disease, which present an inflammatory origin.

Biologics of selectively blocking proinflammatory cytokine signaling by targeting TNFα receptors [42] or IL-1 activities [43] have been developed and used clinically. Infliximab and adalimumab are two examples of anti-TNFα biologics for treating inflammatory bowel disease [44]. The IL-1 receptor antagonist anakinra, the soluble decoy receptor rilonacept, and the neutralizing monoclonal anti–IL-1β antibody canakinumab are three examples of anti–IL-1 biologics used to treat classic rheumatic diseases. The anti-TNFα therapy can be combined with other anti-cytokine or anti–T cell biologicals or with small molecules targeting NF-κB and p38 mitogen-activated protein kinases (MAPKs) [42].

However, costs to produce biologics are usual much higher than small molecules and it is a big burden to patients. Just to cite a latest example, Spinraza, which uses a survival motor neuron–directed antisense oligonucleotide and is marketed by Biogen for spinal muscular atrophy, would cost US$750,000 in the first year of treatment [45].

In summary, steroidal antiinflammatory drugs are very effective and have an obvious role in the treatment of inflammatory diseases, but they can be used for only short periods of time due to their systematic toxicity. NSAIDs, when used over a long time, tend to cause gastrointestinal side effects. Clinical use of COX-2 selective drugs also causes serious cardiovascular problems. Emerging antiinflammatory biologics are yet to be produced economically.

Since the successful story of salicylate and aspirin, structurally diverse natural products continue to attract research interest and serve as alternatives for therapeutic intervention of many inflammatory diseases. Many plant-derived natural products are known to directly or indirectly interact with inflammation mediators, and they are discussed in detail next.

3. NATURAL PRODUCTS TARGETING THE COMMON INFLAMMATORY MEDIATORS

3.1 Introduction and the Salicylate Story

Throughout recorded history in all cultures, plant remedies for the control of inflammation have been used all over the world and provided a rich source for the development of antiinflammatory agents, such as chewing willow bark for reducing fever and relieving pains until the beginning of the 19th century, when the active ingredient, salicin (1), was discovered. Aspirin (2), one of the most widely used NSAIDs today, is an analogue of salicin with a chemical name of acetyl salicylic acid. Salicylic acid (3), or its sodium salt salicylate (4), is widely distributed in plants. In comparing the antiinflammatory actions of aspirin and sodium salicylate, it was observed that aspirin directly acetylates COXs and irreversibly blocks the conversion of AA to prostanoids, whereas salicylate's inhibition of prostanoid biosynthesis is weak and other mechanisms such as intracellular signaling pathways with NF-κB and MAPK may be involved [46]. For example, it has been reported that aspirin inhibits IKKβ with an approximate IC_{50} of 50 μM [47], which is consistent with its reported inhibitory effect on the NF-κB pathway [48].

On June 2, 1763, the Reverend Edmund Stone of Chipping Norton in Oxfordshire read a report to the Royal Society on the use of willow bark against fever. He believed in the doctrine that the cures for diseases are often found in the same location that the malady occurs. He gathered a pound of willow bark, because "the willow delights in a moist and wet soil, where agues chiefly abound," and dried it in a baker's oven for 3 months, and then pulverized it. He reported using 1-g doses with about 50 patients with safety and great success. He concluded in his paper [49] that "I have no other motives for publishing this valuable specific, than that it may have a fair and full trial in all its variety of circumstances and situations, and that the world may reap the benefits occurring from it."

In 1860, salicylic acid, found in the bark of the willow tree, was chemically synthesized in Germany and its ready supply led to even more extended use as an external antiseptic, as an antipyretic, and in the treatment of rheumatism. In 1893, the German chemist Felix Hoffman synthesized the acetyl derivative of salicylic acid in response to the urging of his father, who took salicylic acid for rheumatism. This more palatable form of salicylate is aspirin (acetylsalicylic acid). This important new drug was soon recognized by Dr. Heinrich Dreser,

Bayer's research director, and introduced to market in 1899, accompanied by a paper suggesting that aspirin was a convenient way of supplying the body with the active substance of salicylate [50].

Annual world production of aspirin is many thousands of tons with an average consumption in developed countries of about 100 tablets per person per year. The dream of Edmund Stone in 1763 that "the world may reap the benefits occurring from willow bark" has been fulfilled.

Nowadays, many types of natural product molecules are used clinically in western medicine. Plants and plant products are present in 14 of the 15 therapeutic categories of pharmaceutical preparations that have been recommended to medical practitioners in the United Kingdom, and they form an important part of the health care system in the Western world [51].

Although natural products are not dominant in Western therapy, they do account for the active ingredients of 35%–50% of preparations and the future drug development [52]. Between 2000 and 2005, 23 new drugs launched on the market were originated from natural products, among which five (apomorphine, tiotropium, nitisinone, galanthamine, and arteether) were derived from plants or plant metabolites and derivatives [53].

Next, we discuss several groups of natural products with distinguish antiinflammatory actions (for their molecular diversity, see Fig. 10.2).

3.2 Phenolics

Phenolic compounds are ubiquitous in plants, a major group of plant secondary metabolites and an essential part of our vegetables and foods. Major groups of phenolic compounds from dietary plants and medicinal herbs are:

- simple phenolic acids including hydroxybenzoic acids (e.g., salicylic acid) and hydroxycinnamic acid;
- flavonoids including flavone, flavonol, flavanone, flavanol, and anthocyanidin;
- tannins;
- stilbenes;
- curcuminoids;
- coumarins;
- lignans;
- quinones, and
- xanthones.

Since the discovery of salicylic acid, antiinflammatory action of phenolic compounds has been well researched and documented. Phenolic compounds are found to interfere with many mediators or pathways including AA metabolites and NO [54], cytokines [55], and NF-κB signaling pathways [56].

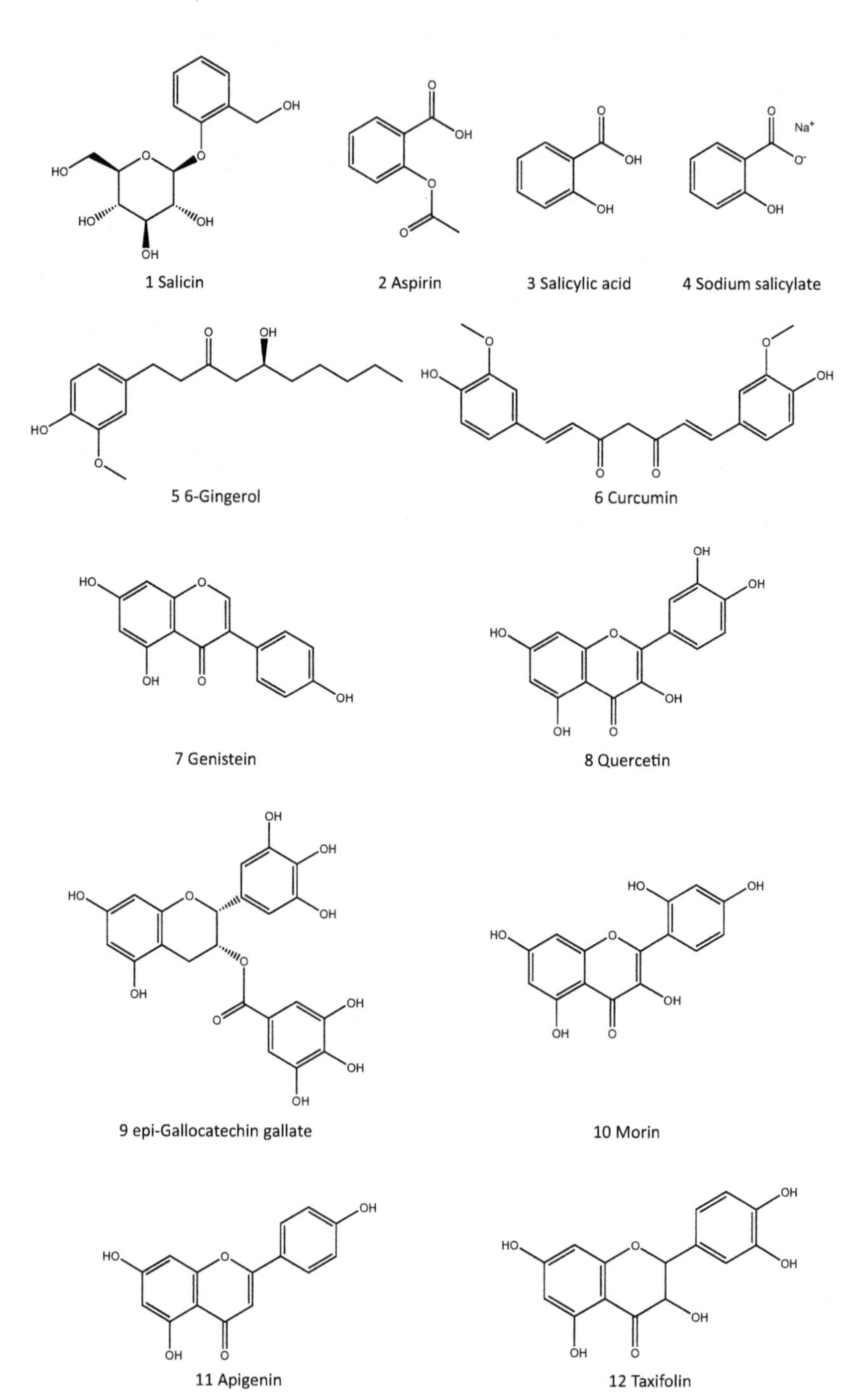

FIGURE 10.2 Molecular diversity of natural products with antiinflammatory action.

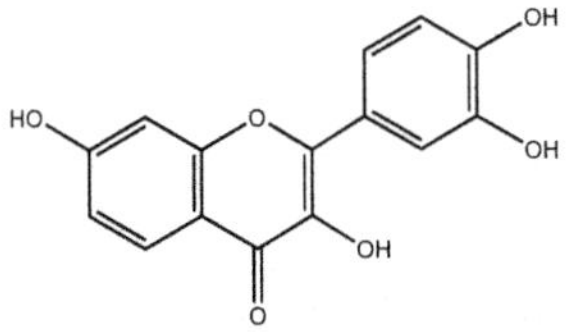

13 Fisetin

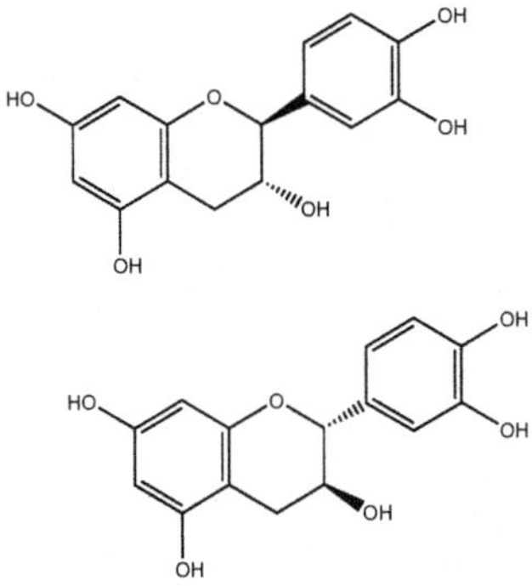

14 Catechin

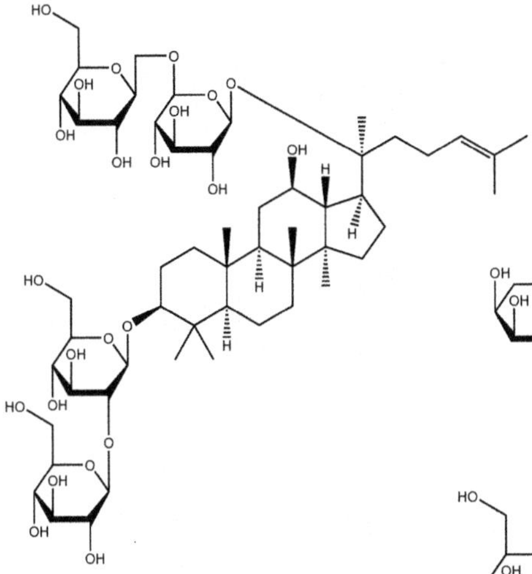

15 Ginsenoside Rb1

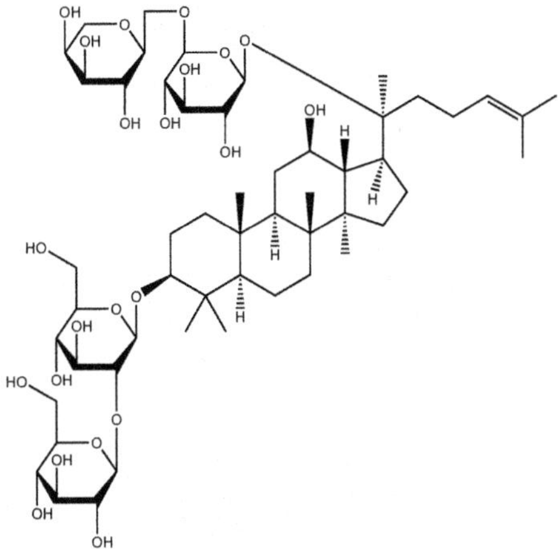

16 Ginsenoside Rb2

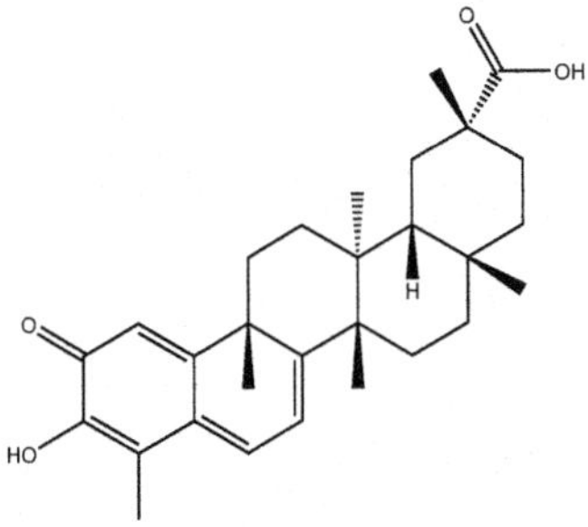

17 Tripterine

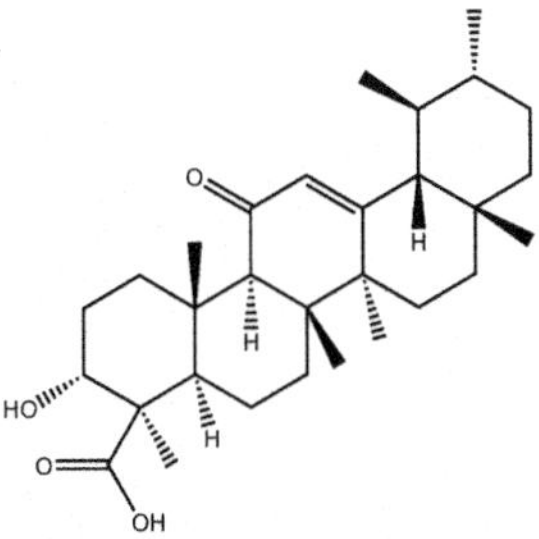

18 11-Keto-β-boswellic acid

FIGURE 10.2 Continued

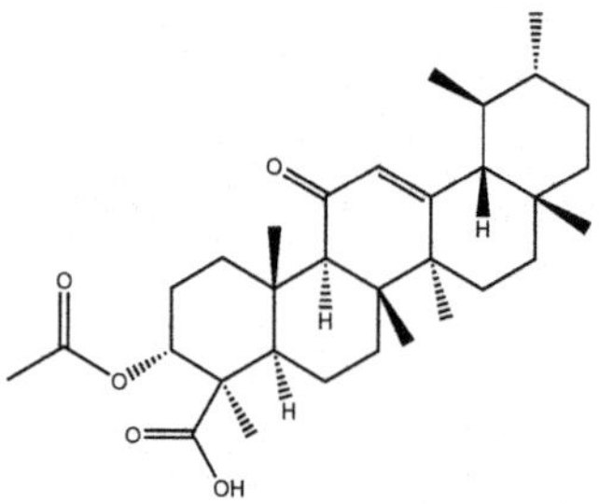

19 3-*O*- Acetyl-11-keto-β-boswellic acid

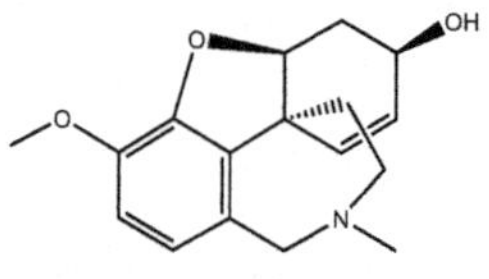

20 Galanthamine

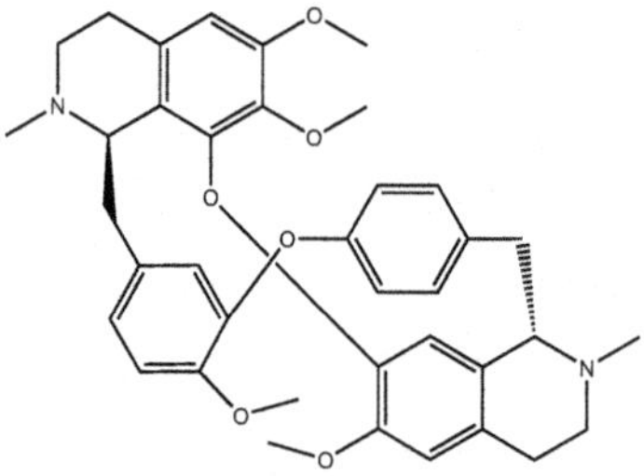

21 Tetrandrine

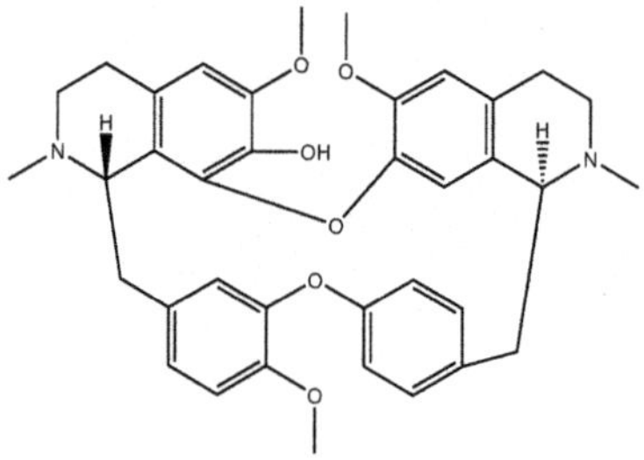

22 Fangchinoline

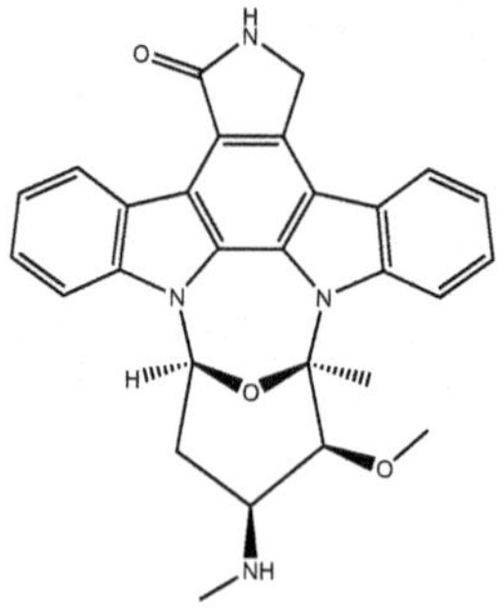

23 Staurosporine

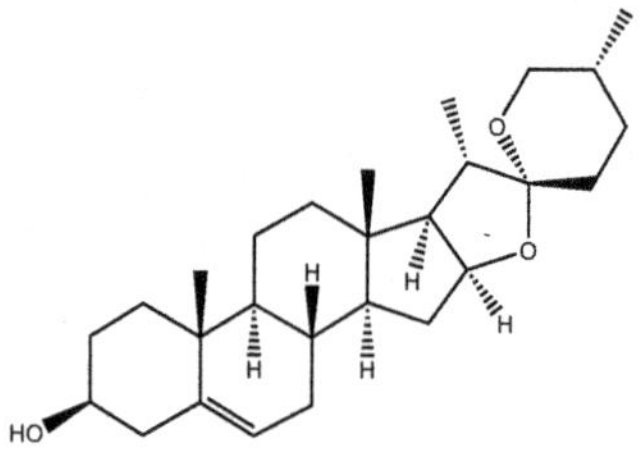

24 Diosgenin

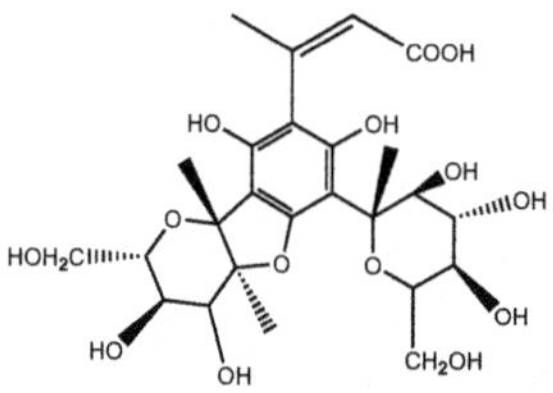

25 Racemosic acid

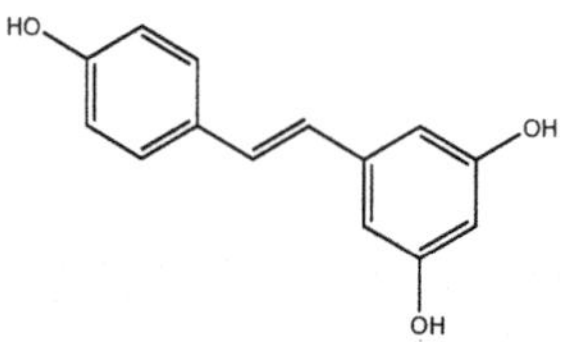

26 trans-resveratrol

FIGURE 10.2 Continued

6-Gingerol (5) is found in rhizomes of *Zingiber officinale* (ginger) and is a potent inhibitor against a number of prostaglandin biosynthesizing enzymes including 5-LO and LT synthetases.

Curcumin (6), a natural diphenyl pentanoid from dried root and rhizome of turmeric (*Curcuma longa*), possesses chemopreventive, antiproliferative, proapoptotic, antimetastatic, antiangiogenic, and antiinflammatory properties as shown by preclinical and clinical studies [57]. Curcumin is shown to interfere with multiple cell signaling pathways of NF-κB, TNFα, IL-6, IL-1, COX-2, and 5-LO, in addition to the activation of caspases and down-regulation of antiapoptotic gene products.

From 1815 to mid-2009, 728 curcumin analogues were reported, and they were reviewed by Agrawal and Mishra [58] for various pharmacological activities in various models and various cell lines. Some have shown antioxidant, anti-HIV, antimutagenic, antiangiogenic, antimalarial, antitubercular, anti-androgenic, and COX inhibitory activities. Several analogues have shown very potent results and may be considered as clinical candidates for the development of future therapeutics.

Genistein (7), is a PTK inhibitor but also inhibits the generation of NO and the induction of iNOS in murine macrophages [59]. Quercetin (8), epigallocatechin gallate (9), morin (10), apigenin (11), taxifolin (12), fisetin (13), and catechin (14) are shown to attenuate NO production in C6 astrocyte cell cultures [60].

3.3 Terpenes

Terpenes are natural hydrocarbons with isoprene as building blocks and form a key component of essential oils. Terpenes contain less than five units of original carbon skeletons and can be grouped into the first three of the following list:

- monoterpenes (two units, for example, menthol),
- sesquiterpenes (three units, for example, zingiberene),
- diterpenes (four units, for example, retinol),
- triterpenes (six units, for example, squalene), and
- tetraterpenes (eight units, for example, β-carotene).

Terpenoids, also called isoprenoids, have more than five isoprene units and are a large and diverse class of naturally occurring molecules similar to terpenes, but often with altered carbon skeletons (for example, in tetracyclic or pentacyclic) and oxygenated in many ways by nature.

Tetracyclic and pentacyclic terpenoids with significant antiinflammatory properties include ginsenosides Rb1 (15) and Rb2 (16) from Panax ginseng, triperine (17) from *Triperygium wilfordii*, 11-keto-β-boswellic acid (18), and 3-*O*-acetyl-11-keto-β-boswellic acid (19) from *Boswellia serrata*.

Ginsenosides Rb1 and Rb2 inhibit TNFα production in RAW 264.7 cells stimulated with LPS, with mean IC_{50} values 56.5 and 27.5 μM, respectively

[61]. Tripterine, a pentacyclic triterpene isolated from *T. wilfordii* inhibits the production of IL-1 with a mean IC_{50} of 40 nM in human monocytes stimulated by LPS [62]. *T. wilfordii* is one of nine vine plants used in the traditional Chinese medicine to treat inflammatory conditions and studied for the inhibition of isolated enzymes COX-1, COX-2, PLA_2, 5-LO, and 12-LO [63]. Root extracts of *T. wilfordii* show the best inhibition of COX-1, COX-2, and 5-LO with moderate inhibition of PLA_2 and 12-LO among the nine vine extracts.

Two bosswellic acids (18, 19) have been chosen in the monograph of Indian frankincense in European Pharmacopoiea 6.0 as markers to ensure the quality of the air-dried gum resin exudate of *B. serrata*. In addition to inhibiting 5-LO, these two bosswellic acids strongly inhibits two additional molecular targets – microsomal prostaglandin E synthase-1 (mPGES-1) and the serine protease cathepsin G [64].

3.4 Alkaloids

Alkaloids are secondary metabolites containing at least a nitrogen, found primarily in plants for examples nicotine from tobacco and morphine from opium poppy, with very extremely variable structures of various carbon skeletons and nitrogen locations, which also form the basis for classification.

In addition to pyridine alkaloids (for example, nicotine), other medicinally important groups include:

- purine alkaloids (for example, caffeine and theophylline),
- tropane alkaloids (for example, atropine and scopolamine, for an account of chemotaxonomy, see Griffin and Lin [65]),
- indole alkaloids (for example, ergotamine and vincristine),
- quinoline alkaloids (for example, quinine) and
- isoquinoline alkaloids (for example, morphine, galanthamine, tetrandrine, and fangchinoline).

Galanthamine (20), a representative alkaloid isolated from amaryllidaceae, has been approved by FDA for the treatment of mild to moderate Alzheimer's Disease and various other memory and neurocognitive impairments [66]. The cholinesterase inhibitor galantamine is also known to suppress excessive proinflammatory cytokine release from high-fat diet, obese C57BL/6J mice by significantly reducing the plasma level of IL-6 [67]. Similarly, in lipopolysaccharide-induced peritonitis rat models, circulating TNF-α concentration is significantly reduced [68].

Tetrandrine (21) and fangchinoline (22) are the major alkaloids from the Chinese herbal medicine *Stephania tetrandrae* S. Moore. In an in vivo model of mouse ear oedema induced by croton oil, 4 μM of fangchinoline and 6 μM of tetrandrine showed 63% and 86% of inhibitions on IL-6 activity, respectively [69].

Staurosporine (23), a microbial alkaloid isolated from *Streptomyces staurosporeus* inhibits both IKKα and IKKβ in a dose-dependent manner, with an apparent IC_{50} of 0.85 and 1.6 μM for IKKα and IKKβ, respectively [70]. The potent inhibition of IKKα and IKKβ by staurosporine is consistent with its potent inhibition of NF-κB activation in THP-1 monocytic cells stimulated by LPS.

3.5 Others

Diosgenin (24) from the Mexican Yam (*Dioscorea floribunda*) is an example of plant-derived steroidal antiinflammatory drugs and is also largely used as starting materials for the synthesis of many steroidal drugs.

Bromelain, a protein extract of pineapple (*Ananas comosus*), which has been used as a folk medicine to cure various ailments by the aboriginal inhabitants of the Central and South America for centuries, directly inhibited the biosynthesis of prostaglandin E_2 in rats [71]. It is understood that bromelain contains protease enzymes and interferes with the growth of malignant cells, inhibiting platelet aggregation, and thus exerting antiinflammatory action.

In a report by the authors of this book chapter [72], 58 Australian plants and 41 Chinese plants are identified to have potential antiinflammatory activity from cross-referencing 284 Australian bush medicines and 882 Chinese herbal medicines. By further comparing the ethnobotany, ethnopharmacology and phytochemistry of the selected Australian plants with the selected Chinese plants, 14 Australian plants are selected and 13 were collected and studied for inhibitory activities against isolated COXs and LOs. Extracts from 11 of 13 plants (85%) exhibited moderate to high inhibition of COX-1 [73]. Racemosic acid (25) is isolated as a new compound with antiinflammatory activity from *Ficus racemose* [74]. Triterpene-fatty acid esters inhibiting COX-1, COX-2, 5-LO, and PLA_2 are identified in *Tinospora smilacina.* [75].

4. ANTIINFLAMMATORY NATURAL PRODUCTS WITH MULTIPLE TARGETS

4.1 Introduction

Despite great progress within medical research during the past decades, the treatment of inflammatory diseases remains problematic. This is in part, as we have seen in the discussion so far, that pathophysiology of inflammation is defined by the interplay of vast inducers, mediators and effectors in a network of proinflammatory, immunomodulatory, and antiinflammatory pathways.

In principle, drugs need only to attack the cause of the inflammation to solve the problem. Unfortunately, in many cases the cause of inflammation is complex and poorly understood (for instance, the cause of rheumatoid arthritis). It is therefore difficult, at least in the near future, to envisage specific treatments for these conditions.

Obviously classical "one target one disease" approach has its limitation, particularly for complex conditions like inflammation as has discussed above. One direct approach to overcome this limitation is the combination of several drugs that can simultaneously impact multiple targets and this has emerged as a better approach at controlling complex disease systems, with increased efficacy and less prone to drug resistance or side effects [76].

Another approach is to use one molecule that can hit several targets of a complex disease, for example, one compound that inhibits both COX and LO, a duel inhibition concept for suppressing inflammation [77].

Rational design by identifying novel multi-target mechanisms of combined drugs and then developing one molecule that can hit multiple targets is now becoming popular for inflammation [78] and Alzheimer disease [79], just to mention these two complex diseases.

Our mother nature has created many examples of one molecule with a talent of hitting multiple inflammatory mediators or targets, and yet displaying superior safety. This group of unique natural products includes cucurmin (6), (−)-epigallocatechin-3-gallate (EGCG) (9), trans-resveratrol (26), quercetin (8), and racemosic acid (24). Their major molecular targets relevant to common inflammation and their antiinflammatory concentrations are summarized in Table 10.1.

4.2 Curcumin

Curcumin (6), the major yellow pigment in turmeric (*Curcuma longa*), is known for its antiinflammatory, antioxidant, and anticancer properties. Importantly, curcumin has excellent safety profile and lacks side effects in human clinical trials even at daily dosages of 8−12 g [57]. Curcumin acts on many direct molecular targets of inflammation mediators (Table 10.1) including:

- $cPLA_2$: incubation of curcumin with LPS-stimulated RAW264.7 cells decreases the level of phospho-$cPLA_2$ with increasing incubation time and increasing cucurmin concentrations. The significant inhibition is observed at low 5 μM of curcumin [80].
- COX-1 and COX-2: curcumin inhibits isolated ovine COX-1 with an IC_{50} of 50 μM and inhibits COX-2 with 23% at 50 μM [80].
- 5-LO: curcumin inhibits human recombinant 5-LO with an IC_{50} of 0.7 μM [80].
- mPGES-1: curcumin suppresses microsomal prostaglandin E_2 synthase at an IC_{50} of 0.3 μM [81,82]. More detail is discussed below.
- IKK: curcumin suppresses the nuclear translocation of NF-κB signalling pathway by directly blocking the catalytic activity of IKK with an IC_{50} value of 13 μM in TNFα-induced IκB phosphorylation in A549 cells and inhibits recombinant IκB kinase-β with an IC_{50} of 20 μM [83].

TABLE 10.1 Antiinflammatory Natural Products Hitting Multiple Targets

Natural Products	Targets	Results (IC_{50}) or as Stated	References
Cucurmin	$cPLA_2$ (LPS-stimulated RAW264.7 cells)	5 μM (significant)	[80]
	COX-1 (ovine)	50 μM	
	COX-2 (ovine)	50 μM (23%)	
	5-LO (hRecombinant)	0.7 μM	
	mPGES-1 (IL-1β–stimulated A549 lung carcinoma cells)	0.3 μM	[82]
	Nuclear factor (NF)-κB signaling (TNF-α–induced IκB phosphorylation in A549 cells)	13 μM	[83]
	I*k*B kinase-β (recombinant)	20 μM	[83]
EGCG	COX (microsomes from human colon mucosa)	50 μM (42%)	[86]
	• Normal	65 μM (51%)	
	• Tumour	65 μM (40%)	
	5-LO (cytosol from normal human colon mucosa)	50 μM (41%) 65 μM (68%)	[86]
	mPGES-1 (isolated)	1.8 μM	[87]
	NF-κB signaling (LPS-activated mouse peritoneal macrophages)	5 μM (significant)	[88]
trans-Resveratrol	COX-1	15 μM	[91]
	COX-2	85 μM	
	5-LO	2.7 μM	[92]
	IκB kinase	30 μM (significant)	[93] [94]
	iNOs	3 μM (significant)	
Quercetin	iNOs	7.6 μM	[95]
	IκB kinase α	11 μM	
	IκB kinase β	4 μM	
Racemosic acid	COX-1 (ovine)	90 μM	[74]
	5-LO (hRecombinant)	18 μM	

COX, cyclooxygenase; *EGCG*, (−)-Epigallocatechin-3-gallate; *iNOS*, inducible nitric oxide synthase; *LO*, lipoxygenas; *mPGES-1*, microsomal prostaglandin E2synthase 1; *NFκB*, nuclear factor κB; *PLA2*, phospholipase A2.

We know that PGE_2 is a key mediator of proinflammatory prostaglandins for inflammation, fever and pain, and a key metabolite in AA, which is first converted to the peroxide PGH_2 by COXs and then isomerized by PGE_2 synthases (Fig. 10.1). There are three terminal isoforms of PGE_2 synthases: (1). microsomal prostaglandin E synthase (mPGES)-1, (2). cytosolic PGES (cPGES), and (3). mPGES-2, which have been identified so far.

We know that mPGES-1 is functionally coupled to COX-2 and responsible for excessive PGE_2 generation connected to pathologies like inflammation, fever, pain, atherosclerosis and tumorigenesis. Though the consequences of redirection of the COX product and mPGES-1 substrate PGH_2 to other PG synthases following inhibition of mPGES-1 are still a matter for debate, selective inhibition of PGE_2 generation by mPGES-1is considered superior to the overall inhibition of PGs by COX inhibitors, as COXs contribute to homeostatic functions and its inhibition can cause significant side effects [81]. Thus selective suppression of PGE_2 formation via inhibiting mPGES-1 has been considered as an alternative strategy to develop antiinflammatory drugs. Curcumin is a best-known natural product with the most favorite dual inhibition of both prostaglandins (e.g., by inhibiting mPGES-1) and LTs (by inhibiting 5-LO) [81].

Hence multi-target drugs combining potent inhibition of mPGES-1 and 5-LO with moderate inhibition of either additional PG synthases or COX-1 (to buffer redirection to other PGs than PGE_2) might be preferable. Such a pharmacological profile has been demonstrated in two examples of the dual inhibitor of PG and LT biosynthesis licofelone [27,28], and for the natural product curcumin (Table 10.1).

Interestingly, curcumin in a dose-response manner reduces the production of amyloid-β_{42}-induced IL-1β, IL-6, and TNF-α in mRNA and protein levels in microglia [84]. Curcumin also exerts an inhibitory effect on phosphorylation of ERK1/2 and P38, and blockage of ERK1/2 and p38 pathways reduces inflammatory cytokines production from microglia. So curcumin may be useful for Alzheimer's disease.

4.3 (−)-Epigallocatechin-3-gallate

EGCG (9) has gained much attention as an active principle of green tea against inflammation and inflammation-associated disorders like cancer, cardiovascular disease and neurodegeneration [85]. EGCG acts on many direct molecular targets of inflammation mediators (Table 10.1) including:

- COX: in microsomes from human colon mucosa, 50 and 65 μM of EGCG inhibited *normal* mucosa COX at 42% and 51%, respectively; 65 μM of EGCG inhibited *tumor* mucosa COX at 40% [86].
- LO: in cytosol from normal human colon mucosa, 50 and 65 μM of EGCG exhibit an inhibition of at 41% and 68%, respectively [86].
- mPGES-1: $IC_{50} = 1.8$ μM, this concentration can be reached in plasma after oral administration. This inhibition on mPGES-1 is much more significant than inhibiting isolated COX-1 ($IC_{50} > 30$ μM) and isolated COX-2 and $cPLA_2$ (not inhibition up to 30 μM) [87].
- IκB kinase: EGCG at 5 μM significantly blocks disappearance of inhibitor κB from cytosolic fraction by preventing IKK action and inhibits the activation of NF-κB, a transcription factor necessary for iNOS induction,

as evidenced by the amount of nitrite released into the culture medium in peritoneal macrophage cells co-treated with EGCG (5, 10 and 15 μM) and LPS (5 μM) for 6-24 hour [88].

Another study [89] directly assessed IKK activity by detecting changes in phosphorylation of an IκBα–glutathione *S*-transferase (GST) fusion protein. Using IEC-6 cells pretreated with various green tea polyphenols (0–0.4 mg/mL), TNFα-induced IKK and NF-κB activity diminished. EGCG was the most potent inhibitor. In cytosolic extracts of TNFα-stimulated cells, EGCG inhibited phosphorylation of IκBα-GST with $IC_{50} > 18$ μM consistent with inhibition of IKK activity. Also, the gallate group was essential for inhibition, and antioxidants were ineffective in blocking activated IKK. Further, EGCG decreased IKK activity in cytosolic extracts of NF-κB–inducing kinase (NIK) transiently transfected cells indicating the findings were not related to nonspecific kinase activity. EGCG is an effective inhibitor of IKK activity. This may explain why green tea extracts and EGCG exhibit antiinflammatory and anticancer properties.

4.4 *trans*-Resveratrol

trans-Resveratrol (trans-3,4′,5-trihydroxystilbene) is a stilbenoid found in grapes and red wine and has attracted much attention due to its multiple health benefits for inflammation conditions and inflammation-associated disorders like cancer, cardiovascular disease, type 2 diabetes, and neurodegeneration [90].

trans-Resveratrol (resveratrol in short) acts on many direct molecular targets of inflammation mediators (Table 10.1) including:

- COX-1 and COX-2: resveratrol inhibits COX-1 activity of microsomes derived from sheep seminal vesicles at a median effective concentration of 15 μM and inhibits human recombinant COX-2 at a median effective concentration of 85 μM [91].
- 5-LO: Resveratrol at 2.7 μM (IC_{50}) inhibits the formation of 5-HETE catalyzed by rat peritoneal polymorphonuclear leukocyte LO [92].

Resveratrol is a potent inhibitor of both NF-κB activation and NF-κB–dependent gene expression through its ability to inhibit IκB kinase activity. The effective dose of resveratrol for inhibiting activation of NF-κB is 30 μM in macrophage/monocytic cell lines THP-1 and U937, stimulated by LPS. Resveratrol inhibits NF-κB–dependent gene expression through the inhibition of induction of NF-κB DNA binding activity [93]. Also resveratrol is a potent inhibitor of inducible IKK activity in response to TNFα exposure of THP1 cells, but is not an intrinsic IKK inhibitor because it does not block IKK activity when added directly to the in vitro kinase reaction, and treatment of THP-1 cells with 17β-estradiol at concentrations as high as 10^{-5} M or 10 μM does not lead to inhibition of NF-κB. Thus, reseratrol blocks NF-κB activity through the inhibition of induction of IKK activity.

Resveratrol (3–30 μM) significantly inhibits NO generation in macrophages RAW 264.7 cells activated by bacterial lipopolysaccharide (LPS, 50 ng/mL), as measured by the amount of nitrite released into the culture medium, and strongly reduces the amount of cytosolic iNOS protein and steady state mRNA levels [94]. Resveratrol (30 μM) also inhibits the activation of NF-κB induced by LPS for 1 h using an electrophoretic mobility shift assay. Furthermore, in immunoblotting analysis, cells treated with LPS plus resveratrol shows an inhibition of phosphorylation as well as degradation of IκBα, and a reduced nuclear content of NF-κB subunits.

Resvesatrol not only inhibits COX and LO in isolated enzymes, but also suppresses the iNOS expression and NO production by down-regulation of NF-*k*B binding activity via blockade of I*k*Bα degradation in cultured cells.

4.5 Quercetin

Quercetin gains much attention as active principles of vegetables and fruits against inflammation and inflammation-associated disorders like cancer, cardiovascular disease, and neurodegeneration [51]. Quercetin acts on several molecular targets of inflammation mediators (Table 10.1).

Quercetin at 3–100 μM (with 1% ethanol) suppressed iNOS gene expression and NO production (IC_{50} 7.6 μM), as determined by reverse transcription–polymerase chain reaction and nitrite assay in the murine macrophage cell line RAW 264.7. Stimulation of the cells with lipopolysaccharide and interferon-γ leads to expression of the iNOS gene and production of NO [95]. Interestingly, quercetin has a synergic effect with ethanol as evidenced by titrating from 0.75% to 0.1% of ethanol with a fixed concentration of 30 μM quercetin (concentration selected at the log phase of the inhibition curve). When ethanol is at 0.1%, nitrite production is reduced to 58% of the control (cells stimulated with LPS and IFNγ only). When the ethanol content is increased to 0.25%, 0.50%, and 0.75% in the culture, in a concentration-dependent manner, the degree of nitrite production is reduced further to 38%, 30.7%, and 20.7%, respectively.

Quercetin inhibits both IKKα and IKKβ with an apparent IC_{50} value of 11 and 4 μM, respectively [70]. Both IKKα and IKKβ contribute to the activity of the IKK complex and are involved in NF-κB activation.

4.6 Racemosic Acid

Racemosic acid (25), a glycoside isolated Australian bush medicine *Ficus racemosa* (family Moraceae), shows potent inhibitory activities against COX-1 (ovine) and 5-LO (human recombinant) in vitro with IC_{50} values of 90 and 18 μM, respectively [74]. The results of this study provide a scientific explanation for the use of this plant species in Australian indigenous medicine and the traditional system of medicine in other countries, for its use in the treatment of inflammatory diseases.

5-Lipoxygenase is involved in the synthesis of nonresolving inflammatory leukotriene B_4, leukotriene C_4, leukotriene D_4 and leukotriene E_4 but also in the generation of pro-resolving antiinflammatory 15-HETE and lipoxins. (Fig. 10.1). In assessing the efficacy of LT synthesis inhibitors alone and in combination with an NSAID using a mouse model of chronic arthritis, Nickerson-Nutter and Medvedeff [96] observe no inhibition of arthritis with any compound administered alone, but a significant reduction of arthritis animals when a concomitant administration of a LT synthesis inhibitor and an NSAID is used. This clearly demonstrates that inhibitors of COX and 5-LO in combination proved to be more effective than either class of drugs used alone. The dual inhibition of the pathways involved in AA metabolism represents a strong rationale for the development of a new class of agents [77].

From the examples of racemosic acid, quercetin, *trans*-resveratrol, EGCG, and curcumin discussed above, one can see that safe and multitargeting natural products can serve as effective starting points for multitarget drug discovery to tap into the unsatisfactory antiinflammatory drugs and unmet markets for doctors and patients.

5. CONCLUSION

This chapter begins with the discussion of the common inflammation processes and mediators and then moves to a survey of potential natural product lead compounds on targeting the inflammation processes or mediators for the discovery and development of antiinflammatory agents. One can be convinced that natural products particularly phenolics, terpenes and alkaloids play an indispensable role as lead compounds for targeting the inflammatory processes/mediators.

Cucurmin, EGCG, *trans*-resveratrol, quercetin and racemosic acid are good examples of antiinflammatory natural products with multiple targets supporting the current drug discovery trend of "one molecule multiple target several diseases," emerging from the tradition of "one molecule one target one disease."

Mother nature has created these unique molecules, each possessing bioactivities against multiple molecular targets including COXs, LOs, mPGES-1 and NF-κB signaling, all of which are suitable and effective as starting points for discovering single molecule with multiple targets.

Inhibition of mediators involved in the eicosanoid biosynthesis dominates among antiinflammatory natural products with multiple targets. This preference is not surprising because eicosanoids are powerful lipid mediators with pleiotropic, often opposing, activities [28]. These inhibitors combine proinflammatory, antiinflammatory, immunomodulatory, and homeostatic properties depending on their structures, concentrations and responsive tissues, on which they act. Lipids originally known as a cell membrane and energy storing molecules, are now recognized as potent signaling molecules via receptor-mediated pathways related to cell growth and death, inflammation, homeostasis, etc. These signaling molecules include lipids derived from:

- AA: prostanoids, LTs, 5-oxo-6,8,11,14-eicosatetraenoic acid, lipoxins, and epoxyeicosatrienoic acids;
- eicosapentaenoic acid (EPA): E-series resolvins; and
- docosahexaenoic acid (DHA): D-series resolvins, protectins, and maresins.

Side-effects might result from interfering with off-targets but also through compensatory mechanisms and others. Nonselective COX-1 and COX-2 inhibitors, for example, induce gastrointestinal complications, due to the general suppression of COX-1–derived cytoprotective PGE_2 and prostacyclin (PGI_2) in gastroduodenal epithelium, and due to an imbalance between PGs and LTs as a result of redirecting AA from PG to LT biosynthesis.

As safe alternatives to NSAIDs, the dual COX and 5-LO inhibitors may improve gastrointestinal safety profile compared with pure COX blockers alone. Dual inhibitors of proinflammatory PGE_2 synthesis (through mPGES-1) and 5-LO are also promising.

Natural products have a place as valuable NSAIDs but also versatile leads for multi-target drug designs. Our knowledge of the molecular targets of natural products including salicylate, curcumin, catechins, resveratrol, quercetin, and racemosic acid are expanding and would facilitate research and development of multitarget natural products for antiinflammatory therapies.

As Csermely and coworkers [97] advocated, systematic drug design strategies should be directed against multiple targets and the final effect of partial, but multiple, drug actions, which might often surpass that of complete drug action at a single target. As we discussed so far, natural products alone, or in mixture and even in crude extract can be a case for this approach. The future success to show case will depend on:

- quicker computer to identify the correct multiple targets and their multi-fitting, low-affinity drug candidates,
- more-efficient in vivo testing, and
- faster LC-MS techniques to identify and characterise hundreds of active principles and metabolites.

It is anticipated that valuable multitarget leads for structural optimization are often found among natural products with proven preclinical and clinical antiinflammatory potential [98]. High-throughput-compatible chromatographic and mass spectrometric techniques could provide access to complex metabolic profiles, which can lead to a boost in the development of dynamic network models to predict efficacy and safety of multi-target drugs.

REFERENCES

[1] R. Medzhitov, Origin and physiological roles of inflammation, Nature 454 (2008) 428–435.

[2] M. Hansen, Pathophysiology: Foundations of Disease and Clinical Intervention, W.B. Saunders Company, Philadelphia, 1998.

[3] P.W. Baumert, Acute inflammation after injury, Postgrad. Med. 97 (1995) 35–49.

[4] G.M. Rubanyi, The role of endothelium in cardiovascular homeostasis and diseases, J. Cardiovasc. Pharmacol. 22 (Suppl. 4) (1993) S1–S14.

[5] R. Raghow, The role of extracellular matrix in postinflammaory wound healing and fibrosis,, FASEB J. 8 (1994) 823–831.

[6] S.M. Albelda, C.W. Smith, P.A. Ward, Adhesion molecules and inflammatory injury, FASEB J. 8 (1994) 504–512.

[7] R.C. Bone, Immunologic dissonance: a continuing evolution in our understanding of the systemic inflammatory response syndrome (SIRS) and the multiple organ dysfunction syndrome (MODS), Ann. Intern. Med. 125 (1996) 680–687.

[8] T. Lewis, The Blood Vessels of the Human Skin and Their Responses, Shaw and Sons, London, 1927.

[9] J.R. Vane, Inhibition of prostaglandin synthesis as a mechanism of action for aspirin-like drugs, Nat. New Biol. 231 (1971) 232–235.

[10] S.H. Ferreira, S. Moncada, J.R. Vane, Indomethacin and aspirin abolish prostaglandin release from the spleen, Nat. New Biol. 231 (1971) 237–239.

[11] S.H. Ferreira, S. Moncada, J.R. Vane, Further experiments to establish that the analgesic action of aspirin-like drugs depends on the inhibition of prostaglandin biosynthesis, Br. J. Pharmacol. 47 (1973) 629P–630P.

[12] S. Moncada, S.H. Ferreira, R.J. Vane, Inhibition of prostaglandin biosynthesis as the mechanism of analgesia of aspirin-like drugs in the dog knee joint, Eur. J. Pharmacol. 31 (1975) 250–260.

[13] J.R. Vane, R. Botting, Mechanism of action of anti-inflammatory drugs, FASEB J. 1 (1987) 89–96.

[14] J.R. Burke, K.R. Gregor, R.A. Padmanabha, A b-lactam inhibitor of cytosolic phospholipase A_2 which acts in a competitive, reversible manner at the lipid/water interface,, J. Enzyme Inhib. 13 (1998) 195–206.

[15] L.J. Reynolds, L.L. Hughes, E.A. Dennis, Analysis of human synovial fluid phospholipase A_2 on short chain phosphatidylcholine-mixed micelles: development of a spectrophotometric assay suitable for a microtiterplate reader, Anal. Biochem. 204 (1992) 190–197.

[16] T. Nakamura, Y. Morita, M. Kuriyama, K. Ishihara, T. Miyamoto, Platelet-activating factor in late asthmatic response, Int. Archive Allergy Appl. Immunol. 82 (1987) 57–61.

[17] P. Braquet, D. Hosford, Ethnopharmacology and the development of natural PAF antagonists as therapeutic agents, J. Ethnopharmacol. 32 (1991) 135–139.

[18] M. Hamberg, J. Svensson, B. Samuelsson, Thromboxanes: a new group of biologically active compounds derived from prostaglandin endoperoxides, Proc. Natl. Acad. Sci. USA 72 (1975) 2994–2998.

[19] B. Samuelsson, Introduction of a nomenclature: leukotrienes, Prostaglandins 17 (1979) 785–787.

[20] S. Moncada, R. Gryglewski, S. Bunting, J.R. Vane, An enzyme isolated from arteries transforms prostaglandin endopreoxides to an unstable substance that inhibit platelet aggregation, Nature 263 (1976) 663–665.

[21] W.L. Smith, Prostanoid biosynthesis and mechanisms of action, Am. J. Physiol. 263 (1992) 181–191.

[22] K. Seibert, Z. Yan, K. Leahy, Pharmacological and biochemical demonstration of the role of cyclooxygenase 2 in inflammation and pain, Proc. Natl. Acad. Sci. USA 91 (1994) 12013–12017.

[23] J.Y. Jouzeau, B. Terlain, A. Abid, Cyclo-oxygenase isoenzymes: how recent findings affect thinking about nonsteroidal anti-inflammatory drugs, Drugs 55 (1997) 563–582.

[24] D.E. Griswold, J.L. Adams, Constitutive cyclooxygenase (COX-1) and inducible cyclooxygenase (COX-2): rationale for selective inhibition and progress to date, Med. Res. Rev. 16 (1996) 181–206.

[25] S. Yamamoto, Mammalian lipoxygenase: molecular structures and functions, Biochim. Biophys. Acta 1128 (1992) 117–131.

[26] E.S. Silverman, J.M. Drazen, The biology of 5-lipoxygenase: function, structure, and regulatory mechanism, Proc. Assoc. Am. Phys. 111 (1999) 525–536.

[27] B. Samuelsson, S.E. Dahlén, J.A. Lindgren, C.A. Rouzer, C.N. Serhan, Leukotrienes and lipoxins: structures, biosynthesis, and biological effects, Science 237 (1987) 1171–1176.

[28] M.J. Stables, D.W. Gilroy, Old and new generation lipid mediators in acute inflammation and resolution, Prog. Lipid Res. 50 (2011) 35–51.

[29] P.J. Barnes, M. Karin, Nuclear factor-κB — a pivotal transcription factor in chronic inflammatory diseases, N. Engl. J. Med. 336 (1997) 1066–1071.

[30] B.P. Bremner, M. Heinrich, Natural products as targeted modulators of the nuclear factor-kappaB pathway, J. Pharm. Pharmacol. 54 (2002) 453–472.

[31] C.J. Lowenstein, S.H. Snyder, Nitric oxide, a novel biologic messenger, Cell 70 (1992) 705–707.

[32] S. Moncada, A. Higgs, The L-arginine-nitric oxide pathway, N. Engl. J. Med. 329 (1993) 2002–2012.

[33] J.B. Weinberg, Nitric oxide synthase 2 and cyclooxygenase 2 interactions in inflammation, Immunol. Res. 22 (2000) 319–341.

[34] C.A. Dinarello, Proinflammatory cytokines, Chest 118 (2000) 503–508.

[35] S.M. Opal, V.A. DePalo, Anti-inflammatory cytokines, Chest 117 (2000) 1162–1172.

[36] J.J. Haddad, Cytokines and related receptor-mediated signaling pathways, Biochem. Biophys. Res. Commun. 297 (2002) 700–713.

[37] K.D. Rainsford, Anti-inflammatory drugs in the 21st century, Subcell. Biochem. 42 (2007) 3–27.

[38] E.A. Meade, L.S. William, D.L. DeWitt, Differential inhibition of prostaglandin endoperoxide synthase (cyclooxygenase) isozymes by aspirin and other non-steroidal anti-inflammatory drugs, J. Biol. Chem. 268 (1993) 6610–6614.

[39] J.L. Masferrer, B.S. Zweifel, P.T. Manning, S.D. Hauser, K.M. Leahy, W.G. Smith, P.C. Isakson, K. Seibert, Selective inhibition of inducible cyclooxygenase 2 in vivo is anti-inflammatory and nonulcerogenic, Proc. Natl. Acad. Sci. USA 91 (1994) 3228–3232.

[40] W. Xie, D.L. Robertson, D.L. Simmons, Mitogen-inducible prostaglandin G/H synthase: a new target for nonsteroidal anti-inflammatory drugs, Drug Dev. Res. 25 (1992) 249–265.

[41] P. MaGettigan, D. Henry, Cardiovascular risk and inhibition of cycloxydenase: a systematic review of the observational studies of selective and nonselective inhibitors of cyclooxygenase-2, JAMA 296 (2006) 1633–1644.

[42] E.T. Andreakos, B.M. Foxwell, F.M. Brennan, R.N. Maini, M. Feldmann, Cytokines and anti-cytokine biologicals in autoimmunity: present and future, Cytokine Growth Factor Rev. 13 (2002) 299–313.

[43] G. Cavalli, C.A. Dinarello, Treating rheumatological diseases and co-morbidities with interleukin-1 blocking therapies, Rheumatol. Oxf. 54 (2015) 2134–2144.

[44] N. Ghosh, R. Chaki, V. Mandal, G.D. Lin, S.C. Mandal, Mechanisms and efficacy of immunobiologic therapies for inflammatory bowel diseases, Int. Rev. Immunol. 29 (2010) 4–37.

[45] L.M. Jarvis, The year in new drugs, Chem. Eng. News 95 (2017) 28–32.

[46] R. Amann, B.A. Peskar, Anti-inflammatory effects of aspirin and sodium salicylate, Eur. J. Pharmacol. 447 (2002) 1–9.

[47] M.J. Yin, Y. Yamamoto, R.B. Gaynor, The anti-inflammatory agents aspirin and salicylate inhibit the activity of IκB kinase-β, Nature 396 (1998) 77–80.

[48] E. Kopp, S. Ghosh, Inhibition of NF-kappa B by sodium salicylate and aspirin, Science 265 (1994) 956–959.

[49] E. Stone, An account of the success of the bark of the willow in the cure of agues. In a letter to the right honourable George Earl of Macclesfield, President of R. S. from the Rev. Mr. Edmund Stone, of Chipping-Norton in Oxfordshire, Philos. Trans. (1683–1775) 53 (1763) 195–200.

[50] H. Dreser, Pharmakologisches über aspirin (acetyl-salicy-saüre), Pfluegers Arch. 76 (1899) 306–318.

[51] J.D. Phillipson, L.A. Anderson, Ethnopharmacology and western medicine, J. Ethnopharmacol. 25 (1989) 62–73.

[52] N.R. Farnsworth, Ethnopharmacology and future drug development: the North America experience, J. Ethnopharmacol. 38 (1993) 145–152.

[53] Y. Wang, Needs for new plant-derived pharmaceuticals in the post-genome era: an industrial view in drug research and development, Phytochem. Rev. 7 (2008) 395–406.

[54] J.B. Calixto, M.F. Otuki, A.R.S. Santos, Anti-inflammatory compounds of plant origin. Part I. Action on arachidonic acid pathway, nitric oxide and nuclear factor κ B (NF-κB), Planta Med. 69 (2003) 973–983.

[55] J.B. Calixto, M.M. Campos, M.F. Otuki, A.R.S. Santos, Anti-inflammatory compounds of plant origin. Part II. Modulation of pro-inflammatory cytokines, chemokines and adhesion molecules, Planta Med. 70 (2004) 93–103.

[56] N.P. Seeram, H. Ichikawa, S. Shishodia, B.B. Aggarwal, Preventive and therapeutic effects of plant polyphenols through suppression of nuclear factor-kappa B, in: T.Y.M. Hiramatsu, L. Packer (Eds.), Molecular Interventions in Lifestyle-related Diseases, CRC Press LLC, Boca Raton, 2005, pp. 243–275.

[57] T. Esatbeyoglu, P. Huebbe, I.M.A. Ernst, D. Chin, A.E. Wagner, G. Rimbach, Curcumin—from molecule to biological function, Angew. Chem. Int. Ed. 51 (2012) 5308–5332.

[58] D.K. Agrawal, P.K. Mishra, Curcumin and its analogues: potential anticancer agents, Med. Res. Rev. 30 (2010) 818–860.

[59] X.Q.Z. Dong, K. Xie, I.J. Fidler, Protein tyrosine kinase inhibitors decrease induction of Nitric oxide synthase activity in lipopolysaccharide-responsive and lipopolysaccharidenonresponsive murine macrophages, J. Immunol. 151 (1993) 2714–2717.

[60] K.F. Soliman, E.A. Mazzio, In vitro attenuation of nitric oxide production in C6 astrocyte cell culture by various dietary compounds, Proc. Exp. Biol. Med. 218 (1998) 390–397.

[61] J.Y. Cho, E.S. Yoo, K.U. Baik, M.H. Park, B.H. Han, In vitro inhibitory effect of protopanaxadiol ginsenosides on tumor necrosis factor (TNF)-alpha production and its modulation by known TNF-alpha antagonists, Planta Med. 67 (2001) 13–18.

[62] F.C. Huang, W.K. Chan, K.J. Moriarty, D.C. Zhang, M. Chang, W. He, K.T. Yu, A. Zilberstein, Novel cytokine release inhibitors. Part I: triterpenes, Bioorg. Med. Chem. Lett. 8 (1998) 1883–1886.

[63] R.W. Li, G.D. Lin, S.P. Myers, D.N. Leach, Anti-inflammatory activity of Chinese medicinal vine plants, J. Ethnopharmacol. 85 (2003) 61–67.

[64] O.W.M. Abdel-Tawab, M. Schubert-Zsilavecz, *Boswellia serrata*: an overall assessment of in vitro, preclinical, pharmacokinetic and clinical data, Clin. Pharmacokinet. 50 (2011) 349–369.

[65] W.J. Griffin, G.D. Lin, Chemotaxonomy and geographical distribution of tropane alkaloids, Phytochemistry 53 (2000) 623–637.

[66] R.H. Howland, Alternative drug therapies for dementia, J. Psychosoc. Nurs. Ment. Health Serv. 49 (5) (2011) 17–20.

[67] S.K. Satapathy, M. Ochani, M. Dancho, L.K. Hudson, M. Rosas-Ballina, S.I. Valdes-Ferrer, P.S. Olofsson, Y.T. Harris, J. Roth, S. Chavan, K.J. Tracey, V.A. Pavlov, Galantamine alleviates inflammation and other obesity associated complications in high-fat diet–fed mice, Mol. Med. 17 (2011) 599–606.

[68] Z.H. Liu, Y.F. Ma, J.S. Wu, J.X. Gan, S.W. Xu, G.Y. Jiang, Effect of cholinesterase inhibitor galanthamine on circulating tumor necrosis factor alpha in rats with lipopolysaccharide-induced peritonitis, Chin. Med. J. Engl. 123 (2010).

[69] H.S. Choi, H.S. Kim, K.R. Min, Y. Kim, H.K. Lim, Y.K. Chang, M.W. Chung, Anti-inflammatory effects of fangchinoline and tetrandrine, J. Ethnopharmacol. 69 (2000) 173–179.

[70] G.W. Peet, J. Li, IκB kinases α and β show a random sequential kinetic mechanism and are inhibited by staurosporine and quercetin, J. Biol. Chem. 274 (1999) 32655–32661.

[71] V. Cody, E. Middleton, J.B. Harborne (Eds.), Plant Flavonoids and Medicine: Biochemical, Pharmacological and Structure-activity Relationships: Proceedings of a Symposium Held in Buffalo, New York, July 22-26, 1985, vol. 1, Liss, The University of Michigan, 1986.

[72] C.T. Lin, C.J. Chen, T.Y. Lin, J.C. Tung, S.Y. Wang, Anti-inflammation activity of fruit essential oil from *Cinnamomum insularimontanum* Hayata, Bioresour. Technol. 99 (2008) 8783–8787.

[73] R.W. Li, S.P. Myers, D.N. Leach, G.D. Lin, G. Leach, A cross-cultural study: anti-inflammatory activity of Australian and Chinese plants, J. Ethnopharmacol. 85 (2003) 25–32.

[74] R.W. Li, D.N. Leach, S.P. Myers, G.D. Lin, G.J. Leach, P.G. Waterman, A new anti-inflammatory glucoside from *Ficus racemosa* L. Planta Med. 70 (2004) 421–426.

[75] R.W. Li, S.P. Myers, G. Leach, G.D. Lin, D.J. Brushett, P.G. Waterman, Anti-inflammatory activity, cytotoxicity and active compounds of *Tinospora smilacina* Benth, Phytother. Res. 18 (2004) 78–83.

[76] G.R. Zimmermann, J. Lehar, C.T. Keith, Multi-target therapeutics: when the whole is greater than the sum of the parts, Drug Discov. Today 12 (2007) 34–42.

[77] F. Celotti, S. Laufer, Anti-inflammatory drugs: new multitarget compounds to face an old problem. the dual inhibition concept, Pharmacol. Res. 43 (2001) 429–436.

[78] A. Koeberle, O. Werz, Multi-target approach for natural products in inflammation, Drug Discov. Today 19 (2014) 1871–1882.

[79] N. Guzior, A. Wieckowska, D. Pannek, B. Malawska, Recent development of multifunctional agents as potential drug candidates for the treatment of Alzheimer's disease, Curr. Med. Chem. 22 (2014) 373–404.

[80] J. Hong, M. Bose, J. Ju, J.H. Ryu, X. Chen, S. Sang, M.J. Lee, C.S. Yang, Modulation of arachidonic acid metabolism by curcumin and related beta-diketone derivatives: effects on cytosolic phospholipase A(2), cyclooxygenases and 5-lipoxygenase, Carcinogenesis 25 (2004) 1671–1679.

[81] A. Koeberle, O. Werz, Inhibitors of the microsomal prostaglandin E2 Synthase-1 as alternative to non steroidal anti-inflammatory drugs (NSAIDs) a critical review, Curr. Med. Chem. 16 (2009) 4274–4296.

[82] A. Koeberle, O. Werz, Curcumin blocks prostaglandin E2 biosynthesis through direct inhibition of the microsomal prostaglandin E2 synthase-1, Mol. Cancer Ther. 8 (2009) 2348–2355.

[83] Y.D.A.L. Kasinski, S.L. Thomas, J. Zhao, S.Y. Sun, F.R. Khuri, C.Y. Wang, M. Shoji, A. Sun, J.P. Snyder, D. Liotta, H. Fu, Inhibition of IκB kinase-nuclear factor-κB signaling pathway by 3,5-bis(2-flurobenzylidene)piperidin-4-one (EF24), a novel monoketone analog of curcumin, Mol. Pharmacol. 74 (2008) 654–661.

[84] X. Shi, Z. Zheng, J. Li, Z. Xiao, W. Qi, A. Zhang, Q. Wu, Y. Fang, Curcumin inhibits Aβ-induced microglial inflammatory responses in vitro: involvement of ERK1/2 and p38 signaling pathways, Neurosci. Lett. 594 (2015) 105–110.

[85] B.N. Singh, S. Shankar, R.K. Srivastava, Green tea catechin, epigallocatechin-3-gallate (EGCG): mechanisms, perspectives and clinical applications, Biochem. Pharmacol. 82 (2011) 1807–1821.

[86] J. Hong, T.J. Smith, C.T. Ho, D.A. August, C.S. Yang, Effects of purified green and black tea polyphenols on cyclooxygenase- and lipoxygenase-dependent metabolism of arachidonic acid in human colon mucosa and colon tumor tissues, Biochem. Pharmacol. 62 (2001) 1175–1183.

[87] A. Koeberlea, J. Bauera, M. Verhoffa, M. Hoffmannb, H. Northoffc, O. Werza, Green tea epigallocatechin-3-gallate inhibits microsomal prostaglandin E2 synthase-1, Biochem. Biophys. Res. Commun. 388 (2009) 350–354.

[88] Y.L. Lin, J.K. Lin, (-)-epigallocatechin-3-gallate blocks the induction of nitric oxide synthase by down-regulating lipopolysaccharide-induced activity of transcription factor nuclear factor-κB, Mol. Pharm. 52 (1997) 465–472.

[89] F. Yang, H.S. Oz, S. Barve, W.J.S.D. Villiers, C.J. McClain, G.W. Varilek, The green tea polyphenol (-)-epigallocatechin-3-gallate blocks nuclear factor-κB activation by inhibiting IκB kinase activity in the intestinal epithelial cell line IEC-6, Mol. Pharm. (2001).

[90] W. Yu, Y.C. Fu, W. Wang, Cellular and molecular effects of resveratrol in health and disease, J. Cell Biochem. 113 (2012) 752–759.

[91] M. Jang, L. Cai, G.O. Udeani, K.V. Slowing, C.F. Thomas, C.W.W. Beecher, H.H.S. Fong, N.R. Farnsworth, A.D. Kinghorn, R.G. Mehta, R.C. Moon, J.M. Pezzuto, Cancer chemopreventive activity of resveratrol, a natural product derived from grapes, Science 275 (1997) 218–220.

[92] Y. Kimura, H. Okuda, S. Arichi, Effects of stilbenes on arachidonate metabolism in leukocytes, Biochim. Biophys. Acta 834 (1985) 275–278.

[93] M. Holmes-McNary, A.S. Baldwin Jr., Chemopreventive properties of trans-resveratrol are associated with inhibition of activation of the IκB kinase, Cancer Res. 60 (2000) 3477–3483.

[94] S.H. Tsai, S.Y. Lin-Shiau, J.K. Lin, Suppression of nitric oxide synthase and the down-regulation of the activation of NFκB in macrophages by resveratrol, Br. J. Pharmacol. 126 (1999) 673–680.

[95] M.M. Chan, J.A. Mattiacci, H.S. Hwang, A. Shah, D. Fong, Synergy between ethanol and grape polyphenols, quercetin, and resveratrol, in the inhibition of the inducible nitric oxide synthase pathway, Biochem. Pharmacol. 60 (2000) 1539–1548.
[96] C.L. Nickerson-Nutter, E.D. Medvedeff, The effects of leukotriene synthesis inhibitors in models of acute and chronic inflammation,, Arthritis Rheum. 39 (1996) 515–521.
[97] P. Csermely, V. Ágoston, S. Pongor, The efficiency of multi-target drugs: the network approach might help drug design, Trends Pharmacol. Sci. 26 (2005) 178–182.
[98] P. Csermely, T. Korcsmáros, H.J.M. Kissa, G. London, R. Nussinove, Structure and dynamics of molecular net works: a novel paradigm of drug discovery: a comprehensive review, Pharmacol. Ther. 138 (2013) 333–408.

Chapter 11

Biologically Functional Compounds From Mushroom-Forming Fungi

Hirokazu Kawagishi
Shizuoka University, Shizuoka, Japan

There is an expression that states that "plants act as producer, animals as consumer, and fungi as restorer and decomposer." In other words, plants create organic compounds by means of photosynthesis and animals consume such plants. Then fungi, including mushrooms, play an important role in restoring the plants and animals back to the land. There are a number of differences in the structures of metabolic products by mushroom-forming fungi compared to those by plants and animals. These differences sometimes indicate biological activities indigenous to mushroom-forming fungi. This chapter presents some of our studies on biologically functional molecules isolated from mushroom-forming fungi.

1. ANTIDEMENTIA COMPOUNDS

Hericium erinaceus is a well-known edible and medicinal mushroom known in Japan as Yamabushitake, in China as Hou Tou Gu, and in Europe and the United States as Lion's Mane. It showed a number of important bioactivities concerning antidementia such as the promotion of nerve growth factor (NGF) synthesis [1–8] and the reduction of endoplasmic reticulum (ER) stress induced by amyloid β-peptide (Aβ) [9–11].

NGF belongs to a family of neurotrophins that induce survival and proliferation of neurons, and play an important role in the repair, regeneration, and protection of neurons. A remarkable finding was reported: a woman with Alzheimer's dementia improved her symptoms, such as enhancing mental ability, after the administration of NGF directly into her brain using a catheter [12]. It has been suggested by scientists that NGF may be used to treat Alzheimer's disease [13]. However, there is a high risk in such treatment since NGF is a protein that cannot pass through the blood–brain barrier and it

Natural Products and Drug Discovery. https://doi.org/10.1016/B978-0-08-102081-4.00011-3

needs to be injected directly into the brain to be effective. If a compound can be taken by oral administration that can pass through the barrier and stimulate NGF synthesis inside the brain, it may be applied as a safer therapy to prevent the disease. Even if this compound could not go through the barrier, it would be still beneficial to disorders of the peripheral nervous system since NGF has a similar effect on neurons in the system. Based on the foregoing concept, we searched for stimulators of NGF synthesis in nature and found hericenones C to H (**1–6**) from the fruiting bodies from *Hericium erinaceus* in 1991 [1,2]. These compounds were the first NGF stimulators isolated from nature. Later, we also obtained erinacines A to I (**7–15**) from the mycelia of the fungus (Fig. 11.1) [3–7].

Many types of neurodegenerative diseases causing neuronal cell death (associated with Aβ), including Alzheimer's, Parkinson's, Huntington's, and the prion diseases, involve ER stress. 3-Hydroxyhericenone F (**18**) was isolated as an ER stress suppressor from the fruiting bodies of *H. erinaceus* [10]. We also found active compounds (**19–22**) in an extract from the scrap cultivation bed of the mushroom (Fig. 11.2) [11]. The cultivation bed is usually discarded by mushroom growers after harvesting the fruiting bodies. Another purpose of the study was efficient use of scrap cultivation beds. The mushroom *Mycoleptodonoides aitchisonii* also produces suppressors (**23–26**) (Fig. 11.2) [14].

An in vivo study showed the effects of erinacine A on the production of NGF in various regions of the central nervous system of rats [8]. Twenty newborn rats were divided into two groups (10 per group). By oral administration, the control group was given 5% ethanol in a saline phosphate buffer (10 mL/kg body weight), and the treatment group was given 5% ethanol in a saline phosphate buffer with erinacine A (8 mg/kg body weight) for 4 weeks. After the last administration, the rats were decapitated in the anesthetized condition. NGF content was measured in the following rat brain regions: olfactory bulb, locus coeruleus, hippocampus, and cerebral cortex. The results are shown in Fig. 11.3. In the locus coeruleus and hippocampus, NGF contents of the erinacine A-treated group were much higher than that of the control group.

Another in vivo study was done to investigate the effects of hericenone C (**3**) and erinacine A (**7**) on rats with ibotenic acid-induced dementia and rats with artificially induced cerebrovascular dementia. The results demonstrated that these compounds were beneficial to maintaining memory and improving learning skills in these model rats [15].

An in vivo study using the commercially available low-molecular weight fraction (LMF) available in the United States and Japan was followed [16]. The fraction was extracted with ethanol from the fruiting bodies of *H. erinaceus* and the ethanol solution was concentrated under reduced pressure. Water was added to the concentrate and the floating (water-insoluble) fraction was obtained. This fraction (LMF) contained the NGF stimulators and ER stress-protective compounds. Rats were divided into six groups and used in the study. Five

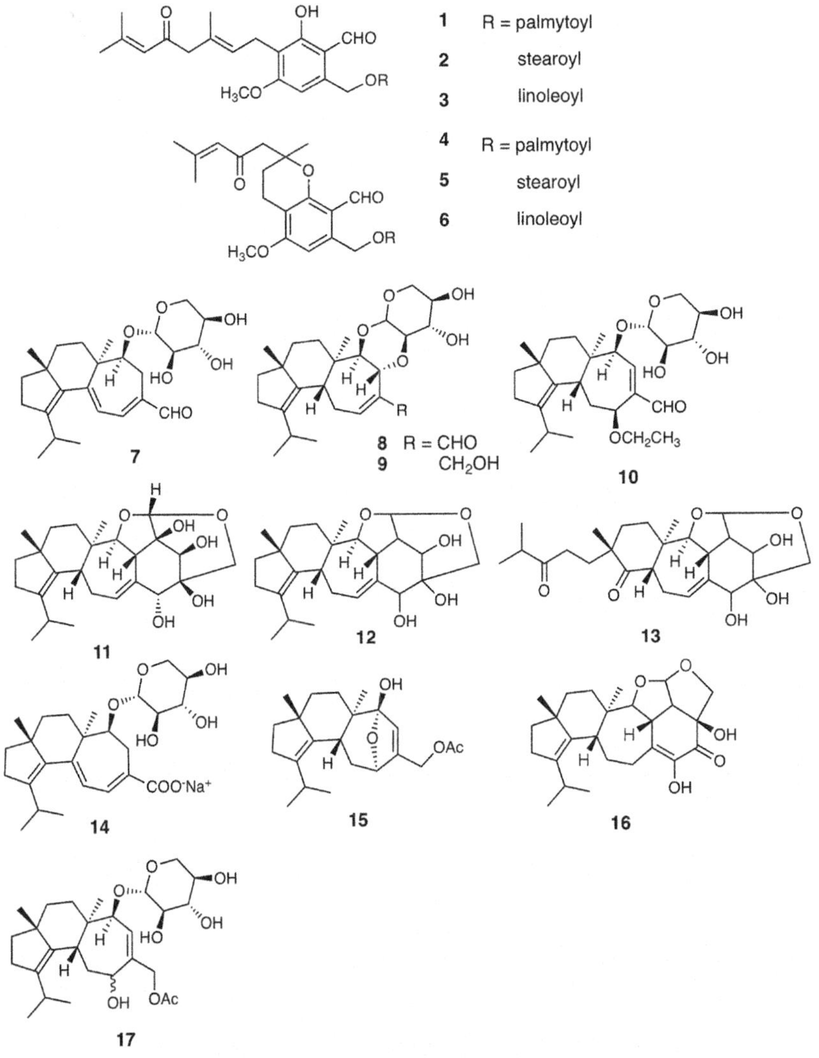

FIGURE 11.1 Simulators of nerve growth factor-synthesis, antimethicillin-resistant *Staphylococcus aureus* compounds, and amyloid-β toxicity-supressing compounds from *Hericium erinaceus*.

groups of rats were separately injected with 5 μL (10 μg) of Aβ1-40 into both sides of the brain hippocampus to cause Alzheimer's disease symptoms (AD rats). Instead of Aβ, the control (normal) group was injected with saline. One group of AD rats was treated with an antidementia drug, donepezil (brand name Aricept). The other three groups of AD rats were treated with LMF-H (high dose), LMF-M (middle dose), and LMF-L (low dose). We compared the effects of LMF with that of donepezil. From the fourth day after injection, the Aβ + D

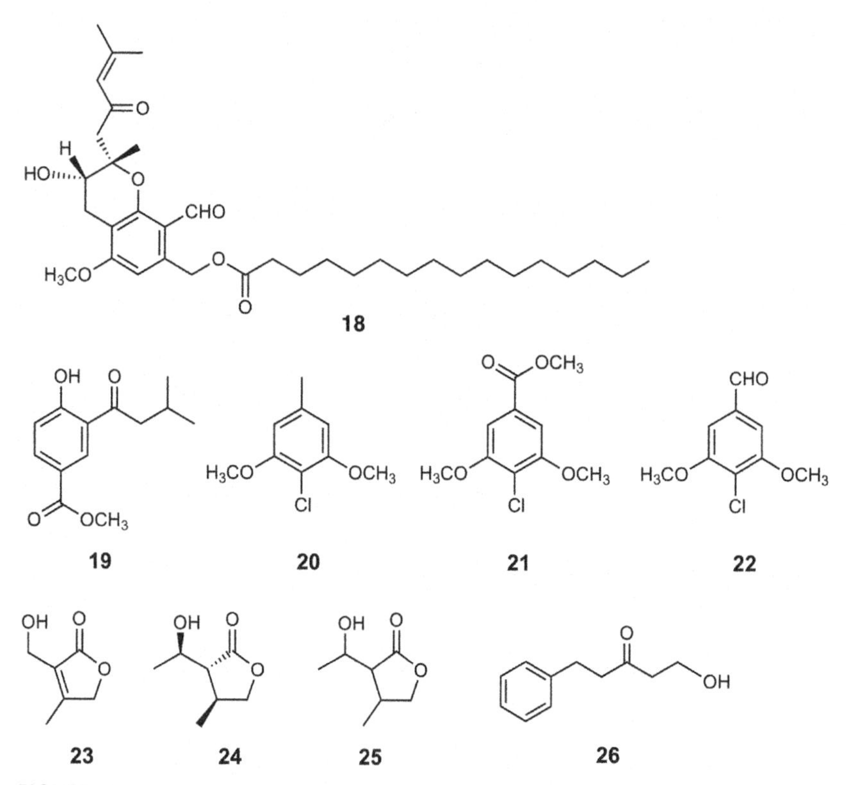

FIGURE 11.2 Endoplasmic reticulum stress protecting compounds from *Hericium erinaceus* and *Mycoleptodonoides aitchisonii*.

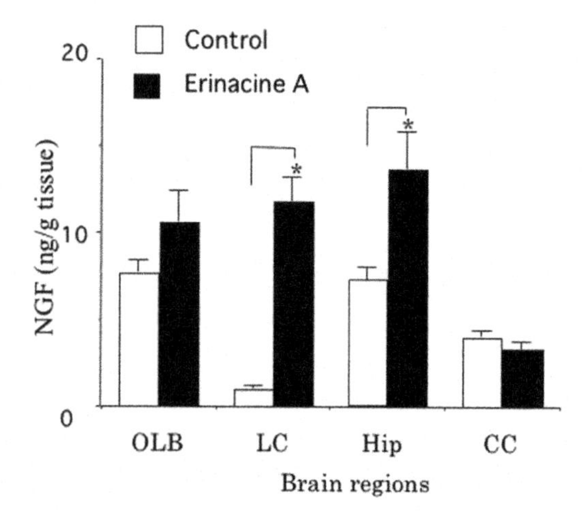

FIGURE 11.3 Effects of erinacine A on nerve growth factor synthesis in the brain of rats. *Hip*, Hippocampus; *LC*, locus coeruleus; *NGF*, nerve growth factor; *OLB*, olfactory bulb. $*P < .05$.

group was administered with donepezil at a dose of 1 mg/kg by stomach tube once per day for 4 weeks. Aβ + LMF-H, Aβ + LMF-M, and Aβ + LMF-L groups were treated with LMF separately at doses of 24, 12, and 6 mg/kg, and the control and Aβ groups were treated with saline. From the fourth week after administration, learning and memory-related behavior was assessed using the Morris water maze test on 10 arbitrarily chosen rats from each group once a day for 1 week. The Morris water maze test is used for investigating the learning and memory-related behavior of animals. A platform was located in one of four quadrants in a pool and hidden below the surface of the water. A rat was put in the water from any one of the other three quadrants. Time of arrival at the platform was measured from day 1 to day 6 in the fourth week after administration. On day 7, the platform was removed and the frequency of crossing the platform location was counted. Fig. 11.4A shows comparison of the arrival time to the platform of Aβ rats with those of donepezil or LMF-treated rats. The Aβ group spent a significantly longer time arriving at the platform from day 1 to day 6 than the control, which correlated with memory impairment of the Aβ group due to the injection of Aβ1-40. Compared with the Aβ group, the arrival times of the Aβ + D, Aβ + LMF-H, and Aβ + LMF-M groups were improved from day 1 to day 6, and in particular were significantly decreased for arriving at the platform from day 4 to day 6 ($P < .01$). Fig. 11.4B shows the frequency of crossing the platform location on day 7. Compared with the Aβ group, the Aβ + D, Aβ + LMF-H, and Aβ + LMF-M groups increased the frequencies to cross the platform location on day 7 ($P < .01$ or $P < .05$). All the results suggested that LMF improved the learning and memory abilities of Alzheimer's model rats.

Researchers in Taiwan showed that a 30-day oral course of erinacines A (**7**) and S (**16**) attenuated Aβ plaque burden in the brains of 5-month-old female APP/PS1 transgenic mice and significantly increased the level of insulin-degrading enzyme in the cerebral cortex [17].

Clinical tests evaluating effects of the fruiting bodies of *H. erinaceus* on dementia have been done [18–21].

In 2001, in one of the clinical studies in a rehabilitation hospital in Japan, two groups of patients were divided as 50 patients in an experimental group (average age 75.0) and 50 patients in a control group (average age 77.2) [18]. All the patients were suffering from cerebrovascular disease, degenerative orthopedic disease, Parkinson's disease, spinocerebellar degeneration, diabetic neuropathy, spinal cord injury, or disuse syndrome. Seven of the patients in the experimental group suffered from different types of dementia. The patients in this group received 5 g of dried fruiting bodies of *H. erinaceus* per day in their soup for a 6-month period. All the patients were evaluated before and after the treatment period for their functional independence measure (FIM), which is an international valuation standard of independence in physical capabilities (eating, dressing, evacuating, walking, bathing/showering, etc.) and in perceptive capabilities (understanding, expression, communication, problem

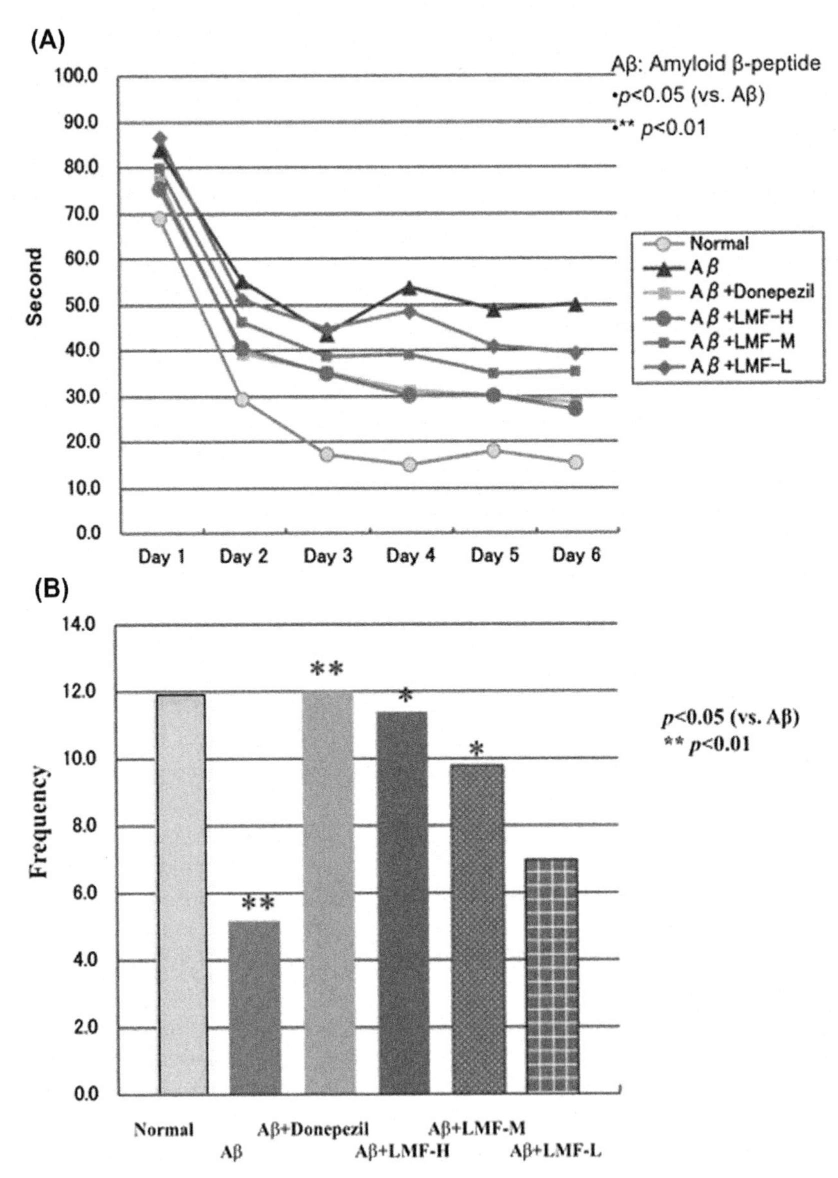

FIGURE 11.4 Antidementia effects of low-molecular weight fraction.

solving, memory). The results showed that after 6 months of taking *H. erinaceus*, six out of seven dementia patients demonstrated improvements in their perceptual capacities, and all seven had improvements in their overall FIM score. Particularly, three bedridden patients were able to get up to eat after administration.

In another clinical test, 18 dementia patients [19] were treated with an antidementia drug, donepezil, before the test, but no positive effect could be observed. The patients ate 1.5 g of the powdered mushroom twice a day after breakfast and supper for 3 months. The effectiveness of the mushroom was evaluated by the mini-mental state examination (MMSE) test. The test is a brief 30-point questionnaire used to screen for cognitive impairment. It is commonly used in medicine to screen for dementia. It is also used to estimate the severity of cognitive impairment at a given point in time and to follow the course of cognitive changes in an individual over time, thus making it an effective way to document an individual's response to treatment. In about 10 min, it samples various functions, including arithmetic, memory, and orientation. The MMSE test includes simple questions and problems in a number of areas: the time and place of the test, repeating lists of words, arithmetic, language use and comprehension, and basic motor skills. For example, one question asks to copy a drawing of two pentagons. MMSE values after the treatment were compared with those before the treatment. As a result, the values of nine people increased by eating the mushroom (Table 11.1). From these data and other observations, the authors concluded that the symptoms of 11 patients improved after mushroom intake [19].

2. ANTIMETHICILLIN-RESISTANT *STAPHYLOCOCCUS AUREUS* COMPOUNDS

In the clinical study in 2001 mentioned earlier, methicillin-resistant *Staphylococcus aureus* (MRSA) in the patients disappeared after taking the mushroom [21]. Therefore we searched for anti-MRSA compounds from the mushroom and obtained erinacines A (**7**), C (**9**), and K (**17**) as the active compounds (Fig. 11.1) [22]. We also purified active compounds **27** and **28** from the mushroom *Sparassis crispa* (Fig. 11.5) [23].

3. OSTEOCLAST-FORMING SUPPRESSING COMPOUNDS

Osteoporosis is caused by an imbalance between bone resorption and bone formation, which results in bone loss and fractures after mineral flux. Hip fracture in senile patients is a very serious problem because it often limits their quality of life. Osteoclasts are multinucleated, giant cells that are primarily responsible for bone resorption. The most characteristic feature of osteoclasts is the presence of ruffled borders and a clear zone. Osteoclast-like multinucleated cells can be differentiated in vitro from cocultures of mouse bone marrow cells and osteoblastic cells by treatment with osteotropic factors, 1α,25-dihydroxyvitamin D_3, and prostaglandin E_2 [24]. During screening for osteoclast-formation suppressing effects of the extracts of various mushrooms by using the assay, we found a known compound (**29**) and five new compounds called chaxines A (**30**) to E (**34**) from the mushroom

TABLE 11.1 Effect of Ingestion of *Hericium erinaceus* on Mini-Mental State Examination (MMSE) Values of Alzheimer Patients[a]

Entry	Patient	Sex	Age	Before	After	Change
1	Female	66	22	24	+2	Improvement
2	Female	69	18	23	+5	Improvement
3	Male	86	21	25	+4	Improvement
4	Male	80	18	22	+4	Improvement
5	Female	72	20	24	+4	Improvement
6	Male	81	23	26	+3	Improvement
7	Female	80	18	21	+3	Improvement
8	Female	69	21	25	+4	Improvement
9	Female	84	17	19	+2	Improvement
10	Male	87	20	20	0	No change
11	Male	79	11	12	+1	No change
12	Female	80	18	19	+1	No change
13	Male	64	26	25	−1	No change
14	Male	84	18	19	+1	No change
15	Male	80	21	22	+1	No change
16	Male	79	20	21	+1	No change
17	Female	81	18	16	−2	Worse
18	Male	73	19	14	−5	Worse

[a] *1.5 g of the powdered mushroom twice a day after breakfast and supper for 3 months.*

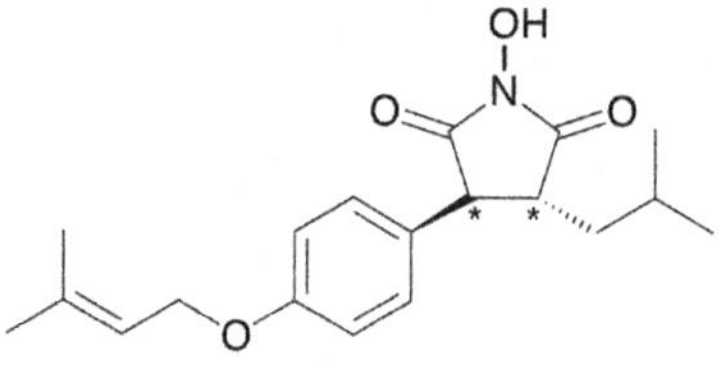

27 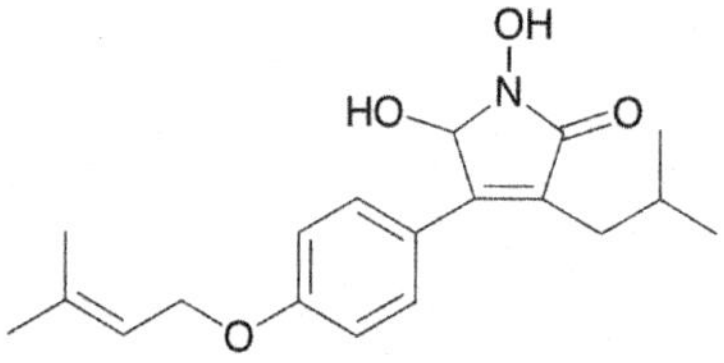 **28**

FIGURE 11.5 Antimethicillin-resistant *Staphylococcus aureus* compounds from *Sparassis crispa.*

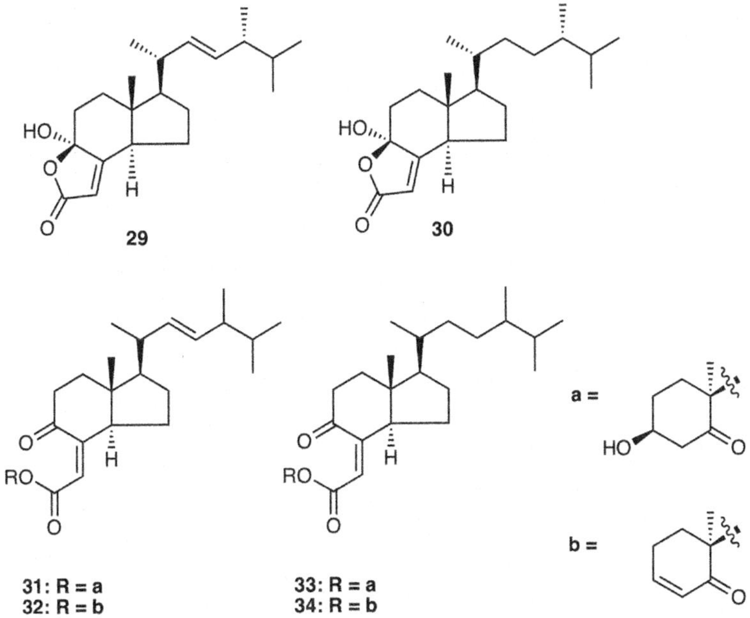

31: R = a
32: R = b

33: R = a
34: R = b

FIGURE 11.6 Osteoclast-forming suppressing compounds from *Agrocybe chaxingu.*

Agrocybe chaxingu (Fig. 11.6). The compounds suppressed the formation of osteoclasts without cytotoxicity [25–27].

4. DIARRHEA-CAUSING COMPOUNDS

Yearly, the three mushrooms responsible for most cases of poisoning in Japan are *Lampteromyces japonicus*, *Rhodophyllus rhodopolium*, and *Tricholoma ustale*. The toxic principles of the former two mushrooms had been isolated and characterized before our study: a hemolytic protein from *R. rhodopolium* and illusion S (**35**) from *L. japonicus* [28–34]. We disclosed the toxic principles of *T. ustale*. This mushroom (Kakishimeji in Japanese) is widespread and common throughout temperate regions in the world, and human ingestion of this mushroom causes gastrointestinal poisoning accompanied by vomiting and diarrhea. The toxins ustalic acid (**35**) and its derivatives (**36–39**) inhibited Na^+, K^+-ATPase (Fig. 11.7). In general, absorption of water from the intestines is suppressed by inhibition of intestinal Na^+, K^+-ATPase, resulting in diarrhea. In fact, rats caused diarrhea by ingesting ustalic acid [35].

Eight isolectins, BVL-1 to -8, were isolated from *Boletus venenatus*. BVLs showed fetal toxicity in mice upon intraperitoneal administration and caused diarrhea upon oral administration in rats [36,37].

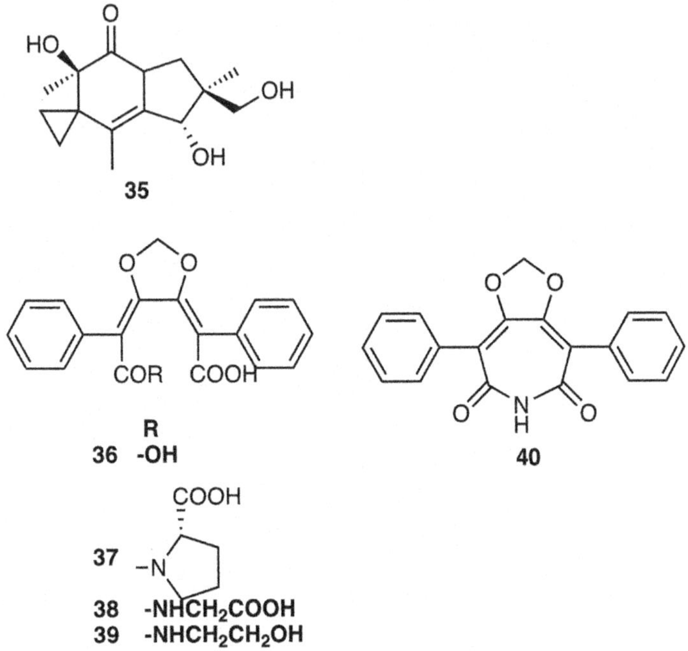

FIGURE 11.7 Diarrhea-causing compounds from *Lampteromyces japonicus* and *Tricholoma ustale*.

5. ACETALDEHYDE DEHYDROGENASE INHIBITORS

Clitocybe clavipes is widespread and common throughout temperate regions of the world. It is a delicious wild mushroom and popular in Japan. However, if ethanol is consumed together with this mushroom, one or more of the following symptoms may be experienced: profound flushing, metallic taste, palpitations, hyperventilation, hypertension, tachycardia, nausea, vomiting, and occasionally collapse. These symptoms are very similar to those caused by an aldehyde dehydrogenase inhibitor, coprine (**41**), isolated from the mushroom *Coprinus atramentarius* [38–41]. In addition, ingestion of the extract of *C. clavipes* in mice increased acetaldehyde concentration in their blood. We found five enone fatty acids (**42–46**) as aldehyde dehydrogenase inhibitors (Fig. 11.8). There are cysteine residues in the active site in the aldehyde dehydrogenase. The active site is formed in adducts with the enones by Michael addition reaction [42]. Nuclear magnetic resonance data of the crude extracts of this mushroom indicated that the mushroom contained many kinds of other enone fatty acids that have not been isolated. The high-abundance enones might be a cause of the symptoms.

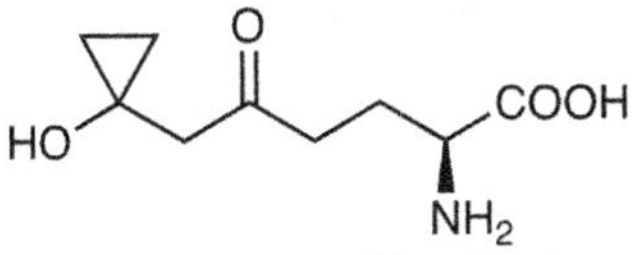

41

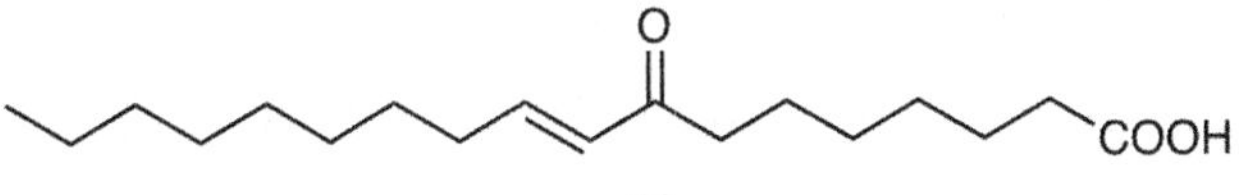

42

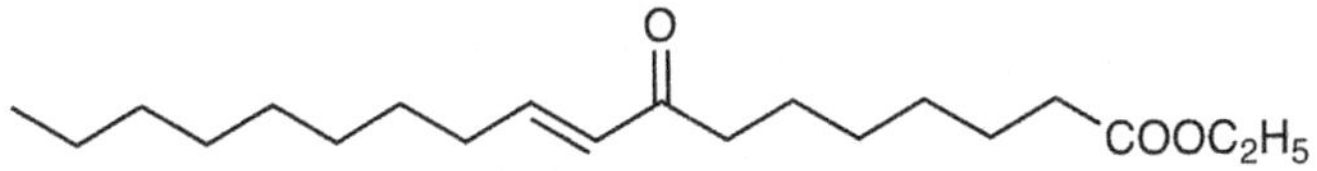

43

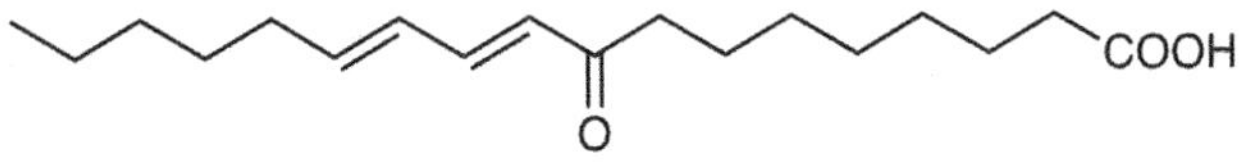

44

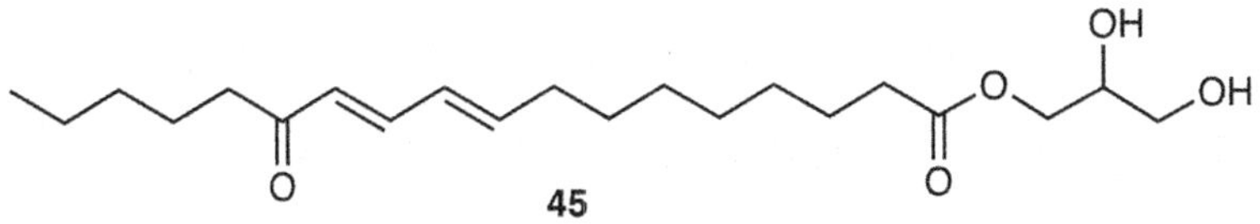

45

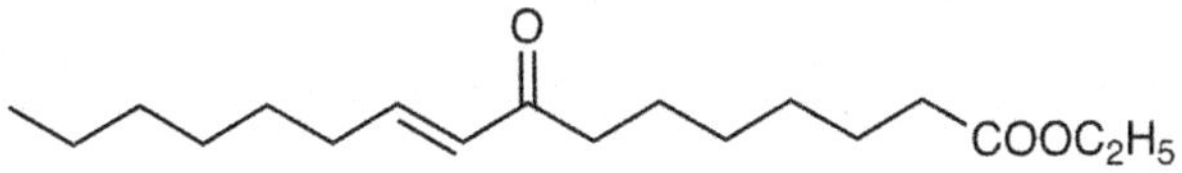

46

FIGURE 11.8 Adehyde dehydrogenese inhibitors from *Coprinopsis atramentaria* and *Clitocybe clavipes*.

6. HYALURONAN-DEGRADATION REGULATING COMPOUNDS

Hyaluronan (HA), a nonsulfated glycosaminoglycan composed of repeating disaccharide units of *N*-acetylglucosamine and glucuronic acid, plays an important role in inflammation, cell locomotion, and wound healing. The skin has more than 50% of total body HA, and the half-life of HA is less than 1 day, suggesting that the rate of HA production and depolymerization (degradation) in the skin is very rapid [43]. Therefore substances that inhibit degradation of

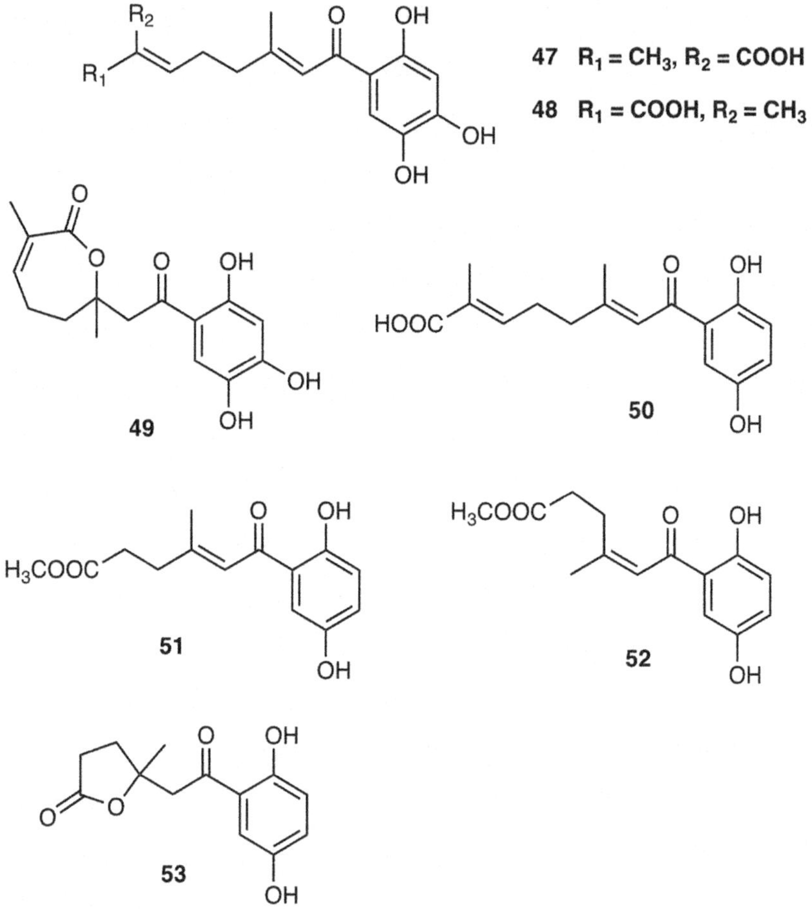

47 $R_1 = CH_3$, $R_2 = COOH$

48 $R_1 = COOH$, $R_2 = CH_3$

FIGURE 11.9 Hyaluronan-degradation inhibitors and those analogs from *Tricholoma orirubens*.

HA can moisturize the skin and prevent infection to the skin. On the other hand, there are many reports showing that HA is elevated in disorders such as fibrosis diseases and tumors. For example, during the progression of hepatitis, accumulated HA causes fibrosis and eventually cirrhosis of the liver [44]. An increase in HA in the endocervical canal at the appropriate stage of pregnancy can result in miscarriage [45]. Overproduction of HA accelerates tumor growth and is associated with cancer metastasis [46,47]. These results show that the balance of production and degradation of HA in tissues is regulated. Therefore regulators targeting HA degradation are useful as drugs. We found the inhibitors against HA degradation, orirubenones A (**47**) to C (**49**), from the extracts of the edible mushroom *Tricholoma orirubens* (Fig. 11.9). IC_{50} of **47**, **48**, and **49** were 15, 21, and 57 μM, respectively [48]. Since orirubenones D (**50**) to G (**53**) obtained from the mushroom have no activity, we concluded

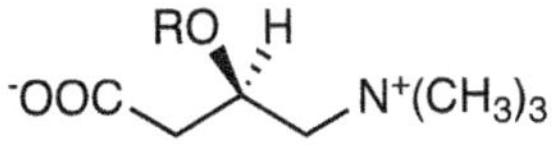

54	R = H
55	CH_3CO-
56	CH_3CH_2CO-
57	(*R*)-$HOCH_2(CH_3)CHCO$-
58	(*R*)-$CH_3(HO)CHCH_2CO$-
59	CH3(HO)CH(CH3)CHCO-
60	$CH_3(CH_3)(HO)CCH_2CO$-
61	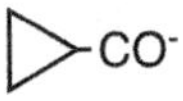

FIGURE 11.10 Hyaluronan-degradation promoter and its analogs from *Suillus laricinus*.

that the catechol skeleton in orirubenones was indispensable for the activity [49]. Carnitine esters (**54–61**) were isolated from the edible mushroom *Suillus laricinus* (Fig. 11.10). Among the eight compounds, only (*R*)-3-hydroxy-2-methylpropanoyl-(*R*)-carnitine (**57**) promoted HA degradation by human skin fibroblasts [50].

7. ACUTE ENCEPHALOPATHY CAUSED BY EATING ANGEL'S WING OYSTER MUSHROOM

Angel's wing mushroom, *Pleurocybella porrigens*, is found worldwide in temperate areas and eaten throughout the world, especially in Japan. Nonetheless, 17 people who ate angel's wing mushroom died of acute encephalopathy in 2004 in Japan [51–53]. Epidemiological investigation of the incident indicated that most patients were undergoing hemodialysis treatment for chronic renal failure and had digested the wild fruiting bodies before the onset of neurological symptoms. Although the mechanism of acute encephalopathy is not yet known, we reported cytotoxic amino acids (**62–67**), three of which (**62–64**) were novel (Fig. 11.11) [54]. All the compounds except for **67** have the β-hydroxyvaline unit attached to endogenous molecules. This led us to believe that the presence of an aziridine amino acid (**68**) was the common precursor, because we thought that attack of a hydroxyl group of each nucleophile to β-carbon of **68** would give **62–66** and attack of water to α-carbon of **68** would give **67** (Fig. 11.11). We hypothesized that the labile feature of **68** should be the reason why it could not be isolated from the mushroom in the previous study. To confirm the existence of aziridine as a component within the mushroom, we synthesized the proposed molecule and proved the existence of the compound in the mushroom [55]. We named the compound (**68**) pleurocybellaziridine. Histological findings of the brain tissues affected by the encephalopathy showed demyelinating symptoms. This indicates that toxic substance(s) in the

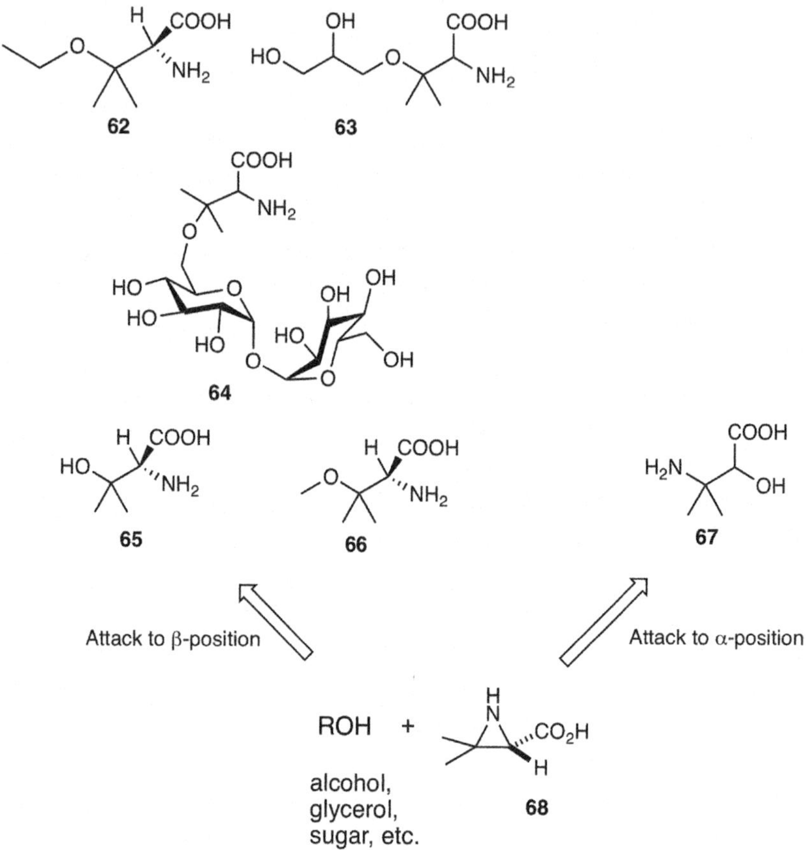

FIGURE 11.11 Cytotoxic compounds from *Pleurocybella porrigens*.

mushroom damaged oligodendrocytes, which constitutes the myelin sheath in the brain. Therefore we examined the toxicity of **68** against rat CG4-16 oligodendrocyte cells. As a result, **68** significantly reduced cell viability at concentrations up to 10 μg/mL (87 μM).

There are more than 140,000 species of mushroom-forming fungi in the world [56]. However, only about 10% of the fungi have been given names. The nameless fungi must also be producing compounds with new functions and structures. Mushrooms are unexplored biological resources.

REFERENCES

[1] H. Kawagishi, M. Ando, H. Sakamoto, S. Yoshida, F. Ojima, Y. Ishiguro, N. Ukai, S. Furukawa, Hericenones C, D and E, stimulators of nerve growth factor (NGF)-synthesis, from the mushroom *Hericium erinaceum*, Tetrahedron Lett. 32 (35) (1991) 4561–4564.

[2] H. Kawagishi, M. Ando, K. Shinba, H. Sakamoto, S. Yoshida, Y. Ishiguro, S. Furukawa, Chromans, hericenones F, G and H from the mushroom *Hericium erinaceum*, Phytochemistry 32 (1) (1993) 175–178.

[3] H. Kawagishi, A. Shimada, R. Shirai, K. Okamoto, F. Ojima, H. Sakamoto, Y. Ishiguro, S. Furukawa, Erinacines A, B and C, strong stimulators of nerve growth factor (NGF)-synthesis, from the mycelia of *Hericium erinaceum*, Tetrahedron Lett. 35 (10) (1994) 1569–1572.

[4] H. Kawagishi, A. Simada, K. Shizuki, H. Mori, K. Okamoto, H. Sakamoto, S. Furukawa, Erinacine D, a stimulator of NGF-synthesis, from the mycelia of *Hericium erinaceum*, Heterocycl. Commun. 2 (1) (1996) 51–54.

[5] H. Kawagishi, S. Furukawa, C. Zhuang, R. Yunoki, The inducer of the synthesis of nerve growth factor from lion's mane (*Hericium erinaceum*), Explore 11 (4) (2002) 46–51.

[6] H. Kawagishi, A. Shimada, S. Hosokawa, H. Mori, H. Sakamoto, Y. Ishiguro, S. Sakemi, J. Bordner, N. Kojima, S. Furukawa, Erinacines E, F, and G, stimulators of nerve growth factor (NGF)-synthesis, from the mycelia of *Hericium erinaceum*, Tetrahedron Lett. 37 (41) (1996) 7399–7402.

[7] E.W. Lee, K. Shizuki, S. Hosokawa, M. Suzuki, H. Suganuma, T. Inakuma, J. Li, M. Ohnishi-Kameyama, T. Nagata, S. Furukawa, H. Kawagishi, Two novel diterpenoids, erinacines H and I from the mycelia of *Hericium erinaceum*, Biosci. Biotechnol. Biochem. 64 (11) (2000) 2402–2405.

[8] M. Shimbo, H. Kawagishi, H. Yokogoshi, Erinacine A increases catecholamine and nerve growth factor content in the central nervous system of rats, Nutr. Res. 25 (2005) 617–623.

[9] K. Nagai, A. Chiba, T. Nishino, K. Kubota, H. Kawagishi, Dilinoleoyl-phosphati dylethanolamine from *Hericium erinaceum* protects against ER stress-dependent Neuro2a cell death via protein kinase C pathway, J. Nutr. Biochem. 17 (2006) 525–530.

[10] K. Ueda, M. Tsujimori, S. Kodani, A. Chiba, M. Kubo, K. Masuno, A. Sekiya, K. Nagai, H. Kawagishi, An endoplasmic reticulum (ER) stress-suppressive compound and its analogues from the mushroom *Hericium erinaceum*, Bioorg. Med. Chem. 16 (2008) 9467–9470.

[11] K. Ueda, S. Kodani, M. Kubo, K. Masuno, A. Sekiya, K. Nagai, H. Kawagishi, Endoplasmic reticulum (ER) stress-suppressive compounds from scrap bed cultivation of the mushroom *Hericium erinaceum*, Biosci. Biotechnol. Biochem. 73 (8) (2009) 1908–1910.

[12] A. Seiger, A. Nordberg, H. Von Holst, Intracranial infusion of purified nerve growth factor to an Alzheimer patients, Behav. Brain Res. 57 (2) (1993) 255–261.

[13] S.A. Scott, E.J. Mufson, J.A. Weingartner, K.A. Kau, K.A. Crutcher, Nerve growth factor in Alzhemer's disease. Increased levels throughout the brain coupled with declines in nucleus basalis, J. Neurosci. 15 (9) (1995) 6213–6221.

[14] J.-H. Choi, M. Horikawa, H. Okumura, S. Kodani, K. Nagai, D. Hashizume, H. Koshino, H. Kawagishi, Endoplasmic reticulum (ER) stress protecting compounds from the mushroom *Mycoleptodonoides aitchisonii*, Tetrahedron 65 (2009) 221–224.

[15] H. Kawagishi, C. Zhuang, Compounds for dementia from *Hericium erinaceum*, Drugs Future 33 (2) (2008) 149–155.

[16] C. Zhuang, H. Kawagishi, L. Zhang, H. Anzai, Anti-dementia Substance from *Hericium erinaceum* and Method of Extraction, United States Patent, US7,214,778 B2/, May 8, 2007.

[17] C.-C. Chen, T.-T. Tzeng, C.-C. Chen, C.-L. Ni, L.-Y. Lee, W.-P. Chen, Y.-J. Shiao, C.-C. Shen, Erinacine S, a rare sesterterpene from the mycelia of *Hericium erinaceus*, J. Nat. Prod. 79 (2) (2016) 438–441.

[18] K. Kasahara, S. Kaneko, K. Shimizu, Effects of Yamabushitake on older disabled people, Gunma Igaku (Suppl. 76) (2001) 77–81 (in Japanese).

[19] E. Ohtomo, M. Shimizu, Y. Komatsu, Clinical effect of Yamabushitake on senile dementia of Alzheimer type, Rounenigaku 47 (2011) 1061–1066 (in Japanese).

[20] K. Mori, S. Inatomi, K. Ouchi, Y. Azumi, T. Tuchida, Improving effects of the mushroom Yamabushitake (*Hericium erinaceus*) on mild cognitive impairment: a double-blind placebo-controlled clinical trial, Phytother. Res. 23 (2009) 367–372.

[21] K. Kasahara, S. Kaneko, Three cases of disappearance of MRSA by tubal feeding of Yamabushitake, Gunma Igaku (Suppl. 78) (2002) 1–5 (in Japanese).

[22] H. Kawagishi, A. Masui, S. Tokuyama, T. Nakamura, Erinacines J and K from the mycelia of *Hericium erinaceum*, Tetrahedron 62 (2006) 8463–8466.

[23] H. Kawagishi, K. Hayashi, S. Tokuyama, N. Hashimoto, T. Kimura, M. Dombo, A novel bioactive compound from the mushroom *Sparassis crispa*, Biosci. Biotechnol. Biochem. 71 (7) (2007) 1804–1806.

[24] T. Suda, N. Takahashi, N. Udagawa, E. Jimi, T.M. Gillespieand, T.J. Martin, Modulation of osteoclast differentiation and function by the new members of the tumor necrosis factor receptor and ligand families, J. Endocr. Rev. 20 (1999) 345–357.

[25] H. Kawagishi, T. Akachi, T. Ogawa, K. Masuda, K. Yamaguchi, K. Yazawa, M. Takahashi, Chaxine A, an osteoclast-forming suppressing substance, from the mushroom *Agrocybe chaxingu*, Heterocycles 6 (2006) 253–258.

[26] J.-H. Choi, A. Ogawa, N. Abe, K. Masuda, T. Koyama, K. Yazawa, H. Kawagishi, Chaxines B, C, D and E from the edible mushroom *Agrocybe chaxingu*, Tetrahedron 65 (2009) 9850–9853.

[27] Y. Hirata, A. Nakazaki, H. Kawagishi, T. Nishikawa, Biomimetic synthesis and structural revision of chaxine B and its analogues, Org. Lett. 19 (3) (2017) 560–563.

[28] K. Suzuki, T. Une, H. Fujimoto, M. Yamazaki, Studies on the toxic components of *Rhodophylllus rhodopolius*. I. The biological activities and screening of the toxic principles, Yakugaku Zasshi 107 (12) (1987) 971–977.

[29] K. Suzuki, T. Une, H. Fujimoto, M. Yamazaki, Studies on the toxic components of *Rhodophylllus rhodopolius*. II. Partial purification and properties of the hemolysin from *Rhodophylllus rhodopolius*: examination on the condition of the hemolysis, Yakugaku Zasshi 108 (3) (1988) 221–225.

[30] K. Suzuki, T. Une, M. Yamazaki, T. Takeda, Purification and some properties of a hemolysin from the poisonous mushroom *Rhodophylllus rhodopolius*, Toxicon 28 (9) (1987) 1019–1028.

[31] K. Nakanishi, M. Tada, Y. Yamada, M. Ohashi, N. Komatsu, H. Terakawa, Isolation of lampterol, an antitumor substance from *Lampteromyces japonicas*, Nature 197 (1963) 292.

[32] T.C. McMorris, M. Anchel, The structures of Basidiomycete metabolites illudin S and illudin M, J. Am. Chem. Soc. 85 (1963) 831–832.

[33] K. Nakanishi, M. Ohashi, M. Tada, Y. Yamada, Illudin S (lampterol), Tetrahedron 21 (1965) 1231–1246.

[34] T. Matsumoto, H. Shirahama, A. Ichihara, Y. Fukuoka, Y. Takahashi, Y. Mori, M. Watanabe, Structure of lampterol (illudin S), Tetrahedron 21 (1965) 2671–2676.

[35] Y. Sano, K. Sayama, Y. Arimoto, T. Inakuma, K. Kobayashi, H. Koshino, H. Kawagishi, Ustalic acid as a toxin and related compounds from the mushroom *Tricholoma ustale*, Chem. Commun. 13 (2002) 1384–1385.

[36] M. Matsuura, M. Yamada, Y. Saikawa, K. Miyairi, T. Okuno, K. Konno, J. Uenishi, K. Hashimoto, M. Nakata, Bolevenine, a toxic protein from the Japanese toadstool *Boletus venenatus*, Phytochemistry 68 (2007) 893–898.

[37] M. Horibe, Y. Kobayashi, H. Dohra, T. Morita, T. Murata, T. Usui, S. Nakamura-Tsuruta, M. Kamei, J. Hirabayashi, M. Matsuura, M. Yamada, Y. Saikawa, K. Hashimoto, M. Nakata, H. Kawagishi, Toxic isolectins from the mushroom *Boletus venenatus*, Phytochemistry 71 (2010) 648–657.

[38] P. Lindberg, R. Bergman, B. Wickberg, Isolation and structure of coprine, a novel physiologically active cyclopropane derivertives from *Coprinus atramentarius* and its synthesis via 1-aminocyclopropanol, J. Chem. Soc. Chem. Commun. (1975) 946–947.

[39] G.M. Hatfield, J.P. Schaumberg, Isolation and structure studies of coprine, the disulfiram-like constituent of *Coprinus atramentarius*, Lloydia 38 (6) (1975) 489–496.

[40] P. Lindberg, R. Bergman, B. Wickberg, Isolation and structure of coprine, the in vivo aldehyde dehydrogenase inhibitor in *Coprinus atramentarius*; synthesis of coprine and related cycloprppane derivatives, J. C. S. Perkin I (1977) 684–691.

[41] J.S. Wiseman, R.H. Abeles, Mechanism of inhibition of aldehyde dehydrogenase by cyclopropane hydrate and the mushroom toxin coprine, Biochemistry 18 (3) (1979) 427–435.

[42] H. Kawagishi, T. Miyazawa, H. Kume, Y. Arimoto, T. Inakuma, Aldehyde dehydrogenase inhibitors from the mushroom *Clitocybe clavipes*, J. Nat. Prod. 65 (11) (2002) 1712–1714.

[43] R. Tammi, A.M. Säämänen, H.I. Maibach, M. Tammi, Degradation of newly synthesized high molecular mass hyaluronan in the epidermal and dermal compartments of human skin in organ culture, J. Invest. Dermatol. 97 (1) (1991) 126–130.

[44] T. Satoh, T. Ichida, Y. Matsuda, M. Sugiyama, K. Yonekura, T. Ishikawa, H. Asakura, Interaction between hyaluronan and CD44 in the development of dimethylnitrosamine-induced liver cirrhosis, J. Gastroenterol. Hepatol. 15 (2000) 402–411.

[45] E. El Maradny, N. Kanayama, H. Kobayashi, B. Hossain, S. Khatun, S. Liping, T. Kobayashi, T. Terao, The role of hyaluronic acid as a mediator and regulator of cervical ripening, Hum. Reprod. 12 (1997) 1080–1088.

[46] N. Itano, T. Sawai, O. Miyaishi, K. Kimata, , Relationship between hyaluronan production and metastatic potential of mouse mammary carcinoma cells, Cancer Res. 59 (1999) 2499–2504.

[47] A. Jacobson, M. Rahmanian, K. Rubin, P. Heldin, Expression of hyaluronan synthase 2 or hyaluronidase 1 differentially affect the growth rate of transplantable colon carcinoma cell tumors, Int. J. Cancer 102 (2002) 212–219.

[48] H. Kawagishi, Y. Tonomura, H. Yoshida, S. Sakai, S. Inoue, Orirubenones A, B and C, novel hyaluronan-degradation inhibitors from the mushroom *Tricholoma orirubens*, Tetrahedron 60 (2004) 7049–7052.

[49] S. Sakai, Y. Tonomura, H. Yoshida, S. Inoue, H. Kawagishi, Orirubenones D to G, novel phenones from the *Tricholoma orirubens* mushroom, Biosci. Biotechnol. Biochem. 69 (2005) 1630–1632.

[50] H. Kawagishi, H. Murakami, S. Sakai, S. Inoue, Carnitine-esters from the mushroom *Suillus laricinus*, Phytochemistry 67 (2006) 2676–2680.

[51] T. Kuwabara, T. Arai, N. Honma, M. Nishizawa, Acute encephalopathy among patients with renal dysfunction after ingestion of "Sugihiratake", angel's wing mushroom – Study on the incipient cases in the northern area of Niigata Prefecture, Rinsho Shinkeigaku 45 (2005) 239–245 (in Japanese with English abstract).

[52] K. Obara, S. Okawa, M. Kobayashi, S. Watanabe, I. Yoyoshima, A case of encephalitis -type encephalopathy related to *Pleurocybella porrigens* (Sugihiratake), Rinsho Shinkeigaku 45 (2005) 253–256 (in Japanese with English abstract).

[53] K. Obara, C. Wada, T. Yoshioka, K. Enomoto, S. Yagishita, I. Toyoshima, Acute encephalopathy associated with ingestion of a mushroom, *Pleurocybella porrigens* (angel's wing), in a patient with chronic renal failure, Neuropathology 28 (2008) 151–156.

[54] T. Kawaguchi, T. Suzuki, Y. Kobayashi, S. Kodani, H. Hirai, K. Nagai, H. Kawagishi, Unusual amino acid derivatives from the mushroom *Pleurocybella porrigens*, Tetrahedron 66 (2010) 504–507.

[55] T. Wakimoto, T. Asakawa, S. Akahoshi, T. Suzuki, K. Nagai, H. Kawagishi, T. Kan, Proof of the existence of an unstable amino acid, pleurocybellaziridine, in *Pleurocybella porrigens* (angel's wing mushroom), Angew. Chem. Int. Ed. 50 (5) (2011) 1168–1170.
[56] D.L. Hawksworth, Mushrooms: the extent of the unexplored potential, Int. J. Med. Mushr. 3 (2001) 333–337.

Chapter 12

Natural Products in Lifestyle Diseases: In Vitro Screening

Anuradha S. Majumdar[1], Sahil J. Somani[2]
[1]*Department of Pharmacology, Bombay College of Pharmacy, Mumbai, India;* [2]*Department of Pharmacology, RK University, Rajkot, India*

1. INTRODUCTION

1.1 Natural Products in Drug Discovery

What exactly defines a natural product varies across the scientific and medicinal communities. Food and Drug Administration (FDA) acceptable definitions largely include: (1) unregulated organisms or natural materials; (2) FDA-regulated, unmodified natural materials or compounds; (3) a naturally occurring compound that has been chemically modified (also called a semi-synthetic); and (4) a purely synthetic medicinal compound inspired by a natural compound [1]. The usefulness of natural products in treating various ailments takes its roots from the Vedic period and even today many modern pharmaceutical drugs are plant-derived natural products or their derivatives.

Applications of these medicinal plants in the healthcare system have been documented in various literature systems such as Mesopotamia, Ebers Papyrus (dating from 1500 BCE), Chinese Materia Medica (1100 BCE), Shennong Herbal (100 BCE), and the Indian Ayurvedic system such as Charaka, Sushruta, and Samhitas (1000 BCE) [2]. Between 1981 and 2010, 1073 new chemical entities had been approved by the FDA, of which only 36% are purely synthetic, 19% correspond to synthetic molecules containing pharmacophores derived directly from natural products, 11% are actually modeled on a natural product inhibitor of the molecular target of interest or mimic (i.e., competitively inhibit) the endogenous substrate of the active site, and the remaining are derived or inspired from nature. Examples of plant-derived natural compounds that are therapeutically available in the market include paclitaxel and its derivatives from *Taxus* species, vincristine and vinblastine from *Catharanthus roseus*, camptothecin from *Camptotheca acuminata*, galanthamine from *Galanthus nivalis*, and the most important antimalarial artemisinin from the traditional Chinese herb *Artemisia annua* [2,3].

Natural Products and Drug Discovery. https://doi.org/10.1016/B978-0-08-102081-4.00012-5

The prime advantage of natural product-based drug discovery is the available ethnopharmacological information concerning the traditional use of the natural product that can endow hints for compound effectiveness in humans. Eighty percent of the 122 plant-derived compounds that are used globally as drugs originate from plants that have ethnomedicinal use. Drugs developed from medicinal plants with traditional use include reserpine from *Rauwolfia serpentina* used for snakebites, ephedrine from *Ephedra sinica* formed on the basis of salbutamol and salmetrol synthesis, tubocurarine from *Chondrodendron* used for arrow poison, guaifenesin from guaiacum resin used for syphilis treatment, and digoxin from *Digitalis lanata* [2].

Sometimes lead compounds of natural products can also be helpful in drug synthesis, which acts as a starting material for various drug reactions. The best example would be Tamiflu (oseltamivir), the single drug available for the treatment of H1N1 swine flu. Shikimic acid, the main active constituent of Chinese star anise, *Calophyllum apetalum*, and *Araucaria excelsa* act as starting materials for the synthesis of oseltamivir. Also many ethnomedicines have formed the basis for several synthetic compounds, thereby guiding the drug discovery process [4]. Pertinent examples include khellin from *Ammi visnaga*, which led to the development of sodium chromoglycate as a bronchodilator, galegine from *Galega officinalis*, which was the model for the synthesis of metformin and other bisguanidine-type antidiabetic drugs, and papaverine from *Papaver somniferum*, which formed the basis for verapamil. Also in recent years the importance of natural product discovery can be evidenced from the increasing amount of scientific literature that is being published in the public domain [5,6].

1.2 Natural Products: Metabolic Disorders (Diabetes, Dyslipidemia, and Obesity)

Metabolic syndrome (MS) is a group of risk factors that affect the ability of the cell to perform critical biochemical reactions that involve the processing or transport of proteins (amino acids), carbohydrates (sugars and starches), or lipids (fatty acids) [7] and occurs primarily as a result of being overweight and obese caused by a sedentary lifestyle, i.e., physical inactivity and the consumption of a diet containing excess calories. Systemic effects of MS influence the metabolism in key metabolic organs, such as liver, muscle, pancreas, and white adipose tissue (WAT) [8]. Alteration of WAT function with subsequent dysfunctional expression and secretion of adipokines plays a key role in the pathogenesis of obesity, diabetes, and other metabolic diseases [9]. **Diabetes mellitus** (DM) is a heterogeneous disorder characterized by abnormally high blood glucose levels [10]. It is one of the most prevalent MSs and has reached epidemic proportions; it affects more than 220 million individuals worldwide. Globally, more than 90%–95% of cases of diabetes are type 2, which is largely associated with diet behavior and excess body weight, while type 1 accounts for only 3%–5% [11] and the number of patients is estimated to increase from currently 150 million to about 300 million by 2025 [12]. The major factors contributing to the hyperglycemia of diabetes are multifactorial but are secondary to the failure of the β-cells of the pancreas to adequately compensate for insulin resistance, characterized by decreased whole body insulin-mediated glucose utilization and elevated hepatic glucose output [13]. Although type 2 diabetes is closely associated with obesity, the mechanisms by which obesity leads to type 2 diabetes remain unclear. Insulin resistance is a common pathogenesis of obesity and type 2 diabetes because obesity leads to hyperlipidemia. A high level of free fatty acids (FFAs) in plasma and tissue reduces insulin sensitivity, impairs insulin signaling, and induces insulin resistance [14]. **Obesity** can be defined as an excess of body fat. A surrogate marker for body fat content is the body mass index (BMI), which is determined by weight (kilograms) divided by height squared (square meters). In clinical terms, a BMI of 25–29 kg/m^2 is called overweight; higher BMIs (30 kg/m^2) are called obesity. The increase in obesity prevalence is due to two major factors: plentiful supplies of inexpensive foods and sedentary jobs [15]. Obesity is characterized at the cellular level by an increase in the number and size of adipocytes (fat storage cells) that have differentiated from preadipocytes in the adipose tissue. A World Health Organization (WHO) report stated that obesity had reached epidemic proportions worldwide. Since then its incidence has continued to rise at an alarming rate in both developed and developing countries and is becoming a major public health concern with incalculable social costs [16]. Over 600 million adults are obese. Overall, about 13% of the world's adult population (11% of men and 15% of women) were obese in 2014. In 2014, 39% of adults aged 18 years and over (38% of men and 40% of women) were overweight. The worldwide prevalence of obesity more than doubled between 1980 and 2014. India has the third-highest number of obese and overweight people (11% of adolescents and 20% of all adults) after the United States and China (WHO). It has been shown that excessive macronutrient intake can contribute to the inflammatory response occurring in MS, whereas some dietary polyphenols are able to reduce the incidence of MS, including diabetes. Many of the metabolites occurring in plants are now recognized as useful for the maintenance of human health; hence the recommended use of plant food supplements [17]. Phytochemicals obtained from plants have been extensively evaluated for MS using in vitro methods to substantiate their molecular mechanisms.

Many scientists have used in vitro methods using cell lines, namely, H4IIE hepatoma cell line, primary human skeletal muscle cell culture (HSMC), mature 3T3-L1 adipocytes, L6 skeletal muscle cells, hepatic stellate cells (HSCs), C2C12 mouse skeletal muscle cells, HepG2 (human hepatoma) cells, FL83B (normal mouse hepatocytes) cells, Chinese hamster ovary/hIR (CHO/hIR) cell lines, and colorectal cancer cell lines such as HT29, HCT116, SW620, and CaCo2 to generate evidence of molecular mechanisms of herbals and herb-derived phytochemicals. A few examples of natural products that have been explored for MS using in vitro tools are extracts of *Artemisia dracunculus*, *Taraxacum officinale*, *Magnolia officinalis*, ginger, barley, *Psoralea corylifolia*, *Euonymus alatus*, salvia, and phytochemicals such as epicatechin gallate (ECG), epigallattocatechin-3-gallate (EGCG), the polyphenols of green tea, curcumin from *Curcuma longa*, tangeretin, active polymethoxy flavones (PMF) from citrus fruits, procyanidins, flavonoids from grape seeds, corosolic acid from *Lagerstroemia speciosa* L. etc.

A. dracunculus L. or Russian tarragon is a species of perennial herb in the sunflower family that has proven antidiabetic action by decreasing the expression of phosphoenolpyruvate carboxykinase (PEPCK), a key enzyme modulating hepatic gluconeogenesis in an animal model of streptozotocin-induced diabetic rats. It is known to be upregulated by glucagon and stress hormones in hepatocytes via a cyclic AMP-dependent pathway. Moreover, insulin strongly represses PEPCK transcription through the activation of the phosphoinositide-3 kinase (PI3K) pathway. The compounds that are able to repress PEPCK expression and overcome insulin resistance are envisaged as promising compounds to treat diabetes. Working on these lines, Govorko et al. explored the polyphenol-rich ethanolic extract of the shoot of this plant in vitro using an H4IIE hepatoma cell line to elucidate its impact on PEPCK expression. It was found that the two active polyphenolic compounds, 6-demethoxycapillarisin and 2′,4′-dihydroxy-4-methoxydihydrochalcone, inhibited PEPCK mRNA levels via activation of the PI3K pathway and the adenosine monophosphate-activated protein kinase (AMPK) pathway, respectively [13].

Wang et al. also demonstrated this activity of **artemisia** using primary HSMC obtained from obese and type 2 diabetic human subjects. The glucose uptake was significantly increased in the presence of increasing concentrations of the extract with restoration of normal glycogen levels and decline in FFA levels. This extract was shown to modulate the levels of a specific protein tyrosine phosphatase (PTP1B), which is expected to enhance insulin receptor signaling [18]. *T. officinale*, commonly known as **dandelion,** is an edible plant distributed worldwide. Leaves and root extracts of dandelion are found to be rich in polyphenols, flavonoids, sesquiterpene lactones, and coumarins. An in vitro study performed by Garcia-Carrosco et al. in mature 3T3-L1 adipocytes demonstrated the inhibition of adipocyte differentiation and lipid accumulation by the extracts of dandelion roots and leaves. Molecular analysis showed that the extract regulated many gene expressions playing a key role in brown fat cell differentiation and diet-induced thermogenesis. The extracts displayed effective antioxidant activity correlating with total flavonoids and polyphenolic contents. The investigation went on to elaborate that the leaf extract and crude powdered roots of dandelion reduced triglyceride accumulation in mature 3T3-L1 adipocytes to a higher extent than the root extract [19].

Green tea is a popular beverage worldwide and its polyphenol **EGCG** has been reported to enhance glucose tolerance in diabetic rodents [20].

Furthermore, to explore the underlying mechanism of EGCG ($\geq$95% pure), Zhang et al. profiled its effect in rat L6 skeletal muscle cells treated with dexamethasone. It was found that EGCG improved glucose uptake by increasing GLUT4 translocation, which was inhibited by dexamethasone. Furthermore, it was shown that EGCG inhibited dexamethasone-induced insulin resistance through the AMPK and PI3K/Akt pathways [21].

Curcumin, which is isolated from the rhizomes of the herb *C. longa*, is known to possess prominent antiinflammatory activities. An important initiator of the inflammatory response to obesity is adipose tissue, which is involved in energy regulation and homeostasis. Adipocytes uniquely secrete adipokines, such as leptin, adiponectin, and resistin as well as inflammatory cytokines such as tumor necrosis factor (TNF) and interleukins 1 and 6 (IL-1, IL-6). These factors are critically involved in obesity-induced insulin resistance and chronic inflammation. Wang et al. investigated the impact of curcumin on the proinflammatory insulin-resistant state in 3T3-L1 adipocytes. Curcumin increased the insulin-stimulated glucose uptake in 3T3-L1 adipocytes and suppressed the transcriptional secretion of TNF-α and IL-6 induced by palmitate in a concentration-dependent manner by inhibition of NF-κB. Moreover, curcumin enhanced the expression of the favorable adipokine, namely, adiponectin, in adipocytes, which in turn inhibits NF-κB activation and negatively controls obesity. It was concluded that curcumin reverses palmitate-induced insulin resistance state in 3T3-L1 cells through the inhibition of NF-κB and c-Jun N-terminal kinase [14].

Nonalcoholic steatohepatitis is commonly found in patients with obesity and is often accompanied by hyperleptinemia with subsequent development of hepatic fibrosis, and even cirrhosis. HSCs are the major effector cells during liver fibrogenesis and could be activated by leptin. Tang et al. demonstrated that curcumin inhibited HSC activation, induced gene expression of endogenous peroxisome proliferator-activated receptor-γ (PPARγ), and suppressed gene expression of αI(I) collagen, α-smooth muscle actin, connective tissue growth factor, receptors for transforming growth factor β (TGF-β), platelet-derived growth factor-β, and epidermal growth factor, thus suppressing hepatic fibrogenesis. Curcumin (purity $\geq$94%) was evaluated using a model of leptin-induced HSC activation using primary HSCs from male Sprague–Dawley rats. It was found that curcumin abrogated the stimulatory effect of leptin on HSC activation by reducing the phosphorylation of leptin receptor (Ob-R), stimulating PPARγ activity, and attenuating oxidative stress, which lead to the suppression of Ob-R gene expression and elimination of leptin signaling [22]. The same research group also reported that curcumin prevents leptin-driven hike in glucose levels in HSCs by blocking translocation of glucose transporter-4 (GLUT4) by interrupting the insulin receptor substrates/phosphatidylinositol 3-kinase/AKT signaling pathway. Furthermore, curcumin stimulated glucokinase activity, increasing conversion of glucose to glucose-6-phosphate [23].

Skeletal muscle is the major site for insulin-stimulated glucose uptake and is involved in energy regulation and homeostasis. To explore the effect of phytopolyphenols such as ECG, EGCG, and curcumin, Deng et al. used 12-*O*-tetradecanoylphorbol 13-acetate (TPA), a protein kinase C (PKC) activator, and palmitate to induce insulin resistance in C2C12 mouse skeletal muscle cells. The study showed that EGCG and curcumin treatment reduced insulin receptor substrate-1 (IRS-1) Ser307 phosphorylation. While curcumin was more effective in increasing Akt phosphorylation after TPA induction, it was also found that with shorter palmitate incubation, ECG can suppress IRS-1 Ser307 phosphorylation and significantly promote Akt, ERK1/2, p38 MAPK, and AMP-activated protein kinase activation, whereas IRS-1 exhibited a dramatic depletion with longer incubation with palmitate. It was noted that EGCG, ECG, and curcumin could reverse IRS-1 expression, Akt phosphorylation, and MAPK signaling cascade activation and improve glucose uptake in C2C12 skeletal muscle cells. This effect was pronounced with ECG and curcumin. Although these polyphenols could suppress acetyl-CoA carboxylase activation, only EGCG could inhibit lipid accumulation in the intracellular site. The scientists concluded that curcumin shows the best capacity to improve FFA-induced insulin resistance than the other two, and ECG was more effective than EGCG in attenuating insulin resistance [11].

M. officinalis is a Chinese herbal medicine, and traditionally the bark of the roots and stems have been used to treat liver and associated gastrointestinal disorders. Nonalcoholic fatty liver disease (NAFLD) is one of the most common causes of chronic liver disease and currently recognized as a manifestation of MS. Earlier evidence indicated that *M. officinalis* extract reverses alcoholic fatty liver disease via the suppression of TNF-α and superoxide anion production, as well as by inhibiting the maturation of sterol regulatory element-binding protein-1c (SREBP-1c) [24]. Min Suk Seo et al. demonstrated the activity of *M. officinalis* in NAFLD using HepG2 (human hepatoma cells) and FL83B (normal mouse hepatocytes) cells. Investigations revealed that *M. officinalis* pretreatment prevented an increase in intracellular lipid accumulation and triglyceride content as well as significantly inhibited SREBP-1c activation. It also abated the rise in fatty acid translocase, fatty acid synthase, and stearoyl CoA desaturase-1 protein expression in FFA-exposed hepatocytes in a dose-dependent manner. *M. officinalis* pretreatment markedly induced AMPK phosphorylation in hepatocytes. *M. officinalis*failed to show an effect in hepatocytes pretreated with compound C (AMPK inhibitor). The scientists concluded that the pharmacological potential of *M. officinalis* in attenuating triglyceride biosynthesis and accumulation in NAFLD is via inhibition of SREBP-1 via AMPK phosphorylation [25].

Corosolic acid or 2α-hydric ursolic acid is a triterpenoid and has been found in many Chinese medicinal herbs, such as *L. speciosa* L., banaba leaves, *Tiarella polyphylla*, etc. Miura T. et al. reported that corosolic acid reduced blood glucose levels and significantly lowered plasma insulin levels in KK-Ay mice by muscle GLUT4 translocation from the low-density microsomal membrane to the plasma membrane. The scientists later worked on the cellular effects of corosolic acid on the glucose metabolism and signaling pathway of CHO/hIR cells (the CHO cell line transfected with an expression plasmid encoding human insulin receptor) using wortmannin as a specific inhibitor of PI3K, which is a key downstream kinase of the insulin pathway. Corosolic acid was able to promote GLUT4 translocation in CHO/hIR cells and also displayed selective inhibition on nonreceptor protein tyrosine phosphatases (PTPs), such as PTP1B, TCPTP, SHP1, and SHP2, which are the diabetes-related nonreceptor proteins responsible for dephosphorylation of phospho-Tyr sites on insulin receptor β. It was concluded that corosolic acid might exert its antidiabetic effects through enhancing insulin receptor β phosphorylation by inhibiting certain PTPs [26].

Ginger, a rhizome of *Zingiber officinale*, a perennial herbaceous plant, has been studied for its benefits in reducing cholesterol and high-density lipoprotein. Deoxyglucose-based studies and blood sugar analysis have confirmed the glucose-lowering effect of ginger extracts. Rani et al. studied this activity of ethyl acetate ginger (EAG) extract in terms of its modulatory effects on glucose uptake, protein glycation, and adipocyte differentiation using L-6 (mouse myoblast cells) and 3T3-L1 (mouse preadipocytes) cell lines. The exploration highlighted the potential of the phenolic fraction (gingerol and shoagol) of ginger in combating reactive oxygen species (ROS). The extract could suppress low-density lipoprotein (LDL) oxidation, enhance glucose uptake and angiotensin-converting enzyme (ACE) inhibition, and inhibit adipocyte differentiation, which points toward its antidiabetic and cardioprotective effect. Antibody-based studies in the treated cells revealed that EAG showed its antidiabetic effect by increasing GLUT4 expression in the cell surface membrane. The scientists concluded that the action of ginger extract was expressed by its antioxidant and antiglycation activity. It also improved expression or transportation of GLUT4 receptors from internal vesicles [27].

E. alatus or burning bush is a species of flowering plant used in folk medicine in China. It has been reported that the extract of *E. alatus* can ameliorate hyperglycemia and hyperlipidemia induced by a high-fat diet in ICR mice. It has shown the effect by increasing the expression of PPARγ in periepididymal fat pad. PPARγ is an essential transcription factor and numerous studies have shown that this isoform participates in biological pathways of adipocyte differentiation and insulin sensitivity, and also plays a critical role in the pathogenesis of type 2 diabetes mellitus. A previous study indicated that a number of flavonoids isolated from the extract of *E. alatus*, including kaempferol and quercetin, improved peripheral glucose uptake into adipocytes [28]. Fang et al. investigated the probable mechanism for the antidiabetic effect of these compounds using 3T3-L1 cell lines. Both actives not only significantly improve insulin-stimulated glucose uptake in mature 3T3-L1 adipocytes but also serve as weak partial agonists in the PPARγ reporter gene assay, which explains the insulin sensitization activity of these components. They both could not induce differentiation of 3T3-L1 preadipocytes as traditional PPARγ agonists but were capable of inhibiting adipocyte differentiation in a dose-dependent manner when administered with PPARγ agonist. In lipopolysaccharide-activated peritoneal macrophages, PPARγ is overexpressed and both the actives proved to be more potent than rosiglitazone, a PPARγ agonist, in inhibiting nitric oxide production. Taken together, these data show that kaempferol and quercetin as multitargeting compounds hold promising therapeutic potential in the treatment of type 2 diabetes mellitus [12].

Salvia is a traditional Jordanian medicine of the family Labiateae (Lamiacae), which is derived from the Latin word salvare, "to heal." For centuries, many salvias have been valued for their medicinal and culinary qualities, enriched with terpenoids, flavonoids, and coumarins. It has been proven that pancreatic triacylglycerol lipase, α-amylase, and α-glucosidase are appealing pharmacological targets for the management of dyslipidemia, atherosclerosis, obesity, and diabetes. An in vitro study performed by Kasabri et al. and conducted on colorectal cancer cell lines such as HT29, HCT116, SW620, and CaCo2 revealed considerable inhibition of these digestive hydrolases by crude aqueous extracts (AEs) of *Salvia* spp. The AEs were identified as potent and efficacious dual inhibitors of α-amylase and α-glucosidase in in vitro experimentation. It has been concluded that *Salvia* spp. can modulate gastrointestinal carbohydrate and lipid absorption and digestion and may be advocated as potential candidates for combinatorial obesity-diabetes (diabesity) prevention and phytotherapy [29].

Citrus fruits contain a large amount of flavonoids, of which hesperetin and naringenin have been proven to have hypolipidemic effects. Human hepatoma cell line HepG2 cells are an important in vitro model system extensively used to study the hypolipidemic effects of citrus flavonoids and catabolism of apolipoprotein B (apoB) containing lipoproteins such as very low-density lipoprotein and LDL. PMF, minor components from citrus fruits, have shown powerful hypolipidemic and apoB-lowering responses. In vitro and in vivo studies have indicated that this particular group of citrus flavonoids might regulate hepatic lipid and apoB metabolism via distinct mechanism(s). Kurowska et al. performed a study to investigate the mechanism(s) by which **tangeretin**, the most abundant and most active PMF from citrus fruits, regulates apoB protein and lipid metabolism using HepG2 cells. The results showed that tangeretin decreased intracellular synthesis of cholesteryl esters and free cholesterol. It also suppressed triacylglycerol (TAG) availability via inhibition of diacylglycerol acyltransferase activity, which partly accounts for the net decrease in apoB secretion. Tangeretin was also found to activate PPAR activity, which was assumed to contribute to the observed drop in cellular accumulation of TAG via β-oxidation. Thus the scientists concluded that tangeretin affects the regulation of hepatic apoB secretion through multiple mechanisms and has potential in the treatment of hypertriglyceridemia associated with obesity and type 2 diabetes [30].

Barley (*Hordeum vulgare* L.) is a member of the grass family and was one of the first cultivated grains. β-Glucan is a major bioactive compound in barley, whereas the other bioactive compounds include phenolic acids such as ferulic acid, coumaric acid, and benzoic acid. These compounds have proven to exert their adipogenetic effects in many cell types, including adipocytes, osteoblasts, and immune cells [31,32]. Various extracting conditions such as storage and heating temperature have shown an influence on bioactive ingredients, resulting in different antioxidative and antiobesity effects [33,34]. An in vitro study performed by Cho-RongSeo et al. on mesenchymal (C3H10T1/2) and preadipocyte (3T3-L1) cells demonstrated the antilipogenic effects of both aqueous and ethanolic extracts of hulled barley, roasted at different temperatures up to 250°C. The phenolic content of aqueous extract of hulled barley (AHB) was observed to be increasing as the roasting temperature increased. However, the phenolic content of AHB roasted at 210°C (AHB210) was found to be higher than that at 250°C (AHB250) and thus was their anti-adipogenic activity. Regardless of all the roasting conditions, ethanolic extract showed the same amount of phenolic content. The lower amount of β-glucan in AHB210 than in AHB250 suggests that it is not solely responsible for antiadipogenesis activity. Furthermore, it was also found that both coumaric acid and ferulic acid dose dependently suppressed lipid accumulation. A correlation between relatively high phenolic contents and the anti-adipogenic effects of AHB210 suggests that polyphenolic compounds found in barley are responsible for antiadipogenic effects, and ferulic and coumaric acid act as the potential mediators [35].

Grapes are a reservoir of many phenolic phytochemicals, including flavonoids, anthocyanins, stilbenes such as resveratrol, etc. **Procyanidins** are flavonoids extracted from grape seed (grape seed proanthocyanidin extract, GSPE), and its impact on adipocyte metabolism in 3T3-L1 cells has shown that it has a protective role in obesity and insulin resistance, with an effect on lipid synthesis, lipid degradation, glucose uptake, and adipocyte differentiation [36]. Chacón et al. investigated the potential antiinflammatory actions of procyanidins on human adipocytes (SGBS) and macrophage-like (THP) cell lines in combating obesity, diabetes, and insulin resistance states. The results showed that the exposure of these cell lines to increasing nontoxic doses of GSPE followed by an inflammatory stimulus modulated IL-6 and monocyte chemoattractant protein 1 expression levels. It was also observed that GSPE alone could modulate in vitro adiponectin gene expression and has a partial inhibitory effect on NF-κB translocation to the nucleus in both the cell lines. It was concluded that GSPE enhances the production of the anti-inflammatory adipokine adiponectin, suggesting that it may have a beneficial effect on low-grade inflammatory diseases such as obesity and type 2 diabetes [37].

P. corylifolia seeds, commonly called "babchi seeds," are a medicinally important in India and have been proved to possess antiinflammatory activity. Suhashini R. et al. demonstrated the antidiabetic activity of different extracts of *P. corylifolia* using in vitro assays such as glucose uptake by yeast cells, glucose diffusion assay, α-amylase inhibition assay, and glycosylation of hemoglobin assay. The data showed that the methanolic extract of *P. corylifolia* possesses higher antidiabetic activity when compared to ethyl acetate and hexane extracts. The plant extract plays a role in improving the formation of glycosylated hemoglobin and hence reduces the concentration of free glucose in the blood. The methanol extract of *P. corylifolia* was found to cause minimum toxicity in L6 cell lines. The scientists concluded that the metabolic extract of *P. corylifolia* possesses a higher antidiabetic activity [38].

1.3 Natural Products: Cardiovascular Disorders

Cardiovascular disease (CVD) as the name suggests encompasses conditions affecting the heart or blood vessels. CVD covers a broad array of diseases, including ischemic heart disease, coronary heart disease (angina, myocardial infarction), cerebrovascular disease (e.g., stroke), hypertension, rheumatic heart disease, heart failure, and peripheral vascular disease. The major causes of CVD are physical inactivity, smoking, unhealthy diets, high cholesterol, high blood pressure, increased stress levels, diabetes, family history of CVD, and excessive consumption of tobacco, alcohol, and sodium. CVD is one of the leading causes of morbidity and mortality in developed as well as developing countries, including India, and its prevention is a major public health challenge. Natural products in the form of neutraceuticals, phytochemicals, or crude plant extracts have shown beneficial results in improving CVD, as substantiated from various epidemiological, experimental, and clinical studies [39]. A few examples of natural products that have been used traditionally for the treatment of CVD include *Panax notoginseng*, saffron extract, berberine, luteolin, *Allium sativum*, rhamnetin, baicalein, tanshinone II, and *Brassica oleracea*.

P. notoginseng is a well-known medicinal herb and functional food. It has a long track record in the prevention and treatment of myocardial infarction. The in vitro cardioprotective effect of standardized notoginseng extract in rat cardiac myoblast H9c2 cell lines was explored by Wang and his colleagues. In this study, notoginseng extract provides cell protection against oxygen and glucose deprivation by inhibiting NLRP3 inflammasome activation and suppresses apoptosis in ROS-associated endoplasmic reticulum stress, thereby protecting the myocardiocytes from cell damage in response to ischemic insult [40]. **Saffron** is a natural plant known for its antioxidant effect and is a rich source of flavonoids. Saffron extract mainly contains saffron, safranal, and corcins. Previous reports have provided evidence of a strong radical scavenging potential of saffron extract. In H9c2 cardiomyocytes, saffron extract has shown novel beneficial outcomes against doxorubicin-induced cardiac toxicity in combination with ischemia/reperfusion conditions. Saffron extract attenuates the expression of inflammatory pathways p-AKT, p-P70S6K, p-ERK1/2, α-actinin, troponin C, apoptotic activity of caspase-3, and improvement of cardiomyocyte survival [41].

Luteolin has shown beneficial effects on rat cardiomyocytes and isolated heart tissue against simulated ischemia/reperfusion injury at least partially through the oxidative (attenuates lactate dehydrogenase, superoxide dismutase, malondialdehyde levels) and inflammatory pathways partially mediated by the PI3K/Akt pathway [42]. Another potential mechanism by which luteolin exhibits its protective action in neonatal cardiomyocytes in myocardial infarction is via inhibiting the mammalian sterile 20-like kinase 1 enzyme, improving mitochondrial biogenesis, upregulating autophagy, and downregulating apoptosis-associated genes. A different novel mechanism by which luteolin alleviates myocardial ischemia/reperfusion injury is by decreasing miR-208b-3p and increasing Ets1 expression levels involved in apoptotic pathways [43]. The role of **berberine** as an M2 muscarinic agonist on the heart has been explored, with its impending actions on ion channels, adrenergic and opioid receptors, and phosphodiesterase enzymes [44].

Tanshinone II, an active constituent from *Salvia miltiorrhiza* (danshen), is a Chinese medicinal plant traditionally used for various cardiovascular diseases such as angina pectoris, atherosclerosis, myocardial infarction, and ischemia/reperfusion injury. In addition to its ethnopharmacological use, the protective effect of tanshinone IIA was explored by Mao et al. in human cardiac fibroblasts in vitro. In this study the phytochemical primarily acts through inhibition of Smad signaling by interfering with Smad-mediated recruitment of CBP1 via the activation of CREB. It also opposes extracellular matrix remodeling through inhibition of MMP9 expression via IκB degradation and p65 nuclear translocation [45]. Furthermore, in hypoxic neonatal cardiomyocytes, tanshinone IIA has shown beneficial effects in myocardial ischemia/reperfusion injury via the PI3K/Akt/mTOR signaling pathway. Additionally, another study demonstrated the beneficial effect of tanshinone IIA on decreasing DOX-induced apoptosis in neonatal rat cardiomyocytes and underlying molecular mechanisms. Tanshinone IIA ameliorated apoptosis and ROS generation induced by DOX in a dose-dependent manner [46]. **Danshen**, a traditional Chinese medicine, has been useful for coronary heart disease such as angina and infarction. Magnesium lithospermate B, a phenolic acid in danshen, protects cadiomyocytes from ischemic injury through specific inhibition of the TGF-β-activated protein kinase 1-binding protein 1−p38 apoptosis signaling pathway. In another study, magnesium lithospermate B prevented myocardial damage in H9c2 cardiomyocytes by modulating apoptosis marker Bcl-2 and Bax protein and phosphorylation of Akt pathways [47].

Garlic (*A. sativum*) is one of the best herbal remedies for various ailments, and a commonly used spice in food. Its ethnopharmacological use has also been ascribed by Charak (around 300 BC) and Egyptian Codex Ebers (approximately 3500 years ago) for maintaining the fluidity of blood and strengthening the heart. In one study, odorless garlic extract activated fibrinolytic activity by accelerating t-PA-mediated plasminogen activation, suppressing the coagulation system by downregulating thrombin formation, and thereby playing an important role in preventing pathological thrombus formation. A recent in vitro study indicated that the chloroform extract of aged black garlic attenuated TNF-α-induced VCAM-1 expression via an NF-κB-dependent pathway in human umbilical vein endothelial cells. In primary human coronary artery endothelial cells, aqueous extract of garlic dose dependently curbs ICAM-1 and VCAM-1 expression induced by IL-1 [48]. Another flavonoid, **rhamnetin**, with antioxidant, anticancer, and antibacterial activities, consists of two benzene rings and an oxygen-containing pyran ring. Park et al. demonstrated the protective effects of rhamnetin against myocardial damage in H9c2 cells. It acts against the oxidative and apoptotic mechanism by modulating caspase-3, Bax, Bcl-2, heme oxygenase-1 (HO-1), SOD, and CAT expressions and activating various inflammatory pathways of Akt/GSK-3β and MAPK, including ERK1/2, p38 MAPK, and JNK/SAPK [49]. **Baicalein**, a compound from *Scutellaria baicalensis* Georgi, exerts beneficial effects from hypoxia-induced embryonic ventricular cardiomyocyte necrosis and apoptosis via μ- and δ-opioid receptors and its downstream signal transduction via PKC and the K_{ATP} channel. Another novel mechanism through which baicalein protects human embryonic stem cell-derived cardiomyocytes is through the NF-E2-related factor 2 and Kelch-like epichlorohydrin-associated protein (Nef2/Keap1) pathway [50].

1.4 Natural Products: Hypertension

Hypertension, also known as the "silent killer," is a disease state in which blood pressure in the arteries is unceasingly elevated with no specific symptoms. Lifestyle factors that increase the disease risk include excessive salt intake, smoking, alcohol, unhealthy diet, and increased body weight. A few examples of natural products that have been explored in the management of hypertension include black tea, cocoa, pomegranate, docosahexaenoic acid, grape seed procyanidin extract, soy pulp, apigenin, etc. **Docosahexaenoic acid** is an essential component of n-3 polyunsaturated fatty acids that may have a role in preventing cardiovascular events, breast cancer, and inflammation. In 2013, Yan reported the evidence of docosahexaenoic acid as an effective inhibitor suppressing hypoxia-induced pulmonary artery smooth muscle cell (PASMC) proliferation, migration, phenotype modulation, and ERK1/2 activation in vitro [51]. Hypoxia-induced oxidative stress and excessive proliferation of PASMCs play important roles in the pathological process of hypoxic pulmonary hypertension. **Grape seed procyanidin extract**, a biologically active polyphenolic flavonoid combination that contains oligomeric proanthocyanidin, has shown beneficial effects in vitro on pulmonary artery smooth muscle cells, playing an important role in hypoxic pulmonary hypertension [52]. **Soy pulp**, called "okara" in Japanese, is known as a by-product of the production of bean curd (tofu) and has been traditionally used in numerous conditions. In a study by Nishibori, okara extract dose dependently inhibited ACE activity in vitro, proposing the likelihood of okara to reduce blood pressure, thereby being effective and useful as a functional food for hypertension [53].

1.5 Natural Products: Stroke

Stroke, also known as "brain attack," is the foremost basis of death and permanent disability in elderly persons worldwide, including in India. Interruption of blood supply to the brain, either in the form of clots or vessel rupture, is the leading cause of brain attack. Lifestyle risk factors such as obesity, diabetes mellitus, hypertension, hyperlipidemia, cigarette smoking, physical inactivity, and excessive consumption of alcohol have all been associated with increasing the likelihood of stroke. During the years 2001–03, 19% of deaths in India were due to cardiovascular disorders, including stroke, which is estimated to rise 36% by 2030. A National Program for Prevention and Control of Cancer, Diabetes, Cardiovascular Diseases and Stroke was launched to implement a 25% reduction in cardiovascular disease and stroke by the year 2025 [54–56].

A growing body of evidence provides beneficial effects of **resveratrol** in diverse in vitro assays, namely, primary neuronal cultures, cortical glial cells, endothelial cultures, and hippocampal slice cultures. Probable mechanisms of neuroprotective effects of resveratrol may be due to its antioxidant and anti-inflammatory potential mediated by hemeoxygenase 1 and AMPK/SIRT1 signaling pathways [57]. In another study, resveratrol also enhanced neurogenesis after stroke in neural stem cells via a novel Sonic hedgehog signaling pathway. In this study, resveratrol (1, 5, and 20 μM) significantly increased cell survival and proliferation and upregulated Patched-1, Smoothened, Gli-1 protein expression [58]. In AD 1596, the compendium of *Materia Medica* by Li Shizhen reported the first medicinal use of *P. notoginseng*. It also finds a place in the Pharmacopoeia of the People's Republic of China, indicating its clinical and ethnopharmacological usefulness in China. *P. notoginseng* saponin's beneficial effects in SH-SY5Y cells exposed to oxygen/glucose deprivation injury may be attributed to its inhibitory effect on NgR1, RhoA, and ROCK2 expressions [59]. The natural phenolic diterpenoid **totarol** is a major constituent isolated from the sap of *Podocarpus totara* and is traditionally used in numerous ailments. Totarol was found to be beneficial against neuronal injury factors in primary rat cerebellar granule neuronal cells and cerebral cortical neurons, i.e., it prevents glutamate-induced excitotoxicity, protects the cells against nutrient deprivation, and suppresses glucose- and oxygen-deprived cell damage. It also increased Akt and GSK-3β phosphorylation, and Nrf2 and HO-1 protein expressions [60]. **Sinomenine**, a bioactive alkaloid of the Chinese plant *Sinomenium acutum*, has been explored clinically for the management of arthritis in China. Wu and colleagues reported the inhibitory mechanism of sinomenine in vitro using whole-cell patch-clamp techniques, calcium imaging in PC12 cells, and rat cortical neurons. In this study, sinominene inhibited the acid-sensing and voltage-gated calcium channels in rat-cultured cortical neurons, and conferred marked cytoprotection against OGD-R and extracellular acidosis-induced cell injury [61].

1.6 Natural Products: Cancer

Cancer results from an uncontrolled growth of abnormal cells in the body. These abnormal cells are called cancer cells, malignant cells, or tumor cells. Cancer types are identified by the name of the tissue from where the abnormal cells originate (i.e., lung cancer, breast cancer, bowel cancer). A causative factor includes exposure to environmental carcinogens that include chemical or toxic compound exposures, ionizing radiation, diet and obesity, viruses, tobacco, stress, lack of physical activity, and genetic predisposition. An approximate one-third of all cancer deaths could be prevented by improving the daily diet, predominantly by increasing the intake of natural products such as fruits, vegetables, and whole grains. It has been clinically also proven that diets rich in fruits and vegetables are connected with reduced risk for cancer development.

Ginger (*Z. officinale* Roscoe) has been widely used as a condiment throughout the world for centuries. The biologically active components contained in ginger are reported to be phenylpropanoid-derived compounds, including gingerols and shogaols. Studies have indicated that shogaols possess anticancer effects. For example, 6-shogaol has been shown to induce apoptosis in human colorectal carcinoma cells via the production of ROS and activation of caspase. In another study, 6-shogaol was reported to reduce gastric cancer viability by impairing tubulin polymerization. Also ginger exhibited antineoplastic effects in SKOV3 ovarian cancer cells via NF-κB inhibition and diminished secretion of angiogenic factors VEGF and IL-8 [62]. In MDA-MB-231 human breast cancer cells, gingerol inhibits cell adhesion, invasion, motility, and activities of MMP-2 and MMP-9 [63]. Also the anticancer potential of 6-shogaol in human lung cancer A549 cells was demonstrated through the AKT/mTOR pathway [64].

Apple extract has caught the interest of the scientific world due to its wide array of polyphenol phytochemicals useful for reducing disease risk and improving health. For example, apple polyphenols induced apoptosis in human colon carcinoma cells HT29 via inhibition of PKC activity, so reducing the protein expression of PKCα, PHCβ, and PKCγ and increasing the proapoptotic PKCδ fragments. In another study, apple extract reported inhibition of

TNF-α-induced NF-κB activation of human breast cancer MCF-7 cells by inhibiting proteosomal activity. Also Boyer demonstrated the antiproliferative activity of apple extract against human HepG2 liver cancer cells, MCF-7 breast cancer cells, and Caco-2 colon cancer cells [65,66].

Cinnamon extract has shown beneficial effects in vitro in a number of cell lines, i.e., B16F10 and Clone M3 (mouse melanoma cell), Hela (human cervical carcinoma cell), and Caco2 (human epithelial colorectal adenocarcinoma cell) via NF-κB inhibition [67]. In another study, cinnamon oil exerts anticancer activity against head and neck squamous cell carcinoma (FaDu, Detroit-562, and SCC-25) by suppressing epidermal growth factor receptor-tyrosine kinase activity.

The reported therapeutic potential of **mangiferin** has been elucidated in myriad malignancies such as lung cancer, colon cancer, and breast cancer. In A549 human lung adenocarcinoma cells, mangiferin inhibited G2/M phase cell cycle proliferation through reduction of the cyclin-dependent kinase 1-cyclin B1 signaling pathway, and induced apoptosis by inhibiting the PKC NF-κB pathway. Also Huang et al. demonstrated the anticancer activity of mangiferin in **breast cancer** cell line-based in vitro models. In cell lines such as human estrogen receptor (ER)-negative (MDA-MB-231 and BT-549) and ER-positive (MCF-7 and T47D), mangiferin treatment decreased the activities of MMP-7 and MMP-9, reversed epithelial mesenchymal transition, and inhibited activation of the β-catenin pathway. Also for the first time in a human cell model of breast adenocarcinoma, mangiferin inhibited the three enzymatic systems, namely, HMG-CoA reductase, the proteasome, and plasmin, respectively in charge of regulating cholesterol homeostasis, protein turnover, and cell adhesion [68,69].

Rosemary is a common culinary/medicinal herb with noteworthy antiproliferative activities against a variety of human cancer cell lines, including breast, leukemia, prostate, lung, and liver. Moore et al. demonstrated the chemotherapeutic role of rosemary extract in human lung A549 cancer cells in vitro. In this study, rosemary extract inhibited Akt/mTOR/p70S6K activities, thereby inhibiting the proliferation, survival, and apoptosis of lung cancer cells. The antitumor potential of rosemary extract has also been proved in human colon cancer cells (SW620 and DLD-1) and pancreatic cancer cells (MIA-PaCa-2 and PANC-1) by upregulating the glycosyltransferase GCNT3 and downregulating the microRNA-15b pathways. In breast cancer cells, its activity may be mediated by modulating the ER-α and HER2 signaling mechanisms. Its potential against prostate and ovarian cancer cell lines has also been explored [70].

Fenugreek (*Trigonella foenum-graecum*) is one of the oldest medicinal plants, originating in India and Northern Africa. In traditional Chinese medicine, fenugreek seeds were used as a tonic, as well as a treatment for weakness and edema of the legs. In India, fenugreek is commonly consumed as a condiment and used medicinally as a lactation stimulant. Shabbeer et al.

demonstrated the anticancer potential of fenugreek against prostate cancer cell lines (DU-145, LNCaP, and PC-3), breast cancer cell lines (MDA-MB-231, MCF-7, T47D, and SKBR3), and pancreatic cancer cell lines (MiaPaCa, HS766T, Panc1, L3.6PL, and BXPC3) mainly by upregulating p21 expression and inhibiting the TGF-β phosphorylation of Akt [71]. In Indian traditional medicine, *Piper nigrum* has been widely used for digestive problems (diarrhea and indigestion) and respiratory difficulties (colds, fever, and asthma). Deng explored the anticancer potential of *Piper nigrum* using breast cancer cell lines MCF-7 and ZR-75-1. In this study, *P. nigrum* at 7.45 mg/mL upregulated p53, downregulated E-cadherin, vascular endothelial growth factor, and c-Myc protein levels, and inhibited tumor size and cell proliferation [72].

The anticarcinogenic potential of **EGCG**, a chief constituent of green tea, was investigated in different tumor cell lines. The mechanism of EGCG intervention in human lung cancer cells (A549 and H1299) involves the novel Wnt/β-catenin pathway [73]. Also the antimitogenic outcome of EGCG on BeWo, JEG-3, and JAR placental choriocarcinoma cells acts via AMPK, p38, and ERK pathways, but not the JNK pathway [74]. The growth inhibitory potential of EGCG in breast cancer progenitor cells (MDA-MB-231 and MDA-MB-231) is seemingly through downregulation of ER-α36 and epidermal growth factor receptor expression. EGCG inhibits the stem cell-like characteristics of nasopharyngeal cancer cells by suppressing STAT3 activation [75].

1.7 Natural Products: Osteoarthritis and Chronic Obstructive Pulmonary Disease

Kaempferol, a kind of flavonol, is derived from the rhizome of *Kaempferia galanga*, which is used in traditional medicine in Asia for hypertension, abdominal pain, headaches, and rheumatism. Recently, the impending role of kaempferol in the management of bone metabolism disorders such as osteoarthritis was explored in MC3T3-E1 cells. In this study, kaempferol increased the expression of osteoblast-activated factors such as RUNX-2, osterix, BMP-2, and collagen I as well as autophagy-related factors beclin-1, SQSTM1/p62, and the conversion of LC3-II from LC3-I. Also another study showed the osteogenic potential of kaempferol by directly modulating the cytokeratin-14 (*Krt-14*) pathway [76]. The aerial part of *Herba epimedii* is commonly used in traditional Chinese medicine for "strengthening the kidney." These Chinese kidney tonifying medicines have also been used for treating bone disorders for thousands of years. Fang Xie and his colleagues reported the antiosteoporotic role in bone cells, i.e., UMR 106 cells, by inhibiting osteoclastogenesis via modulating the OPG/RANKL pathway [77]. Also another Chinese medicinal plant, Korean red ginseng, was found to be of therapeutic importance in glucocorticoid-induced osteoporosis in MC3T3E1 cells by modulating the AKT/JNK and apoptotic pathways [78].

Another lifestyle disorder, chronic obstructive pulmonary disease (COPD), is characterized by airflow limitation that is not fully reversible. Airflow limitation is usually progressive and associated with an abnormal inflammatory response of the lung to noxious particles or gases. Risk factors for COPD include cigarette smoke, fine airborne particles, air pollution, and exposure to particles such as dusts and chemical irritants. Resveratrol (3,5,40-trihydroxystilbene), a natural plant polyphenolic phytoalexin primarily found in grapes, is generating increased attention these days because of its wide applications in cancer, inflammation, and calorie restriction. Because of its potent antioxidant properties, resveratrol induces GSH biosynthesis via Nrf2 activation; protects lung epithelial cells (A549) against CS-mediated oxidative stress; and implicates its usefulness in COPD [79]. Natural products, namely, quercetin, berberine, and glycyrrhizic acid from licorice, have also been useful for lung inflammatory conditions such as COPD.

2. SUMMARY

Over the past seven to eight decades, natural products have been the source for the discovery of many compounds for the management of acute as well as chronic human conditions, indicating their imperative role in drug discovery processes. In this chapter, we discussed the in vitro screening of natural products or natural product-derived compounds with diverse chemical structures for the treatment of lifestyle-related disorders such as diabetes, obesity, stroke, hypertension, and cancer. These natural compounds have also been beneficial for understanding the pathways and targets involved in lifestyle disorders. The positive results from these natural products are preliminary, therefore continued research in the form of in vivo studies and human trials is mandatory before introducing them as new candidate drugs in the market.

REFERENCES

[1] E. Patridge, et al., An analysis of FDA-approved drugs: natural products and their derivatives, Drug Discovery Today 21 (2) (2016) 204–207.

[2] A.G. Atanasov, et al., Discovery and resupply of pharmacologically active plant-derived natural products: a review, Biotechnol. Adv. 33 (8) (2015) 1582–1614.

[3] G.M. Cragg, D.J. Newman, Natural products: a continuing source of novel drug leads, Biochim. Biophys. Acta 1830 (6) (2013) 3670–3695.

[4] L. Katz, R.H. Baltz, Natural product discovery: past, present, and future, J. Ind. Microbiol. Biotechnol. 43 (2–3) (2016) 155–176.

[5] B. David, J.-L. Wolfender, D.A. Dias, The pharmaceutical industry and natural products: historical status and new trends, Phytochem. Rev. 14 (2) (2015) 299–315.

[6] D.J. Newman, G.M. Cragg, Natural products as sources of new drugs from 1981 to 2014, J. Nat. Prod. 79 (3) (2016) 629–661.

[7] G. Enns, Metabolic Disease, Encyclopaedia Britannica, 2016.

[8] C. Carlberg, et al., The Metabolic Syndrome, Nutrigenomics, 2016.

[9] A. Armani, et al., Cellular models for understanding adipogenesis, adipose dysfunction, and obesity, J. Cell. Biochem. 110 (3) (2010) 564–572.

[10] C. Larqué, et al., Early endocrine and molecular changes in metabolic syndrome models, IUBMB Life 63 (10) (2011) 831–839.

[11] Y.-T. Deng, et al., Suppression of free fatty acid-induced insulin resistance by phytopolyphenols in C2C12 mouse skeletal muscle cells, J. Agric. Food Chem. 60 (4) (2012) 1059–1066.

[12] X.-K. Fang, J. Gao, D.-N. Zhu, Kaempferol and quercetin isolated from *Euonymus alatus* improve glucose uptake of 3T3-L1 cells without adipogenesis activity, Life Sci. 82 (11) (2008) 615–622.

[13] D. Govorko, et al., Polyphenolic compounds from *Artemisia dracunculus* L. inhibit PEPCK gene expression and gluconeogenesis in an H4IIE hepatoma cell line, Am. J. Physiol. Endocrinol. Metab. (6) (2007) 293. E1503–E1510.

[14] W. Shao-Ling, et al., Curcumin, a potential inhibitor of up-regulation of TNF-alpha and IL-6 induced by palmitate in 3T3-L1 adipocytes through NF-kappaB and JNK pathway, Biomed. Environ. Sci. 22 (1) (2009) 32–39.

[15] S.M. Grundy, Obesity, metabolic syndrome, and cardiovascular disease, J. Clin. Endocrinol. Metab. 89 (6) (2004) 2595–2600.

[16] D.J. Williams, et al., Vegetables containing phytochemicals with potential anti-obesity properties: a review, Food Res. Int. 52 (1) (2013) 323–333.

[17] C. Di Lorenzo, et al., Metabolic syndrome and inflammation: a critical review of in vitro and clinical approaches for benefit assessment of plant food supplements, Evidence-Based Complementary Altern. Med. 2013 (2013).

[18] Z.Q. Wang, et al., Bioactives of *Artemisia dracunculus* L. enhance cellular insulin signaling in primary human skeletal muscle culture, Metabolism 57 (2008) S58–S64.

[19] B. García-Carrasco, et al., In vitro hypolipidemic and antioxidant effects of leaf and root extracts of *Taraxacum officinale*, Med. Sci. 3 (2) (2015) 38–54.

[20] S. Wolfram, et al., Epigallocatechin gallate supplementation alleviates diabetes in rodents, J. Nutr. 136 (10) (2006) 2512–2518.

[21] Z. Zhang, et al., Epigallocatechin-3-O-gallate (EGCG) protects the insulin sensitivity in rat L6 muscle cells exposed to dexamethasone condition, Phytomedicine 17 (1) (2010) 14–18.

[22] Y. Tang, S. Zheng, A. Chen, Curcumin eliminates leptin's effects on hepatic stellate cell activation via interrupting leptin signaling, Endocrinology 150 (7) (2009) 3011–3020.

[23] Y. Tang, A. Chen, Curcumin prevents leptin raising glucose levels in hepatic stellate cells by blocking translocation of glucose transporter-4 and increasing glucokinase, Br. J. Pharmacol. 161 (5) (2010) 1137–1149.

[24] H.-Q. Yin, et al., *Magnolia officinalis* reverses alcoholic fatty liver by inhibiting the maturation of sterol regulatory element–binding Protein-1c, J. Pharmacol. Sci. 109 (4) (2009) 486–495.

[25] M.S. Seo, et al., *Magnolia officinalis* attenuates free fatty acid-induced lipogenesis via AMPK phosphorylation in hepatocytes, J. Ethnopharmacol. 157 (2014) 140–148.

[26] L. Shi, et al., Corosolic acid stimulates glucose uptake via enhancing insulin receptor phosphorylation, Eur. J. Pharmacol. 584 (1) (2008) 21–29.

[27] M.P. Rani, et al., *Zingiber officinale* extract exhibits antidiabetic potential via modulating glucose uptake, protein glycation and inhibiting adipocyte differentiation: an in vitro study, J. Sci. Food Agric. 92 (9) (2012) 1948–1955.

[28] H. Yang, et al., Effect of extracts from *Euonymus alatus* Sieb. on peripheral glucose uptake of adipocytes, Chin. J. Nat. Med. 2 (6) (2004) 365–368.

[29] V. Kasabri, et al., In vitro modulation of metabolic syndrome enzymes and proliferation of obesity related-colorectal cancer cell line panel by Salvia species from Jordan, Rev. Roum. Chim. 59 (2014) 693–705.

[30] E.M. Kurowska, et al., Modulation of HepG2 cell net apolipoprotein B secretion by the citrus polymethoxyflavone, tangeretin, Lipids 39 (2) (2004) 143–151.

[31] C.-C. Chuang, M.K. McIntosh, Potential mechanisms by which polyphenol-rich grapes prevent obesity-mediated inflammation and metabolic diseases, Annu. Rev. Nutr. 31 (2011) 155–176.

[32] H. Matsuda, et al., Structural requirements of flavonoids for the adipogenesis of 3T3-L1 cells, Bioorg. Med. Chem. 19 (9) (2011) 2835–2841.

[33] E. Gómez-Plaza, et al., Color and phenolic compounds of a young red wine. Influence of wine-making techniques, storage temperature, and length of storage time, J. Agric. Food Chem. 48 (3) (2000) 736–741.

[34] M. Meydani, S.T. Hasan, Dietary polyphenols and obesity, Nutrients 2 (7) (2010) 737–751.

[35] C.-R. Seo, et al., Aqueous extracts of hulled barley containing coumaric acid and ferulic acid inhibit adipogenesis in vitro and obesity in vivo, J. Funct. Foods 12 (2015) 208–218.

[36] M. Pinent, et al., Procyanidin effects on adipocyte-related pathologies, Crit. Rev. Food Sci. Nutr. 46 (7) (2006) 543–550.

[37] M.R. Chacón, et al., Grape-seed procyanidins modulate inflammation on human differentiated adipocytes in vitro, Cytokine 47 (2) (2009) 137–142.

[38] R. Suhashini, S. Sindhu, E. Sagadevan, In vitro evaluation of anti diabetic potential and phytochemical profile of *Psoralea corylifolia* seeds, Int. J. Pharmacogn. Phytochem. Res. 6 (2014) 414–419.

[39] S.K. Shukla, et al., Cardiovascular friendly natural products: a promising approach in the management of CVD, Nat. Prod. Res. 24 (9) (2010) 873–898.

[40] L.-C. Wang, et al., A standardized notoginseng extract exerts cardioprotection by attenuating apoptosis under endoplasmic reticulum stress conditions, J. Funct. Foods 16 (2015) 20–27.

[41] N. Chahine, et al., Saffron extracts alleviate cardiomyocytes injury induced by doxorubicin and ischemia-reperfusion in vitro, Drug Chem. Toxicol. 39 (1) (2016) 87–96.

[42] R.-Q. Zhang, et al., Antioxidative effect of luteolin pretreatment on simulated ischemia/reperfusion injury in cardiomyocyte and perfused rat heart, Chin. J. Integr. Med. (2016) 1–10.

[43] C. Bian, et al., Luteolin inhibits ischemia/reperfusion-induced myocardial injury in rats via downregulation of microRNA-208b-3p, PLoS One 10 (12) (2015) e0144877.

[44] S. Salehi, T.M. Filtz, Berberine possesses muscarinic agonist-like properties in cultured rodent cardiomyocytes, Pharmacol. Res. 63 (4) (2011) 335–340.

[45] S. Mao, et al., Tanshinone IIA inhibits angiotensin II induced extracellular matrix remodeling in human cardiac fibroblasts–Implications for treatment of pathologic cardiac remodeling, Int. J. Cardiol. 202 (2016) 110–117.

[46] Q. Li, et al., Tanshinone IIA protects against myocardial ischemia reperfusion injury by activating the PI3K/Akt/mTOR signaling pathway, Biomed. Pharmacother. 84 (2016) 106–114.

[47] W. Quan, et al., Magnesium lithospermate B improves myocardial function and prevents simulated ischemia/reperfusion injury-induced H9c2 cardiomyocytes apoptosis through Akt-dependent pathway, J. Ethnopharmacol. 151 (1) (2014) 714–721.

[48] T.N. Khatua, R. Adela, S.K. Banerjee, Garlic and cardioprotection: insights into the molecular mechanisms 1, Can. J. Physiol. Pharmacol. 91 (6) (2013) 448–458.

[49] E.-S. Park, et al., Cardioprotective effects of rhamnetin in H9c2 cardiomyoblast cells under H_2O_2-induced apoptosis, J. Ethnopharmacol. 153 (3) (2014) 552−560.
[50] G. Cui, et al., Cytoprotection of baicalein against oxidative stress-induced cardiomyocytes injury through the Nrf2/Keap1 pathway, J. Cardiovas. Pharmacol. 65 (1) (2015) 39−46.
[51] J. Yan, et al., Docosahexaenoic acid inhibits development of hypoxic pulmonary hypertension: in vitro and in vivo studies, Int. J. Cardiol. 168 (4) (2013) 4111−4116.
[52] H. Jin, et al., Grape seed procyanidin extract attenuates hypoxic pulmonary hypertension by inhibiting oxidative stress and pulmonary arterial smooth muscle cells proliferation, J. Nutr. Biochem. 36 (2016) 81−88.
[53] N. Nishibori, R. Kishibuchi, K. Morita, Soy pulp extract inhibits angiotensin I-Converting enzyme (ACE) activity in vitro: evidence for its potential hypertension-improving action, J. Diet. Suppl. (2016) 1−11.
[54] T. Mutoh, Y. Taki, T. Ishikawa, Therapeutic potential of natural product-based oral nanomedicines for stroke prevention, J. Med. Food 19 (6) (2016) 521−527.
[55] T.K. Banerjee, S.K. Das, Fifty years of stroke researches in India, Ann. Indian Acad. Neurol. 19 (1) (2016) 1−8.
[56] J. Kim, et al., Phytochemicals in ischemic stroke, Neuromol. Med. 18 (3) (2016) 283−305.
[57] N. Singh, M. Agrawal, S. Dore, Neuroprotective properties and mechanisms of resveratrol in in vitro and in vivo experimental cerebral stroke models, ACS Chem. Neurosci. 4 (8) (2013) 1151−1162.
[58] W. Cheng, et al., Sonic hedgehog signaling mediates resveratrol to increase proliferation of neural stem cells after oxygen-glucose deprivation/reoxygenation injury in vitro, Cell Physiol. Biochem. 35 (5) (2015) 2019−2032.
[59] X. Shi, et al., *Panax notoginseng* saponins provide neuroprotection by regulating NgR1/RhoA/ROCK2 pathway expression, in vitro and in vivo, J. Ethnopharmacol. 190 (2016) 301−312.
[60] Y. Gao, et al., Totarol prevents neuronal injury in vitro and ameliorates brain ischemic stroke: potential roles of Akt activation and HO-1 induction, Toxicol. Appl. Pharmacol. 289 (2) (2015) 142−154.
[61] W.N. Wu, et al., Sinomenine protects against ischaemic brain injury: involvement of co-inhibition of acid-sensing ion channel 1a and L-type calcium channels, Br. J. Pharmacol. 164 (5) (2011) 1445−1459.
[62] J. Rhode, et al., Ginger inhibits cell growth and modulates angiogenic factors in ovarian cancer cells, BMC Complement. Altern. Med. 7 (2007) 44.
[63] H.S. Lee, et al., [6]-Gingerol inhibits metastasis of MDA-MB-231 human breast cancer cells, J. Nutr. Biochem. 19 (5) (2008) 313−319.
[64] J.Y. Hung, et al., 6-Shogaol, an active constituent of dietary ginger, induces autophagy by inhibiting the AKT/mTOR pathway in human non-small cell lung cancer A549 cells, J. Agric. Food Chem. 57 (20) (2009) 9809−9816.
[65] D.A. Hyson, A comprehensive review of apples and apple components and their relationship to human health, Adv. Nutr. Int. Rev. J. 2 (5) (2011) 408−420.
[66] J. Boyer, R.H. Liu, Apple phytochemicals and their health benefits, Nutr. J. 3 (1) (2004) 1.
[67] H.-K. Kwon, et al., Cinnamon extract induces tumor cell death through inhibition of NFκB and AP1, BMC Cancer 10 (1) (2010) 392.
[68] H. Li, et al., Mangiferin exerts antitumor activity in breast cancer cells by regulating matrix metalloproteinases, epithelial to mesenchymal transition, and β-catenin signaling pathway, Toxicol. Appl. Pharmacol. 272 (1) (2013) 180−190.

[69] M. Cuccioloni, et al., Mangiferin blocks proliferation and induces apoptosis of breast cancer cells via suppression of the mevalonate pathway and by proteasome inhibition, Food Funct. 7 (10) (2016) 4299–4309.
[70] J. Moore, M. Yousef, E. Tsiani, Anticancer effects of Rosemary (*Rosmarinus officinalis* L.) extract and Rosemary extract polyphenols, Nutrients 8 (11) (2016) 731.
[71] S. Shabbeer, et al., Fenugreek: a naturally occurring edible spice as an anticancer agent, Cancer Biol. Ther. 8 (3) (2009) 272–278.
[72] Y. Deng, et al., Anti-cancer effects of Piper nigrum via inducing multiple molecular signaling in vivo and in vitro, J. Ethnopharmacol. 188 (2016) 87–95.
[73] J. Zhu, et al., Wnt/beta-catenin pathway mediates (-)-Epigallocatechin-3-gallate (EGCG) inhibition of lung cancer stem cells, Biochem. Biophys. Res. Commun. (2016).
[74] L.J. Shih, et al., Green tea (-)-epigallocatechin gallate induced growth inhibition of human placental choriocarcinoma cells, Placenta 41 (2016) 1–9.
[75] X. Pan, et al., Estrogen receptor-α36 is involved in epigallocatechin-3-gallate induced growth inhibition of ER-negative breast cancer stem/progenitor cells, J. Pharmacol. Sci. 130 (2) (2016) 85–93.
[76] I.R. Kim, et al., The role of kaempferol-induced autophagy on differentiation and mineralization of osteoblastic MC3T3-E1 cells, BMC Complement. Altern. Med. 16 (1) (2016) 333.
[77] F. Xie, et al., The osteoprotective effect of *Herba epimedii* (HEP) extract in vivo and in vitro, Evidence-Based Complementary Altern. Med. 2 (3) (2005) 353–361.
[78] J. Kim, et al., Protective effect of Korean Red Ginseng against glucocorticoid-induced osteoporosis in vitro and in vivo, J. Ginseng Res. 39 (1) (2015) 46–53.
[79] A. Kode, et al., Resveratrol induces glutathione synthesis by activation of Nrf2 and protects against cigarette smoke-mediated oxidative stress in human lung epithelial cells, Am. J. Physiol. Lung Cell Mol. Physiol. 294 (3) (2008) L478–L488.

Chapter 13

Common Toxic Plants and Their Forensic Significance

Nawal K. Dubey, Abhishek K. Dwivedy, Anand K. Chaudhari, Somenath Das
Centre of Advanced Study in Botany, Institute of Science, Banaras Hindu University, Varanasi, India

1. HISTORICAL ASPECT OF POISONOUS PLANTS

Most children have heard stories about sadhus, who habitually took dhatura seeds, which were considered poisonous. Most people have also read about vishkanyas in fiction and about poison arrows in both fact and fiction. The story of vishkanyas is fascinating and was believed to have been started in *Arthashastra*, composed by Chanakya, adviser and prime minister of the first Maurya emperor, Chandragupta. Vishkanyas, also called poison damsels, were young ladies whose blood was toxic and utilized to execute adversaries [1]. Archeological evidence suggests that primitive men were in search of more efficient weapons for protection from animals or their enemies. In ancient times, poisonous compounds were considered as mysterious substances. Egyptian kings studied poisonous plants around 300 BC. The roman emperor Claudius was murdered by his wife Agripinna by expert administration of *Amanita* or death cap mushroom [2]. Water hemlock containing coniine, another notorious poisonous compound, was used to murder the great Greek philosopher Socrates (c.479–399 BC). Poisoned weapons were also used in ancient India in war tactics. A verse in Sanskrit, "Jalam visravayet sarmava-mavisravyam ca dusayet," means the water of wells was mixed with poison and was thus polluted. Mithridates, the king of Pontus, used to consume small amounts of poison regularly to develop immunity against poisonous compounds [3]. Mistletoe (*Viscum album*) has been known from ancient times for its poisonous properties. It was considered as a holy plant in pagan times, particular by the Celts. In Nordic texts the famous death of the god Baldr was caused by an arrow made of mistletoe [4]. Mistletoe contains many poisonous compounds such as toxic lectins and viscumins [5]. Extracts of mistletoe have been useful in a number of diseases and are still common in alternative medicine, particularly in the treatment of cancer [6]. These extracts are mostly

Natural Products and Drug Discovery. https://doi.org/10.1016/B978-0-08-102081-4.00013-7

comprised of many components and it is difficult to categorize the compounds as having beneficial or harmful effects.

2. COMMON TOXIC PLANTS

Over the past few years a lot of toxins have been used to attempt suicide. Several plant species, poisonous or injurious to the human body, are commonly present in our gardens or planted by the forest department as a roadside tree unknowingly or deliberately without studying their effects on humans. Poisoning can be of different types, namely, by contact (irritation of skin), ingestion (causing internal poisoning), absorption (by the dermal layer), and inhalation (in the respiratory system). Groups of poisonous plants are decided by their poisonous principle. Often, these principles are chemically similar or identical within a single genus or family, especially if the taxa are closely related. Bakain (*Melia azedarach*), a relative of the famous neem tree, also known as chinaberry, is of common occurrence in India either on the roadside or in large gardens; its leaves are used as insect repellents in stored grains and it is advisable to remove all bakain leaves from stored grain before consumption [7]. Generally, all parts of the plant are harmful to animals. However, the fruits are highly poisonous to humans, and birds gorge on them, reaching a level of "intoxication" [8]. The seeds of the castor bean (*Ricinus communis*) and the Jequiriti bean (*Abrus precatorius*) have been recognized for their toxicity since times immemorial. The castor bean originates from Asia and Africa, but has now spread to all temperate and subtropical regions. Ricin is the principal toxin from castor bean having a wide range of biological effects on higher organisms, and more than 700 cases of human intoxication with ricin have been described [9]. Being a protein, to a large extent ricin is destroyed in the intestine. However, the potency of toxin increases 100 times when administered by injection [10]. The seeds of the Jequiriti bean (*A. precatorius*) are highly toxic and often ingested as a means of suicide. Abrin is an immune modulator but inactivates protein synthesis and disturbs the central nervous system [11]. Modeccin is found in the roots of the Southern African plant *Adenia digitata*, which causes serious intoxications due to its resemblance to the root of an edible plant [12]. Volkensin is a toxin closely related to modeccin [13]. *Nerium oleander* is one of the poisonous ornamental plants that can poison food with its fumes during cooking. There is another poisonous plant called Pangi (*Pangium edule*) common in Southeast Asia. Water hemlock (*Cicuta maculata*), a native of the United States and Canada, is also a deadly poison whose small amounts may cause death. In traditional medicine, *Gloriosa superba* is used to treat ailments such as sprains, bruises, colic pain, etc. but an overdose causes poisoning and it is also used for committing suicide. The Manchineel (*Hippomane mancinella*), a native of Florida and the Caribbean, is extremely toxic and causes serious dermatitis even after 30 min [14].

The major poisonous principles present in plants are organic compounds such as alkaloids, diterpenes, flavonoids, tannins, cardiac and cyanogenic glycosides, proanthocyanidins, phenylpropanoids, lignans, nitrogen-containing compounds, resins, oxalates, and certain proteins or amino acids [15]. Some poisonous plants also accumulate inorganic compounds from the soil. Table 13.1 is a list of common toxic plants with their active principles and major symptoms caused in humans after administration.

The synthesis of toxic secondary metabolites follows a number of biochemical pathways, which are interconnected and depend on several biotic and abiotic factors of the regions where the plants flourish. Therefore the nature and toxicity of these compounds vary with respect to locality. This is also evident with the level of poisonous piperidine alkaloids from *Conium maculatum*, which is positively correlated with herbivory by *Agonopterix alstroemeriana*, an oecophorid caterpillar found in New York and Washington, USA [39]. Similarly, the level of saponins is also influenced by surrounding environmental conditions. The saponin content of three *Panax* sp. from Asia (Russia, Korea, China, and Japan), North America, and Asia (China) varies significantly [40]. In addition, the glucosinulate concentration in plants is significantly affected by average temperature and day length [41]. Hence it is advisable to take care of the surrounding conditions as well as the type of toxic plants for social and agroforestry programs.

3. IMPACTS OF POISONOUS PLANTS ON GRAZING ANIMALS

Presently, poisonous plants cause tremendous economic loss for the livestock industry. These economic losses are not only due to the death of livestock but also deterioration in their health, decreased productivity, deformed offspring, and reduced longevity are also leading causes. The nature of these toxic secondary metabolites changes with varying place of origin and environmental conditions. The main toxic components that function in defense systems of plants against herbivores include tannins [42], phenolics [43], alkaloids [44], phytohemagglutinins, terpenes [45], cyanogenic glycosides [46], and oxalates [47], etc. It is relatively clear that the aforementioned compounds are developed for defense. Poisonous compounds of toxic plants such as pyrrolizidine alkaloids of *Crotalaria* spp. [48], piperidine alkaloids of *C. maculatum* [49], and tremetol of *Eupatorium* sp. [50] may cause significant deterioration of milk quality in livestock. Table 13.2 is a list of toxins that affect milk quality if consumed by grazers.

The evolutionary history of poisonous plants, the toxic compounds in the plants, and the animals that graze on these plants have received little attention. Much of the literature dealing with coevolution theory related to poisonous plants has been confined to insects [42,60–63]. All the defense mechanisms proposed are based on the assumption that because these plants are poisonous

TABLE 13.1 Common Toxic Plants With Their Active Principle and Major Symptoms

Plant Species	Family	Active Compounds	Major Symptoms in Humans	References
Conium maculatum	Apiaceae	Coniine, coniceine	Vomiting and diarrhea, inflammation of the gastrointestinal tract, mental confusion, convolution and often death	[16]
Datura stromonium	Solanaceae	Atropine, hyoscamine, scopolamine	Flushed skin, dilated pupils, even death from respiratory failure	[17]
Abrus precatorius	Fabaceae	Abrin abric acid from thoroughly chewed seed	Gastrointestinal disease	[18]
Argemone mexicana	Papaveraceae	Toxic alkaloids sanguinarine, berberine, protopine	Generally causes edema, fruits are contaminated	[19]
Gutierrezia sarothrae	Asteraceae	Saponin in leaves	Severely toxic if eaten in quantity and also causes abortion	[20]
Hippomane mancinella	Euphorbiaceae	Purgative oil milky sap	Severe gastroenteritis	[21]
Manihot esculenta	Euphorbiaceae	Hydrocyanic acid	Death from cyanide poisoning	[22]

Nerium indicum	Apocynaceae	Cardioactive glycosides	Bloody diarrhea, drowsiness, respiratory paralysis	[23]
Nicotiana tabacum	Solanaceae	Nicotine	Vomiting, diarrhea, slow pulse, collapse, respiratory failure	[24].
Passiflora quandrangularis	Passifloraceae	Passiflorine hydrocyanic acid	Symptoms of drunkenness, paralysis, coldness	[25]
Philodendron spp.	Araceae	Tissue contains irritant juice, calcium oxalate crystal	Ingestion results in severe swelling of throat and mouth, possible asphyxiation	[26]
Ricinus communis	Euphorbiaceae	Ricin, cardiac glycoside	Nausea, muscle spasm, purgation, convolution, death	[27]
Urtica dioica	Urticaceae	Formic acid, histamine, acetylcholine in sting	Allergies, eye infection especially in cornea	[28]
Dieffenbachia sp.	Araceae	Oxalic acid, asparagine	Burning in mouth or throat, swelling and blistering of mouth, nausea, damage to cornea of the eye	[29]
Xanthium strumarium	Asteraceae	Hydroquinone in seeds and seedlings	DNA damage, chromosome aberration	[30]

Continued

TABLE 13.1 Common Toxic Plants With Their Active Principle and Major Symptoms—cont'd

Plant Species	Family	Active Compounds	Major Symptoms in Humans	References
Strychnos nux-vomica	Loganiaceae	Strychnine, brucine, vomicine, protostrychnine, chlorogenic acid, *n*-oxystrychnine	Painful seizures, spasms, difficulty breathing, dizziness, confusion	[31]
Calotropis gigantea	Asclepiadaceae	Laurane, saccharose, Β-amyrin, calotroposide; calactin, calotoxin; calotropins DI and DII, gigantin	Diarrhea, vomiting, effects in lactation	[32].
Parthenium hysterophorus	Asteraceae	Germacrene D, *trans*-β-ocimene, β-myrcene, β-caryophyllene	Pollen allergy, dermatitis, inflammation, eczema, asthma, hay fever, burning and blistering around the eye	[33]
Aconitum napellus	Ranunculaceae	Aconitine, mesaconitine, hypaconitine, other Aconitum alkaloids	Weakness or inability to move, sweating, breathing problems, heart problems, ventricular arrhythmias	[34]
Convallaria majalis	Asparagaceae	38 different cardenolides (convallarin, cannogenol-3-*O*-α-L-rhamnoside, neoconvallocide, etc.), saponin, azetidine 2-carboxylic acid	Nausea, vomiting, abnormal heart rhythm, headache, decreased consciousness and responsiveness, visual color disturbances	[35]

Atropa belladonna	Solanaceae	Atropine, hyoscine (scopolamine), hyoscyamine	Dilated pupils, blurred vision, tachycardia, constipation, anticholinergic syndrome	[36]
Cerbera odollam	Apocynaceae	Cerberin, cardiac glycoside	Blocks cellular Na^+/K^+ -ATPase pump, increases heart contractility	[37]
Thevetia peruviana	Apocynaceae	Thevetin A, thevetin B, thevetoxin, neriifolin, peruviside, ruvoside	Sensitive to the interactions of cardenolides with Na^+/K^+ -ATPase	[23]
Colocasia esculenta	Araceae	Insoluble and soluble oxalate	Systemic toxicity, including renal failure, hypocalcemia	[38]

TABLE 13.2 Plant Toxins That Affect Milk Quality

Name of Toxin	Source Plant	References
Pyrrolizidine alkaloids	*Heliotropium europaeum*	[51]
	Crotalaria spp.	[52]
	Festuca sp.	[53]
	Cynoglossum sp.	[54]
	Symphytum sp.	[55]
	Echium sp.	[56]
	Senecio sp.	[57]
Piperidine alkaloids	*Conium maculatum*	[49]
	Nicotiana sp.	[58]
Sesquiterpene lactones	Rubber weed	[58a]
	Bitter weed	[58a]
Tremetol	*Eupatorium* sp.	[50]
	Isocoma sp.	[52]
Cannabinol	*Cannabis sativa*	[59]
Glucosinulates	*Amoracia* sp.	[52]
	Brassica sp.	[52]

the result is a reduction of plants consumed by herbivores, in turn maintaining the vigor of the plants and making them better suited in a natural plant community. Possible ways that poisonous compounds may work include extreme toxicity, poisonous properties correlated with palatability, and adverse conditioning. These poisonous compounds are also known as antinutritional factors (ANFs) or antinutritional compounds [73]. The distribution of ANFs varies with families of plant species with their respective poisonous compounds. Alkaloids are mainly found in Umbelliferae [64], Papaveraceae [65], Leguminosae [66], and Liliaceae [67]. Similarly, cyanogens, saponins, phytohemagglutinins, organic acids, and miscellaneous compounds are mainly distributed in the families Leguminosae [68], Sapotaceae [69], Euphorbiaceae [72], Oxalidaceae [38], and Rhamnaceae [70], respectively. These ANFs are generally not lethal, but they reduce animal productivity, cause toxicity during periods of food scarcity, and consume such substances in large quantities. Table 13.3 is a list of poisonous compounds in the leaves of trees and shrubs.

TABLE 13.3 Poisonous Chemical Compounds Affecting Livestock

Poisonous Compound	Name of the Compound	Occurrence	Family	References
Alkaloids	Coniine	*Conium maculatum*	Umbelliferae	[64]
	Mexicanol and mexicanic acid	*Argemone mexicana*	Papaveraceae	[65]
	Quinolizidine	*Lupinus* sp.	Leguminosae	[66]
	Zygacine and zygadenine	*Zygadenus* sp.	Liliaceae	[67]
Glycosides	Cyanogens	*Acacia giraffae*, *Acacia sieberiana*	Leguminosae	[68]
	Saponins	*Albizia stipulata*	Leguminosae	[71]
		Bassia latifolia	Sapotaceae	[69]
Phytohemagglutinins	Ricin	*Ricinus communis*	Euphorbiaceae	[72]
	Robin	*Robinia pseudoacacia*	Fabaceae	[73]
Organic acid	Oxalic acid	*Halogeton glomeratus*	Chenopodiaceae	[73a]
	Oxalic acid and its salt	*Oxalis* sp.	Oxalidaceae	[38]
Nonprotein amino acid	Mimosine	*Leucaena leucocephala*	Fabaceae	[74]
	Indospecine	*Indigofera spicata*	Fabaceae	[75]
Miscellaneous compounds		*Phyllanthus abnormis*	Euphorbiaceae	[76]
		Karwinskia humboldtiana	Rhamnaceae	[70]
		Baileya multiradiata	Asteraceae	[77]

TABLE 13.4 Common Grass Species Poisonous to Grazing Animals

Common Grass Species	Scientific Name	Susceptible Animal	References
Dallisgrass	*Paspalum dilatatum*	Cattle	[79]
Rye grass	*Lolium* spp.	Sheep	[82]
Green foxtail	*Setaria anceps* var.	Horse	[80]
Foxtail	*Hordeum* spp.	Horse	[83]
Spiny burr grass	*Cenchrus longispinus*	Rat	[81]
Prairie threeawn	*Aristida oligantha*	Sheep	[84]
Black speargrass	*Heteropogon contortus*	All herbivores	[85]

Different species of broomweed, ponderosa pine, Monterey cypress and juniper, Tansy (*Tanacetum vulgare*), and pennyroyal (*Mentha pulegium*) contain essential oils and have been shown to cause abortions. The presence of diterpene acids and germacrene D in *Gutierrezia sarothrae* causes animal diseases [78]. Certain grass species, namely, dallisgrass [79], green foxtail [80], spiny burr grass [81], etc., contain some toxic components that affect livestock. A list of grass species that are poisonous to grazing animals is presented in Table 13.4.

4. TOXIC PLANTS OF FORENSIC SIGNIFICANCE

Plants can be the star witness that provides evidence in suicidal, criminal, burglary, and other criminal offences. Parts of plants can be lodged in the clothing or possessions of the perpetrator of a crime and later provide evidence for a court of law when those trained in taxonomy, molecular taxonomy, anatomy, and ecology can interpret their evidence. The first such incident happened in January 1935 when Bruno Hauptmann was tried for kidnapping the baby son of Charles and Anne Lindbergh, and Arthur Koehler, a wood technologist, used his knowledge of wood anatomy to detect the origin of one of the parts of a wooden ladder used in the crime to reach the baby. Never before in the history of the American judicial system had botanical information been admitted as evidence. Later, Koehler's work and testimony provided an example of the inclusion of botanical evidence in other judiciary cases. Plants producing toxic substances in the form of secondary metabolites for self-defense can also be used as evidence for forensic investigations.

Plant-originated weapons or botanical weapons are frequently used by criminals in murder, burglary, and rape cases. Some poisonous plant species, namely, *Conium* [16], *Cicuta* [86], *Nerium* [75], *Aconitum* [87], *Datura* (Ayuba, 2011), and *Ricinus* [87a], which are very toxic and used for homicidal and suicidal purposes, are used especially for forensic significance to catch criminals. No other plants have such a history of crime as *Datura* and the seeds are charming poisons. Generally, plants that are allergens and less poisonous are used for burglary [88]. These plants are the first choice of professional criminals because they are easily available and have no cost. They have played a large a part in romance as well as in crime. Poisoned weapons were used in ancient times in war tactics [89]. Several modes of poisoning were described in Susruta Samhita, which explained how the poisons were mixed with food, drink, honey, and snuff or sprinkled over bed clothes, couches, shoes, jewelry, garlands, and horse saddles [89a]. Presently, there are many cases where criminals have applied these toxicants in food material to a victim's body during bus and train journeys. A lot of work has been reported on the toxicology of plants but no work has been done on poisonous plants in a forensic context [90]. The toxicological substances of these plants give a prefect database for forensic toxicologists. At a crime scene, forensic experts can retrieve evidence related to deliberate or accidental poisoning by poisonous plants. Similar to parts of plants, crime scenes such as forests/gardens as well as symptoms of poisoning can lead an investigation in the right direction [23]. A list of toxic plants that are of forensic significance is presented in Table 13.5.

Plant toxicity is an emerging science with the development of methods, concepts, and understanding. The field of poisoning related to plant materials is one that is relatively poorly studied. In developed countries, poisoning by plants is more or less due to accidental ingestion of seeds, fruits, pollens, trichomes, etc. [99]. However, in developing countries, lack of knowledge of and forensic expertise in plant-derived toxins has meant that there have been few studies of the extent of poisoning with these toxic plant agents.

Lack of botanical knowledge among most people involved in criminal investigations is the prime cause of underutilization of botanical resources as forensic evidence. However, resourceful investigators and scientists are now taking initiatives to change this perception. Presently, evidence from plant systematics, plant anatomy, palynology, plant ecology, and related areas is acceptable. In the case of poisoning by toxic plants, a fast and easy screening test is required for accurate forensic or medical investigation. Molecular and biotechnological tools and setups are used to detect specific toxins associated with the specified organs of animals or humans. Frequently used techniques for forensic purposes are radioimmunoassay, enzyme-linked immunosorbent assay [100], high-performance liquid chromatography [101], reverse transcription polymerase chain reaction (RT-PCR) [102], gas chromatography-mass spectrometry [103], liquid chromatography-tandem mass spectrometry [104], etc. Nowadays, scientists use modern technologics such as probe

TABLE 13.5 Plants and Associated Toxins of Forensic Significance

SN	Plant Name	Common Name	Toxic Component	Plant Part Used	Mode of Action	Symptoms	References
1	*Conium maculatum*	Poison hemlock	Coniine	Leaves, seeds, and roots	Binds through acetylcholine receptor and blocks neuromuscular junction	Respiratory collapse and death	[16]
2	*Cicuta virosa*	Water hemlock	Cicutoxin and virol A	Mostly roots	Binds to gamma-aminobutyric acid receptor and also blocks the K ion channel in T-lymphocytes	Abdominal pain, seizures, and death	[86]
3	*Nerium oleander*	Oleander	Oleandrin and oleandrigenin	All parts (cardiac glycosides)	Binds with Na, K ion enzymes, and inhibits an increase in intercellular Ca ions	Atrioventricular blockage, xanthopsia (yellow vision), tremors, or even coma	[75]
4	*Aconitum napellus*	Monk's hood or queen of poisons	Aconitine	Roots	Interacts with the voltage-dependent Na-ion channel resulting in long-term depolarization	Nausea, bradycardia, arrhythmias, or death	[87]

5	*Ageratina altissima*	White snakeroot	Tremetol	Leaves	Increases and then suddenly decreases free fatty acids and glucose in blood	Loss of appetite, vomiting, and tremors	[91]
7	*Abrus precatorius*	Crab's eye	Abrin	Seed	Abrin binds to 26S ribosomes and inhibits synthesis of transmembrane protein	Respiratory distress, fever, and tightness in chest	[92]
8	*Cerbera odollam*	Suicide tree	Cerbarin (a cardiocide)	Seed	Binds with the Ca ion channel of heart muscle	Heart beat disruption resulting in heart failure	[93]
9	*Calotropis gigantea*	Crown flower/madar flower	Calotropin/calotoxin and calactin	Latex and roots	Harmful to eyes	Blindness	[94]
10	*Ricinus communis*	Castor bean	Ricin	Seed	Inhibits protein synthesis by binding with 60S RNA	Nausea, pulmonary edema, etc.	[87a]
11	*Datura stramonium*	Devil's trumpet	Atropine and scopolamine	Seed and flower	Acts as an anticholinergic	Trachycardia, sometimes even death	[95]
12	*Argemone mexicana*	Sial kanta	Berberine and protopine	Especially seeds	Appears to depress the action of sympathetic	Dropsy	[96]

Continued

TABLE 13.5 Plants and Associated Toxins of Forensic Significance—cont'd

SN	Plant Name	Common Name	Toxic Component	Plant Part Used	Mode of Action	Symptoms	References
					stimulation and adrenaline		
13	*Nicotiana tabacum*	Tambaku/ tobacco	Nicotine	All parts except seeds	Disrupts normal neurotransmitter activity and increases adrenaline levels in blood	High blood pressure, weak immunity, and cancer	[24]
14	*Parthenium hysterophorus*	Congress weed/chatak chandani	Parthenin and lactones	Mostly pollen and trichomes	–	–	[97]
15	*Dieffenbachia* spp.	Dumb cone	Raphides	Leaves and roots	Causes temporary burning sensation	Erythema and edema of tissue	[98]

–, not known.

binding methodology, which easily binds with specific receptor molecules. Sometimes, to compare particular proteins or genetic samples such as victims' DNA or RNA, RT-PCR is frequently used for multifold amplification to solve cases of rape or sexual harassment. Some common techniques that are generally used for forensic investigations are presented in Table 13.6.

The aforementioned methods are powerful tools for the detection of natural toxins derived from poisonous plants, and a forensic scientist can determine the toxins present in body fluids as well as in tissue both qualitatively and quantitatively based on the principle of these methods. Thus the methods appear to be essential for the resolution of some very delicate forensic cases [93].

5. DETOXIFICATION OF PLANT POISON

Usually, humans and animals except some cattle and insects are prone to toxic compounds. Many insects are coevolved with plants, to be able to sequester the toxic compounds for their own metabolic activity or for feeding purposes [61]. A similar situation arises with native animals coevolved with toxic plants. These animals have a better detoxifying ability than domestic livestock.

Detoxification in broad sense means metabolic inactivation of poisonous compounds [113]. The potency of detoxification varies with the animal species, namely, sheep can consume a large percentage of their own body weight of larkspur (*Delphinium* spp.) without any harmful effects; however, this is not the case with cattle [114]. Sheep and goats are also immune to tansy ragwort (*Senecio jacobaea*), which is highly poisonous to horses even in very smaller doses [115]. Ruminant microflora and microsomal enzymes help to detoxify these poisonous compounds [116]. Sometimes nontoxic plant-based components become toxic in animals after metabolism. For example, pyrrolizidine alkaloids in plants are relatively nontoxic, but they are converted to toxic pyrroles in the liver after metabolism.

6. THERAPEUTIC USE OF POISONOUS PLANTS

We already know that about 70%–80% of the world population, particularly in developing countries, rely on nonconventional medicine for their primary health care, as recommended by the World Health Organization [117]. Homeopathy, a multibillion-dollar industry in the United States, is also based on plant products.

A number of epidemiological studies have regularly shown that dietary habit is one of the most important reasons behind the development of chronic diseases. It is a general perception that the abundant consumption of food from plant origin reduces the risk of chronic diseases. Based on a similar perception, scientists are also trying to resolve the problem of chronic diseases by using toxic plants and their major components. It is estimated that more than 1000

TABLE 13.6 Common Plant-Derived Toxins Detected by Forensic Techniques

Toxin	Plant	Techniques	References
Abrine	*Abrus precatorius*	Radioimmunoassay, enzyme-linked immunosorbent assay (ELISA), and electrochemiluminescence-based assay	[100]
Aconitine	*Aconitum napellus*	Thin-layer chromatography (TLC), high-performance liquid chromatography-mass spectrometry (HPLC-MS), gas chromatography with flame ionization detection (GC-FID)	[105]
Toxalbumin	*Adenia palmata*	Liquid chromatography-electrospray ionization (LC-ESI), GC-FID, and ELISA	[106]
Picrotoxin	*Anamirta cocculus*	TLC, matrix-assisted laser desorption ionization-time of flight (MALDI-TOF), and polymerase chain reaction (PCR)	[106a]
Berberine	*Argemone mexicana*	High-performance capillary electrophoresis, capillary electrophoresis-electrospray ionization, and MALDI-TOF	[107]
Calotropin/ Calotoxin	*Calotropis gigantea*	HPLC	[101]
Cerberin	*Cerbera odollum*	PCR, HPLC, and capillary electrophoresis	[108]
Coniine	*Conium maculatum*	LC-GC-MS, HPLC	[109]
Atropine/ hyoscyamine	*Datura fastuosa*	LC-ESI-MS, PCR HPLC, GC-MS	[103]
Cyanogenic glycosides	*Dieffenbachia* spp.	Raman scattering spectroscopy, Fourier transform infrared and near infrared , and TLC	[110]
Curcin	*Jatropa multifida*	Sonic-spray ionization mass spectroscopy, HPLC-MS, GC-MS, TLC	[111]
Mucunain	*Mucuna prurita*	HPLC-tandem-MS, real time-PCR	[102]

TABLE 13.6 Common Plant-Derived Toxins Detected by Forensic Techniques—cont'd

Toxin	Plant	Techniques	References
Parthenin	*Parthenium hysterophorus*	LC-tandem-MS, LC-MS/MS, sodium dodecyl sulfate polyacrylamide gel electrophoresis	[104]
Ricin	*Ricinus communis*	Avidin/biotin-ELISA, immunochromatography, chemiluminescence-ELISA, and immunopolymerase chain assay	[112]

toxic compounds have been identified from fruits, seeds, leaves, and other parts, but a large fraction of plant-based toxic compounds still remain unknown and efforts are needed for these to be identified and characterized before consumption. In spite of their adverse effects on humans and other animals, poisonous plants also possess medicinal and nutraceutical activities to cure certain life-threatening diseases. However, a lot of evidence suggests that the advantage of plant-derived toxic components may be even greater than is presently understood, because various factors, including protein synthesis, oxidative stress, irregular DNA replication, and inactivation of certain metabolic enzymes, may involve the etiology of a series of chronic diseases such as cardiovascular diseases, cancer, Alzheimer's disease, and functional defects related to aging. Production of metabolic products of plants in the form of toxic components for the formulation of therapeutic drugs, vaccines, and diagnostic agents with the help of recombinant DNA technology is a recent advancement in this area [117a].

The therapeutic use of poisonous plants for curing ailments such as diabetes and cancer, and for use as cytotoxic agents, as well as the principles related to them, is well studied. Some components of plant extracts could have the potential to be immune system stimulators or angiogenesis promoters, which are important processes in the pathogenesis of several diseases and thus can be targeted for the prevention and cure of those disorders. Efforts should be made to improve the bioavailability of these toxic compounds for clinical use. Although toxins produced by plants are hazardous, many plants and their parts have medicinal characteristics, namely, the seeds of *Abrus* [118] and *Conium* [16] possess anticancerous properties, the leaves of *Nerium* [119] and *Argemone* [120] have cardiotonic and antidiabetic properties, and the roots of *Ageratina* [91] and *Aconitum* [121]possess nephrotherapeutic and antileprosic activity. Table 13.7 is a list of toxic plants and their parts commonly in used in the treatment of health hazardous diseases.

TABLE 13.7 Toxic Plants Used in the Treatment of Health Hazardous Diseases

Plant Name	Part Used	Therapeutic use	References
Abrus precatorius	Seeds	Has anticancerous, immunostimulant, and a uterotonic properties	[118]
	Root powder	Used for scorpion bites	
Arum maculatum	Leaves	Cytotoxic	[122]
Gloriosa superba	Alkaloid	Antitumor, antiproliferative	[123]
Nerium oleander	Leaves (koneric acid)	Cardiotonic	[119]
Aconitum ferox	Tuberous roots	Diabetes and leprosy	[121]
Conium maculatum	Seeds	Paralysis, dyspepsia, urinary tract infection, prevents cyst and tumor formation	[16]
Cicuta virosa	Aerial extract	Cerebrospinal meningitis, hemorrhage of ears, epilepsy	[124]
Ailanthus altissima	Roots	Cancer treatment	[124a]
Ricinus communis	Leaf juice	Increases secretion of milk in women when breastfeeding after parturition	[125]
	Seed oil	Applied to lower stomach to relieve pain	
Datura stramonium	Leaf juice	Earache, diabetes, tuberculosis, and filaria	[126]
Argemone mexicana	Leaf extract	Diabetes	[120]
Parthenium hysterophorus	Flower	Muscular rheumatism and antitumor activity	[127]

Based on the aforementioned pharmaceutical properties of toxic plants, the use of plant extracts and phytochemicals in pharmaceuticals has received interest from both homeopathic and allopathic branches. Cancer is a biomedically complex group of diseases involving cell transformation, deregulation of apoptosis, proliferation, invasion, angiogenesis, and metastasis. The developing knowledge of cancer biology suggests that administering cytotoxic drug therapy at very high doses is not always appropriate. An important alternative approach is to administer lower doses of synergistic phytochemicals obtained from various toxic plants. Most of the natural products with anticancer activity modify the activity of one or more protein kinases involved in cell cycle regulation and control. On the other hand, some toxic plants have the potency to treat diabetes mellitus, which is one of the most common and prevalent noncommunicable diseases affecting citizens of both developed and developing countries. Other important risks to humans are associated with neurodegenerative diseases such as Parkinson's disease and Alzheimer's disease, which occur when small groups of brain cells or neurons that control body movement die. There are many drugs available for the treatment of these diseases but none of them are effective. However, *C. maculatum* and other toxic medicinal plants are the only option and there is a need to evaluate or trace the mechanism of action of natural toxic products. Due to increasing demand for plant products containing bioactive compounds as well as nonfood products, the pharmaceutical industry has commercialized its pharmaceutical products containing bioactive compounds in the form of capsules, pills, gels, liquors, and solutions. These chemopreventive toxic substances can act synergistically with chemotherapeutic agents at lower doses and thus minimize chemotherapy-induced toxicity. These compounds can be used not only to treat cancer but also to prevent it because of their pharmacological safety. Application of these chemopreventive phytochemicals has boosted the pharmaceutical and nutraceutical industries to produce large numbers of phytochemical-based nutraceuticals. The nutraceutical industry is a dynamic and evolving industry offering a good opportunity for research in phytochemicals, including both toxic and nontoxic compounds. There are so many disorders having no cure that can be treated with toxic plants and their major metabolic components. Future research should be directed in this area, which should produce fruitful results and better utilization of these toxic plants for the therapy of different diseases.

7. CONCLUSION

Plants contain different types of toxic secondary metabolites, which are lethal to human health and also cause different types of disorders to animals. Most of these toxic principles are of forensic significance. There is a need for proper integration of botanical knowledge of such plants from plant systematics, plant anatomy, palynology, plant ecology, and related areas in criminal investigations as forensic evidence. Still more data on the effects of different environmental and geographical conditions on the formation of toxic principles in plants are desirable to create awareness when planting poisonous trees in social forestry and agroforestry programs.

ACKNOWLEDGMENT

Abhishek Kumar Dwivedy and Anand Kumar Chaudhari are thankful to the Council of Scientific and Industrial Research (CSIR), New Delhi, and Somenath Das is thankful to the Department of Biotechnology, New Delhi, for research fellowship.

REFERENCES

[1] R.S. Chaurasia, History of Ancient India: Earliest Times to 1000 AD, Atlantic Publishers & Dist, 2002.

[2] H.G. Cutler, An historical perspective of ancient poisons, in: Phytochemical Resources for Medicine and Agriculture, Springer, US, 1992, pp. 1–13.

[3] A. Hashim, V. Vaswani, K.L. Pramod, S. Rasheed, Homicidal poisons-past, present and future, Indian J. Forensic Med. Toxicol. 10 (2) (2016).

[4] F.H. Tainter, What Does Mistletoes Have to Do with Christmas, 2002. Online at: http://www.apsnet.org/online/future/mistletoe.

[5] F. Stirpe, K. Sandvig, S. Olsnes, A. Pihl, Action of viscumin, a toxic lectin from mistletoe, on cells in culture, J. Biol. Chem. 257 (22) (1982) 13271–13277.

[6] R. Holtskog, K. Sandvig, S. Olsnes, Characterization of a toxic lectin in Iscador, a mistletoe preparation with alleged cancerostatic properties, Oncology 45 (3) (1988) 172–179.

[7] A.M. Abbasi, M.A. Khan, M. Ahmad, M. Zafar,). Medicinal Plant Biodiversity of Lesser Himalayas-Pakistan, Springer Science & Business Media, 2011.

[8] P. Suhag, Phytochemical Investigation on Melia Azedarach L. And its Bioefficacy against Some Lepidopterous Pests (Doctoral dissertation), Chaudhary Charan Singh Haryana Agricultural University, Hisar, 2001.

[9] G.A. Balint, Ricin: the toxic protein of castor oil seeds, Toxicology 2 (1) (1974) 77–102.

[10] J.F. Morton, Poisonous and Injurious Higher Plants and Fungi [Human Injury], 1977.

[11] R. Sahoo, A. Hamide, S.D. Amalnath, B.S. Narayana, Acute demyelinating encephalitis due to *Abrus precatorius* poisoning—complete recovery after steroid therapy, Clin. Toxicol. 46 (10) (2008) 1071–1073.

[12] S. Olsnes, The history of ricin, abrin and related toxins, Toxicon 44 (4) (2004) 361–370.

[13] F. Stirpe, L. Barbieri, A. Abbondanza, A.I. Falasca, A.N. Brown, K. Sandvig, A. Pihl, Properties of volkensin, a toxic lectin from *Adenia volkensii*, J. Biol. Chem. 260 (27) (1985) 14589–14595.

[14] G.L. Webster, Irritant plants in the spurge family (Euphorbiaceae), Clin. Dermatol. 4 (2) (1986) 36–45.

[15] A. Bernhoft, H. Siem, E. Bjertness, M. Meltzer, T. Flaten, E. Holmsen, Bioactive compounds in plants—benefits and risks for man and animals, in: Proceedings from a Symposium Held at the Norwegian Academy of Science and Letters, Novus Forlag, Oslo, 2010.

[16] J. Vetter, Poison hemlock (*Conium maculatum* L.), Food Chem. Toxicol. 42 (9) (2004) 1373–1382.

[17] P.A. Steenkamp, N.M. Harding, F.R. Van Heerden, B.E. Van Wyk, Fatal Datura poisoning: identification of atropine and scopolamine by high performance liquid chromatography/photodiode array/mass spectrometry, Forensic Sci. Int. 145 (1) (2004) 31–39.

[18] K.J. Dickers, S.M. Bradberry, P. Rice, G.D. Griffiths, J.A. Vale, Abrin poisoning, Toxicol. Rev. 22 (3) (2003) 137–142.

[19] R.R. Dalvi, Sanguinarine: its potential, as a liver toxic alkaloid present in the seeds of *Argemone mexicana*, Experientia 41 (1) (1985) 77–78.
[20] M.H. Ralphs, K.C. McDaniel, Broom snakeweed (*Gutierrezia sarothrae*): toxicology, ecology, control, and management, Invasive Plant Sci. Manage. 4 (1) (2011) 125–132.
[21] K.V. Rao, Toxic principles of *Hippomane mancinella*, Planta Med. 25 (02) (1974) 166–171.
[22] R.D. Cooke, E.M. De La Cruz, The changes in cyanide content of cassava (*Manihot esculenta* Crantz) tissues during plant development, J. Sci. Food Agric. 33 (3) (1982) 269–275.
[23] S.D. Langford, P.J. Boor, Oleander toxicity: an examination of human and animal toxic exposures, Toxicology 109 (1) (1996) 1–13.
[24] N.L. Benowitz, Nicotine Safety and Toxicity, Oxford University Press, USA, 1998.
[25] J.K. Brown, M.H. Malone, "Legal highs"–constituents, activity, toxicology, and herbal folklore, Clin. Toxicol. 12 (1) (1978) 1–31.
[26] R. Mrvos, B.S. Dean, E.P. Krenzelok, Philodendron/dieffenbachia ingestions: are they a problem? J. Toxicol. 29 (4) (1991) 485–491.
[27] K. Sandvig, S. Olsnes, A. Pihl, Kinetics of binding of the toxic lectins abrin and ricin to surface receptors of human cells, J. Biol. Chem. 251 (13) (1976) 3977–3984.
[28] J. Asgarpanah, R. Mohajerani, Phytochemistry and pharmacologic properties of *Urtica dioica* L. J. Med. Plants Res. 6 (46) (2012) 5714–5719.
[29] D.W. Fassett, Oxalates, in: Toxicants Occurring Naturally in Foods, 1973, pp. 346–362.
[30] A. Kamboj, A. Saluja, Phytopharmacological review of *Xanthium strumarium* L. (Cocklebur), Int. J. Green Pharm. 4 (3) (2010) 129.
[31] M. Brown, W. Vale, Central nervous system effects of hypothalamic peptides, Endocrinology 96 (5) (1975) 1333–1336.
[32] Y.M. Shivkar, V.L. Kumar, Anthelmintic activity of latex of *Calotropis procera*, Pharm. Biol. 41 (4) (2003) 263–265.
[33] G.H.N. Towers, P.V. Subba Rao, Impact of the pan-tropical weed, *Parthenium hysterophorus* L. on human affairs., in: Proceedings of the First International Weed Control Congress, vol. 2, February 1992, pp. 134–138.
[34] A. Ameri, The effects of Aconitum alkaloids on the central nervous system, Prog. Neurobiol. 56 (2) (1998) 211–235.
[35] A. Der Marderosian, F.C. Roia, Literature review and clinical management of household ornamental plants potentially toxic to humans, in: Toxic Plants, Columbia University Press, New York, 1979, pp. 103–135.
[36] H. Rajput, Effects of *Atropa belladonna* as an anti-cholinergic, Nat. Prod. Chem. Res. (2013).
[37] S. Senthilkumaran, R. Meenakshisundaram, P. Thirumalaikolundusubramanian, Plant toxins and the heart, in: Heart and Toxins, vol. 151, 2014.
[38] S.C. Noonan, G.P. Savage, Oxalate content of foods and its effect on humans, Asia Pac. J. Clin. Nutr. 8 (1999) 64–74.
[39] E. Castells, M.A. Berhow, S.F. Vaughn, M.R. Berenbaum, Geographic variation in alkaloid production in *Conium maculatum* populations experiencing differential herbivory by *Agonopterix alstroemeriana*, J. Chem. Ecol. 31 (8) (2005) 1693–1709.
[40] A. Szakiel, C. Pączkowski, M. Henry, Influence of environmental abiotic factors on the content of saponins in plants, Phytochem. Rev. 10 (4) (2011) 471–491.

[41] C.S. Charron, A.M. Saxton, C.E. Sams, Relationship of climate and genotype to seasonal variation in the glucosinolate–myrosinase system. I. Glucosinolate content in ten cultivars of *Brassica oleracea* grown in fall and spring seasons, J. Sci. Food Agric. 85 (4) (2005) 671–681.

[42] P. Feeny, Biochemical coevolution between plants and their insect herbivores, Coevol. Anim. Plants 13 (24) (1975) 6. University of Texas Press, Austin.

[43] D.A. Levin, Plant phenolics: an ecological perspective, Am. Nat. 105 (942) (1971) 157–181.

[44] D. McKey, Adaptive patterns in alkaloid physiology, Am. Nat. 108 (961) (1974) 305–320.

[45] P.W. Paré, J.H. Tumlinson, Plant volatiles as a defense against insect herbivores, Plant Physiol. 121 (2) (1999) 325–332.

[46] E.E. Conn, Cyanogenic glycosides, Biochem. Plants 7 (1981) 479–500.

[47] B. Molano-Flores, Herbivory and calcium concentrations affect calcium oxalate crystal formation in leaves of Sida (Malvaceae), Ann. Bot. 88 (3) (2001) 387–391.

[48] M.G. Neuman, L. Cohen, M. Opris, R.M. Nanau, H. Jeong, Hepatotoxicity of pyrrolizidine alkaloids, J. Pharm. Pharm. Sci. 18 (4) (2015) 825–843.

[49] K.E. Panter, R.F. Keeler, Piperidine alkaloids of poison hemlock (*Conium maculatum*), Toxicants Plant Origin 1 (1989) 109–132.

[50] J.A. Elix, Annelated furans. V. Selective Wittig synthesis of euparin and dehydrotremetone, Aus. J. Chem. 24 (1) (1971) 93–97.

[51] J.A. Shimshoni, P.P. Mulder, A. Bouznach, N. Edery, I. Pasval, S. Barel, S. Perl, *Heliotropium europaeum* poisoning in cattle and analysis of its pyrrolizidine alkaloid profile, J. Agric. Food Chem. 63 (5) (2015) 1664–1672.

[52] S. Das, G.P. Mandal, A.K. Tyagi, Milk toxicant of natural plant and feed mold origin: a review, Indian Dairyman 59 (12) (2007) 27.

[53] J.K. Porter, F.N. Thompson, Effects of fescue toxicosis on reproduction in livestock, J. Anim. Sci. 70 (5) (1992) 1594–1603.

[54] Z. Chen, J.R. Huo, Hepatic veno-occlusive disease associated with toxicity of pyrrolizidine alkaloids in herbal preparations, Neth. J. Med. 68 (6) (2010) 252–260.

[55] S.O. Onduso, Determination of Levels of Pyrrolizidine Alkaloids in *Symphytum asperumlepech* Growing in Selected Parts of Kenya (Doctoral dissertation, Kenyatta University), 2015.

[56] P.R. Cheeke, Pyrrolizidine Alkaloid Toxicity and Metabolism in Laboratory Animals and Livestock. Toxicants of Plant Origin Alkaloids, CRC Press, Boca Raton, FL, 1989, p. 1Á22.

[57] S.J. Kramer, L.W. Tsai, S.F. Lee, J.J. Menn, Chemistry of toxic range plants. Variation in pyrrolizidine alkaloid content of Senecio, Amsinckia, and Crotalaria species, J. Agric. Food Chem. 33 (1985) 50–55.

[58] L.F. James, K.E. Panter, B.L. Stegelmeier, R.J. Molyneux, Effect of natural toxins on reproduction, Vet. Clin. North Am. 10 (3) (1994) 587–603;
[58a] K.E. Panter, L.F. James, Natural plant toxicants in milk: a review, J. Anim. Sci. 68 (3) (1990) 892–904.

[59] A. Garry, V. Rigourd, A. Amirouche, V. Fauroux, S. Aubry, R. Serreau, Cannabis and breastfeeding, J. Toxicol. (2009).

[60] P.R. Ehrlich, P.H. Raven, Butterflies and plants: a study in coevolution, Evolution (1964) 586–608.

[61] G.S. Fraenkel, The raison d'etre of secondary plant substances, Science 129 (3361) (1959) 1466–1470.

[62] D.A. Jones, Co-evolution and cyanogenesis, Taxon. Ecol. (1973) 213–242.

[63] R.H. Whittaker, P.P. Feeny, Allelochemics: chemical interactions between species, Science 171 (3973) (1971) 757–770.

[64] B.T. Cromwell, The separation, micro-estimation and distribution of the alkaloids of hemlock (*Conium maculatum* L.), Biochem. J. 64 (2) (1956) 259.

[65] S. Rajvaidhya, B.P. Nagori, G.K. Singh, B.K. Dubey, P. Desai, S. Jain, A review on *Argemone mexicana* linn.-an Indian medicinal plant, Int. J. Pharm. Sci. Res. 3 (8) (2012) 2494.

[66] J.F. Couch, Relative toxicity of lupine alkaloids, Aus. J. Agric. Res. 32 (1926) 51–67.

[67] K.D. Welch, K.E. Panter, D.R. Gardner, B.L. Stegelmeier, B.T. Green, J.A. Pfister, D. Cook, The acute toxicity of the death camas (species) alkaloid zygacine in mice, including the effect of methyllycaconitine coadministration on zygacine toxicity, J. Anim. Sci. 89 (5) (2011) 1650–1657.

[68] J.C. Tanner, J.D. Reed, E. Owen, The nutritive value of fruits (pods with seeds) from four *Acacia* spp. compared with nong (*Guizotia abssinica*) meal as supplements of maize stover for Ethiopian high land sheep, Anim. Prod. 51 (1990) 127–133.

[69] D.C. Joshi, R.C. Katiyar, M.Y. Khan, R. Banerji, G. Misra, S.K. Nigam, Studies on Mahua (*Bassia latifolia*) seed cake saponin (Mowrin) in cattle, Indian J. Anim. Nutr. 6 (1989) 13–17.

[70] C.D. Marsh, A.B. Clawson, G.C. Roe, Coyotillo (*Karwinskia thumboldtiana*) as a Poisonous Plant, vol. 29, US Dept. of Agriculture, 1928.

[71] D.D. Sharma, S. Chandra, S.S. Negi, The nutritive value and toxicity of OHI (*Albizzia stipulata Bovin*) tree leaves, J. Res. Ludhiana 6 (1969) 388–393.

[72] C.R. Behl, M.B. Pande, D.P. Pande, M.S. Radadia, Nutritive value of matured wilted castor (*Ricinus communis* Linn.) leaves for crossbred sheep, Indian J. Anim. Sci. (India) (1986).

[73] P.R. Cheeke, L.R. Shull, Natural Toxicants in Feeds and Poisonous Plants, AVI Publishing Company Inc, 1985;
[73a] A.J. Duncan, P. Frutos, S.A. Young, Rates of oxalic acid degradation in the rumen of sheep and goats in response to different levels of oxalic acid administration, Anim. Sci. 65 (3) (1997) 451–455.

[74] M.P. Hegarty, Toxic amino acids of plant origin, in: Effects of Poisonous Plants on Livestock, vol. 575, 1978.

[75] R. Kumar, Anti-nutritional factors, the potential risks of toxicity and methods to alleviate them, in: Legume Trees and Other Fodder Trees as Protein Source for Livestock. FAO Animal Production and Health Paper, vol. 102, 1992, pp. 145–160.

[76] F.P. Mathews, The toxicity of a spurge (*Phyllanthus abnormis*) for cattle, sheep, and goats, Cornell Vet. 35 (1945) 336–346.

[77] F.P. Mathews, The toxicity of *Baileya multiradiata* for sheep and goats, J. Am. Vet. Med. Assoc. 83 (1933) 673–679.

[78] R.J. Cole, B.P. Stuart, J.A. Lansden, R.H. Coxs, Chemistry of toxic range plants. Volatile constituents of Broomweed (*Gutierrezîa sarothrae*), J. Agric. Food Chem. 1000 (28) (1980) 1332–1333.

[79] S.J. Cysewski, Paspalum staggers and tremorgen intoxication in animals, J. Am. Vet. Med. Assoc. 163 (11) (1973) 1291–1292.

[80] M.A.M. Schenk, T.T. Faria Filho, D.M. Pimentel, L.R.L.S. Thiago, Intoxicac ao por oxalatos em vacas lactantes em pastagem de Setaria, Pesq. Agropec. Bras. 17 (1982) 1403–1407.

[81] J. Randall, R. Authority, Import Risk Analysis, 1999.

[82] C.A. Pimentel, C.L.G.P. Brod, S.M. Pimentel, E.L. Medeiros, C.L.R. Bento, P. Monks, Hiperestrogenismo causado porfito-estroógenos em novilhas da rac a holandesa, Revta. Bras. Reprod. Anim. 1 (1977) 15–20.
[83] E.F. Woodcock, Observations on the poisonous plants of Michigan, Am. J. Bot. 12 (2) (1925) 116–131.
[84] A.P. Knight, R.G. Walter, Plants Affecting the Digestive System, February 19, 2003.
[85] D.G. Parbery, Trophism and the ecology of fungi associated with plants, Biol. Rev. 71 (3) (1996) 473–527.
[86] U. Wittstock, K.H. Lichtnow, E. Teuscher, Effects of cicutoxin and related polyacetylenes from *Cicuta virosa* on neuronal action potentials: a comparative study on the mechanism of the convulsive action, Planta Med. 63 (02) (1997) 120–124.
[87] T.Y. Chan, B. Tomlinson, L.K. Tse, J.C. Chan, W.W. Chan, J.A. Critchley, Aconitine poisoning due to Chinese herbal medicines: a review, Vet. Hum. Toxicol. 36 (5) (1994) 452–455;
[87a] H.M. Coyle (Ed.), Forensic Botany: Principles and Applications to Criminal Casework, CRC Press, 2004.
[88] B.S. Khajja, M. Sharma, R. Singh, G.K. Mathur, Forensic study of Indian toxicological plants as botanical weapon (BW): a review, J. Environ. Anal. Toxicol. 1 (2011) 1–5.
[89] N. Viswanathan, B.S. Joshi, Toxic Constituents of Some Indian Plants, 1983;
[89a] V.K. Raju, Susruta of ancient India, Indian J. Ophthalmol. 51 (2) (2003) 119.
[90] R.W. Byard, R.A. James, P. Felgate, Detecting organic toxins in possible fatal poisonings—a diagnostic problem, J. Clin. Forensic Med. 9 (2) (2002) 85–88.
[91] S.T. Lee, T.Z. Davis, D.R. Gardner, S.M. Colegate, D. Cook, B.T. Green, T.J. Evans, Tremetone and structurally related compounds in white snakeroot (*Ageratina altissima*): a plant associated with trembles and milk sickness, J. Agric. Food Chem. 58 (15) (2010) 8560–8565.
[92] S. Olsnes, K. Refsnes, A. Pihl, Mechanism of action of the toxic lectins abrin and ricin, Nature 249 (1974) 627–631.
[93] Y. Gaillard, G. Pepin, Poisoning by plant material: review of human cases and analytical determination of main toxins by high-performance liquid chromatography–(tandem) mass spectrometry, J. Chromatogr. B 733 (1) (1999) 181–229.
[94] J.N. Seiber, C.J. Nelson, S.M. Lee, Cardenolides in the latex and leaves of seven Asclepias species and *Calotropis procera*, Phytochemistry 21 (9) (1982) 2343–2348.
[95] V.O. Ayuba, T.O. Ojobe, S.A. Ayuba, Phytochemical and proximate composition of *Datura innoxia* leaf, seed, stem, pod and root, J. Med. Plants Res. 5 (14) (2011) 2952–2955.
[96] S.K. Verma, G. Dev, A.K. Tyagi, S. Goomber, G.V. Jain, *Argemone mexicana* poisoning: autopsy findings of two cases, Forensic Sci. Int. 115 (1) (2001) 135–141.
[97] D. Mew, F. Balza, G.N. Towers, J.G. Levy, Anti-tumour effects of the sesquiterpene lactone parthenin, Planta Med. 45 (05) (1982) 23–27.
[98] S.A. Young, Rates of Oxalic Acid Degradation in the Rumen of Sheep and Goats in Response to Different Levels of Oxalic Acid Administration, 1997.
[99] T. Acamovic, C.S. Stewart, T.W. Pennycott, Poisonous Plants and Related Toxins, CABI, 2003.
[100] M.A. Poli, V.R. Rivera, J.F. Hewetson, G.A. Merrill, Detection of ricin by colorimetric and chemiluminescence ELISA, Toxicon 32 (11) (1994) 1371–1377.
[101] H.W. Groeneveld, H. Steijl, B. Berg, J.C. Elings, Rapid, quantitative HPLC analysis of *Asclepias fruticosa* L. and *Danaus plexippus* L. cardenolides, J. Chem. Ecol. 16 (12) (1990) 3373–3382.

[102] K.O. Adebowale, O.S. Lawal, Functional properties and retrogradation behaviour of native and chemically modified starch of mucuna bean (*Mucuna pruriens*), J. Sci. Food Agric. 83 (15) (2003) 1541–1546.

[103] U.R. Cieri, Determination of atropine (hyoscyamine) sulfate in commercial products by liquid chromatography with UV absorbance and fluorescence detection: multilaboratory study, J. AOAC Int. 86 (6) (2003) 1128–1134.

[104] N. Hussain, T. Abbasi, S.A. Abbasi, Transformation of the pernicious and toxic weed parthenium into an organic fertilizer by vermicomposting, Int. J. Environ. Stud. 73 (5) (2016) 731–745.

[105] Q. Liu, L. Zhuo, L. Liu, S. Zhu, A. Sunnassee, M. Liang, Y. Liu, Seven cases of fatal aconite poisoning: forensic experience in China, Forensic Sci. Int. 212 (1) (2011) e5–e9.

[106] C.P. Holstege, T. Neer, R.B. Furbee, Criminal Poisoning: Clinical and Forensic Perspectives, Jones & Bartlett Publishers, 2010;
[106a] A.W. Blyth, Poisons, Their Effects and Detection, vol. 2, 1885.

[107] A. Marston, Thin-layer chromatography with biological detection in phytochemistry, J. Chromatogr. A 1218 (19) (2011) 2676–2683.

[108] D.M. Tian, H.Y. Cheng, M.M. Jiang, W.Z. Shen, J.S. Tang, X.S. Yao, Cardiac glycosides from the seeds of *Thevetia peruviana*, J. Nat. Prod. 79 (1) (2015) 38–50.

[109] J. Beyer, F.T. Peters, T. Kraemer, H.H. Maurer, Detection and validated quantification of toxic alkaloids in human blood plasma—comparison of LC-APCI-MS with LC-ESI-MS/MS, J. Mass Spectrom. 42 (5) (2007) 621–633.

[110] D.M. Jian, H.Y. Cheng, M.M. Jiang, W.Z. Shen, J.S. Tang, X.S. Yao, Cardiac glycosides from the seeds of *Thevetia peruviana*, J. Nat. Prod. 79 (1) (2015) 38–50.

[111] H.P.S. Makkar, K. Becker, F. Sporer, M. Wink, .Studies on nutritive potential and toxic constituents of different provenances of *Jatropha curcas*, J. Agric. Food Chem. 45 (1997) 3152–3157.

[112] A.G. Leith, G.D. Griffiths, M.A. Green, Quantification of ricin toxin using a highly sensitive avidin/biotin enzyme-linked immunosorbent assay, J. Forensic Sci. Soc. 28 (4) (1988) 227–236.

[113] M. Rothschild, Some observations on the relationship between plants, toxic insects and birds, in: Phytochemical Ecology, Academic Press, London, 1972, pp. 1–12.

[114] R.A. James, P. Felgate, Detecting organic toxins in possible fatal poisonings—a diagnostic problem, J. Clin. Forensic Med. 9 (2) (2002) 85–88.

[115] P.R. Cheeke, M.L. Pierson-Goeger, Toxicity of *Senecio jacobaea* and pyrrolizidine alkaloids in various laboratory animals and avian species, Toxicol. Lett. 18 (3) (1983) 343–349.

[116] W.J. Freeland, D.H. Janzen, Strategies in herbivory by mammals, Am. Nat. 108 (1974) 269–289.

[117] N. Tamilselvan, T. Thirumalai, P. Shyamala, E. David, A review on some poisonous plants and their medicinal values, J. Acute Dis. 3 (2) (2014) 85–89;
[117a] K. Chan, Some aspects of toxic contaminants in herbal medicines, Chemosphere 52 (9) (2003) 1361–1371.

[118] N. Garaniya, A. Bapodra, Ethno botanical and Phytophrmacological potential of *Abrus precatorius* L.: a review, Asian Pac. J. Trop. Biomed. 4 (2014) S27–S34.

[119] R.O. Adome, J.W. Gachihi, B. Onegi, J. Tamale, S.O. Apio, The cardiotonic effect of the crude ethanolic extract of *Nerium oleander* in the isolated Guinea pig hearts, Afr. Health Sci. 3 (2) (2003) 77–82.

[120] A. Andrade-Cetto, M. Heinrich, Mexican plants with hypoglycaemic effect used in the treatment of diabetes, J. Ethnopharmacol. 99 (3) (2005) 325–348.

[121] N.D. Prajapati, S.S. Purohit, A.K. Sharma, T. Kumar, Medicinal Plants, third ed., vol. 353, Agrobios Published Company, India, 2003.

[122] M. Nabeel, S. Abderrahman, A. Papini, Cytogenetic effect of Arum maculatum extract on the bone marrow cells of mice, Caryologia 61 (4) (2008) 383–387.

[123] S. Jana, G.S. Shekhawat, Critical review on medicinally potent plant species: *Gloriosa superba*, Fitoterapia 82 (3) (2011) 293–301.

[124] V. Ricotti, N. Delanty, Use of complementary and alternative medicine in epilepsy, Curr. Neurol. Neurosci. Rep. 6 (4) (2006) 347–353;
[124a] S. Madhuri, Govind Pandey, Some anticancer medicinal plants of foreign origin, Current science (2009) 779–783.

[125] A. Scarpa, A. Guerci, Various uses of the castor oil plant (*Ricinus communis* L.) a review, J. Ethnopharmacol. 5 (2) (1982) 117–137.

[126] F.I. Jahan, M.R.U. Hasan, R. Jahan, S. Seraj, A.R. Chowdhury, M.T. Islam, M. Rahmatullah, A comparison of medicinal plant usage by folk medicinal practitioners of two adjoining villages in Lalmonirhat district, Bangladesh, Am. Eurasian J. Sustainable Agriculture 5 (1) (2011) 46–66.

[127] S. Patel, Harmful and beneficial aspects of *Parthenium hysterophorus*: an update, 3 Biotech 1 (1) (2011) 1–9.

Chapter 14

Role of Stress in Diseases and Its Remedial Approach by Herbal and Natural Products in Stress-Related Disease Management: Experimental Studies and Clinical Reports

Dhrubojyoti Mukherjee[1], Partha Palit[2], Shubhadeep Roychoudhury[3], Sukalyan K. Kundu[4], Subhash C. Mandal[1]
[1]*Department of Pharmaceutical Technology, Jadavpur University, Kolkata, India;* [2]*Department of Pharmaceutical Sciences, Assam University, Silchar, India;* [3]*Department of Life Science & Bio-Informatics, Assam University, Silchar, India;* [4]*Department of Pharmacy, Jahangirnagar University, Dhaka, Bangladesh*

In today's 21st century there is a huge generational shift in terms of socio-economic, work, and cultural pressures that can lead to a stressful life. Also there has been a subsequent rise in the range of disorders affecting major populations regarding health and monetary consequences [1]. Stress affects practically everybody on the planet from schoolchildren to the elderly population. The role of modern medical science is constantly updating its parameters to understand the causes of psychological stress-induced disease and its treatment. However, in many cases of chronic lifestyle diseases, one significant point regarding the prognosis of patients is that treatment is focused only on the apparent pathophysiological aspect of the disease, not upon the grassroot level or the underline hidden cause, such as "psychological stress," which remains often unaddressed. As a precursor to many diseases, psychological stress may also worsen in patients suffering from several chronic diseases [2]. According to the World Health Organization and the Global Burden of Disease Survey, it has been estimated that mental disease, including stress-related disorders, will be the second leading cause of disabilities by the year 2020 [3]. Moreover, the United Nations in 1992 affirmed that stress was a 20th

Natural Products and Drug Discovery. https://doi.org/10.1016/B978-0-08-102081-4.00014-9

century disease. Also a proposal has been made to include stress-associated mental disorders and possible outcomes in International Classification of Diseases-11 [4]. The American Psychological Association stated in a survey report that teenagers and younger children are more stressed than adults [5]. In 2011 the Nielsen survey of 6500 women in 21 countries conferred a report showing that women are the most stressed on earth, among them Indian women rated at 87%, which is the highest in the world [6].

Stress is a part of everyday life, which consists of an individual perceptual phenomenon that primarily originates from the imbalance between demand on the individual and his or her ability to cope. Stress is not an illness, rather it is a state. If stress becomes too excessive and prolonged, mental and physical illness may develop. Stress is an adaptive response of the body to psychological and physiological stressors, comprising factors leading to disruption of the normal homeostasis of the body [7]. Physiological stressors are infection, severe illness, fever, blood loss, etc., whereas associated depression, anxiety, fear psychosis, tension due to prime triggering factors such as unemployment, work load, job insecurity, traumatic incidence, family problems, conjugal-life problems, being abused or neglected, conflicts in interpersonal relationships, losing contact with loved ones, etc. are negative personal stressors that lead to psychological stress [8]. Recurrent intense stressful situations are detrimental to the body, resulting in psychosomatic disturbances, which in future may turn into organic illnesses as the responses become pathological in nature [9]. Psychological stress now affects millions of people globally, ultimately leading to increases in health adverse effects such as cognitive and behavioral diseases, heart diseases, metabolic syndrome, infertility, immune problems, psychosomatic disorders, and related neurotic disorders. Several studies reveal that stress reductions in cases of heart diseases, hypertension, diabetes, infertility, etc. give very promising effects on disease progression [10].

1. PATHOPHYSIOLOGY OF STRESS RESPONSE

Stress affects two major body systems: the hormonal system, which is the hypothalamic–pituitary–adrenal (HPA) axis, and the sympathetic adrenomedullary system (SAS). Stress leads to secretion of corticotrophin-releasing hormone (CRH) from the hypothalamus, and then CRH acts on the anterior pituitary gland to release adrenocorticotropic hormone (ACTH). Finally, ACTH acts on the adrenal cortex to release cortisol and epinephrine [11]. Thus activation of the HPA axis ultimately leads to elevated levels of plasma cortisol, a glucocorticoid and end product of the HPA axis that is secreted within a stable and narrow range through the tight regulation of a feedback system that checks excessive and sustained cortisol secretion. Thus a normal response of the HPA axis to stress and subsequent cortisol secretion is maintained by a feedback system, circadian variation, and central regulation. The normal pattern of ACTH and cortisol secretion reaches its peak plasma level in the early morning, decreases during the day, and becomes lowest at midnight. However, at times of chronic exposure to stress, disruption of the HPA axis occurs, which is characterized by downregulation of glucocorticoid receptors in the hippocampus and pituitary gland, causing an increased CRH and cortisol response to stress with inappropriate feedback regulation resulting in disturbed normal diurnal cortisol rhythm [12]. Simultaneously, hyperactivation of SAS results in the release of norepinephrine and epinephrine with an inhibitory effect on GABAergic transmission. The released high concentration of cortisol increases the synthesis of adrenaline from noradrenaline in the adrenal medullary cells through induction of the methylating enzyme [13]. Furthermore, cortisol inhibits extraneuronal uptake of adrenaline by inhibiting the extraneuronal amine transporter and organic cation transporter, thus potentiating further excitatory adrenergic transmission in stressful situations (Fig. 14.1) [14].

In emergency situations or because of acute stressors such as infections or preexams, stress is essential or to some extent physiological. However, chronic and constant hyperactivation of the HPA axis mainly in psychological chronic stress alters the normal homeostatic balance of the body leading to neurohormonal imbalances. In normal conditions in the body, excitatory signaling is counterbalanced by inhibitory signaling, which results in activation of compensatory adaptive responses leading to reestablishment of homeostasis and protection against stress. This stress system is vital in homeostasis maintenance. However, under chronic stressful situations, inhibitory signaling is not able to counterbalance excitatory signaling efficiently leading to chronic and continuous activation of the HPA axis and SAS. This can be manifested by chronic elevation of blood pressure (BP), change in heart rate (HR), etc.,

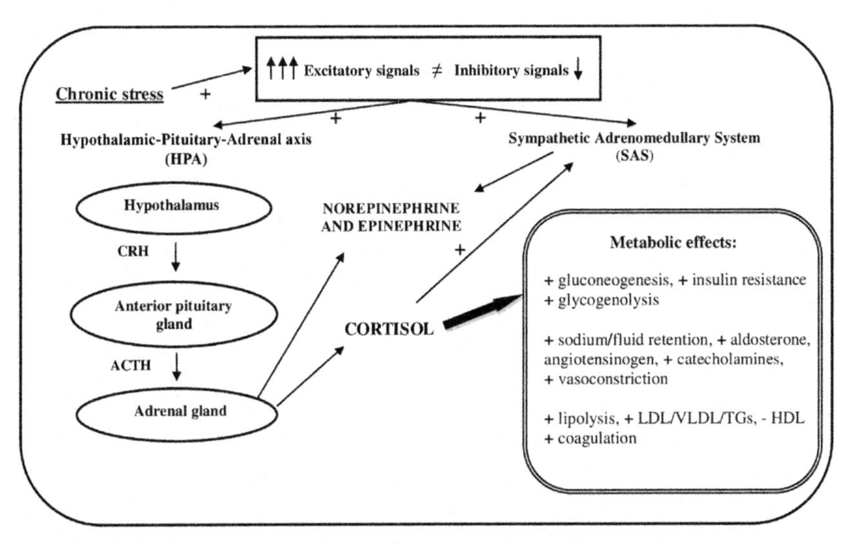

FIGURE 14.1 Physiological stress response in diseased conditions. *HDL*, High-density lipoprotein; *LDL*, low-density lipoprotein; *TGs*, triglycerides; *VLDL*, very low-density lipoprotein.

which affects various systems of the body such as the cardiovascular system, central nervous system (CNS), gastrointestinal (GI) system, immune system, cognitive function, growth and reproductive axis, etc. In the long run these lead to various stress-related disorders, including cardiovascular, metabolic, behavioral, endocrine, autoimmune, etc. disorders [15]. Ultimately, psychological stress disturbs the quality of life and a growing number of illnesses have been found to be associated with it [16]. Apart from stress-induced diseases, patients suffering from chronic diseases are also much more stressed because of their illness, medications, treatment cost, etc., which ultimately hastens their recovery from primary treatment. Patients suffering from metabolic syndrome and heart diseases who are under standard medicine treatment often require dose escalation and additional drugs in the long run because chronic stress-related elevated cortisol levels are poorly managed or neglected. So, a novel antistress product devoid of any side effects may become a gold standard treatment in cases of stress-associated diseases, which will prevent further disease complications.

Long-term psychological stress results in the manifestation of symptoms, mainly:

1. Cognitive, behavioral, and emotional symptoms: memory problems, poor concentration, anxiety, nervousness, negative thoughts, irritability, depression, eating more or less, sleep disturbances, addiction to smoking and drinking, nervous habits such as nail biting, etc. [17].
2. Physical symptoms: back ache, muscle tension, GI disturbances (diarrhea, constipation), nausea, rapid HR, frequent colds, loss of libido, etc.

2. IMPACT OF PSYCHOLOGICAL STRESS ON OCCURRENCE OF DISEASES

2.1 Cardiovascular Diseases

Clinical data points to the fact that long-term psychological stress is associated with increased risk of coronary heart disease (CHD) [18]. The INTERHEART study is the largest study conducted for evaluating the risk factors for myocardial infarctions in over 262 centers in 52 countries. The study included 11,119 patients suffering their first myocardial infarction and 13,648 age-matched and sex-matched controls. Ultimately, very surprising conclusions showed that out of the nine risk factors, psychological stress was the third most prevalent risk factor associated with acute myocardial infarction, after apolipoprotein B/apolipoprotein A ratio and smoking [19]. Interestingly, diabetes and hypertension were lesser risk factors than psychological stress. Furthermore, the Whitehall II prospective cohort study of over 10,308 civil service employees after a 5-year follow-up concluded that psychological stress was associated with an increased incidence of CHD and electrocardiogram abnormalities [20]. Another study of 514 healthy men and women depicted that after experiencing mental stress through behavioral tasks, salivary cortisol levels were elevated and significantly associated with coronary artery calcification (CAC) [21]. However, only 40% of the participants were cortisol responders after stress induction, which reflects the variation in the stress-coping capability of different individuals. Reports showed that elevated urinary levels of the stress hormone cortisol in 861 participants aged 65 year and older suggested that they had a five times increased risk of dying of cardiovascular disease [22]. Elevated cortisol level, which is a biomarker for stress, is related to atherosclerosis of the carotid arteries and is associated with a history of cardiovascular disease [23].

There is evidence that a glucocorticoid receptor gene is related to high proinflammatory activity and CHD risk. It has been shown that blocking cortisol activity prevents stress-induced endothelial dysfunction and baroreflex impairment [24]. A population-based prospective study found an association between cortisol and incident ischemic heart disease, which reflects the effects of chronic stress and behavioral factors [25]. The study also revealed that each change of 10 mmHg in systolic BP in young adults during a video game stressor was associated with increased risk of having CAC in relation to poststress cortisol elevation after 11 years of follow-up [26]. Heightened mental stress response also predicts the possibility of future hypertension, increased low-density lipoprotein/very low-density lipoprotein (LDL/VLDL) triglyceride levels, and subclinical cardiovascular disease progression [27]. Coagulation activity is increased very efficiently through a complex mechanism by cortisol and catecholamines, accompanying plaque rupture and thrombosis leading to future cardiovascular events. Additionally, cortisol has an action on myocardial contractility and tone of arterioles. Circulating elevated cortisol levels are also the initiator of perivascular inflammation and calcification within arteriosclerotic lesions [28].

2.2 Hypertension

Abnormality of the HPA axis and SAS system by psychosocial stress is one of the reasons for the development of hypertension. Studies also indicate that high levels of stress in prehypertensive men were associated with increased risk of progressing to hypertension and incidence of CHD. In a study of 489 healthy normotensive men and women, salivary cortisol levels were measured after facing a 5-min Stroop test (physiological stress) and mirror tracing test (psychological stress). After a 3-year follow-up, the sample that responded to elevated cortisol developed hypertension, suggesting association between psychological stress cortisol reactivity and incidental hypertension [29]. Extensive research indicating the point at which participants were at risk of hypertension showed heightened HPA activity in response to acute stressors. Furthermore, cortisol has a permissive effect in potentiating the pressor activity of adrenaline and angiotensin by direct and indirect effects during simultaneous epinephrine release by SAS activation resulting in vasoconstriction and increased HR [30]. Glucocorticoid receptors present in the heart, smooth muscle cells of the resistance vessels, and the kidneys directly affect BP. Furthermore, cortisol enhances angiotensinogen formation, a substrate of rennin resulting in increased angiotensin II levels, and subsequently affects the rennin–angiotensin system, which is a vital physiological system in maintaining blood volume and electrolyte status in relation to BP regulation [31]. Aldosterone and vasopressin syntheses are also augmented, affecting sodium and fluid retention causing sustained increase in BP due to vascular overload [32]. There is also research regarding the molecular actions of cortisol in the inhibition of vasodilation via decreased levels of plasma kallikrein, prostacyclin, and nitric oxide synthase (NOS) activity [33]. Increased expression of adrenergic and AT1 receptors in vascular smooth muscle cells by chronic elevated cortisol levels is another pathway to BP elevation [34].

2.3 Diabetes

There has been a positive correlation among psychological stress and the incidence of type 2 diabetes mellitus (DM) according to a large number of clinical reports. A major prospective MONICA/KORA Augsburg study declared some interesting outcomes over 5337 of the working population aged from 29 to 66 years without any history of diabetes. After a follow-up of 12.7 years, 291 cases of type 2 DM were observed and participants with high stress levels due to job strain had a 45% higher chance of developing type 2DM than those with a low job strain [35]. So, a high job strain with mental stressors is a vital precursor to type 2 DM independently of traditional risk factors. A relation can be drawn between increasing cases of diabetes among the young adult population and the burden of psychosocial stressors such as career, work load, relationship complexities, etc.

Cortisol and catecholamines, the end products of stress, to some extent have similar metabolic effects. They alter the glucose metabolic pathway by using a plethora of innervating effects. Cortisol promotes glycogen deposition in the liver by inducing hepatic glycogen synthase; it also promotes gluconeogenesis and glycogenolysis in the liver [36]. It inhibits glucose utilization by the peripheral tissues and causes decreased glucose uptake in skeletal muscle [37]. The catabolic action of cortisol causes protein breakdown, and amino acid mobilization from peripheral tissues is used up in gluconeogenesis resulting in increased urea production with negative nitrogen balance in the body. This along with increased glucose release in the liver results in hyperglycemia and causes insulin resistance and a diabetes-like state. In addition, increased adrenergic activity due to SAS activation induces downregulation of insulin receptors. Lipolysis is also stimulated by cortisol, epinephrine, and norepinephrine resulting in increased plasma free fatty acids, which further inhibit release of insulin in response to glucose. Thus all these effects converge to create further insulin resistance and glucose intolerance-like states [38]. Beside chronic stress as a precursor of diabetes, it has been implicated that the further existence of psychological stress in patients with existing type 2 DM results in poor glucose control and future associated complications such as nephropathy, neuropathy, macroangiopathy, etc., which are directly correlated with cortisol secretion [39]. Experimental acute psychological stress is reported to increase postprandial glucose concentrations significantly in patients with type 2 DM, which further justifies the need for stress management in diabetic patients [40].

2.4 Metabolic Syndrome

Metabolic syndrome is characterized by the coexistence of hypertension, diabetes, hyperlipidemia, and abdominal obesity in an individual state, which is recognized as an important risk factor for CHD and mortality. Psychological stress related to depression or environmental stressors is associated with increased cortisol levels, which predispose to features of metabolic syndrome [41]. In a double-blind case–control study of 10,308 people it was found that working people aged between 35 and 55 years with metabolic syndrome had greater psychological stress associated with higher cortisol levels and epinephrine activity compared to healthy controls [42].

As discussed earlier regarding the roles of stress hormones in progression to hypertension and diabetes, there is a prevalence of insulin resistance, which clusters together with increased LDL, VLDL, and triglycerides due to altered lipid metabolism, ultimately leading to features of metabolic syndrome. Insulin resistance is thought to be the underlying abnormality that causes metabolic syndrome-associated complications. Stressor-like depression also predicts insulin resistance, diabetes, and metabolic syndrome [43]. Obesity is a major problem; in fact, visceral obesity, which is a feature of metabolic syndrome, is nowadays found to be very common. People who are greatly exposed to mental stressors such as low social status, anxiety, exhaustion, etc. have more visceral obesity or intraabdominal fat accumulation and constitute dysregulation of cortisol circadian pattern [44]. Stress-induced excess cortisol increases growth and functions of visceral fat cells leading to abdominal adipose tissue accumulation, whereas released catecholamine during SAS activation stimulates inflammatory adipocytokine production such as interleukin (IL)-6 in adipose tissue [45]. This inflammatory cytokine along with increased angiotensin II levels contributes to insulin resistance, thus metabolic syndrome status begins [46].

2.5 Stroke

Research has confirmed that leading a stressful life due to psychological stressors resulting in behavioral symptoms of anxiety, apathy, etc. significantly increases the risk of incidental stroke in humans [47]. In a study among 6749 human subjects, researchers implicated that those who reported higher levels of mental stress were at the greatest risk for having a stroke, compared to those with lower levels of psychological stress and depression [48]. In that particular study, psychological stress significantly increased transient ischemic attack risk by 59%, which is a ministroke, caused by temporary blockage of blood flow to the brain. Mental stress can be related to an increased risk of cerebrovascular disease, as excessive sympathetic activity occurs during stress-induced SAS activation leading to vasoconstriction. The mechanism of chronic psychosocial stress affecting the vascular system, as discussed earlier under hypertension, can also be attributed to the sustained pressor effects of elevated cortisol by potentiating adrenaline and angiotensin levels and lowering plasma kallikrein, prostacyclin, and NOS in vascular endothelium leading to inhibition of vasodilation and ultimately decreased blood flow. Also increased coagulation, vascular calcification, and inflammation are triggered by both cortisol and catecholamines released in stress, which can also be a possible mechanism for strokes.

2.6 Infertility

Perceived mental stress has been regarded as detrimental to successful human reproduction, since a lot of reports are showing a relation between stress and unsuccessful pregnancy outcomes. Data show that psychological stress significantly reduces the probability of conception each day during the fertile window in women [49]. Human data show that chronic stress causes dysmenorrhea and anovulation in women, and lowers sperm count and sperm motility in men [50]. In a study among men from infertile couples, higher amounts of cortisol levels were found in the group of men with low sperm count compared to the control group with normal counts [51]. Couples who have tried to conceive for a longer time always experience frustration and distress if pregnancy is not achieved. Also in infertile couples, much higher levels of psychological stress have been reported among women than men [52]. Stress affects both male and female fertility by causing neurohormonal imbalance due to disruption of normal HPA axis-mediated cortisol secretion. Increased levels of cortisol inhibit the gonadal axis by inhibiting secretion of gonadotropin-releasing hormone (GnRH) resulting in decreased levels of follicular stimulating hormone (FSH) and leutinizing hormone (LH), ultimately causing impaired ovarian and testicular functions [53]. Due to this serum testosterone, estrogen and progesterone levels are lowered and lead to fertility problems. In human males, severe psychological stress causes decreased testosterone synthesis by Leydig cells, which is also due to cortisol-induced apoptosis of Leydig cells [54]. Furthermore, in females, cortisol hampers uterine growth and differentiation [55]. It has been postulated that stress-mediated overactivation of sympathetic nervous transmission into excess released catecholamines alters blood flow in the fallopian tubes and gamete transportation through adrenergic receptors present in the reproductive tract [56].

Ultimately, couples who suffer from infertility frequently turn to artificial reproductive technology (ART) such as in vitro fertilization (IVF), which has increased by 83% [57]. It is required to measure mental stress levels before and during fertilization treatment to achieve success in pregnancy. Studies have shown that psychological stress significantly reduces the chances of obtaining pregnancy with the IVF procedure. Before IVF treatment, women who had mental stress factors such as concerns about aspects of the procedure, treatment cost, etc. had lower pregnancy outcomes after IVF compared to those who were not stressed [58]. Stressful behavior reduces fertilization, implantation, and live birth rates, thus affecting IVF outcomes [59]. Impaired ovarian function due to estradiol inhibition by the stress hormone cortisol leads to decreased number of oocytes to be harvested in IVF [60]. Interruption of menstrual cycle due to mental stress, which is very well documented, is also a causative factor for infertility and unsuccessful IVF outcomes [61].

2.7 Polycystic Ovarian Syndrome

Polycystic ovarian syndrome (PCOS) is the most common gynecological disorder, affecting 12% women of reproductive age with features such as acne due to hyperandrogenism, hirsutism, obesity, irregular menstruation, and anovulation, and is ultimately the leading cause of infertility in women [62]. There is a prevalence of higher levels of behavioral and psychological stresses in patients suffering from PCOS [63]. Reports suggests that PCOS patients have higher risks for developing CHD and type 2 DM (eightfold increased chance), which are manifestations of insulin resistance in metabolic syndrome, thus connection of psychological stress is well established [64].

A study reported that public speaking in PCOS patients resulted in higher plasma ACTH and cortisol levels, with an increased HR and SNS activation compared to normal healthy individuals [65]. Furthermore, they showed that increased levels of baseline serum IL-6 levels in PCOS patients over controls was an indication of the mechanism behind stress-related insulin resistance. Other clinical studies also report higher amounts of perceived mental stress, cortisol levels, and higher HPA axis activation in PCOS women over normal individuals [66]. Thus besides conventional hormonal treatments, PCOS treatment should also cover psychological aspects such as stress reduction to increase the outcomes of infertility, hyperandrogenism, etc..

2.8 Pregnancy Outcomes and Miscarriages

In the previous section the adverse role of psychological stress in causing infertility, menstruation problems, PCOS, etc. was mentioned, but the role of the gestation period ultimately leading to birth defects, miscarriages, etc. has to be discussed. Psychological stress in pregnant women has been clinically established as a vital risk factor for spontaneous abortions [67]. An observational study of pregnant woman confirmed that of those who had increased stress levels with elevated urinary cortisol, 90% resulted in spontaneous abortion, i.e., very early pregnancy loss [68]. As discussed earlier, stressful behavior is related to increased cortisol and decreased FSH, LH, and GnRH release; additionally, cortisol affects luteal progesterone production, and combines to affect uterine maturation and pregnancy maintenance [69]. Studies in humans also confirm the role of maternal psychosocial stress in shortening of the gestation period, which results in preterm delivery and premature babies [70]. So, maternal stress reduction is vital for successful pregnancies and neonatal health.

2.9 Gastric Ulcer

Stress ulcers are commonly encountered in clinical practices, in which mental and behavioral factor-associated stress is greatly responsible. A human observational study indicated the positive association of higher incidence of gastric ulcer in subjects with higher amounts of psychological distress [71]. Cortisol increases the secretion of pepsin and gastric acid in the stomach [72]. Moreover, cortisol decreases the formation of prostaglandins, which are gastroprotective, through negative regulation of the expression of COX-2 [73]. Ulcer-causing effects can also be delineated by cortisol-mediated induction of lipocortins, which are inhibitors of phospholipase A2 and thus result in decreased prostaglandin production [74].

2.10 Irritable Bowel Syndrome

Enhanced mental stress response has been found to be a potential mechanism for pathophysiological changes in causing irritable bowel syndrome (IBS) due to dysregulation of HPA axis activity [75]. A number of studies have depicted that IBS patients perceive greater psychological stress with higher levels of plasma cortisol over healthy controls. In postexperimental mental stress in IBS patients there are also reports of escalation in problematic gastrointestinal symptoms, which also might further establish the relation between perceived stress and the severity of IBS [76]. Crohn's disease and IBS patients have higher anxiety states, depression, and mental exhaustion as reported through a specific stress questionnaire [77]. Research in IBS patients also indicates that a psychological stressor such as public speaking hampers intestinal permeability of IBS patients, and mast cell degranulation is positively correlated with increased salivary cortisol levels [78]. Also research in healthy young women established that increased exposure to psychological stress resulted in prolonged mucosal dysfunction in the jejunum in response to stimuli, which may also represent an initial step in the development of IBS [79].

2.11 Osteoporosis

In a study, depressed young men and women with increased basal cortisol levels, who were without any signs of osteoporosis, were observed for a number of years. Ultimately, observation concluded the significant occurrence of osteoporosis in the depressed subjects, which was also found to be positively associated with elevated cortisol levels [80]. The possible explanation for stress-mediated progression to osteoporosis can be made through cortisol's inhibition of intestinal calcium absorption, with heightened renal calcium excretion, which ultimately leads to calcium loss from bones and decreased osteoid formation and resorption [81].

2.12 Decreased Immunity and Delayed Wound Healing

A number of extensive clinical reports have suggested that mental stress, emotional behavior, etc. decrease the immunity of the body and thereby make the body prone to easily infectious diseases. Reports indicate higher incidences of infectious diseases such as respiratory tract infections, urinary infections, etc. and also neoplastic disease in chronic mentally stressed persons [82]. Clinical cough and cold are frequently reported as common incidents in psychologically stressed students who possess large social networks [83]. A study among medical students revealed that there was significant increase in plasma cortisol levels with a marked decreasing effect on lymphocyte proliferation, IL-2, and lymphocyte CD19 production on the day before examination compared to the beginning of the academic year [84].

Cortisol, which is a glucocorticoid, has a profound pharmacologically immunosuppressive effect through wider suppression of major inflammatory responses. Stress-influencing immunity is attributed to the actions of cortisol, which has a lympholytic action, suppression of the proliferation of T-lymphocyte, and higher rate of destruction of lymphoid cells [85]. Cortisol downregulates the production of IL-1, IL-2, IL-3, tumor necrosis factor-α, and γ-interferon by negative regulation of genes for cytokines in macrophages, endothelial cells, and lymphocytes, resulting in interference of chemotaxis [86]. Cortisol has also been found to decrease the production of cell adhesion molecules such as ELAM-1 and ICAM-1, which affects decreased adhesion and localization of leukocytes in wound healing [87]. Further decrease in the production of acute phase reactants from macrophage and endothelial cells results in interference of the complement function. Thus cortisol favors the spread of infection in the body as the capacity of microorganism killing by defensive immune cells is impaired.

In human subjects it has been established that psychological stress also impairs wound healing, which can reflect poor recovery from surgery [88]. Study also reflects that during examination stress in a group of students, experimentally induced wounds healed after a much longer time when compared to the control ones at summer vacation time [89]. Further studies in humans also clearly show the positive correlation between psychological stress-induced elevated cortisol levels and decreased rate of wound healing [90].

2.13 Mental Diseases

In young insomniac patients, elevated cortisol secretion patterns along with increased sympathetic activity are well established. Chronic stressful events leading to HPA overactivity in insomnia through significant elevated cortisol levels suggests that insomniacs are at risk for anxiety, depression, and psychosis with significant morbidity [91]. Chronic elevated cortisol levels are associated with stressful situations on a daily basis, which ultimately progress to depression and psychosis in the long term, especially in subjects who have impaired coping ability due to impaired HPA axis functioning. Cortisol has been found to increase the expression of serotonin reuptake transporter in humans resulting in higher serotonin uptake, thus directly leading to clinical depression [92]. In a metaanalysis, augmentation of cortisol levels was much higher in clinically depressed persons compared to nondepressed subjects in response to an experimental psychological stressor [93]. Findings also indicated a positive relation between daily life stressors as risk factors for the onset of psychotic symptoms in adolescents. Moreover, an analysis over a baseline and 1 year follow-up among psychotic youths indicated that those who developed psychosis showed significantly higher salivary cortisol levels at the baseline during the first follow-up [94]. Moreover, psychological stress also victimizes short-term memory and increases forgettability to a greater extent. Significant findings have related that the decrease in cortisol levels in the humorous mood of elderly individuals while watching a humorous video caused a greater improvement in learning ability and memory recall [95].

2.14 Need for Herbal and Natural Drugs in the Management of Psychological Stress

Psychological stress management is urgently needed to prevent stress-induced diseases and to increase the quality of life and productivity. However, there are a number of lifestyle modifications and nonpharmacological approaches such as meditation, exercise, music, etc. that may increase the body's adaptation to stress. Nonetheless, shortcomings of this therapy are nonspecific, with fewer clinically proven efficacies, variable acceptances, and questionable approaches and qualities. Speaking of pharmacological approaches, in the modern allopathic system of medicine, antianxiety agents of the benzodiazepines class such as diazepam and antidepressants such as selective serotonin reuptake inhibitors (selective seretonin reuptake inhibitors [SSRIs]) are often prescribed for frank stress management. Though they acutely alleviate some of the behavioral symptoms by affecting GABA transmission and other central neurotransmissions in the body, chronic sustained elevated cortisol by HPA axis involvement remains unaddressed. These synthetic pills also have drawbacks, e.g., they are meant for short-term usage only and there is a risk of dependence, hangover, withdrawal symptoms, cognitive impairments, etc. So far there has been no allopathic or synthetic drug that is specific for antistress situations, which without any adverse effects can safely prevent the elevation of stress-induced cortisol. Today, herbal drug research has reached its peak in terms of active moiety isolation, preclinical studies, and also human clinical trials. As mentioned earlier, some of the potential antistress compounds as isolated molecules with structure elucidation as per modern herbal drug discovery seek genuine standardization of the antistress herbs for therapeutic utility and efficacy. Thousands of years of traditional concepts from Ayurveda, Chinese traditional medicine, etc. have documented valuable herbs for respective vivid indications, which can act as a vital hub for modern drug discovery research. In relation to the management of psychological stress, "adaptogens" that improve an individual's coping ability to stress is also a complex and new drug research area. In situations of increased stress these herbs normalize the physiological process of the body and increase the ability of an organism to adapt to environmental factors and prevent damage from stressors.

Ideally, an antistress candidate should significantly prevent chronic psychological stress-induced cortisol elevation because cortisol is a validated stress biomarker, which seems to be the ideal target for prevention of stress-induced disorders. It should not interfere with normal circadian rhythmic secretion of ACTH, and should only prevent excess sustained cortisol release as in mental stress-associated impaired cortisol release rhythm. Besides this the drug should not impose any adverse effect on the HPA axis, should be free from withdrawal syndromes, and should possess its beneficial effect even if the dosing is tapered. In animal models of stress, extensive study has been done with natural-based compounds, but establishment of an exact replica of human psychological stress in animals is not very practical as compared to controlled human studies. A large number of clinical trials with herbal-based products have been conducted on volunteers for assessing their antistress therapeutic potential, through validation of psychological stress monitoring by mental tasks, computer games, questionnaires, etc. Many of them have shown quite promising effects in human trials by significantly lowering experimental stress-induced plasma and salivary cortisol levels with insignificant side effects. The following are some herbal and natural products that are clinically proven antistress discoveries in human clinical trials.

3. HERBAL THERAPY

3.1 *Withania somnifera*

This is also known as "ashwagandha." It is mentioned in Ayurveda for having rejuvenating power and stamina. The roots of the herb possess clinically and therapeutically confirmed antistress properties as shown in a number of human trials. A double-blind, randomized, and placebo-controlled trial with 64 adults, chronically suffering from mental stress, at a dosage 300 mg of ashwagandha root extract twice a day for 60 days, showed promising results [96]. In that study, stress scores were assessed by using numerous, well-recognized, globally accepted stress questionnaires and a serum cortisol estimation on days 0 and 60. The study concluded that chronic ashwagandha ingestion was significantly responsible for lowering the stress score and serum cortisol levels over placebo and baseline at day 0, with no significant reports of adverse effects during the treatment period.

Several animal studies also validate its clinical utility by significantly ameliorating chronic stress-triggered pathologies such as hyperglycemia, cortisol elevation, immunosuppression, gastric ulceration, and cognitive deficits in chronic foot shock-induced stress models [97]. Withanolides is one of its active constituents, which has been found to be responsible for its adaptogenic activity through improving brain oxidative status in animal models of chronic stress [98]. Another novel withanolide-free aqueous fraction of the root possesses antistress efficacy with a high therapeutic index in mice models, which might indicate other responsible polar antistress compounds [99,100]. The herb is also reported to increase anabolic activity and mice swimming endurance activity, thus nullifying stress-related fatigue [101].

In another double-blind, randomized human study, the antistress properties of standardized ashwagandha root extract were established in a group of mentally stressed subjects. In that study, ashwagandha intake for 2 month significantly prevented the elevation of serum cortisol, serum C-reactive protein, pulse rate, and BP over placebo controls. Improvement was reported in stress-associated symptoms such as fatigue, loss of appetite, headache, palpitations, sleeplessness, irritability, etc. through a Hamilton anxiety scale questionnaire [102]. In relation to therapeutic management of male infertility due to psychological stress, a study result supported the use of 3 g of herb per day for 3 months, as it is reported to increase fertility in normozoospermic psychologically stressed persons [103]. In that particular clinical study, ashwagandha significantly lowered plasma cortisol levels in psychologically stressed infertile men, and there were significant increases in fertility rate, sperm motility, semen quality, and LH levels over untreated individuals. Thus ashwagandha seems to be an effective antistress drug having a therapeutic prospect in managing psychologically stress-induced infertility.

3.2 *Panax ginseng*

Ginseng is an excellent adaptogenic ancient Chinese herb, which contains mainly ginsenosides as bioactive constituents. Ginsenosides are the most active therapeutic moieties possessing antistress properties, as per the reports of various animal studies of acute and chronic stress-induced pathologies affecting metabolism and immune and hormonal status [97,104,105]. Ginsenoside's neuromodulating effects have been found to prevent a chronic stress-mediated decrease in brain-derived neurotropic factor levels in mice, especially in the hippocampus region [106,107]. It is also a therapeutically established antistress herb that has been validated through a number of clinical trials in stressed human beings. In a study among postmenopausal women suffering from long-term psychological stress having symptoms of fatigue, insomnia, depression, etc., ginseng intake for 30 days resulted in a startling improvement as per the results of psychological stress assessments such as the Cornell Medical Index and the State-Trait Anxiety Inventory indicate. Postmenopausal women showed a higher stress score with serum cortisol elevation, but after treatment with ginseng at the 30th day the stress score was in the normal range with a significant decrease in serum cortisol concentration [108].

3.3 *Eleutherococcus senticosus*

This is also known as Siberian ginseng, having similar adaptogenic properties to *P. ginseng*. Its root and stem bark has traditional usage in China, Korea, and Japan, claiming to relieve fatigue and increase vitality. Eleutheroside E is an iridoid glycoside isolated from *E. senticosus* extract and is responsible for its antifatigue property, as evident from various animal studies [109, 110]. Animal data reflect that forced swimming, stress-induced immune downregulation and cortisol elevation are very much inhibited by Eleutheroside E administration [111], which is also evident in human studies. Furthermore, Eleutheroside E also has the propensity to prevent behavioral alterations and cognitive deficits due to stress induced by sleep deprivation in mice [112].

In a clinical trial on a group of club-level endurance athletes under training, which served as a severe stressful situation, *E. senticosus* consumption for 6 weeks was related to lower values of serum cortisol, as reported after posttraining compared to the placebo group [113].

3.4 *Magnolia officinalis* and *Phellodendron amurense* Combination

Magnolia and Phellodendron standardized bark extracts are clinically validated as antistress in a combination form, as various human trials indicate. *M. officinalis* and *P. amurense* combination (MPC) treatment in 56 moderately stressed patients for 4 weeks demonstrated a significant reduction of salivary cortisol levels with an improvement in mood state parameters of anger, tension, confusion, fatigue, etc. in comparison to the placebo group [114]. Among premenopausal women suffering from psychological stress, 6 weeks of oral MPC, extract treatment effectively reduced stress-perceived scores with broader aspects of mood symptoms associated with postmenopausal condition [115]. Magnolol and honokiol are the reported phytomolecules present in magnolia bark, which are being found to interact with the benzodiazepine site of the GABA(A) receptor, which might be responsible for exerting its effect upon controlling anxious and stressful behavior [116,117].

Another human study also indicated the cortisol and stress score-lowering property of MPC upon healthy overweight psychologically stressed women. This study also reported that the extract is free of adverse effects at a dose of 750 mg per day for a period of 6 weeks [118]. Furthermore, in pilot studies in humans, placebo-controlled trial data also suggest the effectiveness of MPC treatment on lowering perceived stress symptoms and cortisol levels, with additional decreased weight gain levels in overweight premenopausal females [119].

3.5 *Rhodiola rosea*

This is a well-known herb in European and Asiatic traditional medicine, widely used for depression, fatigue, stress, etc. since ancient times. Rosavin and salidroside isolated from this plant are largely responsible for its antistress efficacy in both acute and chronic stress models in rodents [120,121]. Salidroside and *R. rosea* extracts are simultaneously reported to decrease the levels of phosphorylated stress-activated protein kinase in immobilized stressed animals, with a modulatory effect in nitric oxide and cortisol release [122].

In a double-blind, randomized, and placebo-controlled study, *R. rosea* extract for 20 days to a group of mentally stressed students during an examination period showed a significant improvement in mental and physical symptoms associated with examination stress compared to placebo treatment [123]. Another clinical study among 161 young subjects under exposure to acute mental stress also showed that *R. rosea* treatment ameliorates stress symptoms such as fatigue, anxiety, etc. with a dual reduction in the physical parameters of BP and pulse rate [124]. Furthermore, another double-blind study showed that 28 days of *R. rosea* extract-repeated intake in psychologically stressed persons with association of fatigue significantly lowered cortisol response and increased mental performance compared to placebo [125]. Salidroside present in *R. rosea* exerts its antistress, adaptogenic, and antidepressive activity through stimulation of neuropeptide-Y with concomitant lower levels of HSP-72 (heat shock protein) release in brain microglial cells, ultimately modulating the HPA axis stress response [126].

3.6 *Lavandula angustifolia*

This is also known as lavender, containing aromatic oils that mainly contain triterpenoids such as linalool. Its aromatic effect is reported to display antistress and mood relaxing activities by biological effects on both stressed animal and humans. In a trial among 24 human subjects experiencing experimental stress, linalool inhalation resulted in a decrease in BP, HR variation, and salivary cortisol levels, thus modulating the stress response [127]. Simultaneously, other studies have also claimed the cortisol and daytime BP-lowering properties of lavender oil aroma in 83 stressed prehypertensive subjects [128]. Lavender oil inhalation in another study among students performing arithmetic tasks as a mental stressor gave a stress-relieving action in aspects of symptoms and lowering of stress biomarkers cortisol and chromogenin levels in saliva [129]. A randomized trial also confirmed the stress-relieving property of lavender oil among female patients suffering from urinary incontinence [130]. A further study of healthy volunteers also clarified that after sniffing lavender oil aroma, significant reductions of stress hormone cortisol are achieved [131]. Linalool's effect upon modulation of hypothalamic gene expression in stressed rats may be one of the pathways to preventing elevation of cortisol levels during stress [132,133].

3.7 *Bacopa monnieri*

Commonly known as brahmi, *B. monnieri* is a widely explored brain tonic with huge amounts of existing scientific evidence in the area of cognitive treatment. Chronic administration of *B. monnieri* in rats ameliorates forced swimming-induced hypothermia, which is a parameter of stress evaluation in rodents [134]; also chronic immobilized stress-related pathological changes upon spleen, adrenal gland, and metabolic enzyme activities in mice are significantly prevented by chronic *B. monnieri* treatment [135]. A study has indicated that chronic stress-induced increase in brain HSP-70 levels, specifically in the hippocampal area, is only prevented by chronic pretreatment with *B. monnieri*, with a simultaneous increase in superoxide dismutase activities in the cortex area [136].

A clinical study with *B. monnieri* has been conducted on human volunteers to validate antistress properties. A double-blind, crossover, placebo-controlled study demonstrated that upon consumption of *B. monnieri* extract (320, 640 mg), volunteers showed a significant positive level of mood with a significant lowering of cortisol levels after postmultitasking experimental mental stressors. In contrast, the placebo group showed a higher cortisol level and negative mood at poststressors [137].

3.8 *Ginkgo biloba*

Animal findings suggest that the flavonoids and terpenoids present in this ancient Chinese herb improve stress adaptation to animals through normalization of alterations in brain catecholamines and plasma cortisol levels [138,139]. The versatile cerebroprotective role of this herb led to research that showed that chronic stress-induced detrimental changes in discrimination learning, cognitive functions, and plasma stress hormones in rats are suppressed by chronic treatment of *G. biloba* (GB) extracts, resulting in facilitation of better behavioral adaptation under stressful situations [140,141].

Standardized extract of GB consumption in a clinical trial upon healthy young volunteers under experimental mental stress has also demonstrated significant lowering of stress-induced BP and salivary cortisol levels, without any effect on HR [142]. Ginkgolide, a bioactive component, demonstrates an effect upon adrenal gene expressions of glucocorticoids, which also causes lowering of circulating cortisol levels in stressful situations in animals [143].

3.9 *Ocimum sanctum*

This is commonly known as holy basil, and one of its isolated phytoconstituents, ocimumosides A from leaves, displays a promising antistress action on rats under stress, with normalization of stress-induced elevation of cortisol, glucose levels, and adrenal hypertrophy [144]. Also chronic restraint stress-induced anxiety and depression in mice is well reversed by oral *O. sanctum* treatment [145,146], with reports of efficacy in swimming models of stress exposure [147].

In a controlled clinical trial among patients suffering from stress-related anxiety disorder, *O. sanctum* capsules of a high dose of 1000 mg per day for 60 days were administered with a control placebo, because a high dose of *O. sanctum* has shown significant antistress effects in a rodent swimming model of stress. The study concluded that *O. sanctum* treatment significantly improved behavioral aspects of psychological stress as evaluated by a stress questionnaire [148]. *Ocimum tenuiflorum* is another species of *O. sanctum* and has also been further clinically tested in 150 psychologically stressed patients in a randomized double-blind trial. The outcomes of the study after 6 weeks depicted that stress-related symptoms such as forgetfulness, exhaustion, and sexual problems of recent origin were significantly improved compared to placebo [149].

3.10 Black Tea

Black tea is enriched with theaflavin, which has good antioxidant properties. Besides this, theaflavin-enriched black tea extract has also shown antistress effects in a number of randomized crossover studies. In a study a group of college students was given black tea extract for 9 days and were gone through stressor, and subsequently serum cortisol levels were significantly lower compared to placebo treatment [150]. Black tea polyphenols exert an antistress and antidepressant action against chronic stressors in mice through a direct alteration of brain monoaminergic responses and antioxidant status [151]. Another parallel group study among 75 healthy nonsmoking men revealed that

6 weeks of continuous black tea consumption led to significant lowering of psychological stress together with lower cortisol levels and reduced platelet activation upon challenge of mental stressors. Moreover, in that study the placebo group received a caffeinated dummy that showed insignificant effect on stress reduction, thus suggesting the role of black tea constituents other than caffeine [152]. Thus apart from the CNS stimulating effects of caffeine from a cup of tea, stress and fatigue-alleviating properties are related to the content of other natural substances present in it.

3.11 Green Tea

Green tea polyphenols prevent cognitive dysfunctions associated with psychological stress [153]. Green tea is fully enriched with the amino acid L-theanine, which is also reported to reduce mental stress levels in humans. L-Theanine consumption in high-stress-response human adults efficiently reduces anxiety and BP elevation specifically after acute psychological stressors [154]. In a randomized crossover design study among students, intake of green tea was responsible for improving stress scores and mood profile, with a decrease in mental stress load-induced salivary chromogranin A levels [155]. A double-blind study concluded that L-theanine consumption in healthy subjects led to reduced HR variation and salivary immunoglobulin-A levels when subjected to a psychological arithmetic task stressor [156].

Catechin, a major constituent like tannins in green tea, is reported to ameliorate corticosterone injection-mediated stressful behavioral alteration in rats through modulation of the HPA axis [157]. Green tea's antistress assessment might be correlated with catechin's, especially epigallocatechin gallate found in green tea, which is a potent inhibitor of the cortisol-producing enzyme 11β-hydroxysteroid dehydrogenase type 1, leading to lowering of cortisol levels [158] (Fig. 14.2).

4. NUTRITIONAL THERAPY

4.1 Vitamin C

Vitamin C or ascorbic acid is abundantly found in various citrus fruits, and has an antistress effect by lowering cortisol levels in stressful situations. Reports indicate that in a double-blind study among 60 human subjects, continuous 2-week high vitamin C consumption prevented psychological stress-induced systolic BP and cortisol elevation in the stressful situations of public speaking and mental arithmetic tests [159]. In clinical studies in marathon athletes supplemented with vitamin C, postracing serum cortisol level elevation was reported to be significantly lower than placebo-treated athletes, which reflects the stress-reducing action of vitamin C during severe stressors [160]. Thus a regular intake of a diet enriched with vitamin C could be a simple strategy for controlling stress cortisol levels.

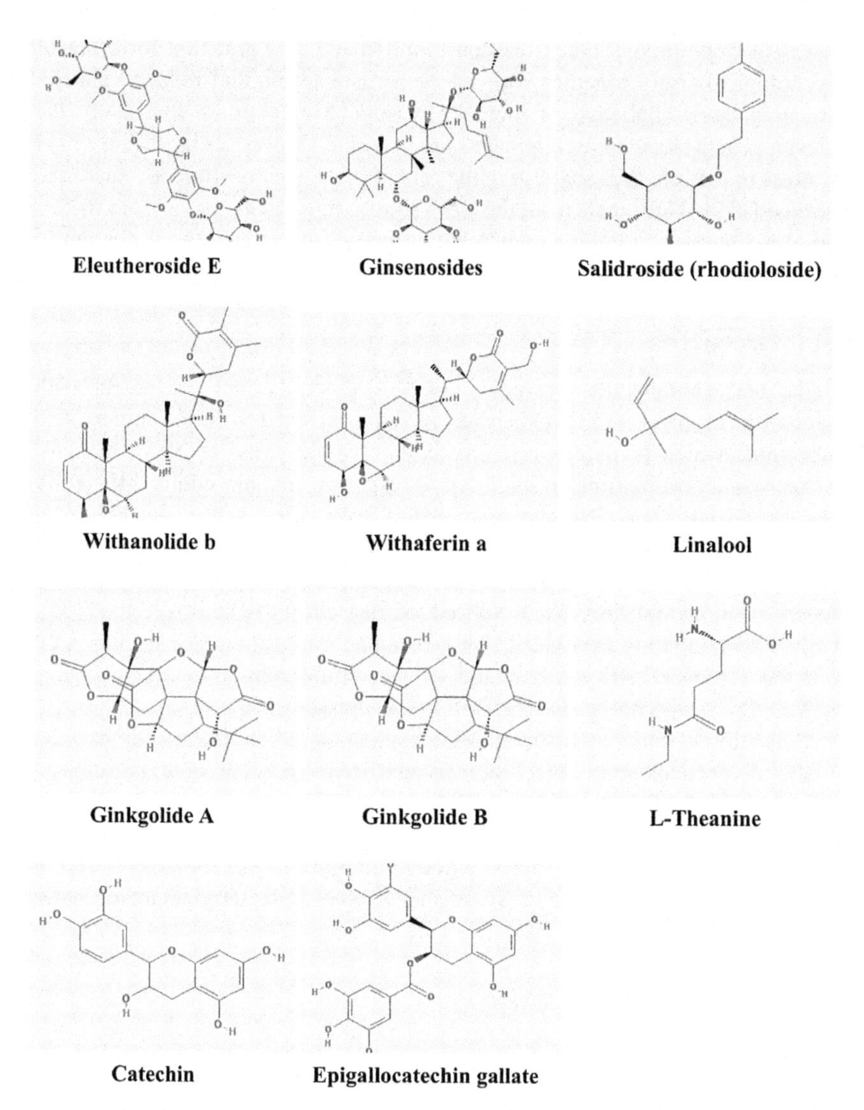

FIGURE 14.2 Respective lead molecules responsible for activity.

4.2 L-Lysine

This is an essential amino acid abundantly found in foods such as spirulina, fenugreek seeds, fish, eggs, meat, etc. The result of a double-blind, placebo-controlled study on 180 adults facing psychological stress confirmed the antistress effect of the amino acid through significant lowering of anxiety and cortisol levels [161]. A 3-month, randomized, double-blind study among a poor Syrian community validated that lysine-fortified wheat consumption resulted in significantly reduced chronic anxiety and lower plasma cortisol response to stress [162].

4.3 L-Ornithine

L-Ornithine is a nonessential amino acid usually contained in protein-rich foods such as fish, meat, eggs, and also nuts. A randomized placebo-controlled trial suggested that 400 mg of L-ornithine consumption in adults led to a marked decrease in negative feelings, stress score, and salivary cortisol levels with a better quality of sleep [163]. In a placebo-controlled trial among 52 women having high levels of mental stress, L-ornithine intake for 8 weeks resulted in a marked improvement in sleep quality and mood states, and also a decrease in serum cortisol levels and perceived mental stress [164].

4.4 Jerte Valley Cherries

These cherries are specially grown in Spain and contain high levels of tryptophan, serotonin, and melatonin. One study claimed that consumption of these cherries in young, middle-aged, and elderly subjects controlled anxiety and perceived stress levels with a significant lowering of urinary cortisol levels and increased serotonergic activity, suggesting that regular consumption of these cherries may reduce stress levels [165].

4.5 Fish Oil

Fish oils (FO) are reported to contain larger amounts of polyunsaturated fatty acids such as omega-3 fatty acids, eicosapentaenoic acid, docosahexaenoic acid, etc., which are widely accepted as beneficial for human health. In alcohol addicts going through rehabilitation programs, 3 weeks of FO supplementation provided a significant reduction in anxiety-distress symptoms and cortisol levels as compared to before treatment and placebo subjects [166]. In another study, reduction of plasma epinephrine, cortisol, and psychological stress response was significant in volunteers taking FO when subjected to experimental mental stress [167]. The stress-related cortisol-lowering action of 6 weeks' FO consumption was also evident in another trial among healthy adults [168].

4.6 Soy Protein

Soy protein is a herbal protein that contains mixtures of lecithin phosphatidylserine and lecithin phosphatidic acid. Several clinical trials with the isolated phosphatidylserine-phosphatidic acid (PAS) mixtures from soy protein had confirmed a protective effect on mood and HRs against severe stress.

Another trial among 75 chronically stressed patients showed that PAS consumption for 42 days effectively reduced the stress-associated increase in serum ACTH and cortisol levels [169]. More trial data also cite that consumption of isolated soy protein complex or PAS in humans gave greater protection against psychological stress and a pronounced decreasing effect on serum ACTH and cortisol concentrations in respect to placebo treatment [170]. Decreased HPA axis-mediated elevation of serum cortisol after stressors was also further evident from another human clinical evaluation with phosphatidylserine consumption [171].

4.7 Casein Tryptic Hydrolysate

Casein is the major milk protein that is hydrolyzed by both pepsin in the stomach and trypsin in the intestines of adults, whereas in infants it is only hydrolyzed by trypsin due to inactive pepsin in a neutral gastric environment. Research has found that the reason behind the calm state of infants after having milk is due to the formation of tryptic hydrolysate of casein in the intestines resulting in the generation of a specific peptide, after cleavage at 91 to 100 number amino acid chain, called α-casozepine, which gives a relaxing effect [172]. In animal models of stressful anxious behavior, the bioactive peptide α-casozepine-enriched casein hydrolysate showed potent activity when compared to benzodiazepines without any side effects such as tolerance and sedation, as seen with diazepam [173,174].

In a randomized, double-blind, placebo-controlled trial among 42 healthy volunteers, two capsules of casein tryptic hydrolysate (CTH), 200 mg three times at 12 h intervals, resulted in better recovery from mental and physical stress. Major findings in that study were that CTH significantly prevented the rise in systolic BP and plasma cortisol levels during the mental stress test [175]. In another crossover, double-blind, placebo-controlled trial among 63 women, CTH 150 mg per day ingested for 30 days significantly reduced the severity of stress-related symptoms. Specifically among 44 stress symptoms, greater improvement was reported in digestive, cardiovascular, intellectual, and emotional aspects of stress-associated manifestations [176].

4.8 Yoghurt

In a double-blind study, consumption of yoghurt enriched with α-lactalbumin, casein tripeptides had been shown to increase the capability to cope with stressful situations. The study pointed out yoghurt's beneficial effect upon psychological stress challenges and resulted in better HR recovery, decreased anxiety, and increased positive mood with lower salivary cortisol levels [177].

4.9 Whey Protein

α-Lactalbumin is a bovine milk-derived protein, also called whey protein. In a randomized human trial it was quite evident that lactalbumin or whey protein consumption in healthy adults increased their coping ability with stressful situations, with a lowering of cortisol levels [178]. Another placebo-controlled human trial also confirmed the stress-relieving effect of whey protein consumption by significantly lowering cortisol levels after stressful exertion [179].

Herbs	Antistress activities
Polygala tenuifolia	Ameliorated chronic mild stress (CMS) and reward insensitivity in rats by reducing serum cortisol, ACTH and CRH levels [180]. Significantly increased sucrose intake and decreased cortisol elevation in chronic mild stress (CMS)-treated rats [181]. Inhibited the decrease in brain derived neurotrophic factor due to CMS [182].
Schisandra chinensis	Treated rats after stressful swimming exercise showed lower values of blood glucose, cortisol, IL-1 and IL-2 levels compared to placebo ones [183]. In isolated neuroglia cells, it modulated in vitro expression of neuropeptide-Y (NPY) and heat shock protein (HSP), the molecular mediators in tolerance and adaptation of stress response [184].
Argyreia speciosa	Significantly lowered the swimming and chronic immobilization stress-induced elevation in cortisol and adrenal gland hypertrophy in treated rats [185]. Pretreatment in rats subjected to cold restraint stress prevented adrenal gland hypertrophy and showed stress ulcer protective and cortisol-lowering properties [186].
Andrographis paniculata	Attenuated the elevation of cortisol, TNF-α, IL-10 and prevented chronic stress-induced pathological changes in rats subjected to chronic unavoidable foot shocks [187]. Also been reported to alter the behavioral pattern, affecting spontaneous motility in rats [188].
Pomegranate peel	In CMS rats treated with extract it showed lower serum cortisol concentrations with an increase in sucrose intake compared to placebo [189]. In rats fed with high-fat diet, extract significantly decreased glucose intolerance, dyslipidemia, TNF-α, IL-6 and cortisol levels compared to control group as metabolic syndrome indicator [190].
Mellisa officinalis	In chronic stressed mice it prevented the physical changes of spleen, thymus, and body weight with a lowered serum cortisol concentration compared to control [191]. Increased neuroblast differentiation and lowered serum cortisol, GABA transaminase levels [192].

Antistress activity of some lead plant extracts and fractions in animal models pending future human trials.

5. CONCLUSION

There is a potential to consider psychological stress management therapeutically by natural substances, which is the safest way of shielding progression to chronic stress-related diseases. Validations exist for pathophysiological connection between stress and various disorders, as proven by globally conducted numerous cohort studies and metaanalyses. As discussed through literature surveys, numerous clinically proven herbal or natural remedies for stress-related disruption of the HPA axis could be advised for increasing patients' adaptability to altered homeostasis for positive outcomes. Of these herbs and dietary interventions, many are effectively confirmed to be devoid of any side effects, with a modulating action on HPA axis-mediated stress hormone levels status; however, a larger amount of clinical data are still awaited for its universal acceptance in specific stress-related diseases. Many of the plant-based herbal drugs and nutritional substances discussed in this chapter have been marketed as antistress, with exclusive patent rights in various countries under dietary supplements or neutraceuticals. Also the active phytomolecules responsible for their antistress activity, as mentioned in Fig. 14.2, could reflect the need for standardization of natural products to be efficacious. Although in future, more randomized controlled trials in large number of subjects with a strong pharmacovigilance-oriented monitoring are required for a FDA approval & recognition. Thus a robust approach must be emphasized for recognition and acceptance by modern prescribing clinicians [193], due to US Food and Drug Administration (FDA)'s and other stringent regulatory authorities' recognition for their acceptance.

Also from a clinician's and prescriber's point of view, more rigorous clinical trials with standardization of these herbal and natural formulations need to be undertaken to validate specific antistress therapies for USFDA approval or be categorizing as antistress nutraceuticals or dietary ingredients, which remains to be resolved. However, traditional usage of natural plant substances and foods with validation through modern studies could be considered for their consumption. Thus along with the prime medications intended to treat various chronic diseases, novel nutritional interventions with plant-based therapies could be considered for effective and safer antistress therapy along with other chronic disease management to be completely beneficial for patients. Besides very commonly and routinely used standard low-dose SSRI antidepressants, anxiolytics, etc., this novel dietary and herbal mode of therapy could replace them and justify their usage for specific modulatory action upon stress biomarkers. We also hereby emphasize the urgent need for antistress herbal drug discovery research programs globally, to design and deliver potential candidates in the future for modulating the human body's ability to adapt to stressors.

ACKNOWLEDGMENTS

The authors are thankful to the University Grants Commission, New Delhi, for providing financial assistance to Dr. Subhash C. Mandal as *UGC Research Awards* (File no.: F.30-1/2013(SA-II)/RA-2012-14-NEW-SC-WES-3684).

REFERENCES

[1] S.S. Lim, et al., A comparative risk assessment of burden of disease and injury attributable to 67 risk factors and risk factor clusters in 21 regions, 1990–2010: a systematic analysis for the Global Burden of Disease Study 2010, Lancet 380 (2012) 2224–2260.

[2] G.E. Miller, E. Chen, K.J. Parker, Psychological stress in childhood and susceptibility to the chronic diseases of aging: moving toward a model of behavioral and biological mechanisms, Psychol. Bull. 137 (2011) 959–997.

[3] C.J. Murray, A.D. Lopez, The Global Burden of Disease: A Comprehensive Assessment of Mortality and Disability from Diseases, Injuries and Risk Factors in 1990 and Projected to 2020, in: Global Burden of Disease and Injury Series, vol. I, Harvard School of Public Health, Cambridge, MA, 1996.

[4] A. Maercker, C.R. Brewin, R.A. Bryant, M. Cloitre, G.M. Reed, M. van Ommeren, A. Humayun, L.M. Jones, A. Kagee, A.E. Llosa, C. Rousseau, D.J. Somasundaram, R. Souza, Y. Suzuki, I. Weissbecker, S.C. Wessely, M.B. First, S. Saxena, Proposals for mental disorders specifically associated with stress in the International Classification of Diseases-11, Lancet 381 (2013) 1683–1685.

[5] http://www.apa.org/news/press/releases/2014/02/teen-stress.aspx.

[6] http://www.nielsen.com/us/en/press-room/2011/women-of-tomorrow.html.

[7] I. Kyrou, C. Tsigos, Stress mechanisms and metabolic complications, Horm. Metab. Res. 39 (2007) 430–438.

[8] L.V. Kessing, E. Agerbo, P.B. Mortensen, Does the impact of major stressful life events on the risk of developing depression change throughout life? Psychol. Med. 33 (2003) 1177–1184.

[9] S. Cohen, T.B. Herbert, Health psychology: psychological factors and physical disease from the perspective of human psychoneuro immunology, Annu. Rev. Psychol. 47 (1996) 113–142.

[10] M. Merkes, Mindfulness-based stress reduction for people with chronic diseases, Aust. J. Prim. Health 16 (2010) 200–210.

[11] E.R. De Kloet, M. Joëls, F. Holsboer, Stress and the brain: from adaptation to disease, Nat. Rev. Neurosci. 6 (2005) 463–475.

[12] A. Gądek-Michalska, J. Spyrka, P. Rachwalska, J. Tadeusz, J. Bugajski, Influence of chronic stress on brain corticosteroid receptors and HPA axis activity, Pharmacol. Rep. 65 (2013) 1163–1175.

[13] E. Viskupic, R. Kvetnansky, E.L. Sabban, K. Fukuhara, V.K. Weise, I.J. Kopin, J.P. Schwartz, Increase in rat adrenal phenylethanolamine N- methyltransferase mRNA level caused by immobilization stress depends on intact pituitary-adrenocortical axis, J. Neurochem. 63 (1994) 808–814.

[14] G. Horvath, Z. Sutto, A. Torbati, G.E. Conner, M. Salathe, A. Wanner, Norepinephrine transport by the extraneuronal monoamine transporter in human bronchial arterial smooth muscle cells, Am. J. Physiol. Lung Cell Mol. Physiol. 285 (2003) 829–837.

[15] G.P. Chrousos, Stress and disorders of the stress system, Nat. Rev. Endocrinol. 5 (2009) 374–381.

[16] N. Schneiderman, G. Ironson, S.D. Siegel, STRESS and health: psychological, behavioral, and biological determinants, Annu. Rev. Clin. Psychol. 1 (2005) 607–628.

[17] M.M. Larzelere, G.N. Jones, Stress and health, Prim. Care 35 (2008) 839–856.

[18] B. Ohlin, P.M. Nilsson, J.A. Nilsson, G. Berglund, Chronic psychosocial stress predicts long-term cardiovascular morbidity and mortality in middle-aged men, Eur. Heart J. 25 (2004) 867–873.

[19] A. Rosengren, S. Hawken, S. Ounpuu, K. Sliwa, M. Zubaid, W.A. Almahmeed, K.N. Blackett, C. Sitthi-amorn, H. Sato, S. Yusuf, Association of psychosocial risk factors with risk of acute myocardial infarction in 11119 cases and 13648 controls from 52 countries (the INTERHEART study): case-control study, Lancet 364 (2004) 953–962.

[20] S.A. Stansfeld, R. Fuhrer, M.J. Shipley, M.G. Marmot, Psychological distress as a risk factor for coronary heart disease in the Whitehall II Study, Int. J. Epidemiol. 31 (2002) 248–255.

[21] M. Hamer, K. O'Donnell, A. Lahiri, A. Steptoe, Salivary cortisol responses to mental stress are associated with coronary artery calcification in healthy men and women, Eur. Heart J. 31 (2010) 424–429.

[22] N. Vogelzangs, A.T. Beekman, Y. Milaneschi, S. Bandinelli, L. Ferrucci, B.W. Penninx, Urinary cortisol and six-year risk of all-cause and cardiovascular mortality, Clin. Endocrinol. Metab. 95 (2010) 4959–4964.

[23] M.J. Dekker, J.W. Koper, M.O. van Aken, H.A. Pols, A. Hofman, F.H. de Jong, C. Kirschbaum, J.C. Witteman, S.W. Lamberts, H. Tiemeier, Salivary cortisol is related to atherosclerosis of carotid arteries, J. Clin. Endocrinol. Metab. 93 (2008) 3741–3747.

[24] A.J. Broadley, A. Korszun, E. Abdelaal, V. Moskvina, C.J. Jones, G.B. Nash, C. Ray, J. Deanfield, M.P. Frenneaux, Inhibition of cortisol production with metyrapone prevents mental stress-induced endothelial dysfunction and baroreflex impairment, J. Am. Coll. Cardiol. 46 (2005) 344–350.

[25] G.D. Smith, Y. Ben-Shlomo, A. Beswick, J. Yarnell, S. Lightman, P. Elwood, Cortisol, testosterone, and coronary heart disease: prospective evidence from the Caerphilly study, Circulation 112 (2005) 332–340.

[26] K.A. Matthews, S. Zhu, D.C. Tucker, M.A. Whooley, Blood pressure reactivity to psychological stress and coronary calcification in the coronary artery risk development in young adults study, Hypertension 47 (2006) 391–395.

[27] A. Steptoe, J. Wardle, Cardiovascular stress responsivity, body mass and abdominal adiposity, Int. J. Obes. (Lond) 29 (2005) 1329–1337.

[28] B.R. Walker, Glucocorticoids and cardiovascular disease, Eur. J. Endocrinol. 157 (2007) 545–559.

[29] M. Hamer, A. Steptoe, Cortisol responses to mental stress and incident hypertension in healthy men and women, J. Clin. Endocrinol. Metab. 97 (2012) 29–34.

[30] K. Pacák, M. Palkovits, Stressor specificity of central neuroendocrine responses: implications for stress-related disorders, Endocr. Rev. 22 (2001) 502–548.

[31] J.P. Henry, C.E. Grim, Psychosocial mechanisms of primary hypertension, Rev. J. Hypertens. 8 (1990) 783–973.

[32] F.E. Marie, M.C. Connell John, Mechanisms of hypertension: the expanding role of aldosterone, J. Am. Soc. Nephrol. 15 (2004) 1993–2001.

[33] S. Yang, L. Zhang, Glucocorticoids and vascular reactivity, Curr. Vasc. Pharmacol. 2 (2004) 1–12.

[34] M.E. Ullian, The role of corticosteroids in the regulation of vascular tone, Cardiovasc Res. 41 (1999) 55–64.

[35] H. Cornelia, T. Barbara, B. Jens, K. Johannes, E.R. Thwing, S. Andrea, M. Christa, L. Karl-Heinz, Job strain as a risk factor for the onset of type 2 diabetes mellitus: findings from the MONICA/KORA Augsburg cohort study, Psychosom. Med. 76 (2014) 562–568.

[36] S. Khani, J.A. Tayek, Cortisol increases gluconeogenesis in humans: its role in the metabolic syndrome, Clin. Sci. (Lond.) 101 (2001) 739–747.

[37] M. Lehrke, U.C. Broedl, I.M. Biller-Friedmann, M. Vogeser, V. Henschel, K. Nassau, B. Göke, E. Kilger, K.G. Parhofer, Serum concentrations of cortisol, interleukin 6, leptin and adiponectin predict stress induced insulin resistance in acute inflammatory reactions, Crit. Care 12 (2008) R157, 1–8.

[38] J.Q. Purnell, S.E. Kahn, M.H. Samuels, D. Brandon, D.L. Loriaux, J.D. Brunzell, Enhanced cortisol production rates, free cortisol, and 11beta-HSD-1 expression correlate with visceral fat and insulin resistance in men: effect of weight loss, Am. J. Physiol. Endocrinol. Metab. 296 (2008) 351–357.

[39] I. Chiodini, G. Adda, A. Scillitani, F. Coletti, V. Morelli, S. Di Lembo, P. Epaminonda, B. Masserini, P. Beck-Peccoz, E. Orsi, B. Ambrosi, M. Arosio, Cortisol secretion in patients with type 2 diabetes: relationship with chronic complications, Diabetes Care 30 (2007) 83–88.

[40] M. Faulenbach, H. Uthoff, K. Schwegler, G.A. Spinas, C. Schmid, P. Wiesli, Effect of psychological stress on glucose control in patients with Type 2 diabetes, Diabet. Med. 29 (2012) 128–131.

[41] B.E. Cohen, P. Panguluri, B. Na, M.A. Whooley, Psychological risk factors and the metabolic syndrome in patients with coronary heart disease: findings from the Heart and Soul Study, Psychiatry Res. 175 (2010) 133–137.

[42] T. Chandola, E. Brunner, M. Marmot, Chronic stress at work and the metabolic syndrome: prospective study, BMJ 332 (2006) 521–525.

[43] L.S. Kinder, M.R. Carnethon, L.P. Palaniappan, A.C. King, S.P. Fortmann, Depression and the metabolic syndrome in young adults: findings from the Third National Health and Nutrition Examination Survey, Psychosom. Med. 66 (2004) 316–322.

[44] P. Björntorp, R. Rosmond, Neuroendocrine abnormalities in visceral obesity, Int. J. Obes. Relat. Metab. Disord. 24 (2000) 80–85.

[45] B.E. Wisse, The inflammatory syndrome: the role of adipose tissue cytokines in metabolic disorders linked to obesity, J. Am. Soc. Nephrol. 15 (2004) 2792–2800.

[46] F. Folli, C.R. Kahn, H. Hansen, J.L. Bouchie, E.P. Feener, Angiotensin II inhibits insulin signaling in aortic smooth muscle cells at multiple levels: a potential role for serine phosphorylation in insulin/angiotensin II crosstalk, J. Clin. Invest. 100 (1997) 2158–2169.

[47] J.A. Egido, O. Castillo, B. Roig, I. Sanz, M.R. Herrero, M.T. Garay, A.M. Garcia, M. Fuentes, C. Fernandez, Is psycho-physical stress a risk factor for stroke? A case-control study, J. Neurol. Neurosurg. Psychiatry 83 (2012) 1104–1110.

[48] S.A. Everson-Rose, N.S. Roetker, P.L. Lutsey, K.N. Kershaw, W.T. Longstreth Jr., R.L. Sacco, A.V. Diez Roux, A. Alonso, Chronic stress, depressive symptoms, anger, hostility, and risk of stroke and transient ischemic attack in the multi-ethnic study of atherosclerosis, Stroke 45 (2014) 2318–2323.

[49] G.M. Louis, K.J. Lum, R. Sundaram, Z. Chen, S. Kim, C.D. Lynch, E.F. Schisterman, C. Pyper, Stress reduces conception probabilities across the fertile window: evidence in support of relaxation, Fertil. Steril. 95 (2011) 2184–2189.

[50] K. Nakamura, S. Sheps, P.C. Arck, Stress and reproductive failure: past notions, present insights and future directions, J. Assist. Reprod. Genet. 25 (2008) 47–62.

[51] M. Klimek, W. Pabian, B. Tomaszewska, J. Kołodziejczyk, Levels of plasma ACTH in men from infertile couples, Neuro. Endocrinol. Lett. 26 (2005) 347–350.

[52] R. Musa, R. Ramli, A.W. Yazmie, M.B. Khadijah, M.Y. Hayati, M. Midin, N.R. Nik Jaafar, S. Das, H. Sidi, A. Ravindran, A preliminary study of the psychological differences in infertile couples and their relation to the coping styles, Compr. Psychiatry 55 (2014) 65–69.

[53] S. Whirledge, J.A. Cidlowski, Glucocorticoids, stress, and fertility, Minerva Endocrinol. 35 (2010) 109–125.

[54] H.B. Gao, M.H. Tong, Y.Q. Hu, Q.S. Guo, R. Ge, M.P. Hardy, Glucocorticoid induces apoptosis in rat leydig cells, Endocrinology 143 (2002) 130–138.

[55] S. Entringer, C. Buss, P.D. Wadhwa, Prenatal stress and developmental programming of human health and disease risk: concepts and integration of empirical findings, Curr. Opin. Endocrinol. Diabetes Obes. 17 (2010) 507–516.

[56] J.G. Schenker, D. Meirow, E. Schenker, Stress and human reproduction, Eur. J. Obstet. Gynecol. Reprod. Biol. 45 (1992) 1–8.

[57] National Board of Health D, IVF-behandlinger I Danmark (2004), 2006, pp. 1–16. Copenhagen.

[58] H. Klonoff-Cohen, L. Natarajan, The concerns during assisted reproductive technologies (CART) scale and pregnancy outcomes, Fertil. Steril. 81 (2004) 982–988.

[59] S.M. Ebbesen, R. Zachariae, M.Y. Mehlsen, D. Thomsen, A. Højgaard, L. Ottosen, T. Petersen, H.J. Ingerslev, Stressful life events are associated with a poor in-vitro fertilization (IVF) outcome: a prospective study, Hum. Reprod. 24 (2009) 2173–2182.

[60] D. Lancastle, J. Boivin, Dispositional optimism, trait anxiety, and coping: unique or shared effects on biological response to fertility treatment? Health Psychol. 24 (2005) 171–178.

[61] J. Boivin, J.E. Takefman, Stress level across stages of in vitro fertilization in subsequently pregnant and nonpregnant women, Fertil. Steril. 64 (1995) 802–810.

[62] W.A. March, V.M. Moore, K.J. Willson, D.I. Phillips, R.J. Norman, M.J. Davies, The prevalence of polycystic ovary syndrome in a community sample assessed under contrasting diagnostic criteria, Hum. Reprod. 25 (2010) 544–551.

[63] F.Z. Zangeneh, M. Jafarabadi, M.M. Naghizadeh, N. Abedinia, F. Haghollahi, Psychological distress in women with polycystic ovary syndrome from imam khomeini hospital, tehran, J. Reprod. Infertil. 13 (2012) 111–115.

[64] T. Apridonidze, P.A. Essah, M.J. Iuorno, J.E. Nestler, Prevalence and characteristics of the metabolic syndrome in women with polycystic ovary syndrome, J. Clin. Endocrinol. Metab. 90 (2005) 1929–1935.

[65] S. Benson, P.C. Arck, S. Tan, S. Hahn, K. Mann, N. Rifaie, O.E. Janssen, M. Schedlowski, S. Elsenbruch, Disturbed stress responses in women with polycystic ovary syndrome, Psychoneuroendocrinology 34 (2009) 727–735.

[66] A. Gallinelli, M.L. Matteo, A. Volpe, F. Facchinetti, Autonomic and neuroendocrine responses to stress in patients with functional hypothalamic secondary amenorrhea, Fertil. Steril. 73 (2000) 812–816.

[67] T. Wainstock, L. Lerner-Geva, S. Glasser, I. Shoham-Vardi, E.Y. Anteby, Prenatal stress and risk of spontaneous abortion, Psychosom. Med. 75 (2013) 228–235.

[68] P.A. Nepomnaschy, K.B. Welch, D.S. McConnell, B.S. Low, B.I. Strassmann, B.G. England, Cortisol levels and very early pregnancy loss in humans, Proc. Natl. Acad. Sci. U.S.A. 103 (2006) 3938–3942.

[69] E.R. Norwitz, D.J. Schust, S.J. Fisher, Implantation and the survival of early pregnancy, N. Engl. J. Med. 345 (2001) 1400–1448.

[70] N. Dole, D.A. Savitz, I. Hertz-Picciotto, A.M. Siega-Riz, M.J. McMahon, P. Buekens, Maternal stress and preterm birth, Am. J. Epidemiol. 157 (2003) 14–24.

[71] S. Levenstein, S. Rosenstock, R.K. Jacobsen, T. Jorgensen, Psychological stress increases risk for peptic ulcer, regardless of *Helicobacter pylori* infection or use of nonsteroidal anti-inflammatory drugs, Clin. Gastroenterol. Hepatol. 13 (3) (2015) 498–506.

[72] I. Gritti, G. Banfi, G.S. Roi, Pepsinogens: physiology, pharmacology pathophysiology and exercise, Pharmacol. Res. 41 (2000) 265–281.

[73] B. García-Bueno, J.L. Madrigal, B.G. Pérez-Nievas, J.C. Leza, Stress mediators regulate brain prostaglandin synthesis and peroxisome proliferator-activated receptor-gamma activation after stress in rats, Endocrinology 149 (2008) 1969–1978.

[74] S.H. Peers, F. Smillie, A.J. Elderfield, R.J. Flower, Glucocorticoid-and non glucocorticoid induction of lipocortins (annexins) 1 and 2 in rat peritoneal leucocytes in vivo, Br. J. Pharmacol. 108 (1993) 66–72.

[75] L. Chang, S. Sundaresh, J. Elliott, P.A. Anton, P. Baldi, A. Licudine, M. Mayer, T. Vuong, M. Hirano, B.D. Naliboff, V.Z. Ameen, E.A. Mayer, Dysregulation of the hypothalamic-pituitary-adrenal (HPA) axis in irritable bowel syndrome, Neurogastroenterol. Motil. 21 (2009) 149–159.

[76] P.J. Kennedy, J.F. Cryan, E.M. Quigley, T.G. Dinan, G. Clarke, A sustained hypothalamic-pituitary-adrenal axis response to acute psychosocial stress in irritable bowel syndrome, Psychol. Med. 44 (2014) 3123–3134.

[77] M. Simrén, J. Axelsson, R. Gillberg, H. Abrahamsson, J. Svedlund, E.S. Björnsson, Quality of life in inflammatory bowel disease in remission: the impact of IBS-like symptoms and associated psychological factors, Am. J. Gastroenterol. 97 (2002) 389–396.

[78] T. Vanuytsel, S. van Wanrooy, H. Vanheel, C. Vanormelingen, S. Verschueren, E. Houben, S. Salim Rasoel, J. Tóth, L. Holvoet, R. Farré, L. Van Oudenhove, G. Boeckxstaens, K. Verbeke, J. Tack, Psychological stress and corticotropin-releasing hormone increase intestinal permeability in humans by a mast cell-dependent mechanism, Gut 63 (2014) 1293–1299.

[79] C. Alonso, M. Guilarte, M. Vicario, L. Ramos, Z. Ramadan, M. Antolín, C. Martínez, S. Rezzi, E. Saperas, S. Kochhar, J. Santos, J.R. Malagelada, Maladaptive intestinal epithelial responses to life stress may predispose healthy women to gut mucosal inflammation, Gastroenterology 135 (2008) 163–172.

[80] M. Vrkljan, V. Thaller, I. Lovricević, P. Gaćina, J. Resetić, M. Bekić, Z. Sonicki, Depressive disorder as possible risk factor of osteoporosis, Coll. Antropol. 25 (2001) 485–492.

[81] F. Manelli, A. Giustina, Glucocorticoid-induced osteoporosis, Trends Endocrinol. Metab. 11 (2000) 79–85.

[82] S.C. Segerstrom, G.E. Miller, Psychological stress and the human immune system: a meta-analytic study of 30 years of inquiry, Psychol. Bull. 130 (2004) 601–630.

[83] N. Hamrick, S. Cohen, M.S. Rodriguez, Being popular can be healthy or unhealthy: stress, social network diversity, and incidence of upper respiratory infection, Health Psychol. 21 (2002) 294–298.

[84] L. Guidi, A. Tricerri, M. Vangeli, D. Frasca, A. Riccardo Errani, A. Di Giovanni, L. Antico, E. Menini, V. Sciamanna, N. Magnavita, G. Doria, C. Bartoloni, Neuropeptide Y plasma levels and immunological changes during academic stress, Neuropsychobiology 40 (1999) 188–195.

[85] R.M. Sapolsky, L.M. Romero, A.U. Munck, How do glucocorticoids influence stress responses? Integrating permissive, suppressive, stimulatory, and preparative actions, Endocr. Rev. 21 (2000) 55–89.

[86] I.J. Elenkov, G.P. Chrousos, Stress hormones, Th1/Th2 patterns, pro/anti-inflammatory cytokines and susceptibility to disease, Trends Endocrinol. Metab. 10 (1999) 359–368.

[87] B.N. Cronstein, S. C Kimmel, R. I Levin, F. Martiniuk, G. Weissmann, A mechanism for the antiinflammatory effects of corticosteroids: the glucocorticoid receptor regulates leukocyte adhesion to endothelial cells and expression of endothelial-leukocyte adhesion molecule 1 and intercellular adhesion molecule 1, Proc. Natl. Acad. Sci. U.S.A. 89 (1992) 9991–9995.

[88] J.K. Kiecolt-Glaser, P.T. Marucha, W.B. Malarkey, A.M. Mercado, R. Glaser, Slowing of wound healing by psychological stress, Lancet 346 (1995) 1194–1196.

[89] P.T. Marucha, J.K. Kiecolt-Glaser, M. Favagehi, Mucosal wound healing is impaired by examination stress, Psychosom. Med. 60 (1998) 362–365.

[90] M. Ebrecht, J. Hextall, L.G. Kirtley, A. Taylor, M. Dyson, J. Weinman, Perceived stress and cortisol levels predict speed of wound healing in healthy male adults, Psychoneuroendocrinology 29 (2004) 798–809.

[91] A.N. Vgontzas, E.O. Bixler, H.M. Lin, P. Prolo, G. Mastorakos, A. Vela-Bueno, A. Kales, G.P. Chrousos, Chronic insomnia is associated with nyctohemeral activation of the hypothalamic-pituitary-adrenal axis: clinical implications, J. Clin. Endocrinol. Metab. 86 (2001) 3787–3794.

[92] G.E. Tafet, V.P. Idoyaga-Vargas, D.P. Abulafia, J.M. Calandria, S.S. Roffman, A. Chiovetta, M. Shinitzky, Correlation between cortisol level and serotonin uptake in patients with chronic stress and depression, Cogn. Affect Behav. Neurosci. 1 (2001) 388–393.

[93] H.M. Burke, M.C. Davis, C. Otte, D.C. Mohr, Depression and cortisol responses to psychological stress: a meta-analysis, Psychoneuroendocrinology 30 (2005) 846–856.

[94] E.F. Walker, P.A. Brennan, M. Esterberg, J. Brasfield, B. Pearce, M.T. Compton, Longitudinal changes in cortisol secretion and conversion to psychosis in at-risk youth, J. Abnorm Psychol. 119 (2010) 401–408.

[95] G.S. Bains, L.S. Berk, N. Daher, E. Lohman, E. Schwab, J. Petrofsky, P. Deshpande, The effect of humor on short-term memory in older adults: a new component for whole-person wellness, Adv. Mind Body Med. 28 (2014) 16–24.

[96] K. Chandrasekhar, J. Kapoor, S. Anishetty, A prospective, randomized double-blind, placebo-controlled study of safety and efficacy of a high-concentration full-spectrum extract of ashwagandha root in reducing stress and anxiety in adults, Indian J. Psychol. Med. 34 (2012) 255–262.

[97] S.K. Bhattacharya, A.V. Muruganandam, Adaptogenic activity of *Withania somnifera*: an experimental study using a rat model of chronic stress, Pharmacol. Biochem. Behav. 75 (3) (June 2003) 547–555.

[98] A. Bhattacharya, S. Ghosal, S.K. Bhattacharya, Anti-oxidant effect of *Withania somnifera* glycow ithanolides in chronic footshock stress-induced perturbations of oxidative free radical scavenging enzymes and lipid peroxidation in rat frontal cortex and striatum, J. Ethnopharmacol. 74 (1) (2001) 1–6.

[99] B. Singh, A.K. Saxena, B.K. Chandan, D.K. Gupta, K.K. Bhutani, K.K. Anand, Adaptogenic activity of a novel, withanolide-free aqueous fraction from the roots of *Withania somnifera* Dun, Phytother. Res. 15 (4) (2001) 311–318.

[100] B. Singh, B.K. Chandan, D.K. Gupta, Adaptogenic activity of a novel withanolide-free aqueous fraction from the roots of *Withania somnifera* Dun. (Part II), Phytother. Res. 17 (5) (2003) 531–536.

[101] A. Grandhi, A.M. Mujumdar, B. Patwardhan, A comparative pharmacological investigation of Ashwagandha and Ginseng, J. Ethnopharmacol. 44 (3) (1994) 131–135.

[102] B. Auddy, J. Hazra, A. Mitra, B. Abedon, S. Ghosal, A standardized *Withania somnifera* extract significantly reduces stress-related parameters in chronically stressed humans: a double-blind, randomized, placebo-controlled study, J. Am. Nutraceutical Assoc. 11 (2008) 50–56.

[103] A.A. Mahdi, K.K. Shukla, M.K. Ahmad, S. Rajender, S.N. Shankhwar, V. Singh, D. Dalela, *Withania somnifera* improves semen quality in stress-related male fertility, Evid. Based Complement. Altern. Med. 576962 (2011) (2011) 1–9.

[104] D. Rai, G. Bhatia, T. Sen, G. Palit, Anti-stress effects of *Ginkgo biloba* and *Panax ginseng*: a comparative study, J. Pharmacol. Sci. 93 (2003) 458–464.

[105] U. Banerjee, J.A. Izquierdo, Antistress and antifatigue properties of *Panax ginseng*: comparison with piracetam, Acta Physiol. Lat. Am. 32 (1982) 277–285.

[106] M. Kim, S.O. Kim, M. Lee, Y. Park, D. Kim, K.H. Cho, S.Y. Kim, E.H. Lee, Effects of ginsenoside Rb1 on the stress-induced changes of BDNF and HSP70 expression in rat hippocampus, Environ. Toxicol. Pharmacol. 38 (1) (2014) 257–262.

[107] B. Jiang, Z. Xiong, J. Yang, W. Wang, Y. Wang, Z.L. Hu, F. Wang, J.G. Chen, Antidepressant-like effects of ginsenoside Rg1 are due to activation of the BDNF signalling pathway and neurogenesis in the hippocampus, Br. J. Pharmacol. 166 (6) (2012) 1872–1887.

[108] T. Tode, Y. Kikuchi, J. Hirata, T. Kita, H. Nakata, I. Nagata, Effect of Korean red ginseng on psychological functions in patients with severe climacteric syndromes, Int. J. Gynaecol. Obstet. 67 (1999) 169–174.

[109] L.Z. Huang, B.K. Huang, Q. Ye, L.P. Qin, Bioactivity-guided fractionation for anti-fatigue property of *Acanthopanax senticosus*, J. Ethnopharmacol. 133 (1) (2011) 213–219.

[110] T. Deyama, S. Nishibe, Y. Nakazawa, Constituents and pharmacological effects of *Eucommia* and *Siberian ginseng*, Acta Pharmacol. Sin. 22 (12) (2001) 1057–1070.

[111] Y. Kimura, M. Sumiyoshi, Effects of various *Eleutherococcus senticosus* cortex on swimming time, natural killer activity and corticosterone level in forced swimming stressed mice, J. Ethnopharmacol. 95 (2004) 447–453.

[112] L.Z. Huang, L. Wei, H.F. Zhao, B.K. Huang, K. Rahman, L.P. Qin, The effect of Eleutheroside E on behavioral alterations in murine sleep deprivation stress model, Eur. J. Pharmacol. 658 (2011) 150–155.

[113] B.T. Gaffney, H.M. Hügel, P.A. Rich, The effects of *Eleutherococcus senticosus* and *Panax ginseng* on steroidal hormone indices of stress and lymphocyte subset numbers in endurance athletes, Life Sci. 70 (2001) 431–442.

[114] S.M. Talbott, J.A. Talbott, M. Pugh, Effect of *Magnolia officinalis* and *Phellodendron amurense* (Relora®) on cortisol and psychological mood state in moderately stressed subjects, J. Int. Soc. Sports Nutr. 10 (2013) 1–6.

[115] D.S. Kalman, S. Feldman, R. Feldman, H.I. Schwartz, D.R. Krieger, R. Garrison, Effect of a proprietary *Magnolia* and *Phellodendron* extract on stress levels in healthy women: a pilot, double-blind, placebo-controlled clinical trial, Nutr. J. 7 (2008) 1–6.

[116] C.R. Chen, X.Z. Zhou, Y.J. Luo, Z.L. Huang, Y. Urade, W.M. Qu, Magnolol, a major bioactive constituent of the bark of *Magnolia officinalis*, induces sleep via the benzodiazepine site of GABA(A) receptor in mice, Neuropharmacology 63 (6) (2012) 1191–1199.

[117] H. Kuribara, W.B. Stavinoha, Y. Maruyama, Honokiol, a putative anxiolytic agent extracted from magnolia bark, has no diazepam-like side-effects in mice, J. Pharm. Pharmacol. 51 (1) (1999) 97–103.

[118] D. Kalman, S. Feldman, D. Krieger, A randomized double blind placebo controlled clinical trial of Relora^{TM} in the management of stress in healthy overweight females, FASEB J. 20 (2006). A-379.

[119] R. Garrison, W.G. Chambliss, Effect of a proprietary *Magnolia* and *Phellodendron* extract on weight management: a pilot, double-blind, placebo-controlled clinical trial, Altern. Ther. Health Med. 12 (2006) 50–54.

[120] M. Perfumi, L. Mattioli, Adaptogenic and central nervous system effects of single doses of 3% rosavin and 1% salidroside *Rhodiola rosea* L. extract in mice, Phytother. Res. 21 (1) (2007) 37–43.

[121] L. Mattioli, C. Funari, M. Perfumi, Effects of *Rhodiola rosea* L. extract on behavioural and physiological alterations induced by chronic mild stress in female rats, J. Psychopharmacol. 23 (2) (2009) 130–142.

[122] A. Panossian, M. Hambardzumyan, A. Hovhanissyan, G. Wikman, The adaptogens rhodiola and schizandra modify the response to immobilization stress in rabbits by suppressing the increase of phosphorylated stress-activated protein kinase, nitric oxide and cortisol, Drug Target Insights 2 (2007) 39–54.

[123] A.A. Spasov, G.K. Wikman, V.B. Mandrikov, I.A. Mironova, V.V. Neumoin, A double-blind, placebo-controlled pilot study of the stimulating and adaptogenic effect of *Rhodiola rosea* SHR-5 extract on the fatigue of students caused by stress during an examination period with a repeated low-dose regimen, Phytomedicine 7 (2000) 85–89.

[124] E.M. Olsson, B. von Schéele, A.G. Panossian, A randomised, double-blind, placebo-controlled, parallel-group study of the standardised extract shr-5 of the roots of *Rhodiola rosea* in the treatment of subjects with stress-related fatigue, Planta Med. 75 (2009) 105–112.

[125] V.A. Shevtsov, B.I. Zholus, V.I. Shervarly, V.B. Vol'skij, Y.P. Korovin, M.P. Khristich, N.A. Roslyakova, G. Wikman, A randomized trial of two different doses of a SHR-5 *Rhodiola rosea* extract versus placebo and control of capacity for mental work, Phytomedicine 10 (2003) 95–105.

[126] A. Panossian, G. Wikman, P. Kaur, A. Asea, Adaptogens stimulate neuropeptide Y and Hsp72 expression and release in neuroglia cells, Front. Neurosci. 6 (2012) 6.

[127] M. Höferl, S. Krist, G. Buchbauer, Chirality influences the effects of linalool on physiological parameters of stress, Planta Med. 72 (2006) 1188–1192.

[128] I.H. Kim, C. Kim, K. Seong, M.H. Hur, H.M. Lim, M.S. Lee, Essential oil inhalation on blood pressure and salivary cortisol levels in prehypertensive and hypertensive subjects, Evid. Based Complement. Altern. Med. 984203 (2012) 1–9.

[129] M. Toda, K. Morimoto, Effect of lavender aroma on salivary endocrinological stress markers, Arch. Oral Biol. 53 (2008) 964–968.

[130] G.H. Seol, Y.H. Lee, P. Kang, J.H. You, M. Park, S.S. Min, Randomized controlled trial for *Salvia sclarea* or *Lavandula angustifolia*: differential effects on blood pressure in female patients with urinary incontinence undergoing urodynamic examination, J. Altern. Complement. Med. 19 (2013) 664–670.

[131] T. Atsumi, K. Tonosaki, Smelling lavender and rosemary increases free radical scavenging activity and decreases cortisol level in saliva, Psychiatry Res. 150 (2007) 89–96.

[132] N. Yamamoto, S. Fujiwara, K. Saito-Iizumi, A. Kamei, F. Shinozaki, Y. Watanabe, K. Abe, A. Nakamura, Effects of inhaled (S)-linalool on hypothalamic gene expression in rats under restraint stress, Biosci. Biotechnol. Biochem. 77 (12) (2013) 2413–2418.

[133] A. Nakamura, S. Fujiwara, I. Matsumoto, K. Abe, Stress repression in restrained rats by (R)-(-)-linalool inhalation and gene expression profiling of their whole blood cells, J. Agric. Food Chem. 57 (12) (2009) 5480–5485.

[134] K.I.W.K. Somarathna, H.M. Chandola, B. Ravishankar, K.N. Pandya, A.M.P. Attanayake, B.K. Ashok, Evaluation of adaptogenic and anti-stress effects of Ranahamsa Rasayanaya-A Sri Lankan classical Rasayana drug on experimental animals, Ayu 31 (1) (2010) 88–92.

[135] D. Rai, G. Bhatia, G. Palit, R. Pal, S. Singh, H.K. Singh, Adaptogenic effect of *Bacopa monniera* (Brahmi), Pharmacol. Biochem. Behav. 75 (4) (2003) 823–830.

[136] D.K. Chowdhuri, D. Parmar, P. Kakkar, R. Shukla, P.K. Seth, R.C. Srimal, Antistress effects of bacosides of *Bacopa monnieri*: modulation of Hsp70 expression, superoxide dismutase and cytochrome P450 activity in rat brain, Phytother. Res. 16 (7) (2002) 639–645.

[137] S. Benson, L.A. Downey, C. Stough, M. Wetherell, A. Zangara, A. Scholey, An acute, double-blind, placebo-controlled cross-over study of 320 mg and 640 mg doses of *Bacopa monnieri* (CDRI 08) on multitasking stress reactivity and mood, Phytother. Res. 28 (2014) 551–559.

[138] Z.A. Shah, P. Sharma, S.B. Vohora, *Ginkgo biloba* normalises stress-elevated alterations in brain catecholamines, serotonin and plasma corticosterone levels, Eur. Neuropsychopharmacol. 13 (5) (2003) 321–325.

[139] C.R. Markus, J.H. Lammers, Effects of *Ginkgo biloba* on corticosterone stress responses after inescapable shock exposure in the rat, Pharmacol. Biochem. Behav. 76 (2003) 487–492.

[140] J.R. Rapin, I. Lamproglou, K. Drieu, F.V. DeFeudis, Demonstration of the "anti-stress" activity of an extract of *Ginkgo biloba* (EGb 761) using a discrimination learning task, Gen. Pharmacol. 25 (5) (1994) 1009–1016.

[141] A. Walesiuk, E. Trofimiuk, J.J. Braszko, *Gingko biloba* extract diminishes stress-induced memory deficits in rats, Pharmacol. Rep. 57 (2) (2005) 176–187.

[142] D. Jezova, R. Duncko, M. Lassanova, M. Kriska, F. Moncek, Reduction of rise in blood pressure and cortisol release during stress by *Ginkgo biloba* extract (EGb 761) in healthy volunteers, J. Physiol. Pharmacol. 53 (2002) 337–348.

[143] H. Amri, S.O. Ogwuegbu, N. Boujrad, K. Drieu, V. Papadopoulos, In vivo regulation of peripheral-type benzodiazepine receptor and glucocorticoid synthesis by *Ginkgo biloba* extract EGb 761 and isolated ginkgolides, Endocrinology 137 (12) (1996) 5707–5718.

[144] P. Gupta, D.K. Yadav, K.B. Siripurapu, G. Palit, R. Maurya, Constituents of *Ocimum sanctum* with antistress activity, J. Nat. Prod. 70 (9) (2007) 1410–1416.

[145] I. Tabassum, Z.N. Siddiqui, S.J. Rizvi, Effects of *Ocimum sanctum* and *Camellia sinensis* on stress-induced anxiety and depression in male albino *Rattus norvegicus*, Indian J. Pharmacol. 42 (5) (2010) 283–288.

[146] L.R. Bathala, ChV. Rao, S. Manjunath, S. Vinuta, R. Vemulapalli, Efficacy of *Ocimum sanctum* for relieving stress: a preclinical study, J. Contemp. Dent. Pract. 13 (6) (2012) 782–786.

[147] S.C. Mandal, T.K. Maity, B.P. Saha, M. Pal, Effect of *Ocimum sanctum* root extract on swimming performance in mice, Phytother. Res. 14 (2000) 120–121.

[148] D. Bhattacharyya, T.K. Sur, U. Jana, P.K. Debnath, Controlled programmed trial of *Ocimum sanctum* leaf on generalized anxiety disorders, Nepal Med. Coll. J. 10 (2008) 176–179.

[149] R.C. Saxena, R. Singh, P. Kumar, M.P. Singh Negi, V. Saxena, P. Geetharani, J.J. Allan, K. Venkateshwarlu, Efficacy of an extract of *Ocimum tenuiflorum* (OciBest) in the management of general stress: a double-blind, placebo-controlled study, Evid. Based Complement. Altern. Med. 894509 (2012) 1–7.

[150] S.M. Arent, M. Senso, D.L. Golem, K.H. McKeever, The effects of theaflavin-enriched black tea extract on muscle soreness, oxidative stress, inflammation, and endocrine responses to acute anaerobic interval training: a randomized, double-blind, crossover study, J. Int. Soc. Sports Nutr. 7 (2010) 1–7.

[151] Y. Liu, G. Jia, L. Gou, L. Sun, X. Fu, N. Lan, S. Li, X. Yin, Antidepressant-like effects of tea polyphenols on mouse model of chronic unpredictable mild stress, Pharmacol. Biochem. Behav. 104 (2013) 27–32.

[152] A. Steptoe, E.L. Gibson, R. Vuononvirta, E.D. Williams, M. Hamer, J.A. Rycroft, J.D. Erusalimsky, J. Wardle, The effects of tea on psychophysiological stress responsivity and post-stress recovery: a randomised double-blind trial, Psychopharmacol. Berl. 190 (2007) 81–89.

[153] W.Q. Chen, X.L. Zhao, Y. Hou, S.T. Li, Y. Hong, D.L. Wang, Y.Y. Cheng, Protective effects of green tea polyphenols on cognitive impairments induced by psychological stress in rats, Behav. Brain Res. 202 (1) (2009) 71–76.

[154] A. Yoto, M. Motoki, S. Murao, H. Yokogoshi, Effects of L-theanine or caffeine intake on changes in blood pressure under physical and psychological stresses, J. Physiol. Anthropol. 31 (2012) 28.

[155] A. Yoto, S. Murao, Y. Nakamura, H. Yokogoshi, Intake of green tea inhibited increase of salivary chromogranin A after mental task stress loads, J. Physiol. Anthropol. 33 (1) (2014) 20.

[156] K. Kimura, M. Ozeki, L.R. Juneja, H. Ohira, L-Theanine reduces psychological and physiological stress responses, Biol. Psychol. 74 (2007) 39–45.

[157] B. Lee, B. Sur, S. Kwon, M. Yeom, I. Shim, H. Lee, D.-H. Hahm, Chronic administration of catechin decreases depression and anxiety-like behaviors in a rat model using chronic corticosterone injections, Biomol. Ther. Seoul. 21 (4) (2013) 313–322.

[158] J. Hintzpeter, C. Stapelfeld, C. Loerz, H.J. Martin, E. Maser, Green tea and one of its constituents, Epigallocatechine-3-gallate, are potent inhibitors of human 11β-hydroxysteroid dehydrogenase type 1, PLoS One 9 (1) (2014) e84468.

[159] S. Brody, R. Preut, K. Schommer, T.H. Schürmeyer, A randomized controlled trial of high dose ascorbic acid for reduction of blood pressure, cortisol, and subjective responses to psychological stress, Psychopharmacol. Berl. 159 (2002) 319–324.

[160] E.M. Peters, R. Anderson, A.J. Theron, Attenuation of increase in circulating cortisol and enhancement of the acute phase protein response in vitamin C-supplemented ultramarathoners, Int. J. Sports Med. 22 (2001) 120–126.

[161] M. Smriga, T. Ando, M. Akutsu, Y. Furukawa, K. Miwa, Y. Morinaga, Oral treatment with L-lysine and L-arginine reduces anxiety and basal cortisol levels in healthy humans, Biomed. Res. 28 (2007) 85–90.

[162] M. Smriga, S. Ghosh, Y. Mouneimne, P.L. Pellett, N.S. Scrimshaw, Lysine fortification reduces anxiety and lessens stress in family members in economically weak communities in Northwest Syria, Proc. Natl. Acad. Sci. U.S.A. 101 (2004) 8285–8288.

[163] T. Kokubo, E. Ikeshima, T. Kirisako, Y. Miura, M. Horiuchi, A. Tsuda, A randomized, double-masked, placebo-controlled crossover trial on the effects of L-ornithine on salivary cortisol and feelings of fatigue of flushers the morning after alcohol consumption, Biopsychosoc. Med. 7 (2013) 1–6.

[164] M. Miyake, T. Kirisako, T. Kokubo, Y. Miura, K. Morishita, H. Okamura, A. Tsuda, Randomised controlled trial of the effects of L-ornithine on stress markers and sleep quality in healthy workers, Nutr. J. 13 (2014) 1–8.

[165] M. Garrido, J. Espino, D. González-Gómez, M. Lozano, C. Barriga, S.D. Paredes, A.B. Rodríguez, The consumption of a Jerte Valley cherry product in humans enhances mood, and increases 5-hydroxyindoleacetic acid but reduces cortisol levels in urine, Exp. Gerontol. 47 (2012) 573–580.

[166] P. Barbadoro, I. Annino, E. Ponzio, R.M. Romanelli, M.M. D'Errico, E. Prospero, A. Minelli, Fish oil supplementation reduces cortisol basal levels and perceived stress: a randomized, placebo-controlled trial in abstinent alcoholics, Mol. Nutr. Food Res. 57 (2013) 1110–1114.

[167] J. Delarue, O. Matzinger, C. Binnert, P. Schneiter, R. Chioléro, L. Tappy, Fish oil prevents the adrenal activation elicited by mental stress in healthy men, Diabetes Metab. 29 (2003) 289–295.

[168] E.E. Noreen, M.J. Sass, M.L. Crowe, V.A. Pabon, J. Brandauer, L.K. Averill, Effects of supplemental fish oil on resting metabolic rate, body composition, and salivary cortisol in healthy adults, J. Int. Soc. Sports Nutr. 7 (2010) 1–7.

[169] J. Hellhammer, D. Vogt, N. Franz, U. Freitas, D. Rutenberg, A soy-based phosphatidylserine/phosphatidic acid complex (PAS) normalizes the stress reactivity of hypothalamus-pituitary-adrenal-axis in chronically stressed male subjects: a randomized, placebo-controlled study, Lipids Health Dis. 13 (2014) 1–11.

[170] J. Hellhammer, E. Fries, C. Buss, V. Engert, A. Tuch, D. Rutenberg, D. Hellhammer, Effects of soy lecithin phosphatidic acid and phosphatidylserine complex (PAS) on the endocrine and psychological responses to mental stress, Stress 7 (2004) 119–126.

[171] M.A. Starks, S.L. Starks, M. Kingsley, M. Purpura, R. Jäger, The effects of phosphatidylserine on endocrine response to moderate intensity exercise, J. Int. Soc. Sports Nutr. 5 (2008) 1–6.

[172] L. Miclo, E. Perrin, A. Driou, V. Papadopoulos, N. Boujrad, R. Vanderesse, J.F. Boudier, D. Desor, G. Linden, J.L. Gaillard, Characterization of α-casozepine, a tryptic peptide from bovine αs1-casein with benzodiazepine-like activity, FASEB J. 15 (2001) 1780–1782.

[173] N. Violle, M. Messaoudi, C. Lefranc-Millot, Ethological comparison of the effects of has bovine alpha s1-casein tryptic hydrolysate and diazepam on the behaviour of rats in two models of anxiety, Pharmacol. Biochem. Behav. 84 (2006) 517–523.

[174] M. Messaoudia, R. Lalonde, H. Schroederd, D. Desor, Anxiolytic-like effects and safety profile of a tryptic hydrolysate from bovine alpha s1-casein in rats, Fundam. Clin. Pharmacol. 23 (2009) 323–330.

[175] M. Messaoudi, C. Lefranc-Millot, D. Desor, B. Demagny, L. Bourdon, Effects of a tryptic hydrolysate from bovine milk αS1-casein on hemodynamic responses in healthy human volunteers facing successive mental and physical stress situations, Eur. J. Nutr. 44 (2005) 128–132.

[176] J.H. Kim, D. Desor, Y.T. Kim, W.J. Yoon, K.S. Kim, J.S. Jun, K.H. Pyun, I. Shim, Efficacy of as1-casein hydrolysate on stress-related symptoms in women, Eur. J. Clin. Nutr. 61 (2007) 536–541.

[177] N. Jaatinen, R. Korpela, T. Poussa, A. Turpeinen, S. Mustonen, J. Merilahti, K. Peuhkuri, Effects of daily intake of yoghurt enriched with bioactive components on chronic stress responses: a double-blinded randomized controlled trial, Int. J. Food Sci. Nutr. 65 (2014) 507–514.

[178] C.R. Markus, B. Olivier, G.E. Panhuysen, J. Van Der Gugten, M.S. Alles, A. Tuiten, H.G. Westenberg, D. Fekkes, H.F. Koppeschaar, E.E. de Haan, The bovine protein alpha-lactalbumin increases the plasma ratio of tryptophan to the other large neutral amino acids, and in vulnerable subjects raises brain serotonin activity, reduces cortisol concentration, and improves mood under stress, Am. J. Clin. Nutr. 71 (2000) 1536–4154.

[179] W.J. Kraemer, G. Solomon-Hill, B.M. Volk, B.R. Kupchak, D.P. Looney, C. Dunn-Lewis, B.A. Comstock, T.K. Szivak, D.R. Hooper, S.D. Flanagan, C.M. Maresh, J.S. Volek, The effects of soy and whey protein supplementation on acute hormonal reponses to resistance exercise in men, J. Am. Coll. Nutr. 32 (2013) 66–74.

[180] Y. Hu, et al., A bioactive compound from *Polygala tenuifolia* regulates efficiency of chronic stress on hypothalamic-pituitary-adrenal axis, Pharmazie 64 (9) (September 2009) 605–608.

[181] Y. Hu, et al., Possible mechanism of the antidepressant effect of 3,6′-disinapoyl sucrose from *Polygala tenuifolia* Willd, J. Pharm. Pharmacol. 63 (6) (June 2011) 869–874.

[182] Y. Hu, et al., Antidepressant-like effects of 3,6′-disinapoyl sucrose on hippocampal neuronal plasticity and neurotrophic signal pathway in chronically mild stressed rats, Neurochem. Int. 56 (3) (February 2010) 461–465.

[183] J. Li, et al., Effect of *Schisandra chinensis* on interleukins, glucose metabolism, and pituitary-adrenal and gonadal axis in rats under strenuous swimming exercise, Chin. J. Integr. Med. 21 (1) (January 2015) 43–48.

[184] A. Panossian, et al., Adaptogens stimulate neuropeptide y and hsp72 expression and release in neuroglia cells, Front. Neurosci. 6 (February 1, 2012) 6.

[185] P.V. Habbu, et al., Adaptogenic and in vitro antioxidant activity of flavanoids and other fractions of Argyreiaspeciosa (Burm.f) Boj. in acute and chronic stress paradigms in rodents, Indian J. Exp. Biol. 48 (1) (January 2010) 53–60.

[186] N.B. Patel, et al., Antistress activity of *Argyreia speciosa* roots in experimental animals, J. Ayurveda Integr. Med. 2 (3) (July 2011) 129–136.

[187] A.K. Thakur, et al., Protective effects of Andrographis paniculata extract and pure andrographolide against chronic stress-triggered pathologies in rats, Cell Mol. Neurobiol. 34 (8) (November 2014) 1111–1121.

[188] S.C. Mandal, et al., Studies on psychopharmacological activity of *Andrographis paniculata* extract, Phytother. Res. 15 (3) (May 2001) 253–256.

[189] S. Naveen, et al., Anti-depressive effect of polyphenols and omega-3 fatty acid from pomegranate peel and flax seed in mice exposed to chronic mild stress, Psychiatry Clin. Neurosci. 67 (7) (November 2013) 501–508.

[190] M. Dushkin, et al., Effects of rhaponticum carthamoides versus glycyrrhiza glabra and punica granatum extracts on metabolic syndrome signs in rats, BMC Complement. Altern. Med. 14 (2014) 33.

[191] K. Feliú-Hemmelmann, et al., *Melissa officinalis* and Passiflora caerulea infusion as physiological stress decreaser, Int. J. Clin. Exp. Med. 6 (6) (2013) 444–451.

[192] D.Y. Yoo, et al., Effects of *Melissa officinalis* L. (lemon balm) extract on neurogenesis associated with serum corticosterone and GABA in the mouse dentate gyrus, Neurochem. Res. 36 (2) (February 2011) 250–257.

[193] S.A. Nirmal, S.C. Pal, S.C. Mandal, Pharmacovigilance of herbal medicines, Pharma Times 46 (2014) 19–21.

Chapter 15

Antiinflammatory Medicinal Plants: A Remedy for Most Disease Conditions?

Sunday O. Otimenyin
Department of Pharmacology, University of Jos, Plateau State, Nigeria

1. INFLAMMATION

Mammalian bodies are constantly being injured internally or externally. Injury to the body often refers to the destruction of body cells. It can be moderate or serious, simple or complicated, noticed or unnoticed. Destruction of body cells may result from toxic chemical interaction, sharp objects, invading organisms, heat, and pressure. The destroyed cells release their contents into the interstitial space. Released contents may be toxic (i.e., destroy another cell), inert, or stimulate the production and/or release of inflammatory mediators. The body may not be able to withstand harsh conditions, such that stress, the presence of toxic agents, and extreme temperatures and pressures (from cancer cells), which may inflict injury on it, for example, sunburn or contents released by adjacent cells. These events are unavoidable, since neither the cells nor the body can live in isolation in higher animals, they must interact with their environment for them to function as part of the body.

The mammalian body has mechanisms for responding to noxious insults and effecting repair/healing of the affected parts. These mechanisms are often referred to as inflammation and healing. Inflammation is the body's response to injury. It involves a well-organized cascade of fluidic and cellular changes. It is recognizable grossly and histologically, and has both beneficial and detrimental effects locally (leads to the destruction of cells resident in the affected areas) and systemically (releases agents that may have an effect on distant cells in the blood stream or/and other parts of the body). It involves simple and at times complicated processes. It can be lifesaving or life threatening (leading to instantaneous death) in some instances. Simple reactions include rashes, followed by immediate healing, while anaphylactic reactions to antigen invasion can be deadly. Inflammation can therefore be

Natural Products and Drug Discovery. https://doi.org/10.1016/B978-0-08-102081-4.00015-0

defined as a physiological process that is intended to neutralize injurious stimulus, limit tissue injury, and initiate healing and repair [1]. It can occur in acute or chronic phases; in the acute phase, a shift or mobilization of transvascular fluid, plasma proteins, and immune cells occurs in the microvasculature of the injured tissue. Vascular injury induces the activation of both platelets and coagulation proteins, which work to maintain homeostasis. We can conclude that inflammation is a potentially harmful process; it is directed to destroy/neutralize the invading agents/organisms, halt their activities, and effect repair. Organisms are living cells; the mechanism put in place to destroy these living cells (invading organisms) via inflammation can also destroy mammalian cells and kill the host. The implication of this is that inflammation must be checked or controlled if the body is to survive. The body also has mechanisms for controlling inflammation and putting a halt to it when it has accomplished its task and healing sets in.

1.1 Agents That Trigger and Sustain Inflammation

Generally, stimuli that cause cellular injury include immunological reactions (hypersensitivity reaction to foreign agents, autoimmune reactions, immune deficiency), nutritional imbalances (protein calorie malnutrition, excessive intake of fats, carbohydrates, and proteins), genetic defects (inborn errors in metabolism, gross malformations), hypoxia (shock, localized areas of inadequate blood supply, hypoxemia), microorganisms (viruses, bacteria, fungi, protozoa), chemical agents (drugs, poisons, food, irritating substances), and physical agents (trauma, thermal agitation of electrical charges, irradiation). When tissue injury occurs, numerous substances are released from the injured tissues, which cause changes to the surroundings of uninjured cells. Some of the tissue products that trigger inflammatory reaction include: (1) plasma-derived mediators of inflammation: Hageman factor, kinins, complement system, and the membrane attack complex; (2) cell-derived mediators of inflammation: arachidonic acid and platelet activating factor, prostanoids, leukotrienes and lipoxins, cytokines, reactive oxygen species (ROS), histamine (increases permeability of epithelial cells, causes contraction of smooth muscle, and constricts the bronchioles), serotonin, and bradykinin; and (3) cells of inflammation: neutrophils, endothelial cells, monocytes/macrophages, mast cells and basophils, eosinophils, and platelets. These substances are the mediators of inflammation, and have been viewed as areas of therapeutic intervention, since blockade of any of these agents will truncate inflammatory processes. Collectively, they are called autacoids. Autacoids are substances released from the cells in response to various stimuli and elicit normal physiological responses locally. An imbalance in the synthesis and release of the autacoids contributes significantly to pathological conditions such as chronic inflammation, allergy, hypersensitivity, cardiovascular diseases, and ischemia/reperfusion. Imbalance may result from nutritional imbalances and genetic defects.

Drugs and herbal products that interfere with these stimuli and/or triggers of inflammation may be beneficial in halting or controlling inflammation. Such agents may be of benefit in the management of most disease conditions, since inflammation has been implicated in the pathogenesis of disease conditions. *Chenopodium ambrosioides* (Amarantaceae) was reported to inhibit the influx of neutrophils and leukocytes to inflamed sites, thereby inhibiting inflammatory processes [2]. It also inhibits bradykinin, nitric oxide (NO) and tumor necrosis factor (TNF)-α levels, and the activities of myeloperoxidase and adenosine-deaminase during inflammation induced by carrageenan.

1.2 Mechanism of Inflammation

It is established that substances released from injured cells promote inflammation. These chemical agents (released from the cells) can also induce inflammation if introduced (injected) into the body. There are specialized cells that store high concentrations of the mediators of inflammation: typical examples are mast cells and white blood cells (not as much as those produced by mast cells). The degranulation of the mast cell results in the release of its contents (potent inflammatory mediators). A host of agents can trigger the degranulation of mast cells, but the most common are allergens. Allergens refer to a wide range of chemical substances that can bind to the receptors on the mast cell surface and sensitize it to a state that triggers its degranulation at subsequent exposure.

The mechanism that underlies inflammatory processes varies; it depends on the insult or the initiator of inflammation. Irrespective of the initiator, mediators are released in the process, it is these mediators that initiate and sustain inflammation. These mediators also make inflammatory processes similar irrespective of the initiator, but the progress and end result may differ depending on the tissue and the available mediators in that tissue. These mediators can be released by any form of injury: excessive pressure (which may result in tissue damage and subsequent microorganism infiltration), microorganism invasion, antigen introduction, burns, radiation, nutritional imbalance, etc. When excessive pressure is applied to the body, especially via sharp objects, cells are ruptured and cell contents are released, for example, the vascular endothelial cell if damaged allows the movement of cells that release mediators of inflammation into the interstitial space. The released agents may trigger inflammatory processes, leading to the movement of cells and chemical agents that sustain inflammation to the site, and "gating" of the exposed site to prevent the invasion of microorganisms and to arrest further loss of body fluids and cells. Similarly, when there is microorganism invasion, macrophages are mobilized to the site of invasion. At the site, macrophages engulf the microorganism and release toxic chemicals that are harmful to the bacterial and host cells. Rupture of adjacent macrophage/adjacent cells then results in the release of chemical mediators of inflammation. Antigen invasion may lead to the sensitization of mast cells, such that subsequent exposure will lead to their degranulation and release of their contents, which are mediators of inflammation.

Inflammation occurs in three phases: silent, vascular, and cellular phases. The presence of chemical mediators of inflammation at the site of provocation leads to the mass movement of cells, and the production and release of other mediators of inflammation to the inflamed site. Inflammation starts unnoticed, "silent phase," with a series of events. In the "silent phase" the body mobilizes resident inflammatory cells present at the site of injury and stimulates them to produce and secrete inflammatory mediators, most of which attract other inflammatory cells (macrophages and mast cells) to the site of injury. The resident cells, mast cells, and macrophages play important roles in alerting the body to tissue injury. This is achieved by their ability to release their contents (inflammatory mediators, such as NO, histamine, kinins, cytokines, leukotrienes, or prostaglandins) into the interstitial space. The presence of these mediators in the interstitial space leads to further degranulation of mast cells, epithelial cells, macrophages, and monocytes and further release of mediators (inflammatory cytokines such as interleukin (IL)-1, −6, −8, TNF-α, and prostanoids). As time progresses, vasomotor mediators and other mediators are released from the resident cells. These mediators attract and mobilize cells to the inflamed site, leading to the second phase of the inflammatory reaction: the vascular phase.

In the vascular phase, the released NO [3], cytokines [4], and prostanoids increase membrane permeability of endothelial cells [5], resulting in leakage of vascular content into the interstitial space. Adjacent cells also release substance P, whose primary effect is to increase the permeability of the vascular membrane (resulting in the leakage of plasma and cells from the blood vessels into the inflamed region), cause vasodilatation, and relay pain signals. The weakened integrity of the vascular membrane facilitates the passage of leukocytes to the damaged tissues, resulting in the third phase of the inflammatory reaction known as the cellular phase, a phase that is characterized by the movement of leukocytes and fluid circulating in the blood into the damaged part of the body. This movement is preceded by the events of rolling, adhesion, and transmigration of leukocytes on the endothelial membrane: a process induced by thrombin [6]. These events lead to degranulation and superoxide production in eosinophils [7]. Neutrophils degranulate and release cytotoxic agents into the environment from their cytoplasmic granules, a process known as respiratory burst, which requires the consumption of glucose and oxygen. Destructive substances released from the neutrophils include highly reactive oxygen and nitrogen species and various proteinases. These substances destroy both pathogens and hosts cells and essentially induce cell damage.

1.3 Healing of Injured Tissue

Inflammation is often follow by the healing process. During acute inflammation, some inflammatory cells produce proinflammatory prostaglandins and leukotrienes, and later lipoxins. Lipoxins block further neutrophil recruitment and favor enhanced infiltration of monocytes into the site of injury. It is these monocytes that are important in wound healing and actually trigger wound healing and halt inflammation.

1.4 Active Antiinflammatory Constituents in Plants

Plants have been known to ameliorate inflammation and promote wound healing (both internal and external wounds) since the existence of humans. These effects have promoted the use of plants and certain animal products for the management of inflammation. Scientific data supporting the use of plants in folkloric medicine have been accumulated over time. Many plants have been documented to have antiinflammatory properties [8,9]. These properties are thought to be mediated by the bioactive constituents in the plant. Documented research has shown that plant bioactive substances inhibit some inflammatory mediators (interleukins, inducible nitric oxide synthase [iNOS], bradykinins, TNF-α, etc.), thereby arresting inflammation and promoting wound healing. These findings have made plants useful in the treatment of most disease conditions, since disease presentations are preceded by inflammation of the relevant tissue or organ.

1.5 Inflammatory Mediator Inhibitors in Plants

Extracts or juice obtained from plants contain many bioactive principles, which are formed in the plants from simple minerals obtained from the soil. Plants use these bioactive constituents for different purposes ranging from enzymes to protective agents or as waste products. One plant may have batteries of bioactive constituents, with each having different bioactivities. Some are toxic, while others are beneficial. Some of the bioactions of plant constituents include the inhibition of bioactive compounds in human/mammalian body. Emphasis here is on mediators of inflammation and these are discussed later.

1.5.1 Interleukins

Interleukins are a group of chemical agents called cytokines. They are produced by leukocytes, hence the name interleukins, a word derived from leukin (from leukocytes) and inter (communication). Interleukins occur in different forms and types, and are differentiated by attaching specific numerals to the interleukin: interleukin 1, interleukin 2, etc. Interleukins are produced by several other cells in the body. They are potent mediators of inflammation, though some may reverse inflammation and promote wound healing. IL-6, IL-10, and IL-19 have been reported to have antiinflammatory activities [10,11]. Some of the interleukins (IL-6) induce the production and secretion of other inflammatory mediators. IL-6 induces acute phase inflammatory response, which may result in fever, secondary amyloidosis, anemia, and elevations in acute-phase proteins, such as C-reactive protein (CRP). It participates in the pathogenesis of several disease conditions, for example, systemic and local rheumatoid arthritis. The ability of IL-6 to induce B-cell differentiation may lead to the formation of rheumatoid factor and other autoantibodies. In joints, IL-6 promotes osteoclast activation and induces the release of matrix metalloproteinases, thus contributing to joint damage [12]. IL-17A plays a significant role in many inflammatory diseases [13] and cancers. IL-17A could promote the invasion of colorectal cancer cells [14]. Inhibition of IL-17 halts inflammation in angiotensin II-induced hypertension, leading to the lowering of blood pressure [15].

A number of plants, *Pseuderanthemum palatiferum* [16], *Scirpus yagara* [17], *Elephantopus scaber* [18], *Emilia sonchifolia* [73], and *Fructus xanthii* [19], inhibit the production and release of interleukins. Fu-ling, a medicinal plant, increases IL-10 production via an increase in the expression of IL-10 mRNA. This then impacts (increases) on IgG and IgA secretion by the spleen and boosts immunity [20]. Herbal melanin, a constituent of some antiinflammatory medicinal plants, modulates cytokine production. Melanin modulates the production of TNF-α, IL-6, and vascular endothelial growth factor mRNA expression in monocytes. *Neuclea sativa* is an example of a plant that contains melanin [21]. *Andrographis paniculata*, a plant with antiinflammatory effects, inhibits the expression of IL-6, a major mediator of inflammatory response [22]. Reduction in the production of IL-1β, IL-6, IL-8, and TNF-α by antiinflammatory constituents of medicinal plants has been reported [23–25]. Their inhibition plays a significant role in the resolution of many disease conditions, and underlies the rationale for using medicinal plants to manage more than one disease condition.

1.5.2 Bradykinins

Bradykinin is a potent mediator of pain and inflammation. It is released during mast cell degranulation. It mediates inflammation by causing vasodilatation and increasing blood microvasculature permeability, thereby promoting the movement of fluids and cells from the blood vessels [26] into the interstitial space. Bradykinin also induces the release of other inflammatory mediators (e.g., histamine) from mast cells. Bradykinin is a nonapeptide generated from kininogen by the action of protease during tissue injury. It is implicated in the etiology of several inflammatory diseases, including arthritis, allergic rhinitis, asthma, anaphylaxis, pancreatitis, and inflammatory bowel diseases [27]. Its actions are mediated by specific membrane receptors and involve a complex signal transducer and also second message mechanisms [27]. Inhibition of bradykinin will truncate inflammatory processes and halt progression of inflammation and further tissue destruction. Many medicinal plants (e.g., *C. ambrosioides*, *Mandevilla velutina*, and *Helminthostachys zeylanica*) inhibit the effects of bradykinin [27–29]. Amentoflavone from *Ginkgo biloba*, cupressuflavone from *Cupressus tarulosa*, vitexin from *Ochrocarpus longifolius* and *Arnebia hispidissima*, jotrophone from *Jatropha elliptica*, and plants rich in vitamin K have been found to be potent antagonists of bradykinin [27]. They have also been found useful in the management of certain disease conditions and promote wound healing.

It has also been documented that some medicinal plants (*Gynura procumbens* and *Echinodorus grandiflorus*) increase bradykinin concentration in vessels [30,31]. This action is often elicited by medicinal plants that cause vasodilatation (via an increase in bradykinin concentration) and reduce blood pressure.

1.5.3 Histamine

Histamine is produced and released from mast cells and basophils (found in connective tissues) during cell degranulation. Histamine increases the permeability of the capillaries to fluid, white blood cells, and certain proteins. This results in the leakage of plasma and inflammatory mediators, and cells into the interstitial space, a process that fuels inflammation. It modulates T-lymphocyte migration and cytotoxicity [32], and induces endothelial hyperpermeability [33]. The pharmacology of histamines and antihistamines has been well studied and its benefits in the management of many disease conditions documented and harnessed. Drugs that block the action of histamine now form major therapies in orthodox medicine. The benefit of such drugs ranges from the management of allergies to the management of cold to its use in other disease conditions. Medicinal plants (e.g., *Peucedanum praeruptorum* and *Peucedanum decursivum*) have been reported to decrease the production of histamine [34]. *Viola yedoensis* inhibits histamine release from mast cells and basophils [25], while *Bixa orellana* suppresses histamine production and release [33]. The majority of plants that produce antihistaminic effects elicit their effects by blocking the histamine receptors (H1). Closely tied to inflammatory response to histamine are the H1 histamine receptors. *Dichrostachys cinerea* [35], *Aegle marmelos* [36], and *Diplosolenodes occidentalis* [37] are reported to antagonize the effect of histamine on H1 receptors. These effects of plants will truncate inflammatory processes fueled by histamine and halt the progression of disease conditions. Some antihistamines have been reported to inhibit the proliferation of hepatocellular carcinoma cells [38].

1.5.4 Mast Cell Stabilizers

Mast cells are found in virtually all tissues in the human body. Their degranulation results in the release of their contents (stored in granules in the cell), which are potent inflammatory mediators. Mast cell degranulation will therefore promote inflammatory processes, cause tissue damage, and result in a number of diseases. Any substance that prevents the degranulation (stabilization) of mast cells will halt inflammatory processes and improve the prognosis of many disease conditions. Mast cell stabilization has been credited to some medicinal plants (e.g., *V. yedoensis*). Such plants inhibit B-hexosaminidase and histamine release from mast cells by stabilizing (preventing degranulation of) mast cells [25], and preventing the release of all the contents of mast cells: histamine, bradykinin, prostaglandins, etc. Many Chinese medicinal plants have been shown to elicit their pharmacological effect by stabilizing mast cells and inhibiting the release of their contents [39].

1.5.5 Inducible/Endothelial Nitric Oxide Synthase and Nitric Oxide

NO regulates a number of biological processes, including inflammation, by binding to a wide range of molecules, such as free radicals, metal centers of enzymes (e.g., guanylyl cyclase and cytochrome C oxidase), tyrosine or cysteine residues of proteins, guanine nucleotides, and polyunsaturated fatty acids [39a]. It is a by-product of the oxidation of L-arginine to citrulline by the action of three isoforms of NO synthase, including iNOS. Inhibition of iNOS modulates inflammatory processes [25]. Inducible forms of iNOS have been implicated in the pathogenesis of liver fibrosis [40]. Nitric oxide (NO) upregulates cyclooxygenase-2 (COX-2) and the synthesis of several other inflammatory cytokines. *Stellera chamaejasme* is an example of medicinal plants that inhibit the release of NO in certain tissues [41]. In the endothelium, NO is a potent endogenous antiatherogenic molecule that suppresses key processes in atherosclerosis. NO is produced through the action of the enzyme NO synthase on the amino acid arginine to produce NO and citrulline. The availability of arginine determines the amount of NO that can be produced. Arginine modulates the development of atherosclerotic cardiovascular disease, improves immune function in healthy and ill patients, stimulates wound healing in healthy and ill patients, and modulates carcinogenesis and tumor growth. Arginine, the source of NO (conversion is catalyzed by iNOS found in the endothelial cell) is found in proteins, which are present in some medicinal plants, implying that plants rich in arginine will increase the turnover of NO in the endothelium. This will manifest as improvement in the prognosis of some disease conditions. Other nutrients, antioxidants (vitamins C and E, lipoic acid, and glutathione) and enzyme cofactors (vitamins B2 and B3, folate, and tetrahydrobiopterin) help to elevate NO levels and play an important role in the management of cardiovascular disease [42].

1.5.6 Tumor Necrosis Factor Alpha

TNF-α is a pleiotropic cytokine and a cell-signaling protein (cytokine) involved in systemic inflammation [43]. TNF is one of the major stress-induced proinflammatory cytokines. It functions in homeostasis and disease pathogenesis. It is one of the cytokines that are involved in acute phase inflammatory reaction. TNF-α induces the production and release of other inflammatory mediators, such as IL-6 [44]. It drives vascular and lymphatics remodeling in sustained airway inflammation [45]. It elicits its effects by binding to its receptors (TNF receptors) on target cells, leading to the modification of cellular response to other stimuli [46]. TNF and its receptors are at the crossroads of inflammation, survival, apoptosis, and necroptosis [47]. TNF has homeostatic functions in addition to its immune and inflammatory roles [48]; it is required for defense against pathogens, proper lymphoid—organ architecture and germinal—center formation, development of granulomas, resolution of inflammation, and induction of tissue repair [43]. Inhibition of TNF will inhibit inflammatory processes and halt disease (rheumatoid arthritis, ankylosing spondylitis, Crohn's disease, psoriatic arthritis, and other inflammatory conditions) progression [43]. Medicinal plants (e.g., *P. palatiferum*, *Corydalis impatiens*, *Scutellaria baicalensis*) have been reported to inhibit the release and production of TNF-α [16,25, 49–51]. Dihydrotanshinone I from *Salvia miltiorrhiza* exhibited anticancer activity via the blockade of TNF-α [52].

1.5.7 Prostaglandins

Prostaglandins are produced from arachidonic acid by the action of cyclooxygenases. Prostaglandin E2 is involved in the mediation of certain inflammatory processes and in the pathogenesis of certain disease conditions (e.g. osteoarthritis). The contractile effects of some prostaglandins (prostaglandin E2) results in premature abortion, cramping pain, and blood vessel occlusion (lack of blood supply to cells leads to tissue necrosis or ischemia). Medicinal plants (e.g., *Persea americana* and *Glycine max*) that inhibit prostaglandin E2 (through the inhibition of COX2) are helpful in resolving some of these disease conditions [41,53]. Inhibition of proinflammatory mediators that increase the turnover of prostaglandin will improve disease conditions fueled by such prostaglandins. *Sophora japonica* and *G. biloba* inhibit bradykinin and lead to the inhibition of bradykinin-enhanced production of prostaglandin F in cultured endothelial cells [54].

1.5.8 Leukotrienes

Leukotrienes are products of arachidonic acid. They are produced by the action of lipoxygenases and are inflammatory mediators. They have been implicated in many pathological conditions. Leukotrienes are produced by immune system cells. They facilitate the production of immune system signaling molecules: interleukins, interferon, and substances involved in anaphylactic reactions. They are found culpable in allergies, bronchospasms, drug disease interactions, and autoimmune diseases. They are implicated in asthma, cancer, kidney disease, obesity, and many other diseases [14,55,56]. Plants that inhibit leukotrienes have been of benefit in the management of inflammatory disease conditions. *Pantago lanceolata* and Oren-gedoku-to (a mixture of plant constituents) [2] are a few of the many plants that inhibit the actions of leukotrienes, and have been used for the management of disease conditions.

1.5.9 Thromboxanes

Thromboxanes, a substance produced by platelets, lead to occlusion of blood vessels by fueling blood clots inside the vascular system. This has been implicated in many cardiovascular conditions, ranging from hearth attack to stroke. Thromboxanes are also implicated in asthma, but its primary role during tissue injury and inflammation is to reduce vascular content (plasma and blood) loss.

The last three products discussed are produced from arachidonic acid simultaneously. Blockade of the production of one will lead to the overproduction of the other (leukotriene concentration increases when the production of prostaglandins is inhibited by nonsteroidal antiinflammatory drugs [NSAIDs]). This may impact negatively on disease conditions such as asthma. This explains why NSAIDs are contraindicated in asthma.

Inflammatory responses mediated by these eicosanoids have beneficial effects, but persistent inflammation has a degrading effect that can lead to debility and death [57].

It is evident that chronic inflammation is involved in the pathogenesis of most disease conditions and in the production of free radicals that may be involve in cell gene mutation. Substances that inhibit any or some of the mediators of inflammation may contribute to the resolution of disease conditions, and may trigger healing of the organs affected. A number of medicinal plants have been demonstrated to have antiinflammatory properties, and this may explain why they are effective in the management of several disease conditions.

1.6 Medicinal Plants That Prevent Cell/Tissue Injury

Damage to cells or tissues is often caused by excess or inadequate chemicals in the body. These important chemicals that are needed for normal function of cells are supplied in food, and when the chemicals are not supplied in food as required, the result is nutritional imbalance.

1.6.1 Nutritional Imbalances

Cells can be injured when the requirement for their normal function or activity is oversupplied, undersupplied, or lacking. Nutritional imbalances result in shortfall in the delivery of amino acids for the synthesis of essential proteins, metals (calcium, potassium, etc.), glucose, etc. Shortfall in the supply of essential nutritional substances causes cellular dysfunction and cell death. On the other hand, excessive supply of these essential nutrients leads to an imbalance and cell injury through the production of excessive lipids in the cells. Excessive fat intake is associated with cancer, cardiovascular diseases, and respiratory and gastrointestinal disorders [58]. Some medicinal plants provide concentrated nutrients that resolve the deficiencies, making them the treatment of choice in the management of such disorders.

1.6.2 Free Radicals

Free radicals are species that have a single unpaired electron in an outer orbit. This makes the free radical extremely reactive and unstable, making it possible for it to react with any available organic or inorganic chemicals, including lipids, proteins, carbohydrates, nucleic acids, etc. They convert the molecules they have reacted with to free radicals (autocatalytic reaction), thereby further propagating their destructive effects on the cells. Free radicals are often produced during enzymic metabolism of exogenous chemicals or drugs, exposure to absorption radiant energy, or endogenous oxidative processes (inflammation). Some medicinal plants contain substances (antioxidants) that neutralize free radicals. They are useful in mopping up the free radicals and halting disease progression.

1.7 Cosmetic Effects of Analgesic and Antiinflammatory Medicinal Plants

Plants with analgesic and/or antiinflammatory effects are often useful in ameliorating the suffering of sick people. Relief of pain or inflammation does not often translate to resolution of the disease condition if the underlying cause of the disease is not addressed. For example, if the cause of inflammation is the invasion of microorganisms, it will be necessary to use antimicrobial agents alongside analgesics or antiinflammatory agents.

Patients are often satisfied with the relief of pain and inflammation, but this does not always translate to total cure. The inability of some medicinal plants to take care of the underlying cause of a disease, but silence the alarm system of the body (pain and inflammation), is referred to as a plant's cosmetic effect. This is dangerous and can lead to permanent disability and sudden death. A typical example is a patient who has malaria (caused by malaria plasmodium parasites). The parasites destroy the red blood cells and cause pain and weakness of the body. Analgesics and antiinflammatory agents are very helpful in curtailing the symptoms (pain, weakness, and inflammation) of the disease. Patients who are on analgesics or antiinflammatory drugs will appear treated, but their red blood cells are being destroyed by the parasites. Failure to manage this will lead to anemia and sudden death. It is therefore imperative to attend to the underlying cause of the disease conditions and to educate traditional medical practitioners on the need to carry out proper diagnosis before the commencement of treatment. Another crucial repercussion of using analgesics/antiinflammatory drugs without proper diagnosis is that the inflammatory aspect of the disease condition may be drastically reduced, while the disease progresses unnoticed and kills the host. Death may not always be the end result as it can result in permanent disability of the affected organ (for example, end-stage kidney disease).

1.8 The Role of Inflammation in Disease Conditions

Inflammation is the major causative factor of different diseases: cardiovascular disease, diabetes, obesity, osteoporosis, rheumatoid arthritis, inflammatory bowel disease, asthma, central nervous system disorders, diabetes, Alzheimer's disease, stroke, infectious diseases, and cancer. Antiinflammatory drugs are often the first step in treatment in many of these diseases. They are often prescribed alongside specific drugs for managing specific conditions, with the aim of reducing pain and controlling the symptoms of these conditions.

In inflammation, many mediators are released, namely, chemokines, prostaglandins, histamines, and vasodilators. The overall effect of these mediators is an increase in vascular permeability. The compromised integrity of the endothelial membrane of the blood vessels leads to local edema, during which there is massive movement of inflammatory cells (phagocytes) and mediators (humoral factors) into the extracellular space. Depending on the part of the body where these mediators are released and the type of mediators released, inflammation may proceed to chronic disease condition, complete or partial healing, or cell death. Prostaglandins, thromboxanes, and leukotrienes act as mediators in dysmenorrhea, colitis, arthritis, heart attack, asthma, and recurrent headaches.

Recent reports showed that in many disease conditions there is upregulation of inflammatory mediators. CRP is upregulated in colon cancer, age-related macular degeneration, gum disease, diabetes, cardiovascular disease, myocardial infarction, coronary artery disease, angina pectoris, and Alzheimer's disease [59,60]. NO is upregulated in sepsis, arthritis, thrombotic thrombocytopenic purpura, and antiphospholipid syndrome. Cytokines are upregulated in oxidative stress. TNF-α is upregulated in cancer. Prostaglandin E2 is upregulated in dysmenorrhea and arthritis. Inflammatory cells have been observed to have a wide gap from control during disease progression, thereby making it almost impossible for the tissues involved to halt the process of disease progression. In normal mammalian body, inflammation does occur frequently, due to stress, tear, and trauma. The body has mechanisms put in place for the regulation of inflammatory processed. These mechanisms are absent or repressed in diseased tissues, creating a gap in the normal control processes.

2. INFLAMMATION IN DISEASE CONDITIONS

Strong evidence of the role of inflammation in disease conditions has been substantiated by scientific reports. Only a few of these disease conditions will be discussed.

2.1 Alzheimer's Disease

In Alzheimer's disease, patients show increased levels of proinflammatory cytokines in the brain. Neuroscientists believe the brain's immune cells rally to attack a form of plaque that signals Alzheimer's disease.

2.2 Asthma

In asthma, activated eosinophils and neutrophils are associated with increased levels of IL-5, IL-8, and other proinflammatory mediators. Allergen exposure sensitizes the IgE on mast cells such that subsequent exposure triggers mast cell degranulation and release of inflammatory mediators (histamine, leukotrienes, and other mediators that perpetuate airway inflammation). Neutrophils are then recruited via IL-8 production by activated macrophages or epithelial cells. The released inflammatory mediators are responsible for the diffuse bronchial inflammation and subsequent airway narrowing seen in asthma. Persistent release of inflammatory cell mediators in bronchi, particularly neutrophils, may explain the epithelial damage, the extensive mucus plugging, and the abnormalities of epithelial and endothelial permeability seen in severe acute asthma [61]. The continual presence of allergens in the airways sustains asthma by mobilizing more mast cells and causing their degranulation. Leukocytes from the blood stream are also mobilized and inflammatory mediators from eosinophils and lymphocytes are released. Further degranulation of mast cells (releasing histamines, leukotrienes, and other mediators that perpetuate airway inflammation), release of cytokines, and the recruitment and activation of eosinophils sustain asthma and cause perpetuation of inflammation in the airway smooth muscles. This process results in narrowing of the airway and reduction in the quantity of oxygen delivered to the lungs. It leaves the patient gasping for air (asthma) and if not quickly resolved can lead to death.

2.3 Cancer

Chronic inflammation often results in the production and release of inflammatory mediators, free radicals, ROS, metabolites (malondialdehyde), and DNA-reactive aldehydes from lipid peroxidation. These agents deregulate cellular homeostasis, induce DNA damage, and activate normal cells to malignancy. Therapeutic intervention aimed at inhibiting the progression of inflammation, reducing angiogenesis, and stimulating cell-mediated immune responses may have a major role in reducing the incidence of common cancer.

IL-17A is an example of inflammatory mediators that play a significant role in cancers, because it promotes cancer invasiveness [62]. Some medicinal plants' constituents inhibit mediators of inflammation that are implicated in cancer. Betulinic acid from *Inonotus obliquus* (Chaga) selectively inhibits some of these mediators to suppress melanoma and other cancers.

2.4 Cardiovascular Disease

Cardiovascular disease begins with an injury or change in the integrity of the endothelial walls of blood vessels, leading to an alteration in the intimal layer. This increases leukocyte, low-density lipoprotein , and platelet adhesion to the endothelium. Endothelial dysfunction is caused by the presence of free radicals (produced during chronic inflammation). Hypertension often induces stress on the vascular walls. Such stress may lead to a tear of the vascular wall and kick-start inflammation, leading to the production and release of batteries of inflammatory mediators. This may result in the formation of atherosclerotic plaque and blockade of coronary vessels, which may result in myocardial ischemia. During the development of atherosclerosis, macrophages interact with vascular endothelial cells, medial smooth muscle cells, and infiltrated inflammatory cells. Unquenched intracellular ROS induce monocytes to differentiate into macrophages. These cells then accumulate cholesterol esters in the cytoplasm, which leads to foam cell (fatty cells that together form the fatty streak) formation. Further recruitment of monocytes and macrophages can occur with the release of cytokines from the endothelium and vascular smooth muscle as part of the inflammatory cycle. As a result, the atheromatous plaque core becomes rich in macrophages as the plaque ages. A vicious cycle of endothelial cell activation ensues that induces the expression of vascular cell adhesion molecule-1 and monocyte chemoattractant protein-1, leading to increased monocyte/macrophage recruitment into the intima. This results in a soft plaque that increases the risk of unstable angina, thrombosis, and acute myocardial infarction [42,51]. Antiinflammatory agents are assumed to mitigate atherosclerosis formation by stalling the release or activity of inflammatory mediators.

2.5 Inflammatory Bowel Disease

Inflammatory bowel disease is a chronic inflammatory disease that results from the invasion of microorganisms. The invasion of microorganisms results in the initiation of inflammatory processes that recruit macrophages to kill the microorganisms. Inflammatory bowel disease results if there is an imbalance of proinflammatory and inhibitory cytokines. The intestine remains in a state of controlled inflammation because of the presence of aggravating substances in the intestine. The imbalance caused by the presence of the microorganism and mediators in the bowel progresses to inflammatory bowel disease [63].

2.6 Rheumatoid Arthritis

Rheumatoid arthritis may set in when there is a slight injury or stress to the muscles (ligament) at the joint, resulting in the production and release of inflammatory mediators into the synovial fluid. Harmful mediators include monocytes and macrophages, and have an important role in joint inflammation and destruction [64]. Cytokines, IL-1 and TNF-α, play important roles in cartilage and bone degradation and upregulate production of other cytokines, including IL-1, granulocyte-macrophage colony-stimulating factor, IL-6, IL-8, and IL-10 [65]. These inflammatory mediators recruit more mast cells, leukocytes, and macrophages into the synovial fluid: a process that leads to the further release of inflammatory mediators and incapacitation of the joint.

2.7 Infection

Infection refers to the presence of foreign living cells (microorganisms) in the organs or parts of the body. Depending on the organs, invasion by these microorganisms destroys the organ cells, and may result in organ failure and loss of function of the said organ. Human tissue reaction to the presence of microorganisms is to recruit macrophages that will destroy the microorganism. The process of recruitment is via the release of inflammatory mediators. Macrophages engulf the microorganisms and kill them with toxic substances produced in the macrophages. The toxic substance is also toxic to the host (human) cell. If this process is not controlled, massive recruitment of macrophages will result in death of the host cell and the loss of function of the organ involved.

2.8 Metabolic Syndrome

Metabolic syndrome describes a wide range of diseases: atherogenic dyslipidemia, insulin resistance, and elevated blood pressure (cardiovascular diseases). Obesity predisposes to these disease conditions and influences their gravity through the actions of proinflammatory mediators and prothrombotic state. Metabolic syndrome is mediated by uncontrolled production and release of biologically active molecules (adipokines and cytokines/chemokines) by fat cells, which results in the infiltration of other cells into fat tissue. Proinflammatory mediators involved in metabolic syndrome include TNF-α, adiponectin, resistin, visfatin, monocyte chemoattractant protein-1, IL-6, and plasminogen activator inhibitor. Herbal products that modulate these mediators are useful in the correction of the syndrome.

3. MEDICINAL PLANTS WITH ANTIINFLAMMATORY PROPERTIES

Medicinal plants are widely used in communities for the management of health-compromised conditions. Many plants with antiinflammatory properties have been cited under relevant sections of this text. Perilla seed (*Perilla frutescens*), flaxseed (*Linum usitatissimum*), and coconut (*Cocos nucifera*) are rich in omega-6 and omega-3. Omega-6 and omega-3 have been shown to inhibit inflammatory mediators [66], thereby preventing the progression of certain disease conditions (cardiovascular disease, cancers, prostate cancer, and colon cancer). Omega-3 oil in fish reduces T-cell stimulation in the intestinal tract resulting in delay of inflammation in the gut.

Curcuma longa (containing curcumin) is a useful antiinflammatory medicinal plant. It inhibits iNOS, COX, and/or lipoxygenase (LOX) enzymes, the production of IL-8, monocyte inflammatory protein-1α, monocyte chemotactic protein-1, and TNF-α. The use of curcumin in different pathological conditions is credited to its antioxidant and antiinflammatory properties.

Greek sage (*Salvia triloba*), oleander (*Nerium oleander*), rosemary, lavender (*Lavandula angustifolia*), apple (*Malus pumila*), and thyme (*Thymus vulgaris*) are rich in ursolic acid and inhibit both COX-2 and 5-LOX. Ursolic acid is a potent antiinflammatory agent found in plants that are used in the management of certain disease conditions.

Alpha-linolenic acid, a substance found in a number of medicinal plants, decreases IL-6 levels in dyslipidemic patients. Arginine, a precursor of NO, modulates the development of atherosclerotic cardiovascular disease, improves immune function in healthy and ill patients, stimulates wound healing in healthy and ill patients, and modulates carcinogenesis and tumor growth. Plants (peanuts, *Arachis hypogaea*, soybeans, *G. max*, walnuts, *Juglans nigra*) rich in arginine increase the synthesis of NO and improve the integrity of vascular endothelial cells. Compromised vascular endothelial cells facilitate the movement of inflammatory mediators into the extravascular space to favor inflammation and the progression of certain diseases and formation of atherosclerotic plaque.

Boswellic acid (found in frankincense, *Boswellia carterii*, and *Boswellia serrata*) inhibits 5-LOX and is used as an antiinflammatory for arthritis, Crohn's disease, and asthma. Boswellic acid has been reported to be effective in the management of cancer [67], psoriasis [68], arthritis [69], and asthma [70]. *S. baicalensis* (an anticancer, antiviral, antibacterial plant) and *Tylophora* spp. (antitumor, antiangiogenic, and antiasthmatic agents) exhibit antiinflammatory properties by suppression of immune modulators and kinase signaling molecules. These plants reduce IL-1β, IL-2, IL-6, IL-12, and TNF-α concentration [71–73].

REFERENCES

[1] P. Tan, F.W. Luscinskas, S. Homer-Vanniasinkam, Cellular and molecular mechanisms of inflammation and thrombosis, Eur. J. Vasc. Endovasc. Surg. 17 (5) (1999) 373–389.

[2] J.K. Triantafillidis, A. Triantafyllidi, C. Vagianos, A. Papalois, Favorable results from the use of herbal and plant products in inflammatory bowel disease: evidence from experimental animal studies, Ann. Gastroenterol. 29 (3) (2016) 268–281.

[3] M.D. Hollenberg, S.J. Compton, International union of pharmacology, XXVIII. Protein-activated receptors, Pharmacol. Rev. 54 (2) (2002) 203–207.

[4] G. Cirino, C. Cicala, M.R. Bucci, L. Sorrentino, J.M. Maraganore, S.R. Stone, Thrombin function as an inflammatory mediator through activation of its receptor, J. Exp. Med. 183 (3) (1996) 821–827.

[5] M. Fresno, M.D. Diaz-Munoz, N. Cuesta, C. Cacheiro-llaguno, M.A. IniGuez, Prostanoid actions in cardiovascular physiolopathology, R. Acad. Nac. Farm. 74 (4) (2008) 1–23.

[6] N. Vergnolle, C.K. Derian, M.R. DAndrea, M. Steinhoff, P. Andrade-Gordon, Characterization of thrombin-induced leukocyte roling and adherence: a potential proinflammatory role for proteinase-activated receptor-4, J. Immunol. 169 (3) (2002) 1467–1473.

[7] S. MiiKe, H. Kita, Human eosinophils are activated by cysteine proteases and release inflammatory mediators, J. Allergy Clin. Immunol. 111 (4) (2003) 704–713.

[8] S.O. Otimenyin, M.O. Uguru, B.L. Atang, Antiinflamatory and analgesic activities of *Ficus thonningii* and *Pseudocedrela kotschyi* extracts, Niger. J. Pharm. Res. 3 (1) (2004) 82–85.

[9] S.Y. Sabo, S.O. Otimenyin, Anti-nociceptive and anti-inflammatory properties of the aqueous leave extract of *Sterculia setigera* Del (Sterculiaceae), IJPI's J. Pharmacol. Toxicol. 1 (5) (2011) 35–41.

[10] K. Gabunia, S. Ellison, S. Kelemen, F. Kako, W.D. Cornwell, T.J. Rogers, P.K. Datta, M. Ouimet, K.J. Moore, M.V. Autieri, IL-19 halts progression of atherosclerotic plaque, polarizes, and increases cholesterol uptake and efflux in macrophages, Am. J. Pathol. 186 (5) (2016) 1361–1374.

[11] Y. Tian, L.J. Sommerville, A. Cuneo, S.E. Kelemen, M.V. Autieri, Expression and suppressive effects of interleukin-19 on vascular smooth muscle cell pathophysiology and development of intimal hyperplasia, Am. J. Pathol. 173 (3) (2008) 901–909.

[12] B.N. Cronstein, Interleukin-6—a key mediator of systemic and local symptoms in rheumatoid arthritis, Bull NYU Hosp. Jt. Dis. 65 (Suppl. 1) (2017) S11–S15.

[13] Q. Tan, H. Yang, E.M. Liu, H. Wang, Establishing a role for Interleukin-17 in atopic dermatitis-related skin inflammation, J. Cutan. Med. Surg. (2017). https://doi.org/10.1177/1203475417697651.

[14] I. Ray, S.K. Mahata, R.K. De, Obesity: an immunometabolic perspective, Front Endocrinol. (Lausanne) 7 (2016) 157.

[15] M.A. Saleh, A.E. Norlander, M.S. Madhur, Inhibition of interleukin 17-a but not Interleukin-17F signaling lowers blood pressure and reduces end-organ inflammation in angiotensin II-induced hypertension, JACC Basic Transl Sci. 1 (7) (2016) 606–616.

[16] P. Sittisart, B. Chitsomboon, N.E. Kaminski, *Pseuderanthemum palatiferum* leaf extract inhibits the proinflammatory cytokines, TNF-α and IL-6 expression in LPS-activated macrophages, Food Chem. Toxicol. 97 (2016) 11–22.

[17] S.H. Dong, J.F. Zhang, Y.M. Tang, J. Li, Y.R. Xiang, Q.L. Liang, Chemical constituents from the tubers of *Scirpus yagara* and their anti-inflammatory activities, J. Asian Nat. Prod. Res. 18 (8) (2016) 791–797.

[18] A.P. Abhimannue, M.C. Mohan, B PK, Inhibition of tumor necrosis factor-α and Interleukin-1β production in lipopolysaccharide-stimulated monocytes by methanolic extract of *Elephantopus scaber* Linn and identification of bioactive components, Appl. Biochem. Biotechnol. 179 (3) (2016) 427–443.

[19] M.Y. Song, E.K. Kim, H.J. Lee, J.W. Park, D.G. Ryu, K.B. Kwon, B.H. Park, *Fructus xanthii* extract protects against cytokine-induced damage in pancreatic beta-cells through suppression of NF-kappa B activation, Int. J. Mol. Med. 23 (4) (2009) 547–553.

[20] C.J. Liou, J. Tseng, A Chinese herbal medicine, fu-ling, regulates interleukin-10 production by murine spleen cells, Am. J. Chin. Med. 30 (4) (2002) 551–560.

[21] A. El-Obeid, S. Al-Harbi, N. Al-Jomah, A. Hassib, Herbal melanin modulates tumor necrosis factor alpha (TNF-α), interleukin 6 (IL-6) and vascular endothelial growth factor (VEGF) production, Phytomedicine 13 (5) (2006) 324–333.

[22] J.Y. Chun, R. Tummala, N. Nadiminty, W. Lou, C. Liu, J. Yang, C.P. Evans, Q. Zhou, A.C. Gao, Andrographolide, an herbal medicine, inhibits Interleukin-6 expression and suppresses prostate cancer cell growth, Genes Cancer 1 (8) (2010) 868–876.

[23] X. Lu, Y. Pu, W. Kong, X. Tang, J. Zhou, H. Gou, X. Song, H. Zhou, N. Gao, J.C. Shen, Antidesmone, a unique tetrahydroquinoline alkaloid, prevents acute lung injury via regulating MAPK and NF-κB activities, Int. Immunopharmacol. 45 (2017) 34–42.

[24] F. Meng, S. Yang, X. Wang, T. Chen, X. Wang, X. Tang, R. Zhang, L. Shen, Reclamation of Chinese herb residues using probiotics and evaluation of their beneficial effect on pathogen infection, J. Infect Public Health S1876-0341 (17) (2016) 30025–30034.

[25] H.R. Zeng, B. Wang, Z. Zhao, Q. Zhang, M.Y. Liang, Y.Q. Yao, K. Bian, W.R. Zhang, Effects of *Viola yedoensis* Makino anti-itching compound on degranulation and cytokine generation in RBL-2H3 mast cells, J. Ethnopharmacol. 189 (2016) 132–138.

[26] A. Siltari, R. Korpela, H. Vapaatalo, Bradykinin -induced vasodilatation: role of age, ACE1-inhibitory peptide, mas- and bradykinin receptors, Peptides 85 (2016) 46–55.

[27] J.B. Calixto, R.A. Yunes, Effect of a crude extract of *Mandevilla velutina* on contractions induced by bradykinin and [des-Arg9]-bradykinin in isolated vessels of the rabbit, Br. J. Pharmacol. 88 (4) (1986) 937–941.

[28] H.L. Hsieh, S.H. Yang, T.H. Lee, J.Y. Fang, C.F. Lin, Evaluation of anti-inflammatory effects of *Helminthostachys zeylanica* extracts via inhibiting bradykinin-induced MMP-9 expression in brain astrocytes, Mol. Neurobiol. 53 (9) (2016) 5995–6005.

[29] L. TrivellatoGrassi, A. Malheiros, C. Meyre-Silva, S. Buss Zda, E.D. Monguilhott, T.S. Fröde, K.A. da Silva, M.M. de Souza, From popular use to pharmacological validation: a study of the anti-inflammatory, anti-nociceptive and healing effects of *Chenopodium ambrosioides* extract, J. Ethnopharmacol. 145 (1) (2013) 127–138.

[30] T.F. Poh, H.K. Ng, S.Z. Hoe, S.K. Lam, *Gynura procumbens* causes vasodilation by inhibiting angiotensin II and enhancing bradykinin actions, J. Cardiovasc. Pharmacol. 61 (5) (2013) 378–384.

[31] T.B. Prando, L.N. Barboza, O. Araújo Vde, F.M. Gasparotto, L.M. de Souza, E.L. Lourenço, A. Gasparotto Junior, Involvement of bradykinin B2 and muscarinic receptors in the prolonged diuretic and antihypertensive properties of *Echinodorus grandiflorus* (Cham. & amp; Schltdl, Micheli. Phytomed. 23 (11) (2016) 1249–1258.

[32] K. Truta-Feles, M. Lagadari, K. Lehmann, L. Berod, S. Cubillos, S. Piehler, Y. Herouy, D. Barz, T. Kamradt, A.A. Maghazachi, J. Norgauer, Histamine modulates γδ-T lymphocyte migration and cytotoxicity, via Gi and Gsprotein-coupled signalling pathways, Br. J. Pharmacol. 161 (6) (2010) 1291–1300.

[33] Y.K. Yong, H.S. Chiong, M.N. Somchit, Z. Ahmad, *Bixa orellana* leaf extract suppresses histamine-induced endothelial hyperpermeability via the PLC-NO-cGMP signaling cascade, BMC Complement Altern. Med. 15 (2015) 356.

[34] A.R. Lee, J.M. Chun, A.Y. Lee, H.S. Kim, G.J. Gu, B.I. Kwon, Reduced allergic lung inflammation by root extracts from two species of Peucedanum through inhibition of Th2 cell activation, J. Ethnopharmacol. 196 (2017) 75–83.

[35] R.R.R. Aworet-Samseny, A. Souza, F. Kpahé, K. Konaté, J.Y. Datté, *Dichrostachys cinerea* (L.) Wight et Arn (Mimosaceae) hydro-alcoholic extract action on the contractility of tracheal smooth muscle isolated from Guinea-pig, BMC Complement Altern. Med. 11 (2011) 23.

[36] A.E. Nugroho, D.D. Agistia, M. Tegar, H. Purnomo, Interaction of active compounds from *Aegle marmelos*CORREA with histamine-1 receptor, Bioinformation 9 (8) (2013) 383–387.

[37] A.S. Jacob, O.R. Simon, D. Wheatle, P. Ruddock, K. McCook, Antihistamine effect of a pure bioactive compound isolated from slug (*Diplosolenodes occidentalis*) material, West Indian Med. J. 63 (5) (2014) 401–407.

[38] Yu-M. Feng, C.-W. Feng, S.-Y. Chen, H.-Y. Hsieh, Yu-H. Chen, C.-D. Hsu, Cyproheptadine, an antihistaminic drug, inhibits proliferation of hepatocellular carcinoma cells by blocking cell cycle progression through the activation of P38 MAP kinase, BMC Cancer 15 (2015) 134.

[39] H.-C. Chang, C.-C. Gong, C.-L. Chan, Oi-T. Mak, A nebulized complex traditional Chinese medicine inhibits Histamine and IL-4 production by ovalbumin in Guinea pigs and can stabilize mast cells in vitro, BMC Complement Altern. Med. 13 (2013) 174;
[39a] Y. Iwakiri, M.Y. Kim, Nitric oxide in liver diseases, Trends Pharmacol. Sci. 36 (8) (August 2015) 524–536.

[40] I. Yasuko, Nitric oxide in liver fibrosis: the role of inducible nitric oxide synthase, Clin. Mol. Hepatol. 21 (4) (2015) 319–325.

[41] M. Kim, H.J. Lee, A. Randy, J.H. Yun, S.R. Oh, C.W. Nho, *Stellera chamaejasme* and its constituents induce cutaneous wound healing and anti-inflammatory activities, Sci. Rep. 7 (2017) 42490.

[42] H. Osiecki, The role of chronic inflammation in cardiovascular disease and its regulation by nutrients, Altern. Med. Rev. 9 (1) (2004) 32–53.

[43] C.T.K. Le, G. Laidlaw, C.A. Morehouse, B. Naiman, P. Brohawn, T. Mustelin, J.R. Connor, D.M. McDonald, Synergistic actions of blocking Angiopoietin-2 and tumor necrosis factor-α in suppressing remodeling of blood vessels and lymphatics in airway inflammation, Am. J. Pathol. 185 (11) (2015) 2949–2968.

[44] H. Matsuno, K. Yudoh, R. Katayama, F. Nakazawa, M. Uzuki, T. Sawai, T. Yonezawa, Y. Saeki, G.S. Panayi, C. Pitzalis, T. Kimura, The role of TNF-alpha in the pathogenesis of inflammation and joint destruction in rheumatoid arthritis (RA): a study using a human RA/SCID mouse chimera, Rheumatology (Oxford) 41 (3) (2002) 329–337.

[45] P. Baluk, L.C. Yao, J. Feng, T. Romano, S.S. Jung, J.L. Schreiter, L. Yan, D.J. Shealy, D.M. McDonald, TNF-alpha drives remodeling of blood vessels and lymphatics in sustained airway inflammation in mice, J. Clin. Invest. 119 (10) (2009) 2954–2964.

[46] C. Sohn, A. Lee, Y. Qiao, K. Loupasakis, L.B. Ivashkiv, G.D. Kalliolias, Prolonged tumor necrosis factor α primes fibroblast-like synoviocytes in a gene-specific manner by altering chromatin, Arthritis Rheumatol. 67 (2015) 86–95.

[47] D. Brenner, H. Blaser, T.W. Mak, Regulation of tumour necrosis factor signalling: live or let die, Nat. Rev. Immunol. 15 (2015) 362–374.

[48] J. Liu, et al., TNF is a potent anti-inflammatory cytokine in autoimmune-mediated demyelination, Nat. Med. 4 (1998) 78–83.

[49] B.O. Henriques, O. Corrêa, E.P.C. Azevedo, R.M. Pádua, V.L.S. de Oliveira, T.H.C. Oliveira, D. Boff, A.C.F. Dias, D.G. de Souza, F.A. Amaral, M.M. Teixeira, R.O. Castilho, F.C. Braga, In vitro TNF-α inhibitory activity of Brazilian plants and anti-inflammatory effect of *Stryphnodendron adstringens* in an acute arthritis model, Evid. Based Complement Alternat Med. (2016) 9872598.

[50] W. Li, H. Huang, Y. Zhang, T. Fan, X. Liu, W. Xing, X. Niu, Anti-inflammatory effect of tetrahydrocoptisine from *Corydalis impatiens* is a function of possible inhibition of TNF-α, IL-6 and NO production in lipopolysaccharide-stimulated peritoneal macrophages through inhibiting NF-κB activation and MAPK pathway, Eur. J. Pharmacol. 715 (1–3) (2013) 62–71.

[51] J. Li, J. Ma, K.S. Wang, C. Mi, Z. Wang, L.X. Piao, G.H. Xu, X. Li, J.J. Lee, X. Jin, Baicalein inhibits TNF-α-induced NF-κB activation and expression of NF-κB-regulated target gene products, Oncol. Rep. 36 (5) (2016) 2771–2776.

[52] F. Wang, J. Ma, K.S. Wang, C. Mi, J.J. Lee, X. Jin, Blockade of TNF-α-induced NF-κB signaling pathway and anti-cancer therapeutic response of dihydrotanshinone I, Int. Immunopharmacol 28 (1) (2015) 764–772.

[53] R. Goudarzi, J.F. Taylor, P.G. Yazdi, B.A. Pedersen, Effect of Arthrocen, an avocado/soy unsaponifiables agent, on inflammatory mediators and gene expression in human chondrocytes, FEBS Open Bio 7 (2) (2017) 187–194.

[54] H.T. Nguyen, H.T. Nguyen, M.Z. Islam, T. Obi, Antagonistic effects of *Gingko biloba* and Sophora japonica on cerebral vasoconstriction in response to histamine, 5-hydroxytryptamine, U46619 and bradykinin, Am. J. Chin. Med. 44 (8) (2016) 1607–1625.

[55] G.M. Nassar, J.D. Morrow, L.J. Roberts 2nd, F.G. Lakkis, K.F. Badr, Induction of 15-lipoxygenase by interleukin-13 in human blood monocytes, J. Biol. Chem. 269 (1994) 27631–27634.

[56] R. Mashima, T. Okuyama, The role of lipoxygenases in pathophysiology; new insights and future perspectives, Redox Biol. 6 (December 2015) 297–310.

[57] S. Dharmananda, Reducing Inflammation With Diet and Supplements: The Story of Eicosanoid Inhibition, Institute for Traditional Medicine, Portland, OR, May 2003.

[58] Y. Zhang, J. Liu, J. Yao, G. Ji, L. Qian, J. Wang, G. Zhang, J. Tian, Y. Nie, Y.E. Zhang, M.S. Gold, Y. Liu, Obesity: pathophysiology and intervention, Nutrients 6 (11) (2014) 5153–5183.

[59] S.S. Bassuk, N. Rifai, P.M. Ridker, High-sensitivity C-reactive protein: clinical importance, Curr. Probl. Cardiol. 29 (8) (2004) 439–493.

[60] A. Weil, The truth about inflammation, Dr. Andrew Weil's Self Healing (2004) 4–5.

[61] I. Tillie-Leblond, P. Gosset, A.B. Tonnel, Inflammatory events in severe acute asthma, Allergy 60 (1) (2005) 23–29.

[62] D. Liu, R. Zhang, J. Wu, Y. Pu, X. Yin, Y. Cheng, J. Wu, C. Feng, Y. Luo, J. Zhang, Interleukin-17A promotes esophageal adenocarcinoma cell invasiveness through ROS-dependent, NF-κB-mediated MMP-2/9 activation, Oncol. Rep. 37 (3) (2017) 1779–1785.

[63] T. Kanai, M. Watanabe, Regulatory T cells and inflammatory bowel diseases, Nihon Rinsho Meneki Gakkai Kaishi. 27 (5) (2004) 302–308.

[64] E.N. Van Roon, T.L. Jansen, L. Mourad, P.M. Houtman, G.A. Bruyn, E.N. Griep, B. Wilffert, H. Tobi, J.R. Brouwers, Leflunomide in active rheumatoid arthritis: a prospective study in daily practice, Br. J. Clin. Pharmacol. 58 (2) (2004) 201–208.

[65] D.E. Woolley, L.C. Tetlow, Mast cell activation and its relation to proinflammatory cytokine production in the rheumatoid lesion, Arthritis Res. 2 (1) (2000) 65–74.

[66] H. Yu, J.F. Qiu, L.J. Ma, Y.J. Hu, P. Li, J.B. Wan, Phytochemical and phytopharmacological review of *Perilla frutescens* L. (Labiatae), a traditional edible-medicinal herb in China, Food Chem. Toxicol. S0278–6915 (16) (2016) 30439–30442.

[67] V.R. Yadav, S. Prasad, B. Sung, J.G. Gelovani, S. Guha, S. Krishnan, B.B. Aggarwal, Boswellic acid inhibits growth and metastasis of human colorectal cancer in orthotopic mouse model by downregulating inflammatory, proliferative, invasive, and angiogenic biomarkers, Int. J. Cancer 130 (9) (2012) 2176–2184.

[68] A. Bader, F. Martini, G.R. Schinella, J.L. Rios, J.M. Prieto, Modulation of Cox-1, 5-, 12- and 15-Lox by popular herbal remedies used in Southern Italy against Psoriasis and other skin diseases, Phytother. Res. 29 (1) (2015) 108–113.

[69] S.H. Venkatesha, B.M. Berman, K.D. Moudgil, Herbal medicinal products target defined biochemical and molecular mediators of inflammatory autoimmune arthritis, Bioorg. Med. Chem. 19 (1) (2010) 21–29.

[70] Z. Liu, X. Liu, L. Sang, H. Liu, Q. Xu, Z. Liu, Boswellic acid attenuates asthma phenotypes by downregulation of GATA3 via pSTAT6 inhibition in a murine model of asthma, Int. J. Clin. Exp. Pathol. 8 (1) (2015) 236–243.

[71] J. Stephen, P.L. Vijayammal, Anti-tumor activity of *Tylophora asthmatica*, Anc. Sci. Life 20 (1–2) (2000) 88–91.

[72] J.B. Calixto, R.A. Yunes, Natural bradykinin antagonist, Mem. Inst. Oswaldo Cruz 86 (2) (1991) 195–202.

[73] C.S. Nworu, P.A. Akah, F.B. Okoye, C.O. Esimone, Inhibition of pro-inflammatory cytokines and inducible nitric oxide by extract of *Emilia sonchifolia* L. aerial parts, Immunopharmacol. Immunotoxicol. 34 (6) (2012) 925–931.

Section III

Herbal Drug Research

Chapter 16

Techniques and Technologies for the Biodiscovery of Novel Small Molecule Drug Lead Compounds From Natural Products

Phurpa Wangchuk, Alex Loukas
Centre for Biodiscovery and Molecular Development of Therapeutics, Australian Institute of Tropical Health and Medicine, James Cook University, QLD, Australia

1. INTRODUCTION

Millions of years of testing by natural selection have made organisms chemists of superhuman skill, champions of defeating most of the kinds of biological problems that undermine human health…

E.O. Wilson

Nature is the best chemist – over millennia of evolution it has synthesized many primary and secondary metabolites. The primary metabolites, including proteins, fats, nucleic acids, and carbohydrates, are the biochemicals that are essential for the growth, development, and reproduction of an organism. The secondary metabolites are often called natural products and include alkaloids, glucosinolates, terpenoids, and phenolics; these natural products are biosynthesized as a result of an organism adapting to its environment either to assist in its survival or as a defensive mechanism against predators [1]. The natural products have unusual structures and possess an array of biological functions and activities. For example, the alkaloids have the ability to block ion channels, inhibit enzymes, block neurotransmission, and cause hallucinations and even death. Terpenoids, phenolics, and glucosinolates provide an organism with a characteristic flavor, aroma, and color – some of which can interfere with digestion, block cell division, provide an organism an awful smell and taste, and even arrest growth and aging of the consumers. These

Natural Products and Drug Discovery. https://doi.org/10.1016/B978-0-08-102081-4.00016-2

characteristics of natural products are being exploited and manipulated by humans for treating various ailments as antiparasiticides, anticancer, and digestive agents. Both traditional and modern medicines use natural products. In traditional medicine (TM), natural products are used as tinctures, teas, herbal supplements, and formulated products. As a result of complex mixtures of chemicals present in the crude extracts used in TM, maintaining standard efficacy, quality, and safety and elucidating the mechanisms of action are often obscured or difficult. These hindrances dictate the isolation of single-component pure compounds from natural products.

Generally, isolation of pure compounds and identification of drug leads from natural products represent the first critical stage of the drug development process, and the successful development of a newly identified drug lead/candidate depends on how each technique/step is handled through research and development processes. The drug lead discovery phase involves intensive exploration to find drug-like small molecules or biological therapeutics with the potential to progress into preclinical development, clinical trials, and ultimately a marketable medicine. However, this stage is a demanding task requiring multidisciplinary approaches, including material identification, collection, extraction, isolation, purification, characterization, structure elucidation, identification, biological activity screening, lead optimization, and preclinical evaluation in animal models. The oldest isolated compound is morphine (an analgesic drug), which was isolated from opium in 1804 [2]. The isolation and identification of antibacterial penicillin in 1929 marked the dawn of the antibiotic era and revolutionized medicine. Later, extensive screening of microbes for other antibiotic compounds and their commercialization potential laid the foundation of the modern pharmaceutical industry. Currently, there are 270,000 isolated pure compounds recorded in the *Dictionary of Natural Products* – an authoritative and comprehensive database of natural products [3]. Many of these compounds have formed the lead candidates or skeletal frameworks of modern drugs. Of 1562 drugs approved between 1981 and 2014, 73% of these drugs were developed from lead candidates sourced from natural products, including botanicals, biologics, vaccines, synthesized natural products, and natural product mimics [4].

Life science manufacturers have developed large numbers of powerful tools and technologies to support all these lead identification processes. In addition, advances in '-omics' fields (genomics, transcriptomics, proteomics, and metabolomics), medicinal and computational chemistry, molecular biology, automation, detection, and bioinformatics have led to new paradigms in high-throughput (HTP) drug discovery. These techniques are powerful tools in chemometric profiling, creation of natural product libraries, biomolecular target development, and early hit characterization from crude extracts, semi-pure mixtures, or single-component natural products. However, none of them in isolation can provide a complete picture of the metabolites in a living sample, and, hence, combinations of two or more of these techniques are highly recommended for characterizing and identifying therapeutically useful compounds from different sources of natural products. While screening for biologically active natural products, in-depth understanding of the habitat and environment, ecological adaptations, morphological and physiological responses, host–pathogen interactions, and chemical properties of natural products would assist researchers in framing strategic extraction, isolation, and biological assay protocols. This chapter discusses the approaches, techniques, and technologies that are currently available for the identification of novel nature-based drug lead compounds and provides a checklist and tips for strategically engaging in the identification of novel drug lead compounds from different biological resources.

2. BIOLOGICAL RESOURCES AND THE SEARCH STRATEGIES FOR NOVEL DRUG LEAD COMPOUNDS

2.1 Biological Resources With Chemotherapeutic Compounds

Synthetic chemical libraries, combinatorial chemicals, and natural product compounds provide abundant resources for target-based screening – often facilitating HTP screening (HTS) platforms. The advantage of natural product compounds over the other two resources is that they provide unmatched chemical structural diversity, which has been selected by nature for specific and meaningful biological interactions. These natural product compounds are obtained mainly from plants, animals, and microorganisms. Among these resources, the plants—especially the medicinal plants—present distinct advantages because of their bulk and easy access. More than 350,000 accepted plant species names, which belong to 642 plant families and 17,020 plant genera, are recorded in "The Plant List" [5]. Only 10%–15% of these terrestrial plants have been studied for their phytochemical properties and biological activities [6]. The main categories of plant-derived drugs that are available today are terpenes (34%), glycosides (32%), alkaloids (16%), and others (18%) [7]. For example, an antimalarial drug, artemisinin (discovered by Youyou Tu, a Chinese scientist, who was awarded half of the 2015 Nobel Prize in in Physiology or Medicine for her discovery), is a sequiterpene drug. Digitoxin is a cardiac glycoside, and an anticancer drug (taxol) is an alkaloid. The alkaloids are the most fascinating group of secondary metabolites, yet only a few have been identified as drug leads and even less have become actual drugs. There are currently 40,000 isolated alkaloids [8].

In the animal kingdom, 953,434 species have been described and cataloged [9], and the lower groups of animals, especially insects, amphibians, reptiles, butterflies, bees, snails, scorpions and spiders, produce interesting novel bioactive compounds. For example, captopril, an antihypertensive agent [10], was derived from a snake venom compound called teprotide. Similarly, epibatidine, isolated from the frog *Epipedobates tricolor* [11], was shown to be a novel nonopioid analgesic agent with 200 times the analgesic potency of morphine. The parasitic helminths' excretory-secretory products, which are composed of proteins, glycans, and small molecules, were recently identified as an entirely new generation of pharmacopoeia of novel anti-inflammatory therapeutics [12], especially effective against inflammatory bowel disease and asthma [13]. Another important resource that is rich in unusual novel bioactive compounds is marine resources. Almost 250,000 species in the oceans have been described and entered into central databases [9]. However, *The Dictionary of Marine Natural Products* recorded only 30,000 known natural products [14], and even fewer have been studied for medicinal applications. Approximately 300 patents on marine natural products were issued between 1969 and 1999 [15], and few compounds have progressed to clinical trials in recent years. One of these compounds is bryostatin-1, which was isolated from a marine bryozoan, *Bugula neritina,* and was found effective against ovarian carcinoma and non-Hodgkin's lymphoma [16]. Marine resources like mollusks, jellyfish, and sea sponges have recently caught the attention of academic researchers and pharmaceutical companies. Fungi have been a frequent source of antibiotics, ever since the discovery of penicillin from the filamentous fungus *Penicillium notatum* by Alexander Fleming in 1928. For the discovery of penicillin and its curative effect in various infectious diseases, the Nobel Prize in Physiology or Medicine was awarded jointly to Fleming, Ernst Boris Chain, and Howard Walter Florey in 1945. Currently, more than 80,000 species of fungi with highly specialized nutritional niches have been recorded. Insect-fungi, often known as entomopathogenic fungi, constitute 700 known fungal species from 100 genera of insects and present niche chemical diversity with potential to create new waves of therapeutics. Microbial resources including virus and bacteria also host colossal chemical diversity. The antimycobacterial drug streptomycin was discovered from *Streptomyces griseus,* and Selman Abraham Waksman was awarded the Nobel Prize in Physiology or Medicine for this discovery. Lately, the antiparasitic drug (ivermectin) was isolated from *Streptomyces avermitilis*, and William C. Campbell and Satoshi Ōmura were conferred the Nobel Prize in Physiology or Medicine in 2015 for this discovery. Other microbial resources, which remain underexplored, include extremophiles such as alkalophiles, halophiles, barophiles, thermophiles, and psychrophiles. As genetic information from microbial sources are currently being amassed, there is likelihood for discovering more novel natural product pharmacophoric skeletons in the future. Overall, natural resources have provided many life-saving de novo drugs despite the path for natural products drug discovery having faced several challenges, including access to biological diversity, collection of sufficient material, extraction of complex fractions, long purification processes via chromatograms that often yield minute amounts, and, finally, testing of each fraction and compound for biological activities. Recent advancements in screening strategies, sampling techniques, spectroscopy, and bioassay-guided fractionation and isolation techniques have revamped the interest of many researchers and pharmaceutical companies.

2.2 Search Strategies for Novel Drug Lead Compounds

In natural product drug discovery, it is critical to know what strategies and approaches are available in the pharmaceutical world and then devise one's own strategic approach by considering the available expertise, equipment, and financial resources. The most common approaches or tactics used for accessing and identifying biologically active natural products are biorational, chemorational, and random approaches [17]. The biorational approach includes ethnomedically directed screening—based on the ethnopharmacological uses of plants and HTS—based on the observation of pest–plant analysis, plant characteristics, and their ecological adaptations and functions. An anti-HIV agent—conocurvone—was isolated from an Australian smokeweed bush plant using an HTS approach. The chemorational approach is based on chemotaxonomical considerations including chemoinformatics of a plant, specific alkaloid surveys, and investigation of specific families that are known to contain similar compounds and showed prior biological activities or had been identified as drug leads or had become actual drugs [18]. It has been found that species in a given genus or related genera yield the same or structurally related compounds. For examples, 164 genera of 47 families produce both isoquinoline [19] and indole alkaloids, and seven different genera of plants in the Solanaceae family contain the same tropane alkaloid—hyoscyamine [20]. Acquiring this information beforehand would facilitate generating good bioactivity hit rates and shorten the screening procedures. The limitation of this approach is that it would minimize the window of screening to what is already known and may not capture the biological activity spectrum of unknown scaffolds. Finally, the random approach involves studying plant extracts and compounds based on personal interest and easy accessibility of the materials and has no prior ethnopharmacological uses or chemotypical rationality whatsoever. This approach, sometimes referred to as "find and grind," relies mostly on random HTS. Having access to collections of huge numbers of extracts and large libraries of pure compounds, which are generated from natural resources or engineered through combinatorial or computational chemistry, are therefore highly desirable in this approach.

Among these three search strategies, the ethno-directed biorational approach has proved to be the shortest and the most effective search strategy for discovering drugs from nature. This is mainly because their long history of clinical uses enhances the hit rate of a new drug lead candidate. For example, the National Cancer Institute of the National Institutes of Health reported that medicinal plant extracts gave greater bioactivity hit rates than did randomly studied plant extracts [21]. A recent study involving Bhutanese medicinal plants also supports this viewpoint of bioactivity hit rates and scientific findings. From nine Bhutanese medicinal plant species investigated for their phytochemical and biological activities between 2002 and 2015, 181 phytochemicals including five new phytochemicals were identified by using gas chromatography–mass spectrometry (GC-MS) and nuclear magnetic resonance (NMR) spectroscopy. Of 55 compounds that were isolated from these nine plants, 32 compounds were tested for their biological activities and 15 of them exhibited promising antimicrobial, antimalarial, anti-inflammatory, and/or anthelmintic properties. While five of these compounds were identified as new antimalarial drug lead compounds against multidrug-resistant strains of *Plasmodium falciparum* [22–26], one compound showed strong acetylcholinesterase activity with a minimum inhibitory concentration of 0.0015 nmol, which is twofold better than galanthamine (0.003 nmol), a drug currently used for treating Alzheimer's disease [27]. Another four compounds showed strong dual anthelmintic activities against major human helminth parasites, the blood flukes and whipworms [28,29]. 14-*O*-Acetylneoline, which was isolated from the Bhutanese *Aconitum laciniatum*, was identified as a novel anti-inflammatory drug lead compound [30]. There exists vast scientific literature on the phytochemicals and biological activities of medicinal plants with new drug lead compounds being identified and published in many natural product journals on a monthly basis. Such published information helps or inspires researchers and the pharmaceutical companies to further develop these as drug lead candidates into more amenable and palatable drugs. An analysis by Fabricant and Farnsworth [31] on the origin of the drugs developed between 1981 and 2001 showed that 80% of 122 plant-derived drugs were discovered as a result of chemical studies directed at isolating the biologically active substances from the plants used in TM—the most popular being artemisinin, morphine, quinine, and ephedrine. Between 2000 and 2005, five medicinal plant-based drugs were introduced in the United States market, and another seven plant-derived compounds are currently in clinical trials around the world [32,33]. However, in using this ethno-directed search strategy, it is crucial to have intimate understanding of the disease concepts of the culture whose pharmacopeia is under examination. It is also important to consider the convention on biodiversity, intellectual rights– and benefit-sharing mechanisms, memoranda of understanding, and collection permits for accessing the biological materials.

3. LOGICAL FRAMEWORK APPROACHES FOR THE BIODISCOVERY OF SMALL MOLECULE DRUG LEAD COMPOUNDS

In general, identifying novel small molecule drug lead compounds involves using carefully considered search strategies. Using the right approach is crucial for isolating compounds with desirable bioactivities. Devising one's own strategy and logical framework approaches could accelerate the process in

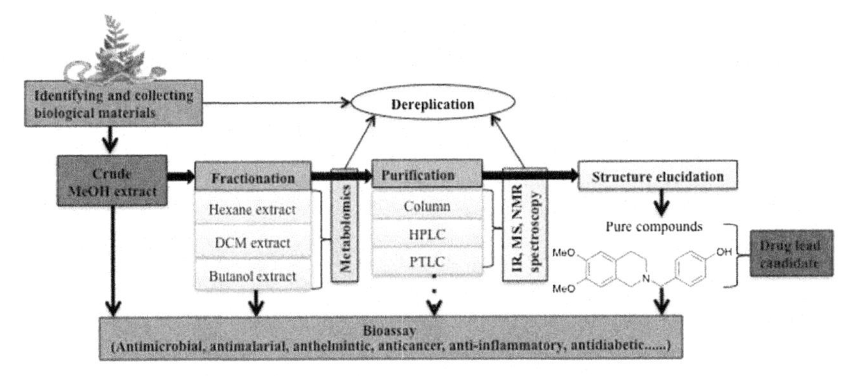

FIGURE 16.1 General schematic logical framework approaches for identifying novel drug lead compounds from medicinal plants.

which bioactive compounds and new drug lead candidates are discovered. Usually, the logical framework approaches include selecting right natural product resources, sample preparation, extraction methods, biological activity screening of crude extracts, fractionation and bioassays, metabolomics studies, purification, structure elucidation, biological activity screening of pure compounds, and finally, identifying the novel drug compounds based on their activity index (Fig. 16.1). Dereplication can be performed at three tiers: (1) while selecting a candidate for the biodiscovery project, (2) metabolomics studies, and (3) spectroscopic investigation. Each of these important stages is described in details in the ensuing sections.

3.1 Selecting Biological Materials: Their Identification and Collection Processes

The most important thing to consider before undertaking such a process is to conduct SWOT (strength, weakness, opportunities, and threat) analysis for the project samples and then devise appropriate selection criteria. Situating the selecting criteria and critical questions beforehand would avoid many undesirable issues in the later stages of biological activity screening processes. Answering these three basic research questions is critical: (1) What natural resources or materials are relevant from a disease perspective? (2) Have the samples or materials been previously identified and studied against the target diseases (dereplication)? (3) Are the materials easily accessible and collectable in sufficient quantities for the biodiscovery project? If the biodiscovery project samples involve medicinal plants used by indigenous communities privately or in scholarly TM, surveying the plants, botanically identifying them, and correctly translating their ethnopharmacological uses (including mode of use) are critical steps in the ethno-directed biorational approach to natural product screening. This information helps researchers to retrieve accurate scientific

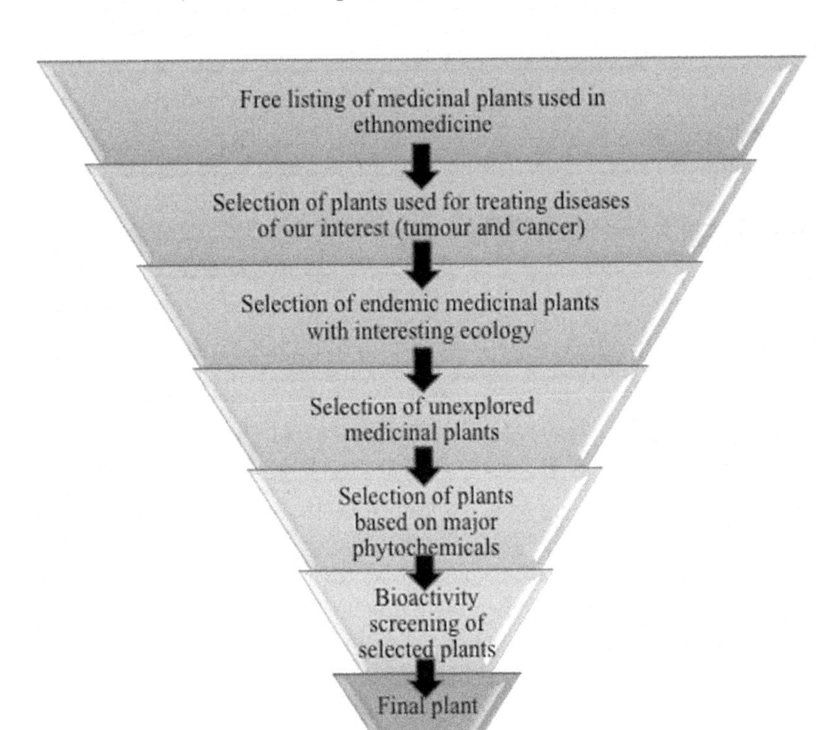

FIGURE 16.2 Schematic selection of medicinal plants for phytochemical and pharmacological studies.

literature from databases to formulate an informed literature review on their biodiscovery project and guides them to select appropriate bioassay targets as well as to devise proper modes of compound administration for in vivo studies. For example, in an ethno-directed biorational approach, the selection criteria as depicted in Fig. 16.2 can be followed.

While every step is important for successfully isolating bioactive compounds from medicinal plants, the botanical identification and collection of plant materials play a critical role in downstream processes leading to phytochemical and bioactivity analyses. When it involves medicinal plants, it is necessary to be mindful of the collection season and how they are collected and dried. Different scholarly traditional medicinal systems have very specific collection seasons, times, and drying methods for every medicinal plant, and these are important factors in determining the quality and efficacy of medicines. For example, the quality of a medicinal plant used in the scholarly Bhutanese TM largely depends on strict adherence to the seven ethnoquality parameters: right identification, right natural habitat, right collection (including plant parts, season and time), right preprocessing and detoxification, right drying methods, appropriate storage conditions, and the spiritual empowerment [34]. Following these seven traditional guidelines of medicinal plant collection would potentially render better bioactivity hit rates than not adhering to the ancient practices. This claim and the associated practices have a sound scientific underpinning since the secondary metabolites produced by a plant very much depend on the available nutrients sustained by the habitat or the environment in which it grows. Time, season, and temperature also affect the quality and the quantity of bioactive metabolites produced by medicinal plants.

3.1.1 Screening for Major Classes of Biochemicals From Crude Extracts

Plants and animals contain complex matrices of metabolites. It is critical to understand their nature of existence and how best to extract the biological molecules. Organic solvents with different polarity indices are used for extraction of biological samples, aiming to efficiently extract most metabolites in their natural forms. The choice of solvents and conditions depends on the spectrum of small molecules desired, and the norm is that most drug-like molecules of intermediate polarity can be extracted using a 1:1 v/v mixture of chloroform/methanol. When screening for biological activities, heat-assisted or microwave-assisted extraction techniques—even though they could increase the yield of extracts—should be avoided unless the desired compounds are known to be heat stable. While plant materials are usually dried after collection from their natural habitat, animal materials are generally freeze-dried or snap-frozen to stop deterioration of metabolites. Plants contain different classes of small molecules, often called phytochemicals, which includes alkaloids, flavonoids, tannins, terpenoids, and saponins. It is often found that plants containing diverse classes of phytochemicals are of superior biological activities [35,36]. It may be also deduced that the higher the diversity and quantity of chemical classes that a plant contains, the stronger and broader are the spectrum of biological activities that plant may exhibit. Thus, screening the biological samples from the biodiscovery project for their major classes of biochemicals provides a snapshot of possible biological activities present in the samples under investigation. The presence or absence of different classes of phytochemicals in the crude extracts of plants can be detected by adding appropriate chemicals or reagents to the concentrated methanol extracts of a plant (Fig. 16.3).

The changes in appearance of crude extracts foretell the presence or absence of different classes of phytochemicals [37]. Standards (commercially available) can be used for confirming their physiochemical tests. The physiochemical test can detect specific biochemical entities present in biological samples by visualizing spots or bands on thin layer chromatography (TLC) plates loaded with the crude extracts or samples [38]. This method involves

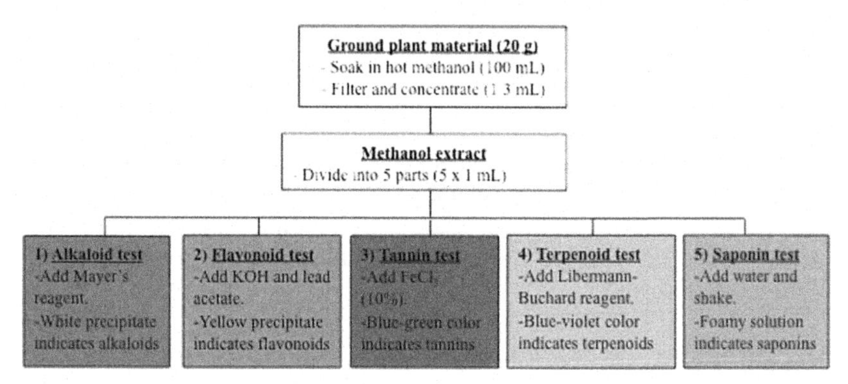

FIGURE 16.3 Test methods for screening plants for major classes of phytochemicals.

developing a TLC plate for the crude extracts under appropriate mobile phases constituted by organic solvents. The developed TLC plates are dried in a fume hood, observed under an ultraviolet (UV) lamp, and then moderately sprayed with an appropriate reagent to obtain endpoint reactions, which are indicated by color changes in the TLC spots. Care must be taken not to oversoak/overspray the TLC plates or overheat it, which may result in loss or decomposition of the components. These experiments require appropriate safety precautions that are listed in the material safety data sheets of each reagent. Iodine vapor is a nonspecific universal reagent that can be used for detecting many organic compounds present in samples. There are specific spray reagents for each class of chemicals. We have given representative examples of biochemical-specific visualization reagents in Table 16.1. Slowly heating the TLC plates with a heat gun or heating in an oven helps in achieving prominent TLC spot color. Recently, automated high-performance TLC (HPTLC)—an enhanced form of TLC—has been developed to achieve many analytical parameters including the precise sample application, efficient separation, reproducible chromatogram, accurate quantitative measurements, and generation of digital image. Recent advances in metabolomics studies have made it even easier to identify biochemicals from crude biological mixtures, which enables separation scientists to perform targeted molecular ion-based natural product isolation.

3.2 Metabolomics Studies of Crude Extracts: A Recent Development

Metabolomics is a recent advance in the postgenomics revolution that studies small molecules within a mass range of 50–1500 Da. It combines analytical chemistry, biochemistry, and bioinformatics and enables the detection and identification of metabolites from biological samples in large numbers. The metabolites represent the molecular phenotype, which directly reflects the

TABLE 16.1 Representative TLC Visualization Reagents for Selected Biochemicals

Classes of Biochemical	Spray Reagent to Detect Biochemicals in a TLC Plate	Visualization
Amino acids, amines, and amino sugars	Ninhydrin (0.2 g) in ethanol (100 mL). Heat it at 110°C until spots appear	Reddish spots
Aldehydes and ketones	2,2′-Diphenylpicryhydrazyl (15 mg) in chloroform (25 mL)	Yellow spots on purple background
Carbohydrates and sugars	3% *p*-Anisidine hydrocholride in *n*-butanol. Heat at 100°C for 2–10 min	Green/yellow/brown/red spots
Glycosides and glycolipids	10% Diphenylamine in ethanol (10 mL) and add HCl (100 mL) and glacial acetic acid (80 mL). Heat 30–40 min at 110°C	Blue spots
Halogenated compounds	To silver nitrate solution (0.1 g in 1 mL water), add 2-phenoxyethanol (10 mL), acetone (200 mL), and 1 drop of hydrogen peroxide (30%) and irradiate with unfiltered UV light for 15 min	Dark spots formed
Flavonoids	1% Ethanolic solution of aluminum chloride and observe under UV lamp (360 nm)	Yellow fluorescence
Iodinated compounds	50% Acetic acid and irradiate TLC with unfiltered UV light. Dry TLC plate at 100°C prior to spraying	Weakly violet-brown spots
Lipids	0.1% Gentian violet in methanol and place in a bromine vapor tank	Blue spots on yellow background
Organic acids	Bromocresol green (0.1 g) in ethanol (500 mL) with 1 mol/L NaOH (5 mL)	Yellow spots on blue background
Sterols, steroids, and bile acids	85% Phosphoric acid with water (1:1, v/v) and heat for 10–15 min at 120°C. Observe under visible and UV light	Produce colourful spots

underlying biochemical activity and state of cells or tissues. There exists a wide range of highly sophisticated and hyphenated metabolomics tools; three main instruments that are widespread in use today are GC-MS, NMR, and liquid chromatography–mass spectrometry (LC-MS) [39–41]. These instruments provide more robust identification for both targeted and untargeted metabolomics analyses of the samples where patterns can be matched with standards or previously known spectra. While targeted metabolomics studies are hypothesis-driven experiments and are characterized by predefined sets of metabolites with high levels of precision and accuracy, untargeted metabolomics studies are characterized by the simultaneous measurement of a large number of metabolites from each sample with complex data that often require bioinformatics tools for data analyses and reliable interpretation [42]. General processes in metabolomics studies include collecting samples and preparing at least three to six biological replicate samples, data acquisition, spectral processing, analyzing the data using bioinformatics software, and identifying the metabolites (Fig. 16.4). The methods for biological sample preparation vary depending on the types of instruments to be used for data acquisition, types of biological materials, and the research questions that need to be answered. Usually, universal solvents (water) or other organic solvents (methanol or

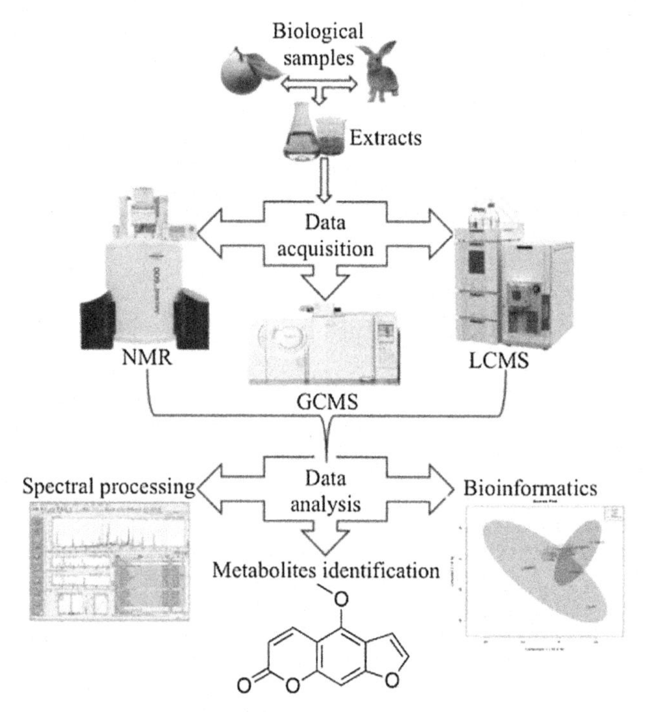

FIGURE 16.4 Metabolomics workflow.

ethanol or chloroform) are used for extracting whole ranges of metabolites present in both plant and animal samples. A review by Mushtaq et al. [43] describes comprehensively the preextraction techniques, the choice of extraction solvents and their main features, and the description of the most used extraction techniques for a variety of biological materials. Their review of more than 200 studies showed that sample collection, homogenization, grinding, and storage could affect the yield and reproducibility of results.

The data acquisition phase of metabolomics studies involves using analytical instruments including NMR, LC-MS, and GC-MS. Although individual instruments are often used for metabolomics, cross-platform techniques that engage all three types of instruments (GC-MS, LC-MS, and NMR) are highly recommended to comprehensively detect and identify large numbers of metabolites with reproducible data, and this is more appropriate for analyzing crude extracts that are very complex. NMR is based on the energy absorption and reemission of the atom nuclei when subjected to an external magnetic field. Owing to the natural abundance in the biological samples, hydrogen is the most commonly targeted nucleus in ^{1}H-NMR spectroscopy [42]. Each molecule of a metabolite generates a unique one-dimensional NMR spectral pattern referenced by the chemical shift expressed in parts per million (ppm). One-dimensional NMR is the most commonly used method in HTP studies, but two-dimensional NMR, including correlation spectrometry (COSY), total correlation spectroscopy (TOCSY), and nuclear Overhauser effect spectroscopy (NOESY), are also used when the compounds are difficult to identify using only one-dimensional NMR spectra. LC-MS and GC-MS instruments use MS approaches that first ionize compounds, separate the ions, and elute them as different peak signals and patterns in the form of a mass-to-charge ratio (*m/z*) with reference to retention time and relative intensity of the measured compounds. Both of these instruments use CG separation techniques based on the interaction of different metabolites in the sample with the adsorbent materials in the chromatographic column. There are varieties of GC columns that can separate different metabolites based on their polarity indices. A review by Parasuraman et al. [44] describes different types of LC-MS instrumentations, including different types of ionization sources, mass analyzers, and detectors. Comparatively, GC-MS has higher sensitivity and resolution and can identify hundreds of compounds. The disadvantage of a GC-MS metabolomics study is that the instrument can detect only the volatile components. The nonvolatile components, especially the highly polar metabolites, have to be derivatized before injection into the GC-MS instrument; as a result, the samples are prone to artifact development.

The data analyses consist of raw spectral preprocessing, peak alignment, normalization, statistical analyses, and compound identification. A review by Alonso et al. [42] gives insightful and detailed information on untargeted metabolomics studies including a list of spectral databases for metabolomics studies and different bioinformatics tools/software required for spectral processing, statistical data analyses, and metabolite identification. The spectral databases including NMR spectra, EIMS, and MS/MS are essential for metabolite identification. NMR-based metabolite identification is carried out either by matching the reference peak positions in ppm against the list of detected peak positions [45] or, more recently, by using the valid cluster concept approach [46]. The GC-MS–based metabolite identification uses ionization pattern–matching techniques, in which each MS peak that gives ionization pattern is matched with the reference mass spectral library/database of the National Institute of Standards and Technology (NIST) [47]. This NIST database contains the mass spectra for 242,466 chemical compounds. Many of these bioinformatics software programs, including MetaboAnalyst 3.0, a web-based analytical pipeline, support the calculation of compound concentration in samples and data processing with statistical analysis including multivariate principal component analysis, partial least squares–discriminant analysis (PLS-DA), and orthogonal partial least squares (orthogonal PLS-DA). This statistical tool allows researchers to discriminately identify the group of active compounds from inactive compounds. In addition, MetaboAnalyst 3.0 has 1600 metabolic pathways that could facilitate metabolic and gene–metabolite pathway analyses, and the interpretation of their chemobiological functions.

Despite the wide range of metabolomics applications, there are only a few reports that have combined natural product chemistry with metabolomics approaches to identify novel bioactive extracts and compounds, and even these studies have focused mostly on plants. However, it is increasingly being applied to drug discovery, disease diagnosis, precision medicine, and biomarker identification [48]. Metabolomics studies provide a snapshot of the constituents in the extract, which will facilitate rational decisions on choosing the best method of fractionation or whether to further pursue the isolation. Combining these metabolomics data with corresponding biological activity can potentially furnish new drug lead compounds. For example, NMR-based metabolomics studies of a Mexican plant, *Galphimia glauca*, demonstrated that galphimine was responsible for the strong sedative and anxiolytic activities exhibited by its two crude extracts [49]. Metabolomics studies have the ability to identify only the known component signals of a mixture, and the unknown signals (which can constitute 60%–80% of all detected signals), often require detailed isolation and structure elucidation techniques to identify them.

3.3 Techniques for Separation, Isolation, and Structure Elucidation of Natural Products

3.3.1 Purification Tools and Techniques

Crude extracts of biological origin are complex mixtures of various types of biochemicals, so their mechanisms of action are difficult to determine. Isolated bioactive molecules or natural product compounds are, therefore, the ideal targets for drug discovery. Once the crude plant extract has been confirmed as a hit in a biological assay and metabolomics data have shown the presence of compound peaks, the bioactive compounds in the extract must be isolated and identified. The classic approach is that the bioactive crude extract is first separated into several fractions that are tested for biological activities. As described earlier, one alternative is to conduct bioassay-guided separation techniques in which each fraction is screened for its biological properties. The advantage of this route is that the desired fractions can be identified and the likelihood of isolating pure compounds with bioactivities is high. However, a single separation step is rarely sufficient to obtain purified bioactive compounds, and often repeated isolation of fractions and repeated testing is required to locate the bioactive individual compounds, making this approach expensive and laborious. Another alternative, which is less expensive, is to separate all the pure compounds first and then test them for their biological activities. The disadvantage of this route is the potential to miss or lose the bioactive agents during the separation process. Often, the unknown components and the bioactive compounds are present in low concentrations and are prone to loss or degradation in the process of separation. For these reasons, selection of appropriate tools is essential for obtaining the target compounds at high yield.

During the past decades, many isolation and purification techniques, such as membrane filtration, HPLC, countercurrent CG (CCC), supercritical fluid CG (SFC), column CG (CC), and preparative TLC (PTLC) have been introduced. However, it must be stressed that none of these techniques provide in itself a comprehensive solution to all separation problems, and the best approach is usually a combination of different techniques. For example, a crude plant extract is first fractionated using organic solvents and then separated using flash CC before isolating the compounds by using HPLC or PTLC. While many polar and nonpolar organic solvents exist, the most commonly used or essential solvents for the fractionation and isolation of pure compounds are water, methanol, butanol, acetonitrile, dichloromethane, chloroform, hexane, and petroleum ether. Combining these solvents in proportionate ratios can form good fractionating solvents or mobile phases for CG systems. A solvent miscibility table with polarity indices [50] can facilitate tailored fractionating solvent phases and CG mobile phases.

Before mass separation of crude extracts, it is common practice to develop a good TLC profile or analytical HPLC profile, which would facilitate the choice of separation techniques. For visualizing separated compound spots on the developed TLC plate, a UV lamp (254 and 354 V) and Dragendorff's reagent and many other spray reagents described previously can be used. Dragendorff's reagent can be used specifically for detecting alkaloids, and it can be prepared using the methods described by Munier and Mache-Boeuf [51]. When this reagent is sprayed on TLC plates, alkaloids adopt an orange-red coloration. HPLC comes with the attached UV/visible (UV/VIS) or diode array detector (DAD) or photo diode array detector (PDA). HPLC PDA offers autoprogrammed multiple spectral displays and is suitable for isolating different compounds present in crude mixtures. Purification process is the most tedious stage of natural product isolation. It requires repeated cycles of CC, PTLC, or HPLC until one TLC spot or one HPLC spot with a single mass peak (*m/z*) is obtained for the isolated compound. Achieving single TLC spots or HPLC peaks is often construed as pure compounds and thus can be deceiving for the untrained chemist. However, this is not necessarily correct, especially when the compounds of similar or closer rates of flow (Rf) or retention time (Rt) represented by TLC spots or HPLC peaks, respectively, are present in what appears to be pure one spot/peak compounds. This is often observed with isomeric compounds. The mass spectrum is the best approach to detect such anomalies. Changing the solvent system and repeated development of PTLC by using the same plate and mobile phases also helps to determine the purity of compounds and in isolating hard-bound isomeric compounds. Some compounds can be easily purified using a liquid–liquid crystallizing technique wherein one or two solvent systems (most commonly used) with different volatility indices can be used to grow the crystal from the mixture. This single crystal can be used for determining the structure of a compound. Depending on the nature of the target compounds, the crude bioactive fraction is subjected to repeated large-scale purification rituals using CC or preparative HPLC. For example, fractionation and isolation of pure alkaloids can be achieved using the acid-base fractionation technique followed by CC or HPLC purification processes as outlined in Fig. 16.5. In order to prevent decomposition or the

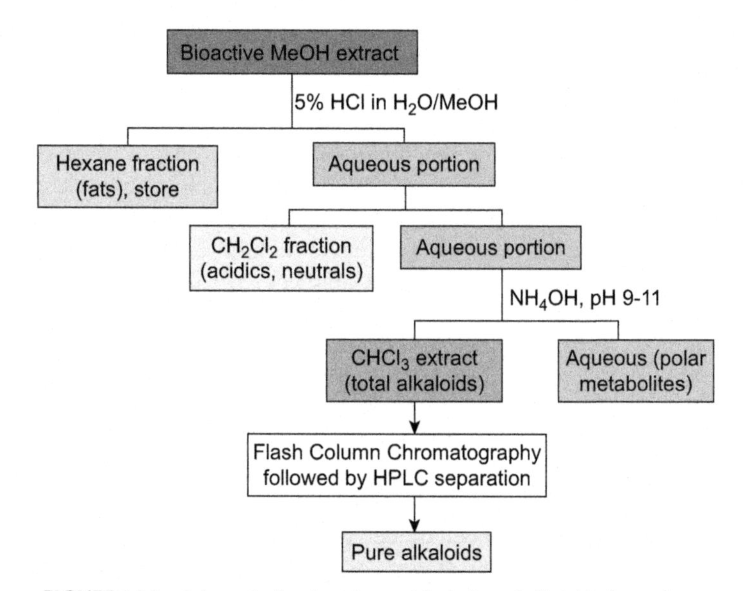

FIGURE 16.5 Schematic fractionation and isolation of alkaloids from plants.

formation of artifacts during separation processes, the alkaloids or compounds must be recovered from the silica gel, columns, and the solvents as soon as possible. Depending on the nature of the compound, aqueous ammonia solution or trifluoroacetic acid (TFA) is often used for enhancing separation of compounds.

There are four main types of HPLC: normal-phase, reverse-phase, ion-exchange, and size-exclusion CG. Different HPLC columns come in different sizes and lengths embedded with various column chemistries, particle sizes and porosity. Both Agilent and Shimadzu HPLC instruments have their own standard protocols and column specifications. It is important to discuss column requirements with the suppliers as they can provide custom-made preparation and development methods. A review by Tistaert et al. [52] and a book by Hostettmann et al. [53] provide excellent overviews of the applications of HPLC separation techniques in natural product isolation.

3.3.2 Tools and Techniques for the Identification and Structure Elucidation of Compounds

Once pure compounds are obtained, the next crucial step is to identify them using MS, infrared spectroscopy (IR), one-dimensional NMR, and X-ray crystallography. There are only limited techniques to identify the isolated compounds. Identification of a compound serves as a dereplication process and can prevent unnecessary and complex structure elucidation experiments. In the absence of GC-MS or NMR instruments, either TLC co-spotting or HPLC coelution with reference standards can be used. A book by Mandal et al. [54] provides an overview of the identification strategies of phytocompounds both with reference standards and without reference standards and describes the stages in structure elucidation of unknown compounds. The disadvantages of identifying compounds using reference standards are that it can only identify known compounds, it can be expensive while identifying large numbers of isolated compounds, and it is impossible to obtain all the required reference standards for the compounds, which may be isolated from the biological samples. A GC-MS library matching technique is the simplest and quickest method for identifying large numbers of known compounds. When the isolated compound is injected into the GC-MS system, which is supported with the NIST EI MS compounds library, an ionized spectrum (*m/z*) with a specific ion pattern is generated. This ion pattern is then matched to hits in the NIST library, and the compound can be identified. A review by Kind and Fiehn [55] highlights the advances in structure elucidation of small molecules using MS. The compounds identified using MS can be confirmed by comparing the physiochemical properties, IR and one-dimensional NMR (^{1}H and ^{13}C-NMR) of the isolated compound with the reported literature. The physiochemical properties include molecular weight, molecular formula, melting point, and optical rotation (for chiral compounds). If the references of the reported

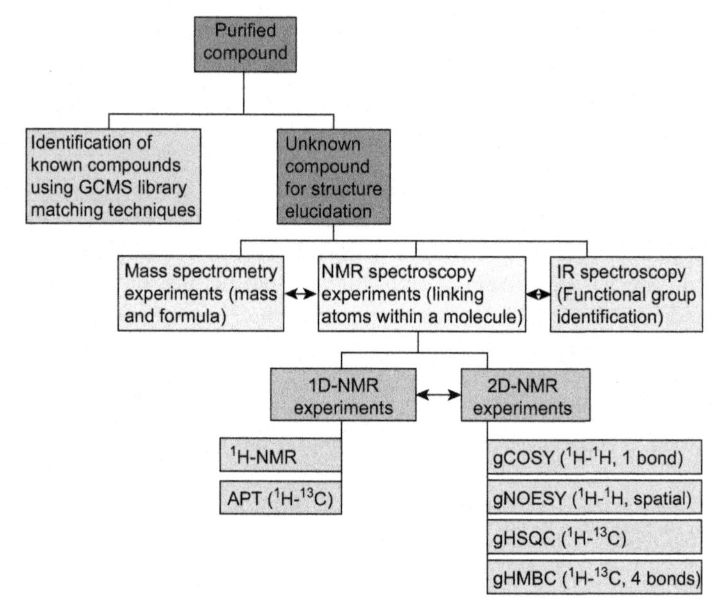

FIGURE 16.6 Identification and structure elucidation of a compound using mass spectroscopy, infrared spectroscopy (IR), and nuclear magnetic resonance (NMR).

compounds predate the 1970s, it is possible that the literature lacks complete data on NMR. This sometime misleads researchers into thinking that the compounds are new and novel in structure. Fig. 16.6 shows the standard approach to identify and elucidate the structure of an isolated compound using GC-MS, IR, and NMR.

To identify unknown compounds, stepwise and detailed experiments including MS, IR, and NMR are necessary. The first step is to obtain information on the molecular weight (mass), molecular formula, and functional group of the compound under investigation. The molecular weights and ion fragmentation patterns can be determined from a low-resolution chemical ionization mass spectra (LR-CIMS) and electron impact mass spectra (at 70 eV, LR-EIMS). The Shimadzu QP-5000 spectrometer (isobutane as the carrier gas, direct insertion technique) is a popular mass spectrometer used for obtaining these spectra. While LR-CIMS provides information on the nature of the compound (m/z MH^+), LR-EIMS provides information on the ionisation pattern of a compound (m/z M^+). Electrospray ionization (ESI) MS performed on a micro-mass Q-TOF-2 (acetonitrile as solvent) can also give reliable information on the mass of a compound. On the basis of its LR-CIMS/ESI-MS spectra, the molecular formula of a compound can be determined using a high-resolution mass spectrometer including A Fison/VG Autospec oa-TOF mass spectrometer (methane as the carrier gas). This instrument generates a high-resolution chemical ionization mass spectra (HR-CIMS) and the high-resolution electron ionization mass spectra (HR-EIMS), which determines the molecular formula based on the abundance and the elemental composition (commonly observed in nature are C, H, N, and O) of a compound. Once this information is obtained, the melting points and optical rotation (chiral compounds) can be determined using a Reichert hot stage apparatus and polarimeter, respectively. The specific rotation can be calculated using the equation: $[\alpha]_D^t = \alpha / l \cdot c$ where α = observed rotation at temperature t°C, l = length of polarimeter cell (dm), and c = concentration of sample (g/100 mL). An average of at least 10 optical rotation readings should be taken for the rotation α-value.

Simultaneously, the unknown compounds can be prepared in deuterated solvents (most commonly used are *d*-chloroform, methanol, and water) for obtaining one- and two-dimensional NMR spectra associated with specific ranges of chemical shifts in ppm (δ) expressed by a compound. Recently published books (volumes I and II) by Williams et al. [56,57] discuss modern NMR approaches to the structure elucidation of natural products in detail and give comprehensive instrumentation and experimental techniques. There are many NMR experiments required for the structure elucidation of a new compound, and the approaches differ among researchers and institutions. Mainly, two types of NMR exist: continuous-wave NMR (Cw-NMR) and pulsed or Fourier-transform (FT-NMR). The latter type is used in modern NMR spectroscopy. NMR is basically used to determine the relative location of atoms within a molecule and construct a complete structure through correlations of carbon and proton signals. One-dimensional NMR experiments for small molecules include ^{1}H-NMR, ^{13}C-NMR, attached proton test (APT), distortionless enhancement by polarization transfer experiment (DEPT), COSY, NOESY and TOCSY. While ^{1}H-NMR provides information on the total number of hydrogens (integrated proton numbers) and the chemical or functional groups present in a compound, scalar one-dimensional NOESY (^{1}H-^{1}H) is most effective at determining a rough estimate of relative proximity of the proton atoms based on the intensity of the nuclear Overhauser effect (NOE). One-dimensional TOCSY spectra are derived from a long-range (^{1}H-^{1}H, maximum four-five bond away) spin–spin coupling experiment, and it is useful for dividing the proton signals into groups or coupling networks, especially when the multiplets overlap. These proton experiments should be then followed with ^{13}C-NMR, APT, and DEPT experiments. ^{13}C-NMR gives overall information on a total number of carbon atoms present in a compound. This experiment is often replaced by APT, which can generate all possible carbon-hydrogen (^{13}C-^{1}H) configurations in one experiment, yielding methane (CH) and methyl (CH_3) groups with positive signals, while the methylene (CH_2) and quaternary carbons (C) with negative signals. DEPT is more sensitive compared to APT but it suppresses quaternary carbons and requires up to three different acquisitions to yield full results. Thus, APT is a good carbon experiment that is often used for structure elucidation. Once the total number of protons, carbons, and different functional groups (CH, CH_2, CH_3, C=O) and aromatic protons are confirmed by one-dimensional NMR and their combined atomic weights correspond to the molecular weight and molecular formula of a compound (determined by MS), the networks of carbon-proton arrangement should be established through two-dimensional NMR spectroscopy. Two-dimensional NMR, which is generally displayed as a contour plot (similar to a topographical plot), includes gCOSY, gTOCSY, gNOESY, gradient heteronuclear single quantum correlation (gHSQC) and gradient heteronuclear multiple bond correlation (gHMBC). While gCOSY generates information on correlations between coupled proton signals (represented by *J* coupling through a bond), gTOCSY is sometimes helpful to obtain cross-peaks through uninterrupted pattern of coupling constants. gNOESY spectra provide information on the alignment of protons through space effects, and it is useful for establishing conformation and stereochemistry of the structure of a compound. gHSQC is used for obtaining information on paired narrower resonances for ^{1}H-^{13}C correlations (CH, CH_2, and CH_3). Finally, gHMBC spectra are used for establishing the long-range chemical shift correlation of ^{1}H-^{13}C that are typically two to four bonds away from a proton in the structure of a compound. Generally, in both the gHSQC and gHMBC contour plots, ^{1}H-NMR spectra are aligned on the *x*-axis and APT spectra are aligned on the *y*-axis. gHMBC is sometimes helpful to spot the carbon atoms, especially the quartenary carbons that may not have been picked up by APT. Only 1−10 mg of compound is needed to carry out all these experiments. The less sample available, the longer it takes to run these experiments, especially ^{13}C-NMR and two-dimensional NMR. For smaller quantities of samples, nanoprobe experiments are available. For frequent small molecule structure elucidation, a 500-MHz system is sufficient. However, for HTP and for obtaining enhanced spectra, higher-capacity NMR machines such as Bruker's next-generation ADVANCE NEO 900-MHz NMR are available. Emsley and Feeney [58] have diligently highlighted the last 40 years of progress in NMR applications.

Structure elucidation can be easy if crystals can be grown from a pure compound. Crystals can be grown using either one or two miscible (diffusible) solvent systems with varying volatility indices. Various sources [59−61] provide hints on how to successfully grow crystals from organic compounds. X-ray crystallography or more advanced Synchrotron technology can be used for determining the crystal structure of a compound from its single crystal.

3.4 Biological Activity Screening of Crude Extracts and Pure Compounds

In natural product−based drug lead identification, biological activity screening is a crucial step. To conduct the bioassays, productive collaborations with pharmacologists, immunologists, physicians, plant pathologists and biologists may be crucial, and there are relatively few simple laboratory-based in vitro methods that could be performed by chemists alone [62]. Biological activity screening for natural product samples is performed at various stages including crude extracts, fractions, and the pure compounds. Based on the types of major classes of biochemicals present in crude extracts, strategic biological screening and isolation protocols should be developed. While researchers with access to advanced laboratory facilities can choose to conduct bioassay-guided biochemical analyses to identify drug lead compounds, researchers with compromised access to such technologies often test crude extracts first for their biological activities and, based on the bioactivity hits, they proceed with the isolation of pure compounds from bioactive crude extracts and, finally, test the isolated compounds for their biological activities. A semibioassay-guided isolation model, in which crude extracts, major fractions, and pure compounds are targeted for biological activity screening, is cost-effective for identifying new drug lead compounds. For resource-limited countries, testing all the HPLC peak-based fractions is often prohibitive due to the need for large numbers of fractions requiring sophisticated HTP technological platforms, which would increase with the number of purification steps. This can be even more challenging when screening is carried out in an in vivo setting.

Crude extracts are a complex mixture of different biochemicals, which are present in various concentrations. For this reason, it is essential to concentrate the crude extracts to reach the biological target in the assay. Despite concentrating the extracts, it is possible that some components may be in concentrations that are too low to have noticeable effects on the target organisms or cells if not enriched in some form. Sometimes, "nuisance" and additive compounds (examples are tannins and resins) or the synergistic effects of several compounds can confound the desired bioactivity data generated by the bioassays. Prefractionating crude extracts using liquid–liquid fractionation techniques is a first step toward simplifying complex mixtures and making them more suitable for use in bioassays, as this step can remove "false-positive" compounds or artifacts from the mixture. This can be followed by liquid–solid fractionation of the crude fractions, which involves precipitation, crystallization, or CG techniques. These techniques can generate large number of fractions that can be easily mined only by using automated HTS platforms.

Bioassays involve quantitation of the response that follows the application of a stimulus to a biological system [63]. Mainly, three bioassay platforms (in vitro, ex vivo, and in vivo approaches) are used for the discovery of new drug lead compounds. All these platforms are based on two approaches: (a) phenotypic screening (classic pharmacology or forward pharmacology), in which extracts and compounds are screened in cellular or animal disease models to identify desirable candidates that can change a phenotype of an organism; and (b) reverse pharmacology screening (target-based drug discovery), in which compounds or extracts are tested against certain isolated biological targets believed to be disease modifying/modulating [64]. Generally, in vitro screening precedes in vivo animal model testing to mainly minimize the cost of research and reduce the need for animal-based studies. In vitro testing is usually carried out on cultured cells to understand the nature of biological samples including how the cells behave when treated with test samples. For example, an in vitro dendritic cell (DC) assay is often used to look for biological samples that could potentially upregulate or downregulate inflammatory (or anti-inflammatory) mediators. DC immunomodulator screening assay, which uses high-content flow cytometry platform with 12 or more independent fluorescent readouts and polymerase chain reaction, is often applied in the search for adjuvants, anti-inflammatories, anticancer, allergies, and infectious diseases [65]. A DC-based extracts/compound screening assay scheme is given in Fig. 16.7. It is argued that cellular-based systems are unable to adequately model human disease processes that involve many different cell types across many different organ systems and that this type of complexity can only be emulated in model organisms [66,67]. In in vivo animal-based systems, phenotypic high-content screening uses model organisms, which mostly include fruit fly (*Drosophila melanogaster*) and mice (*Mus musculus*), to evaluate the effects of a test sample. Each approach has an array of advantages and disadvantages to its use, and therefore a comprehensive study examining multiple aspects of target diseases requires using more than one platform to assess the efficacies of the test samples. What is critical is that any test models chosen for the biological activity studies must mimic common human diseases and suit the scope and relevance of outcomes achievable based on the experimental design. Some of the common examples of in vitro, ex vivo, and in vivo screening assays are described next.

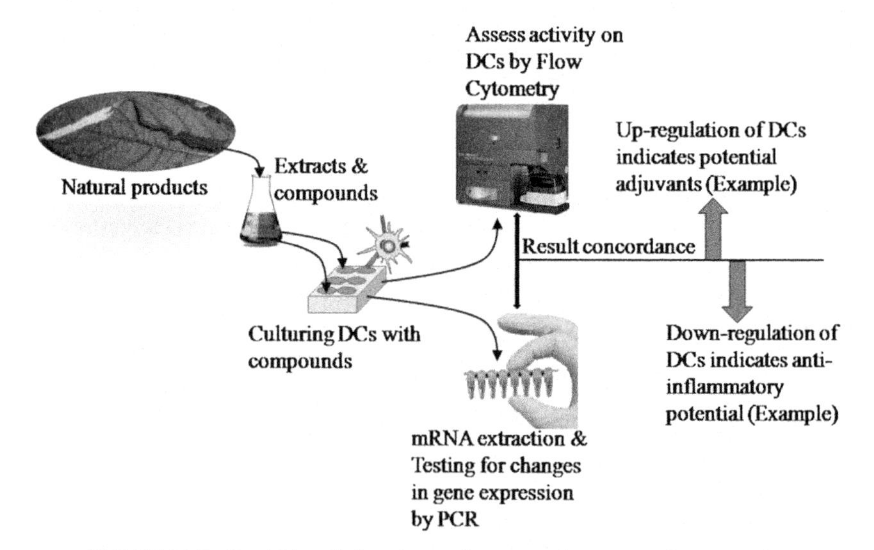

FIGURE 16.7 Dendritic cell–based screening assay of extracts and compounds.

3.4.1 In Vitro and In Vivo Antimicrobials Screening Assays

Microbiological bioassays involve the detection of antibacterial and antifungal activities and potencies of compounds using different end-point methods like plate assays (agar diffusion assay) and tube assays (agar dilution streak assays and broth dilution minimum inhibition concentration assays) [63,68]. Due to the requirement of a long incubation time, the agar diffusion assay (plate assay) cannot be used for volatile and unstable compounds and, therefore, agar-overlay methods (agar dilution methods or liquid media methods) have normally been followed. In the agar-overlay methods, the degree of antimicrobial activity and the minimum inhibition concentration (MIC) of a compound are greatly influenced by factors like "compound concentration, carbon dioxide and oxygen concentration, media compatibility with the test compound, solubility and stability of the test compound, pH of the medium, types nutrient of media, temperature, the duration of the exposure to the compound and the types of strains (bacterial, fungal or viral) used in the testing" [69,70]. For these reasons, growth-promoting or -inhibiting potencies of test samples have to be standardized and compared with standard preparations of the test compound [63]. Thus, the choice of antimicrobial bioassay method has to be selectively adapted taking these factors into consideration. Such methods should be sensitive enough to detect active principles, the MIC values, and the dose−response of the microorganisms (bacteria and fungi). A comparison of the disk diffusion method, the hole-plate diffusion method, and the bioautographic TLC assay showed that the nature and amount of substance influence to a great extent the selection of the test method [71]. Using these assays, more than 300 alkaloids have been shown to possess antimicrobial activity [72]. The antibacterial activity assay is distinguishable into bacteriostatic activity and bactericidal activity. Bacteriostatic activity of the compounds refers to stopping or inhibiting bacterial growth and multiplication. Bactericidal activity is the property of an antibacterial agent responsible for killing bacteria. Both disk diffusion assay and broth methods can be used to examine bacteriostatic activities, but the bactericidal activities can be detected only by microscopic analysis or by the assessment of bacterial cell recovery in fresh media [70]. Susceptibility of *Mycobacterium tuberculosis* to antimicrobial agents has been tested by the bioluminescent assay (ATP) method, the Mycobacteria growth indicator tube (MGIT) method, the MIC method, and the M24-T agar proportion method. It was found that the ATP method was the most rapid and reliable method for the assessment of drug (antibacterial) susceptibility testing [73]. The fluorescein diacetate (FDA) assay, based on the enzymatic hydrolysis of fluorescein diacetate to fluorescein, is suitable for use with a wide range of different microorganisms and was recommended for determining the MIC of crude extracts and purified compounds [74]. This method is commonly used in antibacterial testing, although problems can arise with fluorescent components in the growth media or fluorescence of the test compound. Kamia et al. [75] reported that the turbidometric assay was rapid, reproducible, and a very useful method for quantitative estimation of antibacterial activity.

3.4.2 In Vitro and Ex Vivo Antiparasitic Screening Assays

Parasites affect billions of people worldwide. Searches for new antiparasitic drugs to combat parasitic infections is ongoing, and various bioassay protocols have been developed with a renewed interest in whole-organism screening (ex vivo) as part of the drug discovery process. Commonly used in vitro and animal models in antiparasitic drug discovery is highlighted in a review article by Pink et al. [76]. A more recent review by Buckingham et al. [77] reported on the progress of the development of automation of measuring nematode behaviors using machine vision, statistical imaging, and tracking approaches. Monitoring and recording motility with micromotility meters, micromotility recorders, video motilitimeters, larval migration assays, microfluidic chips, and sinusoidal movement techniques offer direct readouts of neuromuscular activity and the health of animals during the screening process. However, these approaches can be laborious and subjective, present bias, are difficult to standardize, and are low-throughput in nature. The xCELLigence motility assay method, which we have applied in our anthelmintic drug lead identification from natural products [28,29], overcomes some of these low-throughput bottlenecks in anthelmintic drug screening and resistance detection pipelines. This method uses a 96-well E-plate cell-monitoring device (xCELLigence) to monitor the motility of parasitic worms in a fully automated label-free manner and facilitates objective assessment of real time motility of anti-parasite efficacy of drugs on different developmental stages of the parasites. It enables HTS of large numbers of compounds. Peak et al. [78] developed an HTP fluorescent-based bioassay to detect *Schistosoma* viability, and this assay has been recently applied to natural product drug screening against the schistosomulum stage of *Schistosoma mansoni* [28], to which the frontline drug praziquantel is ineffective. For antimalarial screening, different types of test methods, like biochemical and molecular techniques, colorimetric enzyme assays, flow cytometry, and microdilution assays, have been developed during the past years [79]. The colorimetric enzyme assay that was described by Basco et al. [80] is based on the measurement (using spectrophotometry) of the activity of the enzyme lactate dehydrogenase in living *Plasmodium falciparum* and *Plasmodium ovale*. A semiautomated micro-dilution assay [81] involves the inhibition of uptake of a nucleic acid precursor, ^{3}[H]hypoxanthine, into the malarial parasite. A simple and reproducible method has also been described by Pandey et al., and modified microdilution assay methods have been routinely used in the identification of potential antimalarial compounds [69].

3.4.3 In Vivo Screening Assays for Inflammatory Bowel Diseases

Inflammatory bowel diseases (IBDs) includes Crohn's disease and ulcerative colitis and is often referred to as a disease of the developed world. The search for anti-inflammatory agents are ongoing due to only limited frontline drugs being available in the market. Various animal models that mimic human IBD have been developed for screening and identifying new anti-inflammatory drug lead compounds. A recent review by Goyal et al. [82] comprehensively described 34 available models and classified them into seven major groups: chemical-induced models, bacteria-induced models, spontaneous models, genetically engineered models, transgenic mouse models, mutation knock-in models, and adoptive transfer models. Presently, there is no "perfect" animal model that can address all the mechanisms involved in intestinal inflammation. Each animal model has an array of advantages and disadvantages to its use, and therefore a comprehensive study examining multiple aspects of intestinal inflammation requires multiple models. What is important in an IBD animal model is that a successful animal model should have the ability to induce inflammation that mimics the major pathologic sequelae in humans. Due to its short time frame and relative ease of implementation, chemical-induced animal models of IBD are widely adopted; as many as 10 chemical-induced mouse models have been developed. According to Wirtz et al. [83], the most commonly used models are 2,4,6-trinitrobenzene sulfonic acid (TNBS), oxazolone, and dextran sodium sulfate (DSS) colitis models. While the TNBS and oxazolone models use intrarectal administration of the covalently reactive reagents TNBS/oxazolone, the DSS model uses drinking water supplemented with DSS to induce colitis in mice. There are relative strengths and limitations in using these three animal models. The TNBS model is short and effective for large-scale anti-inflammatory screening projects, especially for natural products. TNBS induces acute and resolving colitis rather than the chronic inflammation experienced by many people with IBD. However, a review by Jiminez et al. [84] highlighted the suitability of TNBS as an incitant of both acute and chronic inflammation. Moreover, genes with known crucial roles in human Crohn's disease such as *nod2* and *il-17* are essential for pathology in TNBS colitis in mice (reviewed in [85]). TNBS elicits cell-mediated mixed T helper (Th)1 and Th2 cytokine-based immune responses and induces transmural inflammation in the gut with clinical, morphological, and histopathological features similar to those of human IBD. Clinically, IBD is characterized by severe diarrhea, bleeding, abdominal pain, and fluid and electrolyte loss, which underlies the initial acute nature of IBD. Similarly, patients with ulcerative colitis experience significant colonic thickening and dense infiltration of neutrophils, monocytes, macrophages, and T cells, typically in the mucosal layer of the bowel wall. All these clinical and histological signs are exhibited in the TNBS colitis model. We have previously identified a novel anticolitis drug lead compound from a Bhutanese medicinal plant using this model [30] but recommend using additional models of colitis to confirm the translatability of these findings to human IBD, particularly the T-cell transfer model to $RAG^{-/-}$ mice.

4. CONCLUSIONS AND FUTURE DIRECTIONS

Natural products, including plants, animals, and microbes, continue to be rich sources for new drug lead discovery. However, a majority of these sources remain underexplored for medicinal applications. While this is partly due to overemphasis on the development of combinatorial and synthetic chemistry, the complexity in isolation and structure elucidation of pure compounds was the major hurdle in natural product–based drug discovery. The advancement in spectroscopy, separation technologies, omics sciences, medicinal chemistry, molecular biology, automation, bioassay techniques, and bioinformatics have led to new paradigms in HTP biodiscovery from natural products. Most recently developed "-omics" techniques are powerful tools in chemometric profiling, creation of natural product libraries, biomolecular target development, dereplication, and early hit characterization from crude extracts, semi-pure mixtures, or single-component natural products.

Although vast arrays of these advanced screening techniques and technologies are available today, the success of a natural product drug lead compound depends largely on how efficiently they are used and how strategic the researchers are in their respective development pathways. It also depends on the choice of the biodiscovery source materials. In-depth understanding of the habitat and environment, ecological adaptations, morphological and physiological responses, host–pathogen interactions, traditional uses, and chemical properties of the natural products all help researchers to develop tailored strategic extraction, isolation, and biological assay protocols. The general consensus is that ethno-medically directed and biorational search strategies supported by bioassay-guided isolation techniques provide high biological activity hit rates, many of which could potentially become new drug lead compounds. It must also be understood that discovery of a new drug lead compound is not a one-person show. It requires a multidisciplinary approach involving botanists, zoologists, chemists, immunologists, pharmacologists, and clinicians. Close horizontal collaboration among these experts is essential for the success of any biodiscovery projects.

REFERENCES

[1] D.A. Dias, S. Urban, U. Roessner, A historical overview of natural products in drug discovery, Metabolites 16 (2) (2012) 303–336.

[2] J.A. Beutler, Natural products as a foundation for drug discovery, Curr. Protoc. Pharmacol. 46 (9) (2009) 1–9.

[3] Dictionary of Natural Products, Taylor & Francis Group, CRC Press, 2017. CHEMnetBASE, Available from: http://www.chemnetbase.com/.

[4] D.J. Newman, G.M. Cragg, Natural products as sources of new drugs from 1981 to 2014, J. Nat. Prod. 79 (3) (2016) 629–661.

[5] The Plant List: A Working List of All Plant Species, Version 1.1, 2013. Available from: http://www.theplantlist.org/.

[6] M.F. Balandrin, A.D. Kinghorn, N.R. Farnsworth, Plant-derived natural products in drug discovery and development: an overview, in: A. Douglas Kinghorn, M.F. Balandrin (Eds.), Human Medicinal Agents from Plants, ACS Symposium Series 534, American Chemical Society, Washington, DC, 1993, pp. 1–12 (Chapter 1).

[7] J.A. Wilkinson, M.L. Wahlqvist, J. Clark, New Food and Pharmaceutical Products from Agriculture: Papers from Outlook, RIRDC Publication No 02/015, Rural Industries Research and Development Corporation, Kingston, ACT, 2002. Available from: http://www.rirdc.gov.au.

[8] Dictionary of Alkaloids, Taylor & Francis Group, CRC Press, 2017. CHEMnetBASE, Available from: http://www.chemnetbase.com/.

[9] C. Mora, D.P. Tittensor, S. Adl, A.G.B. Simpson, B. Worm, How many species are there on Earth and the Ocean? PLoS Biol. 9 (8) (2011) e1001127.

[10] G. Patrick, Medicinal chemistry-instant notes, in: B.D. Hames (Ed.), The Instant Notes Series, first ed., Bios Scientific Publishers Limited, Leeds, UK, 2001, pp. 77–83.

[11] J.W. Daly, H.M. Garaffo, T.F. Spande, M.W. Decker, J.P. Sullivan, M. Williams, Alkaloids from frog skin: the discovery of epibatidine and the potential for developing novel non-opiod analgesics, Nat. Prod. Rep. 17 (2000) 131–135.

[12] C. Shepherd, S. Navarro, P. Wangchuk, D. Wilson, N. Daly, A. Loukas, Identifying the immunomodulatory components of helminths, Parasite Immunol. 37 (6) (2015) 293–303.

[13] S. Navarro, D. Pickering, I. Ferreira, L. Jones, S. Ryan, S. Troy, et al., Hookworm recombinant protein promotes regulatory T cell responses that suppress experimental asthma, Sci. Transl. Med. 26 (3) (2016).

[14] Dictionary of Marine Natural Products, Taylor & Francis Group, CRC Press, 2017. CHEMnetBASE, Available from: http://www.chemnetbase.com/.

[15] N. Fusetani, Introduction, in: N. Fusetani (Ed.), Drugs from the Sea, 2000, pp. 1–5.

[16] G. Cragg, D. Newman, Chemists' toolkit, Nature's bounty, Chem. Brit. (2001) 22–26.

[17] R. Verpoorte, Chemodiversity and the biological role of secondary metabolites, some thoughts for selecting plant material for drug development, in: L. Bohlin, J.G. Bruhn (Eds.), Bioassay Methods in Natural Product Research and Drug Development, Proceedings of the Phytochmical Society of Europe, vol. 43, 1999, pp. 11–23.

[18] K. Hostettmann, A. Marston, J.L. Wolfender, Strategy in the search for new biologically active plant constituents, in: K. Hostettmann, A. Marston, A. Maillard, M. Hamburger (Eds.), Phytochemistry of Plants Used in Traditional Medicine, Clarendon Press, Oxford, 1993, pp. 17–45.

[19] G.A. Cordell, M.L. Quinn-Beattie, N.R. Farnsworth, The potential of alkaloids in drug discovery, Phytother. Res. 15 (2001) 183–205.

[20] S.W. Pelletier, Introduction, in: S.W. Pelletier (Ed.), Chemistry of the Alkaloids, first ed., Van Nostrand Reinhold Inc., U.S., 1970, pp. 1–10.

[21] G.T. Prance, D.J. Chadwick, J. Marsh, Ethnobotany and the search for new drug discovery, in: Conference Proceedings Ed. Ciba Foundation Symposium Series 185, New York, 1994, pp. 1–280.

[22] P. Wangchuk, J.B. Bremner, Samten, B.W. Skelton, A.H. White, R. Rattanajak, S. Kamchonwongpaisan, Antiplasmodial activity of atisinium chloride from the Bhutanese medicinal plant, *Aconitum orochryseum*, J. Ethnopharmacol. 130 (2010) 559–562.

[23] P. Wangchuk, J.B. Bremner, Samten, R. Rattanajak, S. Kamchongpaisan, Antiplasmodial agents from the Bhutanese medicinal plant, *Corydalis calliantha*, Phytother. Res. 24 (2010) 481–485.

[24] P. Wangchuk, P.A. Keller, S.G. Pyne, W. Lie, A.C. Willis, R. Rattanajak, S. Kamchonwongpaisan, A new protoberberine alkaloid from *Meconopsis simplicifolia* (D. Don) Walpers with potent antimalarial activity against a multidrug resistant *Plasmodium falciparum* strain, J. Ethnopharmacol. 150 (2013) 953–959.

[25] P. Wangchuk, S.G. Pyne, P.A. Keller, M. Taweechotipatr, S. Kamchonwongpaisan, Phenylpropanoids and furanocoumarins as antibacterial and antimalarial constituents of the Bhutanese medicinal plant, *Pleurospermum amabile*, Nat. Prod. Commun. 9 (2014) 957–9660.

[26] P. Wangchuk, P.A. Keller, S.G. Pyne, A.C. Willis, S. Kamchonwongpaisan, Antimalarial alkaloids from a Bhutanese traditional medicinal plant *Corydalis dubia*, J. Ethnopharmacol 143 (2012) 310–313.

[27] P. Wangchuk, T. Tobgay, Contributions of medicinal plants to the gross national happiness and biodiscovery in Bhutan, J. Ethnobiol. Ethnomed. 11 (2015) 48.

[28] P. Wangchuk, P.R. Giacomin, M.S. Pearson, M.J. Smout, A. Loukas, Identification of lead chemotherapeutic agents from medicinal plants against blood flukes and whipworms, Sci. Rep. 6 (2016) 32101.

[29] P. Wangchuk, M.S. Pearson, P.R. Giacomin, L. Becker, J. Sotillo, D. Pickering, M.J. Smout, A. Loukas, Compounds derived from the Bhutanese daisy, *Ajania nubigena*, demonstrate dual anthelmintic activity against *Schistosoma mansoni* and *Trichuris muris*, PLoS Negl. Trop. Dis. 10 (2016) e0004908.

[30] P. Wangchuk, S. Navarro, C. Shepherd, P.A. Keller, S.G. Pyne, A. Loukas, Diterpenoid alkaloids of *Aconitum laciniatum* and mitigation of inflammation by 14-*O*-acetylneoline in a murine model of ulcerative colitis, Sci. Rep. 5 (2015) 12845.

[31] D.S. Fabricant, N.R. Farnsworth, The value of plants used in traditional medicine for drug discovery, Environ. Health. Perspect. 109 (2001) 69–75.

[32] M.J. Balunas, A.D. Kinghorn, Drug discovery from medicinal plants, Life Sci. 78 (2005) 431–441.

[33] Y.W. Chin, M.J. Balunas, H.B. Chai, A.D. Kinghorn, Drug discovery from natural sources, AAPS J. 8 (2) (2006) 239–253.

[34] P. Wangchuk, S.G. Pyne, P.A. Keller, An assessment of the Bhutanese traditional medicine for its ethnopharmacology, ethnobotany and ethnoquality: textual understanding and the current practices, J. Ethnopharmacol. 148 (1) (2013) 305–310.

[35] M.M. Cowan, Plant products as antimicrobial agents, Clin. Microbiol. Rev. 12 (4) (1999) 564–582.

[36] A. Geyid, D. Abebe, A. Debella, Z. Makonnen, F. Aberra, F. Teka, et al., Screening of some medicinal plants of Ethiopia for their anti-microbial properties and chemical profiles, J. Ethnopharmacol. 97 (2005) 421–427.

[37] P. Wangchuk, P.A. Keller, S.G. Pyne, M. Taweechoitpatr, A. Tonsomboon, R. Rattanajak, S. Kamchonwongpaisan, Evaluation of an ethnopharmacologically selected Bhutanese medicinal plants for their major classes of phytochemicals and biological activities, J. Ethnopharmacol. 137 (1) (2011) 730–742.

[38] H. Jork, W. Funk, W. Fishcer, H. Wimmer, Thin Layer Chromatography: Reagents and Detection Methods: Physical and Chemical Detection Methods: Fundamentals, Reagents I, vol. 1a, Wiley, New York, 1989, pp. pp.1–465.

[39] J. Xia, I.V. Sinelnkov, B. Han, D.S. Wishart, MetaboAnalyst 3.0-making metabolomics more meaningful, Nucleic Acids Res. 43 (W1) (2015) W251–W257.

[40] B. Guo, B. Chen, A. Liu, S. Yao, A tandem liquid chromatography–mass spectrometry (LC–MS) method for profiling small molecules in complex samples, Curr. Drug Metab. 13 (9) (2012) 1226–1243.

[41] Y. Zhou, Q. Liao, M. Lin, X. Deng, P. Zhang, M. Yao, et al., Combination of ^{1}H NMR- and GC-MS-based metabonomics to study on the toxicity of Coptidis Rhizome in rats, PLoS One 9 (2) (2014) e88281.

[42] A. Alonso, S. Marsal, A. Julia, Analytical methods in untargeted metabolomics: state of the art in 2015, Front. Bioeng. Biotechnol. 3 (23) (2015) 1–20.

[43] M.Y. Mushtaq, Y.H. Choi, R. Verpoorte, E.G. Wilson, Extraction for metabolomics: access to the metabolome, Phytochem. Anal. 25 (2013) 291–306.

[44] S. Parasuraman, R. Anish, S. Balamurugan, S. Muralidharan, K.J. Kumar, V. Vijayan, An overview of liquid chromatography-mass spectroscopy instrumentation, Pharm. Methods 5 (2) (2014) 47–55.

[45] D. Tulpan, S. Leger, L. Belliveau, A. Culf, M. Cuperlovic-Culf, Metabo-Hunter: an automatic approach for identification of metabolites from 1H-NMR spectra of complex mixtures, BMC Bioinform. 12 (2011) 400.

[46] D. Jacob, C. Deborde, A. Moing, An efficient spectra processing method for metabolite identification from 1H-NMR metabolomics data, Anal. Bioanal. Chem. 405 (2013) 5049–5061.

[47] NIST Standard Reference Database 1A V14, Software Version 2.2 G, Mass Spectrometry Data Center, National Institute of Standards and Technology, Gaithersburg, 2017. Accessed from, https://www.nist.gov/srd/nist-standard-reference-database-1a-v14.

[48] D.S. Wishart, Emerging applications of metabolomics in drug discovery and precision medicine, Nat. Rev. Drug Discov. 15 (2016) 473–484.

[49] A.T. Cardoso-Taketa, R. Pereda-Miranda, Y.H. Choi, R. Verpoorte, M.L. Villarreal, Metabolic profiling of the Mexican anxiolytic and sedative plant *Galphimia glauca* using nuclear magnetic resonance spectroscopy and multivariate data analysis, Planta Med. 74 (2008) 1295–1301.

[50] Solvent Miscibility Table, Phenogel: No-aqueous GPC/SEC Columns, Phenomenex, 2017. Accessed from, https://phenomenex.blob.core.windows.net/documents/75795f06-c624-456a-8fa11761ea980bda.pdf?.

[51] A.B. Svendsen, R. Verpoorte, Chromatography of alkaloids, Part A, in: A.B. Svendsen, R. Verpoorte (Eds.), Thin-Layer Chromatography, first ed.J. Chromatogr. Lib., vol. 23A, 1983, p. 531.

[52] C. Tistaert, B. Dejaegher, Y.V. Heyden, Chromatographic separation techniques and data handling methods for herbal fingerprints: a review, Anal. Chim. Acta 690 (2) (2011) 148–161.

[53] K. Hostettmann, A. Marston, M. Hostettmann, Preparative Chromatography Techniques: Applications in Natural Product Isolation, Springer, Berlin, 1998.

[54] S. Mandal, V. Mandal, A. Das, Essentials of Botanical Extraction: Principles and Applications, Academic Press, Elsevier, U.S.A., 2015.

[55] T. Kind, O. Fiehn, Advances in structure elucidation of small molecules using mass spectrometry, Bioanal. Rev. 1 (1–4) (2010) 23–60.

[56] A. Williams, G. Martin, D. Rovnyak (Eds.), Modern NMR Approaches to the Structure Elucidation of Natural Products: Data Acquisition and Applications to Compound Classes, vol. 2, 2016, pp. 1–516.

[57] Royal Society of Chemistry, in: A. Williams, G. Martin, D. Rovnyak (Eds.), Modern NMR Approaches to the Structure Elucidation of Natural Products: Instrumentation and Software, vol. 1, 2015, pp. 1–329.

[58] J.W. Emsley, J. Feeney, Forty years of progress in nuclear magnetic resonance spectroscopy, Prog. Nucl. Magn. Reson. Spectrosc. 50 (2007) 179–198.

[59] P.G. Jones, Crystal growing, Chem. Brit. 17 (1981) 222–225.

[60] P. Slus, A.M. Hezemans, J. Kroon, Crystallization of low-molecular weight organic compounds for X-ray crystallography, J. Appl. Crystal 22 (1989) 340–344.

[61] R.J. Staples, Growing and Mounting Crystals Your Instrument Will Treasure, Department of Chemistry, Michigan Sate University, Michigan, 2017, pp. 1–17.

[62] A. Marston, K. Hostettmann, Biological and chemical evaluation of plant extracts and subsequent isolation strategy, in: L. Bohlin, J.G. Bruhn (Eds.), Bioassay Methods in Natural Product Research and Drug Development, 1999, pp. 67–80.

[63] D. Hawcroft, T. Hector, F. Rowell, in: A.M. James (Ed.), Quantitative Bioassay, Analytical Chemistry by Open Learning, John Wiley & Sons, London, 1987, pp. 1–300.

[64] J.A. Lee, M.T. Uhlik, C.M. Moxham, D. Tomandl, D.J. Sall, Modern phenotypic drug discovery is a viable, neoclassic pharma strategy, J. Med. Chem. 55 (10) (2012) 4527–4538.

[65] M. Spitzer, G. Haaima, Dendritic Cell Immunomodulator Screening Assay, Bussiness Development Office, QIMR Berghofer Medical Research Institute, Brisbane, Australia, 2014. Available from:www.qimrberghofer.edu.au.

[66] M.K. Hellerstein, Exploiting complexity and the robustness of network architecture for drug discovery, J. Pharmacol. Exp. Ther. 325 (1) (2008) 1–9.

[67] M.K. Hellerstein, A critique of the molecular target-based drug discovery paradigm based on principles of metabolic control: advantages of pathway-based discovery, Metab. Eng. 10 (1) (2008) 1–9.

[68] M.E. Venturini, D. Blanco, R. Oria, *In vitro* antifungal activity of several antimicrobial compounds against *Penicillium expansum*, J. Food Protect. 65 (5) (2002) 834–839.

[69] S. Hadi, Bioactive Alkaloids from Medicinal Plants of Lombok (Ph.D. thesis), Department of Chemistry, University of Wollongong, Wollongong, 2002.

[70] K. Benkendorf, Bioactive Molluscan Resources and Their Conservation: Biological and Chemical Studies on Egg Masses of Marine Molluscs (Ph.D. thesis), Department of Chemistry and Department of Biological Sciences, University of Wollongong, Wollongong, 1999.

[71] A. Brantner, K.P. Pfeiffer, H. Brantner, Applicability of diffusion methods required by the pharmacopoeas for testing antibacterial activity of natural compounds, Pharmazie 49 (7) (1994) 512–516.

[72] R. Verpoorte, Antimicrobially active alkaloids, in: M.F. Roberts, M. Wink (Eds.), Alkaloids: Biochemistry, Ecology, and Medicinal Applications, 1998, pp. 397–434.

[73] Y. Toshio, S. Naoki, Y. Kenya, O. Yutaka, A. Hajime, M. Akinari, Reliability of bioluminescent assay (ATP method) for testing antimicrobial susceptibility of *Mycobacterium tuberculosis*, Rinsho Byori (Jpn J. Clin. Pathol.) 51 (3) (2003) 194–200.

[74] S. Chand, I. Lusunzi, D.A. Veal, L.R. Williams, P. Karuso, Rapid screening of the antimicrobial activity of extracts and natural products, J. Antibiot. 47 (11) (1994) 1295–1304.

[75] H. Kamiya, K. Muramoto, K. Ogata, Antibacterial activity in the egg mass of a sea hare, Experientia 40 (1984) 947.

[76] R. Pink, A. Hudson, M.A. Mouries, M. Bendig, Opportunities and challenges in antiparasitic drug discovery, Nat. Rev. Drug Discov. 4 (2005) 727–740.

[77] S. Buckingham, P.A. Frederick, D. Sattelle, Automated high-throughput motility analysis in *C. elegans* and parasitic nematodes: applications in the search for new anthelmintics, Int. J. Parasitol. Drugs Drug Res. 4 (2014) 226–232.

[78] E. Peak, I.W. Chalmers, K.F. Hoffmann, Development and validation of a quantitative, high-throughput, fluorescent-based bioassay to detect schistosoma viability, PLoS Negl. Trop. Dis. 4 (7) (2010) e759.

[79] S.L. Croft, C.R. Weiss, Natural products with antiprotozoal activity, in: L. Bohlin, J.G. Bruhn (Eds.), Bioassay Methods in Natural Product Research and Drug Development, 1999, pp. 81–100.

[80] L.K. Basco, F. Marquet, M.M. Makler, J. Le Bras, *Plasmodium falciparum* and *Plasmodium vivax*-lactate dehydrogenase activity and its application for *in vitro* drug susceptibility assay, Exp. Parasitol. 80 (2) (1995) 260–271.

[81] R.E. Desjardins, C.J. Canfield, J.D. Haynes, J.D. Chulay, Quantitative assessment of antimalarial activity *in vitro* by a semiautimated microdilution technique, Antimicrob. Agents Chemother. (1979) 710–718.

[82] N. Goyal, A. Rana, A. Ahlawat, K.R.V. Bijjem, P. Kumar, Animal models of inflammatory bowel disease: a review, Inflammopharmacol 22 (2014) 219–233.

[83] S. Wirtz, C. Neufert, B. Weigmann, M.F. Neurath, Chemically induced mouse models of intestinal inflammation, Nat. Protoc. 2 (3) (2007) 541–546.

[84] J.A. Jiminez, T.C. Uwiera, I.G. Douglas, R.R. Uwiera, Animal models to study acute and chronic intestinal inflammation in mammals, Gut Pathog. 7 (2015) 29.

[85] E. Antoniou, G.A. Margonis, A. Angelou, A. Pikouli, P. Argiri, I. Karavokyros, et al., The TNBS-induced colitis animal model: an overview, Ann. Med. Surg. (Lond) 11 (2016) 9–15.

Chapter 17

Herb and Drug Interaction

Nilanjan Ghosh[1], Rituparna C. Ghosh[1], Anindita Kundu[2], Subhash C. Mandal[2]
[1]*Dr. B.C. Roy College of Pharmacy and Allied Health Sciences, Durgapur, India;* [2]*Department of Pharmaceutical Technology, Jadavpur University, Kolkata, India*

1. INTRODUCTION

Herbal medicines have been used for a long time to promote health and treat common diseases. The use of natural products as herbal medicines has increased steadily over the last few years, and a very significant proportion of the population in both developing and developed countries depends on traditional medicines for its primary source of healthcare. Recent literature surveys and data suggest that they are increasingly used worldwide as alternative medicines to manage various chronic diseases and also constitute an important group of multicomponent therapeutics, which are used to manage various chronic diseases [1]. Many health ailments such as common colds, inflammation, hyperalgesia, heart diseases, liver cirrhosis, diabetes, and central nervous system diseases respond well to herbal treatments [1–3]. Such widespread use of herbal medicinal products throughout the world has raised serious questions concerning the quality, safety, and efficacy of these products. Another very important aspect that arises is that patients with chronic diseases are likely to be treated with multiple drugs, which in many cases include herbs as well. This coadministration of herbal drugs with therapeutic drugs increases the risk of herb–drug interactions (HDIs) [3,4]. Despite the popularity of herbal medicines, their use is largely not evidence based because there is a lack of clinical evidence for the efficacy, target, and safety of most herbal medicines. Coadministration of herbs with drugs may mimic, increase, or decrease the effect of either component, resulting in clinically important HDIs. These interactions are generally pharmacokinetic or pharmacodynamic in nature. Synergistic or additive therapeutic effects may lead to unfavorable toxicities and complicate the dosage regimen of long-term medications, while antagonistic interactions will result in decreased efficacy and failure of therapy [5].

The potential interaction of herbal medicines with drugs is a major safety concern, especially for drugs with narrow therapeutic indices, and may lead to life-threatening adverse effects. Interactions of several commonly used herbal

Natural Products and Drug Discovery. https://doi.org/10.1016/B978-0-08-102081-4.00017-4

medicines with therapeutic drugs, including anticoagulants (warfarin and aspirin), sedatives and antidepressants (midazolam, alprazolam, amitriptyline, and trazodone), antihuman immunodeficiency virus (HIV) agents (indinavir and saquinavir), cardiovascular drugs (digoxin, nifedipine, and propranolol), immunosuppressants (cyclosporine and tacrolimus), and anticancer drugs (irinotecan and imatinib) in humans have been extensively reviewed. Therefore HDIs have important clinical significance [5–7].

Understanding the pharmacodynamics and pharmacokinetics of herbal medicine presents a difficult situation in itself. Herbal medicines are usually a mixture of structurally diverse chemical compounds, and therefore the type of bioavailability study to be conducted for herbal medicines is based on the information concerning active constituents and the availability of the analytical methods. A complete understanding of all the phytoconstituents in a herbal agent is very difficult and identification of the specific constituent responsible for activity is a significant challenge in itself. Thus the assessment of bioavailability, metabolic pathways, and the kinetics and routes of elimination are very complex and highly variable. Content uniformity of bioactive constituents within and across marketed products is also a major restraint and the fact that there is no regulatory need to study the pharmacokinetics and absorption, distribution, metabolism, and excretion of herbal medicines further complicates the scenario [8]. The bioactivity of herbal medicines originates from the interaction of the biological system with multiple biologically active phytocompounds present in them. The quantitative knowledge of phytocompounds in most of herbal medicines is not very clear and thus the subsequent quantification of the pharmacodynamic activities of herbal medicine becomes difficult. Variable amounts of active compounds in herbal preparations and limited activity data from clinical studies present a major challenge in the standardization of herbal products [9,10].

2. PHARMACOKINETIC HERB–DRUG INTERACTIONS

Pharmacokinetic HDIs are the result of altered absorption, metabolism, distribution, and excretion of drugs. Herbal drugs are a mixture of phytochemicals that are eliminated by the same system of xenobiotic metabolism and transport mechanism that removes synthetic drugs from the human body. Thus pharmacokinetic HDIs are inevitable. The modern drug development process of synthetic drugs involves the assessment of drug–drug interactions that involves many in vitro and in vivo investigations. The basic objective of these studies is to identify or predict possible interactions with in vitro screening models, which are then followed up by in vivo studies. Pharmacokinetic modeling based on physiological parameters is widely used to guide in vivo drug–drug interaction trials, but is difficult to apply to herbal drugs because of factors discussed earlier. Also there is a dearth of well-designed in vitro studies that can exemplify the interaction potential of phytoconstituents. Additionally, poor knowledge regarding the systemic bioavailability and human pharmacokinetic parameters of bioactive phytoconstituents presents a big challenge in understanding and predicting pharmacokinetic HDIs [9–11].

The underlying mechanisms for the altered drug concentrations by concomitant herbal medicines are of many types. Induction or inhibition of hepatic and intestinal drug-metabolizing enzymes (e.g., cytochrome P450 [CYP]) and/or drug transporters such as P-glycoprotein (P-gp) are the most significant factors involved in pharmacokinetic interactions [12]. Herbal constituents often modulate gastrointestinal pH and motility, which may lead to interactions. Phytoconstituents have the potential to alter renal elimination of other drugs and thus may lead to reduced elimination of drugs [10,13].

2.1 Metabolism of Herbal Drugs by Intestinal Microflora

The intestinal microbiota plays a critical role not only in human health and disease but also in the pharmacokinetics of xenobiotics. Various enzymes produced by intestinal microflora such as β-glucuronidase, sulfatase, and glycosidases could affect the absorption and metabolism of xenobiotics in the intestine [14,15]. Xenobiotic metabolism by intestinal microbiota could alter the pharmacodynamic as well as the pharmacokinetic profile of administered drugs. The variation of gut microbiota also contributes to the subjective differences toward drug therapy and can result in drug-induced toxicity or lack of efficacy. Glycosidic compounds are highly metabolized by those enzymes before absorption from the gastrointestinal tract to the blood. This leads to changes in pharmacokinetics, pharmacodynamics, and toxicity of glycosidic compounds [16]. Baicalin, a main constituent of the rhizome of *Scutellaria baicalensis*, is metabolized to baicalein and oroxylin A in the intestine before its absorption [17]. Baicalin is metabolized to baicalein by β-glucuronidase in the intestine. Ginsenoside Rb1 is metabolized by β-D-glucosidase produced by *Bifidobacterium longum* to a pharmacologically active compound K [18]. α-L-Rhamnosidase produced by *Bifidobacterium dentium* was found to hydrolyze phytochemicals such as rutin, poncirin, naringin, and ginsenoside Re [19]. What is very important is that individual changes in the intestinal flora are caused by the use of many commonly used medicaments such as antibiotics. This may affect the effectiveness of other coadministered therapeutic agents.

2.2 Hepatic Metabolism of Herbal Medicines

The elimination of xenobiotics from the body is an essential defensive process that aims to protect against potential toxicity from the foods we eat. The food broken down in the stomach is absorbed by the small intestine and is then diverted to the liver by the portal circulatory system. The metabolism takes place in a biphasic manner involving phase I and phase II metabolism. Phase I metabolism results in chemical changes that make a compound more hydrophilic, so it can be easily eliminated by the kidneys. These reactions usually involve either adding or unmasking a hydroxyl group, or some other hydrophilic group such as an amine or sulfydryl group, and usually involve hydrolysis, oxidation, or reduction mechanisms. CYP enzymes are the most important enzymes involved in phase I metabolism, and are primarily responsible for the metabolism of most of the drugs. Phase II metabolism takes place if phase I is insufficient to clear a compound from circulation, or if phase I generates a reactive metabolite. These reactions usually involve adding a large polar group (conjugation reaction), such as glucuronide, to further increase the compound's aqueous solubility. Often, the functional groups generated in phase I reactions are required for attachment of the phase II polar groups. As such, the metabolism and clearance of herbal medicines may be altered when the enzymes that metabolize them are modulated by coadministered drugs or herbal medicines, thus leading to HDIs [20,21].

2.3 Phase I: Metabolism by the Cytochrome System

The CYP superfamily is the largest group of enzymes involved in xenobiotic metabolism. They carry out oxidative, peroxidative, and reductive biotransformation of xenobiotics and endogenous compounds. These enzymes are found in the endoplasmic reticulum of hepatocytes and mucosal cells of the gastrointestinal tract. This superfamily is conventionally divided into families and subfamilies based on nucleotide sequence homology [22,23]. There is a high degree of substrate specificity among the various families. CYP belonging to families 1, 2, and 3 are principally involved in xenobiotic metabolism, while others play a major role in the metabolism of endogenous compounds such as hormones, bile acids, and fatty acids [24]. The most important CYP subfamilies responsible for drug metabolism in humans are 1A2, 2A6, 2C9, 2C19, 2D6, 2E1, 3A4, and 3A5 [25,26]. CYP1A1 and 1A2 are the two major members of the human CYP1A subfamily. CYP1A1 is mainly expressed in extrahepatic tissues such as the kidneys, the intestines, and the lungs, while CYP1A2 constitutes about 15% of total hepatic CYP. CYP2B6 is the major enzyme from the 2B family, which is involved in drug metabolism [27]. The subfamily 2C is the second most abundant CYP after 3A, representing over 20% of the total CYP in the human liver. 2C8, 2C9, 2C19, and 2E1 are the active members of this family. The CYP3A subfamily constitutes over 40% of the total CYP in the human body with CYP3A4 being the most abundant of all isoforms. It is highly expressed in the liver and the intestines and participates in the metabolism of most of the drugs in use today [28–30]. Many phytochemicals are also metabolized by CYP enzymes. Magnolin is a natural compound abundantly found in *Magnolia flos*, which has been traditionally used in traditional medicine as an antiinflammatory medicament. Magnolin is metabolized by CYP and the metabolites produced are *O*-desmethyl magnolin, didesmethylmagnolin, and hydroxymagnolin. CYP2C8,

CYP2C9, CYP2C19, and CYP3A4 are responsible for the formation of the two *O*-desmethyl magnolins, whereas the formation of hydroxymagnolin involves CYP2C8 [31].

2.4 Induction and Inhibition of Metabolic Enzymes

2.4.1 Phase I

Induction is the increase in intestinal and hepatic enzyme activity that leads to a corresponding increase in the rate of drug metabolism affecting both oral bioavailability and systemic disposition. The formulation and dosage design of medications by the oral route involves an adjustment for this presystemic metabolism to achieve predictable levels of the drug in systemic circulation. Naturally occurring therapeutic agents can modulate hepatic and intestinal CYP activity. There are basically three types of mechanisms by which herbal drugs or constituents interact with CYP enzymes. An herbal constituent can be an inducer of one or several CYP isoforms, thereby lowering plasma concentrations due to higher metabolism. Such interactions may produce subtherapeutic plasma drug concentrations. A compound can also be an inhibitor of CYP enzymes resulting in reduced metabolic activity. This would lead to higher concentrations of the coadministered drugs. An herbal component can be a substrate of one or several isoforms of CYP enzymes, which leads to competition between it and another coadministered substrate for metabolism by the same CYP isozyme resulting in higher plasma concentrations due to competitive inhibition [10,32].

2.4.2 Phase II Reactions

Phase II biotransformation reactions are also called conjugation reactions. They generally serve as a detoxifying step in drug metabolism. Phase II metabolic enzymes include uridine diphosphoglucuronosyl transferase (UGT), *N*-acetyl transferase, glutathione *S*-transferase, sulfotransferase, and methyl transferase [33]. They catalyze the attachment of polar and ionizable groups to phase I metabolites, which increases their renal elimination. While there has been a very high degree of scientific enterprise on evaluating and predicting CYP-mediated HDIs, there is sufficient evidence to suggest that the effects of herbal extracts on phase II enzymes cannot be overlooked.

UGTs include 18 different endoplasmic reticulum-bound enzymes that catalyze glucuronidation, which accounts for approximately 35% of all drugs metabolized via phase II metabolism [33]. UGTs are divided into two families, UGT1 and UGT2, and three subfamilies, UGT1A (1A1, 1A3, 1A4, 1A5, 1A6, 1A7, 1A8, 1A9, and 1A10), 2A (2A1 and 2A2), and 2B (2B4, 2B7, 2B10, 2B11, 2B15, 2B17, and 2B28), based on sequence homology. UGT enzymes are widely and differentially expressed throughout the human body. Although the majority of UGT enzymes are expressed in the liver, UGT1A7, 1A8, and

1A10 are expressed in the intestine. Many phytochemicals are metabolized by UGTs [34,35]. UGT1A1 is involved in the glucuronidation of many phytochemicals, including anthraquinones, coumarins, flavonoids, and tea catechins.

Eupatilin, a pharmacologically active flavone derived from *Artemisia* plants, is extensively metabolized to eupatilin glucuronide, 4-*O*-desmethyleupatilin, and 4-*O*-desmethyleupatilin glucuronide in human liver microsomes. *O*-Demethylation of eupatilin is mediated by CYP1A2 (a major action) and CYP2C8 (a minor action) and leads to the formation of 4-*O*-desmethyleupatilin, which is further glucuronidated by UGT1A1, UGT1A3, UGT1A7, UGT1A8, UGT1A9, and UGT1A10 [36,37]. In a similar study, metabolism of jatrorrhizine (an alkaloid from *Coptis chinensis*) was studied in human liver microsomes. Demethyleneberberine (a demethylated product) and jatrorrhizine glucuronide were identified as the phase I and phase II metabolites, respectively. Demethylation was catalyzed by CYP1A2, whereas glucuronidation was catalyzed by UGT1A1, UGT1A3, UGT1A7, UGT1A8, UGT1A9, and UGT1A10 [38,39]. Demethylation was inhibited by furafylline (a methylxanthine derivative), which is a potent and selective inhibitor of CYPIA2 [39].

2.5 Efflux of Drugs Through Efflux Transporters

Efflux transporters are proteinaceous transporters localized in the cell membrane of all kinds of cells. Efflux systems function via an energy-dependent mechanism (active transport) to pump out unwanted toxic substances through specific efflux pumps and are considered to be a vital part of xenobiotic metabolism. These transporters are important determinants in the absorption, distribution, and elimination of drugs. Some efflux systems are drug specific, whereas others may accommodate multiple drugs and thus contribute to multidrug resistance (MDR). P-gp, a member of the ATP-binding cassette (ABC) transporter superfamily, is a well-known drug transporter that is highly expressed in the apical membrane of several pharmacologically important epithelial barriers such as the kidneys, liver, intestines, and blood–brain barrier [40]. ABC transporters are efflux transporters and they use energy derived from ATP hydrolysis to mediate the primary active export of drugs from the intracellular to the extracellular compartment against a concentration gradient. P-gp is a 170 kDa phosphorylated glycoprotein encoded by human MDR1 gene and has very extensive substrate specificity. It recognizes compounds ranging from small molecules of 350 Da up to polypeptides of 4000 Da [41]. Two other forms of ABC transporters participate in the MDR of tumors. These are MDR protein 1 (MRP1, ABCC1) and breast cancer resistance protein (BCRP or ABCG2). The interplay between ABC transporters and drug-metabolizing enzymes makes the intestinal and hepatic barrier to absorption of xenobiotics even more effective. A very significant perspective in this context is that on many occasions P-gp and CYP3A4 share the same substrates and thus appear to be functionally related [42,43].

Numerous studies have reported that the modulation of P-gp can affect the oral bioavailability, biliary or renal clearance, and brain uptake of drugs. In addition, P-gp serves as a key factor in conferring the MDR phenotype to cancer cells. Thus the modulation of P-gp by herbal agents may result in significant changes in the pharmacokinetics of prescription drugs and increase the risk of HDIs, and its importance is increasingly being identified [44]. P-gp has been reported to affect the pharmacokinetics of numerous structurally and pharmacologically diverse substrate drugs. The genetic variability in the MDR1 gene further influences absorption and tissue distribution of the drugs transported. Inhibition or induction of P-gp by coadministered drugs or food as well as herbal constituents may result in pharmacokinetic interactions leading to unexpected toxicities or treatment failures [42,45].

Like CYPs, P-gp is vulnerable to inhibition, activation, or induction by herbal constituents. Curcumin, ginsenosides, piperine, some catechins from green tea, and silymarin from milk thistle were found to be inhibitors of P-gp, while some catechins from green tea increased P-gp-mediated drug transport by a heterotropic allosteric mechanism, and St. John's wort (SJW) induced the intestinal expression of P-gp in vitro and in vivo [26]. Some components (e.g., bergamottin and quercetin) from grapefruit juice were reported to modulate P-gp activity [22]. Studies suggest that many of these herbal constituents, in particular flavonoids, modulated P-gp by directly interacting with the ATP-binding site or the substrate-binding site. Some herbal constituents (e.g., hyperforin and kava) were shown to activate the pregnane X receptor, an orphan nuclear receptor acting as a key regulator of MDR1 and many other genes [45]. The modulation of P-gp activity and expression by these herb constituents may result in altered absorption and bioavailability of drugs that are P-gp substrates. This is exemplified by increased oral bioavailability of phenytoin and rifampin by piperine and decreased bioavailability of indinavir, tacrolimus, cyclosporine, digoxin, and fexofenadine by coadministered SJW [46].

2.6 Organic Anion-Transporting Polypeptide

The solute carrier superfamily (SLC) includes hundreds of proteins involved in the transmembrane transport of small molecules or solutes of various degrees of hydrophilicity and lipophilicity. Among the SLC superfamily members, organic anion transporter proteins (OATPs) are the primary influx transporters that play an important role in transporting endo- as well as xenobiotics (including numerous drugs) across plasma membranes. Of the 11 human OATPs, OATP1B1, OATP1B3, and OATP2B1 are expressed on the sinusoidal membrane of hepatocytes and can facilitate liver uptake of their substrate drugs. Among the well-characterized substrates are numerous drugs, including statins, angiotensin-converting enzyme inhibitors, angiotensin receptor blockers, antibiotics, antihistaminics, antihypertensives, and anticancer drugs

[47,48]. The OATP1B1 transporter exhibits broad substrate selectivity that includes anionic (e.g., statins such as pravastatin, pitavastatin, and rosuvastatin), zwitterionic (e.g., rifampicin), and neutral lipophilic (e.g., paclitaxel) drugs [48,49]. OATP1B1 also transports endogenous substances such as bile acids, thyroid hormones, steroid sulfates, glucuronide conjugates, and peptides. Downregulation of OATP expression by a drug may lead to pharmacokinetic alterations and an increase in the chances of drug–drug interactions. Cyclosporin, rifampicin, and some HIV protease inhibitors (lopinavir, ritonavir) are OATP1B1 inhibitors and distinctly increase the plasma concentrations of OATP1B1 substrates [50]. Cyclosporine has been reported to significantly increase the levels of statins [51].

Herbal medicines such as hawthorn and salvia are frequently combined with statins in the Chinese system of medicine. Ursolic acid present in these medicines presents a probability of HDIs as it is an inhibitor of OATP1B1. In a study using HEK 293T cells, ursolic acid was found to significantly inhibit the uptake of rosuvastatin in HEK 293T cells. This study suggests that ursolic acid can affect the uptake of rosuvastatin in hepatocytes by inhibiting the transport of OATP1B1 [52]. DA-9801 (*Dioscorea* extract) is a new standardized extract currently being evaluated for diabetic peripheral neuropathy in a phase II clinical study. DA-9801 inhibited the in vitro transport activities of OCT1, OCT2, OAT3, and OATP1B1 [50,53]. Aqueous decoction of *Hypoxis hemerocallidea* (Hypoxidaceae) is widely consumed in Southern Africa by people suffering from AIDS. The effect of *H. hemerocallidea* on OATP1B1 and OATP1B3 was evaluated using recombinant HEK293 cells overexpressing OATP1B1 and OATP1B3. The inhibition of the uptake activity of OATP1B1 and OATP1B3 was also observed with IC_{50} values of 93.4 and 244.8 μg/mL, respectively [54]. Aristolochic acids, rhein, and gallic acid are phytoconstituents that are found to inhibit OATP1 in in vitro systems [55,56].

3. PHARMACODYNAMIC INTERACTIONS

HDIs can occur through the synergistic or additive actions of herbal products with conventional medications as a result of affinities for common receptor sites. This can precipitate pharmacodynamic toxicity or antagonistic effects. Like most other herbs, SJW contains a complex mixture of phytochemicals, including phenylpropanes, naphthodianthrones, acylphloroglucinols, flavonoids, flavanol glycosides, and biflavones. Hyperforin is known to inhibit the reuptake of neurotransmitters (dopamine, serotonin, noradrenalin) and is believed to be the bioactive responsible for the antidepressant activity of SJW [57,58]. A pharmacodynamic interaction between cranberry and warfarin has been elucidated. Warfarin acts as an anticoagulant by opposing the procoagulant effect of vitamin K by inhibiting the vitamin K epoxide reductase complex (VKORC) enzyme. This enzymatic protein complex is responsible for reducing vitamin K 2,3-epoxide to its active form, which is important for effective clotting. Cranberry has been shown to inhibit VKORC, which could act as an additive factor to the pharmacological action of warfarin. Cranberry can thus alter the pharmacodynamics of warfarin with the potential to increase its effects significantly [59]. Drugs such as quinidine, verapamil, and captopril reduce the plasma protein binding and renal clearance of cardiac glycosides such as digoxin, which leads to plasma accumulation of digoxin leading to toxicity. Quinidine, an alkaloid, used in the treatment of arrhythmia, leads to a marked fall in blood pressure when used in patients on vasodilator therapy. It also produces very severe cardiac depression when used along with verapamil and β-blockers [8].

Allium sativum (garlic) has been used as a food and remedy of various ailments for centuries. In the form of oils, liniments, poultices, and powders, garlic has been extensively used in the Indo-Pak subcontinent, in indigenous systems of medicine for the treatment of hypertension, atherosclerosis, rheumatism, bronchial asthma, and as a spasmolytic, vermifuge, and antiseptic. Several epidemiological studies suggest an antihypertensive effect of garlic and of many of its bioactive components [60,61]. The primary bioactive constituents are *S*-allyl cysteine and allicin. The exact mode of action is not known, but there are studies that indicate that the blood pressure-lowering action could be due to the inhibition of angiotensin and nitric oxide expression. This is of particular importance if there is consistently high consumption of garlic or its dosage forms in hypertensive patients who are on a maintenance therapy consisting of angiotensin converting enzyme (ACE) inhibitors. ACE inhibitors are very widely used drugs in the long-term maintenance of hypertension and of late have assumed first choice drug status in the treatment of hypertension. Thus simultaneous use of garlic and ACE inhibitors could exert superadditive interaction with respect to a fall in blood pressure and ACE inhibition, indicating that hypotension could be a potential adverse reaction [62,63].

Ginseng has been used for a long time for its immunomodulatory benefits. Many findings suggest that ginseng displays beneficial effects in the treatment of diabetes, which could be due to stimulation of insulin release in a glucose-independent manner. Ginseng extracts were able to enhance ATP production and in turn increase insulin production, as insulin deficiency is often linked to a lack of ATP produced [64]. Along with an increase in ATP production, ginseng reduced mitochondrial protein UCP-2, which negatively regulates insulin secretion [65]. Rh2, a ginsenoside present in ginseng, was found to lower plasma glucose. This effect is due to an increase in β-endorphin secretion that activates opioid μ-receptors thereby resulting in an increased expression of GLUT4, a glucose transporter in fat and muscle tissue. Protopanaxatriol, a ginsenoside metabolite, also increased GLUT4 and improved insulin resistance [66]. The upregulation of GLUT4 signals that ginseng has an effect on fat/muscle tissue, possibly decreasing insulin resistance. Thus ginseng could act to reduce plasma glucose levels by multiple mechanisms.

Thus there is the possibility of synergistic effects when used along with insulin or oral hypoglycemic (particularly sulfonylureas) drugs. This synergistic effect could lead to acute hypoglycemia, leading to hypoglycemic episodes.

4. SELECTED CLINICAL HERB–DRUG INTERACTIONS

4.1 Grapefruit Juice

One of the earliest reported clinically relevant drug interactions was observed with grapefruit juice. Simultaneous consumption of grapefruit juice with a number of therapeutic agents resulted in higher plasma levels with subsequent adverse effects [67,68]. Grapefruit juice acts through inhibition of intestinal CYP3A4, which leads to inhibition of metabolism of coadministered therapeutic substances and consequently to toxicity. This was observed with a coadministered felodipine (calcium channel blocker). Felodipine is a widely used drug in the treatment of hypertension, ischemic heart disease, and congestive heart failure. Repeated consumption of grapefruit juice can result in a cumulative increase in the area under the curve (AUC) and C_{max} of felodipine [69]. Similar interactions for grapefruit juice have been reported for terfenadine, saquinavir, cyclosporin, midazolam, triazolam, verapamil, lovastatin, cisapride, and astemizole [70]. Grapefruit juice is rich in a number of phytochemicals, including flavonoids and furanocoumarins. There are numerous in vitro reports that suggest that the major phytoconstituents of grapefruit juice are inhibitors of CYP3A4. The most abundant flavonoid in the juice is naringin. The furanocoumarins are a structurally distinct class of compounds found in grapefruit juice. In vitro experiments have confirmed that furanocoumarins from grapefruit juice are competitive inhibitors of CYP3A4. Recent evidence indicates that furanocoumarins related to bergamottin [5-[(3′,7′-dimethyl-2′,6′-octadienyl)oxy]psoralen] and 6′,7′-dihydroxybergamottin are primarily responsible for CYP3A4 inhibition [71–73]. Grapefruit juice is also an inhibitor of P-gp, and thus could lead to accumulation of P-gp substrates. In addition, grapefruit juice has also been shown to be a potent in vitro inhibitor of OATP 1A2 [74].

4.2 St. John's Wort

SJW (*Hypericum perforatum*) is a widely used herbal medicine mainly used to relieve conditions such as depression, anxiety, and inflammatory skin disorders. The active phytoconstituents present are hypericin, quercetin, isoquercetin, biflavonoids, hyperforin, procyanidin, catechin, tannin, and chlorogenic acid. Among these, hypericin is the major antidepressant component and it inhibits the reuptake of neurotransmitters such as 5-hydroxytryptamine, norepinephrine, and dopamine [75–79]. In vivo and in vitro studies in both animals and humans have shown that SJW is a strong inducer of CYP3A4. Due to its inducing effects on CYP3A4, it significantly reduces the plasma levels of CYP3A4 substrates. A number of clinically significant interactions of SJW have been identified with prescription drugs, which include anticancer agents (e.g., imatinib and irinotecan), anti-HIV agents (e.g., indinavir, lamivudine, and nevirapine), antiinflammatory agents (e.g., ibuprofen and fexofenadine), antimicrobial agents (e.g., erythromycin and voriconazole), cardiovascular drugs (e.g., digoxin, ivabradine, warfarin, verapamil, nifedipine, and talinolol), central nervous system agents (e.g., amitriptyline, buspirone, phenytoin, methadone, midazolam, alprazolam, and sertraline), hypoglycemic agents (e.g., first- and second-generation sulfonylureas such as tolbutamide and gliclazide, respectively), immunomodulatory agents (e.g., cyclosporine and tacrolimus), oral contraceptives, proton pump inhibitors (e.g., omeprazole), bronchodilators (e.g., theophylline), and statins (e.g., atorvastatin and pravastatin) [46,80–83].

The group of drugs with the highest potential for clinically significant pharmacokinetic drug interaction with SJW is antidepressants because SJW itself is consumed by patients with depression. Amitriptyline, a widely used antidepressant, is a substrate to CYP3A4. The risk of therapeutic failure is thus high due to induction of CYP3A4-dependent metabolism activities resulting in poor oral bioavailability. In a clinical study, a 22% and 41% decrease in the AUC of amitriptyline and nortriptyline (an active metabolite of amitriptyline), respectively, was observed in 12 depressed patients who were concomitantly administered with extracts of SJW and amitriptyline for 2 weeks [84]. It has been shown to reduce the efficacy of oral contraceptives [85]. In a clinical study the effect of SJW on the pharmacokinetics and therapeutic efficacy of an oral contraceptive pill (containing ethinyl estradiol and norethindrone) has been evaluated. Twelve healthy premenopausal women using oral contraception were involved in the study that was conducted over a period of three consecutive 28-day menstrual cycles. During the second and third cycles, the participants received 300 mg SJW three times a day. The results indicate that SJW decreased the therapeutic efficiency of the contraceptive combination. This is due to induction of ethinyl estradiol-norethindrone metabolism resulting in increased CYP3A activity [86]. Cyclosporine is an immunosuppressant that is widely used in organ transplantation procedures. Due to its immunosuppressant activities, it decreases graft rejections and increases the chances of organ transplantation. Cyclosporine is metabolized by CYP3A4. Thus SJW interacts with cyclosporine, causing a decrease of cyclosporine blood levels, and leads to transplant rejection as reported by numerous clinical studies [87–89]. Similar activities are observed with tacrolimus, another commonly used immunosuppressant. Increase in metabolism of tacrolimus is most likely through induction of CYP3A4 [90].

A clinical study investigated the effect of SJW on the pharmacokinetics and pharmacodynamics of warfarin. The study was an open-label, three-way crossover randomized study involving 12 healthy male subjects, who received warfarin after 14 days' pretreatment with SJW. Dosing with SJW was continued for 7 days after administration of the warfarin dose. SJW significantly induced the clearance of warfarin, leading to a significant reduction in the pharmacological effect of warfarin [91]. In a different open-label clinical study involving antiretroviral drugs, SJW significantly reduced the AUC of the HIV-1 protease inhibitor indinavir (approximately 57%) in healthy volunteers, and could lead to the development of drug resistance and treatment failure [92]. Imatinib is a potent inhibitor of the Bcr-Abl and c-kit tyrosine kinases and is approved for the treatment of chronic myelogenous leukemia. An open-label, fixed-sequence study involving 12 healthy subjects was carried out to evaluate the effect of SJW administration on imatinib. SJW administration increased imatinib clearance by 43%. Also, $T_{1/2}$ and C_{max} of imatinib were significantly decreased [93]. The results of an unblinded, randomized crossover study suggest that SJW inhibits the production of SN-38, an active metabolite of irinotecan, in cancer patients, thereby reducing its clinical efficiency [94].

SJW is a very potent inducer of P-gp. Therefore concomitant use of SJW with P-gp substrates always has the tendency to lower the concentration of the substrate drug and reduce its therapeutic effect. SJW extracts containing increased concentration of hyperforin were found to induce MDR1 mRNA expression to greater extents [95]. The effect of SJW on the metabolism of fexofenadine was studied in an isolated perfused liver. It was found that SJW administration was responsible for increased elimination of fexofenadine (approximately 71%). This effect could be mediated by increased P-gp levels [96]. In a clinical study, the administration of SJW extract to eight healthy male volunteers during 14 days resulted in an 18% decrease in digoxin concentration after a single 0.5 mg dose of digoxin. This was due to increased expressions (1.4 times) of duodenal P-gp [97].

4.3 *Ginkgo biloba*

Ginkgo biloba leaf extracts (GLEs) are popular herbal remedies for the treatment of Alzheimer's disease, tinnitus, vertigo, and peripheral arterial disease. As GLEs are taken regularly by older people who are likely also to use multiple other drugs for the treatment of diseases such as hypertension, diabetes, rheumatism, or heart failure, potential HDIs are of interest [98]. Ginkgo extracts contain flavonoid glycosides (quercetin, kaempferol, isorhamnetin) and terpene lactones (ginkgolides A, B, C, J, and bilobalides) [99]. Oral standardized dry extracts of *G. biloba* usually contain between 22% and 27% flavone glycosides and 5%–7% terpene lactones. Many studies have been conducted using EGb 761, a well-defined extract of *G. biloba*. *G. biloba* exerts a significant inductive effect on CYP2C19 activity and there is an increased probability of interactions with CYP2C19 substrates. *G. biloba* has been reported to induce CYP2C19-dependent omeprazole metabolism in healthy human subjects. A study investigated the potential HDI between *G. biloba* and omeprazole, a widely used CYP2C19 substrate. Plasma concentrations of omeprazole and omeprazole sulfone were significantly decreased and 5-hydroxyomeprazole significantly increased following *G. biloba* administration [100].

Clopidogrel is a platelet aggregation inhibitor and is used in a prophylactic manner to reduce the risk of thrombus formation in patients who have already had a heart attack or stroke. *G. biloba* extract (GBE) is often used with clopidogrel for the treatment of cerebrovascular diseases. One study explored the effect of GBE on the metabolism and pharmacokinetics of clopidogrel. The in vitro study using rat liver microsomes revealed that GBE significantly induced the conversion of clopidogrel into its active metabolite. The effect of GBE on the pharmacokinetics of clopidogrel was also investigated in vivo. Administration of 4, 20, and 100 mg/kg of GBE significantly decreased the C_{max} and AUC of clopidogrel in a dose-dependent manner [101]. An open-label study evaluated the effect of 2 weeks of standardized GBE administration on the levels of midazolam in 14 healthy volunteers. Single oral doses of midazolam were administered prior to and after 4 weeks of GBE to assess the influence of GBE on CYP3A. GBE decreased midazolam AUC and C_{max} by 34% and 31%, respectively. The results suggest that GBE induces CYP3A metabolism, which is seen with the lowering of midazolam concentrations [102].

A study attempted to investigate the effect of GBE on the pharmacokinetics of theophylline, a CYP1A2 substrate and an important therapeutic agent with a narrow therapeutic window used for the treatment of asthma. Commercial GBE was given to rats for five consecutive days and on the 6th day theophylline was administered. The results showed that pretreatment of rats with GBE resulted in a dose-dependent increase in the total clearance of theophylline. AUC of theophylline was reduced by 40% following pretreatment with GBE. These results demonstrate that GBE pretreatment increased CYP1A2 metabolic activity and the clearance of theophylline in rats [103].

GBE has been shown to attenuate the hypoglycemic action of tolbutamide by induction of CYP2C9 in rats [104]. A study was undertaken to examine the influence of repeated oral administration of GBE on CYP2C9. The CYP2C9 substrate tolbutamide, 125 mg, was orally administered to 10 male healthy volunteers before and after GBE intake (360 mg/day) for 28 days. GBE intake resulted in lower AUC for tolbutamide. Concomitantly, GBE reduced the blood glucose-lowering effect of tolbutamide [105]. Ginkgo is an inducer of P-gp. Regular intake of ginkgo has been shown to increase the plasma levels of talinolol, a substrate of P-gp. Repeated ingestion of GBE increased the talinolol maximum plasma concentration C_{max} by 36% and the AUC by 26% [106,107].

4.4 Garlic

Garlic phytochemicals and garlic supplements influence the pharmacokinetic and pharmacodynamic behavior of concomitantly ingested drugs. Phytochemicals derived from aged garlic extract modify the activities of transporters in both the intestines and liver and competitively inhibit CYP3A4 enzyme. A study evaluated the effect of garlic supplements on the pharmacokinetics of saquinavir in which 10 healthy volunteers received 10 doses of saquinavir at a dosage of 1200 mg three times daily and 41 doses of garlic taken two times for 20 days. In the presence of garlic, the AUC decreased by 51% and the C_{max} decreased by 54%. This could be attributed to CYP3A4 induction by garlic [108]. Ingestion of garlic extract for 21 days in healthy human volunteers increased expression of duodenal P-gp to 131% [109]. An in vitro study that aimed to characterize the effects of garlic extract on CYP2C9 activity using immortalized human hepatocytes (Fa2N-4 cells) validated the effect of garlic on liver microsomal enzymes. Exposure to increasing concentrations of garlic extract led to progressive reduction in Fa2N-4 CYP2C9 activity. CYP2C9 mRNA expression also revealed a concentration-dependent reduction. Greater than 90% reduction in CYP2C9 activity was observed following 4 days of exposure to 50 μg/mL garlic extract [110]. In vitro permeability studies through rat jejunum and Caco-2 cell monolayers studied the impact of first-pass intestinal metabolism and therapeutic efficacy of antiretrovirals in the presence of garlic supplements. Saquinavir and darunavir efflux from enterocytes into gastrointestinal lumen significantly increased in the presence of aged garlic extract, whereas their CYP3A4 metabolism was inhibited [111].

4.5 Berberine

Berberine is an isoquinoline alkaloid mainly isolated from plants of the Berberidaceae family; it is extensively used to treat gastrointestinal infections. Berberine has a variety of pharmacological functions and has been widely used in clinical conditions. The metabolic activities of berberine were studied with specific emphasis on CYP1A2, CYP2D6, and CYP3A4 in an in vitro study. Berberine showed a mechanism-based inhibition of CYP3A4 and CYP2D6 along with nonmechanistic inhibition of CYP1A2. Berberine demonstrated both reversible and irreversible inhibition of CYP2D6 and CYP3A4 [112]. In an in vitro study the metabolism of berberine and its effect on the main metabolic enzymes in HepG2 cells was investigated. Berberine significantly increased the metabolism of midazolam, phenacetin, and tolbutamide by inducing the CYP1A2 and 3A4 enzyme in a dose-dependent manner and the simultaneous expression of CYP1A2 and 3A4 was increased by berberine at 1000 ng/mL [113]. The interaction of lovastatin and berberine was studied in an in vivo and in vitro study using pharmacokinetic and hepg2 cell metabolism studies. The results of the in vivo study indicate that the AUC and C_{max} are significantly decreased in the berberine-treated group, and in the HepG2 cells the metabolic activity was increased by berberine. It appears from the results that berberine could be an inducer for the metabolism of lovastatin, which could be attributed to CYP3A4 stimulation by berberine [114]. Losartan and berberine are often simultaneously used for the treatment of senile diabetic nephropathy and hence the pharmacokinetic interaction between them is very important. In a different study the pharmacokinetic profiles of losartan and EXP3174 (an active metabolite of losartan) with berberine (20 mg/kg) were determined in animal models. The C_{max} and AUC of losartan were significantly increased by berberine, while the C_{max} and T_{max} of EXP3174 were significantly decreased compared to the control. These effects could be due to modulation of CYP3A4 or CYP2C9 by berberine [115].

It has been reported that berberine can block the human ether-a-go-go-related gene (hERG) potassium channel and inhibit its membrane expression. The hERG channel plays a crucial role in cardiac repolarization and is the target of diverse proarrhythmic drugs [116]. Dysfunction of the hERG channel can cause long QT syndrome. Inhibition of the hERG channel could be due to decreased caveolin-1 expression by berberine [117,118]. Berberine is widely used as an antibacterial agent and often combined with macrolides to treat gastropathy. A study was conducted to assess cardiac toxicity during the combined use of berberine with macrolides. Azithromycin reduced hERG currents by accelerated channel inactivation and thus the combination of berberine with azithromycin reduced hERG currents and produced an inhibitive effect much stronger than use of a single drug alone. The combined administration of berberine with clarithromycin also had a powerful inhibitive effect on CYP3A activities than the use of a single drug alone. Collectively, these results demonstrated that concomitant use of berberine with macrolides may require close monitoring because of potential drug toxicities, especially cardiac arrhythmias [119].

4.6 Licorice

Licorice is one of the most frequently used herbal medicines worldwide, along with being a significant food additive. Many studies have claimed that licorice and its important bioactive substances tend to promote HDI by their action on drug-metabolizing enzymes. A study evaluated the effect of licorice on the pharmacokinetics of methotrexate. The results revealed that the AUC and mean residence time of methotrexate were significantly increased by licorice [120]. A study investigated the effects of licorice and its major ingredient glycyrrhizin on the pharmacokinetics of cyclosporine in rats. The results indicated that licorice and glycyrrhizin significantly decreased the peak blood concentration and the AUC of cyclosporine in rats. Mechanism studies revealed that glycyrrhetic acid, the major metabolite of glycyrrhizin, significantly activated the functions of P-gp and CYP3A4. In conclusion, licorice significantly reduced the oral bioavailability of cyclosporine through activating P-gp and CYP3A4 [121]. Licorice increases cortisol availability in patients with Addison's disease, as the median AUC for serum cortisol increased with licorice treatment [122].

The potent antioxidants present in licorice root extract and glabridin, an isoflavan purified from licorice root extract, were tested for their ability to modulate the activities of several cytochrome enzymes. CYP3A4 was inactivated by licorice root extract and by glabridin in a time- and concentration-dependent manner. The inactivation was NADPH dependent and was not reversible. Similarly, CYP2B6 was also inactivated by glabridin in a time- and concentration-dependent manner [123].

5. APPROACHES TO IDENTIFY HERB–DRUG INTERACTIONS

As the use of herbal medicine is progressively increasing, development of models that can help to identify and predict such types of drug interactions are the need of the hour to avoid or minimize such unfavorable interactions. Various in vitro and in vivo models have been used to identify herbal medicines and prescription drugs that have the potential to interact with each other. The jury is still out on the accuracy of the results, but there is the opinion that these results can be beneficial in clinical practice since warnings and proper professional advice may be elicited and shared with patients to prevent adverse outcomes [124]. The various methods and studies that are being used are structure–activity relationships, pharmacophore modeling, molecular docking, homology modeling, and quantitative structure–activity relationships. These help in understanding the effects of the interaction of CYPs, phase II enzymes, and P-gps with herbal medicine. A better understanding of the effect of phytochemicals on the xenobiotic metabolic system would lead to accumulation of knowledge that would be beneficial in avoiding dangerous interactions [23,125].

Also multiple in vitro systems involving isolated/cultured cell lines, subcellular fractions (liver microsomes, cytosols, and homogenates), and cDNA-expressed enzymes have been used to explore the potential for HDIs. The effect on transporter molecules is carried out by using models that employ Caco-2, MDCKII cells, oocytes, membrane vesicles, and cDNA-expressed drug transporters [9]. Combination of those methods will provide very valuable insights into the mechanisms and potential of HDIs. Another strategy for identifying HDIs involves high-throughput screening (HTS) assays. HTS assays enable the testing of large numbers of chemical substances for activity in diverse areas of biology [126,127]. It is a drug-discovery process widely used in the pharmaceutical industry. It uses automation technology to quickly assay the biological or biochemical activity of a large number of potential drug candidates. HTS is of great utility in identifying ligands or substrates for receptors, enzymes, ion channels, or any other specific pharmacological target.

It is being used to generate data regarding in vitro inhibition or induction that can be used to predict HDIs and guide clinical studies [126].

For example, a rapid and comprehensive mass spectrometry-based in vitro high-throughput P450 cocktail inhibition assay has been developed that uses 10 substrates simultaneously against nine CYP isoforms. This cocktail assay simultaneously evaluates many P450 enzymes and includes probe substrates for CYP1A2, CYP2A6, CYP2B6, CYP2C8, CYP2C9, CYP2C19, CYP2D6, and CYP2E1 and two probes targeting different binding sites of CYP3A4/5. It is among the fastest methods due to short incubation times and rapid analysis using ultrahigh-pressure liquid chromatography-tandem mass spectrometry [128]. Another rapid simultaneous analytical method uses a cocktail approach for measuring the activities of seven CYP enzymes (CYP1A2, CYP2A6, CYP2C9, CYP2C19, CYP2D6, CYP2E1, and CYP3A4) derived from human liver microsomes and uses gas chromatography-mass spectrometry for analyzing their activities. The method was validated using kinetic studies and inhibition assays of CYP enzymes using known substrates and inhibitors [129].

6. CONCLUSION

The potential for HDIs is a major safety concern and may lead to alterations in pharmacokinetic and/or pharmacodynamic parameters. This issue assumes greater importance for drugs having narrow therapeutic indices, such as as warfarin and digoxin. Changes in plasma concentrations of these and similar drugs could initiate lethal adverse events. The problem with HDIs is that they are multifactorial and depend on numerous variables such as patient-specific characteristics, concurrent prescription drug regimen(s), and the herbal agent. These multiple points of concern make it very important for both patients and clinicians to exercise caution and diligence. Herbal medicine components can alter drug absorption and metabolism by altering the activities of drug-metabolizing enzymes and transporter proteins. Drug development strategies are using methodologies that specifically aim to identify drugs and herbal agents that are substrates for CYP enzymes and/or P-gp. Understanding the effect of herbal medicine and prescription medicine on metabolizing enzymes and transporter proteins is of great value in predicting and avoiding adverse interactions. With an increase in investigations of the interactions between herbal medicines and prescription drugs, very valuable knowledge is being accumulated that will pave the way for rational clinical drug usage, reducing treatment failure, and preventing the occurrence of undesirable effects. Familiarity with this knowledge would be critical in identifying interaction mechanisms between herbal and prescription medicines and application of rational strategies to minimize HDIs.

REFERENCES

[1] Y. Jaiswal, Z. Liang, Z. Zhao, Botanical drugs in Ayurveda and Traditional Chinese Medicine, J. Ethnopharmacol. 194 (2016) 245–259.

[2] E. van Galen, Traditional herbal medicines worldwide, from reappraisal to assessment in Europe, J. Ethnopharmacol. 158 (Pt B) (2014) 498–502.

[3] B. Patwardhan, R.A. Mashelkar, Traditional medicine-inspired approaches to drug discovery: can Ayurveda show the way forward? Drug Discov. Today 14 (2009) 804–811.

[4] K.K. Bhutani, V.M. Gohil, Natural products drug discovery research in India: status and appraisal, Indian J. Exp. Biol. 48 (2010) 199–207.

[5] D.X. Kong, X.J. Li, H.Y. Zhang, Where is the hope for drug discovery? Let history tell the fut2ure, Drug Discov. Today 14 (2009) 115–119.

[6] J.W. Li, J.C. Vederas, Drug discovery and natural products: end of an era or an endless frontier? Science 325 (2009) 161–165.

[7] J.G. Choi, S.M. Eom, J. Kim, S.H. Kim, E. Huh, H. Kim, Y. Lee, H. Lee, M.S. Oh, A comprehensive review of recent studies on herb-drug interaction: a focus on pharmacodynamic interaction, J. Altern. Complement. Med. 22 (2016) 262–279.

[8] X.W. Chen, E.S. Serag, K.B. Sneed, J. Liang, H. Chew, S.Y. Pan, S.F. Zhou, Clinical herbal interactions with conventional drugs: from molecules to maladies, Curr. Med. Chem. 18 (2011) 4836–4850.

[9] X.W. Chen, K.B. Sneed, S.Y. Pan, C. Cao, J.R. Kanwar, H. Chew, S.F. Zhou, Herb-drug interactions and mechanistic and clinical considerations, Curr. Drug Metab. 13 (2012) 640–651.

[10] Z. Hu, X. Yang, P.C. Ho, S.Y. Chan, P.W. Heng, E. Chan, W. Duan, H.L. Koh, S. Zhou, Herb-drug interactions: a literature review, Drugs 65 (2005) 1239–1282.

[11] B.L. Ma, Y.M. Ma, Pharmacokinetic herb-drug interactions with traditional Chinese medicine: progress, causes of conflicting results and suggestions for future research, Drug Metab. Rev. 48 (2016) 1–26.

[12] J.J. Wu, C.Z. Ai, Y. Liu, Y.Y. Zhang, M. Jiang, X.R. Fan, A.P. Lv, L. Yang, Interactions between phytochemicals from traditional Chinese medicines and human cytochrome P450 enzymes, Curr. Drug Metab. 13 (2012) 599–614.

[13] N.C. Brazier, M.A. Levine, Drug-herb interaction among commonly used conventional medicines: a compendium for health care professionals, Am. J. Ther. 10 (2003) 163–169.

[14] M.J. Kang, H.G. Kim, J.S. Kim, D.G. Oh, Y.J. Um, C.S. Seo, J.W. Han, H.J. Cho, G.H. Kim, T.C. Jeong, H.G. Jeong, The effect of gut microbiota on drug metabolism, Expert Opin. Drug Metab. Toxicol. 9 (2013) 1295–1308.

[15] H.I. Swanson, Drug metabolism by the host and gut microbiota: a partnership or rivalry? Drug Metab. Dispos. 43 (2015) 1499–1504.

[16] M.J. Kang, G.S. Ko, D.G. Oh, J.S. Kim, K. Noh, W. Kang, W.K. Yoon, H.C. Kim, H.G. Jeong, T.C. Jeong, Role of metabolism by intestinal microbiota in pharmacokinetics of oral baicalin, Arch. Pharm. Res. 37 (2014) 371–378.

[17] K. Noh, Y. Kang, M.R. Nepal, K.S. Jeong, D.G. Oh, M.J. Kang, S. Lee, W. Kang, H.G. Jeong, T.C. Jeong, Role of intestinal microbiota in baicalin-induced drug interaction and its pharmacokinetics, Molecules 21 (2016) 337.

[18] I.H. Jung, J.H. Lee, Y.J. Hyun, D.H. Kim, Metabolism of ginsenoside Rb1 by human intestinal microflora and cloning of its metabolizing β-D-glucosidase from *Bifidobacterium longum* H-1, Biol. Pharm. Bull. 5 (2012) 573–581.

[19] S.H. Bang, Y.J. Hyun, J. Shim, S.W. Hong, D.H. Kim, Metabolism of rutin and poncirin by human intestinal microbiota and cloning of their metabolizing α-L-rhamnosidase from *Bifidobacterium dentium*, J. Microbiol. Biotechnol. 25 (2015) 18–25.

[20] H. Remmer, The role of theliver in drug metabolism, Am. J. Med. 49 (1970) 617–629.

[21] R.T. Williams, Hepatic metabolism of drugs, Gut 13 (1972) 579–585.

[22] C.X. Liu, X.L. Yi, D.Y. Si, X.F. Xiao, X. He, Y.Z. Li, Herb-drug interactions involving drug metabolizing enzymes and transporters, Curr. Drug Metab. 12 (2011) 835–849.

[23] D.H. Na, H.Y. Ji, E.J. Park, M.S. Kim, K.H. Liu, H.S. Lee, Evaluation of metabolism-mediated herb-drug interactions, Arch. Pharm. Res. 34 (2011) 1829–1842.

[24] J.F. Wang, K.C. Chou, Molecular modeling of cytochrome P450 and drug metabolism, Curr. Drug Metab. 11 (2010) 342–346.

[25] M. Hiratsuka, In vitro assessment of the allelic variants of cytochrome P450, Drug Metab. Pharmacokinet. 27 (2012) 68–84.

[26] S.F. Zhou, J.P. Liu, B. Chowbay, Polymorphism of human cytochrome P450 enzymes and its clinical impact, Drug Metab. Rev. 41 (2009) 89–295.

[27] B. Wang, J. Wang, S.Q. Huang, H.H. Su, S.F. Zhou, Genetic polymorphism of the human cytochrome P450 2C9 gene and its clinical significance, Curr. Drug Metab. 10 (2009) 781–834.

[28] D. Singh, A. Kashyap, R.V. Pandey, K.S. Saini, Novel advances in cytochrome P450 research, Drug Discov. Today 16 (2011) 793–799.

[29] V.Y. Martiny, M.A. Miteva, Advances in molecular modeling of human cytochrome P450 polymorphism, J. Mol. Biol. 425 (2013) 3978–3992.

[30] D.W. Nebert, D.W. Russell, Clinical importance of the cytochromes P450, Lancet 360 (2002) 1155–1162.

[31] D.K. Kim, K.H. Liu, J.H. Jeong, H.Y. Ji, S.R. Oh, H.K. Lee, H.S. Lee, In vitro metabolism of magnolin and characterization of cytochrome P450 enzymes responsible for its metabolism in human liver microsomes, Xenobiotica 41 (2011) 358–371.

[32] D. Pal, A.K. Mitra, MDR- and CYP3A4-mediated drug-herbal interactions, Life Sci. 78 (2006) 2131–2145.

[33] C. Guillemette, Pharmacogenomics of human UDP-glucuronosyltransferase enzymes, Pharmacogenomics J. 3 (2003) 136–158.

[34] S.M. He, C.G. Li, J.P. Liu, E. Chan, W. Duan, S.F. Zhou, Disposition pathways and pharmacokinetics of herbal medicines in humans, Curr. Med. Chem. 17 (2010) 4072–4113.

[35] S.M. He, E. Chan, S.F. Zhou, ADME properties of herbal medicines in humans: evidence, challenges and strategies, Curr. Pharm. Des. 17 (2011) 357–407.

[36] H.S. Lee, H.Y. Ji, E.J. Park, S.Y. Kim, In vitro metabolism of eupatilin by multiple cytochrome P450 and UDP-glucuronosyltransferase enzymes, Xenobiotica 37 (2007) 803–817.

[37] W.Y. Song, H.Y. Ji, N.I. Baek, T.S. Jeong, H.S. Lee, In vitro metabolism of jaceosidin and characterization of cytochrome P450 and UDP-glucuronosyltransferase enzymes in human liver microsomes, Arch. Pharm. Res. 33 (2010) 1985–1996.

[38] H. Zhou, R. Shi, B. Ma, Y. Ma, C. Wang, D. Wu, X. Wang, N. Cheng, CYP450 1A2 and multiple UGT1A isoforms are responsible for jatrorrhizine metabolism in human liver microsomes, Biopharm. Drug Dispos. 34 (2013) 176–185.

[39] R. Shi, H. Zhou, B. Ma, Y. Ma, D. Wu, X. Wang, H. Luo, N. Cheng, Pharmacokinetics and metabolism of jatrorrhizine, a gastric prokinetic drug candidate, Biopharm. Drug Dispos. 33 (2012) 135–145.

[40] J.A. Silverman, Multidrug-resistance transporters, Pharm. Biotechnol. 12 (1999) 353–386.

[41] S.V. Ambudkar, C. Kimchi-Sarfaty, Z.E. Sauna, M.M. Gottesman, P-glycoprotein: from genomics to mechanism, Oncogene 22 (2003) 7468–7485.

[42] J.A. Silverman, S.S. Thorgeirsson, Regulation and function of the multidrug resistance genes in liver, Prog. Liver Dis. 13 (1995) 101–123.

[43] S.F. Zhou, Structure, function and regulation of P-glycoprotein and its clinical relevance in drug disposition, Xenobiotica 38 (2008) 802–832.

[44] S. Marchetti, R. Mazzanti, J.H. Beijnen, J.H. Schellens, Concise review: clinical relevance of drug drug and herb drug interactions mediated by the ABC transporter ABCB1 (MDR1, P-glycoprotein), Oncologist 12 (2007) 927–941.

[45] T.J. Opperman, S.T. Nguyen, Recent advances toward a molecular mechanism of efflux pump inhibition, Front. Microbiol. 6 (2015) 421.

[46] S. Zhou, E. Chan, S.Q. Pan, M. Huang, E.J. Lee, Pharmacokinetic interactions of drugs with St John's wort, J. Psychopharmacol. 18 (2004) 262–276.

[47] A. Kalliokoski, M. Niemi, Impact of OATP transporters on pharmacokinetics, Br. J. Pharmacol. 158 (2009) 693–705.

[48] Y. Shitara, K. Maeda, K. Ikejiri, K. Yoshida, T. Horie, Y. Sugiyama, Clinical significance of organic anion transporting polypeptides (OATPs) in drug disposition: their roles in hepatic clearance and intestinal absorption, Biopharm. Drug Dispos. 34 (2013) 45–78.

[49] K. Maeda, Organic anion transporting polypeptide (OATP)1B1 and OATP1B3 as important regulators of the pharmacokinetics of substrate drugs, Biol. Pharm. Bull. 38 (2015) 155–168.

[50] Y. Shitara, Clinical importance of OATP1B1 and OATP1B3 in drug-drug interactions, Drug Metab. Pharmacokinet. 26 (2011) 220–227.

[51] S.G. Simonson, A. Raza, P.D. Martin, P.D. Mitchell, J.A. Jarcho, C.D. Brown, A.S. Windass, D.W. Schneck, Rosuvastatin pharmacokinetics in heart transplant recipients administered an antirejection regimen including cyclosporine, Clin. Pharmacol. Ther. 76 (2004) 167–177.

[52] W.J. Hua, W.X. Hua, F.Y. Nan, W.A. Jiang, C. Yan, The influence of herbal medicine ursolic acid on the uptake of rosuvastatin mediated by OATP1B1*1a and *5, Eur. J. Drug Metab. Pharmacokinet. 39 (2014) 221–230.

[53] I.S. Song, T.Y. Kong, H.U. Jeong, E.N. Kim, S.S. Kwon, H.E. Kang, S.Z. Choi, M. Son, H.S. Lee, Evaluation of the transporter-mediated herb-drug interaction potential of DA-9801, a standardized dioscorea extract for diabetic neuropathy, in human in vitro and rat in vivo, BMC Complement. Altern. Med. 14 (2014) 251.

[54] P.S. Fasinu, H. Gutmann, H. Schiller, P.J. Bouic, B. Rosenkranz, The potential of *Hypoxis hemerocallidea* for herb-drug interaction, Pharm. Biol. 51 (2013) 1499–1507.

[55] L. Wang, X. Pan, D.H. Sweet, The anthraquinone drug rhein potently interferes with organic anion transporter-mediated renal elimination, Biochem. Pharmacol. 86 (2013) 991–996.

[56] E. Babu, M. Takeda, R. Nishida, R. Noshiro-Kofuji, M. Yoshida, S. Ueda, T. Fukutomi, N. Anzai, H. Endou, Interactions of human organic anion transporters with aristolochic acids, J. Pharmacol. Sci. 113 (2010) 192–196.

[57] A. Singer, M. Wonnemann, W.E. Müller, Hyperforin, a major antidepressant constituent of St. John's Wort, inhibits serotonin uptake by elevating free intracellular Na+1, J. Pharmacol. Exp. Ther. 290 (1999) 1363–1368.

[58] W.E. Müller, Current St John's wort research from mode of action to clinical efficacy, Pharmacol. Res. 47 (2003) 101–109.

[59] M.I. Mohammed Abdul, X. Jiang, K.M. Williams, R.O. Day, B.D. Roufogalis, W.S. Liauw, H. Xu, A.J. McLachlan, Pharmacodynamic interaction of warfarin with cranberry but not with garlic in healthy subjects, Br. J. Pharmacol. 154 (2008) 1691–1700.

[60] K. Ried, Garlic lowers blood pressure in hypertensive individuals, regulates serum cholesterol, and stimulates immunity: an updated meta-analysis and review, J. Nutr. 146 (2016) 389S–396S.

[61] I.A. Sobenin, I.V. Andrianova, I.V. Fomchenkov, T.V. Gorchakova, A.N. Orekhov, Time-released garlic powder tablets lower systolic and diastolic blood pressure in men with mild and moderate arterial hypertension, Hypertens. Res. 32 (2009) 433–437.

[62] R. Shouk, A. Abdou, K. Shetty, D. Sarkar, A.H. Eid, Mechanisms underlying the anti-hypertensive effects of garlic bioactives, Nutr. Res. 34 (2014) 106–115.

[63] S. Brankovic, M. Radenkovic, D. Kitic, S. Veljkovic, V. Ivetic, D. Pavlovic, B. Miladinovic, Comparison of the hypotensive and bradycardic activity of ginkgo, garlic, and onion extracts, Clin. Exp. Hypertens. 33 (2011) 95–99.

[64] J.Z. Luo, L. Luo, Ginseng on hyperglycemia: effects and mechanisms, Evid. Based Complement. Altern. Med. 6 (2009) 423–427.

[65] C.Y. Zhang, G. Baffy, P. Perret, S. Krauss, O. Peroni, D. Grujic, T. Hagen, A.J. Vidal-Puig, O. Boss, Y.B. Kim, X.X. Zheng, M.B. Wheeler, G.I. Shulman, C.B. Chan, B.B. Lowell, Uncoupling protein-2 negatively regulates insulin secretion and is a major link between obesity, beta cell dysfunction, and type 2 diabetes, Cell 105 (2001) 745–755.

[66] H.J. Lee, Y.H. Lee, S.K. Park, E.S. Kang, H.J. Kim, Y.C. Lee, C.S. Choi, S.E. Park, C.W. Ahn, B.S. Cha, K.W. Lee, K.S. Kim, S.K. Lim, H.C. Lee, Korean red ginseng (*Panax ginseng*) improves insulin sensitivity and attenuates the development of diabetes in Otsuka Long-Evans Tokushima fatty rats, Metabolism 58 (2009) 1170–1177.

[67] D.G. Bailey, J. Malcolm, O. Arnold, J.D. Spence, Grapefruit juice-drug interactions, Br. J. Clin. Pharmacol. 46 (1998) 101–110.

[68] D.G. Bailey, J.H. Kreeft, C. Munoz, D.J. Freeman, J.R. Bend, Grapefruit juice-felodipine interaction: effect of naringin and 6′,7′-dihydroxybergamottin in humans, Clin. Pharmacol. Ther. 64 (1998) 248–256.

[69] D.G. Bailey, G.K. Dresser, J.R. Bend, Bergamottin, lime juice, and red wine as inhibitors of cytochrome P450 3A4 activity: comparison with grapefruit juice, Clin. Pharmacol. Ther. 73 (2003) 529–537.

[70] U. Fuhr, Drug interactions with grapefruit juice. Extent, probable mechanism and clinical relevance, Drug Saf. 18 (1998) 251–272.

[71] J.A. Manthey, B.S. Buslig, Distribution of furanocoumarins in grapefruit juice fractions, J. Agric. Food Chem. 53 (2005) 5158–5163.

[72] L.Q. Guo, Y. Yamazoe, Inhibition of cytochrome P450 by furanocoumarins in grapefruit juice and herbal medicines, Acta Pharmacol. Sin. 25 (2004) 129–136.

[73] M.J. Hanley, P. Cancalon, W.W. Widmer, D.J. Greenblatt, The effect of grapefruit juice on drug disposition, Expert Opin. Drug Metab. Toxicol. 7 (2011) 267–286.

[74] T. Uno, N. Yasui-Furukori, Effect of grapefruit juice in relation to human pharmacokinetic study, Curr. Clin. Pharmacol. 1 (2006) 157–161.

[75] N. Galeotti, *Hypericum perforatum* (St John's wort) beyond depression: a therapeutic perspective for pain conditions, J. Ethnopharmacol. 200 (2017) 136–146.

[76] A.I. Oliveira, C. Pinho, B. Sarmento, A.C. Dias, Neuroprotective activity of *Hypericum perforatum* and its major components, Front. Plant Sci. 7 (2016) 1004.

[77] R.C. Shelton, St John's wort (*Hypericum perforatum*) in major depression, J. Clin. Psychiatry 70 (2009) 23–27.

[78] V. Butterweck, Mechanism of action of St John's wort in depression: what is known? CNS Drugs 17 (2003) 539–562.

[79] J. Barnes, L.A. Anderson, J.D. Phillipson, St John's wort (*Hypericum perforatum* L.): a review of its chemistry, pharmacology and clinical properties, J. Pharm. Pharmacol. 53 (2001) 583–600.

[80] A.A. Sprouse, R.B. van Breemen, Pharmacokinetic interactions between drugs and botanical dietary supplements, Drug Metab. Dispos. 44 (2016) 162–171.

[81] B.J. Gurley, S.F. Gardner, M.A. Hubbard, D.K. Williams, W.B. Gentry, Y. Cui, C.Y. Ang, Clinical assessment of effects of botanical supplementation on cytochrome P450 phenotypes in the elderly: St John's wort, garlic oil, *Panax ginseng* and *Ginkgo biloba*, Drugs Aging 22 (2005) 525–539.

[82] R. Rahimi, M. Abdollahi, An update on the ability of St. John's wort to affect the metabolism of other drugs, Expert Opin. Drug Metab. Toxicol. 8 (2012) 691–708.

[83] Y.M. Di, C.G. Li, C.C. Xue, S.F. Zhou, Clinical drugs that interact with St. John's wort and implication in drug development, Curr. Pharm. Des. 14 (2008) 1723–1742.

[84] A. Johne, J. Schmider, J. Brockmöller, A.M. Stadelmann, E. Störmer, S. Bauer, G. Scholler, M. Langheinrich, I. Roots, Decreased plasma levels of amitriptyline and its metabolites on comedication with an extract from St. John's wort (*Hypericum perforatum*), J. Clin. Psychopharmacol. 22 (2002) 46–54.

[85] E.N. Berry-Bibee, M.J. Kim, N.K. Tepper, H.E. Riley, K.M. Curtis, Co-administration of St. John's wort and hormonal contraceptives: a systematic review, Contraception 94 (2016) 668–677.

[86] S.D. Hall, Z. Wang, S.M. Huang, M.A. Hamman, N. Vasavada, A.Q. Adigun, J.K. Hilligoss, M. Miller, J.C. Gorski, The interaction between St John's wort and an oral contraceptive, Clin. Pharmacol. Ther. 74 (2003) 525–535.

[87] E. Ernst, St John's Wort supplements endanger the success of organ transplantation, Arch. Surg. 137 (2002) 316–319.

[88] S.M. Turton-Weeks, G.W. Barone, B.J. Gurley, B.L. Ketel, M.L. Lightfoot, S.R. Abul-Ezz, St John's wort: a hidden risk for transplant patients, Prog. Transpl. 11 (2001) 116–120.

[89] G.W. Barone, B.J. Gurley, B.L. Ketel, S.R. Abul-Ezz, Herbal supplements: a potential for drug interactions in transplant recipients, Transplantation 71 (2001) 239–241.

[90] M.F. Hebert, J.M. Park, Y.L. Chen, S. Akhtar, A.M. Larson, Effects of St. John's wort (*Hypericum perforatum*) on tacrolimus pharmacokinetics in healthy volunteers, J. Clin. Pharmacol. 44 (2004) 89–94.

[91] X. Jiang, K.M. Williams, W.S. Liauw, A.J. Ammit, B.D. Roufogalis, C.C. Duke, R.O. Day, A.J. McLachlan, Effect of St John's wort and ginseng on the pharmacokinetics and pharmacodynamics of warfarin in healthy subjects, Br. J. Clin. Pharmacol. 57 (2004) 592–599.

[92] S.C. Piscitelli, A.H. Burstein, D. Chaitt, R.M. Alfaro, J. Falloon, Indinavir concentrations and St John's wort, Lancet 355 (2000) 547–548.

[93] R.F. Frye, S.M. Fitzgerald, T.F. Lagattuta, M.W. Hruska, M.J. Egorin, Effect of St John's wort on imatinib mesylate pharmacokinetics, Clin. Pharmacol. Ther. 76 (2004) 323–329.

[94] R.H. Mathijssen, J. Verweij, P. de Bruijn, W.J. Loos, A. Sparreboom, Effects of St. John's wort on irinotecan metabolism, J. Natl. Cancer Inst. 94 (2002) 1247–1249.

[95] H. Gutmann, B. Poller, K.B. Büter, A. Pfrunder, W. Schaffner, J. Drewe, *Hypericum perforatum*: which constituents may induce intestinal MDR1 and CYP3A4 mRNA expression? Planta Med. 72 (2006) 685–690.

[96] J. Turkanovic, S.N. Ngo, R.W. Milne, Effect of St John's wort on the disposition of fexofenadine in the isolated perfused rat liver, J. Pharm. Pharmacol. 61 (2009) 1037–1042.

[97] D. Dürr, B. Stieger, G.A. Kullak-Ublick, K.M. Rentsch, H.C. Steinert, P.J. Meier, K. Fattinger, St John's Wort induces intestinal P-glycoprotein/MDR1 and intestinal and hepatic CYP3A4, Clin. Pharmacol. Ther. 68 (2000) 598–604.

[98] J. Birks, J. Grimley Evans, *Ginkgo biloba* for cognitive impairment and dementia, Cochrane Database Syst. Rev. (1) (2009) CD003120.

[99] B. Singh, P. Kaur, S.R.D. Gopichand, P.S. Ahuja, Biology and chemistry of *Ginkgo biloba*, Fitoterapia 79 (2008) 401–418.

[100] O.Q. Yin, B. Tomlinson, M.M. Waye, A.H. Chow, M.S. Chow, Pharmacogenetics and herb-drug interactions: experience with *Ginkgo biloba* and omeprazole, Pharmacogenetics 14 (2004) 841–850.

[101] Y. Deng, Y.F. Mo, X.M. Chen, L.Z. Zhang, C.F. Liao, Y. Song, C. Xu, Effect of *Ginkgo biloba* extract on the pharmacokinetics and metabolism of clopidogrel in rats, Phytother. Res. 30 (2016) 1886–1892.

[102] S.M. Robertson, R.T. Davey, J. Voell, E. Formentini, R.M. Alfaro, S.R. Penzak, Effect of *Ginkgo biloba* extract on lopinavir, midazolam and fexofenadine pharmacokinetics in healthy subjects, Curr. Med. Res. Opin. 24 (2008) 591–599.

[103] J. Tang, J. Sun, Y. Zhang, L. Li, F. Cui, Z. He, Herb-drug interactions: effect of *Ginkgo biloba* extract on the pharmacokinetics of theophylline in rats, Food Chem. Toxicol. 45 (2007) 2441–2445.

[104] T. Sugiyama, Y. Kubota, K. Shinozuka, S. Yamada, J. Wu, K. Umegaki, *Ginkgo biloba* extract modifies hypoglycemic action of tolbutamide via hepatic cytochrome P450 mediated mechanism in aged rats, Life Sci. 75 (2004) 1113–1122.

[105] S. Uchida, H. Yamada, X.D. Li, S. Maruyama, Y. Ohmori, T. Oki, H. Watanabe, K. Umegaki, K. Ohashi, S. Yamada, Effects of *Ginkgo biloba* extract on pharmacokinetics and pharmacodynamics of tolbutamide and midazolam in healthy volunteers, J. Clin. Pharmacol. 46 (2006) 1290–1298.

[106] L. Fan, G.Y. Tao, G. Wang, Y. Chen, W. Zhang, Y.J. He, Q. Li, H.P. Lei, F. Jiang, D.L. Hu, Y.F. Huang, H.H. Zhou, Effects of *Ginkgo biloba* extract ingestion on the pharmacokinetics of talinolol in healthy Chinese volunteers, Ann. Pharmacother. 43 (2009) 944–949.

[107] M. Unger, Pharmacokinetic drug interactions involving *Ginkgo biloba*, Drug Metab. Rev. 45 (2013) 353–385.

[108] S.C. Piscitelli, A.H. Burstein, N. Welden, K.D. Gallicano, J. Falloon, The effect of garlic supplements on the pharmacokinetics of saquinavir, Clin. Infect. Dis. 34 (2002) 234–238.

[109] J. Hajda, K.M. Rentsch, C. Gubler, H. Steinert, B. Stieger, K. Fattinger, Garlic extract induces intestinal P-glycoprotein, but exhibits no effect on intestinal and hepatic CYP3A4 in humans, Eur. J. Pharm. Sci. 41 (2010) 729–735.

[110] B.E. Ho, D.D. Shen, J.S. McCune, T. Bui, L. Risler, Z. Yang, R.J. Ho, Effects of garlic on cytochromes P450 2C9- and 3A4-mediated drug metabolism in human hepatocytes, Sci. Pharm. 78 (2010) 473–481.

[111] K. Berginc, T. Trdan, J. Trontelj, A. Kristl, HIV protease inhibitors: garlic supplements and first-pass intestinal metabolism impact on the therapeutic efficacy, Biopharm. Drug Dispos. 31 (2010) 495–505.

[112] Y. Zhao, B.H. Hellum, A. Liang, O.G. Nilsen, Inhibitory mechanisms of human CYPs by three alkaloids isolated from traditional Chinese herbs, Phytother. Res. 29 (2015) 825–834.

[113] H.M. Cui, Q.Y. Zhang, J.L. Wang, J.L. Chen, Y.L. Zhang, X.L. Tong, In vitro studies of berberine metabolism and its effect of enzyme induction on HepG2 cells, J. Ethnopharmacol. 158 (Pt A) (2014) 388–396.

[114] H. Cui, J. Wang, Q. Zhang, M. Dang, H. Liu, Y. Dong, L. Zhang, F. Yang, J. Wu, X. Tong, In vivo and in vitro study on drug-drug interaction of lovastatin and berberine from pharmacokinetic and HepG2 cell metabolism studies, Molecules 21 (2016) 464.

[115] H. Li, L. Liu, L. Xie, D. Gan, X. Jiang, Effects of berberine on the pharmacokinetics of losartan and its metabolite EXP3174 in rats and its mechanism, Pharm. Biol. 54 (2016) 2886–2894.

[116] D. Thomas, C.A. Karle, J. Kiehn, The cardiac hERG/IKr potassium channel as pharmacological target: structure, function, regulation, and clinical applications, Curr. Pharm. Des. 12 (2006) 2271–2283.

[117] M. Yan, K. Zhang, Y. Shi, L. Feng, L. Lv, B. Li, Mechanism and pharmacological rescue of berberine-induced hERG channel deficiency, Drug Des. Devel. Ther. 9 (2015) 5737–5747.

[118] K. Zhang, D. Zhi, T. Huang, Y. Gong, M. Yan, C. Liu, T. Wei, Z. Dong, B. Li, B. Yang, Berberine induces hERG channel deficiency through trafficking inhibition, Cell Physiol. Biochem. 34 (2014) 691–702.

[119] D. Zhi, P.F. Feng, J.L. Sun, F. Guo, R. Zhang, X. Zhao, B.X. Li, The enhancement of cardiac toxicity by concomitant administration of Berberine and macrolides, Eur. J. Pharm. Sci. 76 (2015) 149–155.

[120] S.P. Lin, S.Y. Tsai, Y.C. Hou, P.D. Chao, Glycyrrhizin and licorice significantly affect the pharmacokinetics of methotrexate in rats, J. Agric. Food Chem. 57 (2009) 1854–1859.

[121] Y.C. Hou, S.P. Lin, P.D. Chao, Liquorice reduced cyclosporine bioavailability by activating P-glycoprotein and CYP 3A, Food Chem. 135 (2012) 2307–2312.

[122] P. Methlie, E.E. Husebye, S. Hustad, E.A. Lien, K. Løvås, Grapefruit juice and licorice increase cortisol availability in patients with Addison's disease, Eur. J. Endocrinol. 165 (2011) 761–769.

[123] U.M. Kent, M. Aviram, M. Rosenblat, P.F. Hollenberg, The licorice root derived isoflavan glabridin inhibits the activities of human cytochrome P450S 3A4, 2B6, and 2C9, Drug Metab. Dispos. 30 (2002) 709–715.

[124] S.F. Zhou, Z.W. Zhou, C.G. Li, X. Chen, X. Yu, C.C. Xue, A. Herington, Identification of drugs that interact with herbs in drug development, Drug Discov. Today 12 (2007) 664–673.

[125] X.X. Yang, Z.P. Hu, W. Duan, Y.Z. Zhu, S.F. Zhou, Drug-herb interactions: eliminating toxicity with hard drug design, Curr. Pharm. Des. 12 (2006) 4649–4664.

[126] J. Inglese, R.L. Johnson, A. Simeonov, M. Xia, W. Zheng, C.P. Austin, D.S. Auld, High-throughput screening assays for the identification of chemical probes, Nat. Chem. Biol. 3 (2007) 466–479.

[127] J. Inglese, C.E. Shamu, R.K. Guy, Reporting data from high-throughput screening of small-molecule libraries, Nat. Chem. Biol. 3 (2007) 438–441.

[128] G. Li, K. Huang, D. Nikolic, R.B. van Breemen, High-throughput cytochrome P450 cocktail inhibition assay for assessing drug-drug and drug-botanical interactions, Drug Metab. Dispos. 43 (2015) 1670–1678.

[129] H.A. Oh, H. Lee, D. Kim, B.H. Jung, Development of GC-MS based cytochrome P450 assay for the investigation of multi-herb interaction, Anal. Biochem. 519 (2017) 71–83.

Chapter 18

Toxicity Studies Related to Medicinal Plants

Kavimani Subramanian, Divya Sankaramourthy, Mahalakshmi Gunasekaran

Department of Pharmacology, Mother Theresa Post Graduate and Research Institute of Health Sciences, Puducherry, India

LIST OF ABBREVIATIONS

AKR/J mice Mice are viremic from birth, and express the ecotropic retrovirus AKV in all tissues. Widely used in cancer research for their high leukemia incidence (60%–90%) and in immunology
BMD Bench mark dose
C57BL/6 Often referred to as "C57 black 6," "C57," or "black 6," a common transgenic strain of laboratory mouse widely used as models for human disease
CNS Central nervous system
LD_{50} Median lethal dose or lethal dose, 50%
NOAEL No observed adverse effect level
OECD Organization for Economic Cooperation and Development
RBC Red blood corpuscles
Sc Subcutaneous
TG Test guidelines
US FDA US Food and Drug Administration
WBC White blood corpuscles
WHO World Health Organization

1. INTRODUCTION

Plants that possess secondary metabolites of therapeutic potential are broadly referred to as medicinal plants. The use of plant parts for combating various diseases has been practiced by humankind from time immemorial. Every civilization had identified the healing properties of specific medicinal plants and the knowledge was subsequently passed onto successive generations. According to the World Health Organization (WHO), 75% of the world's population depends on herbs for basic healthcare needs [1]. However, in the late 19th and early 20th centuries, interest in the use of medicinal plants for therapeutic purposes declined as there were many shortcomings in the

Natural Products and Drug Discovery. https://doi.org/10.1016/B978-0-08-102081-4.00018-6

processing methods used and a paucity of information on their side effects. In addition, the availability of techniques for the isolation of pure compounds and synthesis of new compounds with proven biomedical applications also had contributed to a major extent to the reduced interest in the use of plant parts. Moreover, knowledge of the toxicity of medicinal plants can be evidenced from ancient inscriptions such as Ebers Papyrus (1500 BC) with 800 different drug preparations from 700 plant species for therapy [2,3]. In this context, there is an impending requirement to compile information on the probable adverse reactions of the medicinal plants through appropriate scientific studies before formulating and marketing them as herbal drugs.

2. TOXICITY STUDIES ARE INDISPENSABLE FOR MEDICINAL PLANTS

Medicinal plants that have been used in various traditional practices lack valid experimental evidence from meticulously planned research that overrules their adverse effects. In addition, those herbal drugs prepared from combinations of different medicinal plants also fail to predict their toxic outcomes. In recent times, several adverse effects have been observed with medicinal plants. An appealing example of unassuming toxicity had been reported in rural Africa where a traditional eye medicine containing toxic substances instilled into the conjunctival sac had resulted in corneal ulceration and childhood blindness [4,5].

Ephedra, used for the treatment of anxiety, insomnia, pain, muscle tension, and as an energy boost, has been banned in Europe and Canada for its adverse effects such as stroke, heart attacks, heart rate irregularities, liver toxicity, seizures, psychoses, and sometimes death [6]. *Ginkgo biloba*, the commonly used herb for dementia, tinnitus, vertigo, asthma, allergies, etc. [7], has been found associated with increased risk of intracranial hemorrhages, bilateral subdural hematomas [8], and bleeding [9,10] and these effects were attributed to interactions with platelet activating factor and collagen culminating in decreased platelet aggregation [11,12]. *Ginseng*, the most widely used Chinese herb for improving immunity [13] and psychomotor functions [14,15], has been reported for its adverse effects such as nausea, diarrhea, euphoria, insomnia, headaches, hypertension, mastalgia, and vaginal bleeding [16]. Aristolochic acid, one of the active constituents present in *Aristochia* species that is used in Chinese herbal medicines for its antineoplastic, antiinflammatory, bactericidal, antiseptic, and phospholipase A_2 inhibitory properties [17,18], has been shown to be associated with carcinogenic [19,20], genotoxic [21], and kidney damaging properties [22,23]. In addition, some herbal drugs have been reported for corticosteroid-like adverse effects due to the presence of trace amounts of betamethasone [24]. Earlier reports of the WHO quoted nearly 5000 herb-related adverse effects and the US Food and Drug Administration reported 2621 adverse events, including 101 deaths by using dietary supplements. Of that, *Ephedra* containing dietary supplements used for the purpose of weight reduction had been the subject of 1200 complaints, including 10 cardiac arrests and 70 sudden death reports [25]. Another troubling example of dietary supplements is β-carotene taken at higher doses, along with vitamin A or E, which increased the incidence of lung cancer in people with higher risk of this disease [26]. Thereby, the WHO accentuated the safety of herbal medicines [27] by issuing guidelines for assessment of herbal medicines [28] and emphasized the significance of toxicity studies.

3. PREPARATION OF A TEST SUBSTANCE FOR TOXICITY STUDIES

The medicinally active plant part to be tested needs to be identified, collected, air dried in shade, ground coarsely, and extracted with appropriate solvent using Soxhlet apparatus [29] or by a simple cold maceration process. The obtained plant extracts should be prepared as solutions, suspensions, or emulsions with a suitable vehicle, preferably water or sodium carboxymethyl cellulose or carboxymethyl cellulose, for administration to animals.

4. TOXICITY STUDIES: GENERAL CONSIDERATIONS

Toxicity testing for medicinal plants can be carried out with reference to the Organization for Economic Cooperation and Development (OECD) guidelines where animals of the same species and sex, preferably female rodents (rats or mice), nulliparous and nonpregnant, 8–12 weeks old with a mean body weight ±20% are maintained under ideal husbandry conditions and reared under standard conditions of temperature (22 ± 3°C), humidity (30%–70%), and exposure to a 12:12 h light/dark cycle. Prior acclimatization for a period of 5 days to laboratory conditions under a standard pellet diet with a supply of drinking water ad libidum is generally preferred and the animals are divided into caged groups of five in each according to doses of test substance to be administered. The test substances of varying concentrations are prepared prior to administration, considering the stability of preparation as a solution, suspension, or emulsion with a suitably inert vehicle. The administration of test doses should be at a constant volume not exceeding 1 mL/100 g body weight and 2 mL/100 g body weight (for aqueous solutions) using oral gavage tubes or intubation cannula to overnight fasted animals (withheld from food but not water).

4.1 Gross Behavioral Studies [30]

After a single dose administration of test material (plant extract) the behavioral profile such as awareness (alertness or stupor), stimulation or depression of the central nervous system, stereotype, passivity, grooming, vocalization, restlessness, irritability, fearfulness, spontaneous motor activity touch response, pain response and neurological profile such as startle response, Straub's response, tremor, convulsions, body position, limb position, gait (staggering/abnormal), muscle tone, pinna reflex, corneal reflex and autonomic profile such as pupil size, palpebral opening, exophthalmos, urination, salivation, writhing, piloerection, hypothermia, skin color, heart rate, and respiratory rate have to be determined.

4.2 Conventional Methods for LD_{50} Determination

The methods conventionally adopted for determining LD_{50} values are the graphical method of Miller and Tainter [31], the arithmetic method of Reed and Muench [32], the arithmetic method of Karber [33], and the classic method of Litchfield and Wilcoxon [34].

The Miller and Tainter method derives the LD_{50} value from a graphical representation of log dose versus probit scale where the percent mortalities are converted to probits using a table of probits. In the Reed and Muench method, the cumulative values of dead and survived animals and the percent survival of two doses adjacent to the LD_{50} value are estimated. The proportional distance of two doses from 50% is calculated with an appropriate formula and the log LD_{50} is calculated from the obtained values of log dilution factor and proportional distance. The antilog of the calculated log LD_{50} gives the LD_{50} value. In the Karber method, the interval mean values between number of dead and the dose difference between each group are initially calculated. Excluding the values of least and highest dose, the product of interval mean and dose difference of all other doses is computed. The LD_{50} value in mg/kg is estimated using the formula where (LD_{50} = [least lethal dose] – [sum of product values/no of animals per group]).

5. ACUTE TOXICITY TESTING

As per the definition of the Globally Harmonized System, acute toxicity testing was defined as "those adverse effects occurring oral or dermal administration of a single dose of a substance or multiple doses given within 24 hours or as an inhalation exposure of 4 hours." With further modification of testing methods minimizing the use of experimental animals for determination of LD_{50}, acute toxicity studies were broadly classified as fixed dose procedure (test guideline [TG] 420), acute toxic class method (TG 423), up and down procedure (TG 425), acute dermal toxicity (TG 402), and acute inhalational toxicity (TG 403) as per OECD guidelines.

5.1 Fixed Dose Procedure [35]

This procedure involves the administration of a test substance in a stepwise manner using fixed doses of 5, 50, 300, and 2000 mg/kg (an additional dose of

5000 mg/kg may be used with proper justification and demands of the study) and the starting dose of the test substance is selected based on previous in vivo data producing signs of toxicity without causing mortality or a dose of 300 mg/kg is taken as starting dose if no information is available. Following the starting dose, other groups of animals were dosed either with higher or lower fixed doses and the experiment continues until the appearance of evident signs of toxicity.

Following drug administration in a single dose by oral gavage not exceeding 24 h, the experimental animals were observed for a period of 14 days for changes in skin, fur, eyes, mucous membranes, respiratory, circulatory, autonomous, central activity, and behavioral pattern. Abnormal signs such as tremors, convulsions, salivation, diarrhea, lethargy, sleep, and coma must be noted carefully and taken into consideration. Body weight changes of each animal before and after drug administration must be recorded. At the end of the study, all the animals were euthanized and subjected to gross necropsy to observe signs of toxicity through microscopic pathological changes in organs; hence toxicity is interpreted.

5.1.1 Limitations

The doses fixed from 2000 to 5000 mg/kg and above are generally not encouraged by the Globally Harmonized Classification System (category 5) considering animal welfare unless justified with specific regulatory needs.

5.2 Acute Toxic Class Method [36]

This test procedure consists of sequential administration of a test substance to a minimum of three experimental animals at each step. In this method, four predefined starting doses may be used and the test dose is selected according to the Globally Harmonized Classification System [37]. The starting dose is selected from one of four doses of 5, 50, 300, and 2000 mg/kg, or 300 mg/kg is taken as a starting dose if no information on the test substance is available. The test starts with the administration of the selected dose to experimental animals in such a way that the dose-related mortality decides the next step of whether further testing is needed. All other observations are similar to fixed dose procedure OECD TG 420.

5.3 Up and Down Procedure [38]

This method is recommended over the others since use of vertebrate animals is minimized. The test procedure is also known as staircase design and deals with the administration of a test substance with one of the defined doses to one of the experimental animals at each step with an interval of 48 h. Doses are selected below the best estimated level of LD_{50} as per in vivo data or 175 mg/kg is taken as a starting dose if no information on the test substance is available. The test starts with administration of the selected dose to the first animal and proceeds based on signs of mortality of the animal. The dose is increased by a factor of 3.2 times the original dose to the next animal if the animal survived or vice versa if the animal died. Thus the presence or absence of dose-related mortality determines the need for further tests and the observations are similar to those of the fixed dose procedure.

5.4 Acute Inhalational Toxicity [39]

Preparations taken as inhalation such as aerosols can be tested for their hazardous effects upon short-term exposure by acute inhalational toxicity study. Animals acclimatized to laboratory conditions are maintained with an air flow of 12–15 air changes/h and an adequate oxygen supply (19%). Following the period of acclimatization, the test substance in the form of inhalation is exposed to the experimental animal for a period of 4 h followed by observation for 14 days for abnormal changes such as tremors, convulsions, salivation, diarrhea, lethargy, sleep, and coma. Mortality during exposure time or observation period is also recorded. At the end of the study, all animals were sacrificed and examined for gross pathological changes.

5.5 Acute Inhalational Toxicity [40]

This method is an alternative to acute inhalational toxicity test OECD TG 403 with fewer modifications to OECD TG 403. In this method, a similar testing procedure is adopted except for endpoint toxicity determination where evident signs of toxicity are considered as the endpoint rather than death of test animals.

5.6 Acute Dermal Toxicity [41]

For testing the toxicity of topical preparations either in the form of solutions or incorporated with ample ointment base, acute dermal toxicity studies using OECD TG 402 can be employed. The test animals are subjected to depilation of the dorsal area of the trunk 24 h before the start of the study. The test substance is applied to the skin covering nearly 10% of the total body surface (except highly toxic substances where the covering area can be less) and covered with a porous gauze and adhesive tape to ensure the substance is in contact with the dermal surface for up to 24 h. Precautions must be taken in the preparation of the test substance considering its physical properties, where solids are pulverized well and moistened with a suitable vehicle (0.5–1 mL), while liquids are prepared in undiluted form. The starting dose is selected from one of three fixed doses of 5, 200, or 1000 mg/kg body weight, based on the Globally Harmonized System category for acute dermal toxicity [42] or taken as 2000 mg/kg if no information is available. The doses are administered in a sequential manner with not more than two to three animals in each step. Testing of the substance on each animal, considering its onset, duration, and signs of toxicity, is allowed for a period of 24–48 h and, based on the survival of the animals, the next dose level is administered. The overall period of observation is 14 days.

Observations are made immediately after dosing from 30 min followed by periodical observations during the first 24 h with special attention during the first 2–6 h and for a total of 14 days. All the observations are individually recorded for each animal, e.g., changes in skin, fur, eyes, mucous membranes, respiratory, circulatory, autonomous, central activity, and behavioral pattern. Abnormal signs such as tremors, convulsions, salivation, diarrhea, lethargy, sleep, and coma must be noted carefully and taken into consideration. Body weight changes of each animal before and after drug administration must be recorded. At the end of the study, all the animals were euthanized and subjected to gross necropsy for observing the signs of toxicity through microscopic, pathological changes in organs; hence toxicity is interpreted.

6. SUBACUTE TOXICITY STUDIES [43]

Evaluation of toxicity of a test substance with repeated dose administration to experimental animals with a short-term exposure of 14–20 days can be studied by subacute toxicity test. The toxic effects of a test substance on target organs, including nervous, endocrine, and immunological systems, are studied. A total of 10 animals of each sex (five female and five male) are used at each group with rats as the preferred rodent species (other rodents can also be used under proper justification). Administration of a test substance to experimental animals is through the oral route in single graduated doses daily for a period of 28 days. A minimum of three test groups along with a control group should be used. Additional use of animals as satellite groups (five per sex) can also be planned before the study for observing the reversibility of toxic outcomes, sudden intermittent toxic effects, or mortality and are maintained for a minimum period of 14 days after drug treatment. Doses are selected based on the toxicokinetic particulars of the test substance such that the highest dose selected must be able to induce severe toxic effects without causing mortality and the lowest dose selected is devoid of adverse effects. Doses exceeding 1000 mg/kg not producing adverse effects need a limit test. If the test substance is administered via diet, precaution should be taken to avoid interference of the test substance with the animal's regular nutritional pattern.

Animals are observed for a period of 28 days (satellite group observed for 14 days) with everyday clinical observation after drug administration followed by mortality observation twice a day. Apart from noting changes in skin, fur, eyes, mucous membranes, secretions, excretions, and autonomous activity, changes in posture, gait, tonic clonic movements, unusual behaviors such as backward walking, and circling movements must be recorded [44].

Observations of sensory responses such as auditory, visual, and proprioceptive stimuli [45–47], grip strength, and motor activity must also be investigated by the fourth week. Body weight changes and food consumption weight are recorded weekly. Water consumption can also be measured weekly in cases of drugs administered via drinking water. At the end of the study period, the animals are sacrificed and examined for hematological, clinical, biochemical, and pathological changes. Hematological examinations such as hemoglobin concentration, hematocrit, red blood corpuscle count, reticulocytes, total and differential white blood corpuscle count, and platelet count are measured from blood samples withdrawn from euthanized animals. Major toxic effects on target organs can be detected by estimating clinical biochemical parameters with isolated plasma or serum samples prepared from blood samples of euthanized animals. Estimation of sodium, potassium, glucose, total cholesterol, urea, creatinine, total protein, albumin, enzymes such as alanine aminotransferase, aspartate aminotransferase, alkaline phosphatase, glutamyltranspeptidase, glutamyl dehydrogenase, bilirubin, bile acids, etc. must be done to analyze the effect of the test substance. In the last week of study, analysis of urine samples for urine volume, appearance, specific gravity, pH, protein, glucose, blood, or blood cells are done.

7. SUBCHRONIC TOXICITY STUDIES (OECD TG 408, 409, 411, 413)

Following subacute toxicity studies, the test substance's toxicity is assessed with repeated dose exposure for a period of 90 days via oral [48,49], dermal [50], or inhalational [51] administration. The preferred species is rat though other rodents such as mice can also be used. When conducting toxicity studies in nonrodents, dogs of a definite breed especially beagles are used and the use of other nonrodents such as swine and minipigs is also encouraged. Toxicity studies in rodents allow a total of 20 animals (10 male and 10 female) per group, whereas in nonrodents only eight animals (four male and four female) at each dose level is allowed. The study of inhalational or dermal toxicity allows an exposure time of 6 h per day for 5 days per week up to a period of 90 days. All the considerations and observations of parameters are similar to that of the subacute toxicity studies.

8. CHRONIC TOXICITY STUDIES [52,53]

Toxicity effects produced on long-term exposure of a test substance can be studied by chronic toxicity studies. The study chiefly aims to identify the toxic nature of a test substance, effect on target organs, establishment of dose–response relationship, no observed adverse effect level, and benchmark dose with a long-term exposure of 6–12 months.

The principle of the testing procedure is to study the cumulative toxicity produced on repeated dosing of the test substance for a period of 12 months (deviations from 12 months requires justification). Selection of the animal species depends on the nature of the study since use of rodents and nonrodents is encouraged. However, in general, rodents (rats and mice) are preferred more for their short lifespan and wide usage in toxicological studies. Doses are selected based on repeated 28- and 90-day exposure toxicity studies. A total of 40 animals (20 male and 20 female) per group can be used at each dose level. In the case of mice, an additional number of animals per group can also be planned before starting the study as satellite and sentinel groups (five animals per sex) to understand the reversibility of toxicological outcomes and monitor the disease status, respectively. With proper scientific justification, provision for interim kills may also be allowed for the purpose of studying the progression of toxicological changes in between the study. All the additional animals included as satellite groups, sentinel groups, and for interim kills must also be observed for parameters of body weight, food/water consumption, clinical, biochemical, hematological, and pathological changes during the phase of study similar to experimental animals in the test groups. A minimum of three groups with varying dose levels along with a control group is generally preferred for the study. The highest dose selected for the study must be able to elicit effective toxicological outcomes without causing severe suffering and mortality and should not exceed 1000 mg/kg (dose limit). The dose spacing intervals must be optimal between each group for interpreting stepwise increased dose response. An additional fourth test group and large dose spacing intervals (more than a factor of 6—10) between the groups are always advantageous.

Route of administration of a test substance based on the purpose of the study must be rationalized. Test substances administered via drinking water or diet should not interfere with nutritional balance and water intake, concentration of the test substance must not exceed 5% of total diet, and a constant dose level (mg/kg or ppm) must be maintained. Dosing of the test substance is usually carried out for 7 days per week for 1 year via the oral or dermal route, whereas dosing for 5 days per week with an exposure time of 6 h per day is carried out via the dermal route. All animals are observed for abnormal signs of toxicological outcomes, behavioral changes, and mortality during the study period twice a day.

Clinical observations and ophthalmological examinations are carried out in all animals before the first exposure of a test substance followed by the first week of exposure and every month thereafter. Changes in skin, fur, eyes, mucous membranes, secretions, excretions and autonomous activity (lacrimation, piloerection, pupil size), unusual respiration, gait, postural changes, tonic clonic movements, grooming, repetitive circling, and abnormal behaviors must also be recorded.

Test substances assessed for neurotoxic effects have to be observed for behavioral parameters such as grip strength, motor activity tests, auditory, visual, and proprioceptive stimuli response, etc. before the study and thereafter at 3, 6, 9, and 12 months. Parameters such as body weight and food and water consumption are recorded weekly from the beginning of the study for the first 13 weeks and then monthly thereafter. Hematological, clinical, and biochemical estimations and urine analysis in rodents require at least 10 male and 10 female animals as representatives of each group and are tested at the 3rd, 6th and 12th month during the study, whereas in nonrodents only a small number of animals are required (e.g., four animals per sex and per group in dogs). Blood samples are withdrawn after overnight fasting (except mice) via retro-orbital sinus or through cardiac puncture. At the end of the study period, all the animals are subjected to gross necropsy with careful examination of body surface, orifices, cavities, and their contents.

Individual organ weights and paired organ weights (kidneys, adrenal glands) of all the animals are recorded after trimming off the extraneous tissues as early as possible after the dissection process. All the tissues are subjected to histopathological analysis along with blood and bone marrow smear examinations.

9. SPECIAL TOXICITY STUDIES

The special toxicity studies carried out with reference to OECD guidelines are eye irritation tests, skin sensitization tests, prenatal developmental toxicity study, neurotoxicity study, carcinogenicity study, and reproductive toxicity such as one-generation reproduction toxicity and two-generation reproduction toxicity study.

9.1 Acute Eye Irritation/Corrosion Test [54]

The ocular corrosivity of a substance can be evaluated in vivo by using the acute eye irritation test in albino rabbits. With prior treatment of systemic analgesia (buprenorphine 0.01 mg/kg subcutaneously) and topical ocular anesthesia (0.5% proparacaine hydrochloride or 0.5% tetracaine hydrochloride), the test substance is instilled as a single dose into the conjunctival sac of one eye, while the other eye serves as a control. One hour after administration of the test substance, the eye is evaluated for the presence or absence of clear ocular lesions followed by 24-h, 48-h, 72-h, 7-day, 14-day, and 21-day observation periods. Clinical signs of pain or distress such as repeated rubbing of eyes and excessive blinking are also recorded twice a day. At the end of the stipulated observation period, the animals are euthanized and graded for eye irritation response in conjunction with the ocular lesions and corneal perforations to assess the corrosive nature of the test substance.

9.2 Skin Sensitization Test [55]

The dermal sensitization response of the test substance can be investigated by using the guinea pig maximization test (adjuvant test) and Buehler's test (nonadjuvant test) by measuring the hypersensitivity reactions in test animals compared with control animals.

The adjuvant test is involved with the administration of three pairs of injections, namely, an intradermal challenge dose of Freund's complete adjuvant (1:1 mixture of Freund's complete adjuvant/water or saline), a test substance suspended in a suitable vehicle, and a test substance suspended in Freund's complete adjuvant followed by topical exposure of the test substance over a depilated dermal surface for a period of 24 h. In the case of the non-adjuvant test, only the topical induction phase is used in guinea pigs with an exposure of 6 h. After 24 h, on removal of the patch test, erythema has to be evaluated by the Magnusson and Klingman grading scale (0—no visible change; 1—discrete or patchy erythema; 2—moderate erythema; 3—intense erythema and swelling) [56].

9.3 Prenatal Developmental Toxicity [57]

The prenatal developmental toxic effects of a substance can be studied in female rodents (rats) or nonrodents (rabbits) by continuous administration of the test substance by oral gavage tube (intubation) after the implantation of fertilized ovum (i.e., day 5 of postmating) until the day prior to cesarean sectioning. Observations for mortality, behavioral changes, body weight, and food and water consumption are generally performed before and after exposure of the test substance. At the end of the study period, animals are euthanized and examined for uterine contents and number of live and dead fetuses. The body weight, sex, soft tissue, and skeletal abnormalities of fetuses are also recorded.

9.4 Neurotoxicity Studies [58]

Neurotoxicity studies are performed for detecting major neurotoxicological and neuropathological changes of a substance upon its exposure in rodents (rats) via oral, dermal, or inhalational routes. Based on the need of the study, duration of the study period varies from 28 days to 1 year or more. Before and after exposure of the test substance, clinical observations and neurobehavioral function tests for auditory, proprioceptive stimuli, motor activity, changes in body weight, food and water consumption, and ophthalmological examination are frequently performed and the study is terminated after the specified duration by sacrificing the animals and examining neuropathological changes by histopathology.

9.5 Carcinogenicity Studies [59]

This study is used for assessing the oncogenicity of a test substance in rodents (preferably rats or mice). The test substance is generally administered via the oral route, while other routes such as topical and dermal are also encouraged if the study demands an exposure period of 6 h per day. Based on selection of the animal, duration of the study is determined from 18 months (specific strains of mice: AKR/J, C_{57} BL/6J) to 24 months (rats). At the end of the study period, the animals are sacrificed and observed for abnormal signs of toxicity in terms of neoplastic developmental lesions. The time of onset, location, dimensions, appearance, and progression of each tangible tumors must be recorded.

9.6 Reproduction Toxicity Studies

One-generation [60] and two-generation reproduction toxicity studies [61] are performed either in mice (8 weeks of treatment for one generation and 56 days of treatment for two generation) or rats (10 weeks of treatment for one generation and 70 days of treatment for two generation). The body weight of prenatal and parental animals, the duration of gestation of prenatal females, estrus cycle length, number and sex of pups, number of dead pups, and physical and behavioral abnormalities of pups are measured in female animals whereas sperm mobility, weight of testis, epididymis, and prenatal and F_1 animals are measured in male animals.

REFERENCES

[1] S.Y. Pan, G. Litscher, S.H. Gao, S.F. Zhou, Z.L. Yu, H.Q. Chen, S.F. Zhang, M.K. Tang, J.N. Sun, K.M. Ko, Historical Perspective of Traditional Indigenous Medical Practices: The Current Renaissance and Conservation of Herbal Resources Evidence Based Complementary and Alternative Medicine, 2014, pp. 1–21.

[2] L. Glesinger, Medicine through Centuries, Zora, Zagreb, 1954, pp. 21–38.

[3] J. Tucakov, Pharmacognosy, Beograd: Institute for Text Book Issuing in SR, Srbije, 1964, pp. 11–30.

[4] P. Courtright, S. Lewallen, S. Kanyaloti, D.J. Diwala, Traditional eye medicine use among patients with corneal disease in rural Malawi, Br. J. Ophthalmol. 78 (11) (1994) 810–812.

[5] D. Yorston, A. Foster, Traditional eye medicines and corneal ulceration in Tanzania, J. Trop. Med. Hyg. 97 (4) (1994) 211–214.

[6] Health Canada Online, Advisory: Health Canada Is Advising Consumers Not to Use Any Products Containing Kava, 2002. http:www.hc.sc.gc.ca/english/protection/warnings/2002/2002_ø2e.html.

[7] E. Ernst, M.H. Pittler, *Ginkgo biloba* extract for dementia: a systematic review of double-blind, placebo-controlled trials, Clin. Drug Invest. 17 (4) (1999) 301–308.

[8] J. Rowin, S.L. Lewis, Spontaneous bilateral subdural haematomas associated with chronic *Ginkgo biloba* ingestion, Neurology 46 (6) (1996) 1775–1776.

[9] E. Ernst, The risk-benefit profile of commonly used herbal therapies: *Ginkgo*, St. John's Wort, Ginseng, Echinacea, Saw Palmetto, and Kava, Ann. Intern. Med. 136 (1) (2002) 42–53.

[10] A.M.K. Lee, J. Moss, C.S. Yuan, Herbal medicines and perioperative care, JAMA 286 (2) (2001) 208–216.

[11] K.F. Chung, G. Dent, M. McCusker, P. Guinot, C.P. Page, P.J. Barnes, Effect of a ginkgolide mixture (BN 52063) in antagonising skin and platelet responses to platelet activating factor in man, Lancet 1 (1987) 248–251.

[12] G.B. Kudolo, S. Dorsey, J. Blodgett, Effect of the ingestion of *Ginkgo biloba* extract on platelet aggregation and urinary prostanoid excretion in healthy and type 2 diabetic subjects, Thromb. Res. 108 (2–3) (2002) 151–160.

[13] F. Scaglione, G. Cattaneo, M. Alessandria, R. Cogo, Efficacy and safety of the standardized Ginseng extract G115 potentiating vaccination against the influenza syndrome and protection against the common cold, Drugs Exp. Clin. Res. 22 (2) (1996) 65–72.

[14] L. D'Angelo, R. Grimaldi, M. Caravaggi, M. Marcoli, E. Perucca, S. Lecchini, G.M. Frigo, A. Crema, A double-blind, placebo-controlled clinical study on the effect of a standardized ginseng extract on psychomotor performance in healthy volunteers, J. Ethnopharmacol. 116 (1) (1986) 15–22.

[15] J.M. Ellis, P. Reddy, Effects of *Panax ginseng* on quality of life, Ann. Pharmacother. 36 (3) (2002) 375–379.

[16] J.T. Coon, E. Ernst, *Panax ginseng*: a systematic review of adverse effects and drug interactions, Drug Saf. 25 (5) (2002) 323–344.

[17] J. Buckingham, Dictionary of Natural Products on CD-ROM, CRC Press, Chapman & Hall/CRC, Boca Raton, FL, 2001.

[18] J.P. Cosyns, Aristolochic acid and Chinese herbs nephropathy: a review of the evidence to date, Drug Saf. 26 (1) (2003) 33–48.

[19] M. Cui, Z.H. Liu, Q. Qiu, H. Li, L.S. Li, Tumour induction in rats following exposure to short-term high dose aristolochic acid I, Mutagenesis 20 (1) (2005) 45–49.

[20] H.H. Schmeiser, J.W. Janssen, J. Lyons, H.R. Scherf, W. Pfau, A. Buchmann, C.R. Bartram, M. Wiessler, Aristolochic acid activates ras genes in rat tumors at deoxyadenosine residues, Cancer Res. 50 (17) (1990) 5464–5469.

[21] V.M. Arlt, M. Stiborova, H.H. Schmeiser, Aristolochic acid as a probable human cancer hazard in herbal remedies: a review, Mutagenesis 17 (4) (2002) 265–277.

[22] J.P. Cosyns, M. Jadoul, J.P. Squifflet, F.X. Wese, C. Van Ypersele de Strihou, Urothelial lesions in Chinese-herb nephropathy, Am. J. Kidney Dis. 33 (1999) 1011–1017.

[23] J.L. Nortier, M.C. Martinez, H.H. Schmeiser, V.M. Arlt, C.A. Bieler, M. Petein, M.F. Depierreux, L. De Pauw, D. Abramowicz, P. Vereerstraeten, J.L. Vanherweghem, Urothelial carcinoma associated with the use of a Chinese herb (*Aristolochia fangchi*), N. Engl. J. Med. 342 (23) (2000) 1686–1692.

[24] WHO, Guidelines on Safety Monitoring of Herbal Medicines in Pharmacovigilance Systems, World Health Organization, Geneva, 2004, p. 19.

[25] C.A. Haller, N.L. Benowitz, Adverse cardiovascular and central nervous system events associated with dietary supplements containing ephedra alkaloids, N. Engl. J. Med. 343 (25) (2000) 1833–1838.

[26] C. Marwick, Trials reveal no benefit, possible harm of beta carotene and vitamin A for lung cancer prevention, JAMA 37 (1996) 1197–1200.

[27] WHO, Reasearch Guidelines for Evaluating the Safety and Efficacy of Herbal Medicines, 1993. Manila.

[28] WHO, Annex II, Guidelines for the Assessment of Herbal Medicines, WHO Technical Report Series No. 863, 1996. Geneva.

[29] N.N. Azwanida, A review on the extraction methods use in medicinal plants, principle, strength and limitation, Med. Aromat. Plants 4 (2015) 196.
[30] S. Irwin, Drug screening and evaluative procedures, Science 136 (1962) 123–128.
[31] L.C. Miller, M.L. Tainter, Estimation of LD_{50} and its error by means of log-probit graph paper, Proc. Soc. Exp. Biol. Med. 57 (1944) 261.
[32] L.J. Reed, H. Muench, A simple method of estimating fifty percent endpoints, Am. J. Hyg. 27 (1938) 493–497.
[33] G. Karber, Contribution to the collective treatment of pharmacological serial experiments, Arch. Exp. Pathol. Pharmacol. 162 (1931) 480–483.
[34] J.T. Litchfield, F.A. Wilcoxon, A simplified method of evaluating dose-effect experiments, J. Pharmacol. Exp. Ther. 96 (1949) 99–113.
[35] OECD, OECD Guideline for Testing of Chemicals: Acute Oral Toxicity – Fixed Dose Procedure (No. 420), Section 4, OECD Publishing, 2001, pp. 1–14.
[36] OECD, OECD Guideline for Testing of Chemicals: Acute Oral Toxicity – Acute Toxic Class Method (No. 423), Section 4, OECD Publishing, 2001, pp. 1–14.
[37] N. Stallard, A. Whitehead, Reducing animal numbers in the fixed dose procedure, Hum. Exp. Toxicol. 14 (4) (1995) 315–323.
[38] OECD, OECD Guideline for Testing of Chemicals: Acute Oral Toxicity – Up and Down Procedure (No. 425), Section 4, OECD Publishing, 2001, pp. 1–14.
[39] OECD, OECD Guideline for Testing of Chemicals: Acute Inhalational Toxicity (No. 403), Section 4, OECD Publishing, 2009, pp. 1–19.
[40] OECD, OECD Guideline for Testing of Chemicals: Acute Inhalation Toxicity – Fixed Concentration Procedure (No: 433), Section 4, OECD Publishing, 2004, pp. 1–24.
[41] OECD, OECD Guideline for Testing of Chemicals: Acute Dermal Toxicity (No. 402), Section 4, OECD Publishing, 1987, pp. 1–7.
[42] UN, United Nations Globally Harmonized System of Classification and Labelling of Chemicals (GHS), Fifth Revised Edition, 2013. UN New York and Geneva. Available at: www.unece.Org/trans/danger/public/ghs/ghs_rev05/05files_ehtml.
[43] OECD, OECD Guideline for Testing of Chemicals: Repeated Dose 28 Day Oral Toxicity Study in Rodents (No. 407), Section 4, OECD Publishing, 1995, pp. 1–8.
[44] ICPS, Principles and Methods for Assessment of Neurotoxicity Assosiated with Exposure to Chemicals, Environmental Health. Criteria Document No. 60, 1986.
[45] S.C. Gad, A neromuscular screen for use in industrial toxicology, J.Toxicol. Environ. Health 9 (5–6) (1982) 691–704.
[46] V.C. Moser, K.M. McDaniel, P.M. Philips, Rat strain and stock comparisons using a functional observational battery, baseline values and effects of *Amitraz*, Toxicol. Appl. Pharmacol. 108 (1991) 267–283.
[47] D.E. Tupper, R.B. Wallace, Utility of the neurologic examination in rats, Acta Neurobiol. Exp. 40 (1980) 999–1003.
[48] OECD, OECD Guideline for Testing of Chemicals: Repeated Dose 90 Day Oral Toxicity Study in Rodents (No. 408), Section 4, OECD Publishing, 1998, pp. 1–10.
[49] OECD, OECD Guideline for Testing of Chemicals: Repeated Dose 28 Day Oral Toxicity Study in Non Rodents (No. 409), Section 4, OECD Publishing, 1998, pp. 1–9.
[50] OECD, OECD Guideline for Testing of Chemicals: Sub Chronic Dermal Toxicity: 90 Day Study (No. 411), Section 4, OECD Publishing, 1981, pp. 1–9.
[51] OECD, OECD Guideline for Testing of Chemicals: Sub Chronic Inhalational Toxicity: 90 Day Study (No. 413), Section 4, OECD Publishing, 2009, pp. 1–15.

[52] OECD, OECD Guideline for Testing of Chemicals, Section 4, Chronic Toxicity Studies (No. 452), Section 4, OECD Publishing, 2009, pp. 1–16.
[53] OECD, OECD Guideline for Testing of Chemicals, Section 4, Combined Chronic Toxicity/Carcinogenicity Studies (No. 453), Section 4, OECD Publishing, 2009, pp. 1–20.
[54] OECD, OECD Guideline for Testing of Chemicals: Acute Eye Irritation/Corrosion (No: 405), Section 4, OECD Publishing, 2012, pp. 1–19.
[55] OECD, OECD Guideline for Testing of Chemicals: Skin Sensitisation (No. 406), Section 4, OECD Publishing, 1992, pp. 1–9.
[56] B. Magnusson, A.M. Kligman, The identification of contact allergens by animal assay. The Guinea pig maximization test, J. Invest. Dermatol. 52 (3) (1969) 268–276.
[57] OECD, OECD Guideline for Testing of Chemicals: Prenatal Developmental Toxicity Study (No. 414), Section 4, OECD Publishing, 2001, pp. 1–11.
[58] OECD, OECD Guideline for Testing of Chemicals: Neurotoxicity Study in Rodents (No. 424), Section 4, OECD Publishing, 1997, pp. 1–15.
[59] OECD, OECD Guideline for Testing of Chemicals, Section 4, Carcinogenicity Studies (No. 451), Section 4, OECD Publishing, 2009, pp. 1–15.
[60] OECD, OECD Guideline for Testing of Chemicals: One-generation Reproduction Toxicity Study (No. 415), Section 4, OECD Publishing, 1983, pp. 1–8.
[61] OECD, OECD Guideline for Testing of Chemicals: Two – Generation Reproduction Toxicity Study (No. 416), Section 4, OECD Publishing, 2001, pp. 1–13.

Chapter 19

Prebiotics: A Functional Food in Health and Disease

Dharmik Joshi, Somdatta Roy, Sugato Banerjee
Department of Pharmaceutical Sciences and Technology, Birla Institute of Technology, Ranchi, India

1. GUT MICROBES

The four dominant phyla of the human gut are Actinobacteria, Bacteroidetes, Firmicutes, and Proteobacteria. Most bacteria belong to the genera *Bacteroides*, *Bifidobacterium*, *Clostridium*, *Eubacterium*, *Faecalibacterium*, *Peptococcus*, *Peptostreptococcus*, and *Ruminococcus* [1]. The currently known fungi of the gut flora are *Aspergillus*, *Candida*, *Saccharomyces*, and *Penicillium*.

Types of Intestinal Flora [1]:

1. Essential or beneficial flora: These are indigenous friendly bacteria. The main members of this group are Lactobacteria (*Lactobacillus rhamnosus*, *Lactobacillus acidophilus*, *Lactobacillus plantarum*), Bifidobacteria (*Bifidobacterium bifidum*), *Enterococci*, *Propionobacteria*, and *Peptostreptococci*. Beneficial flora is made up of probiotics, which are beneficial or good bacteria. They are the housekeepers of the gut, fulfilling myriad vital functions in the body and providing a natural barrier protecting us against all sorts of invaders, parasites, fungi, viruses, bacteria, toxins, etc. They produce antibiotic-like substances that are antiviral and antifungal, dissolving viruses and "bad" bacteria.
2. Opportunistic flora: This is a large group comprising Bacteriodes, Bacilli, Clostridia, Enterobacteria, Fuzobacteria, Eubacteria, Actenobacteria, Peptococci, Staphylococci, Streptococci, Yeasts, and many others.

The effect of gut microbes on obesity and the immune system is illustrated in Figs. 19.1 and 19.2.

2. FACTORS INFLUENCING THE COMPOSITION OF GUT FLORA

Age: The first microbiota profile is largely determined by the mode of delivery, i.e., infants born vaginally or by cesarean process [2]. After birth the gut

Natural Products and Drug Discovery. https://doi.org/10.1016/B978-0-08-102081-4.00019-8

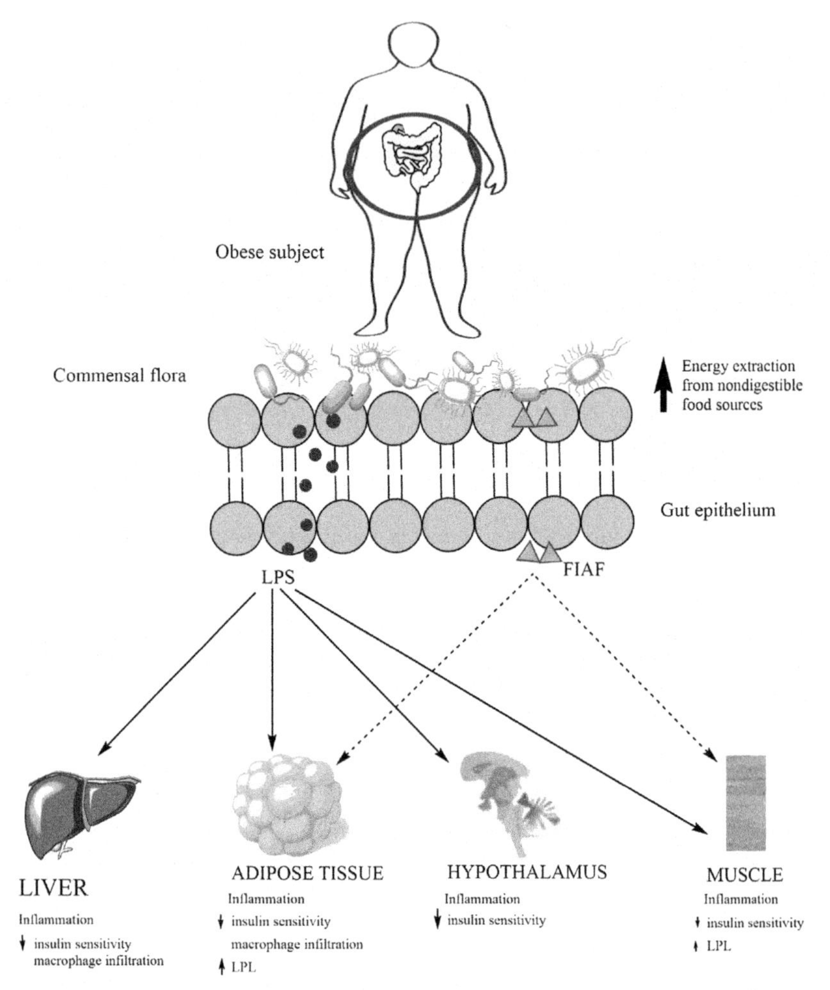

FIGURE 19.1 Gut bacteria and obesity. Gut bacteria may affect the release of liposaccharides (LPS) and fasting-induced adipose factor (FIAF), which are responsible for the inflammatory response of liver, adipose tissue, hypothalamus, and muscles by affecting insulin sensitivity, macrophage infiltration, and increasing lipoprotein lipase (LPL).

microbiota is more or less stable until 70 years. However, gut microbiota diminishes in diversity with aging and host health. Functional impacts of this alteration include reduction in the synthesis of vitamin B12, reduced microbial reductase activity, increased chances of DNA damage, altered stress response, and immune suppression [3].

Diet: The earliest effect on gut microbiota after birth is from the early infant diet, i.e., breast milk and formula feeds. Breast milk contains several bioactive compounds that enrich the intestinal flora of neonates, while formula

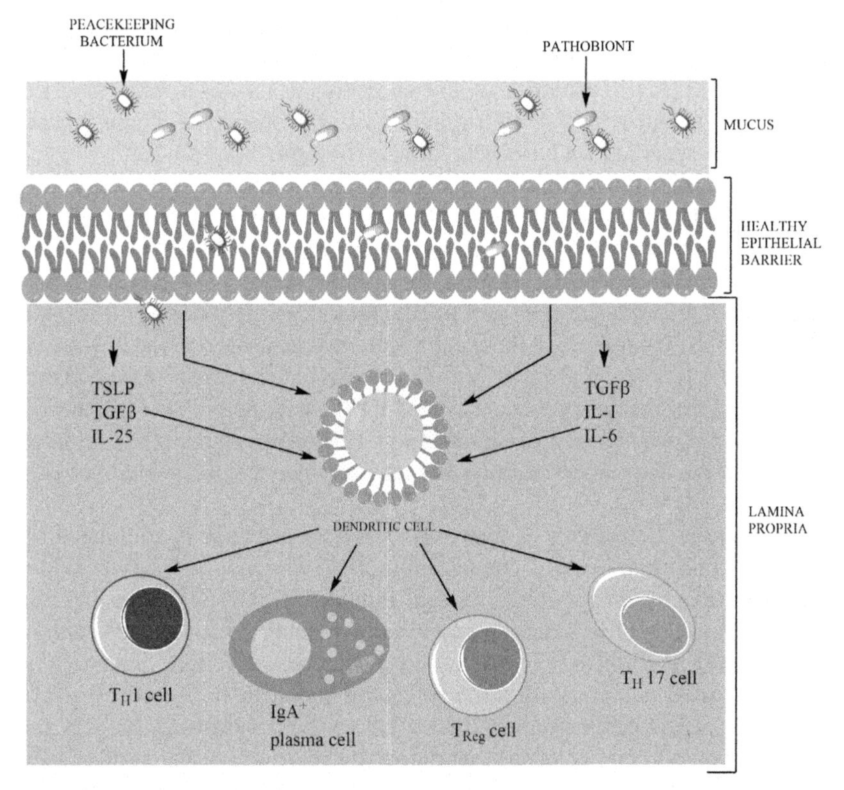

FIGURE 19.2 Effect of gut bacteria on the immune system. Gut bacteria-mediated thymic stromal lymphopoietin (TSLP), transforming growth factor beta (TGFβ), and interleukins (IL-25, IL-1, IL-6) may affect the differentiation of dendritic cells into T_H1 helper cells, regulatory T cells (T_{Reg}cell), T helper 17 (T_H17) cells, and immunoglobulin A (IgA^+).

feeds, which lack them, reduce the diversity of intestinal flora [4,5]. In general, intake of a diet rich in fruits, vegetables, legumes, and fibers is associated with healthy gut microbiota. Dietary polyphenols also play an important role in the promotion of beneficial intestinal bacteria [6].

Antibiotics: The use of antibiotics has major implications in shaping gut microbiota. Frequent use of antibiotics diminishes the diversity of enteric flora. Another concern from the use of broad-spectrum antibiotics is the propagation of resistant strains of organisms through horizontal gene transfer [7,8].

Lifestyle and environment: Smoking may alter intestinal microflora resulting in increased risk of Crohn's disease. Airborne toxins, including air pollutants, may reach the intestine via mucociliary clearance from the lungs, leading to increased chances of developing irritable bowel syndrome (IBS) [9]. Stress is one of the major regulators of gut microbiota. A sedentary lifestyle or lack of exercise may influence gut microbial populations. Poor sanitary conditions and poor personal hygiene leading to spread of pathogenic infections may disrupt gut microbial homeostasis. Altered sleep cycles, due to overseas travel, work schedule, or stress, may also alter the enteric flora of individuals [10].

Probiotics [11]: "Probiotics are live microorganisms which, when ingested in adequate amounts as a single strain or as a combination of strains, confer one or more specified health benefits to the consumer" (WHO 2002).

Synbiotics [12]: Synbiotics by definition is a combination of probiotics and prebiotics that act in synergism. The synbiotic improves the overall intestinal flora by promoting growth and by activating the metabolism of a limited number of health-promoting bacteria, thus improving host welfare. Examples include kefir, *Bifidobacterium*, and fructooligosaccharides (FOS).

Prebiotics: Prebiotics are those substances that contribute to the well-being of their host by inducing the growth or activity of microbiota (e.g., bacteria and fungi) [13]. Prebiotics that act as feed for the gut microbiome, which may alter their composition, is a common notion. However, in principle, it refers to any part of the body; hand moisturizers may improve the composition of skin microbiota.

The growth or activity of intestinal beneficial bacteria is stimulated by prebiotics that are typically nondigestible fiber compounds. They pass undigested through the upper part of the gastrointestinal (GI) tract by acting as a substrate for them [13]. Marcel Roberfroid first identified and named them in 1995 [13,14]. Prebiotics, like probiotics, fall in between foods and drugs as functional food components. Roberfroid first stated that: "A prebiotic is a selectively fermented ingredient that allows specific changes, both in the composition and activity of the gastrointestinal microflora that confers benefits upon host well-being and health" [15]. In 2007, only two prebiotics, *trans*-galactooligosaccharide and inulin, fully met the definition of Roberfroid [16]. Larch arabinogalactan [17], resistant starch [18], pectin [19], β-glucans [20], and xylooligosaccharides (XOS) [21] also fall under the same category of prebiotics. Prebiotics are short-chain carbohydrates that are nondigestible by human digestive enzymes. They selectively enhance the activity of some beneficial bacteria. Beneficial intestinal bacteria ferment prebiotics into short-chain fatty acids. Prebiotics also reduce cancer risk and increase calcium and magnesium absorption. Prebiotics, considered as functional food components, are found in several vegetables and fruits, which present significant technological advantages. The addition of prebiotics improves sensory characteristics such as taste and texture. It enhances the stability of foams, emulsions, and mouth feel in a large range of food applications such as dairy products and bread [22].

The criteria for nondigestible carbohydrates to be considered as prebiotic are:

1. They must be resistant to gastric acidity and mammalian enzymes;
2. They must be susceptible to fermentation by gut bacteria; and
3. They must be able to enhance the viability and activity of beneficial microorganisms.

The most commonly known prebiotics are galactooligosaccharide (GOS), FOS, and inulin. GOS occurs naturally in mammalian milk. It is nondigestible, derived from lactose, and consists of chains of galactose monomers. According to Roberfroid [23], inulin and inulin-type fructans are known soluble dietary fibers. There are dietary fibers that contain several nonstarch polysaccharides, such as cellulose, dextrins, pectins, β-glucans, waxes, and lignin. They can adjust the transfer time through the gut, and offer the same useful effects as those of inulin-type fructans. Asparagus, chicory, tomatoes, and wheat are naturally occurring prebiotics found in various foods, and it is a natural constituent of breast milk. Prebiotics along with their sources are summarized in Tables 19.1 and 19.2, while their chemical structures are elucidated in Fig. 19.3.

3. HEALTH BENEFITS OF PREBIOTICS

3.1 Acute Gastroenteritis

Ingested food or contaminated water may cause acute gastroenteritis, which affects everyone at some point in life especially in the developing nations. Gastroenteritis may be primarily caused by Shigellae, Salmonellae, *Yersinia enterocolitica*, *Campylobacter jejuni*, *Escherichia coli*, *Vibrio cholera*, and *Clostridium perfringens* [24]. The presence of beneficial bacteria in increased levels along with high immunity provides protection from these pathogens. The antiadhesive property of prebiotics adds significant functionality to the approach of altering gut pathogenesis. Many intestinal pathogens utilize monosaccharides or short oligosaccharide sequences as receptors by binding to them, which is the first step of colonization [25]. Prebiotic oligosaccharide may block and dislodge the adherent pathogen [26]. Prebiotics may also reduce the virulence of certain pathogens. For example, pathogenicity of *Listeria monocytogenes* can be repressed through the downregulation of virulence factors by cellobiose [27]. *Listeria* is virulent in the soil, due to the presence of cellobiose. In the human body, cellobiose is absent, which reduces its virulence, while additional incorporation of disaccharide *Listeria*-contaminated food could reduce this virulence even further.

3.2 Cancer

Prebiotics may decrease genotoxicity. GOS administration in humans reduces nitroreductase indole and isovaleric acid, which are shown to have mutagenic and carcinogenic potential [28]. Other genotoxic enzymes such as β-glucosidase, β-glucuronidase, and arylsulfatase have also been shown to be strongly inhibited by GOS [29]. Bifidobacteria produces lower levels of GOS than lactobacilli and thus bifidobacteria and lactobacilli decrease genotoxic enzyme production [30]. In a study on human flora-associated rats [31], it

TABLE 19.1 Sources for Prebiotics [65]

Foods Rich in Prebiotics	
Food	**Prebiotic Fiber Content by Weight (%)**
Gum arabic	85
Raw dry chicory root	64.6
Raw dry Jerusalem artichoke	31.5
Raw dry dandelion greens	24.3
Raw dry garlic	17.5
Raw dry leek	11.7
Raw dry onion	8.6
Raw asparagus	5
Raw wheat bran	5
Whole wheat flour, cooked	4.8
Raw banana	1
Food	**Amount of Food to Obtain 6 g Serving of Prebiotics**
Raw chicory root	9.3 g (0.33 oz)
Raw Jerusalem artichoke	19 g (0.67 oz)
Raw dandelion greens	24.7 g (0.87 oz)
Raw garlic	34.3 g (1.21 oz)
Raw leek	51.3 g (1.81 oz)
Raw onion	69.8 g (2.46 oz)
Cooked onion	120 g (4.2 oz)
Raw asparagus	120 g (4.2 oz)
Raw wheat bran	120 g (4.2 oz)
Whole wheat flour, cooked	125 g (4.4 oz)
Raw banana	600 g (1.3 lb)

was found that β-glucosidase increased, whereas β-glucuronidase and ammonia levels decreased, upon feeding resistant starch to these animals. Butyrate, one of the primary sources of energy for colonocytes, may also help in preventing cancer [32]. It may be proapoptotic, thus helping in the elimination of cancer cells [33].

TABLE 19.2 Types and Corresponding Sources of Prebiotics

Type of Prebiotic	Sources of Prebiotic
Fructooligosaccharides	Asparagus, sugar beet, garlic, chicory, onion, Jerusalem artichoke, wheat, honey, banana, barley, tomato, and rye
Isomaltulose	Honey, sugarcane juice
Xylooligosaccharides	Bamboo shoots, fruits, vegetables, milk, honey, and wheat bran
Galactooligosaccharides	Human's milk and cow's milk
Cyclodextrins	Water-soluble glucans
Raffinose oligosaccharides	Seeds of legumes, lentils, peas, beans, chickpeas, mallow composite, and mustard
Soybean oligosaccharide	Soybean
Lactulose	Lactose (milk)
Lactosucrose	Lactose
Isomaltulose	Sucrose
Palatinose	Sucrose
Maltooligosaccharides	Starch
Isomaltooligosaccharides	Starch
Arabinoxylooligosaccharides	Wheat bran
Enzyme-resistant dextrin	Potato starch

3.3 Mineral Absorption

Increasing intake of calcium and magnesium, which are important components of bone structure, may prevent the development of osteoporosis. Chonan et al. [34] have shown that GOS may increase calcium and magnesium absorption in rats. This report says that colonic flora may be required for GOS to have this effect. FOS have also been shown to affect mineral absorption; mainly, calcium absorption is increased by 40 g inulin or 15 g of oligofructose daily [35]. Magnesium intake has been found to be increased following ingestion of FOS.

3.4 Lipid Metabolism

Prebiotics may also influence lipid metabolism. XOS may reduce serum cholesterol and triglyceride levels [23]. FOS has been shown to inhibit lipid metabolizing enzymes, thus affecting serum lipids. The mechanism may

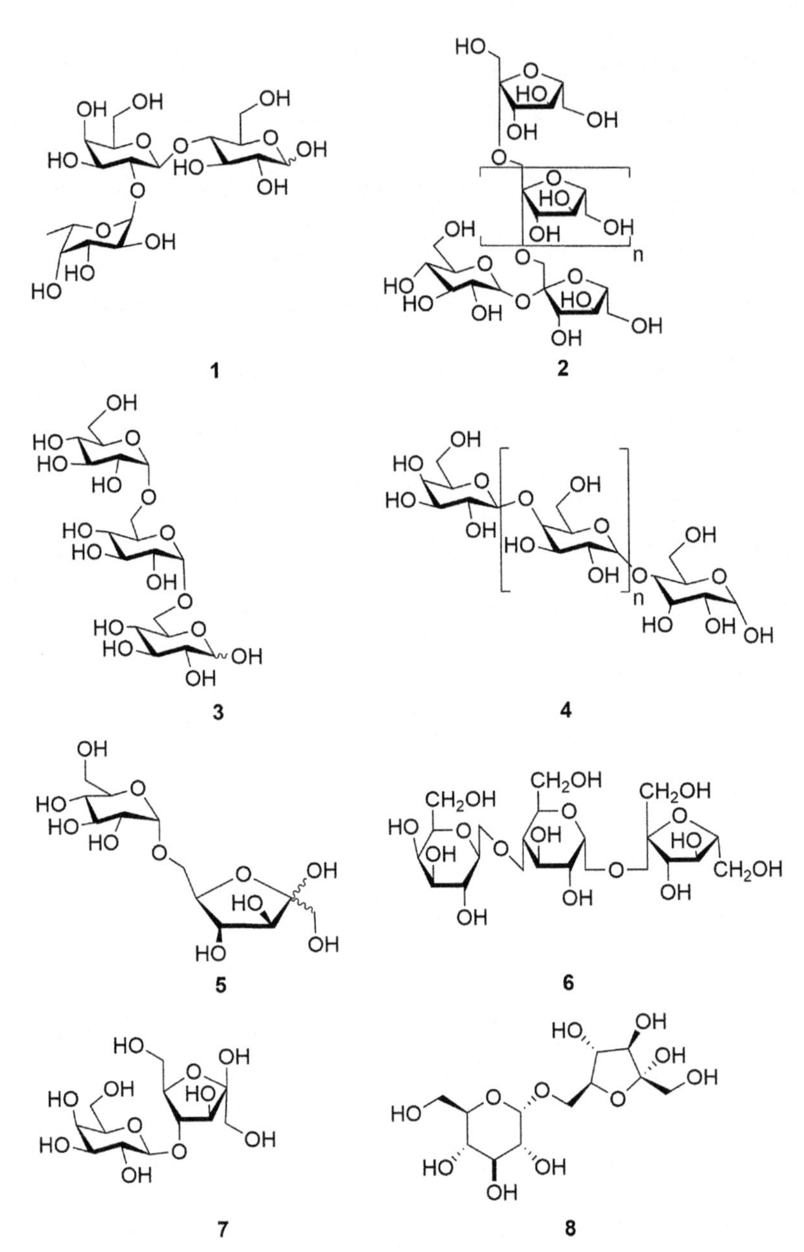

FIGURE 19.3 Chemical structures of natural prebiotics. (**1**) 2′-Fucosyllactose; (**2**) inulin; (**3**) isomaltotriose; (**4**) galactooligosaccharide; (**5**) isomaltulose; (**6**) lactosucrose; (**7**) lactulose; (**8**) palatinose; (**9**) panose; (**10**) raffinose; (**11**) amylopectin (starch); (**12**) stachyose; (**13**) xylooligosaccharides; (**14**) amylose (starch).

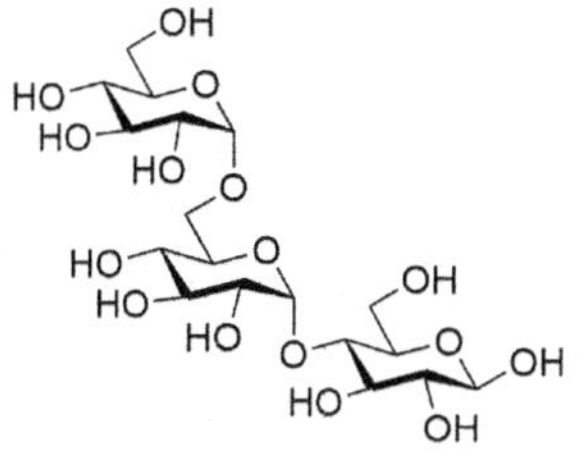

FIGURE 19.3 cont'd

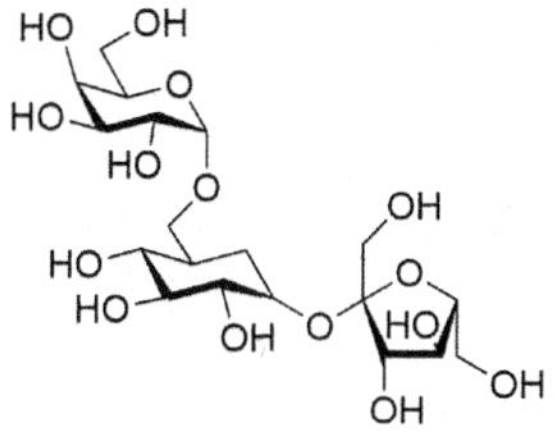

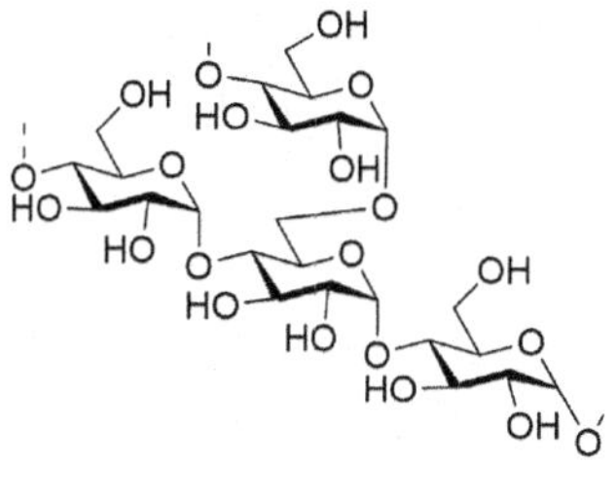

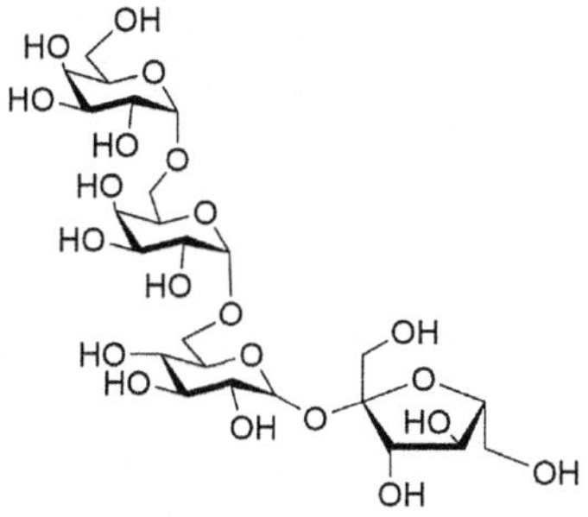

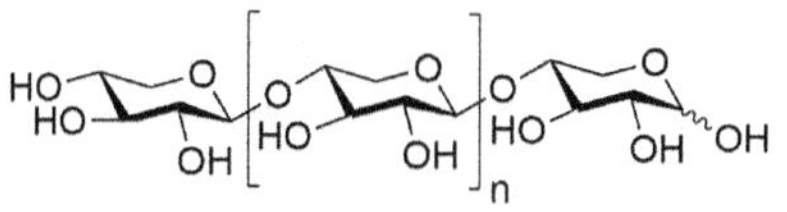

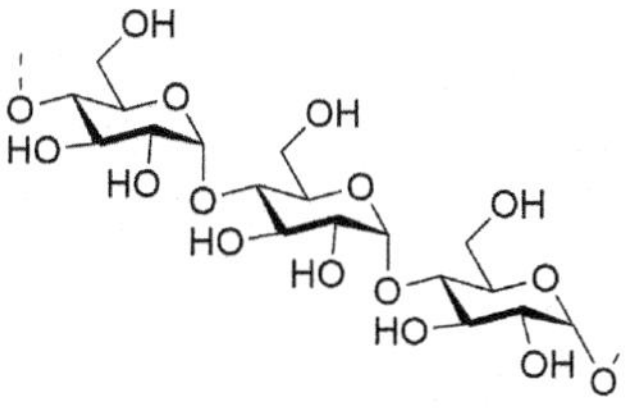

involve propionates, which are fermented prebiotics produced by gut bacteria. While prebiotics can correct hyperlipidemia associated with diabetes, they may not decrease lipids in healthy subjects. Prebiotics increase viscosity in the upper intestinal tract. This viscosity may act as a physical barrier decreasing the (re)absorption of fats, including cholesterol and bile acids leading to increased fecal output of cholesterol and bile acids resulting in greater liver cholesterol metabolism [36]. Inulin may reduce cecal pH [36] resulting in reduced levels of soluble bile acids leading to decreased lipid absorption. Prebiotics may alter hepatic triacylglycerol synthesis. The lipid-lowering action of inulin is possible because it influences hepatic triacylglycerol synthesis, very low-density lipoprotein cholesterol secretion, and decreases circulating bile acids by increasing its excretion [37]. This increased excretion makes liver cholesterol synthesize bile acids. Furthermore, hepatic low-density lipoprotein (LDL) receptor expression increases to replenish hepatic cholesterol leading to decreased serum LDL [38]. Prebiotics ferment to lower cholesterol. The hypocholesterolemic effect of prebiotics is derived from a metabolic effect [39], because these compounds are fermented in the lower intestinal tract [40]. Oligosaccharides are rapidly fermented [41], thus increasing propionates, which reach liver and inhibit cholesterol pathways by preventing the conversion of acetyl-coenzyme A to mevalonate [40], similar to the mechanism of action of statins. Prebiotics reduce serum insulin and glucose. Oligofructose and inulin coordinately reduce hepatic fatty acid and triacylglycerol synthesis. The triacylglycerol-lowering effect of oligofructose is primarily due to its antilipogenic action [42] by reducing the expression of lipogenic enzymes [43]. LDL receptor expression increases in serum total cholesterol occurring in rats receiving resistant starch extracted from beans [44]. Moreover, studies have also reported a decrease in serum total cholesterol after dietary supplementation with prebiotics in vivo [43,44].

3.5 Distant Effects of Prebiotics

Effect on the brain: The gut has been dubbed the "second brain" since it has been shown to communicate directly with the brain. Neural, endocrine, and immunological signaling regulates gut–brain communication. Since the 1970s it has been known that stress and gut flora have a bidirectional relation, while the mechanism of such communication has recently been elucidated. A major factor is the composition of the microbiota, which may be regulated by age, diet, geographical location, diseases, etc. Growth substrates, such as conversion of sugar into fermentation products and by-products, may have an inhibitory effect on good versus pathogenic microbes [12]. Other factors include bacteriocins, competition between microbes, and immune activation, which may affect gut–brain crosstalk. Microbiotic influence on the innate immune system indirectly mediates pro- or antiinflammatory cytokine circulation, which may modulate brain function. Vagus nerve mediates may be responsible for transmitting luminal environmental signals such as carbohydrate levels, hyperosmolarity, and levels of intestinal bacterial products [45]. Various neurotransmissions are produced by microbes. Examples include γ-aminobutyric acid by *Lactobacillus* spp. and *Bifidobacterium* spp., serotonin by *Candida* spp., *Streptococcus* spp., *Escherichia* spp., and *Enterococcus* spp., noradrenaline by *Escherichia* spp., *Bacillus* spp., and *Saccharomyces* spp., acetylcholine by *Lactobacillus* spp., and dopamine by *Bacillus* spp. They have a direct action on the enteric nervous system, and act through GI nerve endings or via enterochromaffin cells accessible to microbes. Microbial metabolites may modulate the metabolism of host short-chain fatty acids, choline, and bile acids. Microbes also break down carbohydrates from dietary fibers and produce acetate, hydrogen sulfide, *n*-butyrate, and propionate, which may affect the host nervous system. Alterations or overproduction of these metabolites may be associated with autism. Tryptophan is an essential amino acid that must be consumed or generated by gut bacteria. Altered tryptophan metabolism has been associated with various digestive and brain disorders. Tryptophan is a precursor for serotonin, the primary neurotransmitter in mediating mood and well-being [46]. A potential strategy for modulating the microbiome–gut–brain axis is the use of prebiotics, nondigestible food ingredients that promote the growth of gut microorganisms as well as probiotics. Specific prebiotics have been shown to increase brain-derived neurotrophic factor levels [47]. In dementia, diabetes has shown to be a risk factor. A trial of a synbiotic (*L. acidophilus* plus prebiotic) demonstrated that consumption of the synbiotic food for 6 weeks had significant positive effects on serum insulin, C-reactive protein, and uric acid. Gut microbiota has been shown to modulate functions of the hypothalamic–pituitary–adrenal (HPA) axis [48]. Microbiota-mediated antiinflammatory and immune activation following probiotic administration has been suggested to be a potential mechanism [49]. The vagus nerve has been shown to be the primary gut–brain connection [50] (Fig. 19.4). In 2015, the *Journal of Agricultural and Food Chemistry* published that brain function can be improved in controlled human, animal, and in vitro studies after oral, systemic, and localized administration of varieties of polysaccharides. For example, isolichenan is an α-glucan from the lichen *Cetrariella islandica*, whose ingestion reversed ethanol-induced memory impairment in mice [51]. In healthy middle-aged adults, cognitive function and mood improved significantly when supplemented with a mixed polysaccharide product (Ambrotose complex). Memory in a mouse model of vascular dementia was preserved by oral administration of arabinoxylan from the yeast *Triticum aestivum* and β-glucan from barley [52]. Results from a study exploring the neuroendocrine and affective effects of FOS and Bimuno-galactooligosaccharides (B-GOS) in healthy human volunteers revealed that B-GOS lowered cortisol awakening reactivity and increased attentional vigilance in the dot-probe task [53]. B-GOS administration has also been shown to have an antidepression effect by modulating the HPA axis [54]. Therefore the question arises: how do prebiotics act on the gut microbiota-brain axis? Arabinoxylan stimulates the growth of known butyrate producers (*Roseburia intestinalis*, *Eubacterium rectale*, *Anaerostipes caccae*) [55]; fucoidan reduces the Enterobacteriaceae population in the newly weaned pig [56]; and glucan, which is highly fermentable by the intestinal microbiota in the cecum and colon, enhances the growth of *Lactobacillus* strains in the human intestine [19]. On the other hand, prebiotics directly influence signaling molecules in the brain. In a study it was revealed by Savignac et al. [47] that prebiotics (FOS and GOS) increased hippocampal BDNF and NR1 subunit expression but did not alter amino acids associated with glutamate neurotransmission compared to controls. GOS has also been shown to increase hippocampal NR2A subunits, frontal cortex NR1, and plasma peptide YY. [47]. This suggests that the prebiotics-mediated altered microbiota's effect on brain chemistry is similar to that of probiotics that increases brain BDNF expression probably through the involvement of gut hormones [47]. Compared with

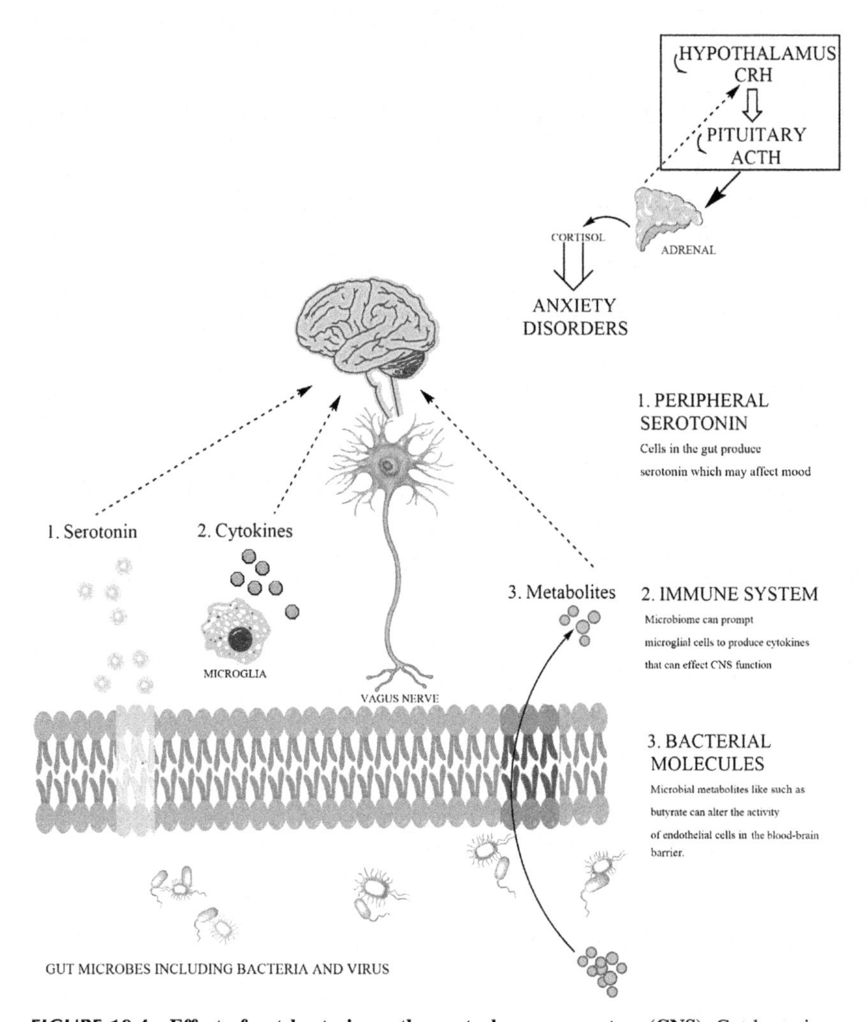

FIGURE 19.4 Effect of gut bacteria on the central nervous system (CNS). Gut bacteria as well as their metabolites affect the release of neurotransmitters such as serotonin and facilitate the release of cytokines by microglia. Gut flora also modulates the hypothalamic–pituitary–adrenal (HPA) axis. The vagus nerve is responsible for gut–brain communication. *ACTH*, Adrenocorticotropic hormone; *CRH*, corticotropin-releasing hormone.

probiotics-based modulation, prebiotics could be advantageous to some extent due to the presence of a survival problem in the GI tract for probiotics. In particular, genetic factors could have major impacts such that a single host gene, FUT2, is strongly associated with the diversity and composition of the human bifidobacterial population. Specifically, several genotypes, including *B. bifidum*, *Bifidobacterium adolescentis*, and *Bifidobacterium catenulatum*/*pseudocatenulatum*, were absent or rarely colonized in individuals without the

FUT2 gene. On the other hand, probiotics are usually supplemented by one or several species at one time, whereas prebiotics supplementation could stimulate beneficial species simultaneously, which is good for their synergistic actions.

IBS: Hypothetically, prebiotics' ability to enrich gut microbiota may correct the composition of the intestinal microbial population associated with IBS. A study by Olesenand and Gudmand-Hoyer [57] contradicts this theory. This group showed 20 g of inulin treatment for 12 weeks worsened IBS symptoms, whereas placebo symptoms improved after 4–6 weeks of treatment. However, a clinical trial by Paineau et al. [58] on 105 IBS subjects, upon treatment with inulin-type fructan, improved quality of life due to the reduction in intensity of GI symptoms. Another study tested the effect of GOS on IBS for 6 weeks. The prebiotic groups showed improvement in flatulence, bloating, and other IBS-associated symptoms, which were correlated with an increase in the intestinal Bifidobacteria count [59]. Thus clinical studies testing prebiotics may suggest that specific prebiotics (inulin and GOS) could improve IBS symptoms. A GOS and polydextrose combination has been shown to reduce anxiety-related behavior and improve positive social interactions in animals.

Antiviral action: Oligofructose administered to infants for several weeks may elevate their measles-specific IgG response, thus leading to measles vaccination [60]. Fructan prebiotics promotes adaptive responses against influenza immunization [61].

Type IV hypersensitivity response is improved by an oligosaccharide mixture (9:1 short-chain GOS to long-chain FOS + pectin acidic-oligosaccharides). Prebiotics have a protective effect against the onset of eczema by combatting allergic diseases. Partially hydrolyzed whey protein can minimize allergic skin reactions masking the effect of supplementation of prebiotics [62]. Human milk oligosaccharides were expressing the sugar secretory protein FUT2 causing allergic disease and eczema in breastfed children due to the inhibitory effect on IgE by fructan.

Skin diseases: Prebiotics can attenuate allergic skin diseases such as atopic dermatitis [63]. Phenolic compounds may cause keratin depletion that can be prevented by GOS alone or with *Bifidobacterium* in women [64].

REFERENCES

[1] J. Bienenstock, W. Kunze, P. Forsythe, Microbiota and the gut–brain axis, Nutr. Rev. 73 (Suppl. 1) (2015) 28–31.

[2] M.G. Dominguez-Bello, E.K. Costello, M. Contreras, M. Magris, G. Hidalgo, N. Fierer, R. Knight, Delivery mode shapes the acquisition and structure of the initial microbiota across multiple body habitats in newborns, Proc. Natl. Acad. Sci. USA 107 (2010) 11971–11975.

[3] Y. Lan, A. Kriete, G.L. Rosen, Selecting age-related functional characteristics in the human gut microbiome, Microbiome 1 (2013) 2.
[4] L.G. Albenberg, G.D. Wu, Diet and the intestinal microbiome: associations, functions, and implications for health and disease, Gastroenterology 146 (2014) 1564–1572.
[5] E.M. Brown, M. Sadarangani, B.B. Finlay, The role of the immune system in governing host-microbe interactions in the intestine, Nat. Immunol. 14 (2013) 660–667.
[6] M. Dueñas, I. Muñoz-González, C. Cueva, A. Jiménez-Girón, F. Sánchez-Patán, C. Santos-Buelga, M.V. Moreno-Arribas, B. Bartolomé, A survey of modulation of gut microbiota by dietary polyphenols, Biomed. Res. Int. 2015 (2015) 850902.
[7] L.S. Frost, R. Leplae, A.O. Summers, A. Toussaint, Mobile genetic elements: the agents of open source evolution, Nat. Rev. Microbiol. 3 (2005) 722–732.
[8] H. Ochman, J.G. Lawrence, E.A. Groisman, Lateral gene transfer and the nature of bacterial innovation, Nature 405 (2000) 299–304.
[9] J.L. Benjamin, C.R.H. Hedin, A. Koutsoumpas, S.C. Ng, N.E. McCarthy, N.J. Prescott, P. Pessoa-Lopes, C.G. Mathew, J. Sanderson, A.L. Hart, et al., Smokers with active Crohn's disease have a clinically relevant dysbiosis of the gastrointestinal microbiota, Inflamm. Bowel Dis. 18 (2012) 1092–1100.
[10] R.M. Voigt, C.B. Forsyth, S.J. Green, E. Mutlu, P. Engen, M.H. Vitaterna, F.W. Turek, A. Keshavarzian, Circadian disorganization alters intestinal microbiota, PLoS One 9 (2014) e97500.
[11] G.B. Nair, Y. Takeda, Probiotics Food in Health and Disease, CRC Press, 2011, pp. 1–11.
[12] M. Anandharaj, B. Sivasankari, R. Parveen Rani, Effects of probiotics, prebiotics, and synbiotics on hypercholesterolemia: a review, Chin. J. Biol. 2014 (2014) 1–7.
[13] R.W. Hutkins, J.A. Krumbeck, L.B. Bindels, P.D. Cani, G. Fahey, Y.J. Goh, B. Hamaker, E.C. Martens, D.A. Mills, R.A. Rastal, E. Vaughan, Prebiotics: why definitions matter, Curr. Opin. Biotechnol. 37 (2016) 1–7.
[14] G.R. Glenn, M.B. Roberfroid, Dietary modulation of the human colonic microbiota: introducing the concept of prebiotics, J. Nutr. 125 (1995) 1401–1412.
[15] M. Roberfroid, Prebiotics: the concept revisited, J. Nutr. 137 (3) (March 1, 2007) 830S–837S.
[16] G.S. Kelly, Larch arabinogalactan: clinical relevance of a novel immune-enhancing polysaccharide, Altern. Med. Rev. J. Clin. Ther. 4 (2) (1999) 96–103.
[17] S.A. Zaman, S.R. Sarbini, The potential of resistant starch as a prebiotic, Crit. Rev. Biotechnol. 36 (3) (2016) 578–584.
[18] B. Gomez, B. Gullon, C. Remoroza, H.A. Schols, J.C. Parajo, J.L. Alonso, Purification, characterization, and prebiotic properties of pectic oligosaccharides from orange peel wastes, J. Agric. Food Chem. 62 (40) (2014) 9769–9782.
[19] M.P. Arena, G. Caggianiello, D. Fiocco, P. Russo, M. Torelli, G. Spano, V. Capozzi, Barley β-glucans-containing food enhances probiotic performances of beneficial bacteria, Int. J. Mol. Sci. (2) (2014) 3025–3039.
[20] I. Jain, V. Kumar, T. Satyanarayana, Xylooligosaccharides: an economical prebiotic from agroresidues and their health benefits, Indian J. Exp. Biol. 53 (2015) 131–142.
[21] S.H. Al-Sheraji, A. Ismail, M.Y. Manap, S. Mustafa, R.M. Yusof, F.A. Hassan, Prebiotics as functional foods: a review, J. Funct. Foods 5 (4) (2013) 1542–1553.
[22] A.J. Moshfegh, J.E. Friday, J.P. Goldman, J.K. Ahuja, Presence of inulin and oligofructose in the diets of Americans, J. Nutr. 129 (7) (1999) 1407S–1411S.
[23] M. Roberfroid, Functional food concept and its application to prebiotics, Dig. Liver Dis. 34 (2002) S105–S110.

[24] J. Slavin, Fiber and prebiotics: mechanisms and health benefits, Nutrients 5 (4) (April 22, 2013) 1417–1435.

[25] Y.H. Hui, J.R. Gorham, K.D. Murrell, D.O. Cliver, Food Borne Diseases Handbook – Diseases Caused by Bacteria, Marcel Dekker Inc. New York, New York, 1994.

[26] K.A. Karlsson, Animal glycosphingolipids as membrane attachment sites for bacteria, Annu. Rev. Biochem. 58 (1) (1989) 309–350.

[27] N. Jayaraman, S.A. Nepogodiev, J.F. Stoddart, Synthetic carbohydrate-containing dendrimers, Chem. A Eur. J. 3 (8) (1997) 1193–1199.

[28] S.F. Park, R.G. Kroll, Expression of listeriolysin and phosphatidylinositol-specific phospholipase C is repressed by the plant-derived molecule cellobiose in *Listeria monocytogenes*, Mol. Microbiol. 8 (4) (1993) 653–661.

[29] M. Ito, Y. Deguchi, A. Miyamori, K. Matsumoto, H. Kikuchi, K. Matsumoto, Y. Kobayashi, T. Yajima, T. Kan, Effects of administration of galactooligosaccharides on the human faecal microflora, stool weight and abdominal sensation, Microb. Ecol. Health Dis. 3 (6) (1990) 285–292.

[30] A.J. McBain, G.T. Macfarlane, Modulation of genotoxic enzyme activities by non-digestible oligosaccharide metabolism in in-vitro human gut bacterial ecosystems, J. Med. Microbiol. 50 (9) (2001) 833–842.

[31] A.J. Burns, I.R. Rowland, Anti-carcinogenicity of probiotics and prebiotics, Curr. Issues Intest. Microbiol. 1 (1) (2000) 13–24.

[32] S. Silvi, C.J. Rumney, A. Cresci, I.R. Rowland, Resistant starch modifies gut microflora and microbial metabolism in human flora-associated rats inoculated with faeces from Italian and UK donors, J. Appl. Microbiol. 86 (3) (1999) 521–530.

[33] D.L. Topping, P.M. Clifton, Short-chain fatty acids and human colonic function: roles of resistant starch and nonstarch polysaccharides, Physiol. Rev. 81 (3) (2001) 1031–1064.

[34] O. Chonan, R. Takahashi, M. Watanuki, Role of activity of gastrointestinal microflora in absorption of calcium and magnesium in rats fed β1-4 linked galactooligosaccharides, Biosci. Biotechnol. Biochem. 65 (8) (2001) 1872–1875.

[35] F.R. Bornet, F. Brouns, Y. Tashiro, V. Duvillier, Nutritional aspects of short-chain fructooligosaccharides: natural occurrence, chemistry, physiology and health implications, Dig. Liver Disease 34 (2002) S111–S120.

[36] K. Imaizumi, Y. Nakatsu, M. Sato, Y. Sedarnawati, M. Sugano, Effects of xylooligosaccharides on blood glucose, serum and liver lipids and cecum short-chain fatty acids in diabetic rats, Agric. Biol. Chem. 55 (1) (1991) 199–205.

[37] K. Vanhoof, R. De Schrijver, Effect of unprocessed and baked inulin on lipid metabolism in normo-and hypercholesterolemic rats, Nutr. Res. 15 (11) (1995) 1637–1646.

[38] E.A. Trautwein, D. Rieckhoff, H.F. Erbersdobler, Dietary inulin lowers plasma cholesterol and triacylglycerol and alters biliary bile acid profile in hamsters, J. Nutr. 128 (11) (1998) 1937–1943.

[39] L. Ellegård, H. Andersson, Oat bran rapidly increases bile acid excretion and bile acid synthesis: an ileostomy study, Eur. J. Clin. Nutr. 61 (8) (2007) 938–945.

[40] M.A. Levrat, M.L. Favier, C. Moundras, C. Rémésy, C. Demigné, C. Morand, Role of dietary propionic acid and bile acid excretion in the hypocholesterolemic effects of oligosaccharides in rats, J. Nutr. 124 (4) (1994) 531–538.

[41] A. Adam, M.A. Levrat-Verny, H.W. Lopez, M. Leuillet, C. Demigné, C. Rémésy, Whole wheat and triticale flours with differing viscosities stimulate cecal fermentations and lower plasma and hepatic lipids in rats, J. Nutr. 131 (6) (2001) 1770–1776.

[42] G.R. Gibson, Dietary modulation of the human gut microflora using the prebiotics oligofructose and inulin, J. Nutr. 129 (7) (1999) 1438S−1441S.

[43] N. Kok, M. Roberfroid, A. Robert, N. Delzenne, Involvement of lipogenesis in the lower VLDL secretion induced by oligofructose in rats, Br. J. Nutr. 76 (06) (1996) 881−890.

[44] N.M. Delzenne, N.N. Kok, Biochemical basis of oligofructose-induced hypolipidemia in animal models, J. Nutr. 129 (7) (1999) 1467S−1470S.

[45] P. Forsythe, W.A. Kunze, J. Bienenstock, On communication between gut microbes and the brain, Curr. Opin. Gastroenterol. 28 (6) (2012) 557−562.

[46] P. Forsythe, N. Sudo, T. Dinan, V.H. Taylor, J. Bienenstock, Mood and gut feelings, Brain Behav. Immun. 24 (1) (2010) 9−16.

[47] H.M. Savignac, G. Corona, H. Mills, L. Chen, J.P. Spencer, G. Tzortzis, P.W. Burnet, Prebiotic feeding elevates central brain derived neurotrophic factor, N-methyl-d-aspartate receptor subunits and d-serine, Neurochem. Int. 63 (8) (2013) 756−764.

[48] J.F. Cryan, T.G. Dinan, Mind-altering microorganisms: the impact of the gut microbiota on brain and behaviour, Nat. Rev. Neurosci. 13 (10) (2012) 701−712.

[49] A. Ait-Belgnaoui, H. Durand, C. Cartier, G. Chaumaz, H. Eutamene, L. Ferrier, E. Houdeau, J. Fioramonti, L. Bueno, V. Theodorou, Prevention of gut leakiness by a probiotic treatment leads to attenuated HPA response to an acute psychological stress in rats, Psychoneuroendocrinology 37 (11) (2012) 1885−1895.

[50] J.A. Bravo, P. Forsythe, M.V. Chew, E. Escaravage, H.M. Savignac, T.G. Dinan, J. Bienenstock, J.F. Cryan, Ingestion of Lactobacillus strain regulates emotional behavior and central GABA receptor expression in a mouse via the vagus nerve, Proc. Natl. Acad. Sci. 108 (38) (2011) 16050−16055.

[51] M. Smriga, J. Chen, J.T. Zhang, T. Narui, S. Shibata, E. Hirano, H. Saito, Isolichenan, an α-glucan isolated from lichen *Cetrariella islandica*, repaires impaired learning behaviors and facilitates hippocampal synaptic plasticity, Proc. Jpn. Acad. Ser. B 75 (7) (1999) 219−223.

[52] H.S. Han, J.H. Jang, J.H. Jang, J.S. Choi, Y.J. Kim, C. Lee, S.H. Lim, H.K. Lee, J. Lee, Water extract of *Triticum aestivum* L. and its components demonstrate protective effect in a model of vascular dementia, J. Med. Food 13 (3) (2010) 572−578.

[53] K. Schmidt, P.J. Cowen, C.J. Harmer, G. Tzortzis, S. Errington, P.W. Burnet, Prebiotic intake reduces the waking cortisol response and alters emotional bias in healthy volunteers, Psychopharmacology 232 (10) (2015) 1793−1801.

[54] Z.N. Mannie, C.J. Harmer, P.J. Cowen, Increased waking salivary cortisol levels in young people at familial risk of depression, Am. J. Psychiatry 164 (4) (2007) 617−621.

[55] P. Van den Abbeele, P. Gérard, S. Rabot, A. Bruneau, S. El Aidy, M. Derrien, M. Kleerebezem, E.G. Zoetendal, H. Smidt, W. Verstraete, T. Van de Wiele, Arabinoxylans and inulin differentially modulate the mucosal and luminal gut microbiota and mucin-degradation in humanized rats, Environ. Microbiol. 13 (10) (2011) 2667−2680.

[56] A.M. Walsh, T. Sweeney, C.J. O'Shea, D.N. Doyle, J.V. O'Doherty, Effect of dietary laminarin and fucoidan on selected microbiota, intestinal morphology and immune status of the newly weaned pig, Br. J. Nutr. 110 (09) (2013) 1630−1638.

[57] M. Olesen, E. Gudmand-Hoyer, Efficacy, safety, tolerability fructooligosaccharides treatment irritable bowel syndrome, Am. J. Clin. Nutr 72 (6) (2000) 1570−1575.

[58] D. Paineau, F. Payen, S. Panserieu, G. Coulombier, A. Sobaszek, I. Lartigau, et al., The effects of regular consumption of short-chain fructo-oligosaccharides on digestive comfort of subjects with minor functional bowel disorders, Br. J. Nutr. 99 (2008) 311−318.

[59] D.B. Silk, A. Davis, J. Vulevic, G. Tzortzis, G.R. Gibson, Clinical trial: the effects of a trans-galactooligosaccharide prebiotic on faecal microbiota and symptoms in irritable bowel syndrome, Alimen. Pharmacol. Ther. 29 (5) (2009) 508–518.

[60] A. Firmansyah, G. Pramita, A. Carrie Fassler, F. Haschke, H. Link-Amster, Improved humoral response to measles vaccine in infants receiving infant cereal with fructo-oligosaccharides, J. Pediatr. Gastroenterol. Nutr. 31 (2001) A521.

[61] A.R. Lomax, L.V. Cheung, P.S. Noakes, E.A. Miles, P.C. Calder, Inulin-type β2-1 fructans have some effect on the antibody response to seasonal influenza vaccination in healthy middle-aged humans, Nutr. Immunol. 6 (2015) 490.

[62] A.P. Vos, J. Knol, B. Stahl, L. M'rabet, J. Garssen, Specific prebiotic oligosaccharides modulate the early phase of a murine vaccination response, Int. Immunopharmacol. 10 (2010) 619–625.

[63] G. Moro, S. Arslanoglu, B. Stahl, J. Jelinek, U. Wahn, G. Boehm, A mixture of prebiotic oligosaccharides reduces the incidence of atopic dermatitis during the first six months of age, Arch. Dis. Child. 91 (2006) 814–819.

[64] M. Kano, N. Masuoka, C. Kaga, S. Sugimoto, R. Iizuka, K. Manabe, T. Sone, K. Oeda, C. Nonaka, K. Miyazaki, et al., Consecutive intake of fermented milk containing *Bifidobacterium* breve strain Yakult and galacto-oligosaccharides benefits skin condition in healthy adult women, Biosci. Microbiota Food Health 32 (2013) 33–39.

[65] Prebiotic Nutrition. https://en.wikipedia.org/wiki/Prebiotic_(nutrition).

Chapter 20

Cultivation of Medicinal and Aromatic Plants

Ajay G. Namdeo
Poona College of Pharmacy, Bharati Vidyapeeth Deemed University, Pune, India

1. INTRODUCTION

Since time immemorial, people have gathered plant and animal resources for their needs. Examples include edible nuts, mushrooms, fruits, herbs, spices, gums, game, fodder, fibers used for the construction of shelter and housing, clothing or utensils, and plant or animal products for medicinal, cosmetic, or cultural uses. Even today, hundreds of millions of people, mostly in developing countries, derive a significant part of their subsistence needs and income from gathered plant and animal products. Gathering of high-value products such as mushrooms and medicinal plants also continues in developed countries for cultural and economic reasons [1].

Among these uses, medicinal plants play a central role, not only as traditional medicines used in many cultures but also as trade commodities that meet the demand of often distant markets. Medicinal plants can be defined as the plants that possess therapeutic properties or exert beneficial pharmacological effect on the human or animal body. Demand for a wide variety of wild species is increasing with growth in human needs, numbers, and commercial trade. With the increased realization that some wild species are being overexploited, a number of agencies are recommending that wild species be brought into cultivation systems. Cultivation can also have conservation impacts, however, and these need to be better understood. Medicinal plant production through cultivation, for example, can reduce the extent to which wild populations are harvested, but it also may lead to environmental degradation and loss of genetic diversity as well as loss of incentives to conserve wild populations. The relationship between in situ and ex situ conservation of species is an interesting topic with implications for local communities, public and private land owners and managers, entire industries, and, of course, wild species. Identifying the conservation benefits and costs of the different production systems for medicinal and aromatic plants (MAPs) should help

Natural Products and Drug Discovery. https://doi.org/10.1016/B978-0-08-102081-4.00020-4

guide policies as to whether species conservation should take place in nature or the nursery, or both [2–5].

2. SUSTAINABLE DEVELOPMENT

The cultivation of medicinal and aromatic crops provides sustainable means of a natural source of high-value industrial raw material for pharmaceutical, agrichemical, food, and cosmetic industries and opens up new possibilities for higher level of gains for farmers with a significant scope for progress in rural economy. Though these plants have been known and used since ancient times to heal and cure diseases, recently, technological advancements and validation of traditional knowledge and usage are leading to a consumer inclination toward natural products and high market value for these crops. Such crops are finding a much higher place in international agribusiness, with an estimated annual growth rate of 10%–15%.

In modern medicine, too, plants occupy a very significant place as raw material for some important drugs, although synthetic drugs and antibiotics brought about a revolution in controlling different diseases. But these synthetic drugs are out of the reach of millions of people. Those who live in remote places depend on traditional healers, whom they know and trust judicious use of medicinal herbs can even cure deadly diseases that have long defied synthetic drugs.

In India, the use of different parts of several medicinal plants to cure specific ailments has been in vogue since ancient times. The indigenous systems of medicine, namely "Ayurvedic," "Siddha," and "Unani," have been in existence for several centuries. These systems of medicine cater the needs of nearly 70% of the population residing in the villages. Apart from India, these systems of medicines are prevalent in Korea, China, Singapore, west Asia, and many other countries. Besides the demands made by these systems as their raw material, the demand for medicinal plants made by the modern pharmaceutical industries has also increased fold. Thus, these constitute a group of industrially important crops that bring appreciable income to the country through export.

The medicinal and aromatic crops are used in Ayurveda, Chinese, Siddha, Unani, and other traditional health systems mainly due to the holistic approach, cost of treatment, and least side effects. These herbs and plants were collected from the natural habitat and under a minimally supervised environment. As a result, the density of MAPs in the natural habitat started declining at a faster rate. The overexploitation of these plant species has led to the cultivation of these under field conditions. Finally, medicinal and aromatic crops have better economic opportunities than do the traditional field crops. The price of these crops as raw material to the pharmaceutical industries has increased substantially, which provides a higher income for the cultivators and collectors. This is also encouraged by the increasing demand of these crops in the world trade. All these factors have led to the emergence of medicinal and aromatic crops as alternatives to some of the traditional uneconomic crops.

According to various reports, the estimated number of plant species used for medicinal purposes is greater than 70,000 (Table 20.1).

Today, the pharmaceutical industry is fairly advanced and sophisticated. The plant-based drugs, however, have shortened the life span of the source of material. There is a continuous search for more-potent and less-expensive raw materials to feed the industry; with a concerted research and development effort, many medicinal plants could provide raw material in abundance to the indigenous pharmacies and local herbalists. Strong linkages should be developed between medicinal plant growers, health experts and pharmaceutical industries for developing scientific basis on which these systems of medicines are working.

India possesses a huge wealth of medicinal and aromatic crops and is traditionally associated with the use of MAPs for human, livestock, and plant health. With the present world "going herbal," trends at the consumer level favoring the use of natural products are considered more harmonious with human metabolism. With increasing global population and consequent rise in demand of pharma, perfumery, and flavor industries for MAPs, farming of these crops becomes an attractive alternative to conventional crops.

In the coming years, land and water resources are sure to face acute degradation; the supply of nonrenewable fossil fuel and phosphates will be constrained while agrichemical inputs gradually will have to be substituted. Under such conditions, MAPs with inherent capabilities to grow under natural stress with better yields of secondary metabolites (active constituents) would be the crops of choice for sustainability. Isolation of biomolecules from MAPs and improving their contents in plants would further enhance the demand, and, consequently, the profits as these molecules may find use in a wide range of applications such as antibacterials, antifungals, antiprotozoans, anticancer agents, hepatoprotective agents, antiobesity agents, antidiabetics, antidepressants, and cardiovascular and nerving stimulators and bioenhancers.

Introduction of new rotation based on newly developed short-duration genotypes have enabled growers to achieve high profits, transforming the socioeconomic status of poor farmers with all-round improvement in the quality of their lives with visible impacts in the forms of housing, nutrition in balanced meals for the family, mechanization of their farms and machinery including postharvest processing capabilities, and visible move toward literacy with their children going to school.

Many medicinal and aromatic crops do not require intensive agri-inputs and grow well under natural stress conditions, whether water, light, nutrients, or salts. A number of aromatic grasses such as lemongrass, palmarosa, and vetiver and plants like chamomile are the potential crops suited for salt-affected soils providing subsistence profits. The great importance of collecting good herbarium material for taxonomic identification of the collected

TABLE 20.1 Worldwide Use of Medicinal Plants

Country	Plant Species	Medicinal Plant Species	%
Bulgaria	3,567	750	21.0
China	26,092	4,941	18.9
France	4,630	900	19.4
Hungary	2,214	270	12.2
India	18,664	3,000	16.1
Jordan	2,100	363	17.3
Rep. of Korea	2,898	1,000	34.5
Malaysia	15,500	1,200	7.7
Nepal	6,973	900	12.9
Pakistan	4,950	1,500	30.3
Philippines	8,931	850	9.5
Sri Lanka	3,314	550	16.6
Thailand	11,625	1,800	15.5
USA	21,641	2,564	11.8
Vietnam	10,500	1,800	17.1
World	**422,000**	**72,000**	17.1 Average

WHO, Medicinal Plants in the Republic of Korea: Information on 150 Commonly Used Medicinal Plants, WHO, Manila, WHO Regional Publications, Western Pacific Series no. 21, 1998. http://www.wpro.who.int/publications/pub_9290611200.htm; J.A. Duke, E.S. Ayensu, Medicinal Plants of China. Reference Publications, Algonac, Medicinal Plants of the World No. 4, 1985, R. Govaerts, How many species of seed plants are there? Taxon 50 (4) (2001) 1085–1090; B. Groombridge, M.D. Jenkins, World Atlas of Biodiversity: Earth's Living Resources in the 21st Century, University of California Press, Berkeley, 2002; B. Groombridge, Biodiversity Data Sourcebook, World Conservation Press, Cambridge, WCMC Biodiversity Series nr. 1, 1994; R. Hardalova, L. Evstatieva, C. Gussev, Wild medicinal plant resources in Bulgaria and recommendations for their long-term development, in: C. Meine, M. Sakalian (Eds.), Bulgarias's Biological Diversity: Conservation Status and Needs Assessment, 1998, pp. 528–561; S.K. Jain, R.A. DeFillipps 1991. Medicinal Plants of India. Vol. 1 & 2. – Algonac, USA. Reference Publications (Medicinal Plants of the World 5); L.S. De Padua, N. Bunyapraphatsara, R.H.M.J. Lemmens, Medicinal and Poisonous Plants, vol. 1, Backhuys Publishers, Leiden, Plants Resources of South-East Asia nr. 12/1, 1999; D. Lange, Europe's medicinal and aromatic plants: their use, trade and conservation. Traffic International, Cambridge, 1998; N.P. Manandhar, S. Manandhar, Plants and People of Nepal, Timber Press, Portland, 2002; D.E. Moerman, An analysis of the food plants and drug plants of native North America, J. Ethnopharmacol. 52 (1) (1996), 1–22; S.A. Oran, D.M. Ali-Eisawi, Checklist of medicinal plants of Jordan. Dirasat. Med. Biol. Sci. 25 (2) (1998), 84–112, and Zahoor Ahmad, Medicinal plants of Pakistan, in: M. Karki, A.N. Rao, V. Ramanatha Rao, et al. (Eds.) The Role of Bamboo, Rattan and Medicinal Plants in Mountain Development. Proceedings of a Workshop Held at the Institute of Forestry, Pokhara, Nepal, 15–17 May 1996. International Development Research Centre, New Delhi, INBAR Technical Report nr. 15, 1997, pp. 207–214.

species must be stressed. There is a need for conservation of all useful plant species, and cultivation, maintenance, and assessment of germ plasm for future use, since among the most vulnerable plant species in India, the most over-exploited are the medicinal plants [18].

A sustainable system for harvesting MAPs is one where fruits, seeds, or other plant parts can be harvested indefinitely from a set area without detrimental impact on the structure and dynamics on the harvested plant populations. Disturbance processes can directly affect sustainable harvesting through their influence on plant populations. Positive links between plant diversity and disturbance factors exist for medicinal plants (Tables 20.2–20.6).

3. CULTIVATION OF MEDICINAL AND AROMATIC PLANTS

Cultivation of MAPs requires intensive care and management. The conditions and duration of cultivation required vary depending on the quality of MAP materials required.

3.1 Methods of Propagation

3.1.1 Vegetative Propagation (Asexual Propagation)

Vegetative propagation can be defined as regeneration or formation of a new individual from any vegetative part of the plant body. The method of vegetative propagation involves separation of a part of plant body, which develops into a new plant (Fig. 20.1).

3.1.1.1 Methods of Vegetative Propagation

They are two types:

1. Methods of natural vegetative propagation
2. Methods of artificial vegetative propagation

3.1.1.1.1 Methods of Natural Vegetative Propagation Vegetative propagation by stem (examples):

1. **Bulb**: Allium, squill
2. **Corms**: Colchicum
3. **Tuber**: Potato, aconite
4. **Offset**: Valerian
5. **Rhizome**: Ginger, Turmeric

Vegetative propagation by root (example): Asparagus

TABLE 20.2 Major Medicinal and Aromatic Crops of India and Their Uses

S. No.	Medicinal Crops	Useful Part	Uses
1	*Ashwagandha* (*Withania somnifera*)	Root	Skin disease, Blood pressure, Swelling, Wounds, filler, Joint pain.
2	*Sarpgandha* (*Rauvolfia serpentina*)	Root	High blood pressure, hysteria
3	*Kalmegh* (*Andrographis paniculata*)	Plant	Skin disease, Malaria, fever, blood purifier
4	*Safed Musali* (*Chlorophytum borivillinum*)	Rhizome root	Ayurvedic medicine- Chavanprash, making diabetic medicine
5	*Satawar* (*Asparagus racemosus willd*)	Root	Acidity, Ulcer, to increase milk production in Cow & buffalo, skin disease, eye disease, develop resistance power.
6	*Sanai* (*Cassia angustifolia*)	Leaf	Stomach disease
7	*Gudmar* (*Gymnema sylvestre*	Leaf	Liver tonic, diabeties, heart disease, fever, white spot, snake bites, stomach pain, eye pain.
8	*Chandrasur* (*Lepidium sativum* L.)	Leaf, Seed	Seed- as vegetable, salad, gum, increase milk production in mother, and in dairy cow & buffalow, digestion, eye disease, loose motion, ladies disease, child development, Asthama, piles, siflice, **Leaf**- anti scarbutic
9	*Ratanjot* (*Jatropha curcus* L.)	Plant Branch	Exima, Dad Used as Daton, Bio diesel, skin disease
10	*Isabgol* (*Plantago ovata Forsk*)	Husk	Piles, Loose motion, Stomach disease
11	*Tulsi* (*Ocimum sanctum*)	Leaf, Seed	Cosmetics, Cough seerup, digestion, ear pain, oil
12	*Bhui aounla* (*Phyllanthus amarus*)	Plant	Urinary disease, Jaundice, stomach pain

13	*Mulaithi* (*Glycyrriza glabra* L.)	Under ground stem	Heart disease, Prepare tasteful medicine
14	*Kalihari* (*Gloriosa superba Linn*)	Rhyzomes	Medicine of anticancer, antijaundice, Piles, Asthama
15	*Giloe* (*Tinospora cordifolia Willd*)	Root, Stem, Leaf, Fruit	Root- Laprocy, Stem- Jaundice, Cough fever, white discharge, control of heart beeting, control blood pressure Leaf- Jaundice, Chicken pox Fruit- Jaundice, tonic
16	*Brahmi* (*Bacopa monnieri* L.) *Penn.*	Plant	Increase memory, Nerve tonic, Histeria
17	*Pattharchur* (*Coleus aromaticus*)	Leaf	Stomach pain, Karminative, Urine disease, kidney stone
18	*Makoy* (*Solanum nigrum* L.)	Plant	Fruit- Fever, Loose motion, eye disease Plant- Piles, Liver disease Leaf – Urinary disease
19	*Bia vidung* (*Embelia ribes Brum* F.)	Fruit	Anti worms, Loose motion, skin disease, tonic snake & crabs bites.
20	*Ajwain* (*Hyoscyamus niger* L.)	Seed	Loose motion, teeth pain relief, eye disease, Asthama, cough, urine, infection, siflice
21	*Pan* (*Piper betle*)	Leaf	Worms, Cough, digestion, heart
22	*Lemongrass* (*Cymbopogon flexuosus steud*)	Oil	Cosmetic creams, soaps, insecticide, joint pain, odomas, gulab jal
23	*Pamarosa* (*Cymbopogon maritini stapf*)	Oil	Soap, Cosmetic, Scent, from its oil Jernoil extracted which used in Aroma industries
24	*Java citronela* (*Cymbopogon winter ianus*)	Oil	Cosmetic, soap, preparation of aromatic jeeraniol, Hydroxisionel chemical, Anti mosqueto ointment deodorant.
25	*Tulsi* (*Ocimum besilium*)	Leaf, Seed	Cosmetics, cough seerup, digestion, earpain, oil

Continued

TABLE 20.2 Major Medicinal and Aromatic Crops of India and Their Uses—cont'd

S. No.	Medicinal Crops	Useful Part	Uses
26	*Mentha* (*Mentha arvensis* L.)	Oil	Karminative, expectorant, stomach disease, cough & cold, throat infection, fever, gas
27	*Pachouli* (*Pogostemon calin Benth*)	Oil	Scent, oil is itself top quality scent, soap, cosmetic tobacco, cream, anti worms, Medicine, its juice used in T.B.
28	*Rajnigandha* (*Polianthes tubrosa* L.)	Oil & Flower	Oil- in Aroma industries Flower- making Bucke, cut flower
29	*Jama rosa* (*Cymbopogon nardas*)	Oil	Cosmetic, knee pain and arthritics
30	*Lavender* (*Lavandula officinalise* L.)	Oil	Soap, shaving cream, powder, Anti-worm, Aromatherapy
31	*Khus* (*Vetiveria zizanioides*)	Oil, Root	Oil- Scent, Cosmetic, Medicine, Root- Used in cooler
32	*Nagarmotha* (*Cyperus scariosus*)	Rhyzomes	Heart disease, Loose motion, Ladies disease, body resistance, siflice
33	*German Chameli* (*Matricaria chamomella*)	Oil	**Oil**- Anti allergy, body resistance, cream, shampoo, **Flower & plant**- Digestion, Cough, hair die, wine industries
34	*Jasmine* (*Jasminum grandiflorum*)	Leaf, Flower, & plant	**Leaf**- mouth ulcer, Ear disease **Flower**- Crab bite, skin disease **Plant**- Loose motion, antiworms, urine disease, and diseases concern with ladies.

TABLE 20.3 Appropriate Temperature for Cultivation of Medicinal and Aromatic Plants

Medicinal Plant	Temperature (°C)
Cinchona	14–21
Tea	23–32
Coffee	12–21
Shatavari	18–20
Aloe	10–18
Kalmegh	18–20
Shatavari	18–20
Tulsi	27–30
Amla	18–20
Saffron	0
Tinospora cardifolia	18–25
Mentha	18–20

TABLE 20.4 Appropriate Altitude for Cultivation of Medicinal and Aromatic Plants

Medicinal Plant	Altitude
Tea	9500–1500 m
Cinnamon	300–1000 m
Saffron	1250 m
Shatavari	800–1500 m
Aloe	1 m
Kalmegh	15 cm^{-1} m
Tulsi	30–60 cm
Amla	700 m
Tinospora cardifolia	10 m
Mentha	10–60 cm
Withania	600–1200 m

TABLE 20.5 Types of Soil

Types of Soil	Geographical	Particle Size (mm)
Fine clay	Islands	Less than 0.002
Coarse clay	Near lake/river	0.002–0.02
Fine sand	Near river	0.02–0.2
Coarse sand	Tropical and subtropical coastal areas	0.2–2.00

TABLE 20.6 Soil Type Suitable for the Cultivation of Some Medicinal Plants

Medicinal Plant	Type of Soil
Aloe	Sandy loam
Kalmegh	Loamy soil
Shatavari	Loamy soil
Tulsi	Loamy soil
Amla	Loamy soil
Saffron	Heavy clay soil
Withania	Sandy loam
Turmeric	Rich loamy soil
Mentha	Loamy soil
Tinospora cardifolia	All types of soil

3.1.1.1.2 Methods of Artificial Vegetative Propagation Various parts developed for natural vegetative propagation have also been used for artificial vegetative propagation, with the following methods used:

1. Cutting:

   The part of the plant that is removed by cutting it from the parent plant is called a "cutting." A plant that is 1 year old is cut from its root at about 20–30 cm. Then, it grows in moist soil. Roots are developed and grow. This method used to obtain new plants (e.g., rose, sugarcane, etc.) (Fig. 20.2).

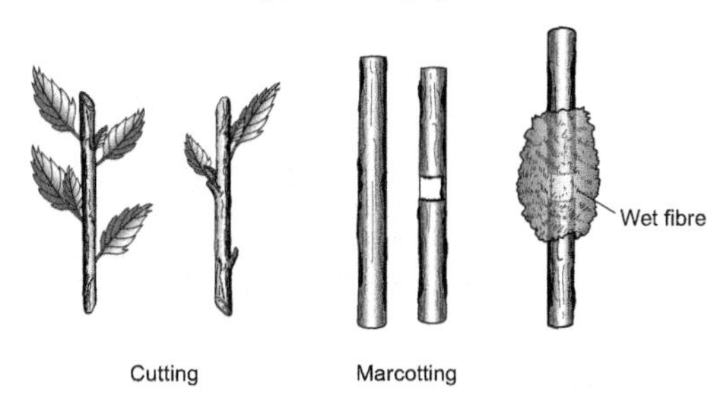

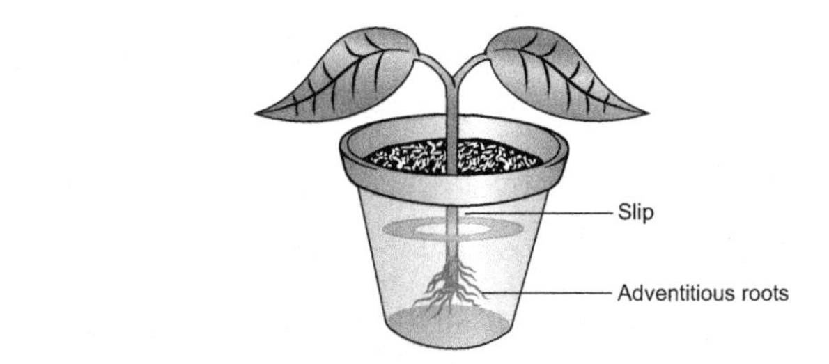

FIGURE 20.1 Vegetative propagation.

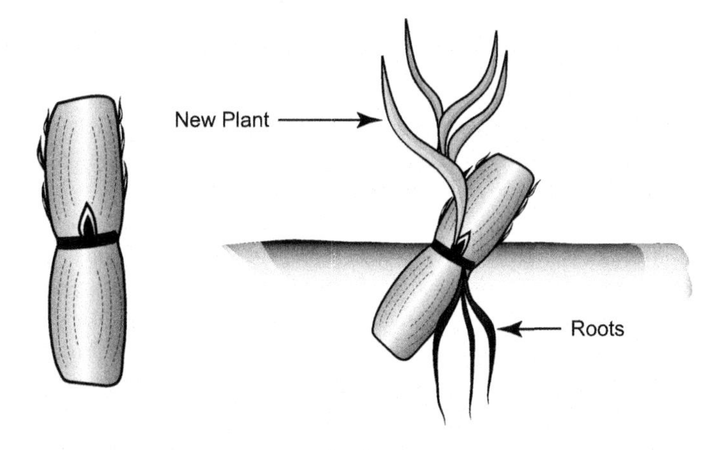

FIGURE 20.2 Vegetative propagation by cutting.

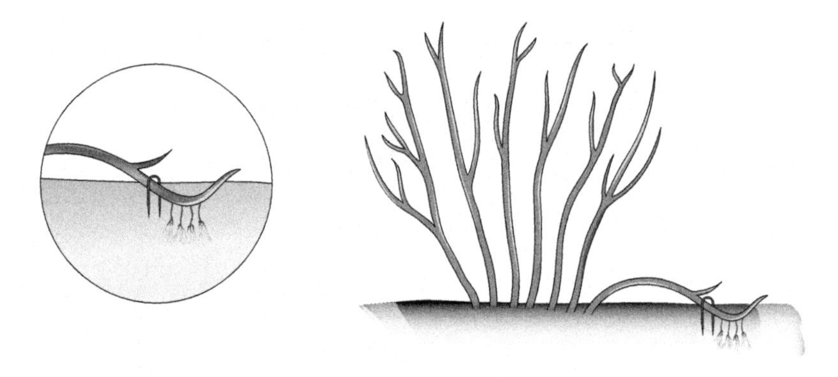

Simple Layering

FIGURE 20.3 Vegetative propagation by simple layering.

2. Layering:

   The soft branches of the plant that are near to the ground are pulled toward the ground and covered with soil. In this case, the tip of a branch leave is above the ground.

   After some time, root development takes place from the branch that was buried in the soil. This part of the plant, along with the roots, is detached from the parent plant and develops into new plant. This method is also used for plants such as jasmine, strawberries, lemon, etc. (Fig. 20.3).
3. Grafting:

   This method is commonly used in horticultural practices. In this method, two different parts of plants are joined together to produce a new plant that has the characteristics of both plants. After some time, the attached cutting becomes a part of the rooted plant. From these two plants, one has a strong root system and the other has a strong flowering and fruit system. The plant from which the shoot system is taken is known as the "scion," and the plant from which the root system is selected is known as the "stock." The root and stem part of plant are placed in such way that no gap remains between them. Finally, they become joined. Wax is applied to cover the place between grafting is done so that the infection is avoid (examples are rose, citrus, rubber, etc.) (Fig. 20.4).
4. Micropropagation:

   This method consists of growing cells, tissues, and organs in culture. Small pieces of plant organs or tissues are grown in a container with suitable nutrient medium, under sterilized conditions. The tissue grows into a mass of undifferentiated cells called callus, which later differentiate into plantlets. These are then transferred into pots or nursery beds and allowed to grow into full plants (Fig. 20.5).

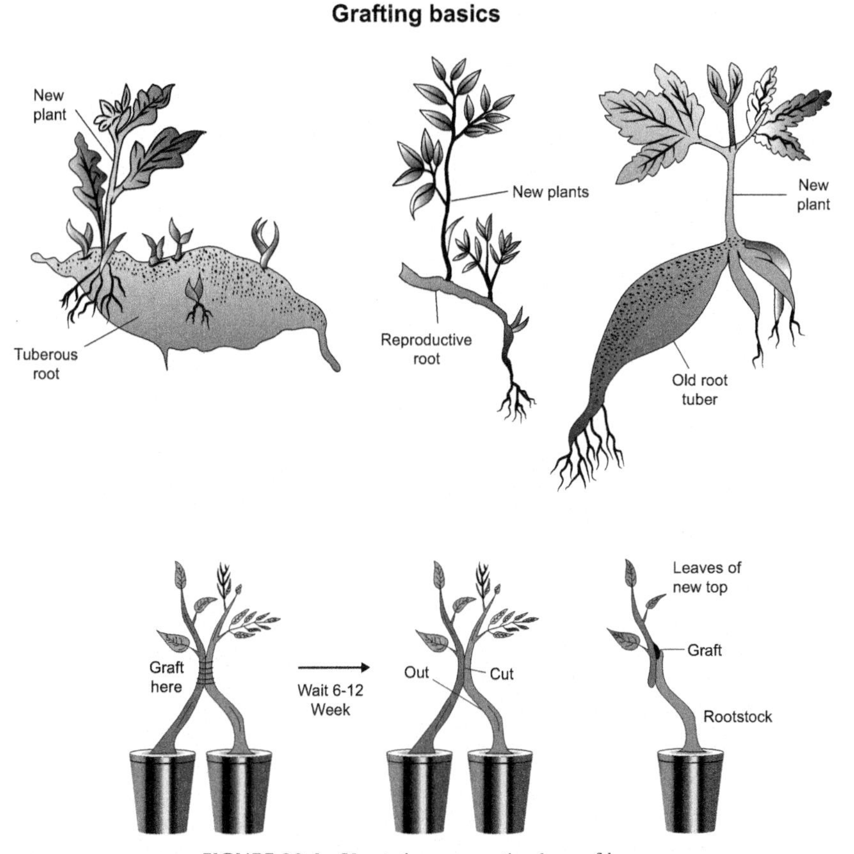

FIGURE 20.4 Vegetative propagation by grafting.

3.1.1.2 Advantages of Asexual Propagation

1. It is a low-cost, easy, and rapid method of multiplication. Many fruit trees require 4–5 years to bear the fruits when developed from seeds. The plants developed by vegetative methods take only 1 year to bear fruits.
2. Plants like roses and chrysanthemum do not form viable seeds. Thus, vegetative propagation is the only method of reproduction and continuation of species in such plants.
3. All the plants developed by these methods will be generally similar to the parent plant.
4. Micropropagation is useful in raising disease-free plants, homozygous diploids, and those without viable seeds.

3.1.1.3 Disadvantages of Asexual Propagation

1. Offspring may find it difficult to survive in a changing environment.
2. If a parent is of "poor quality," the offspring will also be of "poor quality."

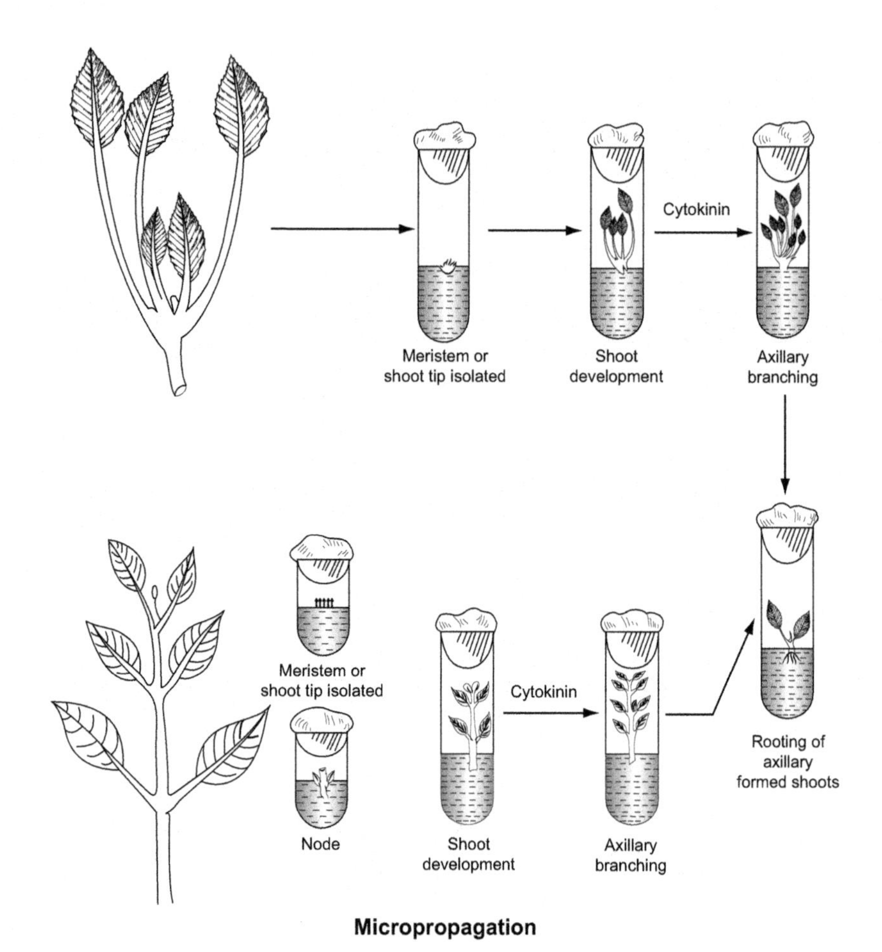

FIGURE 20.5 Micropropagation.

3. Because offspring colonize the same area as parent, competition and overcrowding can easily result.

3.1.2 Seed Propagation (Sexual Propagation)

Propagation by seeds is the major method by which plants reproduce in nature and one of the most efficient and widely used propagation methods for cultivated crops. Seeds are widely available, inexpensive, and easy to handle. Hybrid seeds are more expensive but may have production benefits that offset the cost. Large-scale agriculture (including vegetable crop production) is dependent on seed propagation. Seeds (especially seeds of woody plants) may have complex dormancies that impede germination.

3.1.2.1 Process of Sexual Propagation (Fig. 20.6)

1. **Microsporogenesis**: Microspores are formed from microspore mother cells inside the anther. It is the process of formation of microspores from a pollen mother cell through meiosis division. The cells of sporogenous tissues undergo meiotic division to form microspore tetrad. As the anther mature and dehydrate, the microspore dissociate and develops into pollengrains. The pollen grain represents the male gametophytes. Each pollen grains have two layered wall, the outerexine made up of sporopollenin and inner wall is called intine made up of cellulose and pectin.
2. **Pollination**: Transfer of pollen grains from anther to stigma
   Autogamy – transfer of pollen grain from anther to stigma of same flower
   Cleistogamous – flower that does not open
   Chasmogamou – exposed anther and stigma
   Eitonogamy – transfer of pollen grains from anther to stigma of different flower of same plant
   Xenogamy – ransfer of pollen grain from anther to stigma of different plant's flower of same species

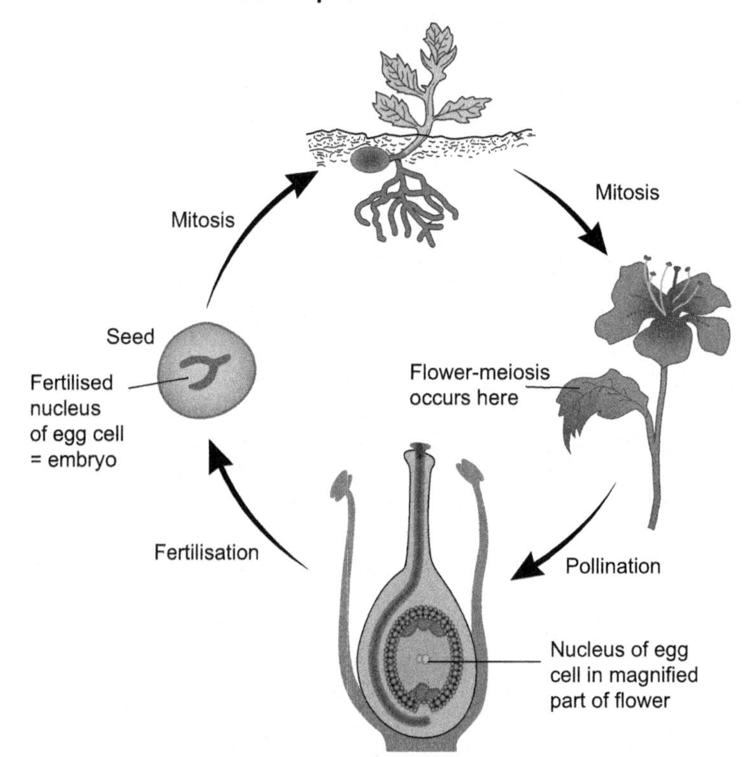

FIGURE 20.6 Sexual reproduction in plants.

3. **Microgametogenesis**: Microgametogenesis is the process in plant reproduction where a microgametophyte develops in a pollen grain to the three-celled stage of its development. These cells are surrounded by a wall of sterile cells called the tapetum, which supplies food to the cell and eventually becomes the cell wall for the pollen grain.
4. **Megasporogenesis**: This process leads to the formation of megaspores from megaspore mother cell, inside the ovule. After megasporogenesis, the megaspore develops into the female gametophyte (the embryo sac) in a process called megagametogenesis. Therefore, the resulting embryo sac is a seven-celled structure consisting of one central cell, one egg cell, two synergid cells, and three antipodal cells.
5. **Megagametogenesis**: The events involving the formation of the embryo sac from megaspores are included in this process. It is the development of a megaspore into an embryo sac, which is the gametophyte – though a highly reduced one – stage in the life cycle of vascular plants.
6. **Fertilization**: Fusion of male and female gametes takes place, resulting in the formation of zygote. In general, male gametes are contained in pollen, which is carried by wind, water, or wildlife (both insects and animals) to reach female gametes. The pollen is deposited on a plant's stigma, which is part of the pistil (the elongated part of a flower extending from the ovary). This process is called pollination.
7. **Embryogeny**: The process involves development of embryo from zygote. This process produces a plant embryo from a fertilized ovule by asymmetric cell division and the differentiation of undifferentiated cells into tissues and organs.

3.1.2.2 Advantages of Sexual Propagation

1. The plant raised by seed is planted lived.
2. They are hardy with deep root system. So they are vigorous in growth.
3. The possibility is there to obtain change in seedling, the performance of which are better than their parents.
4. The polyembryony. The phenomenon of propation of more than one seedling from a single seed, produce true to type, nuclear embryonic seedling that could be used as rootstock for uniform performance.
5. Seed propagation is necessary when vegetative propagation is unsuccessful or expenses (e.g., papaya, coconut, and Areca nut).
6. Exploitation of hybrid, vigor is possible only when the hybrids are multiplied in the first instance through sexual propagation, although subsequent fixing of heterocyst is effected through vegetative propagation.
7. Roots stocks are usually raised by seed.
8. When seedlings are required in large numbers, seed propagation is the only easy method.

3.1.2.3 Disadvantages of Sexual Reproduction

1. When progenies are not true type and so become inferior; in the commercial orchard, it is necessary to have uniform quality, growth, and yielding capacities.
2. Choice tree or any hybrid trees cannot be perpetuated true to type by seed (except in apomixes).
3. Seedling has a long juvenile period in crops like citrus, coca, and rubber. The seeds must be sown afresh (i.e., immediately after extraction). Many varieties are seedless.
4. Seeds lose viability in a short period.

3.2 Factors Affecting Cultivation of Medicinal Plant

3.2.1 Light

Light is an essential factor for plant growth. It helps in photosynthesis, seed germination, flowering and vegetative growth like tuber formation, and opening and closing of stomata. Dry sunny weather increases the proportion of glycosides in digitalis and of alkaloids in belladonna. In some plants, growth will be increased at night.

3.2.2 Temperature

Temperature is very important and a major factor influencing the cultivation of medicinal plants. If the temperature decreases suddenly, then ice crystals are formed in intercellular spaces of the plant. Because of this, water comes out of the cell and plants die due to drought. The ice crystals also cause mechanical injury to the cells and temperature stimulates the growth of seedlings. Water absorption decreases at low temperatures. The rate of photosynthesis is affected by change in temperature. The rate of respiration increases with increase in temperature. Examples;

3.2.3 Atmosphere Humidity

Water is evaporated from the surface of earth, and transpiration from plants is the major effect of humidity on plant life and climate. Humidity is present in the form of water vapors. This is known as atmosphere humidity. Evaporation of water, its condensation, and precipitation depend on relative humidity, and humidity affects structure, form, and transpiration in plants.

3.2.4 Altitude

It is most important factor for the cultivation of medicinal plants. With an increase in the altitude, the temperature and atmospheric pressure decrease while the wind velocity, relative humidity, and light intensity increase.

Vegetation property changes because of change in altitude. The bitter constituents of *Gentiana lutea* increase with altitude, whereas the alkaloids of *Aconitum nacelles* and *Lobelia inflate* and oil content of thyme and peppermint decrease. *Pyrethrum* gives the best yield at high altitude.

3.2.5 Rainfall

Rainfall is a very important factor of cultivation of medicinal plants. Rain water is main source of water for the soil. Rainfall and snowfall have large effects on the climate condition. The water from rainfall flows into the rivers and lakes and percolates into the soil to form ground water; the remaining is evaporated. The minerals in the soil are dissolved in water and then absorbed by plants. Water influences the morphology and physiology of plants. For example, continuous rain can lead to a loss of water-soluble substance from the leaves and root by leaching; this is known to apply to some plants producing glycoside and alkaloids.

3.2.6 Soil

Soil is defined as the surface layer of the earth, formed by the weathering of rocks. The soil is formed as a result of combined action of climate factors like plants and microorganisms. The soil should contain appropriate amounts of nutrients, organic matter, and other elements to ensure optimal medicinal plant growth and quality. Optimal soil conditions, including soil type, drainage, moisture retention, fertility, and pH, will be dictated by the selected medicinal plant species and/or target medicinal plant part.

The soil made of five components:

1. Mineral matter
2. Soil air
3. Soil water
4. Organic matter or humus
5. Soil organisms

Plants depend on soil for nutrients, water supply, and anchorage. Soil influences seed germination, capacity of plant to remain erect, form, vigor and woodiness of the stem, depth of root system, number of flowers on a plant, drought, frost, etc.

3.2.6.1 Classification of Soil Particles

1. Clay
2. Loamy
3. Silt loam
4. Sandy loam
5. Sandy soil
6. Calcareous soil

Clay Soil Particles of clay are very small. These particles fit together very closely and, therefore, leave very little pore space. These spaces are filled with water very easily. Plants growing in this soil are not able to absorb water because the soil has practically no air. This soil known as physiologically dry soil; clay soil is plastic and forms a colloid when moist. When the conditions are dry, it cracks and shrinks the soil rich in nutrient elements and, therefore, acts as a negatively charged colloidal system.

Sandy Soil Particle sizes of sand are very large. These leave large pore spaces, which do not have capillary action an,d therefore, water is not retained by them. Water reaches deep into soil because most of the water quickly drained off. It is poor in nutrient elements; plants growing in this soil have little weight and it is not very fertile.

Loam Soil This is mixture of slit, clay and sand. It is very useful for growing plants. It contains available nutrients in sufficient amounts and is highly fertile soil. The plants growing in loam are vigorous and have very high weight. It has a high water retention capacity and appropriate amount of soil air is also present.

Sandy Loam The amount of sand particles in this soil is greater than in other types of loam.

Silt Loam Silt loam is considered to be the most fertile as it contains more organic substances than other loams.

3.2.7 Fertilizer

The fertilizer types:

1. Biological origin
2. Synthetic
3. Chemical

3.2.7.1 Biological Origin Fertilizer

Soil is generally poor in organic matter and nitrogen. The substances of biological origin used as fertilizer are thus selected if they could provide the elements required. These are two types:

1. Green manures:

   Manure is material, which are mixed with soil. These supply almost all the nutrients required by the crop plants. This results in the increase in crop productivity.

   Manures are of three types:

   Farmyard manure:

   This is a mixture of cattle dung and remaining unused parts of straw and plants stalks fed to cattle.

   Composited manure:

   This consists of a mixture of rotted or decomposed and useless parts of plants and animals.

Green manure:

It is a herbaceous crop ploughed under and mixed with the soil while still green to enrich the soil. The plants used as green manure are often quick growing. These add both organic and nitrogen to the soil. It is also forms a protective soil cover that checks soil erosion and leaching. Thus, the crop yield increases by 30%–50%.

2. Biofertilizer:

   It can be defined as biologically active products or bacteria, algae, and fungi that are useful in bringing about soil nutrient enrichment. These mostly include nitrogen-fixing microorganisms.

Some of the biofertilizers:

1. Legume – *Rhizobium* symbiosis
2. Azolla – *Anabaena* symbiosis
3. Free-living bacteria
4. Loose association of nitrogen-fixing bacteria
5. Cyanobacteria (blue-green algae)
6. Mycorrhiza
   a. Ectomycorrhizae. Increase the interface surface between plant root and soil. Mycorrhizae absorb and store nitrogen, phosphorous, potassium, and calcium.
   b. Endomycorrhizae

3.2.7.2 Chemical Fertilizers

1. **Macronutrients (major elements)**
   a. Nitrogen
   b. Phosphorous
   c. Potassium
   d. Calcium
   e. Magnesium
   f. Sulfur
2. **Micronutrients (minor elements)**
   a. Iron
   b. Manganese
   c. Zinc
   d. Boron
   e. Copper
   f. Molybdenum

Carbon, oxygen, hydrogen, and chorine are provided from water and air.

Examples:

Urea, potash

3.2.8 Greenhouse Effect

When there is an increase in carbon dioxide concentration in the atmosphere, it prevents the heat from being reradiated by forming a thick cover. Consequently, the atmosphere is heated and the temperature increases.

This is called the greenhouse effect. Due to cutting of forests and excessive burning of fossil fuels, the amount of carbon dioxide increased from 290 to 330 ppm. This causes an increase in global temperature.

The global warming by two or three degrees would cause polar ice caps to melt, floods in coastal areas, and changes in the hydrologic cycle, and islands would be submerged. The following gases produce the greenhouse effect: carbon dioxide, sulfur dioxide, oxide of nitrogen, chlorofluorocarbons, etc.

3.2.9 Internal Factors

3.2.9.1 Polyploidy

Plants whose cells contain two sets of chromosomes, derived at fertilization from the union of one set from the pollen and one set from the egg cells, are described as diploids and denoted by "2n." The plants with more than two sets of chromosome in the cell are polyploidy. When four sets are present, the plants are described as tetraploids and denoted by "4n."

On the treatment of colchicines, tetraploidy is induced, which inhibits spindle formation during cell division, so that the divided chromosomes are unable to separate and pass to the daughter cells. The two sets of chromosomes remain in one cell and this develops to give tetraploid plants. Treatment with colchicine may be applied in various ways, but all depend on the effects produced in the meristem. The seeds may be soaked in a dilute solution of colchicine, or the seedlings, the soil around the seedling or the young shoot treated with colchicine solution. Fertile seed and robust, healthy tetraploid plants were obtained, the tetraploid condition being indicated by the increased size of the pollen grains and stomata; chromosome counts in root-tip preparations confirm the tetraploid condition. The average increase in alkaloids content compared with diploid plants of *Datura stromonium* and *Datura tatula* was 68%, with a maximum increase of 211.6%. Similar results were obtained with *Atropa belladonna* and *Hyoscyamus niger*; the average increase in belladonna was 93%. Increased alkaloidal content of tetraploids plants has been confirmed for *Datura stromonium* and *Datura tatula*. The diploid of *Acorus calamus* is 2.1% of volatile oil content, but when they are converted into tetraploid, they produce 6.8% of volatile oil contents.

3.2.9.2 Mutation

Sudden heritable change in the structure of a gene on chromosome or change in the chromosome number.

Type of mutations

1. Spontaneous and induced mutations
2. Recessive and dominant mutations
3. Somatic and germinal mutations
4. Forward, back, and suppressor mutations
5. Chromosomal, genomic, and point mutations

Mutations can be artificially produced by certain agents called mutagens or mutagenic agent.

There are two types.

1. Physical mutagens
   a. Ionizing radiations: X-rays, gamma radiation, and cosmic rays
   b. Nonionizing radiation: ultraviolet (UV) radiation
2. Chemical mutagens
   a. Alkylating and hydroxylating agents: nitrogen and sulfur mustard; methyl and ethylsulfonate, ethylethane sulfonates
   b. Nitrous acid
   c. Acridines

Acridines and proflavins. Ionizing radiation causes breaks in the chromosome. These cells then show abnormal cell divisions. If these include gametes, they may be abnormal and even die prematurely. Nonionizing radiation like UV rays are easily absorbed by purine and pyrimidines. The changed bases are known as photoproducts. UV rays cause two changes in pyrimidine to produce pyrimidine hydrate and pyrimidine dimmers. Thymine dimer is a major mutagenic effect of UV rays that disturbs the DNA double helix and thus DNA replication.

For example, penicillin, as an antibiotic, was first obtained from *Penicillium.* However, the yield was very poor and the preparation was commercially expensive. Since then, mutants with higher yield of penicillin have been selected and produced. *Penicillium chrysogenum* used in the production of penicillin yielded about 100 units of penicillin per mL of culture medium.

By single-spore isolation, strains were obtained that yielded up to 250 units/mL of medium; radiographic treatment of this strain gave mutants that produce 500 units/mL, and ultraviolet mutants of the latter gave strains that produced about 1000 units/mL. Similarly, improvements have been obtained with other antibiotic-producing organisms. Mutant strains of *Capsicum annum* with increasing yields (20%–60%) of capsaicin have been isolated from M3 and M4 generations originating from seed treated with sodium azide and ethyl methane sulfonate.

3.2.9.3 Hybridization

This is the mating or crossing of two genetically dissimilar plants having desired genes or genotypes and bringing them together into one individual called a hybrid. The process through which hybrids are produced is called hybridization. Hybridization, particularly between homozygous strains that have been inbred for a number of generations, introduces a degree of heterozygosis with resultant hybrid vigor often manifest in the dimensions and other characteristics of the plants. A hybrid is an organism that results from crossing of two species or varieties differing at least in one set of characteristics.

The following steps are involved in hybridization of plants:

1. **Choice of parents**:
   Of the two parents selected, at least one should be as well-adopted and proven variety in the area. The other variety should have the characters that are absent in the first chosen variety.
2. **Emasculation**:
   Removal of stamens or anthers or killing the pollen grains of a flower without affecting the female reproductive organs is known as emasculation. Emasculation is essential in bisexual flowers.
3. **Bagging**:
   Immediately after emasculation, the flowers or inflorescences are enclosed in bags of suitable sizes to prevent random cross-pollination.
4. **Pollination**:
   In pollination, mature, fertile, and viable pollens are placed on a receptive stigma. The procedure consists of collecting pollens from freshly dehisced anthers and dusting them onto the stigmata of emasculated flowers.
5. **Raising F1 plants**:
   Pollination is naturally followed by fertilization. It results in the formation of seeds. Mature seeds of the F1 generation are harvested dried and stored; these seeds are grown to produce F1 hybrid. Hybrids of cinchona yield more quinine. A hybrid developed by crossing *Cinchona succirubra* with *Cinchona legering* yields a bark, which contains 11.3% of alkaloids. The parent species produced 3.4% and 5.1% of alkaloids, respectively.
   Pyrethrum hybrids have been used for *Pyrethrum* production; these hybrids are produced by either crossing two clones assumed to be self-sterile or planting a number of desirable clones together and bulking the seed. The hybridization of plants increases pyrethrin contents.

4. OPPORTUNITIES IN DEVELOPING THE MEDICINAL PLANTS SECTOR

4.1 Institutional Support

In India, many government and nongovernment organizations have had the focused attention on improving the medicinal plants sector (Table 20.7).

TABLE 20.7 Major Institutions Involved in Funding Projects for Medicinal Plant Research in India [19]

Institutions	Funding for Major Areas in Medicinal Plant Research
National Medicinal Plants Board, NMPB	Survey, documentation, cultivation, marketing, conservation
Department of Science & Technology, DST	Taxonomy, ecology, pathology, survey, propagation, documentation, cultivation, conservation
Council for Scientific & Industrial Research, CSIR	Ecology, taxonomy, biochemistry, survey, documentation, cultivation, genetics, agrotechnology, conservation
Indian Council of Medical Research, ICMR	Breeding, value addition
All India Council for Technical Education, AICTE	Management technology
Department of Biotechnology, DBT	Agrotechnology, molecular biology, biochemistry, rural biotechnology
Defense Research & Development Organization, DRDO	Agro-technology, survey, documentation, conservation
Indian Council of Agricultural Research, ICAR	Breeding, pathology, molecular biology
Ministry of Environment & Forest, MoEF	Survey, documentation, conservation, management, ecological impact assessment, cultivation
National Bank for Agriculture and Rural Development, NABARD	Cultivation, marketing
University Grant Commission, UGC	Ecology, biochemistry, survey, documentation
Herbal Research and Development Institute, HRDI	Survey, documentation, nursery development
G.B. Pant Institute of Himalayan Environment & Development, GBPIHED	Survey, documentation, cultivation, conservation

Opportunities for funding have been created to assist the person who is willing to work and to build capacity of the medicinal plants sector. According to the mandate of the National Medicinal Plants Board, the projects may be submitted for funding within two major schemes: a promotional scheme and a commercial scheme. The major thrust areas within the promotional scheme are (1) survey and inventory of medicinal plants, (2) in situ conservation and ex situ cultivation of selected medicinal plants, (3) production of quality planting material, (4) diffusion of knowledge through education and communication, (5) promotion of global and domestic market system, and (6) strengthening research, development, and man power. Within the commercial scheme, the major thrust areas are (1) bulk production of medicinal plants and ensuring supply of quality planting material, (2) expansion of selected medicinal plants farming areas, (3) value addition in harvesting, processing, and marketing of medicinal plants, and (4) developing an innovative marketing mechanism [19] (Tables 20.8–20.10).

TABLE 20.8 Export of Major Medicinal Plants From India

S. No.	Medicinal Plant/Part	Quantity (Tons)
1	Liquorice roots (fresh/dried/powdered)	54.9
2	Nux vomica	1.8
3	Galangal (rhizomes and roots)	108.6
4	Ginseng roots	3271
5	Agarwood	169
6	Belladonna leaves	1.7
7	Belladonna roots	2304.6
8	Poppy flowers and unripe heads	9.4
9	Poppy husk	1
10	Isabgol (husk)	19.27
11	Isabgol (seeds)	1000.4
12	Senna (leaves and pods)	7430.25
13	Tukmaria	97.7
14	Catharanthus roseus	522.9
15	Neem (seed)	106.5
16	Neem (leaves/powder)	13.13
17	Gymnema (powder)	19.99
18	Ayurvedic and Unani herbs	9367.12

TABLE 20.9 Over-the-Counter Drugs Used in Western Countries

S. No.	Common Name	Botanical Name	Use
1	Psyllium	*Plantago ovata*	Bulk laxative
2	Ginkgo	*Ginko biloba*	Memory enhancer
3	St. John's wort	*Hyperium perforatum*	Antidepressant
4	Garlic	*Allium sativum*	Hypolipodemic
5	Aloe	*Aloe spices*	Stimulant laxative
6	Peppermint	*Mentha piperita*	Antitussive
7	Saw palmetto	*Senecis repens*	Prostrate hyperplasia
8	Ginseng	*Paraax spices*	Brain
9	Mandhukparani	*Centella asiatica*	Blood circulation
10	Black cohosh	*Cimiccifuga racemosa*	Menopause
11	Kawa	*Piper methysticum*	Permanent syndrome Antidepressant
12	Milk thistle	*Stylybum marianum*	Live protection
13	Valerian	*Valeriana officianalis*	Calmative

TABLE 20.10 Wild-Harvesting Versus Cultivation of Medicinal and Aromatic Plants: A Summary of Advantages and Disadvantages [1]

For ***species and ecosystems*** it is better to…	
Wild-harvest because…	**Cultivate** because…
• it puts wild plant populations in the continuing interest of local people • it provides an incentive to protect and maintain wild populations and their habitats and the genetic diversity of MAP populations **but**… • uncontrolled harvest may lead to the extinction of ecotypes and even species • common access to the resource makes it difficult to adhere to quotas and the precautionary principle • in most cases knowledge about the biology of the resource is poor and the annual sustained yields are not known • in most cases resource inventories and accompanying management plans do not exist	• it relieves harvesting pressure on very rare and slow-growing species that are most susceptible to threat **but**… • it devaluates wild plant resources and their habitats economically and reduces incentive to conserve ecosystems • it narrows the genetic diversity of the gene pool of the resource because wild relatives of cultivated species become neglected • it may lead to conversion of habitats for cultivation • cultivated species may become invasive and have negative impacts on ecosystems • reintroducing plants can lead to genetic pollution of wild populations

The *market* demands...	
Wild-harvest because...	**Cultivated material** because...
• it is cheaper since it does not require infrastructure and investment • many species are only required in small quantities that do not make cultivation economically viable • for some plant parts extra-large cultivation areas are required (e.g., Arnica production for flowers) • successful cultivation techniques do not exist, e.g., for slow-growing, habitat-specific taxa • no pesticides are used • it is often believed that wild plants are more powerful **but**... • there is a risk of adulterations • there is a risk of contaminations • through non-hygienic harvest or post-harvest conditions	• it guarantees continuing supply of raw material • it makes reliable botanical identification possible • genotypes can be standardized or improved • quality standards are easy to maintain • controlled postharvest handling is possible • production volume and price can be agreed for longer periods • resource price is relatively stable over time • certification as organic production is possible **but**... • it is more expensive than wild harvest • it needs substantial investment before and during production
From a perspective of the *people* it is better to...	
Wild-harvest because...	**Cultivated material** because...
• it provides access to cash income without prior investment • it provides herbal medicines for health-care needs • it maintains the resources for rural populations on a long-term basis (if done sustainably) **but**... • unclear land rights create ownership problems • this income and health-care resource is becoming scarce through overharvesting	• it secures steady supply of herbal medicines (home gardens) • it provides in-country value adding **but**... • capital investment for small farmers is high • competition from large-scale production puts pressure on small farmers and on wild harvesters • benefits are made elsewhere and traditional resource users have no benefit return (IPR)

REFERENCES

[1] U. Schippmann, D.J. Leaman, A.B. Cunningham, A comparison of cultivation and wild collection of medicinal and aromatic plants under sustainability aspects, in: R.J. Bogers, L.E. Craker, D. Lange (Eds.), Medicinal and Aromatic Plants, Springer, 2006, pp. 75–95. Printed in the Netherlands.

[2] G. Bodeker, K.K.S. Bhat, J. Burley, et al., Medicinal Plants for Forest Conservation and Health Care, FAO, Rome, 1997. Non-wood Forest Products no. 11, http://www.fao.org/docrep/W7261E/W7261E00.htm.

[3] U. Schippmann, D.J. Leaman, A.B. Cunningham, Impact of cultivation and gathering of medicinal plants on biodiversity: global trends and issues, FAO, in: Biodiversity and the Ecosystem Approach in Agriculture, Forestry and Fisheries. Satellite Event on the Occasion of the Ninth Regular Session of the Commission on Genetic Resources for Food and Agriculture. Rome, 12–13 October 2002, Inter-Departmental Working Group on Biological Diversity for Food and Agriculture, Rome, 2002.
[4] U. Schippmann, D.J. Leaman, C.B. Cunningham, Impact of cultivation and gathering of medicinal plants in biodiversity: global trends and issues, in: FAO (Ed.), Biodiversity and the Ecosystem Approach in Agriculture, Forestry, and Fisheries, FAO, Interdepartmental working group on biological diversity for food and agriculture, Rome, 2002b, pp. 142–167. http://www.fao.org/|DOCREP/005/AA010E/AA010E00.HTM.
[5] U. Schippmann, D.J. Leaman, A.B. Cunningham, et al., Impact of cultivation and collection on the conservation of medicinal plants: global trends and issues, in: A. Jatisatienr, T. Paratasilpin, S. Elliott, et al. (Eds.), Conservation, Cultivation and Sustainable Use of MAPs: A Proceedings of WOCMAP III: the IIIrd World Congress on Medicinal Aromatic Plants, Chiang Mai, Thailand, February 3–7, 2003, ISHS, Leuven, 2005, pp. 31–44. Acta Horticulturae nr. 676, http://www.actahort.org/books/676/.
[6] WHO, Medicinal Plants in the Republic of Korea: Information on 150 Commonly Used Medicinal Plants, WHO, Manila, 1998. WHO Regional Publications, Western Pacific Series no. 21, http://www.wpro.who.int/publications/pub_9290611200.htm.
[7] J.A. Duke, E.S. Ayensu, Medicinal Plants of China, Reference Publications, Algonac. Medicinal Plants of the World No. 4, 1985.
[8] R. Govaerts, How many species of seed plants are there? Taxon 50 (4) (2001) 1085–1090.
[9] B. Groombridge, M.D. Jenkins, World Atlas of Biodiversity: Earth's Living Resources in the 21st Century, University of California Press, Berkeley, 2002.
[10] B. Groombridge, Biodiversity Data Sourcebook, World Conservation Press, Cambridge, 1994. WCMC Biodiversity Series nr. 1.
[11] R. Hardalova, L. Evstatieva, C. Gussev, Wild medicinal plant resources in Bulgaria and recommendations for their long-term development, in: C. Meine, M. Sakalian (Eds.), Bulgarias's Biological Diversity: Conservation Status and Needs Assessment, 1998, pp. 528–561;
[11a] S.K. Jain, R.A. DeFillipps, Medicinal Plants of India. Vol. 1 & 2. – Algonac, USA, Reference Publications (Medicinal Plants of the World 5), 1991.
[12] L.S. De Padua, N. Bunyapraphatsara, R.H.M.J. Lemmens, Medicinal and Poisonous Plants, vol. 1, Backhuys Publishers, Leiden, 1999. Plants Resources of South-East Asia nr. 12/1.
[13] D. Lange, Europe's Medicinal and Aromatic Plants: Their Use, Trade and Conservation, Traffic International, Cambridge, 1998.
[14] N.P. Manandhar, S. Manandhar, Plants and People of Nepal, Timber Press, Portland, 2002.
[15] D.E. Moerman, An analysis of the food plants and drug plants of native North America, J. Ethnopharmacol. 52 (1) (1996) 1–22.
[16] S.A. Oran, D.M. Ali-Eisawi, Checklist of medicinal plants of Jordan, Dirasat. Med. Biol. Sci. 25 (2) (1998) 84–112.
[17] Zahoor Ahmad, Medicinal plants of Pakistan, in: M. Karki, A.N. Rao, V. Ramanatha Rao, et al. (Eds.), The Role of Bamboo, Rattan and Medicinal Plants in Mountain Development. Proceedings of a Workshop Held at the Institute of Forestry, Pokhara, Nepal, 15–17 May 1996, International Development Research Centre, New Delhi, 1997, pp. 207–214. INBAR Technical Report nr. 15.

[18] S.P.S. Khanujia, A. Kalra, A.K. Singh, Gain from Entrepreneurship, The Hindu Survey of Indian Agriculture, 2005, pp. 192–194.

[19] C.P. Kala, P.P. Dhyani, B.S. Sajwan, Developing the medicinal plants sector in northern India: challenges and opportunities, J. Ethnobiol. Ethnomed. 2 (2006) 32. http://www.ethnobiomed.com/content/2/1/32.

Chapter 21

Digitization of Traditional Knowledge

Souvik Basak
Dr. B.C. Roy College of Pharmacy and Allied Health Sciences, Durgapur, India

1. INTRODUCTION

In the last few decades research involving natural products has been a prime focus across the globe since natural products have been found to be toxicologically safer to the human community than synthetic products. The prior utilization of natural products over synthetic or semisynthetic products is the basis for lower toxicity of naturally obtained molecules yet befits the particular receptor for drug action. In addition, natural products are structurally more diverse than synthetic or semisynthetic compounds, hence they are more promising for new drug discovery. Thus increasing attention has been focused on this arena to utilize a broader set of natural products for drug discovery.

2. WHY DIGITIZATION OF NATURAL PRODUCTS IS NECESSARY

Over a time period of 4 billion years evolution has taken place to create complex biodiversity across the globe [1] and to indulge in fundamental life forms over the earth. Of 7–20 million species around the world, only 1–2 million have been identified and scientifically named [2,3]. In addition, tropical countries are the storehouses of biodiversity because of their temperate climate as well as humidity [4]. Concomitant with this biodiversity evolution, plants and animals have also evolved and were used in the past as traditional medicinal plants across the globe [5]. However, with the increase in natural product usage, an emerging need has grown over the years to organize or compile knowledge for the effective utilization of the data. There are almost 300,000 medicinal plants worldwide [6] of which 75,000 plants have been used as medicinal plants [7]. Thus attempts have been made to compile this knowledge in an organized form, which has been initiated through digital

Natural Products and Drug Discovery. https://doi.org/10.1016/B978-0-08-102081-4.00021-6

embodiment of the relevant information. Digitization has occurred predominantly via four approaches:

1. Digital databases on traditional knowledge (web based).
2. Bioinformatics-guided approach of traditional knowledge.
3. Virtual screening of natural products.
4. In-silico approach for natural product-guided drug discovery

2.1 Digital Databases on Traditional Knowledge (Web Based)

To compile, store, and organize medicinal plant knowledge, attempts have been made in several countries to code traditional medicinal plant knowledge in several databases. These databases not only help researchers to find specific information about medicinal plants, it also helps them to locate, identify, and collate data about certain plants in specific places followed by utilization in their system. The most commonly used databases are summarized in Tables 21.1 and 21.2.

2.2 Bioinformatics-Guided Approach for Traditional Knowledge

Although a plethora of digital databases is available for searching medical plants at a particular ecological niche or geographical region, the problem is identifying them based on their evolutionary pattern as well as their genetic makeup, which often determine the plants' properties in an applied field. To help with this, bioinformatics-based approaches have evolved that, based on specific programming-guided coding/decoding/scoring systems, aid in identification and property evaluation of certain plants in a specific database. These approaches often rely on the mutual interaction of two datasets on a common bioinformatics platform either for identification or for revelation of a common sequence space inside a genetic subset for predictive exploration of its functional properties. The general notion in this approach is that plants or animals have evolved due to changes in temperature, humidity, oxygen, or any other geographical factor/s, thus there is a genetic similarity between all these species in spite of biodiversity all over the world. Thus bioinformatics-based approaches can be classified as follows:

1. Identification of traditional components.
2. Digitization of traditional datasets.
3. Creation of biodiversity databases.

2.2.1 Identification of Traditional Components

Identification of traditional components is based on several identification tools that are basically web based. In this format, taxonomic descriptions are coded

TABLE 21.1 Database of Detailed Information on Natural Products

Database	Subdatabase	Summary	Website
African traditional knowledge database	Francophone database or PHARMEL (Pharmacopée et Medecine Traditionnelle)	A database containing 19,691 recipes from 24 African countries, 4000 medicinal records, and 51 references	http://www.ulb.ac.be/sciences/bota/pharmel.htm
	Bilingual database (PRELUDE database)	This database comprises information on traditional veterinary and human medicine plants in sub-Saharan Africa	http://pc4.sisc.ucl.ac.be/prelude.html
	PROTA (Plant Resources of Tropical Africa)	A database containing 7000 useful plants, 200,000 references, 30,000 photographs, and 6000 geographic distribution maps	http://www.prota.org/PROTAstartframes.htm
	The English database TRAMED (Traditional Medicine Database)		http://www.healthnet.org.za/tramed/gen/tramedsearch
	NTRAP	The Database of Natural and Traditional Pesticidal Materials and Pest Control in Sub-Saharan East Africa	http://www.ippc.orst.edu/ipmafrica/db/index.html
	IMRA (Institute Malgache de Recherches Appliquées)	IMRA has a database containing computerized ethnobotanical data of over 4000 Madagascan plants	

Continued

TABLE 21.1 Database of Detailed Information on Natural Products—cont'd

Database	Subdatabase	Summary	Website
	NAPRECA (NAtural Products Research Network for Eastern and Central Africa)	NAPRECA (www.naprecanetwork.net) is a network of natural products research scientists in East and Central Africa. The database tries to discover relevant chemicals used to remediate health problems and other beneficial problems all over the world	http://www.napreca.ne
	WANPRES (West African Network of Natural Products REsearch Scientists)	A sub-Saharan database for the Western Africa Network of Natural Products Research Scientists. It helps coordinate chemists and natural products all over the world	www.wannpres.org
CRISP		Computer-based retrieval of Information on Scientific Projects database, which has been government funded and has been carried out by universities, hospitals, and other research institutions	http://crisp.cit.nih.gov
Indian Medicine		A database for Indian Ayurvedic, Yoga, Siddha, Homeopathy, and Unani systems of medicine	http://indianmedicine.nic.in
NAPRALERT	NAtural PRoducts ALERT from STN International	A relational database of all products including ethnomedical information and other experimental studies including pharmacological in vitro/in vivo bioassays. Currently more than 200,000 scientific reports are in the database	http://info.cas.org/online/DBSS/napralertss.html

UK Crop Net Database			http://ukcrop.net/db.html
Database on medicinal plants, a database formed by the Government of India			http://www.nmpb-mpdb.nic.in/
Medicinal and Aromatic Plants Abstracts (MAPA)		Published by the National Institute of Science Communication, CSIR, New Delhi. 55,000 abstracts published in the first 18 volumes of the journal, which are solely made up of databases on medicinal plants	http://www.fao.org/docrep/w7261e/W7261e09.htm
GLOBinMED		A database on Malaysian traditional and complementary medicine	http://www.globinmed.com/index.php?option=com_content&view=article&id=80849&Itemid=101
Database on Indian medicinal plants			http://www.medicinalplants.in/
An Herbal/Medical Dictionary		A Glossary of terms used in Herbalism, Medicine and Physiology, Descriptions, Explanations, and Implications in Wholistic and Vitalist Therapy	http://www.swsbm.com/ManualsMM/MedHerbGloss2.pdf
Annie's Remedy		General guide to herbal medicine including common and botanical names; some records have citation information (check out the Herb Chart link—chart contains links to records)	http://www.anniesremedy.com/

Continued

TABLE 21.1 Database of Detailed Information on Natural Products—cont'd

Database	Subdatabase	Summary	Website
Arctic Science Portal		A library of links to websites where Arctic data are made publicly available. These websites contain information about indigenous plants and their uses	https://www.uspto.gov/patent/laws-and-regulations/comments-public/traditional-knowledge-and-medicine-dictionariesdatabases
Cacti Guide		Common and Latin names plus photographs	http://cactiguide.com/
Dasherb		Latin, English, PinYin, and Chinese names of medicinal herbs, descriptions, and uses	http://www.dasherb.com/database/index.html
CHEMnetBASE		170,000 natural products with names and synonyms, formulae, chemical structures, CAS Registry Numbers, extensive source data, uses and applications, physical state, melting point, boiling point, pKa, and key literature citations	http://dnp.chemnetbase.com/
Dr. Duke's Phytochemical and Ethnobotanical Databases		All references are to Duke, James A. (1992). The linked records contain plant names, alternative names for plants, and information about chemicals found in plants	https://phytochem.nal.usda.gov/phytochem/search

Eastern Chinese Medicine Export Company		Latin, English, and PinYin names, searchable by medical use or symptoms	http://tcmtreatment.com/images/herb-supply/herb-price/three-lists.htm
Find Me A Cure		Botanical names with common names in English, Chinese, and multiple Indian languages. Includes information about uses of herbs. Searchable via text input box, Herbs Glossary (botanical names), and list of Ailments and Remedies.	http://tcmtreatment.com/images/herb-supply/herb-price/three-lists.htm
Find Wild Flowers		Identification of British flora	http://www.botanicalkeys.co.uk/flora/
Fungal Database		United States Department of Agriculture. Scientific and common names, synonyms, specimens, and literature	https://nt.ars-grin.gov/fungaldatabases/
Germplasm Resources Information Network (GRIN)		United States Department of Agriculture (USDA). Scientific and common names and synonyms	https://nt.ars-grin.gov/fungaldatabases/
Gernot Katzer's Spice Pages		10,500 plant names in more than 60 different languages, with origins, constituents, etymology, images, and uses	http://gernot-katzers-spice-pages.com/engl/index.html?redirect=1
Hawaiian Ethnobotany Online Database		Database of Hawaiian names, species names, and vernacular names of plants. The linked records contain information about the uses of the plants with references.	http://data.bishopmuseum.org/ethnobotanydb/ethnobotany.php?b=list&o=1

Continued

TABLE 21.1 Database of Detailed Information on Natural Products—cont'd

Database	Subdatabase	Summary	Website
Herbal Medicine Materia Medica		Descriptions, constituents, and cited references	http://www.ichineseherbs.com/cross_ref_of_names.html
iChinese Herbs		Latin, common, and PinYin names	http://www.ichineseherbs.com/cross_ref_of_names.html
Interagency Taxonomic Information System (ITIS)		Diverse US government agencies partnership. Standardized nomenclature, taxonomic data, and hierarchical classification. Hyperlinks to diverse off-site resources	https://www.itis.gov/
Korean Traditional Knowledge Portal (KTKP)		Korean Intellectual Property Office's database service for searching traditional knowledge from old and very recent Korean and Chinese medicines, including journal articles and patents. Requires registration/subscription.	http://www.koreantk.com/ktkp2014/
Liber Herbarum		Cross-referenced herbal medicine database based fully on printed sources, inspired by the first known Danish medicine book *Liber Harbarum* written by Henrik Harpestgreng in the 13th century.	http://www.liberherbarum.net/

Malta Wild Plants		Detailed records for wild flowering plants of Malta. By Stephen Mifsud	http://www.maltawildplants.com/
Mushroom Nutrition		Scientific and common names, descriptions, and medicinal uses for mushrooms with citations to literature provided	https://www.mushroomnutrition.com/
NAPRALERT		Database of natural products, including ethnomedical information and pharmacological/biochemical information on extracts of organisms in vitro, in situ, in vivo, in human (case reports, nonclinical trials), and clinical studies. Fee required	https://www.napralert.org/
Native Plants Hawaii		A single, comprehensive, and searchable online knowledgebase of endemic and indigenous plants of Hawaii. Some records name additional reference works	http://nativeplants.hawaii.edu/
Natural Medicines Comprehensive Database		Evidence-based clinical relevance. Searchable by product name or medical condition. Links to PubMed	http://naturaldatabase.therapeuticresearch.com/home.aspx?AspxAutoDetectCookieSupport=1
Northern Ontario Plant Database		Records for some 55,000 herbarium specimens from northern Ontario educational and government institutions	http://naturaldatabase.therapeuticresearch.com/home.aspx?AspxAutoDetectCookieSupport=1

Continued

TABLE 21.1 Database of Detailed Information on Natural Products—cont'd

Database	Subdatabase	Summary	Website
Oro Verde Green Gold from Amazonia		Guide to Amazonian and Andean medicinal plants including common and botanical names; records contain citation information (click on any of the "more info" links to see records)	http://www.oroverde.cz/
Plants For A Future		Latin and common names, uses, constituents, and cited references. Registration required	http://www.pfaf.org/user/plantsearch.aspx
Plantsciencenetbase		CRC collection. Covers individual plants, from historical to modern topics, as well as transgenics and evolutionary biology	http://www.crcnetbase.com/page/plant_science_ebooks
Society for Research and Initiatives for Sustainable Technologies and Institutions		Database of medicinal plants containing uses, botanical names, common names, and Sanskrit names	http://www.sristi.org/hbnew/plant_db.php
Southwest School of Botanical Medicine		Known chemical constituents for over 250 medicinal plants	http://www.swsbm.com/Constituents/Constituents.html
The Encyclopedia of New Zealand: medicinal use of plants		Plants used in traditional methods for healing	http://www.teara.govt.nz/en/rongoa-medicinal-use-of-plants/page-1

The International Plant Names Index		A database of the names and associated basic bibliographical details of seed plants, ferns, and lycophytes	http://www.ipni.org/ipni/plantnamesearchpage.do
The Plant List		From the Royal Botanic Gardens, Kew, and Missouri Botanical Garden. Latin names, synonyms, and unresolved names for vascular plants, mosses, and liverworts. Excludes algae, fungi, and common names	http://www.theplantlist.org/
Traditional Knowledge Digital Library—India (TKDL)		TKDL is based on 148 books of Indian systems of medicine. Requires registration with the Government of India	http://www.tkdl.res.in/
University of Melbourne		Multilingual (21 languages) plant name database with links to sites in various languages	http://www.plantnames.unimelb.edu.au/Sorting/List_bot.html
University of Washington		Hyperlinks to Medline, USDA Plants Database, and Plants For A Future Database	https://staff.washington.edu/boerm/uwmhg//
USDA Plants Database		Searchable via multiple access points	http://plants.usda.gov/java/
Western United States Flora Checklists		Latin names and common names	http://www.swsbm.com/HOMEPAGE/Floras/Checklists.html
Prelude Medicinal Plants Database		The PRELUDE database concerns the use of plants in different traditional veterinarian and human medicines in Africa	http://www.africamuseum.be/collections/external/prelude

Continued

TABLE 21.1 Database of Detailed Information on Natural Products—cont'd

Database	Subdatabase	Summary	Website
Database Anti-Diabetic Medicinal Plants diversity (DADMP)		DADMP is developed and maintained by Gopinath Krishnasamy, Department of Bioinformatics, Alagappa University, Karaikudi. This database contains more than 100 medicinal plants from Valaiyans of Alagarkoil hills to treat diabetes	http://www.mkarthikeyan.bioinfoau.org/dadmp/
HerbMed		Collection of several herbal websites such as AGRICOLA, American Indian Ethnobotany Database, Carotenoid Database for US Foods, IBIDS, etc. It is basically a categorized, evidence-based resource for herbal information, with hyperlinks to clinical and scientific publications and dynamic links for automatic updating; produced by the nonprofit Alternative Medicine Foundation	http://www.herbmed.org/links.html
PLANT		Database on Brazilian medicinal plants	Manha et al. [8]
Medherb		A medicinal plant database with genetic information	Rajoka et al. [9]

AyurMedBase		An Ayurvedic medicinal database for traditional and Ayurvedic medicinal systems	www.grin.com
Cameroon 3D		Botanical database of Cameroon containing 2500 compounds of natural origin and 224 medicinal plants belonging to 55 families	Ntie-Kang et al. [10]
p-ANAPL		A collection of more than 500 natural products from African medicinal plants, which has been subjected to Lipinsky's "Rule of Five" and virtually unleashed relevant compounds with pharmacological properties	Ntie-Kang et al. [11]

TABLE 21.2 Botanical Image Database

Type	Database	Description	Website
General plants worldwide	Albion College Vascular Plant Image Gallery	A database to support organizational botany courses; images mainly collected from North America and the Caribbean basin	http://www.albion.edu/plants/
	An *Array of Botanical Images*	Over 24,000 botanical images, arranged alphabetically by genus	http://www.plantsystematics.org/reveal/pbio/RevealSlides/slideindex.html
	Anthos project	Mainly identifies Spanish flora, including photos, distribution maps, and nomenclature	http://www.anthos.es/
	Atrium (biodiversity information developed by the University of Texas)		http://www.atrium-biodiversity.org/
	Botanical Society of America Online Image Collection	Has photos categorized into a number of groups, including plant anatomy, botany, and plant science (many subcategories), people, places, and events, and plant morphology	http://pix.botany.org/index.php?module=simplemedia&type=user&func=view&ot=collection&tpl=tree
	Botanique.org	Images with botanical and biodiversity information	http://www.botanique.org/
	BotIT (includes fungi too)		http://botit.botany.wisc.edu/
	Digital Flowers	Images of angiosperms	http://www.life.illinois.edu/help/digitalflowers/
	Flora of the World	Flowering plants	http://www.floraoftheworld.org/

	Internet Directory for Botany	Over 100,000 images	http://www.ou.edu/cas/botany-micro/idb-alpha/botany.html
	Life Web Site		http://www.nic.funet.fi/pub/sci/bio/life/plants/magnoliophyta/index.html
	Noble Foundation Plant Image Gallery		https://www.noble.org/imagegallery/
	PhytoImages		http://www.phytoimages.siu.edu/
	PLANTS Database	United States Department of Agriculture National Resources Conservation Service. Searchable database with images, distribution maps, nomenclatural information, and more	http://plants.usda.gov/gallery.html
	Plantillustrations.org		http://plantillustrations.org/
	PlantSystematics.org	Keys, cladograms, and over 45,000 images of vascular plants	http://www.plantsystematics.org/
	Raintree Tropical Plant Database	Images of tropical raintree plants	http://www.rain-tree.com/plantimages.htm#.WGpEIlV97IW
	Scott's botanical links		http://www.ou.edu/cas/botany-micro/bot-linx/subject/sub-pict.shtml

in specific electronic languages, which are open for scientists and accessible for exploration of unknown taxonomic datasets. Such biological identification methodology can be classified into three categories: the field guide method, dichotomous paper keys (which is the mostly used method), and computer-based methods. The latter can also be accomplished by four major techniques: hypertext keys, multiaccess keys, expert systems, and neural networks. Multiaccess keys based on a species–character matrix are used predominantly for the identification of biological databases (http://www.borealis.nu/exjobb/Index_en.html).

A good interactive key bears three fundamental attributes: (1) unrestricted character use, (2) ranking of the best character at any stage of the identification, and (3) opportunity to easily reach explanations of characters or more information about species. In a comparison of 14 identification programs and six interactive keys on the internet, the best keys, according to the three fundamental attributes and other important criteria, were selected: the programs Intkey, Linnaeus II, Lucid, Taxis, XID, and the internet key PollyClave 2 (http://www.borealis.nu/exjobb/Index_en.html). The various identification tools in this regard are summarized in Table 21.3.

In 1988, DELTA was adopted by the International Working Group on Taxonomic Databases for Plant Sciences as a standard language for compilation, analysis, and recognition of taxonomic data. In 2005, FreeDELTA was taken over by sourceforge.net, the largest global web-based platform for developing free software. FreeDELTA is built on program languages such as Python, C++, and Object Pascal libraries and uses an open source code that allows users to develop the program by themselves according to their needs. Currently there are 68 datasets in FreeDELTA and 22 datasets in the NaviKey server.

2.2.2 Digitization Tool

Digitization is used to select the identifying information of the plant under investigation. The identifying information is the detailed pro forma of the plant including nomenclature, genus, species, and other information. These databases are intensively utilizable to search for plants in a particular geographical region and collate information about those plants. The major digitization tools for the creation of databases are summarized in Table 21.4.

2.2.3 Biodiversity-Based Databases

Biodiversity-based databases ensemble and detail information regarding the spectrum of biotic organisms in a particular ecological niche. They actually encompass detailed data about the species, specimen, taxonomic distribution, or phylogenetic hierarchy of the biomass in the particular niche. These kinds of databases collate data regarding either superficial biodiversity of flora or fauna in that particular ecological habitat (e.g., species or taxonomic distribution of biotic lives) or consummate molecular biological information extracted from different species of that particular habitat. The significance of the second algorithm is that it is particularly conducive to searching, exploring,

TABLE 21.3 Various Digitization Tools for Identification of Natural Products

Main Identification Tool	Type of Natural Product	Brief Description	URL
Open Identification API	Plant, animal	A computer-assisted program to identify flora or fauna with free software techniques	http://wwbota.free.fr/Identification/
OpenKey	Plant	A Delta database encrypted by 200 characters to identify any plant in North Carolina Piedmont	http://www.ibiblio.org/openkey/intkey/web/intro.html
PANKEY	Plant, animal	A software program to describe taxonomy of an unknown flora or fauna; also describes numerical taxonomy (clustering and cladistics). Prepared by version 3.0 of Delta	http://www.exetersoftware.com/cat/pankey/pankey.html
PHPKey	Plant	A new interactive key to describe calicioid lichens and fungi of the Nordic countries. The database comprises 83 species, 27 characters, 216 character states, and related information. PHPKey is written in the HTML embedded programming language PHP	http://www.borealis.nu/exjobb/Index_en.html
PollyClave	Plant, animal	A multiple entry identification tool created by the University of Toronto	http://prod.library.utoronto.ca:8090/polyclave/
Rachis	Plant, animal	Software to allocate biological entities in a hierarchical system (both LINUX and MS Windows support) and an interactive key for retrieving data from them	http://rachis.sourceforge.net/
Scratchpads	Plant, animal	A database created by the Natural History Museum (UK)	http://editwebrevisions.info/

Continued

TABLE 21.3 Various Digitization Tools for Identification of Natural Products—cont'd

Main Identification Tool	Type of Natural Product	Brief Description	URL
Stinger's Lightweight Interactive Key Software (SLIKS)	Plant	A Javascript program for biological identification	http://www.stingersplace.com/SLIKS/
Taxy	Fungi		http://www.collectivesource.com/taxy/taxy.html
TeleNature	Plant		http://www3.isrl.uiuc.edu/~TeleNature/projects/telenature.html
X:ID	Plant, animal		http://uio.mbl.edu/services/key.html
Xper	Plant	Free Delta-based software for taxonomical identification and other trait analyses	http://lis-upmc.snv.jussieu.fr/lis/?q=en/resources/software/xper2
Barcode of Life Database (BOLD) (identification by DNA barcoding)	Plant, animal	An identification portal based on sequence search and matching analysis using DNA barcoding	http://www.boldsystems.org/index.php/IDS_OpenIdEngine
EDIT's cybertaxonomy platform	Plant, animal	A common platform for digital analysis of cybertaxonomy, used for data storage and exchange; collections and specimens; descriptions; fieldwork; literature; and geography	http://wp5.e-taxonomy.eu/
Electronic field guide	Plant, animal	A web-based version structured by the Department of Computer Science and Biology at the University of Massachusetts, Boston, with funding from the National Science Foundation	http://wiki.cs.umb.edu/

TABLE 21.3 Various Digitization Tools for Identification of Natural Products—cont'd

Main Identification Tool	Type of Natural Product	Brief Description	URL
FreeDELTA[a]	Plant, animal		http://freedelta.sourceforge.net/
Idenature Guides	Plant, animal	General web-based identification tool for plant, animal, fungi, insects, and other related species	http://www.discoverlife.org/mp/20q
LucID	Plant, animal	An interactive key-based software platform to identify or diagnose biological entities	http://www.lucidcentral.com
Medical fungi identification website	Fungi	Identification tool for filamentous fungi of medical importance except the genera *Aspergillus* and *Penicillium*	http://www.cbs.knaw.nl/medical/DefaultPage.aspx
Meka	Plant		http://ucjeps.berkeley.edu/keys/downloads/Meka31.exe

[a]*FreeDELTA: DELTA stands for DEscription Language for TAxonomy. FreeDELTA is the world's largest software tool that is utilized by taxonomic scientists for the compilation and accumulation of taxonomic data all over the world (http://freedelta.sourceforge.net/). FreeDELTA is a language that comprises both qualitative (binary or multistate, ordered or unordered) or quantitative (integer or real) characters. Although the software was created by Mike Dallwitz at CSIRO Division of Entomology, Canberra, Australia, in the mid-1970s, it was later used by various other taxonomic program developers such as Eric Gouda at the Botanic Gardens of Utrecht University (TAXASOFT) in the Netherlands, Nicholas Lander at the Western Australian Herbarium (DMSWIN) in Australia, Antonio Valdecasas at the Museo Nacional de Ciencias Naturales (EDEL) in Spain, Gregor Hagedorn at the Institute of Microbiology, Federal Biological Research Center (DELTAAccess) in Germany, Michael Bartley and Noel Cross at the Arnold Arboretum of Harvard University (NaviKey) in the United States, Claudio Rivetti and Riccardo Percudani at the University of Parma (WebDelta) in Italy, and Mauro J. Cavalcanti at Museu Nacional/Universidade Federal do Rio de Janeiro (DIANA) in Brazil.*

or interconnecting different biological entities in that specific domain. This unveiling, in turn, helps to predict biological properties of a cluster of entities before performing any operation on them. These can be divided into four major databases

1. Biodiversity databases for all classes.
2. Plant-based biodiversity databases.
3. Animal-based biodiversity databases.
4. DNA barcode-based biodiversity databases.

TABLE 21.4 Major Digitization Tools for the Creation of Databases

Name of the Database	Coverage	Brief Description	Website
Bauble	Flora	Free web interface software. It generates reports through XSL and Mako formatters.	http://bauble.belizebotanic.org/
Bibmaster	Flora, fauna	Database application	http://www.gbif.es/bibmaster/bibmaster_Inphp
Biota	Flora, fauna	Do	http://viceroy.eeb.uconn.edu/biota
Biotica	Flora, fauna		http://www.conabio.gob.mx/biotica/cms/index.php
Brahms	Flora	Digitization tool to collate all the data of botanical species especially in the United States	http://dps.plants.ox.ac.uk/bol/BRAHMS/Home/Index
Herbar	Flora	General digitization program involved in showcasing, storing, and indexing information of traditional products	http://www.gbif.es/herbar/herbar_Inphp
KE Emu-Electronic Museum	Flora, fauna	Do	http://www.kesoftware.com/emu-home.html
Pandora	Flora	Do	http://www.ibiblio.org/pub/academic/biology/ecology+evolution/software/pandora/
Specify 6	Flora, fauna	Do	http://specifysoftware.org/
Zoorbar	Flora, fauna	Do	www.gbif.org/resource/81736

2.2.3.1 Biodiversity Databases for All Classes

The biodiversity databases for all classes include databases where diverse species of a biotic community irrespective of flora or fauna have been created inside the database. A brief description of the databases is summarized in Table 21.5.

TABLE 21.5 Biodiversity Databases for All Classes

Database	Brief Description	Website
Biodiversity Heritage Library (BHL)	Collection of digitized literature including images in open access form	www.biodiversitylibrary.org
Integrated TaxonomicInformation System (ITIS)	Taxonomic information about plants, animals, fungi, and microbes especially in North America	http://www.itis.gov/
Species 2000	A global database of biodiversity of all known species in the world. It is basically a collaborative program between CODATA (International Council for Science: Committee on Data for Science and Technology), IUBS (International Union of Biological Sciences), and the IUMS (International Union of Microbiological Societies) of the early 1990s. It is an associate member in the Global Biodiversity Information Facility (GBIF) , a data provider to EC LifeWatch, and is acknowledged by the United Nations Environment Program (UNEP) together with the Convention on Biological Diversity (CBD)	http://www.sp2000.org/
Tree of Life (TOL)	A biodiversity database compiled by biologists all over the world containing more than 10,000 web pages regarding biodiversity of organisms together with their phylogenetic information	http://tolweb.org/tree/phylogeny.html
TreeBASE	Basically deals with phylogenetic information about trees formed by the Phyloinformatics Research Foundation, Inc.	http://www.treebase.org/treebase-web/home.html
Barcode of Life Databases (BOLD)	Basically integrates biodiversity information all over the world by DNA barcodes. Since the DNA barcode of certain genes (such as *matK*, *rbcL*, mitochondrial cytochrome oxidase c subunit-I [COI], and its genes) shows variation from species to species, it is used as a reference for species identification. BOLD comprises such DNA barcodes to identify unknown species	http://www.barcodinglife.org/views/logInphp

Continued

TABLE 21.5 Biodiversity Databases for All Classes—cont'd

Database	Brief Description	Website
Global Invasive Species Database (GISD)	A database about invasive species across the globe. It was formed by the Invasive Species Specialist Group (ISSG) of the Species Survival Commission of the IUCN (International Union for Conservation of Nature). The development of GISD was initiated by the Global Invasive Species Programme and other statutory bodies such as the National Biological Information Infrastructure, Manaaki Whenua-Landcare Research, the Critical Ecosystem Partnership Fund, the University of Auckland, and private donations	http://www.issg.org/database/welcome/
Invasive and Exotic Species	Deals with invasive and exotic species of North America	http://www.invasive.org/
Invasive Species in Canada	Deals with invasive species of Canada	http://www.invasivespecies.gc.ca/english/view.asp?x=1
Biodiversity of Mexico VN	A Mexican database	http://www.vivanatura.org/About%20VN.html
Natural History Museum	Biodiversity museum, i.e., database of the United Kingdom. Has also been developed in America and other countries	http://www.nhm.ac.uk/nature-online/life/index.html
South African National Biodiversity Institute database	A South African biodiversity database	http://www.sanbi.org/frames/infofram.htm
Arctos	A database for finding information about biodiversity using different identifier tools	http://www.arctos.database.museum
ASEAN Biodiversity Information database (BISS)	A biodiversity system of ASEAN flora and fauna; the database is maintained to be accessible by all the users across the globe	http://www.aseanbiodiversity.org

TABLE 21.5 Biodiversity Databases for All Classes—cont'd

Database	Brief Description	Website
Convention on International Trade in Endangered Species (CITES) species database	A database regarding the Convention on International Trade in Endangered Species of Wild Fauna and Flora. This database maintains the list so that the existence of such kinds of species is not threatened over the world.	All species ever listed in CITES Appendices I, II, and III
iNATURALIST	A site for reporting personal observations of any plants or animals in the world.	www.inaturalist.org
iSpot	Free international database about ecosystems and associated flora and fauna	www.iSpotnature.org
Natural History Information System	A collection of databases of natural products. This is basically a DINA project (DIgital Information System for NAtural History Collections) involving all types of collections such as zoological, botanical, geological, and paleontological collections, living collections, biodiversity inventories, observation records, and molecular data	www.dina-project.net
NatureServe	A natural biodiversity system	www.natureserve.org
WikiSpecies	A free database formed by the Wikimedia foundation	https://species.m.wikimedia.org
Pan European Species directories Infrastructure	A taxonomic database for Europe	www.eu-nomen.eu
NatureDATA	A natural products database in the United Kingdom	naturedata.org.uk
Georgia (country) biodiversity website	A database maintained in Georgia University, USA	biodiversity-georgia.net
Biodiversity Heritage Library (BHL)	A database containing over 2 million volumes of biodiversity literature involving two centuries and is jointly maintained by the United States and the United Kingdom. As of October 2010, BHL had coded 31,397,395 pages from 83,616 volumes, and from 43,140 titles.	www.biodiversitylibrary.org

2.2.3.2 Plant-Based Biodiversity Databases

There are several databases based only on plants or flora as summarized in Table 21.6.

2.2.3.3 Animal-Based Biodiversity Database

There are several databases that contain digitized information about animals, as summarized in Table 21.7.

TABLE 21.6 Digitized Databases Based on Plants or Flora

Database	Brief Description	Website
Algae Base	A botanical database especially focused on algae belonging to aqueous, land, and marine organisms	http://www.algaebase.org/
Australian Biological Resources Study Flora online	A specifically designed database compiling information on Australian biodiversity involving plants	http://www.environment.gov.au/biodiversity/abrs/online-resources/flora/main/
DiaMedBase	A database particularly covering information about the plants to cure diabetes	http://www.progenebio.in/DMP/DMP.html
Encyclopedia of Indian Medicinal Plants	Database containing information on Indian medicinal plants	www.medicinalplants.in
Plants For A Future (edible and medicinal plants)	A database comprising more than 7000 medicinal plants where each plant is categorized on the basis of its edibility and therapeutic use	http://www.pfaf.org/user/plantsearch.aspx
Royal Botanic Garden, Edinburgh	A database containing information about plants inside the Royal Botanical Garden as well as plants involved in specific research projects such as ADIAC Diatom Image Database, DIADIST Website, Apiales Resource Centre, Southeast Asian Begonia Database, etc.	http://www.rbge.org.uk/databases
United States Department of Agriculture (USDA) Plants Database	A database about vascular plants, mosses, liverworts, hornworts, and lichens of the United States and its territories. It contains more than 50,000 images of such plants	http://plants.usda.gov/

TABLE 21.6 Digitized Databases Based on Plants or Flora—cont'd

Database	Brief Description	Website
Fungal Records Database of Britain and Ireland	Mostly focused on the fungal databases of Britain and Ireland; however, it contains information about more than 2 million fungal records	http://www.fieldmycology.net/
Index Fungorum	A project containing information with all formal names (scientific names) of the fungal kingdom. It is a joint collaborative project partnered by the Royal Botanic Gardens, Kew, Landcare Research, and the Institute of Microbiology, Chinese Academy of Sciences	http://www.indexfungorum.org/Names/Names.asp
MycoBank	A fungal database run by the International Mycological Association	http://www.mycobank.org/
USDA fungal database	A database based on US national fungus collections	http://nt.ars-grlngov/fungaldatabases/

TABLE 21.7 Databases About Animals Across the Globe

Database	Brief Description	Website
Amphibian Species of the World	This database contains three types of amphibian species: anura (frogs), salamanders (Caudata), and gymnophiona (caecilians). This database contains 7645 amphibian species, of which 6745 are frogs and toads, 695 are newts and salamanders, and 205 are caecilians	http://research.amnh.org/herpetology/amphibia/index.php
Antbase	Database based on all ant species around the world	http://antbase.org/
Australian Biological Resources Study Fauna Online	A zoological database based on Australian biodiversity.	http://www.environment.gov.au/biodiversity/abrs/onlineresources/fauna/index.html
Biosystematic Database of World Diptera	Knowledge database about dipterian animals. More than 150,000 species of Diptera are described under more than 250,000 names	http://www.sel.barc.usda.gov/Diptera/

Continued

TABLE 21.7 Databases About Animals Across the Globe—cont'd

Butterflies and Moths of the World	Database about butterflies and moths.	http://www.nhm.ac.uk/research-curation/projects/butmoth/
CephBase	Knowledge database about cephalopods around the world	http://www.cephbase.utmb.edu/
Fishbase	Knowledge database about fishes around the world	Fishbase Fauna http://www.fishbase.org/home.htm
Mammal Species of the World	Knowledge database about mammalian species around the world	http://vertebrates.si.edu/mammals/msw/
The Reptile Database	Knowledge database about reptiles around the world	http://www.reptile-database.org/
Universal Chalcidoidea Database	Knowledge database about Chalcidoidea group of wasps around the world	http://www.nhm.ac.uk/research-curation/projects/chalcidoids/index.html
Zoology: Extinct and Endangered database	Organized and integrated information database about extinct and endangered species of the world	http://www.oum.ox.ac.uk/database/zoology/extinct.html

2.2.3.4 DNA Barcode-Based Databases

2.2.3.4.1 Barcode of Life Database Barcode of Life Database (BOLD) is a DNA barcode-based biodiversity database that has four portals. The first is a public data portal that contains 1.7 million DNA barcode sequences, which is freely searchable and categorized by geographical, taxonomical, and depository databases. The second is Barcode Index Numbers, which comprise several numbers signifying specific barcodes, a DNA barcode education portal, which is explorable by students and scientists and could be enriched by the latter through submitting new barcodes, the third portal being Barcode Index numbers and the fourth portal is the workbench that allows scientists to work and analyze DNA barcodes on a common platform. The current coverage of BOLD is given by Table 21.8.

2.2.3.4.2 Korean Barcode of Life This is a barcode database designated to elucidate barcodes of all Korean species. Currently, the database contains 5531 barcode sequences from 2429 Korean species.

TABLE 21.8 Database Having Barcode of Life Database (BOLD) Coverage

Items	Number
Barcode clusters for animals (Barcode Index Numbers)	495,328
All sequences	6,175,187
Barcode sequences	5,339,196
Animals	176,400
Plants	65,732
Fungi and other life forms	20,838

2.3 Metadata Portals

Since the biological databases are interconnected and provide full information upon being integrated in one common data portal, several metadata portals in the digitization of traditional knowledge have been created to access primary databases by searching through secondary search engines. Such metadata engines are provided in Table 21.9.

3. BIODIVERSITY ANALYSIS

Several interdisciplinary approaches have been emerging over the last few decades to organize, narrate, collate, and then use biodiversity data for various purposes such as phylogenetic analysis, evolutionary analysis, metabolic pathway analysis, and many others [11a]. Various bioinformatics, molecular biology, pharmacogenomics, cheminformatics, and other approaches have emerged to process the biodiversity data available across various databases. Thus to cope with this, a plethora of analytical tools has evolved to analyze the biodiversity database. Selected analytical tools based on their usage in this process are summarized in Table 21.10.

4. VIRTUAL SCREENING OF NATURAL PRODUCTS FROM DATABASES

Virtual screening of natural products is the in silico process of screening a large database of natural products obtained in a particular or diverse ecological niche to achieve a specific pharmacological response. The in silico process depends on various bioinformatics, involving docking and network pharmacology as described next.

TABLE 21.9 Metadata Portals Having Barcode of Life Database (BOLD) Coverage

Database	Brief Description	Website
Atlas of Living Australia (ALA)	An online repository of Australian biodiversity including Australian flora, fauna, and fungi. This national database of Australian biodiversity provides a platform to access and search information on specific components	http://www.anbg.gov.au/cpbr/program/hb/index.html
Australian Virtual Herbarium (AVH)	Same as above; however, the major coverage is plant species	http://www.ersa.edu.au/avh/index.jsp
Encyclopedia of Life (EOL)	A metadatabase aimed at collating information about each and every living being of 1.9 million species discovered so far on earth. It is proposed to link information on each species with an infinitely long page containing all the information including images, videos, etc. about the species	http://www.eol.org/home.html
Global Biodiversity Information Facility (GBIF)	Single data management portal through which all taxonomical, biogeographical, hierarchical, and genomic information about various species of the world can be accessed	http://www.gbif.org
iSpecies	A zoological species database created by Glasgow University, Scotland	http://darwlnzoology.gla.ac.uk/~rpage/ispecies/index.php
LifeWatch	A European biodiversity database specially constructed to preserve all environmental biodiversity by accessing such information in Europe	http://www.lifewatch.eu/index.php?id=411
Ocean Biogeographic Information System (OBIS)	This is mainly focused on collecting, compiling, and processing biodiversity present in oceans. Currently OBIS possesses 27.7 million pieces data from 126,000 species from 849 databases. It is jointly run with various other digitization agencies such as GBIF, Consortium for Biodiversity of Life (CBOL), and Taxonomic Database Working Group (TDWG)	http://v2.iobis.org

TABLE 21.9 Metadata Portals Having Barcode of Life Database (BOLD) Coverage—cont'd

Database	Brief Description	Website
SpeciesBase	A species database supported by the Reference Center on Environmental Information (CRIA) to collate and share data regarding various botanical and zoological species across the world. It is structured over Visual Basic for Application (VBA) and Microsoft Access. The user interface is designed on the BONABIO information taxonomic database adopted by the Federal University of Parana system	http://www.speciesbase.org/
Universal Biological Indexer and Organizer	This is a combinatory database where biological data from different resources are collected and presented in a meaningful, legible, and organized way. Web services such as XML and SOAP are used for processing the data. It is basically known as Taxonomic Name Server, interconnected as Name Bank (11,106,374 records) and Classification Bank (90 classifications)	http://www.ubio.org

4.1 Screening Through Network Pharmacology

Different biodiversity databases, in addition to organizing information regarding flora, fauna, or microbiome within a particular area or across the globe, also help in drug discovery with the bioinformatics approach. Over the last few decades attempts have been made to organize the mammoth data of biodiversity in drug discovery processes by the virtual screening method. The most common method of such virtual screening is combining network pharmacology or polypharmacology [12–14] with molecular docking. Since network pharmacology suggests that multiple genes or proteins are involved in a particular phenotype or disease, responsible proteins are searched for first while considering a particular disease. The protein structures are then downloaded from a protein databank and reported compounds from various biodiversity databases are docked onto the particular set of proteins. The "best hit" compounds are then taken as leads for subsequent drug discovery [15,16]. The flow chart for performing virtual screening is summarized in Fig. 21.1.

TABLE 21.10 Tools for Biodiversity Analysis

Software	Use	Category	URL
ADE4	Ecological analysis	S, F	http://cran.rproject.org/src/contrib/Descriptions/ade4.html
ADAPTS	Paleobiological analysis	S, F	http://www.paleodb.org/paleosource/code.php?stage=download&project_no=4
APE	Phylogenetic and diversification analyses	S, F	http://pbil.univ-lyon1.fr/R/ape/
CODA	Nature conservation and planning	S, F	http://members.ozemail.com.au/~mbedward/coda/coda.html
DIVA-GIS	Mapping and ecological modeling	S, F	www.diva-gis.org/
Ecopath with Ecosim (EwE)	Ecological modeling (marine environment, including the effects of fishing)	S, F	http://www.ecopath.org/index.php?name=About
GARP	Ecological modeling	S, F	http://nhm.ku.edu/desktopgarp/
GRASS GIS	GIS is used for geospatial data management and analysis	S, F	http://grass.itc.it/
LAMARC	Population studies	S, F	http://evolution.genetics.washington.edu/lamarc.html
MAXENT	Ecological modeling (species distribution)	S, F	http://www.cs.princeton.edu/~schapire/maxent/
MEGA	Phylogenetic analysis	S, F	http://www.megasoftware.net/
Mesquite	Evolutionary analysis	S, F	http://mesquiteproject.org/mesquite/mesquite.html

Molphy	Phylogenetic analysis	W, F	http://bioweb.pasteur.fr/seqanal/interfaces/prot_nucml.html
MrBayes	Phylogenetic analysis	S, F	http://mrbayes.csit.fsu.edu/index
PAST (PAlaeontological STatistics)	Paleontological statistics	S, F	http://folk.uio.no/ohammer/past/
PATN	Pattern analysis	S, C	http://www.patn.com.au/
PopTools	Population dynamics and ecological model analysis	S, F	http://www.cse.csiro.au/poptools/
PAUP	Phylogenetic analysis	S, C	http://paup.csit.fsu.edu/
PHYLIP	Phylogenetic analysis	S, F	http://evolution.genetics.washington.edu/phylip.html
Rarefaction calculator	Diversity estimation and indices	W, F	http://www2.biology.ualberta.ca/jbrzusto/rarefact.php
TNT	Phylogenetic analysis	S, F	http://www.cladistics.com/aboutTNT.html
TreeAlign	Phylogenetic analysis	W, F	http://bioweb.pasteur.fr/seqanal/interfaces/treealign-simple.html
TreeView X	Phylogenetic tree visualization	S, F	http://darwInzoology.gla.ac.uk/~rpage/treeviewx/

C, Commercial; *F*, free; *S*, standalone application; *W*, web based application.
Reprinted with permission from J. Gaikwad, P.D. Wilson, S. Ranganathan, Ecological niche modeling of customary medicinal plant species used by Australian Aborigines to identify species-rich and culturally valuable areas for conservation, Ecol. Modell. 222 (2011) 3437–3443.

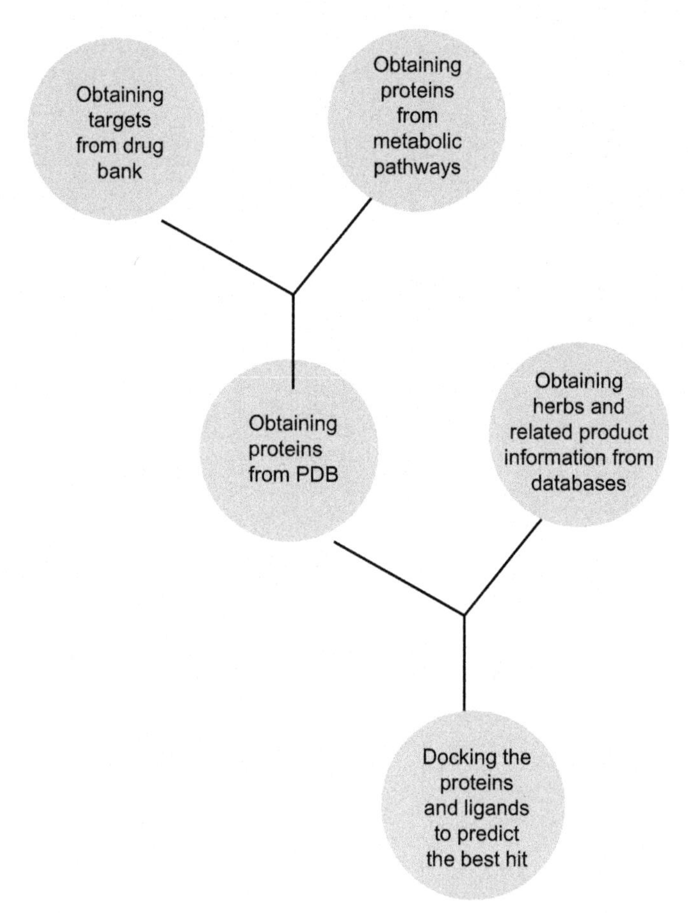

FIGURE 21.1 Tentative methodology for virtual screening of natural products.

For example, Gu et al. [15] reported that Universal Natural Products Database (UNPD)-derived natural products screening yielded five medicinal plants, namely, *Hypericum perforatum*, *Ganoderma lucidam*, *Holarhena antidysenteria*, *Celastrus orbiculatus*, and *Marraya eucherestifolia* as having antidiabetic activity. The authors used the drug target networks (DTN) methodology to explore the new set of plants against the aforementioned disease from a library of 208,000 natural products [15].

4.2 Screening Through Cheminformatics

In a review, Medina-Franco [17] reported various natural products databases, cheminformatics methods of their screening, and ultimately lead findings for various pharmacological responses thereof. For example, he acknowledged that the database ZINC containing more than 19 million molecules, traditional Chinese medicine (TCM) database, UNPD containing 197,201 molecules, UNIIQUIM database (Mexico), and NuBBE database (Brazil) were significantly large databases. These databases have been reported to be used for drug discovery purposes. For example, a web server-based docking of TCM followed by de novo ligand design has been acknowledged by Tsai et al. [18]. Moreover, Chen et al. [19] reported discovery of pancreatic triacylglycerol inhibitors through computational approaches in TCM.

4.2.1 Analysis of Structural Diversity and Complexity

Structural complexity is the hallmark signature of natural product molecules. However, drug discovery in such conditions is aided via two digitized approaches. One is application of structural fingerprints and the other is using chemical scaffolds [20,21]. Apart from benzene and acyclic molecules, flavones, coumarins, and flavanones have been identified as the most frequent scaffolds across the various natural products databases [12].

4.2.2 Structure Promiscuity Index Difference

Dandapani and Marcaurelle [22] in a study reported that the structural diversity of natural products eventually leading to generation of diverse pharmacological activities is due to diverse fraction of unsaturation in various natural products [22]. In continuation, Clemons et al. [23] screened a library of 15,000 compounds, both natural and synthetic, over 100 diverse proteins involved in various metabolic pathways. They later acknowledged that structural diversity actually leads to specificity in substrate-protein binding, finally converging in specific pharmacological activity. To design this in silico, they created an index, namely, Structure Promiscuity Index Difference, to calculate changes in protein binding due to small changes in structure [23].

4.2.3 Chemical Space—Importance and Evaluation

One of the significant approaches to the digitized evaluation of natural products is evaluation of chemical space. It can be defined as defined by Dobson: "the total descriptor space that encompasses all the small carbon-based molecules that could in principle be created" [23a]. In another concept, Lipinsk and Hopkins mentioned that "chemical space can be viewed as being analogous to the cosmological universe in its vastness, with chemical compounds populating space instead of stars." [24]. The evaluation of chemical space has been extensively used by various authors [24a], [15,25–27]. The analysis mainly relies on ChemGPS-NP$_{web}$, an online tool for chemical space analysis. Web analysis is basically reliant on principle component analysis, which divides it into four dimensions and maintains specific compound descriptors in each dimension.

4.2.4 Application of Cheminformatics to Drug Discovery

The cheminformatics approach has been applied to drug discovery to successfully unfold various natural products for a set of pharmacological responses. For example, Cao et al. [28] screened more than 4000 natural products from 100 medicinal plants against estrogen receptors (ERα) and (ERβ), which eventually led to the discovery of 11 selective nonsteroidal estrogen receptor modulators.

Guasch et al. [30] discovered five new drug leads from 89,000 natural products for peroxisome-activated receptors [30]. In continuation, Ngo and Li [31] developed molecules for Alzheimer's disease from a pool of natural products [31]. The authors screened a library of 342 compounds from Vietnamese plants and docked them subsequently against a set of amyloid ($A\beta_{1-40}$ and $A\beta_{1-42}$) peptides to reveal five compounds showing promising potential against Alzheimer's disease. Also Gu et al. [32] performed virtual screening of 676 compounds from a TCM database with 37 proteins related to type 2 diabetes mellitus [32].

5. BIOINFORMATICS APPROACH TO THE DIGITIZATION OF KNOWLEDGE ON NATURAL PRODUCTS

5.1 Quality Control of Herbals Using Next Gen Sequencing

Herbal products are often supplied with supplements from various other natural products. Hence a proper, defined, quality-controlled approach to evaluate these other products is still a difficult task. Ivanova et al. [33] proposed that next gen sequencing followed by DNA barcoding could elucidate the quality of herbal supplements. In this study the authors demonstrated a DNA sequencing approach for taxonomic authentication of herbal supplements from five medicinal plants: *Echinacea purpurea*, *Valeriana officinalis*, *Ginkgo biloba*, *H. Perforatum*, and *Trigonella foenum-graecum*. Using DNA barcoding of *rbcL* and *ITS2* regions the authors successfully accomplished the identification of the foregoing medicinal plants. In addition, the authors also claimed to detect adulterants mixed with the herbal supplements in these food formulations. Interestingly, the amount of contaminants as well as products due to plant–fungi interactions could also be detected by quantitative analysis of next gen sequencing [33].

5.2 Expressed Sequence Tags

In 2012 Sharma and Sarkar described various bioinformatics approaches to discover natural products from various resources, e.g., genomics and transcriptomics data to categorize phylogenetic information about medicinal plants. The authors reported the contribution of "expressed sequence tags" (ESTs) for transcriptomic data organization in universal data portals such as

National Center for Biotechnology Information (NCBI). In addition, they also reported the EGENES database for more authentic information on plant transcriptomic data with better organization of ESTs to correlate genetic information with functional information [34]. In the Medicinal Plants Genomic Resource Database, such complete plant transcriptomic data have been created.

5.3 Simple Sequence Repeats

Apart from ESTs, the authors also acknowledged utilization of simple sequence repeats (SSRs) to compile transcriptomic information of medicinal plants. SSR markers have been shown to be most advantageous because of their multiallelic nature, reproducibility, codominant inheritance, high abundance, and extensive genome coverage [86]. SSRLocator is an example of a computational approach for detection and characterization of SSRs and minisatellite motifs [35].

5.4 Constructing Network Biology Through Chemogenomics

Network biology is an important tool to construct networking maps to unlock the role of various genes in multiple biological functions inside the body. Since body metabolic pathways are usually constructed of various genes or proteins in an orchestrated way, which often involve a spectrum of genes mutually overlapping in nature, perturbing the functional outputs of those genes often elucidates various metabolic pathways inside the body [36]. The Kyoto Encyclopedia of Genes and Genomes (KEGG) is a reliable database that provides information on such metabolic proteins as well as the pathways [37]. In addition, there is a web server called Path Pred [38] that predicts pathways of multistep reaction for a given query compound, starting with a similarity search against the KEGG COMPOUND database. With the help of these chemogenomics databases, network biology is constructed, which on in silico screening leads to the path of drug discovery. For example, the *Catharanthus roseus* gene-metabolite coexpression network was dissected and ultimately led to the discovery of genes associated with the biosynthesis of terpenoid indole alkaloids [39].

5.5 Network Biology Models—Distance-Based Mutual Information Model

In this model, a mutual information entropy and herb distance metric is used to score herb interactions [40] and constitutes an herb network with the combination rules of TCM. Thus, network-lined herb—herb interaction could produce therapeutic activity, which has already been reported to produce angiogenesis activity.

5.6 Quantitative Composition–Activity Relationship Study

Since all herbal components and constituent structures have not yet been discovered or elucidated, quantitative structure–activity relationship (QSAR) studies relating plant constituents and their bioactivities are difficult to perform. Hence the quantitative composition–activity relationship (QCAR) study has been invented to predict plant extract activity in silico [41]. Although this method is still not very accurate and needs full experimental design or bioassays to predict the correct score [42], it could be justifiably used to foretell bioactivity of plant-based extract activity on a computer chip. With approaches such as artificial neuronal network and support vector machines (SVM), prediction with QCAR has been simpler than previously encountered. For example, Nayak et al. [43] reported that adjusting components of a TCM-based herb Qi-Xue-Bing-Zhi-Fang with the aforementioned methods significantly reduced the blood cholesterol level in rats.

5.7 Network Target-Based Identification of Multicomponent Synergy

This methodology of bioinformatics-based digitization of traditional knowledge consists of two components: topology score and agent score. In the first approach, a total network of plant species is analyzed based on their contribution to different diseases and drug actions, and a topology score is assigned. The agent score is given based on their plant phenotypes. Afterward, based on these scores, a synergistic score is given to evaluate the synergistic action between two medicinal plants [44].

5.8 Application of the Bioinformatics Approach for Drug Discovery From Traditional Plants

Phylogenetic analysis of natural products revealed that it has direct correlation with biological activity. Such analysis coupled with gas chromatography-mass spectrometry studies of an alkaloidal fraction of *Phaedranassa dubia* revealed a direct correlation of its acetylcholinesterase inhibitory activity with alkaloids such as galanthamine or lycorine [45]. Anticancer drugs were developed by building network models using a bioinformatics-guided approach [46]. Furthermore, the QSAR-based approach with natural products evolved immunomodulatory compounds; cleomiscosin molecules (A, B, C) were discovered using the QSAR approach [47]. In addition, virtual screening of natural products led to the evolution of peroxisome proliferator-activated receptors (PPARs). Petersen et al. constructed a pharmacophore-based model of 13 PPAR-based partial agonists from *Pistacia lentiscus* from the Chinese Natural Products Database. Virtual screening revealed an oleoresin from the aforementioned plant to have a potential PPAR activator effect [48].

5.9 In Silico Docking

Docking is one of the popular in silico approaches to screen a library of compounds having medicinal interest. This approach is also employed in the case of natural products, because natural products are a diverse set of secondary metabolites and without virtual screening, isolation and subsequent bioactivity estimation are often tedious and complicated jobs. However, docking strategies have evolved a set of newer compounds that can be used as leads for emancipation of medicinal compounds. For example, Zhong et al. reported an inducible nitric oxide synthetase (iNOS) inhibitor of a quinoline derivative, which can be used as scaffold for further designing associated compounds [49]. The author reported docking of more than 90,000 natural products from the ZINC database in silico to evolve one successful compound against iNOS. Again, Li et al. [49a] reported discovery of a potential anticancer compound (breast cancer) through molecular docking by screening of 11,247 compounds from the ZINC database against human epidermal growth factor 2. Likewise, docking has been used to screen a plethora of drugs such as antiinflammatory IKKβ inhibitors [50,51], acid sphingomyelinase inhibitors [52], PPARγ partial agonists, dipeptidyl peptidase inhibitors for antidiabetic drugs [30], STAT 1 and STAT 3 inhibitors [53], multidrug efflux pump inhibitors for reducing antibiotic resistance [54], marine natural products acting on acetylcholine binding protein [55], ellagic acid derivatives on selected enzymes of *Mycobacterium tuberculosis* [56], and others. Thus docking has been a promising alternative approach to drug discovery through digitization of natural products.

6. INVERTNET

InvertNet is a database containing information about invertebrate species across the globe [57]. The database is designed and maintained by the US National Science Foundation's Advancing Digitization of Biological Collections program, and provides digital access to approximately 60 million specimens housed in 22 arthropods (primarily insects). They provide a 3D image of every insect under this category and label them with a unique digital code identifier. This database provides a unique solution to the digital database monitoring system mainly focusing on insects.

7. SCREENING FROM ACTINOBACTERIA

Doroghazi et al. [58] proposed a newer path for drug discovery with a digitized study of natural products. They undertook Actinobacteria as a model of natural product source. Since organisms classified as actinomycetes are reported for natural product biosynthetic gene clusters [59], the authors used the bioinformatics approach to combine 11,422 gene clusters with 4122 gene cluster families (GCF). Subsequent studies revealed 830 genomes from the microorganism, which exhibited coding for hundreds of future leads. In this process, peptidogenomics tools are used to investigate new peptides from a set of mass spectrometry-based peptide fragmentation datasets [60]. Various other bioinformatics tools have been used such as NaPDoS, a natural product domain finder that works on clustering phylogenetically correlated secondary metabolite production gene clusters [61], antiSMASH, a rapid identification, annotation, and analysis of secondary metabolites producing genome sequences from bacterial and fungal origins [62], ClusterMine360, a database for microbial polyketide synthetase [63], SEARCHPKS, a program for investigating polyketide synthetase domains [64], ASMPKS, an analysis program for molecular polyketide synthase domains [85], DoBISCUIT, a database for secondary metabolite producing gene clusters [65], NORINE, a database for nonribosomal peptides [66], PKMIner, a database for exploring type-II polyketide synthetase [67], and others.

Ikram et al. [16] used a digitization screening model to isolate several compounds as neuraminidase inhibitors, i.e., active against influenza. The authors used a docking approach to hit 3000 compounds from the Malaysia Natural Products Database to find the best 12 hits as leads for antiinfluenza drugs. Lead compounds, their docking scores, and IC_{50} values are provided in Table 21.11.

8. PREDICTION INFORMATICS FOR SECONDARY METABOLOMES

For the prediction of secondary metabolites from genetic subsets in natural products, Skinnider et al. published a report of a new web-based software design [68] that they called Prediction Informatics for Secondary Metabolomes. In this web user interface, the software has several components. One is the BLAST search program to find the homologous sequence of the gene subspace under investigation, it is then applied to hidden Markov models to identify different protein domains such as polyketide synthetase domains, transacting acyl transferase and adenylation domains, deoxysugar biosynthesis domains, β-lactam-specific domains, etc. The other components tools are HMMER (version 3.1) for hidden Markov model searches, the Chemistry Development Kit (version 1.4.19) for chemical abstractions, BioJava (version 3.0.7) for sequence translation, RDKit (version 2014.03.1) for Tanimoto coefficient computation, and Apache Batik (version 1.7) for vector image generation [68].

9. BIOINFORMATICS TO NATURAL PRODUCTS THROUGH SYNTHETIC BIOLOGY

Bioinformatics is a subject that creates a bridge between genomic data and natural product discovery. Several tools have been discovered that have been

TABLE 21.11 Neuraminidase Inhibitors From Malaysia Natural Products Database

Compounds	Autodock Score (kcal/mol)	IC_{50} (μM)	% Inhibition (at 250 μg/mL)
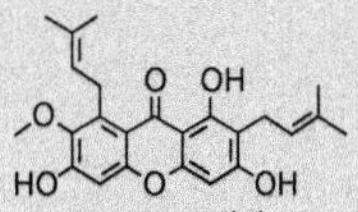 α-mangostin*	−8.87	91.95 ± 0.09	93.08 ± 0.04 (at 609 μM)
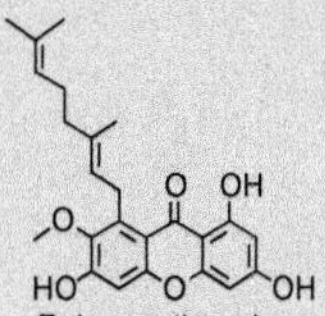 Rubraxanthone*	−9.85	89.71 ± 0.08	92.42 ± 0.12 (at 609 μM)
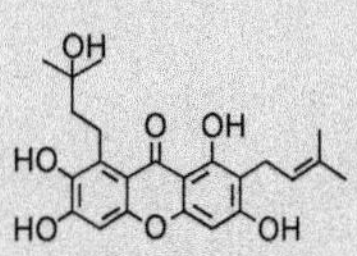 Garcinone C*	−8.85	95.49 ± 0.08	90.13 ± 0.02 (at 603 μM)
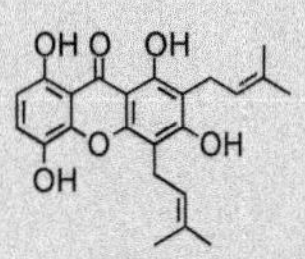 Gartanin*	−11.07	126.64 ± 0.13	80.25 ± 0.32 (at 631 μM)
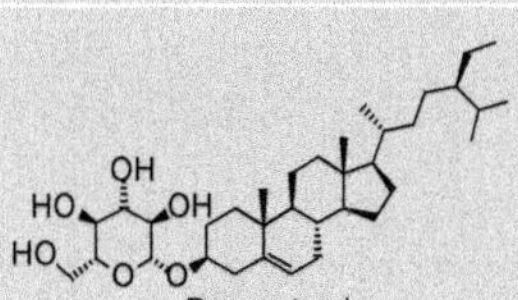 Daucosterol	−8.99	275.45 ± 0.03	60.65 ± 0.29 (at 433 μM)

Continued

TABLE 21.11 Neuraminidase Inhibitors From Malaysia Natural Products Database—cont'd

Compounds	Autodock Score (kcal/mol)	IC_{50} (μM)	% Inhibition (at 250 μg/mL)
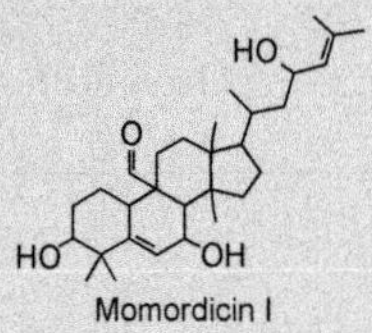 Momordicin I	−11.49	≥250	15.58 ± 0.36 (at 529 μM)
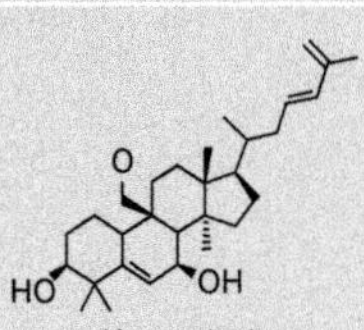 Kuguacin J	−10.21	≥250	21.42 ± 0.50 (at 550 μM)
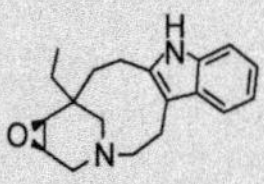 Voaphylline	−10.49	≥250	20.95 ± 0.09 (at 800 μM)
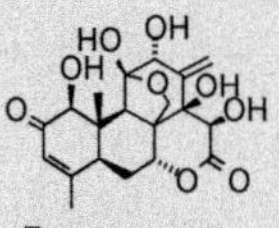 Eurycomanone	−10.89	≥250	20.84 ± 0.67 (at 612 μM)
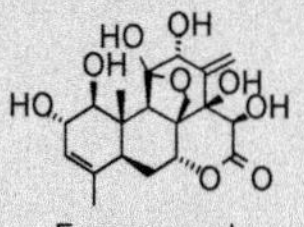 Eurycomanol	−9.83	≥250	21 ± 02.18 (at 609 μM)

TABLE 21.11 Neuraminidase Inhibitors From Malaysia Natural Products Database—cont'd

Compounds	Autodock Score (kcal/mol)	IC_{50} (μM)	% Inhibition (at 250 μg/mL)
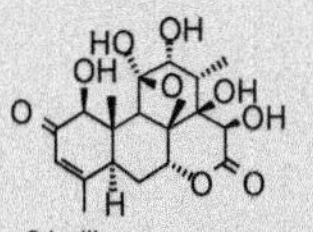 13α,21-dihyroeurycomanone	−9.92	≥250	2.90 ± 0.34 (at 631 μM)
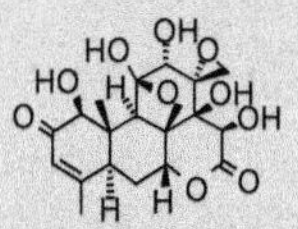 13α,21-epoxyeurycomanone	−10.45	≥250	34.50 ± 0.27 (at 589 μM)

Reprinted from N.K.K. Ikram, J.D. Durrant, M. Muchtaridi, A.S. Zalaludin, N. Purwitasari, N. Mohamed, et al., A virtual screening approach for identifying plants with Anti H5N1 neuraminidase activity. J. Chem. Inf. Model 55 (2015) 308–316.

useful to predict the coding of natural products from a set of genes or protein clusters. The tools and their applications are shown in Table 21.12.

Bioinformatics tools are also used to study natural products using synthetic biology tools. The design of natural products from biosynthetic gene clusters depends on searching relevant sequence space from a database of millions of gene sequences. Afterward, domains are located in the gene using a domain search tool, which often relies on homology match of the unknown gene sequence with established genes of known function. The most putative or conserved sequences are found and synthesizable natural products are then designed based on pharmacophore matches against a known database of secondary metabolites. The tools related to natural product discovery from genomic clusters are summarized in Table 21.13.

10. ESNAPD, A NOVEL WEB-BASED BIOINFORMATICS TOOL

Environmental Surveyor of Natural Products Diversity (eSNaPD) is a web-based bioinformatics-based platform to discover gene clusters for the discovery of natural products. This database first relies on construction of the

TABLE 21.12 Bioinformatics Tools for Natural Products

	Enzymes	Pathways	Regulatory Components	Chassis
Selection	*Mining*	*Rankling*	*Characterization*	*Genome scale modeling*
	antiSMASH	FindPath	Registry of standard biological	
		RetroPath		
		GemPath		
		Metabolic Tinker		
Prediction	*Annotation*	*Search*	*Tuning*	*Optimization*
	antiSMASH	BNICE	RBS Calculator	Optknock
	Enzymes	Pathways	Regulatory Components	Chassis
	CanOE	Route Search		EMILIO
	Enzyme Function Initiative	PathPred		SIMUP
	SymZime	RetroPath		
		GEM-Path		

Reprinted with permission from P. Carbonell, A. Currin, A.J. Jervis, N.J.W. Rattray, N. Swainston, Y. Cunyu, E. Takano, R. Breitling, Bioinformatics for the synthetic biology ofnatural products: integrating across the Design-build-test cycle. Nat. Prod. Rep. 16 (2016) 925–932.

database and then searches for any unknown gene sequence within the database. For construction of the database, first, amplification of different natural product biosynthetic gene clusters by polymerase chain reaction (PCR) is done where various biosynthetic gene clusters such as acyl carrier protein, polyketide synthetase, adenylation, acyltransferase, condensation, dehydratase, epimerization, enoyl reductase, ketoreductase, methyltransferase, peptidyl carrier protein, and thioesterase are involved. After amplification, 95% sequence identity of the PCR-amplified genes is mined and saved as consensus sequence as a unique sequence read. In search space, once an unknown gene sequence is placed after PCR and thereafter sequencing, the sequence is searched for the highest hit in the database by the NCBI BLAST algorithm and

TABLE 21.13 Computational Tools for Natural Products Discovery

Software Program or Database	URL	Last Publication or Document Update	Main Content/Function
Database Focusing on Gene Clusters			
Bactibase	http://bactibase.pfba-lab-tun.org	2011	Web accessible database of bacteriocins
ClusterMine360	http://www.clustermine360.ca/	2013	Web accessible database of biosynthetic gene clusters
ClustScanDatabase	http://csdb.bioserv.pbf.hr/csdb/ClustScanWeb.html	2013	Web accessible database of polyketide synthetase/natural product biosynthetic gene clusters
DoBISCUIT	http://www.bio.nit_e.g_o_jp/pks/	2015	Web accessible database of polyketide synthetase/natural product biosynthetic gene clusters
Integrated Microbial Genome-Atlas of Biosynthetic Gene Clusters	http://img.jgi.doe.gov/abc	2015	Web accessible database of biosynthetic gene clusters
MIBiG	http://mibig.secondarymetabolites.org	2015	Web accessible repository of biosynthetic gene clusters
Recombinant ClustScan Database	http://csdb.bioserv.pbf.hr/csdb/R CSDB.html	2013	In silico recombinant database
Database Focusing on Bioactive Compounds			
Antibioticome	http://magarveylab.ca/antibioticome	2015	Web accessible database of compounds, compound families, and mode of action

Continued

TABLE 21.13 Computational Tools for Natural Products Discovery—cont'd

Software Program or Database	URL	Last Publication or Document Update	Main Content/Function
Database Focusing on Gene Clusters			
ChEBI	https://www.ebi.ac.uk/c hebi/	2015	Web accessible database of compounds, compound families, and mode of action
ChEMBL	https://www.ebi.ac.uk/c hembl/	2015	Web accessible database of bioactive compounds with drug-like properties
Chem Spider	http://www.chemspider.com/	2015	Web accessible database of structures and properties
KNAPSAcK database	http://kanaya.aist-nara.ac.jp/ KN ApSAcK/	2015	Web accessible database of bioactive compounds with KNAPSAcK standalone database
NORINE	http://bioinfo.lifl.fr/norine	2015	Web accessible database of natural products
Novel Antibiotics Database	http://www.antibiotics.or.jp/ journal/database/database-top. htm	2008	Web accessible database of compounds
PubChem	http://pubchem.ncbi.nlm.nih. gov/	2015	Web accessible database of compounds and bioactives
StreptomeDB	http://www.pharmaceutical-bioinformatics.de/s tr ep t ome db	2015	Web accessible database of compounds produced by streptomycetes; download of compounds and metadata

Metabolomics Tools			
Cycloquest	http://cyclo.ucsd.edu	2011	Web application to correlate tandem MS data of cyclopeptides with gene clusters
GNPS	http://gnps.ucsd.edu/	2015	Generic metabolomics portal to analyze tandem mass spectrometry data (dereplication and molecular networking)
GNP/iSNAP	http://magarveylab.ca/gnp/	2015	Web application to automatically identify tandem mass spectrometry data based on genomics data
NRPquest	http://cyclo.ucsd.edu	2014	Web application to automatically identify tandem mass spectrometry data based on genomics data
Pep2Path	http://pep2path.sourceforge.net	2014	Standard replication data relating peptide sequence tags with biosynthetic gene clusters
RiPPquest	http://cyclo.ucsd.edu	2014	Web application to correlate ribosomally and posttranslationally modified peptide tandem data with gene clusters

Reprinted with permission from T. Weber, H.U. Kim, The secondary metabolite bioinformatics portal: computational tools to facilitate synthetic biology of secondary metabolite production, Syn. Sys. Biotechnol. 1 (2016) 69–79.

the most matched hit is calculated by e-value as convened by the NCBI BLAST algorithm. The search hits so far are further processed by hierarchical clustering and a phylogenetic relationship is established. Thus relevant domains cloned in the gene cluster are mined and established [70].

11. DNA BARCODING IN NATURAL PRODUCTS

A DNA barcode is a short segment of genomic DNA (<1000 bp), which is highly variable in sequence and used to determine hierarchical and evolutionary relationships between plants and animals [71]. This is used for species identification through sequence alignment by a series of sequence alignment algorithms [72]. For DNA barcoding, the standard genomic spaces used are chloroplast ribulose 1,5-bisphosphate carboxylase/oxygenase large subunit (rbcL) and maturase K (matK) as core barcodes [73]. Together with this, other regions are also used as DNA barcodes such as the spacer between photosystem II protein D1, tRNA-His (psbA-trnH spacer), the nuclear ribosomal internal transcribed spacer 2 (ITS2) in plants, and cytochrome oxidase c subunit-I (COI) for animals [74–80]. DNA barcoding has been applied to identify the contamination of natural products such as identifying consumer relevant mushrooms [81] among poisonous and nonedible mushrooms, detecting contamination and substitution of herbal products [79], herbal medicines, and dietary supplements [82], and many others. DNA barcoding is also used for phylogenetic evolution of plants [83].

12. DISCUSSION AND CONCLUSION

The knowledge of usage of traditional medicinal plant databases has become a paradigm of immense importance due to intense utilization of natural products across the globe over the last few decades. However, attempts have been made to perturb the dataset of natural products digitally due to complexity and difficulty of exploring millions of natural products by physical sorting. In accordance with this, digitization of natural products has crept in via four major approaches. Data preservation is found in various web databases wherefrom data can be mined according to the user's demands, providing virtual screening of different DTNs or databases for drug discovery, bioinformatics-guided approaches for proper utilization of natural products knowledge for lead optimization in discovery processes, and in silico approaches such as docking or molecular modeling for drug discovery. In the first approach, several UNPDs have been created such as UNPD, CMKb, ebDB, ZINC, TCM, UNIIQUIM, NuBBE, pANAPL, InvertNet, CamMED NP, and DIVERSet, where information on 560 to more than 19 million compounds has been stored based on the database. In addition, different digitization tools have been created for various purposes such as identification tools (EDIT's cybertaxonomy platform, Electronic Field Guide, Medical Fungi Identification Website, Free Delta, and Meka), digitization tools (Bauble, Bibmaster, Biota, and Biotica), and biodiversity together with ecological modeling tools (ADE4, APE, DIVA-GIS, GARP, LAMARC, Molphy, and others). Virtual screening-based natural product search is based on in silico chemical space and docking analyses where similar property harnessing compounds are searched for based on molecular descriptors so that new leads can be discovered from those analogous natural products. For example, using chemical space analysis and subsequent docking on estrogen receptors (ERα and ERβ) led to the discovery of 11 nonsteroidal estrogen modulators. Furthermore, screening 89,000 natural compounds from the ZINC database, five compounds as PPARs have been revealed. Bioinformatics-guided drug discovery from natural products analyzes sequence space to investigate the phylogenetic relationship, biodiversity, and ecological modeling. Some of the popular approaches for bioinformatics investigation use ESTs or SSR locators, restriction fragment length polymorphism, randomly amplified polymorphic DNA, and single nucleotide polymorphism to investigate the phylogenetic relationship between and potential gene clusters among the species in the database. For example, in one study, 11,422 natural product gene clusters from Actinobacteria were grouped into 4122 GCF. This ultimately led to the revelation of 830 genome datasets encompassing the potential for biosynthesizing newer drug leads. In another study, distance-based mutual information model and network target-based identification of multicomponent synergy approaches have been undertaken to generate synergy scores for ranking synergistic effects of agent combinations in a specific database. Statistical learning methods such as probabilistic neuronal network, k-nearest neighbor method, SVM, and decision tree have also been undertaken to elucidate similar gene clusters for new drug discovery. Most importantly, the DNA barcoding approach has also been used over the last few years to search phylogenetic and neighborhood relationships together with synergistic likeliness among diverse natural products. As a rule of thumb, *matK*, *rbcL*, and *ITS2* sequences have been undertaken as DNA barcodes for plant investigation, while for animals, mitochondrial COI has been considered as a DNA barcode. Lastly, the in silico docking approach has been a popular tool for predictive approaches in drug discovery. This approach has been successfully implemented for drug discovery in iNOS inhibitors, antineoplastic compounds such as HER-2 inhibitors in female breast cancer, and many more. Most interestingly, a study has been performed to encode entire medicinal and aromatic plants in Africa in a digital database by using programming language C++. Thus digitization of traditional knowledge is an updated, time-economic, highly investigative, and efficient strategy for studying natural products as well as for drug discovery based on these products.

REFERENCES

[1] W. Altermann, J. Kazmierczak, Archean microfossils: a reappraisal of early life on Earth, Res. Microbiol. 154 (2003) 611–617.

[2] R.T. Watson, V.H. Heywood, I. Baste, B. Dias, R. Gamez, T. Janetos, W. Reid, R. Ruark (Eds.), Global Biodiversity Assessment, 1995 ed., Panama Cambridge University Press, 1995.

[3] E.O. Wilson, The encyclopedia of life, Trends Ecol. Evol. 18 (2003) 77–80.

[4] J. Gaikwad, V. Chavan, Open access and biodiversity conservation: challenges and potentials for the developing world, Data Sci. J. 5 (2006) 1–17.

[5] C.D. Becker, K. Ghimire, Synergy between traditional ecological knowledge and conservation science supports forest preservation in Ecuador, Conserv. Ecol. 8 (2003) 1.

[6] R. Govaerts, How many species of seed plants are there? Taxon 50 (2001) 1085–1090.

[7] N.R. Farnsworth, Computerized data base for medicinal plants, in: World Health Forum, vol. 5(4), World Health Organization, Geneva, 1984, pp. 373–376.

[8] E.M. Manha, M.C. Silva, M.G.C. Alves, M.G.L. Brandão, M.B. Almeida, PLANT – a bibliographic database about medicinal plants, Braz. J. Pharmacogn. 18 (2008) 614–617.

[9] M.I. Rajoka, I. Sobia, K. Sana, E. Beenish, Medherb: an interactive bioinformatics database and analysis resource for medicinally important herbs, Curr. Bioinform. 9 (2014) 23–27.

[10] F. Ntie Kang, J.A. Mbah, L.M. Mbaz, et al., CamMedNP: building the Cameroonian 3D structural natural products database for virtual screening, BMC Complement. Altern. Med. 13 (2013) 88.

[11] F. Ntie-Kang, P.A. Onguene, G.W. Fotso, et al., Virtualizing the p-ANAPL library: a step towards drug discovery from African medicinal plants, PLoS One 9 (2014a) e90655.

[11a] J. Gaikwad, P.D. Wilson, S. Ranganathan, Ecological niche modeling of customary medicinal plant species used by Australian Aborigines to identify species-rich and culturally valuable areas for conservation, Ecol. Modell. 222 (2011) 3437–3443.

[12] A.L. Hopkins, Network pharmacology, Nat. Biotechnol. 25 (2007) 1110–1111.

[13] A.L. Hopkins, Network pharmacology: the next paradigm in drug discovery, Nat. Chem. Biol. 4 (2008) 682–690.

[14] S. Li, B. Zhang, Traditional chinese medicine network pharmacology: theory, methodology and application, Chin. J. Nat. Med. 11 (2013) 110–120.

[15] J. Gu, L. Chen, G. Yuan, X. Xu, A drug-target network-based approach to evaluate the efficacy of medicinal plants for type II diabetes mellitus, Evid. Based Complement. Altern. Med. (2013), 203614, 7 pages.

[16] N.K.K. Ikram, J.D. Durrant, M. Muchtaridi, A.S. Zalaludin, N. Purwitasari, N. Mohamed, et al., A virtual screening approach for identifying plants with Anti H5N1 neuraminidase activity, J. Chem. Inf. Model 55 (2015) 308–316.

[17] J.L. Medina-Franco, Chemoinformatics characterization of the chemical space and molecular diversity of compound libraries, in: T. Andrea (Ed.), Diversity-Oriented Synthesis: Basics and Applications in Organic Synthesis, Drug Discovery, and Chemical Biology, John Wiley & Sons Inc., 2013, pp. 325–352.

[18] T.-Y. Tsai, K.-W. Chang, C. Chen, iScreen: world's first cloud-computing web server for virtual screening and de novo drug design based on TCM database @Taiwan, J. Comput. Aided Mol. Des. 25 (2011) 525–531.

[19] K.-Y. Chen, S.-S. Chang, C.Y.-C. Chen, *In Silico* identification of potent pancreatic triacylglycerol lipase inhibitors from traditional Chinese Medicine, PLoS One 7 (2012) e43932.

[20] M.A. Koch, A. Schuffenhauer, M. Scheck, S. Wetzel, M. Casaulta, A. Odermatt, P. Ertl, H. Waldmann, Charting biologically relevant chemical space: a structural classification of natural products (SCONP), Proc. Natl. Acad. Sci. U.S.A. 102 (2005) 17272–17277.
[21] N. Singh, R. Guha, M.A. Giulianotti, C. Pinilla, R.A. Houghten, J.L. Medina-Franco, Chemoinformatic analysis of combinatorial libraries, drugs, natural products, and molecular libraries small molecule repository, J. Chem. Inf. Model. 49 (2009) 1010–1024.
[22] S. Dandapani, L.A. Marcaurelle, Accessing new chemical space for 'undruggable' targets, Nat. Chem. Biol. 6 (2010) 861–863.
[23] P.A. Clemons, N.E. Bodycombe, H.A. Carrinski, J.A. Wilson, A.F. Shamji, B.K. Wagner, A.N. Koehler, S.L. Schreiber, Small molecules of different origins have distinct distributions of structural complexity that correlate with protein-binding profiles, Proc. Natl. Acad. Sci. U.S.A. 107 (2010) 18787–18792.
[23a] C.M. Dobson, Chemical space and biology, Nature 432 (2004) 824–828.
[24] C. Lipinski, A. Hopkins, Navigating chemical space for biology and medicine, Nature 432 (2004) 855–861.
[24a] J.L. Medina-Franco, Interrogating Novel Areas of Chemical Space for Drug Discovery using Chemoinformatics, Drug Dev. Res. 73 (2012) 430–438.
[25] J. Larsson, J. Gottfries, L. Bohlin, A. Backlund, Expanding the ChemGPS chemical space with natural products, J. Nat. Prod. 68 (2005) 985–991.
[26] J. Larsson, J. Gottfries, S. Muresan, A. Backlund, ChemGPS-NP: tuned for navigation in biologically relevant chemical space, J. Nat. Prod. 70 (2007) 789–794.
[27] T.I. Oprea, J. Gottfries, Chemography: the art of navigating in chemical space, J. Comb. Chem. 3 (2001) 157–166.
[28] X. Cao, J. Jiang, S. Zhang, L. Zhu, J. Zou, Y. Diao, W. Xiao, L. Shan, H. Sun, W. Zhang, J. Huang, H. Li, Discovery of natural estrogen receptor modulators with structure-based virtual screening, Bioorg. Med. Chem. Lett. 2013 (23) (2013) 3329–3333.
[29] Deleted in review.
[30] L. Guasch, E. Sala, J. Iwaszkiewicz, M. Mulero, M.-J. Salvado, M. Pinent, V. Zoete, A. Grosdidier, G. Pujadas, S. Garcia-Vallve, Identification of PPARgamma partial agonists of natural origin (I): development of a virtual screening procedure and in vitro validation, PLoS One 7 (2012) e50816.
[31] S.T. Ngo, M.S. Li, Top-leads from natural products for treatment of Alzheimer's disease: docking and molecular dynamics study, Mol. Simul. 39 (2013) 279–291.
[32] J. Gu, H. Zhang, L. Chen, S. Xu, G. Yuan, X. Xu, Drug–target network and polypharmacology studies of a Traditional Chinese Medicine for type II diabetes mellitus, Comput. Biol. Chem. 35 (2011) 293–297.
[33] N.V. Ivanova, M.L. Kuzmina, T.W.A. Braukmann, A.V. Borisenko, E.V. Zakharov, Authentication of herbal supplements using next-generation sequencing, PLoS One 11 (5) (2016).
[34] V. Sharm, I. Neil Sarkar, Bioinformatics opprtumities for identification of medicinal plants, Brief. Bioinform. 14 (2012) 238–250.
[35] L.C. da Maia, D.A. Palmieri, V.Q. de Souza, et al., SSR locator: tool for simple sequence repeat discovery integrated with primer design and PCR simulation, Int. J. Plant Genom. (2008) 412696.
[36] K. Mochida, Y. Uehara-Yamaguchi, T. Yoshida, et al., Global landscape of a co-expressed gene network in barley and its application to gene discovery in Triticeae crops, Plant Cell Physiol. 52 (2011) 785–803.

[37] M. Kanehisa, S. Goto, M. Furumichi, et al., KEGG for representation and analysis of molecular networks involving diseases and drugs, Nucleic Acids Res. 38 (2010) D355–D360.

[38] Y. Moriya, D. Shigemizu, M. Hattori, et al., PathPred: an enzyme-catalyzed metabolic pathway prediction server, Nucleic Acids Res. 38 (2010) W138–W143.

[39] H. Rischer, M. Oresic, T. Seppanen-Laakso, et al., Gene-to-metabolite networks for terpenoid indole alkaloid biosynthesis in *Catharanthus roseus* cells, Proc. Natl. Acad. Sci. U.S.A. 103 (2006) 5614–5619.

[40] S. Li, B. Zhang, D. Jiang, et al., Herb network construction and co-module analysis for uncovering the combination rule of traditional Chinese herbal formulae, BMC Bioinform. 11 (Suppl. 11) (2010) S6.

[41] X.P. Zhao, X.H. Fan, J. Yu, et al., A method for predicting activity of traditional Chinese medicine based on quantitative composition-activity relationship of neural network model, China J. China Mater. Med. 29 (2004) 1082–1085.

[42] Y. Wang, X. Wang, Y. Cheng, A computational approach to botanical drug design by modeling quantitative composition-activity relationship, Chem. Biol. Drug Des. 68 (2006) 166–172.

[43] S.K. Nayak, P.K. Patra, P. Padhi, A. Panda, Optimization of herbal drugs using soft computing approach, Int. J. Log. Comput. 1 (2010) 34–39.

[44] S. Li, B. Zhang, N. Zhang, Network target for screening synergistic drug combinations with application to traditional Chinese medicine, BMC Syst. Biol. 5 (Suppl. 1) (2011) S10.

[45] M.M. Larsen, A. Adsersen, A.P. Davis, et al., Using a phylogenetic approach to selection of target plants in drug discovery of acetylcholinesterase inhibiting alkaloids in Amaryllidaceae tribe Galantheae, Biochem. Syst. Ecol. 38 (2005) 1026–1034.

[46] C.X. Xue, X.Y. Zhang, M.C. Liu, et al., Study of probabilistic neural networks to classify the active compounds in medicinal plants, J. Pharm. Biomed. Anal. 38 (2005) 497–507.

[47] D.K. Yadav, A. Meena, A. Srivastava, et al., Development of QSAR model for immunomodulatory activity of natural coumarinolignoids, Drug Des. Dev. Ther. 4 (2010) 173–186.

[48] R.K. Petersen, K.B. Christensen, A.N. Assimopoulou, et al., Pharmacophore-driven identification of PPARgamma agonists from natural sources, J. Comp. Aided Mol. Des. 25 (2011) 107–116.

[49] H.-J. Zhong, L. Li-J, C.-M. Chong, et al., Discovery of a natural product-like iNOS inhibitor by molecular docking with potential neuroprotective effects in vivo, PLoS One 9 (2014) e92905.

[49a] J. Li, H. Wang, J. Li, J. Bao, C. Wu, Discovery of a Potential HER2 Inhibitor from Natural Products for the Treatment of HER2-Positive Breast Cancer, Int. J. Mol. Sci. 17 (2016) 1055.

[50] C.-H. Leung, D.S.-H. Chan, Y.-W. Li, et al., Hit identification of IKKβ natural product inhibitor, BMC Pharmacol. Toxicol. 14 (2013) 1–8.

[51] E. Sala, L. Guasch, J. Iwaszkiewicz, M. Mulero, M.-J. Salvado, Pinent M., V. Zoete, A. Grosdidier, S. Garcia-Vallve, O. Michielin, G. Pujadas, Identification of human IKK-2 inhibitors of natural Origin (Part I): modeling of the IKK-2 kinase domain, virtual screening and activity assays, PLoS One 6 (2011) e16903.

[52] J. Kornhuber, M. Muelbacher, S. Trapp, S. Pechmann, A. Friedl, M. Reichel, C. Muhle, L. Terfloth, T.W. Groemer, G.M. Spitzer, R.L. Klaus, E. Gulbins, P. Tripal, Identification of novel functional inhibitors of acid sphingomyelinase, PLoS One 6 (2011) e23852.

[53] M. Szelag, A. Czerwonlec, J. Wesoly, H.A.R. Bluyssen, Identification of STAT1 and STAT3 specific inhibitors using comparative virtual screening and docking validation, PLoS One 10 (2) (2015) e0116688.

[54] V. Aparna, K. Dineshkumar, N. Mohanalakshmi, D. Velmurugan, W. Hopper, Identification of natural compound inhibitors for multidrug efflux pumps of *Escherichia coli* and *Pseudomonas aeruginosa* using in silico high throughput virtual screening and in vitro validation, PLoS One 9 (2014) e101840.

[55] D. Kudryavtsev, T. Makarieva, N. Utkina, E. Santalova, E. Kryukova, C. Methfessel, V. Tsetlin, V. Stonik, I. Kasheverov, Marine natural products acting on the cetylcholine-binding protein and nicotinic receptors: from computer modeling to binding studies and electrophysiology, Mar. Drugs 12 (2014) 1859–1875.

[56] J.A. Shilpi, M.T. Ali, S. Saha, S. Hasan, A.I. Gray, V. Seidel, Molecular docking studies on InhA, MabA and PanK enzymes from *Mycobacterium tuberculosis* of ellagic acid derivatives from *Ludwigia adscendens* and *Trewia nudiflora*, In Silico Pharmacol. 3 (2015) 1–7.

[57] C. Dietrich, J. Hart, D. Raila, U. Ravaioli, N. Sobh, O. Sobh, C. Taylor, InvertNet: a new paradigm for digital access to invertebrate collections, ZooKeys 209 (2012) 165–212.

[58] J.R. Doroghazi, J.C. Albright, A.W. Goering, K.-S. Ju, R.R. Haines, K.A. Tchalukov, D.P. Labeda, N.L. Kelleher, W.W. Metcalf, A roadmap for natural product discovery based on large-scale genomics and metabolomics, Nat. Chem. Biol. 10 (2014) 963–968.

[59] S.D. Bentley, et al., Complete genome sequence of the model actinomycete *Streptomyces coelicolor* A3(2), Nature 417 (2002) 141–147.

[60] R.D. Kersten, et al., A mass spectrometry-guided genome mining approach for natural product peptidogenomics, Nat. Chem. Biol. 7 (2011) 794–802.

[61] N. Ziemert, et al., The natural product domain seeker NaPDoS: a phylogeny based bioinformatic tool to classify secondary metabolite gene diversity, PLoS One 7 (2012) e34064.

[62] M.H. Medema, et al., AntiSMASH: rapid identification, annotation and analysis of secondary metabolite biosynthesis gene clusters in bacterial and fungal genome sequences, Nucleic Acids Res. 39 (2011) W339–W346.

[63] K.R. Conway, C.N. Boddy, ClusterMine360: a database of microbial PKS/NRPS biosynthesis, Nucleic Acids Res. 41 (2013) D402–D407.

[64] G. Yadav, R.S. Gokhale, D. Mohanty, SEARCHPKS: a program for detection and analysis of polyketide synthase domains, Nucleic Acids Res. 31 (2003) 3654–3658.

[65] N. Ichikawa, et al., DoBISCUIT: a database of secondary metabolite biosynthetic gene clusters, Nucleic Acids Res. 41 (2013) D408–D414.

[66] S. Caboche, et al., NORINE: a database of nonribosomal peptides, Nucleic Acids Res. 36 (2008) D326–D331.

[67] J. Kim, G.-S. Yi, PKMiner: a database for exploring type II polyketide synthases, BMC Microbiol. 12 (2012) 169.

[68] M.A. Skinnider, C.A. Dejong, P.N. Rees, C.W. Johnston, H. Li, A.L. Webster, M.A. Wyatt, N.A. Magarvey, Genomes to natural products prediction informatics for secondary metabolomes (PRISM), Nucleic Acids Res. 43 (2015) 9645–9662.

[69] P. Carbonell, A. Currin, A.J. Jervis, N.J.W. Rattray, N. Swainston, Y. Cunyu, E. Takano, R. Breitling, Bioinformatics for the synthetic biology of natural products: integrating across the Design-build-test cycle, Nat. Prod. Rep. 16 (2016) 925–932.

[70] B.V.B. Reddy, A. Milshtein, Z. Charlop-Powers, S.F. Brady, eSNaPD: a versatile, web-based bioinformatics platform for surveying and mining natural product biosynthetic diversity from Metagenomes, Chem. Biol. 21 (8) (2014) 1023–1033.

[71] N.B. Zahra, Z.K. Shinwari, M. Qaiser, DNA barcoding: a tool for standardization of herbal medicinal products (HMPS) of lamiaceae from Pakistan, Pak. J. Bot. 48 (2016) 2167–2174.

[72] Z.K. Shinwari, K. Jamil, N.B. Zahra, Molecular systematics of selected genera of subfamily Mimosoideaefabaceae, Pak. J. Bot. 46 (2014) 591–598.

[73] P.M. Hollingsworth, L.L. Forrest, J.L. Spouge, A DNA barcode for land plants, Proc. Natl. Acad. Sci. 106 (2009) 12794–12797.

[74] S. Chen, H. Yao, J. Han, et al., Validation of the ITS2 region as a novel DNA barcode for identifying medicinal plant species, PLoS One 5 (2010) e8613.

[75] T. Gao, H. Yao, J. Song, Y. Zhu, C. Liu, S. Chen, Evaluating the feasibility of using candidate DNA barcodes in discriminating species of the large Asteraceae family, BMC Evol. Biol. 10 (2010) 324–330.

[76] H. Yao, J. Song, C. Liu, K. Luo, J. Han, Y. Li, X. Pang, H. Xu, Y. Zhu, P. Xiao, S. Chen, Use of ITS2 region as the universal DNA barcode for plants and animals, PLoS One 5 (2010) e13102.

[77] Y.M. Fu, W.M. Jiang, C.X. Fu, Identification of species within Tetrastigma (Miq.) Planch. (Vitaceae) based on DNA barcoding techniques, J. Syst. Evol. 49 (2011) 237–245.

[78] J.P. Han, L.C. Shi, X.C. Chen, Y.L. Lin, Comparison of four DNA barcodes in identifying certain medicinal plants of Lamiaceae, J. Syst. Evol. 50 (2012) 227–234.

[79] S.G. Newmaster, M. Grguric, D. Shanmughanandhan, S. Ramalingam, S. Ragupathy, DNA barcoding detects contamination and substitution in North American herbal products, BMC Med. 11 (2013) 222.

[80] C.I. Michel, S.M. Rachel, T. Yanille, M. Jeanmaire, The nuclear internal transcribed spacer (ITS2) as a practical plant DNA barcode for herbal medicines, J. Appl. Res. Med. Aromat. (2016). https://doi.org/10.1016/j.jarmap.2016.02.002.

[81] H.A. Raja, T.R. Baker, J.G. Little, N.H. Oberlies, DNA barcoding for identification of consumer-relevant mushrooms: a partial solution for product certification? Food Chem. 214 (2017) 383–392.

[82] I. Parveen, S. Gafner, N. Techen, S.J. Murch, I.A. Khan, DNA barcoding for the identification of botanicals in herbal medicine and dietary supplements: strengths and limitations, Planta Med. 82 (2016) 1225–1235.

[83] Q.-Y. Dai, Q. Gao, C.-S. Wu, D. Chesters, C.-D. Zhu, Ai-B. Zhang, Phylogenetic reconstruction and DNA Barcoding for closely related pine moth species (Dendrolimus) in China with multiple gene markers, PLoS One 7 (2012) e32544.

[84] A.T. Peterson, D.A. Vieglais, A.G.N. Sigüenza, M. Silva, A global distributed biodiversity information network: building the world museum, Bull. Br. Ornithol. Club 123A (2003) 186–196.

[85] H. Tae, E.-B. Kong, K. Park, ASMPKS: an analysis system for modular polyketide synthases, BMC Bioinform. 8 (2007) 327.

[86] R.K. Varshney, A. Graner, M.E. Sorrells, Genic microsatellite markers in plants: features and applications, Trends Biotechnol. 23 (2005) 48–55.

Chapter 22

Good Agricultural Practices: Requirement for the Production of Quality Herbal Medicines

Supradip Saha, Abhishek Mandal, Anirban Dutta
ICAR-Indian Agricultural Research Institute, New Delhi, India

1. INTRODUCTION

Unwanted and/or tacit materials, time and again, have been present or been claimed to be present in medicinal or herbal plant medicines around the globe. The substances that have been in the news included microbes such as pathogens, pesticides, mycotoxins, radioactive particles, and heavy metals such as arsenic. The incremental demand and usage of herbal medicines around the world, coupled with the vigorous expansion of the global market demand for the medicinal plants or medicinal plant–derived active ingredients, and quality control (QC) of medicinal plant materials as well as the finished herbal medicinal products have taken center stage as issues of major concern for health agencies, herbal pharmaceutical industries, and the general public, as a whole [1].

National rules for registration and regulation of herbal medicines vary from country to country. Herbal medicines are categorized as prescription medicines or nonprescription medicines, wherever they are regulated. Herbal products as a group along with medicines, may coexist in a certain country. Due to lacunae in regulation, poor QC systems, and faulty distribution channels (which includes Internet-based sales), herbal products categorized other than as medicines and foods are inclined toward increasing potential for drastic consequences. There is a belief that GAP standards are restrictive and obstruct farmers and their agriculture processes. However, the fundamental guiding principle of GAP is the achievement of a safe and sustainable food production system for both growers and consumers. This safe production system is necessary to ensure the right of consumers to hygienic, nutritious, and affordable

Natural Products and Drug Discovery. https://doi.org/10.1016/B978-0-08-102081-4.00022-8

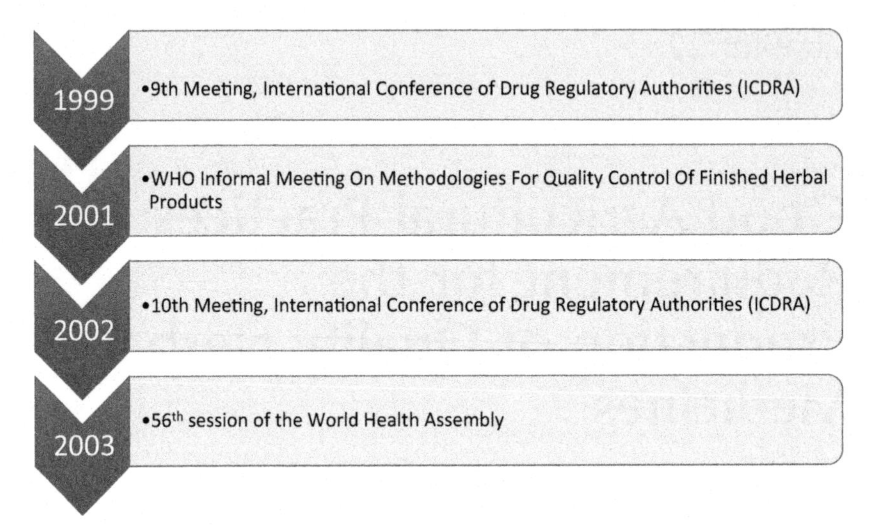

FIGURE 22.1 Advent of good agricultural practices.

food. In addition, it is also essential for food production to safeguard the health, hygiene, and welfare of growers and farm workers. They must not be exposed to hazards and dangers during input applications.

Farmers are most concerned about growing successful crops that allow them to sell quality products at a favorable price. However, the power of the market rests heavily in favor of consumers. Consumers now demand and expect that GAP standards are applied for market access for many food crops. Crops from farms that are noncompliant with GAP standards have to be traded in lower market destinations, which also means they are to be sold at lower prices.

In a resolution on traditional medicine, the 56th session of the World Health Assembly in May 2003, member states were urged to ensure the safety, efficacy, and quality of herbal medicines by determining national standards for, or issuing monographs on, herbal raw materials and traditional medicine formulae, where found to be appropriate. The chronological advent of GAPs can be seen in Fig. 22.1.

1.1 What Are Good Agricultural Practices?

Discussions, as held in the early stages, on "common principles of good agricultural practices," [1] centered on the work of David Connor, a visiting scientist. The above-mentioned principles describe GAPs as a farming practice that makes optimal utilization of the presently available technology to promote sustainable agricultural production of safe as well as healthy food, to achieve economic feasibility as well as environmental sustainability and social responsibility, although the social perspective remains somewhat inadequately addressed in the grid lines. The process in question is one of knowledge, perception, planning, assessment, documentation, and managing in order to recognize social, environmental, and economic aspects of production goals and monitoring of associated impacts. A sound and comprehensive farming approach is imperative along with the capability for decisive tactical adjustments for any circumstantial fluctuations for the afore-mentioned process to be successfully implemented. Development of expertise and knowledge on continuous documentation, performance analytics, and timely required technical advice are also parameters on which the success counts on.

Application of available knowledge for the sustainable utilization of the natural resources for the production of healthy food as well as nonfood agricultural products, in a safe, humane manner, vis-à-vis taking into consideration the economic viability and social stability factors, is the GAP, in a nutshell. The term "GAP" refers to a large plethora of components, from regulation of pesticides use vis-à-vis monitoring of pesticide residues in the produce, to additional surrounding aspects of primary as well as postproduction processes, such as environmental impact assessment or labor working conditions.

The FAO Committee on Agriculture proposed that the GAP framework provides an insight into the scope and wide-ranging objectives of GAPs [2].

The FAO describes GAPs as follows:

> *The concept of Good Agricultural Practices is the application of available knowledge to the utilization of the natural resource base in a sustainable way for the production of safe, healthy food and non-food agricultural products, in a humane manner, while achieving economic viability and social stability. The underlying theme is one of knowing, understanding, planning, measuring, recording, and managing to achieve identified social, environmental and production goals… which requires a sound and comprehensive management strategy and the capability for responsive tactical adjustments as circumstances change. Success depends upon developing the skill and knowledge bases, on continuous recording and analysis of performance, and the use of expert advice as required.*

The FAO concept/framework identifies 10 generic components of GAPs: soil management, water management, crop and fodder production, crop protection, animal production, animal health and welfare, harvest and on-farm processing and storage, energy and waste management, human welfare, health and safety, and wildlife and landscape conservation.

Working on a pledge to ensure the quality and safety of herbal medicines, the national authorities in many member states of the World Health Organization as well as other stakeholders in purview of herbal medicines are likely to face a plethora of challenges, including quality standard setting and their adoption, monitoring, and implementation. Relevant and apt national and location-specific standards, if absent, pose a risk of losing these herbal medicines to traditional users, rendering them useless to newer end users due to various factors. Some of the factors include failure to meet various trade, registration, import, and export norms; loss of confidence in these herbal products pertaining to the presence of actual or factual/perceived health risks; and increased reports related to an adverse aftermath pertaining to use/abuse of these herbal medicines.

1.2 Why Good Agricultural Practices?

A majority of farmers and agricultural practitioners throughout the world have difficulty in having a clear perception of the GAP standards that have been set by national regulatory bodies, international bodies, or local retailers. The farmers' first impression of GAP standards is simply of debarment on many of their on-farm activities. In addition to that, many farmers who still use traditional methods of farming hold an opinion that GAPs are a humiliation to their indigenous practices. In Asia, a vast majority of farmers are either completely or partially illiterate, which is why cultivation skills learned by farmers is purely through their exposure to crude field situations or from their elders.

However, with the adoption of science into farm practices, crop cultivation is gradually turning into a scientific process, which in every aspect of plant growth can be logically accounted for. Crop cultivation practices, which are scientifically augmented, have far outridden traditional cultivation practices, in terms of both productivity and overall production, and are pivotal for the food and nutritional security of the ever-increasing world population. The advent of science in food production practices is now on the verge to become the conventional approach to cultivation.

The farmers now in majority have turned toward the modern-day agrochemicals in every aspect of their farm practices, including crop production chemicals, plant growth regulators, crop protection chemicals, postharvest treatments, and storage chemicals, in order to enhance as well sustain their yield, productivity, and quality vis-à-vis protecting their crops from various pests and diseases. Dubious practices, however, in terms of chemical use can result in the creation of a "Frankenstein," better known as "toxic waste." This toxic waste subsequently affects the quality and safety of crop products, making them contaminated with harmful residues. This in turn destabilizes the ecological equilibrium of flora as well as fauna pertaining to farm and the adjoining areas. The drastic aftermath of low-quality practices comes full circle to subsequently affect the food production, livelihood of farmers, and the consumers as well.

GAPs aim to bring balance into the food production equation. GAPs help all stakeholders of the food production chain to understand the importance of food safety, the necessity of a sustainable food production system, and the fact that we must not produce waste. GAP does not prescribe techniques to increase crop productivity. It does, however, help farmers to effectively produce profitable and sustainable crops, creating benefits that directly affect them.

The GAP guidelines are principally intended with the idea to provide general technical guidance to the member states, keeping in mind the holistic backdrop of quality assurance to the end users as well as in the assessment of quality pertaining to the safety of herbal medicine/product usage with regard to both major and usual contaminants/residues. There may also arise a need to adjust the following guidelines per se according to each country's situation.

The objectives of these guidelines are to provide guiding principles:

- For assessment of the quality in relation to the safety of herbal medicines, with specific reference to both biotic and chemical contaminants/residues,
- For some model criteria to be generated in order to be used in identifying the possible contaminants/residues,
- For establishing ideal examples of methods and protocols, and
- For bringing to the forefront examples of practical technical procedures in order to control the quality of the final/packaged herbal products.

2. BASIC COMPONENTS OF GOOD AGRICULTURAL PRACTICES

2.1 Hygiene and Cleanliness

Medicinal plant material is tested for bacterial and fungal residues before being bought by most of the medicine manufacturers; if the product contains levels greater than permitted, they are rejected. Microbes like *Escherichia coli* or *Salmonella* are a very common contamination found in medicinal plant materials, which are usually caused by lack of basic cleanliness and hygiene during harvest and primary processing of the farm produce. It is therefore a central theme throughout the good agricultural collection practices (GACPs) guidelines to prevent microbial contamination by means of improved hygiene and cleanliness [3].

2.2 Prevention of Contamination

During its journey from field to shelf, medicinal plant material may come into close contact with many sources of contamination. The risks of contamination with heavy metals, pesticides, and other agrochemicals are a matter of particular concern for growers as well as buyers of medicinal plants. If contaminants of any of these sorts find their way into herbal medicine products, they are likely to cause an adverse health aftermath to the consumer.

2.3 Identification

Adverse health consequences have been reported in a number of cases pertaining to consumers consuming herbal medicines devised from incorrect species of medicinal plants. Guidelines on GACPs for starting materials of herbal origin place serious emphasis on ensuring that medicinal plants are correctly identified so that systems are in place in order to provide herbal medicine manufacturers with the vital assurance that they are buying the intended species material.

2.4 Efficacy

If herbal/medicinal plants not cultivated in a suitable environment, not harvested at the right time of the calendar year, or if not processed correctly, the consequences are straightforward; the active ingredient content of the medicinal plant material is likely to be lesser in amount, leading to the manufacture of partially effective/ineffective herbal medicines. GACP guidelines for starting materials of herbal origin have explicitly laid down the main principles that should be followed by farmers and collectors to ensure that medicinal plant materials are produced with optimum/maximum levels of active ingredients.

2.5 Production and Income

"Casual" agricultural and collection practices not only reduce the end quality of the herbal medicines but also affect the production and income of the farmers as well the collectors drastically. Knowhow on the optimization of both quality and extent of production of medicinal plant material vis-à-vis maximization of farmers' income has been provided by GACP guidelines for starting materials of herbal origin.

2.6 Sustainability

Sustainable collection methods in the wild should be encouraged and taken up as an increasing number of medicinal plant species are on the verge of being endangered or have become endangered in recent times. Medicinal plants collected in a manner that allow the plants to regenerate perennially represents one of the pivotal themes presented or laid out in GACP guidelines for starting materials of herbal origin. Sustainable availability of the plant species may be ensured if this path is followed. In addition, the benefactors of the talisman are not singular in nature as it is not only the medicinal plant collectors who shall have a regular source of income but also the herbal medicine industry.

2.7 Documentation and Traceability

Proper and scientific documentation of activities of the medicinal plant growers as well as collectors is required to trace the medicinal plant material back to its origin. Traceability is a key factor of GACP guidelines; it ensures that problems can be traced back to the root place of occurrence so that timely measures may be taken in response to any recurrence. Traceability is possible only through documentation, which is why documentation is a recurring theme throughout the GACP guidelines for starting materials of herbal origin.

2.8 Social and Legal Concerns

The source of a vast majority of medicinal plants is still the wild, which is why collection of plant material is from land that is either owned by the government or a local community (or both the parties are involved) and is subject to local, national, and international codes and regulations. With competition for limited resources increasing, there is increased pressure on collection of medicinal plants in the afore-mentioned areas as well. The GACP guidelines stress the importance of adherence to both traditional and relevant government collection regulations so as to ensure management of common resources in a sustainable mode. The compliance with these norms will also ensure the availability of medicinal plants for household usage of the local communities (Fig. 22.2).

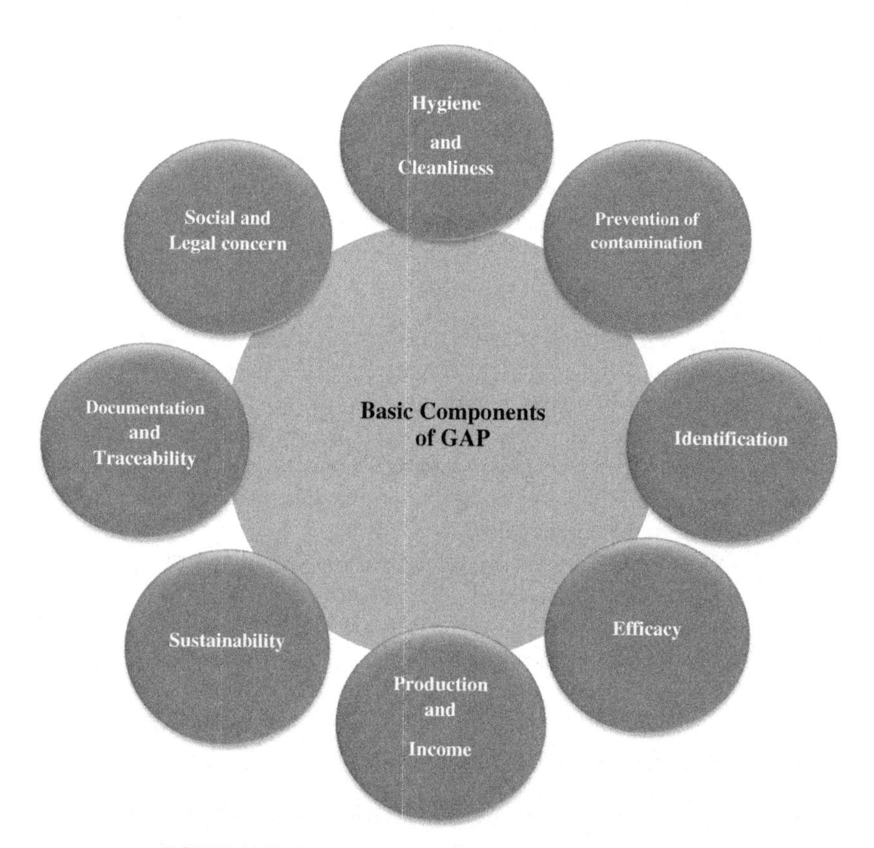

FIGURE 22.2 Basic components of good agricultural practices.

3. GOOD AGRICULTURAL PRACTICES FOR MEDICINAL PLANTS

3.1 Identification

The identity (species or botanical variety) of a medicinal plant selected for cultivation in a location should be specified in the national pharmacopoeia of that country [4,5]. In the countries where such kinds of documents are not available, authoritative documents from other countries should be considered for identification purposes. If the medicinal plant is newly introduced in a country, the identity should be recorded as the source material described or used in traditional medicine of the country of origin.

The botanical identity of each medicinal plant under cultivation should be verified and recorded. This should include a complete listing of the scientific name, family, genus, species, subspecies, variety, and author. Common and local names of the plant should also be recorded for convenience of identification. Additional information regarding cultivar, ecotype, chemotype, phenotype, hybrid, and so on may be recorded. If a commercially available cultivar is being grown, the name(s) of its supplier(s) should be mentioned in the record. If a landrace is being grown, the local name of the line, origin, and source of seeds/propagules should be kept in record.

If a medicinal plant is being registered for the first time in a country or in case of ambiguity over the identity of a species, a properly collected specimen should be submitted to a regional or national herbarium for identification. The specimen should be compared with an authentic specimen and, if possible, genetically.

3.2 Healthy Propagation Materials

First, the mode and method of propagation of an identified plant should be specified. A record should be kept whether the plant is propagated through seeds or any other specific propagules (e.g., root, rhizome, vegetative cuttings, and so on). It should be authenticated that quality seeds or propagating materials are free from any contamination and pests or diseases to ensure healthy plant growth. Traits like resistance or tolerance to biotic and/or abiotic stresses are preferred in propagules.

The suppliers of seeds or propagating materials should ensure the identity, quality, purity, and performance of the products and should provide their breeding history, if possible. The quality of the planting materials should be ensured through proper certification, labeling, and documentation, particularly in the case of organic farming and genetically modified germplasms. Regional and/or national regulations must be complied to use genetically modified medicinal plants or seeds. The use of substandard and adulterated propagation materials must be avoided.

Prescribed seed treatment protocol, if any, should be followed well in advance of cropping season. Seedlings should be raised in nursery reasonably well before transplanting schedule following recommended agronomic package of practices. Propagating materials in terms of vegetative cuttings should be of uniform dimensions and maturity. Only healthy seedlings should be transplanted to the main field.

3.3 Agronomic Practices

Assured quality of medicinal plant materials can be produced only by following recommended agronomic practices that demand intensive care and management. Scientifically documented cultivation practices should be followed. If such documents are not available, traditional method of cultivation can be followed with reinforcement through scientific research. Whatever standard operating procedure (SOP) is established, the principles of good crop husbandry must be followed including conservation agriculture and appropriate crop rotation.

3.3.1 Soil

Geographical and ecological variables have a huge impact on the quality of medicinal crops produced. Physical appearance and relative content of constituents may vary significantly depending on extrinsic environmental conditions when the same herb is grown at different locations. Thus, it is important to know about any specific agricultural requirements of a particular medicinal crop and to assess whether the site at which the crop is to be planted is able to provide those requirements.

In general, information regarding the following should be obtained to ensure selection of a suitable site for cultivation of medicinal plants:

1. The site should be free from any potential risks of contamination by hazardous chemicals through soil, water, or air pollution (e.g., industrial wastes and effluents).
2. A thorough history of past land uses of the cultivation site, including information of previous crops grown, any applications of plant protection products, any recent use of the site as a feedlot, any corrective measures taken, should be evaluated.
3. Soil sampling and analysis should be conducted to assess soil fertility and nutrient status.
4. In case of unknown soil history, tests should be conducted with proper sampling procedures to determine presence of any contaminants such as. residual pesticides or potentially toxic heavy metals at unacceptably high levels. Site selection should be based on knowledge of a crop's potential to accumulate these contaminants.
5. Relevant information regarding the site location to be recorded, such as annual and seasonal rainfall, cardinal direction, slope, access to reliable source of irrigation, crops grown in adjoining sites and their management practices, and the presence of any industries, mines, parking lots, waterways, underground storage tanks, and so on at the immediate vicinity of the site.

Soil is one of the most important factors for successful cultivation of any crop. For optimal plant growth, appropriate amounts of macronutrients, micronutrients, and organic matter should be present in soil. Physicochemical properties of soil including soil type, texture, fertility, pH, water-holding capacity, drainage, and so on determine the medicinal plant to be grown at the location. A soil map of the selected site should be considered as the determining factor for cultivation of medicinal crops.

Growers should adapt tillage practices depending on the requirement of the medicinal plants. However, practices that contribute to soil conservation and minimize erosion should be encouraged and implemented.

3.3.2 Climate

Quality of medicinal plants is significantly influenced by the climatic condition of a location. Meteorological data of at least the preceding 3 years should be available to judge the suitability of a site for growing a specific crop. Prior knowledge regarding average rainfall, average temperature, diurnal temperature variation, day length, humidity, and so on should be considered as important parameters before cultivating medicinal plants.

3.3.3 Supply

Basically, three major inputs are essential for the crop husbandry: nutrients, water, and protective chemicals. The quality of these supplies determines the quality of harvest.

To obtain better crop yield, the application of fertilizers and manures is indispensable. However, need-based applications should be ensured with due considerations to the effects such use may have in the environment. Regulations of the country's regulatory authorities may apply when using chemical fertilizers on commercially grown herbal crops. In case of organically grown medicinal crops, such chemical fertilizers must be avoided and naturally sourced amendments should be used. However, it should be ensured that correct types and quantities of fertilizers are used through documented research. As general guidelines, the following points should be considered:

1. Mineral supplements for a target crop must be based on complete soil analysis report by a competent laboratory.
2. Fertilizers and manures should be applied sparingly and in accordance with the needs of the particular medicinal plant species.
3. Use of fertilizers should be in accordance with the label directions.
4. Sewage sludge, industrial wastes, medical refuse, and human excreta must not be used as fertilizers in medicinal plants.
5. Thoroughly composted manures should be used to meet safe sanitary standards of acceptable microbial limits.
6. Periodic testing through approved procedures should be carried out to monitor any undesirable microbial pathogens.

7. Records must be maintained for application of any fertilizers/manures with their sources.
8. It should be ensured that only properly trained personnel apply the fertilizers.
9. Fertilizers should be applied at sufficiently early phase of the crop growth period so as to maintain a considerable interval between application and harvest.
10. They must be applied in such a manner that should not contribute to water contamination through leaching.

Water is another basic supply for crop growth. Access to sufficient quantity of quality irrigation water is essential for supplementing the requirement of the plants as received from normal rainfall. So, irrigation should be controlled as per the need of the medicinal plant at different growth stages. Cultivated crops should not suffer moisture stress or water logging. Water used for irrigation purpose should comply with the national/regional quality standards. All sources of irrigation water should be identified and recorded. Quality of irrigation water must be checked on a regular basis through standard procedures for any possible contamination of heavy metals, pesticides, pathogens, and so on. The water management plan should be optimized to reduce wastage, ensure conservation, and facilitate drainage. Records should be maintained for irrigation schedule and water usage.

To obtain a healthy crop, application of plant protective chemicals is a necessity. Different medicinal plants have typical growth and developmental characteristics. Depending on their individual needs and requirements, management practices should be planned. Timely implementation of intercultural operations such as weeding, hoeing, pruning, topping, bud nipping, shading, and so on should be followed for better quality and controlled growth of target plant parts being produced.

In general, application of any pesticides should be avoided for medicinal herbs as far as possible. Emphasis should be given on alternatives such as integrated pest management practices. Any agrochemicals to protect plants or to promote crop growth should be applied at a minimum level only when no alternative measures are available. The following points must be considered in case of a necessary application of agrochemicals:

1. Only approved plant protection chemicals should be applied at a minimum effective level.
2. Applications should be done in accordance with the label instructions of the product.
3. Only trained and qualified personnel should carry out the application with approved equipment.
4. All agrochemical applications should be properly documented (time, dose, and mode of application) and such documents should be made available to the buyers on request.

5. A minimal interval between the application of pesticides and the harvest of plant parts should be maintained according to the guidelines.
6. All such treatments should be carried out in consultation or by agreement with the buyers of the plant materials.
7. Growers should comply with the maximum residue limits of pesticides in the harvested plant parts as fixed by the regulatory authorities of respective countries of both the grower and the consumer.
8. In case of any skepticism regarding pesticide use and residues, guidelines of international agreements such as the Codex Alimentarius Commission should be consulted.

3.3.4 Harvest

To ensure quality, medicinal plants should be harvested at the right developmental stage of the target plant part during optimal weather conditions with proper techniques. The best time of harvest has to be determined depending on the optimum quality and quantity of the biologically active constituents in the targeted plant parts rather than the vegetative yield. Considerations should be given to the following factors to ensure quality harvest:

- Detailed information about appropriate timing of harvest is often available in national pharmacopoeias, official monographs, published standards, and major reference books. However, harvest should be scheduled when the crop is in appropriate condition to meet the quality requirements in terms of high concentration of biologically active constituents and low concentration of nontargeted ingredients.
- Harvesting should be done under the best possible weather conditions, avoiding rain, dew, or exceptionally high humidity. Evaluation of weather condition at the actual time of harvest can help in avoiding such conditions to prevent any negative effect of excessive moisture on the quality of the harvest. Consideration of weather forecasting for several days immediately following harvest is also necessary to prevent any microbial growth on the harvested products due to a sudden increase in moisture level. If harvesting must occurs in wet conditions, extra care should be taken to transport the harvested materials quickly to an indoor drying facility to expedite drying.
- Harvest equipment (mechanical equipment, containers, hand tools, and so on) should be made of nontoxic and noncorrosive materials that can be easily and thoroughly cleaned. Cutting devices, harvesters, and other machines should be in proper working condition and kept clean. Care should be taken to ensure that the parts of the equipment that come in direct contact with the crop during harvest are clean and free from any potential contaminants. Remnants of any previous harvest should be cleaned from the equipment to prevent cross-contamination. Clean containers that are free from any contamination by previously harvested crops

and other foreign materials should be used. All the harvest equipment and containers, when not in use, should be stored in an uncontaminated, dry place that is protected from any pests and inaccessible to livestock and domestic animals. Personnel should be properly trained in use of mechanized harvest equipment to ensure safety while operation and to avoid damage to the harvested produce.

- Harvesting should be conducted with care to ensure minimum presence of foreign matter such as soil, weeds, trash, and so on in harvested produce. Special attention should be given to remove any potentially toxic weed species inadvertently collected with harvested plant parts. Harvested medicinal plant parts should be ensured to have minimum contact with soil to minimize microbial contamination. Soil adhering to the underground plant parts should be immediately removed after harvesting.
- Harvested raw plant parts should be transported in clean, dry, well-aerated containers to the processing facility as soon as possible. They must be protected from moisture and unnecessary exposure to insects, birds, rodents, or other animals. Any mechanically damaged or decomposed medicinal plant materials resulting from compacting or stacking should be monitored and discarded at any stage during harvest and processing, to avoid loss of product quality due to microbial contamination.

3.4 Good Collection Practices

Not all plants/plant parts used for medicinal purpose are cultivated. A good number of medicinal plants/plant parts are being collected from the wild habitats. Setting a framework for good collection practices is therefore necessary to ensure long-term survival of wild species and their associated habitats [6]. Though the set of guidelines may vary from region to region, general comprehensive management practices should be maintained to protect wild species from unsustainable harvest while addressing various complex environmental and social issues.

The collection, possession, transit, and sale of medicinal plants should be carried out in compliance with the existing laws and legislation of the country from where they are being collected. Prior permission should be obtained from the competent government authorities and landowners for collecting any plant materials from the wild. Copy of the documentary proof of such permissions should always be kept while collecting.

Regulations as per the Convention on International Trade in Endangered Species of Wild Fauna and Flora (CITES) must be adhered to while collecting medicinal plant parts from the wild. Unless there is proper authorization from a competent authority, plant parts from endangered species must not be collected. The national "red" list should be consulted and respected in this regard.

If the collected medicinal plant materials are intended for export from the country of collection, export permits, phytosanitary certificates, CITES permits, and relevant certificates must be obtained as per requirement. Existing laws and legislation of importing country are to be regarded.

3.5 Postharvest Processing

Immediately after medicinal plant parts are harvested either from a cultivated field or from the wild habitat, postharvest processing must start. This has considerable impact on the product quality as it should be maintained from the points of harvest to manufacture. The steps involved primary processes like sorting, protection of the materials from moisture, washing, drying, and so on to specialized processes like boiling, roasting, fermenting and so on. Care should be taken at every step to maintain the quality of the product in compliance with both GAP and good collection practice.

Harvested raw medicinal plant parts should be inspected carefully for any cross-contamination and foreign materials before primary processing. Any damaged material should be sorted out, and screening may be done depending on appearance, size, color, and so on.

SOP should be followed for primary processing of medicinal raw materials in compliance with the national/regional regulations of producer and purchaser countries. All steps should be carried out carefully without compromising with the quality of the product. The following points must be considered during primary processing:

- Immediately following harvest, the plant parts should be stored in clean containers. If necessary, refrigeration or other measures of conservation should be undertaken. Use of any preservatives should be avoided.
- Raw materials that are to be used fresh should be quickly transported to the processing unit to prevent any degradation.
- Immediately on arrival of harvested raw materials at the processing facility, they should be unpacked and unloaded.
- The raw materials should be protected from moisture, rain, or other means that could deteriorate them.
- Some plant materials like roots are needed to be washed properly to remove dirt, soil, and foreign materials. Only potable water should be used for this purpose. Washing/cleaning should be done in an area with good drainage facility where buildup of mud can be prevented. Washed material should be adequately dried before further processing.
- Exposure to direct sunlight should be done only if there is such specific requirement for drying.
- If the materials are to be prepared in dry form, moisture content must be kept as low as possible. Drying can be done by various ways as per requirements like open air drying in the shade, oven drying, frame drying,

indirect drying by lyophilization, microwave, infrared, and so on. If possible, the temperature and humidity during drying should be controlled to protect the active ingredient from degradation.

- Uniform drying of the materials is desirable. Efforts should be made by turning or stirring of bulk materials while drying. Drying directly on bare ground must be avoided.
- At any stage, protection from any contamination, decomposition, and pest infestation of the plant materials must be ensured.

Apart from primary processing, a few specific processes are also required for specific plant materials. Certain plant materials need special treatment to meet quality specifications. Such specific processing may include removal of skin or peeling, chopping, milling, steaming, boiling, soaking, distillation, fumigation, roasting, frying, acidic/basic treatment, fermentation, and so on. The purpose of these specific treatments aims toward enhancing purity percentage of active ingredient, detoxifying inherent toxic chemicals, improving therapeutic efficiency, protecting from microbes and other pests, and so on. All the treatments should be carried out by trained personnel following standard protocol using correct type of equipment and according to the regulations of producer and purchaser countries.

3.6 Packaging and Labeling

The next step immediately after processing is appropriate packaging of the materials to prevent deterioration, pest infestation and contamination. Before packaging, in-process QC measures should be followed to eliminate any foreign matter, contaminant and sub-standard materials.

Processed medicinal plant materials should be packed in clean, dry, undamaged and preferably new bags, sacks, boxes, cans or other containers as prescribed by the regulatory authorities of producer and consumer countries. Packaging material should meet the criteria of the product's packaging specifications. For example, food-grade or pharmaceutical-grade packaging materials should be used for this purpose. Recycled packaging materials are acceptable as long as they maintain the status of food-grade or pharmaceutical grade. Reusable packaging materials should be properly cleaned, disinfected and well-dried before reusing. They should not have any contaminants present from the previously packed materials. Packaging materials should be stored in clean, dry place which is free from any pests to ensure no contamination through them. While packaging care should be taken for not causing any mechanical damage or undue compaction to deteriorate the packed products.

Proper labeling of the packed material is very important to identify and obtain valuable information about the product without opening the package. The label should be clear, permanently fixed on the package and in compliance with the regulations of producer and end-user countries. It should clearly indicate information like: common English and botanical name of the medicinal plant, used plant part, form of the material (whole, powder, and so on), place of cultivation/collection, date of harvesting, names and contacts of producer/collector and processor, quantitative information, conservation techniques, danger indication, and packaging and transport modalities. The label should also contain production lot/batch number. Additional information regarding quality and production parameters may be given as a separate certificate along with the package linked by the lot/batch number. All the information linked to the lot/batch number, assignment number, and date must be documented and the record should be kept for a period of 3 years or as specified by the legislative authorities.

3.7 Storage and Transportation

Packaged herbal materials should be stored in a cool, dry place away from direct sunlight with good aeration and limited temperature fluctuation. Fresh plant materials should be stored at 2–8°C, while frozen products should be stored at −20°C. Storehouses should have concrete flooring for easy cleaning. Bulk packages should be stored on pallets at sufficient distance from the walls. Segregated storage should be exercised for different kind of material to avoid cross-contamination. Organically grown materials should be stored separately from nonorganic products. Highly aromatic plant materials must be stored separately from other materials to prevent aroma absorption. Storage area should be protected against any insects, birds, rodents, and domestic animals. Regular on-spot monitoring is required to check any pest infestation, fungal attack, or rotting. Fumigation should be carried out against pest infestation only in case of necessity. Only registered products should be used and licensed personnel should carry out fumigation. Such activities should be documented properly.

For bulk transport, the containers should be suitable for transporting food-grade or pharmaceutical-grade materials, clean and dry. Fresh materials must be transported in well-aerated containers, while air-tight containers must be used for dehydrated products. Transport vehicles should be well ventilated and must be cleaned between loads. Proper security measures should be taken while transporting potentially toxic or poisonous materials.

3.8 Sanitation

All the phases of production, collection, and postharvest processing of medicinal plants involve a large number of personnel at different stages. Successful execution of GAP or good collection practice of medicinal herbs depends on training, safety, and hygiene of the personnel involved. Quality products can be ensured if these three issues are properly addressed.

Training by trained personnel is of utmost importance for the maintenance of proper sanitation.

- Each individual assigned to various tasks of growing, collecting, and postharvest processing must receive relevant and adequate training.
- Personnel with the responsibility of plant identification should have adequate training and experience for performing their task with certainty.
- Producers and collectors should possess sufficient knowledge about proper harvesting techniques and maintenance of medicinal plants.
- Personnel assigned the task of agrochemical application must be well-trained about their use.
- Persons involved in collection of plant materials from the wild should have adequate knowledge about environmental protection, soil management, and species conservation. Long-term sustainable harvest from the wild depends on their expertise and level of understanding.
- All the personnel who handle medicinal plant materials should have training about hygienic practices.

Proper safety measurement is also integral part of sanitation in GAP of medicinal plants. The following safety measures need to be considered during the execution of GAP.

- All personnel should wear clothing appropriate for the working environment.
- Protective clothing like gloves, masks, eye coverings, boots, and so on should be used when required.
- Safety of the workers at any stage should be ensured. Workers should be protected from situations such as extreme heat and cold, noxious insects, weeds and animals present at the collection area, and work environments where they may encounter excess noise, dust, or other factors due to specific operations.
- All equipment and tools should be maintained properly to make them reasonably safe for the operator when in use.

Hygiene is also another component of sanitation that requires special mention. Several critical points are to be considered in this regard.

- All personnel handling medicinal plants should ensure personal hygiene and cleanliness to prevent any accidental microbial contamination.
- Toilet facilities should be provided at the working places with proper sanitation measures.
- Personnel should wash their hands always at the beginning of work, after using the toilet, and after handling contaminated materials.
- Eating and smoking should not be permitted at the processing area. Coughing, sneezing, and spitting over unprotected materials should be prohibited.

- Sick personnel who have known communicable diseases must not be allowed to work in the processing unit. Immediate medical care should be given to the worker falling ill during work.
- Workers having open wounds, burns, or skin infections should not be allowed in the handling operations. Minor cuts and burns should be covered by appropriate protective clothing before work.

Furthermore, national/regional regulations regarding labor should be respected while employing staffs and the overall welfare of the workers should be a priority.

4. QUALITY CONTROL AND GOOD AGRICULTURAL PRACTICES

QC is a combination of procedures that ensure that a particular product or service adheres to a defined set of quality criteria, set by the client or customer or organization. Considerations need to keep in mind for maintaining quality of herbal medicines includes authenticity of the species, botanical authenticity of new plant, healthy collection protocol, harvesting at right stage, sorting of produce, presence/absence of foreign matter, and admixture of toxic weed species. Quality attributes that will deteriorate due to malpractices of harvesting and handling are discussed in an earlier section. Major impetus lies on analytical and biological tools to authenticate botanical species and identifiy adulterants/admixtures.

A major factor that governs the quality of the herbal medicine is the authenticity of the plant species. There is every chance of misidentification of the plant species, intentional/unintentional contamination with other species, and adulteration with extraneous materials. In today's context, authentication of plant species can be done based on botanical/morphological characteristics and genetic and chemical approaches. Although it is known to be trivial in nature, classic botanical tools were used extensively to identify the samples, and the same must be enlisted in the herbarium of a certified research/teaching institute. Apart from the vegetative part, mostly reproductive parts are commonly used for the confirmation of the species. Authentication of species for a particular genus is important because apart from the concentration of a particular bioactive compound, there might be other unwanted compounds that might be toxic in nature. Chinese star anise (*Illicium verum*) is known to have properties as a stimulant, diuretic, and digestion aid and thus it was treated as generally recognized as safe material approved by the US Food and Drug Administration. But due to adulteration with Japanese star anise (*Illicium anisatum*), several reports of health-related issues especially in children were reported from the United States and Europe [7,8]. Adulteration-related toxicity was reported to be due to the presence of a few toxic secondary metabolites in the species, namely, anisatin, neoanisatin, and 2-oxoneoanisatin [9]. The most commonly used tool to identify the authentic species is "chemical fingerprinting" using marker compounds. In this case, using gas chromatography (GC), the admixture can be identified by the detection of eugenol, methoxy eugenol, and 2,6-dimethoxy-4-allylphenol.

Genetic fingerprinting is another powerful tool to differentiate the authentic sample from the adulterated samples [10]. Using different molecular tools, identification at the DNA level is also possible. Genetic profiling using high throughput analysis has proved to be valuable new tool for authentication. Best-quality DNA can be extracted from the fresh samples, whereas powdered samples provide relatively poor-quality DNA. So, appropriate QC measures need to be taken. It is such a powerful tool that it helps in confirmatory identification of variety, species, and even contamination of other species. For powdered samples, where DNA extraction is not feasible, or extract-based formulations, it is difficult to authenticate the source material with this tool. In that case, chemical fingerprinting is the preferred protocol to follow.

With the advancement in chromatographic techniques, high-performance liquid chromatography (HPLC)-GC and capillary electrophoresis (CE) were found to be reliable and authentic techniques for the identification of plant samples. In chemical techniques, marker compounds are identified and used for fingerprinting. For an ideal analytical method, criteria like precision, accuracy, robustness, specificity, reproducibility, and ruggedness should meet the standard values. Citrus fruits were known to contain thermogenic constituents like adrenergic amines and flavonoids [11]. These amines include synephrine, octopamine, hordenine, *n*-methyltyramine, and tyramine. One such source of these compound is *Citrus aurantium*, bitter orange, and it has been used for weight loss. Synephrine is used as a marker compound in a CE method for the identification of authentic samples [12]. *Citrus* sp. constitute several flavonoids, which are known to be active as health-benefitting agents. The developed CE was unable to estimate this class of compounds, so HPLC with UV detection was developed to estimate and identify these flavonoids for authentication of citrus plants. The method could analyze six amines apart from 20 flavonoids in a single run of 1 hour. Electro light scattering detection was used in HPLC for those compounds where UV detection was normally unable to recognize. For examples, saponins and other glycosides are in general less sensitive in HPLC-UV detection, due to lack of unsaturation or conjugation. Ginsenosides, the major constituent of ginseng roots, are known to be active as health-benefitting molecules. These glycosides were detected and analyzed by ultra-performance liquid chromatography coupled with evaporative light scattering detection (UPLC–ELSD) for the identification of authentic samples [13].

Instead of a single-marker compound, multiple-marker compounds were preferred for the obvious reason of less chance of adulteration. Use of multiple-maker compounds for authentication eliminates the unsavory practice of fortifying the product with other bioactive compounds (Fig. 22.3).

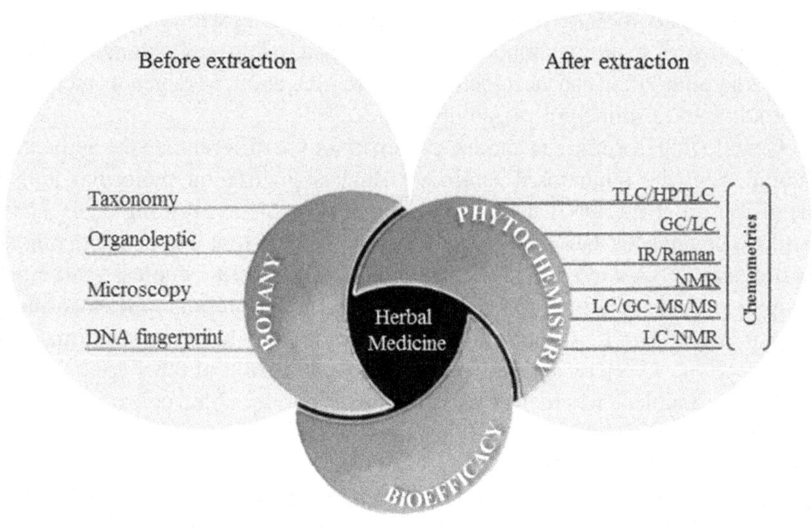

FIGURE 22.3 Techniques (macroscopic, microscopic, and analytical) that helps in authentication of herbal plant species techniques.

With the advent of further sophisticated and highly sensitive tandem techniques like liquid chromatography−mass spectrometry (LC-MS)^n/high-resolution mass spectrometry (HRMS), one can identify and estimate the marker compound(s) with confidence. Because of their morphological similarities, *Lonicera japonica* flower buds were misidentified with six other species of the same genus. The flower bud is used as herbal medicine according to Chinese medicine. Metabolomics tools were used for the identification and differentiation of one species from another [14]. It is a powerful tool for classification and characterization of plant species for QC purposes, and it is more useful when the active ingredients/marker compounds are known.

Simpler techniques like high-performance thin-layer chromatography (HPTLC) are also being used for the authentication of plant species. The use of different silica gels as a stationary phase coupled with automation is helpful for researchers. This technique provides better resolution and separation for the enhancement of theoretical plates over TLC. Ginseng is a powerful candidate as a nutraceutical compound and commands a high price across the globe; thus, the chances of adulteration are high with cheaper plant products. American and Asian ginseng was usually adulterated with sawdust and *Platycodon grandiflorum* [15]. *Panax ginseng* (ginseng) and *Panax quinqeefolius* (American ginseng) are two species of ginseng that are close to each other in terms of their morphological characteristics. Dereplication using single ginsenoside led to erroneous results most of the time. HPTLC profiling distinguished these two species as it considered number of secondary metabolites [16]. Fourier-transform infrared spectroscopy (FTIR) was used for the confirmation of these two species with their fingerprint regions using principal component analysis (PCA) [17].

Among vibrational spectroscopy, IR spectroscopy is also used for the authentication purposes. Another Chinese traditional medicine, *Sedum sarmentosumbunge,* is differentiated from other species: *Sedum lineare* Thunb, *Sedum aizoon*, and *Sedum erythrostictum migo*. Specific frequency regions were used as a tool to differentiate these species [18]. Pattern recognition and PCA were performed to differentiate the species. Another vibrational spectroscopic technique (Raman spectroscopy) was also used to identify the presence of American ginseng, the endangered species in Asian ginseng samples. Country of origin and effect of processing also can be determined with Raman spectroscopy [15]. A peak at 980 cm^{-1} specifically is an attribute of Chinese ginseng and is otherwise absent in American ginseng, which is the Raman spectroscopic application is so widespread.

Chemometrics, a science of extracting information by using statistical and mathematical methods, improves the cognizance of chemical information [19]. Of late, this technique is being used for the authentication of plant samples. As there are limitations of using a single-marker compound for the identification of authentic samples, the chemometric approach could be the better method for the QC purpose. With the help of statistical tools, chemical information derived from analytical instruments could be further processed for its effective utilization. Hyphenated techniques (coupling of two techniques with the help of proper interface) are used as the analytical tool to generate chemical data. To discriminate and evaluate the quality of *Curcuma* sp., chemometric approaches were deployed using GC-MS and HPLC–diode-array detection (DAD) data [20,21]. *Actea racemosa*, a North American plant known to have medicinal properties related to hormone substitution, was ranked among top the five best-selling herbal supplements in 2014 [22]. Adulteration of *A. racemosa* with other species was prevented by the analysis using chemometric tools [23]. RP-PDA-HPLC along with PCA provided reliable results for the proper authentication of the exact species (Fig. 22.4).

Herbal dietary supplements and medicinal products made from the North American plant *Actae aracemosa* L. (Ranunculaceae, syn. *Cimicifuga racemosa* L. Nuttal, black cohosh) are widely known to be used in the treatment of climacteric complaints as a safer and more “natural” alternative to hormone substitution.

For QC, selection of method of analysis is very important, but there is no single analytical method that can authenticate every plant sample in raw or processed conditions. Complete understanding of the type of secondary metabolites present in a particular sample is very much required for the identification of authentic sample. The process may vary from simple organoleptic or microscopic investigation to HRMS or chemometric analysis.

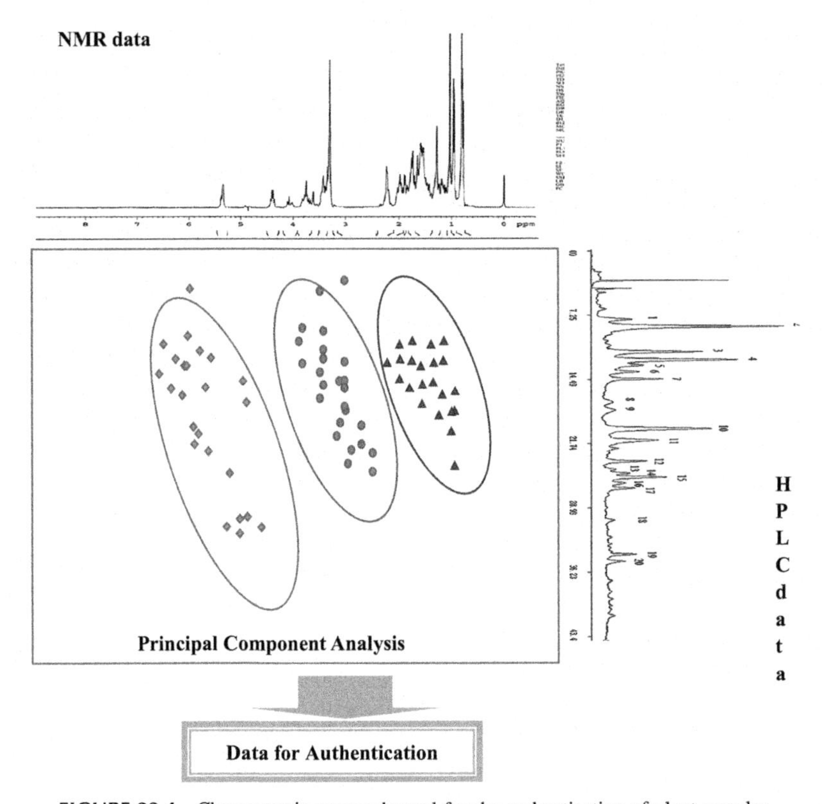

FIGURE 22.4 Chemometric approach used for the authentication of plant samples.

With the continuous advancement of science, if a plant identified as a source of bioactive molecules for treating human ailments, then the plant species needs to be identified using different molecular as well as chemical tools before placing it into a herbarium of a reputable institution or university. Identification of marker compounds and full metabolome profiling also need to be done for future references.

Apart from authentication of plant species, there are other several important points where QC and quality assurance need to be maintained [24]. Introduction of new plant species in the herbarium or enlisting the name in the list of medicinal plants should be done with utmost importance. Future identification and QC-related issues will be executed based on the initial information. QC measures are also important while collecting as well as storing the plants. Suitable measures need to be taken while harvesting the produce after following GAPs for the production of medicinal plant materials. Harvested material strictly should be free from foreign matter as well as any other plant/weed.

5. GOOD AGRICULTURAL PRACTICES: ETHICAL AND LEGAL CONSIDERATIONS

GAP for the production of medicinal plants involves multifaceted inputs and considerations to ensure that the produce is fit for human consumption. Production of healthy materials, which involves cultivation, harvesting, and postharvest processing, must be executed following legal and environmental considerations. The package of practices must also be carried out in accordance with the ethical norms of the community as well as the country without affecting the regular functions of the society in large. Due respect must be given to Convention on Biological Diversity in order to maintain the sustainability; this is more important for medicinal plants as it fetches foreign currency in distant markets.

Intellectual property rights and benefit sharing out of the produce are of utmost importance in the context of international trade. Written agreement on the return of short- and long-term benefits as well as compensation for the use of source material should be executed before cultivation of the planting material. Property rights of a planting material used from indigenous sources of a country need to be clarified before execution of the project. In general, issues related to access to genetic resources are complex, in particular when cross-border movement of plant/propagation material is required. Thus, considerations must be followed according to the international law and regulations.

Ethical considerations are more important throughout the production process including its postharvest processing steps. Starting from the identification of the particular plant species and production of planting material to its cultivation package of practices, ethical issues needs to be dealt in a suitable manner to obtain healthy and safe finished products.

Conservation status or maintenance of natural resources, especially medicinal plants, has been a global challenge, and it is more important for medicinal plants for its increasing use as human medicine [25]. Saving biodiversity issue and severity of the loss of different plant species are indicated under the categories of "critically endangered," "endangered," or "vulnerable" [26].

6. EPILOGUE

Following the protocols of GAPs for the production of quality herbal drugs is essential. The protocol involves site selection to labeling and packaging of the produce. GAP, in general, has four integral structural components, commonly known as the 4Ss of GAP (clean soil, supply, sanitation, and surfaces) apart from other sustainability, social, and legal issues. Some regulations need to be followed during processing as well as using equipment in the pilot- and industrial-scale productions.

Although various components under GAP dictate the quality of herbal material, QC of the finished product remains the mainstay as far as quality assurance of the material is concerned. Selection of analytical/biological methods for the authentication and QC is very important. There is no single method that can ensure in the authentication process. Thus, the use of multiple markers for the authentication as well as searching of adulteration is very much required.

When phytochemical analysis is used for authentication purposes, the analyst should use scientifically validated analytical methods such as those methods developed by the Association of Official Analytical Chemists, the herbal pharmacopoeia of the country/international standard, and academicians/researchers to determine whether the selected method(s) can accurately detect and quantify the marker compounds of the sample material.

REFERENCES

[1] K. Bone, S. Mills, Principles and Practice of Phytotherapy: Modern Herbal Medicine, vol. 2000, Churchill Livingstone, London, 2000, pp. 350–353.

[2] FAO, Development of a Framework for Good Agricultural Practices, Committee on Agriculture, 17th Session, Food and Agriculture Organization of the United Nations (FAO), Rome, 2003. COAG/2003/6.

[3] WHO Guidelines on Good Agricultural and Collection Practices (GACP) for Medicinal Plants, World Health Organization, Geneva, 2003.

[4] AHPA, Good Agricultural and Collection Practice for Herbal Raw Materials, American Herbal Products Association, American Herbal Pharmacopoeia, December 2006.

[5] NMPB, Good Agricultural Practices for Medicinal Plants, National Medicinal Plants Board, Department of AYUSH, Ministry of Health and Family Welfare, Government of India, New Delhi, 2009.

[6] EUROPAM, Guidelines for Good Agricultural and Wild Collection Practice (GACP) of Medicinal and Aromatic Plants, EUROPAM, the European Herb Growers Association, GACP-subcommittee, Brussels, April 2006. EUROPAM GACP Working Copy No. 7.3.

[7] D. Ize-Ludlow, S. Ragone, I.S. Bruck, J.N. Bernstein, M. Duchowny, B.M. Pena, Neurotoxicities in infants seen with the consumption of star anisetea, Pediatrics 114 (2014) e653–e656.

[8] Y. Kudo, J.I. Oka, K. Yamada, Anisatin, a potent GABA antagonist, isolated from *Illicium anisatum*, Neurosci. Lett. 25 (1981) 83–88.

[9] K. Yamada, S. Takada, S. Nakamura, Y. Hirata, Structures of anisatin and neoanisatin toxic sesquiterpenes from *Illicium anisatum*, Tetrahedron 24 (1968) 199–229.

[10] S. Chen, X. Pang, J. Song, L. Shi, H. Yao, J. Han, C. Leon, A renaissance in herbal medicine identification: from morphology to DNA, Biotechnol. Adv. 32 (7) (2014) 1237–1244.

[11] K. Kubo, C. Kiyose, S. Ogino, M. Saito, Suppressive effect of *Citrus aurantium* against body fat accumulation and its safety, J. Clin. Biochem. Nutr. 36 (2005) 11–17.

[12] B. Avula, S.K. Upparapalli, I.A. Khan, Enantiomeric separation of adrenergicamines in citrus species, related genera and dietary supplements by capillary electrophoresis, Chromatographia 62 (2005) 151–157.

[13] D.Y. Lee, J.G. Cho, M.K. Lee, J.W. Lee, Y.H. Lee, D.C. Yang, N.I. Baek, Discrimination of Panax ginseng roots cultivated in different areas in Korea using HPLC-ELSD and principal component analysis, J. Ginseng Res. 35 (1) (2011) 31–38.

[14] W. Gao, H. Yang, L.W. Qi, E.H. Liu, M.T. Ren, Y.T. Yan, J. Chen, P. Li, Unbiased metabolite profiling by liquid chromatography–quadrupole time-of-flight mass spectrometry and multivariate data analysis for herbal authentication: classification of seven *Lonicera* species flower buds, J. Chrom. A 1245 (2012) 109–116.

[15] A. Rohman, A. Nugroho, E. Lukitaningsih, Sudjadi, Application of vibrational spectroscopy in combination with chemometrics techniques for authentication of herbal medicine, Appl. Spect. Rev. 49 (8) (2014) 603–613.

[16] M. Ram, M.Z. Abdin, M.A. Khan, P. Jha, HPTLC Fingerprint Analysis: A Quality Control for Authentication of Herbal Phytochemicals, In High-Performance Thin-Layer Chromatography (HPTLC), Springer Berlin Heidelberg, 2011, pp. 105–116.

[17] K.Y.-L. Yap, S.Y. Chan, C.S. Sing Lim, Infrared-based protocol for the identification and categorization of ginseng and its products, Food Res. Int. 40 (2007) 643–652.

[18] R. Xu, Y. Chen, D. Wan, J. Wang, Identification of four *Sedum* plant medicines by fourier transform infrared spectra, Pharmacogn. Mag. 8 (30) (2012) 107–110.

[19] P.J. Gemperline, Principal component analysis, in: P. Gemperline (Ed.), Practical Guide to Chemometrics, second ed., CRC Press, London, 2006, pp. 69–104.

[20] Y. Ni, M. Mei, S. Kokot, One-and two-dimensional gas chromatography–mass spectrometry and high performance liquid chromatography–diode-array detector fingerprints of complex substances: a comparison of classification performance of similar, complex Rhizoma curcumae samples with the aid of chemometrics, Anal. Chim. Acta 712 (2012) 37–44.

[21] Z. Xiang, X.Q. Wang, X.J. Cai, S. Zeng, Metabolomics study on quality control and discrimination of three Curcuma species based on gas chromatograph–mass spectrometry, Phytochem. Anal. 22 (5) (2011) 411–418.

[22] T. Smith, Herbal dietary supplement sales in US increase 6.8 percent in 2014, Herb. Gram. 107 (2015) 52–59.

[23] M. Bittner, R. Schenk, A. Springer, M.F. Melzig, Economical, plain, and rapid authentication of *Actaearacemosa* L. (syn. *Cimicifuga racemosa*, black cohosh) herbal raw material by resilient RP-PDA-HPLC and chemometric analysis, Phytochem. Anal. 27 (6) (2016) 318–325.

[24] T.J. Smillie, I.A. Khan, A comprehensive approach to identifying and authenticating botanical products, Clin. Pharmacol. Ther. 87 (2) (2010) 175.

[25] J.M. Affolter, A. Pengelly, Conserving medicinal plant biodiversity, in: S.G. Wynn, B.J. Fougère (Eds.), Veterinary Herbal Medicine, Mosby, Saint Louis, Missouri, USA, 2007, pp. 257–263.

[26] IUCN, The IUCN Red List of Threatened Species, Version 2017-1, 2017. Downloaded on June 8, 2017.

Chapter 23

Fundamentals of Microwave-Based Sample Preparation for Plant-Based Drug Discovery

Roshni Tandey, Kavi B.S. Chouhan, Vivekananda Mandal
Institute of Pharmacy, Guru Ghasidas Central University, Bilaspur, India

1. INTRODUCTION

Extraction is as simple as making a cup of tea, but rarely do we consider the steps involved in making of a cup of tea significant because the entire focus is on enjoying the flavor and taste of the tea prepared. Similarly, in natural product research rarely do we tend to realize the importance of selecting a judicious extraction process and the entire research is isolation or bioactivity driven. However, a few minutes of careful thinking can make us understand that an inappropriately selected extraction method is sufficient to jeopardize the entire objective of natural product research because any mistake can then make subsequent steps of isolation and bioactivity determination suffer heavily, thus leading to a shaky foundation. It is similar to optimizing the heating time for making tea. Imagine how tea would taste if heat were not applied at all but the ingredients were just macerated; equally, what would have been the taste if prolonged heating of a few hours were given for making tea. In today's modern technological era, no production line in the field of food and nutraceuticals is devoid of an extraction unit. After the recently concluded Paris Convention for Climate Change, strict protocols were drafted for limiting carbon contributions from each country, with heavy restriction being imposed on developed countries. In such a situation, technology and environment should be in tandem so that the greenness of planet earth is sustained by reducing emissions of greenhouse gases and the carbon load. This forms the backdrop for this chapter.

In recent times (from 2011) sufficient research has been done on designing and developing newer and greener extraction methods. In this regard, microwave-assisted extraction (MAE) needs no introduction. In one of our

Natural Products and Drug Discovery. https://doi.org/10.1016/B978-0-08-102081-4.00023-X

recent articles published in *Trends in Analytical Chemistry* we presented a critical analysis report on the publication trends in MAE of botanicals that clearly stated a sharp rise in research interest in MAE of botanicals from 2011. Plenty of review articles have already been written, including by the authors of this chapter, on different operational and technical aspects of MAE of botanicals, probably leaving no corner untouched. Some of the best review articles on MAE of botanicals have also been summarized briefly by us in the article mentioned earlier. Henceforth, it becomes a challenge to draft another review that provides readers with newer information without repeating earlier published facts. The basic objective of this chapter is to answer technical queries posed by "first-timers" in the field of research in developing green extraction methodologies for botanicals with special emphasis on the MAE technique. All earlier published reviews are literature oriented, consisting of a collection of past research reports, but this specially drafted review in the form of FAQs is purely opinion based for an easy understanding of first-timers and industries trying to take up similar ventures for large-scale production. FAQs have been framed according to the chronological order of events occurring in MAE of botanicals, which shall help readers to map the sequence of work flow. No such review of a similar pattern in the area of natural products has so far been attempted. All past research related to MAE of botanicals has been targeted toward developing optimized conditions for maximum yield of a target compound, and to fulfill this task, investigators have made use of different optimization tools. This review shall provide its prospective readers (natural product researchers, biochemists, chemists, botanists) with a highly organized map of events occurring in MAE of botanicals, which will help others to conceptualize the entire project with a vivid understanding of the significance of each step involved. To simplify the literature search in understanding different aspects of MAE, we have also provided a one-stop center as shown in Table 23.1 to address all issues related to MAE.

2. FREQUENTLY ASKED QUESTIONS

2.1 What Shall Be the Strategy for Plant Selection in the Case of Microwave-Assisted Extraction of Botanicals?

In the case of MAE of botanicals, generally a potential bioactive with commercial and therapeutic significance is treated as a target analyte so that a large-scale extraction technique for the said bioactive may be explored. Therefore more emphasis should be given to specific or multiple bioactive compounds. In this regard, newer bioactive compounds whose investigation is still ongoing should be avoided and instead well-known potential bioactive(s) should be targeted, which are commercially viable, so that something meaningful and economical regarding their large-scale production can be discovered through such research. Once the bioactive that is to be treated as target

TABLE 23.1 One-Stop Solution Center for Different Issues Related to Microwave-Assisted Extraction (MAE)

Issues Related to MAE	One-Stop Solution
Scenario of research and research trends in MAE of botanicals	Trends Anal. Chem. 82 (2016) 100–108
Guide to MAE of phenolics	Trends Anal. Chem. 85 (2016) 140–152
Optimization strategy in MAE	Phytochem. Anal. 25 (2014) 1–12
Elucidation of mechanism of action	Biochem. Eng. J. 50 (2010) 63–70
Degradation profiling	Food Chem. 187 (2015) 507–516
MAE of essential oil and solvent-free MAE	Trends Anal. Chem. 71 (2015) 157–168
Value addition in MAE	J. Chromatogr. A. 1218 (2011) 6213–6225
Industrial scale of MAE	Chem. Eng. J. 299 (2016) 420–430
Factors governing MAE	Pharmacogn. Rev. 1 (2007) 7–18
Energy audit	J. Food Eng. 167 (2015) 51–58

analyte is decided, then the plant from which it is largely reported should be selected, keeping in mind biodiversity issues. A list of bioactive compounds extracted using MAE can be found in a review article published by the authors, which happens to be the first review article on MAE solely dedicated to botanicals. It is important to understand that a particular secondary metabolite may be present in multiple plants of different species. In light of such a situation, attempting MAE of a bioactive whose MAE reports already exist but from a different plant will not be considered a good idea unless the plant part treated as the raw material is different. MAE conditions of a particular bioactive, even though attempted from different plants but from the same plant part, will not differ significantly. However, conditions may vary significantly if the same bioactive is extracted from a different plant part. To simplify, if MAE conditions for a particular alkaloid compound have been developed using leaves as the raw material, then the same MAE conditions shall be deemed to be valid for a different plant as long as leaves are used as the raw material. However, MAE conditions will markedly vary if the same alkaloid is being extracted using MAE from a plant part other than leaves. Microwaves have different penetration power for different plant parts, with the softer parts such as leaves and flowers being more susceptible to microwaves. On the other hand, harder parts such as roots and barks may require a longer time in the case of MAE. However, when a specific volatile oil is selected as the target analyte, its operating conditions will differ totally from other nonvolatile bioactives as the basic operational setup of microwave extraction will then change.

2.2 Is Any Special Preextraction Treatment Necessary for the Raw Material Before Subjecting It to Microwave-Assisted Extraction?

The sample pretreatment protocol is as simple as that applicable to conventional Soxhlet extraction. One striking difference is that for plants bearing volatile principles, fresh plant material is used for MAE, whereas for others, dried material after necessary size reduction is used. Fresh material offers sufficient hydration status and subsequent absorption of microwaves leading to effective cell rupture of the oil glands followed by oozing of the oil from the ruptured glands. In the case of nonvolatile bioactive(s) it may be advisable to allow sufficient preleaching time before actual exposure to microwaves. Preleaching time is defined as the time of contact of the plant matrix with the extraction solvent before irradiation with microwaves. Preleaching time has been reported by multiple authors with positive effect on the yield of bioactives. This preleaching is expected to improve the hydration status of the matrix resulting in effective absorption of microwaves by the plant matrix, which will then make the absorbed solvent evaporate and generate thermal stress from inside the plant cell wall, which ultimately leads to cell wall rupture.

2.3 How Is Microwave-Assisted Extraction to Be Performed and Which Factors Need to Be Optimized and How?

MAE should be strictly done with a commercially available microwave extractor and use of domestic microwave ovens should be strictly discouraged. Microwave extractors are available both as closed and open types. Closed vessel systems operate under pressure and also enjoy the additional benefit of a pressurized system causing forced quicker entry of solvent into the plant matrix whose cell wall is already compromised. However, such systems are costly and require higher laboratory setup cost compared to the open system. In a recently published article of ours we surveyed that the rate of occurrence of use of open vessel systems is almost twice that of closed vessel systems. Ideally, an open vessel system should comprise variable power levels, time controller, temperature controller, stirring device, powerful exhaust, beam reflectors for focusing the microwaves at the center of the extractor cavity, condenser, and alarm system. In the case of open vessel systems it is advisable to perform microwave firing in an intermittent pattern allowing sufficient cooling time between two subsequent irradiations to reduce the risk of rapid solvent evaporation. In the event of the solvent level inside the extractor drastically decreasing, this may raise fear of thermal degradation, because in such a case the plant matrix is exposed to microwaves directly, which may cause intense thermal stress inside the plant cell wall leading to thermal degradation of the target analyte. Henceforth, it is mandatory to ensure that the plant matrix is sufficiently immersed inside the extraction solvent throughout the course of extraction.

Smaller sample loading is attempted for lab-scale MAE ranging from 500 mg to 2 g; of course, a lot depends on the actual capacity of the microwave. A serious concern that is often raised by industrial researchers is that scaling-up becomes a major issue when something from 1 g needs to be scaled up to 100 kg or more. However, scaling-up technology is a separate area of research and is not included in the preview of this chapter. MAE technology definitely conforms to "push a button technology" but to achieve this comfort of "pushing a button" a rigorous, intensive optimization study has to be carried out to map the exact operational parameters for obtaining maximum yield. Mandal et al., Huei et al., and others have talked vividly about the different extraction parameters that can have significant influence on the extraction performance of MAE. However, from our own personal research experience it can be opined that microwave power and extraction time are very critical factors and may be considered as the rate limiting steps of the entire machinery. In the past, several investigators have reported that a slight exposure to microwaves can be extremely detrimental for the target analyte.

As far as the method of optimization is considered, using chemometric tools such as factorial designs, fractional factorial designs, and response surface methodology has been extremely effective in mapping exact optimum operational conditions. These tools allow results to be obtained by performing minimum experimental runs and also such methods take into account the interaction of other participating factors. A review article by Mandal et al. vividly explains the different chemometric options available and their usage strategy in developing a strong optimization protocol for a robust process.

2.4 How Can the Performance of the Microwave-Assisted Extraction Technique Be Monitored and Decisions Taken on the Optimum Set of Conditions?

The performance of the process can be monitored by quantifying the yield of the target analyte through chromatographic techniques such as high-performance liquid chromatography/high-performance thin-layer chromatography, or gas chromatography (in the case of volatile principles) by using external standards of the target analyte and constructing a linear calibration curve over a fixed concentration range. Results can be expressed in terms of relative extraction efficiency by comparing the results of MAE with that of Soxhlet or reflux extraction. After MAE, the sample may be centrifuged or filtered and the weight of the dried extract can be recorded. The extract may then be reconstituted in a suitable solvent and subjected to solid phase extraction (SPE) purification for producing an enriched version of the analyte under investigation. During SPE treatment the interfering compounds are retained inside the cartridge and the analyte of interest is eluted using a suitable mobile phase. However, this step may not be essential if the overall content and quality of the extract is such that it shall not affect the quantification of the target analyte during chromatography. However, past reports have clearly shown that the quality of extract produced by Soxhlet maceration is very similar to that produced by microwave extraction. This can be easily ascertained by overlaying the chromatogram of the extract obtained from MAE with that obtained from other conventional methods. This indicates that the sample purification and enrichment protocol after obtaining the extract is the same for all methods and use of a microwave in this case does not offer additional advantage. The results obtained can then be analyzed by construction of a response surface graph that clearly indicates the intensity of interaction and level of influence of each factor on the performance of the method under investigation.

2.5 How Can It Be Ensured That No Thermal Degradation Takes Place at the Optimum Operating Conditions?

To ensure that no thermal degradation takes place at the optimum operating conditions a fixed concentration of the target analyte can be subjected to microwave firing at the optimum power and exposure time and the analyte can then be requantified to check the recovery. Also there should not be any change in the retention time of the analyte. Alternatively, the plant sample can be spiked and then subjected to microwave extraction under the derived operating conditions and the recovery of the compound shall then be the indicative marker for degradation-related issues. Microwave exposure may also raise fears that the biological activity of the target analyte might be compromised. To abolish such fears the extract may be checked on an in vitro testing system for the concerned biological activity for which the target analyte is mainly responsible. To simplify, if the target compound has well-established antioxidant activity, then the extract may be checked for the same using a 1,1-diphenyl-2-picrylhydrazyl model and the results in terms of IC50 values may be compared with the extract obtained from Soxhlet or reflux (conventional methods). If no thermal degradation has taken place and maximum or exhaustive extraction has occurred in the case of a microwave-treated sample, then it should show a significantly lower IC50 value than the samples obtained from conventional methods. A lower IC50 value indicates higher scavenging potency. Improved exhibition of biological potency from the samples obtained from the MAE technique is also an indirect marker for "no thermal degradation."

2.6 How Is the Mechanism of Accelerated Extraction Phenomenon in the Case of Microwave Treatment Elucidated and How Can Its Complete Supremacy Over Conventional Methods Be Proved?

To elucidate the mechanistic pathway behind accelerated extraction yield, scanning electron micrographs of marc obtained from various extraction methods should be analyzed. Microwave-treated samples will clearly show large perforations and cell rupture, which form the direct exit pathway for the target analytes in a much shorter time. The marc left over after MAE can be reextracted to check for the presence of any trace amount of target analyte. Based on the quantity of the leftover analyte in the marc it can be decided whether it is judicious to run a second cycle of MAE on the marc. The precision level may be compared with conventional methods by performing five sets of extraction separately and comparing the relative standard deviation.

3. KEY REMINDERS FOR MICROWAVE-ASSISTED EXTRACTION OF BOTANICALS

- Energy audit of the entire MAE can be done and compared with conventional methods, which will then give a true picture of its impact on the environment. To obtain 1 kWh from coal or fuel, 800 g of CO_2 will be ejected into the atmosphere during the combustion of fossil fuel. Based on this standard assumption the amount of CO_2 released from the microwave extractor and Soxhlet operation can easily be compared.
- Dual heating mechanism as proposed by Mandal et al. may be adopted whereby the sample is pretreated separately with a better microwave absorbing solvent (water) without affecting the actual quality of the extract.
- Use of a microwave transparent solvent strategy as proposed by Mandal et al. may be adopted for extremely thermolabile principles whereby the sample is pretreated with methanol/water and then extracted using a microwave transparent solvent.
- The kinetic extraction pattern for MAE must be established so that the peak extraction period may be identified followed by the decline stage.
- MAE may not be used for new drug discoveries from medicinal plants as each bioactive principle needs a separate optimization of operational parameters and there exist no universal standard operational settings for microwaves that can be applied to any plant material. Therefore if a plant material needs to be investigated for its biological activity (ethnopharmacology) followed by isolation of bioactive principles, then probably conventional methods need to be adopted to start with. Once the bioactive principle has been isolated and established, then the MAE technique can be optimized for the large-scale extraction of that particular bioactive. The performance of MAE needs to be closely monitored so that the starting point of degradation can be mapped for which a target analyte is required and whose yield shall be the indicative marker for measuring the performance of the said process. When we start with a new plant material in ethnopharmacology research we have no such ready available target analyte and hence MAE cannot be applied. Henceforth, every research on MAE is unique and its novelty cannot be questioned.

- Different innovations have been suggested by Chan et al., which can be applied on a basic MAE model. To get a microwave to work in tandem with a vacuum, an oxygen-free environment and sonication are some of the innovations or value additions mentioned by Chan et al. However, it is not mandatory because it is evident from past results that such innovations will definitely yield positive results when compared to basic MAE. A detailed illustrated analysis of such innovations has been done by us in one of our recent review articles dedicated to MAE of phenolics.
- Use of ionic liquids and eutectic solvents could also be investigated. However, as stated earlier, confirmed positive reports through the use of such solvents cannot be guaranteed.
- A close watch on the degradation profile should be undertaken as the fact that degradation products may be toxic cannot be ruled out.

4. CONCLUSION

Through the presentation of the sequence of events occurring in chronological order this chapter aimed to deliver the "must know" strategic facts in working with MAE. The chapter is not a collection of past reports that review articles generally tend to be, but makes a novel effort to present a critical analysis report in the form of FAQs to its readers so that a vivid description may be presented, which shall be helpful for first-timers in this area of research to strategize their work. A list of do's and don'ts regarding MAE of plant bioactives has also been provided (Fig. 23.1) for quick assessment.

The chapter through its FAQs highlighted the basic mandatory strategic steps that have to be performed for MAE of bioactives. Nevertheless, it shall depend upon the critical thinking and innovative capacity of the researcher to make any value addition over and above the basic steps. The opinions presented through different FAQs are only applicable to MAE. Other methodologies such as ultrasound-assisted extraction and pressurized liquid extractions are governed by different operating principles and shall have different work strategies. Considering the increasing concern over climate change it can be opined that in the near future environment friendly-driven technology shall be the key for economic development for any country and hence this review is an attempt to attract more researchers and industrialists to such areas through strategizing and prioritizing extraction of botanicals. In other words this review can be treated as a promotional review for wider acceptance of MAE of plant bioactives through prioritizing such research.

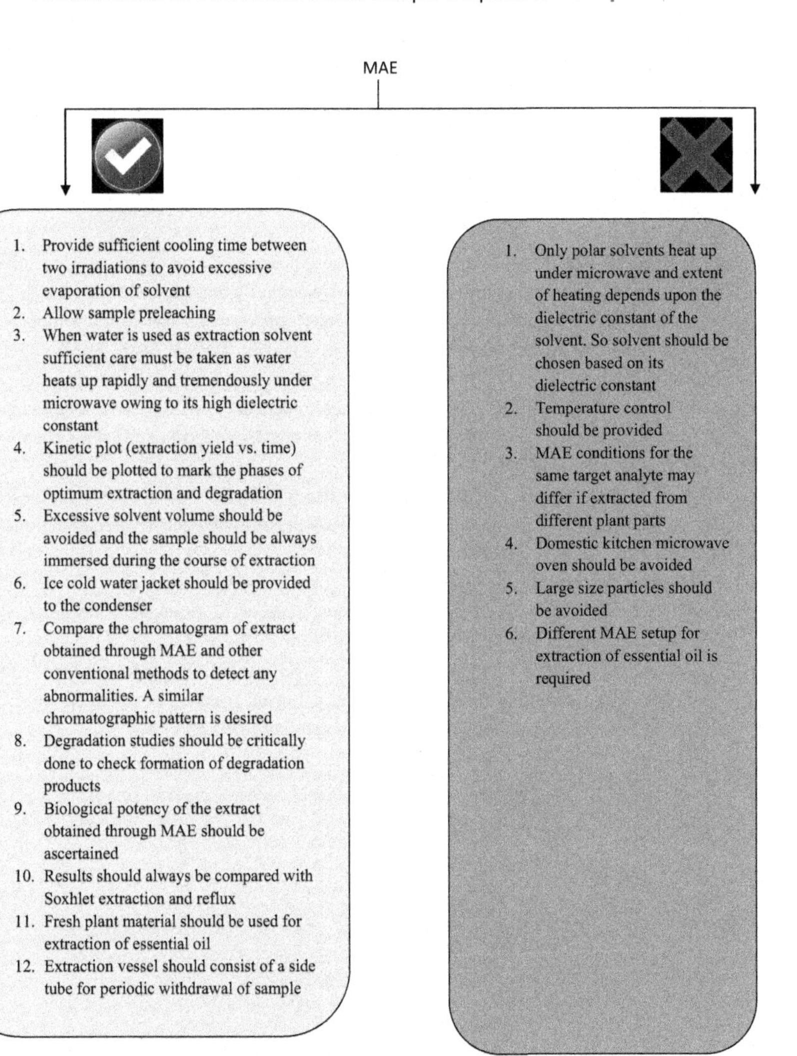

FIGURE 23.1 Do's and don'ts for microwave-assisted extraction.

ACKNOWLEDGMENTS

The authors are thankful to their funding agency SERB, New Delhi, India (Grant Number: EEQ/2016/000067). Infrastructural and research support provided by home university is also duly acknowledged. Kind support and cooperation from our industry research partner Mr. Shrikant Potdar, Catalyst Systems, Pune (India), is also duly acknowledged.

FURTHER READING

[1] S.C. Mandal, V. Mandal, A. Das, Essentials of Botanical Extraction-principles and Applications, first ed., Academic Press (Elsevier), London, 2015.

[2] V. Mandal, R. Tandey, A critical analysis of publication trends from 2005–2015 microwave assisted extraction of botanicals: how far we have come and the road ahead, Trends Anal. Chem. 82 (2016) 100–108.

[3] V. Mandal, Y. Mohan, S. Hemalatha, Microwave assisted extraction: an innovative and promising extraction tool for medicinal plant research, Pharmacogn. Rev. 1 (2007) 7–18.

[4] V. Mandal, R. Tandey, Critical analysis of research trends and issues in microwave assisted extraction of phenolics: have we really done enough, Trends Anal. Chem. 85 (2016) 140–152.

[5] A.K. Das, V. Mandal, S.C. Mandal, A brief understanding of process optimization in microwave assisted extraction of botanicals: options and opportunities with chemometric tools, Phytochem. Anal. 25 (2014) 1–12.

[6] V. Mandal, S.C. Mandal, Design and performance evaluation of a microwave based low carbon yielding extraction technique for naturally occurring bioactive triterpenoid: oleanolic acid, Biochem. Eng. J. 50 (2010) 63–70.

[7] V. Mandal, Y. Mohan, S. Hemalatha, Microwave assisted extraction of curcumin by sample–solvent dual heating mechanism using Taguchi L_9 orthogonal design, J. Pharm. Biomed. Anal. 46 (2008) 322–327.

[8] A.K. Das, V. Mandal, S. C Mandal, Design of experiment (DOE) approach for the process optimization of microwave assisted extraction of Lupeol from Ficus racemosa leaves using Response surface methodology, Phytochem. Anal. 24 (2013) 230–247.

Author Index

'*Note:* Page numbers followed by "f" indicate figures and "t" indicate tables.'

A

B

C

D

E

F

G

H

I

J

K

L

M

N

O

P

Q

R

T

X

Y

Z

Subject Index

'*Note*: Page numbers followed by "f" indicate figures, "t" indicate tables.'

A

B

C

E

F

H

I

J

K

L

M

N

P

Q

R

S

T

U